The Science of Biology

Sixth Edition

Life

Sixth Edition

The Science of Biology

William K. Purves
Emeritus, Harvey Mudd College
Claremont, California

David Sadava
The Claremont Colleges
Claremont, California

Gordon H. Orians
Emeritus, The University of Washington
Seattle, Washington

H. Craig Heller
Stanford University
Stanford, California

Sinauer Associates, Inc.

W. H. Freeman and Company

The Cover

Giraffes (*Giraffa camelopardalis*) near Samburu, Kenya.
Photograph © BIOS/Peter Arnold, Inc.

The Opening Page

Soap yucca (*Yucca elata*), White Sands National Monument, New Mexico.
Photograph © David Woodfall/DRK PHOTO.

The Title Page

The endangered Florida panther (*Felis concolor coryi*).
Photograph © Thomas Kitchin/Tom Stack & Associates.

Life: The Science of Biology, Sixth Edition

Address editorial correspondence to:
Sinauer Associates, Inc., 23 Plumtree Road, Sunderland, Massachusetts 01375 U.S.A.
www.sinauer.com

Email: publish@sinauer.com

Address orders to:
VHPS/W. H. Freeman & Co. Order Department, 16365 James Madison Highway,
U.S. Route 15, Gordonsville, VA 22942 U.S.A.
www.whfreeman.com

Examination copy information: 1-800-446-8923
Orders: 1-888-330-8477

Library of Congress Cataloging-in-Publication Data

Life, the science of biology / William K. Purves...[et al.]--6th ed.
p. cm.
Includes index.
ISBN 0-7167-3873-2 (hardcover) – ISBN 0-7167-4348-5 (Volume 1) –
ISBN 0-7167-4349-3 (Volume 2) – ISBN 0-7167-4350-7 (Volume 3)
1. Biology I. Purves, William K. (William Kirkwood), 1934–

QH308.2 .L565 2000
570--dc21 00-048235

Printed in U.S.A.

Fourth Printing 2003 Courier Companies Inc.

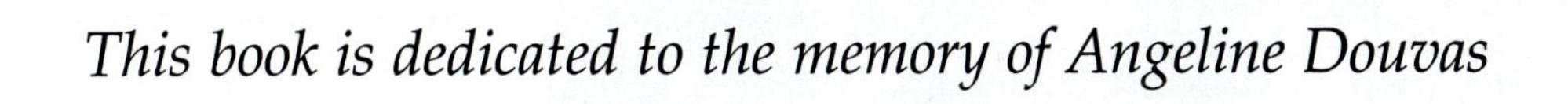

This book is dedicated to the memory of Angeline Douvas

About the Authors

Gordon Orians Craig Heller Bill Purves David Sadava

William K. Purves is Professor Emeritus of Biology as well as founder and former chair of the Department of Biology at Harvey Mudd College in Claremont, California. He received his Ph.D. from Yale University in 1959 under Arthur Galston. A fellow of the American Association for the Advancement of Science, Professor Purves has served as head of the Life Sciences Group at the University of Connecticut and as chair of the Department of Biological Sciences, University of California, Santa Barbara, where he won the Harold J. Plous Award for teaching excellence. His research interests focused on the chemical and physical regulation of plant growth and flowering. Professor Purves elected early retirement in 1995, after teaching introductory biology for 34 consecutive years, in order to turn his skills to writing and producing multimedia for introductory biology students. That year, he was awarded the Henry T. Mudd Prize as an outstanding member of the Harvey Mudd faculty or administration.

David Sadava is now responsible for *Life*'s chapters on the cell (2–8), in addition to the chapters on genetics and heredity that he assumed in the previous edition. He is the Pritzker Family Foundation Professor of Biology at Claremont McKenna, Pitzer, and Scripps, three of the Claremont Colleges. Professor Sadava received his Ph.D. from the University of California, San Diego in 1972, and has been at Claremont ever since. The author of textbooks on cell biology and on plants, genes, and agriculture, Professor Sadava has done research in many areas of cell biology and biochemistry, ranging from developmental biology, to human diseases, to pharmacology. His current research concerns human lung cancer and its resistance to chemotherapy. Virtually all of the research articles he has published have undergraduates as coauthors. Professor Sadava has taught a variety of courses to both majors and nonmajors, including introductory biology, cell biology, genetics, molecular biology, and biochemistry, and he recently developed a new course on the biology of cancer. For the last 15 years, Professor Sadava has been a visiting professor in the Department of Molecular, Cellular, and Developmental Biology at the University of Colorado, Boulder, and is currently a visiting scientist at the City of Hope Medical Center.

Gordon H. Orians is Professor Emeritus of Zoology at the University of Washington. He received his Ph.D. from the University of California, Berkeley in 1960 under Frank Pitelka. Professor Orians has been elected to the National Academy of Sciences and the American Academy of Arts and Sciences, and is a Foreign Fellow of the Royal Netherlands Academy of Arts and Sciences. He was President of the Organization for Tropical Studies, 1988–1994, and President of the Ecological Society of America, 1995–1996. He is chair of The Board on Environmental Studies and Toxicology of the National Research Council and a member of the board of directors of World Wildlife Fund–US. He is a recipient of the Distinguished Service Award of the American Institute of Biological Sciences. Professor Orians is a leading authority in ecology, conservation biology, and evolution, with research experience in behavioral ecology, plant–herbivore interactions, community structure, the biology of rare species, and environmental policy. He elected early retirement to be able to devote more time to writing and environmental policy activities.

H. Craig Heller is the Lorry Lokey/Business Wire Professor of Biological Sciences and Human Biology at Stanford University. He has served as Director of the popular interdisciplinary undergraduate program in Human Biology and is now Chairman of Biological Sciences. Professor Heller received his Ph.D. from Yale University in 1970 and did postdoctoral work at Scripps Institute of Oceanography on how the brain regulates body temperature of mammals. His current research focuses on the neurobiology of sleep and circadian rhythms. Professor Heller has done research on a great variety of animals ranging from hibernating squirrels to exercising athletes. He teaches courses on animal and human physiology and neurobiology.

Preface

Biologists' understanding of the living world is growing explosively. This isn't the world that the four authors of this book were born into. We never dreamed, as we began our research careers as freshly minted Ph.D.'s, that our science could move so rapidly. Biology has now entered the post-genomic era, allowing biologists and biomedical scientists to tackle once-unapproachable challenges. We are also at the threshold of some experiments that raise ethical concerns so great that we must stand back and participate with others in determining what is right to do and what is not.

The enormous growth and changes in biology create a special challenge for textbook authors. How can a biology textbook provide the basics, keep up with the exciting new discoveries, and not become overwhelming. The increasing bulk of textbooks is of great concern to authors as well as to instructors and their students, who blanch at the prospect of too many pages, too many term papers, and too little sleep. Some reconsideration of what is essential and how that is best presented needs to be made if the proliferation of facts is not to obscure the fundamental principles.

Our major goals were brevity, emphasis on experiments, and better ways to help students learn

In writing the Sixth Edition of *Life*, we committed ourselves to reversing the pattern of ever increasing page lengths in new editions. We wanted a shorter book that brings the subject into sharper focus. We tried to achieve this by judicious reduction of detail, by more concise writing, and by more use of figures as primary teaching sources. It worked! Our efforts were successful. This edition is 200 pages shorter than its predecessor, yet it covers much exciting new material.

While working to tighten and shorten the text, we were also determined to retain and even increase our emphasis on *how* we know things, rather than just *what* we know. To that end, the Sixth Edition inaugurates 72 specially formatted figures that show how experiments, field observations, and comparative methods help biologists formulate and test hypotheses (the figure at right is an example). Another 26 figures highlight some of the many field and laboratory methods created to do this research. These Experiment and Research Methods illustrations are listed on the endpapers at the back of the book.

In the Fifth Edition, we introduced "balloon captions" that guide the reader through the illustrations (rather than having to wade through lengthy captions). This feature was widely applauded and we have worked to refine the balloons' effectiveness. In response to suggestions from users

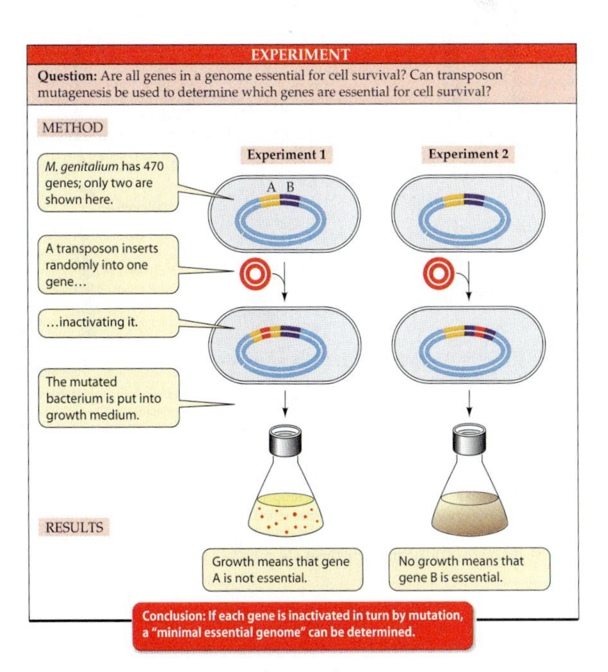

13.22 Using Transposon Mutagenesis to Determine the Minimal Genome
By inactivating genes one by one, scientists can determine which ones are essential for the cell's survival.

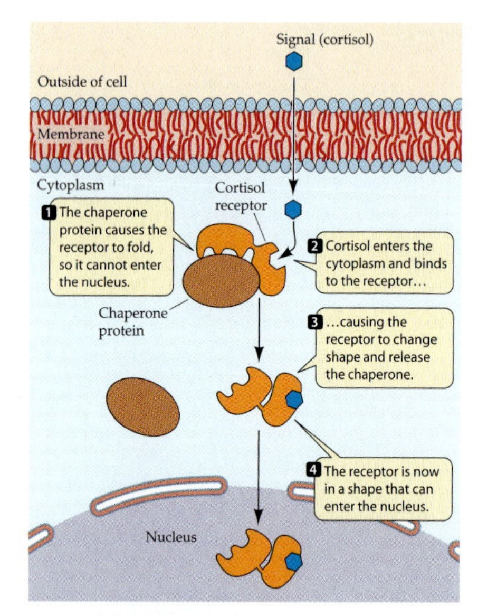

15.9 A Cytoplasmic Receptor
The receptor for cortisol is bound to a chaperone protein. Binding of the signal (which diffuses directly through the membrane) releases the chaperone and allows the receptor protein to enter the cell's nucleus, where it functions as a transcription factor.

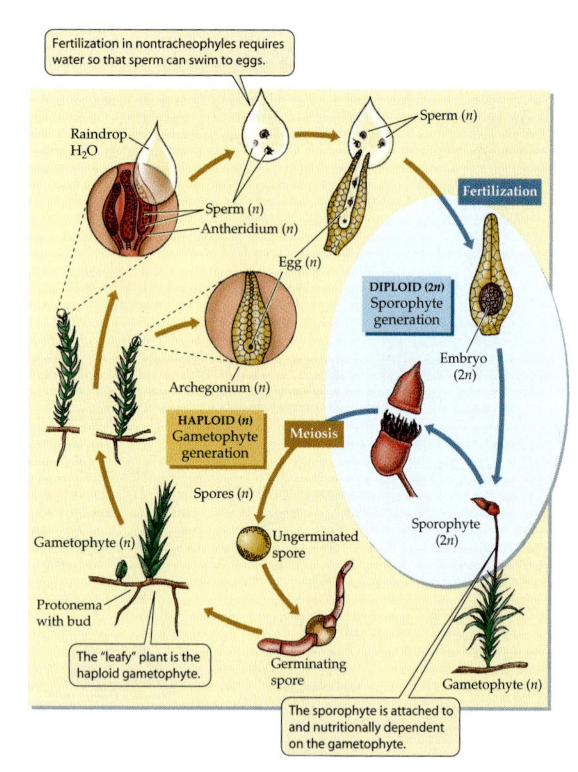

28.3 A Nontracheophyte Life Cycle
The life cycle of nontracheophytes, illustrated here by a moss, is dependent on an external source of liquid water. The visible green structure of nontracheophytes is the gametophyte; in nontracheophyte plants, the "leafy" structures are sporophytes.

of the Fifth Edition, in the Sixth Edition we now number many of the balloons, emphasizing the flow of the figure and making the sequence easier to follow (Figure 15.9 at left is an example).

This edition is accompanied by a comprehensive website, www.thelifewire.com (and an optional CD-ROM that contains the same material) that reinforces the content of every chapter. A key component of the website is a combination of animated tutorials and activities for each chapter, all of which include self-quizzes. Within each book chapter, this icon refers students to a tutorial or an activity. An index of the icons begins in the front endpapers of the book. Figure 28.3 (left, below) shows a typical web icon placement.

As part of the ongoing challenge of keeping the writing and illustrations as clear as possible, we frequently employ bulleted lists. We think these lists will help students sort through what is, even after pruning, a daunting amount of material. And we have continued to provide plenty of interim summaries and bridges that link passages of text.

In all the introductory textbooks, the chapters end with summaries. In ours, we have organized the material within the chapter's main headings. In most cases, we tie key concepts to the figure (or figures) that illustrate it. For visual learners, this provides an efficient mode of reviewing the chapter.

From our many decades in the classroom, we know how important it is to motivate students. Each chapter begins with a brief description of some event, phenomenon, or idea that we hope will engage the reader while conveying a sense of the significance and purpose of the chapter's subject.

Evolution Continues to be the Dominant Theme

Evolution continues to be the most important of the themes that link our chapters and provide continuity. As we have written the various editions of the book, however, the emergence of *genomics* as a new paradigm in the late twentieth century has developed, revolutionizing most areas of biology. In this new century, understanding the workings of the genome is of paramount importance in almost any biological discussion.

In this edition, we have moved further toward updating the evolutionary theme to encompass the postgenomic era. Just two examples are the addition of a section on genomic evolution to our coverage of molecular evolution, and a section on "evo/devo" in the chapter on molecular biology of development. In addition, the chapters on the diversity of life reflect the vast changes in our understanding of systematics and phylogenetic relationships thanks to the genomic perspective.

In fact, each chapter of the book has undergone important changes.

The Seven Parts: Content, Changes, and Themes

In Part One, The Cell, the emphasis in the discussions of biological molecules and thermodynamics has shifted more decisively toward biological aspects and away from pure chemistry. We have made our discussions of enzymes, cell respiration, and photosynthesis less detailed and more focused on the biological applications.

A major addition to Part Two, Information and Heredity, is a new chapter (Chapter 15) on cell signaling and communication, introduced at a place where the students have the necessary grounding in cell biology and molecular genetics. That chapter leads logically into an updated chapter (Chapter 16) on the molecular biology of development, which includes a new section on the intersection of evolutionary and developmental biology—"evo-devo" in the modern jargon. Several chapters incorporate the exciting new work in genomics of prokaryotes, humans, and other eukaryotes.

We have updated all the chapters in Part Three, Evolutionary Processes. In particular, Chapter 24 ("Molecular and Genomic Evolution") reflects the rapid advances in this exciting field. The section on genomic evolution (on pages 446–447) is brand new and includes Figure 24.9 (shown at right).

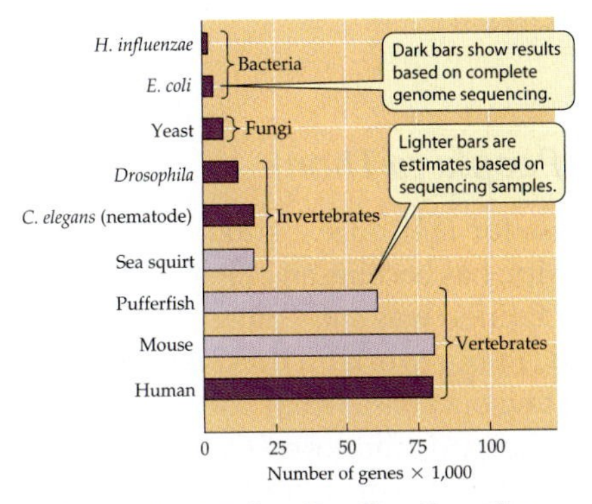

24.9 Complex Organisms Have More Genes than Simpler Organisms
Genome sizes have been measured or estimated in a variety of organisms, ranging from single-celled prokaryotes to vertebrates.

Part Four, The Evolution of Diversity, now reflects some exciting changes. The chapter on the protists—which can no longer be treated as a single "kingdom"—reflects the continuing uncertainty over the origin and early diversification of eukaryotes. The equally great uncertainty over prokaryote phylogeny, as we deal with the implications of extensive lateral transfer of genes, is evident in the chapter on prokaryote phyla.

We have extended the coverage of the evolution and diversity of plants to two chapters, and that of the animals to three. Recent findings stemming largely from molecular research have led to modifications of the phylogenies of angiosperms and of the animal kingdom. These changes are reflected in the many simplified "trees" that give a broad overview of systematic relationships. Key evolutionary events that separate and unite the different groups are highlighted with red "hot spots" (see Figure 33.1 at right).

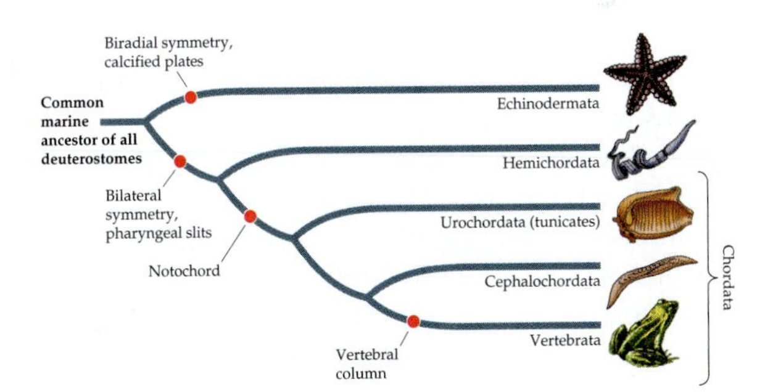

33.1 A Probable Deuterostomate Phylogeny
There are fewer major lineages and many fewer species of deuterostomes than of protostomes.

We have rearranged Part Five, "The Biology of Flowering Plants," to allow Chapter 39 ("Plant Responses to Environmental Challenges") to serve as a capstone to the whole part, drawing together some of the major threads. We have added sections on hormones and photoreceptors discovered in recent years, and on their signal transduction pathways. The opening chapter (Chapter 34) on "The Flowering Plant Body" has an increased emphasis on meristems.

Part Six, The Biology of Animals, continues to be a broad, comparative treatment of animal physiology with an emphasis on mechanisms of control and regulation. Much new material has been added, including a major revision of Animal

Development (Chapter 43) to complement and extend the earlier Chapter 16 (Development: Differential Gene Expression). Some other new topics are the role of melatonin in photoperiodism, the role of leptin in the control of food intake, and the discovery in fruit flies of a gene that controls male mating behavior. The extensive coverage of the fast moving field of neurobiology has been substantially updated.

Throughout Part Seven, Ecology and Biogeography, we have added examples of experimental approaches to understanding the dynamics of ecological systems. Some of the examples illustrate the use of experimental and comparative methods. As before, we conclude the book with a chapter on conservation biology (Chapter 58), emphasizing the use of scientific principles to help preserve Earth's vast biological diversity.

There Are Many People to Thank

The reviewing process for *Life*, once a single pass at the stage of draft manuscript, has become an ongoing phenomenon. When the Fifth Edition was still young, we received critiques that influenced our work on this Sixth Edition. The two most penetrating ones came from Zach Gertz, then an undergraduate at Harvard, and Joseph Vanable, a veteran introductory biology professor at Purdue.

Next, still during the Fifth Edition run, 18 instructors recorded their suggestions for improvements in *Life* while teaching from the book. We call these reviews Diary Reviews. The third stage was the Manuscript Reviews. Seventy-three dedicated teachers and researchers read the first-draft chapters and gave us significant and cogent advice. Still another stage has been added to the process and it turned out to be invaluable. We are indebted to 16 Accuracy Reviewers, colleagues who carefully reviewed the almost final page proofs of each chapter to spot lingering errors or imprecisions in the text and art that inevitably escape our weary eyes. Finally, we appreciate the advice given by several experts who reviewed the animations and activities that our publishers developed for the student Web Site/CD-ROM that accompanies this edition of *Life*. We thank all these reviewers and hope this new edition measures up to their expectations. They are listed after this Preface.

J/B Woolsey Associates has again worked closely with each of us to improve an already excellent art program. They helped to refine the very successful "balloon captions" that were introduced in the Fifth Edition. With their creative input we introduced the Experiment and Research Method illustrations found throughout the text.

James Funston joined us again as the developmental editor for the Sixth Edition. As always, James enforced a rigorous standard for clear writing and illustrating. And he contributed significantly to the process of shortening the book. Norma Roche also suggested cuts, and provided incisive copy editing from beginning to end. Her many astute queries often led to rewrites that enhanced the clarity of the presentation. From first draft to final pages, Susan McGlew was tireless in arranging for expert academic reviews of all of the chapters. Since the First Edition, we have profited immeasurably from the work of Carol Wigg, who again coordinated the pre-production process, including illustration editing and copy editing. She wrote many figure captions, suggested several of the chapter-opening stories, orchestrated the flow of the text and art, kept us mostly on schedule, enforced—sometimes with her red pen—the mandate to be concise, and what's more, did it all with good humor, even under pressure. David McIntyre, photo researcher, found many wonderful new photographs to enhance the learning experience and enliven the appearance of the book as a whole.

We again wish to thank the dedicated professionals in W. H. Freeman's marketing and sales group. Their enthusiasm has helped bring *Life* to a wider audience with each edition. We appreciate their continuing support and valuable input on ways to improve the book. A large share of *Life's* success is due to their efforts in this publishing partnership.

We have always respected Sinauer Associates for their outstanding list of biology books at all levels and we have enjoyed having them lead and assist us through yet another edition. Andy Sinauer has been the guiding spirit behind the development of *Life* since two of us first began to write the First Edition. Andy never ceases helping his authors to achieve our goals, while remaining gentle but firm about his agendas. It has been a very satisfying experience for us to work with him yet again, and we look forward to a continuing association.

Bill Purves   David Sadava   Gordon Orians   Craig Heller

November, 2000

Reviewers for the Sixth Edition

Diary Reviewers

Carla Barnwell, University of Illinois
Greg Beaulieu, University of Victoria
Gordon Fain, University of California, Los Angeles
Ruth Finkelstein, University of California, Santa Barbara
Steve Fisher, University of California, Santa Barbara
Alice Jacklet, SUNY, Albany
Clare Hasenkampf, University of Toronto, Scarborough
Werner Heim, Colorado College
David Hershey, Hyattsville, MD
Hans-Willi Honegger, Vanderbilt University
Durrell Kapan, University of Texas, Austin
Cheryl Kerfeld, University of California, Los Angeles
Michael Martin, University of Michigan, Ann Arbor
Murray Nabors, Colorado State University
Ronald Poole, McGill University
Nancy Sanders, Truman State University
Susan Smith, Massasoit Community College
Raymond White, City College of San Francisco

Manuscript Reviewers

John Alcock, Arizona State University
Allen V. Barker, University of Massachusetts, Amherst
Andrew R. Blaustein, Oregon State University
Richard Brusca, University of Arizona
Matthew Buechner, University of Kansas
Warren Burggren, University of North Texas
Jung Choi, Georgia Institute of Technology
Andrew Clark, Pennsylvania State University
Carla D'Antonio, University of California, Berkeley
Alan de Queiroz, University of Colorado
Michael Denbow, Virginia Tech
Susan Dunford, University of Cincinnati
William Eickmeier, Vanderbilt University
John Endler, University of California, Santa Barbara
Gordon L. Fain, University of California, Los Angeles
Stu Feinstein, University of California, Santa Barbara
Danilo Fernando, SUNY, Syracuse
Steve Fisher, University of California, Santa Barbara
Doug Futuyma, SUNY, Stony Brook
Scott Gilbert, Swarthmore College
Janice Glime, Michigan Technological University
Elizabeth Godrick, Boston University
Robert Goodman, University of Wisconsin, Madison
Nancy Guild, University of Colorado
Jessica Gurevitch, SUNY, Stony Brook
Jeff Hardin, University of Wisconsin, Madison
Joseph Heilig, University of Colorado
David Hershey, Hyattsville, MD
Mark Johnston, Dalhousie University
Walter Judd, University of Florida
Thomas Kane, University of Cincinnati
Laura Katz, Smith College
Elizabeth Kellogg, University of Missouri, St. Louis
Peter Krell, University of Guelph
Thomas Kursar, University of Utah
Wayne Maddison, University of Arizona
William Manning, University of Massachusetts, Amherst
Michael Marcotrigiano, Smith College
Lloyd Matsumoto, Rhode Island College
Stu Matz, The Evergreen State College
D. Jeffrey Meldrum, Idaho State University
Mike Millay, Ohio University (Southern Campus)
David Mindell, University of Michigan, Ann Arbor
Deborah Mowshowitz, Columbia University
Laura Olsen, University of Michigan, Ann Arbor
Guillermo Orti, University of Nebraska
Constance Parks, University of Massachusetts, Amherst
Jane Phillips, University of Minnesota
Ronald Poole, McGill University
Warren Porter, University of Wisconsin, Madison
Thomas Poulson, University of Illinois, Chicago
Loren Rieseberg, Indiana University
Ian Ross, University of California, Santa Barbara
Nancy Sanders, Truman State University
Paul Schroeder, Washington State University
Jim Shinkle, Trinity University
Mitchell Sogin, Marine Biological Laboratory, Woods Hole
Wayne Sousa, University of California, Berkeley
Charles Staben, University of Kentucky
James Staley, University of Washington
Steve Stanley, The Johns Hopkins University
Barbara Stebbins-Boaz, Willamette University
Antony Stretton, University of Wisconsin, Madison
Steven Swoap, Williams College
Gerald Thrush, California State University, San Bernardino
Richard Tolman, Brigham Young University
Mary Tyler, University of Maine
Michael Wade, Indiana University
Bruce Walsh, University of Arizona
Steven Wasserman, University of California, San Diego
Alex Weir, SUNY, Syracuse
Mary Williams, Harvey Mudd College
Jonathan Wright, Pomona College

Accuracy Reviewers

Andrew Clark, Pennsylvania State University
Joanne Ellzey, University of Texas, El Paso
Tejendra Gill, University of Houston, University Park
Paul Goldstein, University of Texas, El Paso
Laura Katz, Smith College
Hans Landel, North Seattle Community College
Sandy Ligon, University of New Mexico
Peter Lortz, North Seattle Community College
Roger Lumb, Western Carolina University
Coleman McCleneghan, Appalachian State University
Janie Milner, Santa Fe Community College
Zack Murrell, Appalachian State University
Ben Normark, University of Massachusetts, Amherst
Mike Silva, El Paso Community College
Phillip Snider, University of Houston, University Park
Steven Wasserman, University of California, San Diego

Media Reviewers

Karen Bernd, Davidson College
Mark Browning, Purdue University
William Eldred, Boston University
Joanne Ellzey, University of Texas, El Paso
Randall Johnson, University of California, San Diego
Coleman McCleneghan, Appalachian State University
Melissa Michael, University of Illinois
Tom Pitzer, Florida International University
Kenneth Robinson, Purdue University

To the Student

Welcome to the study of life! In our student days—and ever since—we have enjoyed studying the fascinating and fast-changing field of biology, and we hope that you will, too.

Getting the Most Out of the Book

There are a few things you can do to help you get the most from this book and from your course. For openers, read the book actively—don't just read passively, but do things that force you to think as you read. If we pose questions, stop and think about them. Ask questions of the text as you go. Do you understand what is being said? Does it relate to something you already know? Is it supported by experimental or other evidence? Does that evidence convince you? How does this passage fit into the chapter as a whole? Annotate the book—write down comments in the margins about things you don't understand, or about how one part relates to another, or even when you find an idea particularly interesting. People remember things they think about much better than they remember things they have read passively. Highlighting is passive; copying is drudge work; questioning and commenting are active and well worthwhile.

"Read" the illustrations actively too. You will find the balloon captions in the illustrations especially useful—they are there to guide you through the complexities of some topics and to highlight the major points.

The chapter summaries will help you quickly review the high points of what you have read. A summary identifies particular illustrations that you should study to help organize the material in your mind. Add concepts and details to the framework by reviewing the text. A way to review the material in slightly more detail after reading the chapter is to go back and look at the boldfaced terms. You can use the boldfaced terms to pose questions—and see if you can answer those questions. The boldfacing will probably be more useful on a second reading than on the first.

Use the "For Discussion" questions at the end of each chapter. These questions are usually open-ended and are intended to cause you to reflect on the material.

The glossary and the index can help you a great deal. When you are uncertain of the meaning of a term, check the glossary first—there are more than 1,500 definitions in it. If you don't find a term in the glossary, or if you want a more thorough discussion of the term, use the index to find where it's discussed.

The Web Site

Use the student Web Site/CD-ROM to help you understand some of the more detailed material and to help you sort out the information we have laid before you. An illustrated guide to the learning resources found on the Web Site/CD-ROM is in the front of this book. Pay particular attention to the activities and animated tutorials on key concepts, and to the self-quizzes. The self-quizzes provide extensive feedback for each correct and incorrect answer, and include hot-linked references to text pages. If you'd like to pursue some topics in greater detail, you'll find a chapter-by-chapter annotated list of suggested readings. We have tried to choose readings from books and magazines, especially *Scientific American*, that should be available in your college library.

What If the Going Gets Tough?

Most students occasionally have difficulty in courses, including biology courses. If you find that you are slipping behind in the course, or if a particular topic is giving you an unreasonable amount of trouble, here are some useful steps you might take. First, the basics: attend class, take careful lecture notes, and read the textbook assignments. Second, note that one of the most important roles of studying is to discover what you *don't* know, so that you can do something about it. Use the index, the glossary, the chapter summaries, and the text itself to try to answer any questions you have and to help you organize the material. Make a habit of looking over your lecture notes within 24 hours of when you take them—find out right away what points are unclear, and get them straightened out in your mind. The web site can help by providing a different perspective.

If none of these self-help remedies does the trick, get help! Other students are often a good source of help, because they are dealing with the material at the same level as you are. Study groups can be very useful, as long as the participants are all committed to learning the material. Tutors are almost always helpful, as are faculty members. The main thing is to *get help when you need it*. It is not a good idea to be strong and silent and drift into a low grade.

But don't make the grade the point of this or any other course. You are in college to learn, to pursue interesting subjects, and to enjoy the subjects you are pursuing. We hope you'll enjoy the pursuit of biology.

Bill Purves David Sadava Gordon Orians Craig Heller

Life's Supplements

For the Student

Web Site/CD-ROM

Student Web Site at www.thelifewire.com
Life 6.0 CD-ROM (optionally bundled with the text)

The Web Site and CD-ROM each support the entire text, offering:

- Over 65 **Animated Tutorials** clarifying key topics from the text
- **Activities**, including flashcards for key terms and concepts, and drag-and-drop exercises
- **Self-quizzes** with extensive feedback, references to the Study Guide, and hot-linked references to *Life: The Science of Biology*, Sixth Edition
- **Glossary** of key terms and concepts
- **End-of-chapter Online Quizzes** (see "Online Quizzing" under "For the Instructor")
- **Lifelines**

  Study Skills (Jerry Waldvogel, *Clemson University*) provides class-tested practical advice on time management, test-taking, note-taking, and how to read the textbook

  Math for Life (Dany Adams, *Smith College*) helps students learn or reacquire basic quantitative skills
- **Suggested Readings** for further study

  Order ISBN 0-7167-3874-0, *Life 6.0* CD-ROM, or ISBN 0-7167-3875-9, Text/CD-ROM bundle

Study Guide

Christine Minor, *Clemson University*, Edward M. Dzialowski and Warren W. Burggren, *University of North Texas*, Lindsay Goodloe, *Cornell University*, and Nancy Guild, *University of Colorado at Boulder*.

For each chapter of the text, the study guide offers clearly defined learning objectives, summaries of key concepts, references to *Life* and to the student *Web/CD-ROM*, and review and exam-style self-test questions with answers and explanations.

Order ISBN 0-7167-3951-8

Lecture Notebook

This new tool presents black and white reproductions of all the Sixth Edition's line art and tables (more than 1000 images, with labels). The *Notebook* provides ample ruled spaces for note-taking.

Order ISBN 0-7167-4449-X

For the Instructor

Instructor's Teaching Kit

This **new** comprehensive teaching tool (in a three-ring binder) combines:

1. **Instructor's Manual**

   Erica Bergquist, *Holyoke Community College*

   The Manual includes:
   - Chapter overviews
   - Chapter outlines
   - A "What's New" guide to the Sixth Edition
   - All the bold-faced key terms from the text
   - Key concepts and facts for each chapter
   - Overviews of the animated tutorials from the Student Web Site/CD-ROM
   - Custom lab ordering information (see "Custom Labs")
2. **Enriched Lecture Notes**, with diagrams

   Charles Herr, *Eastern Washington University*
3. **A PowerPoint® Thumbnail Guide to the PowerPoint®** presentations on the Instructor's CD-ROM

Test Bank

Charles Herr, *Eastern Washington University*

The test bank, available in both computerized and printed formats, offers more than 4000 multiple-choice and sentence-completion questions.

The easy-to-use computerized test bank on CD-ROM includes Windows and Macintosh versions in a format that lets instructors add, edit, and resequence questions to suit their needs. From this same CD-ROM, instructors can access *Diploma* Online Testing from the Brownstone Research Group. *Diploma* allows instructors to easily create and administer secure exams over a network and over the Internet, with questions that incorporate multimedia and interactive exercises. More information about *Diploma* is available at http://www.brownstone.net

Online Quizzing

The online quizzing function is accessed via the Student Web Site at www.thelifewire.com. Using Question Mark's *Perception*, instructors can easily and securely quiz students online using multiple-choice questions for each text chapter and its media resources.

Instructor's Resource CD-ROM

The Instructor's Resource CD-ROM employs **Presentation Manager** and includes:

- All four-color line art and tables from the text (more than 1000 images), resized and reformatted to maximize large-hall projection
- More than 1500 photographic images, including electron micrographs, from the Biological Photo Service collection—all keyed to *Life* chapters
- More than 60 animations from the Student Web Site/CD-ROM
- Exceptional video microscopy from Jeremy Pickett-Heaps and others
- Chapter outlines and lecture notes from the Instructor's Teaching Kit in editable Microsoft® Word documents

PowerPoint® Presentations

The PowerPoint® slide set for *Life* follows the chapter summaries provided in the Instructor's Teaching Kit and can be used directly or customized. Each slide incorporates a figure from *Life*.

PowerPoint® Tutorials
QuickTime™ movies demonstrate how to use PowerPoint®.

Classroom Management

As a service for adopters using WebCT, we will provide a fully-loaded WebCourselet, including the instructor and student resources for this text. The files can then be customized to fit your specific course needs, or can be used as is. Course outlines, pre-built quizzes, activities, and a whole array of materials are included, eliminating hours of work for instructors interested in creating WebCT courses. For more information and a demo of the WebCourselet for this text, please visit our Web Site (http://bfwpub.com/mediaroom/Index.html) and click "WebCT".

Overhead Transparencies

The transparency set includes all four-color line art and tables from the text (more than 1000 images) in a convenient three-ring binder. Balloon captions (and some labels) are deleted to enhance projection and allow for classroom quizzing. Labels and images have been resized for maximum readability.

Slide Set

The slide set includes selected four-color figures from the text. Labels and images have been resized for maximum readability.

Laboratory Manuals

Biology in the Laboratory, Third Edition

Doris Helms, Robert Kosinski, and John Cummings, *all of Clemson University*

The revised edition of this popular lab manual, which includes a CD-ROM, is available to accompany the Sixth Edition of *Life*.
Order ISBN 0-7167-3146-0

Laboratory Outlines in Biology VI

Peter Abramoff and Robert G. Thomson, *Marquette University*
Order ISBN 0-7167-2633-5

The following manuals are available in a bound volume or as separates:
Anatomy and Dissection of the Rat, Third Edition
Warren F. Walker, Jr., *Oberlin College*, and Dominique Homberger, *Louisiana State University*
Order ISBN 0-7167-2635-1
Anatomy and Dissection of the Fetal Pig, Fifth Edition
Warren F. Walker, Jr., *Oberlin College*, and Dominique Homberger, *Louisiana State University*
Order ISBN 0-7167-2637-8
Anatomy and Dissection of the Frog, Second Edition
Warren F. Walker, Jr., *Oberlin College*
Order ISBN 0-7167-2636-X

Custom Labs

Custom Publishing for Laboratory Manuals at www.custompub.whfreeman.com

With this custom publishing option, instructors can build and order customized lab manuals in just minutes, choosing material from Freeman's acclaimed biology laboratory manuals—lab-tested experiments that have been used successfully by hundreds of thousands of students. Instructors determine the manual's content (with the option to incorporate their own material or blank pages), table of contents or index styles, and cover design, and submit the order. A streamlined production process provides a quick turnaround to meet crucial deadlines.

Contents in Brief

Contents

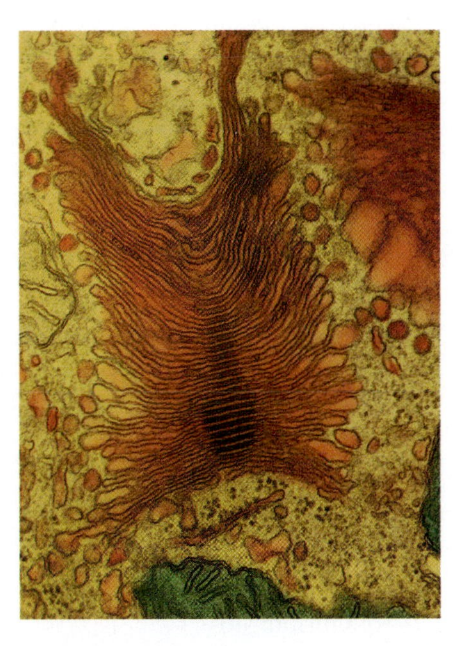

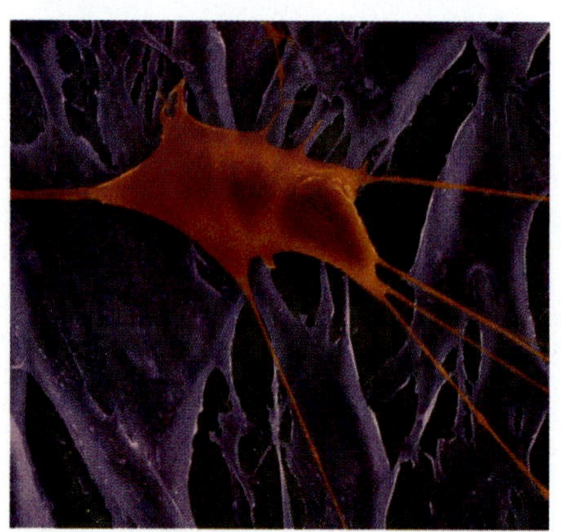

Part Two
INFORMATION AND HEREDITY

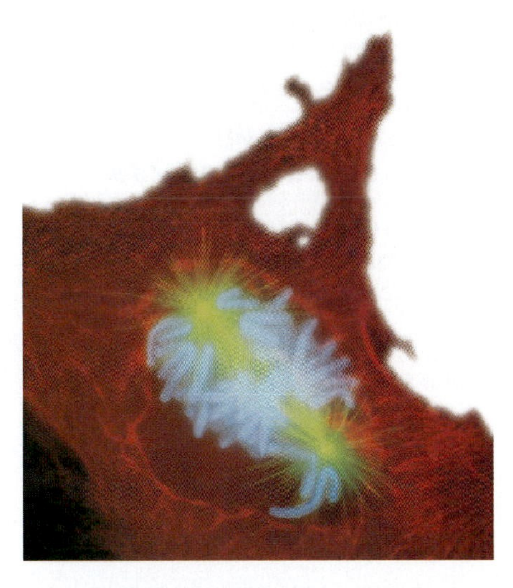

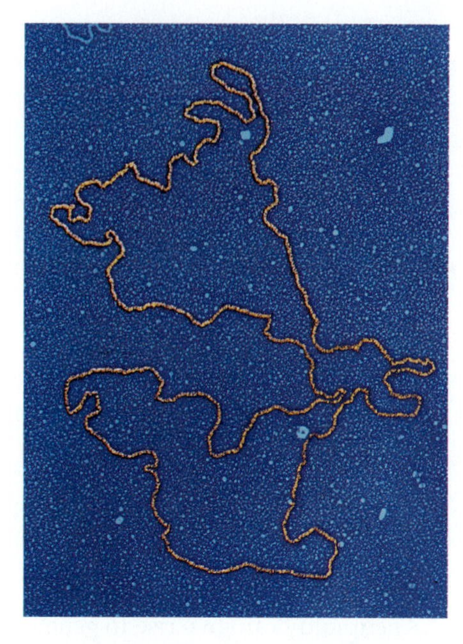

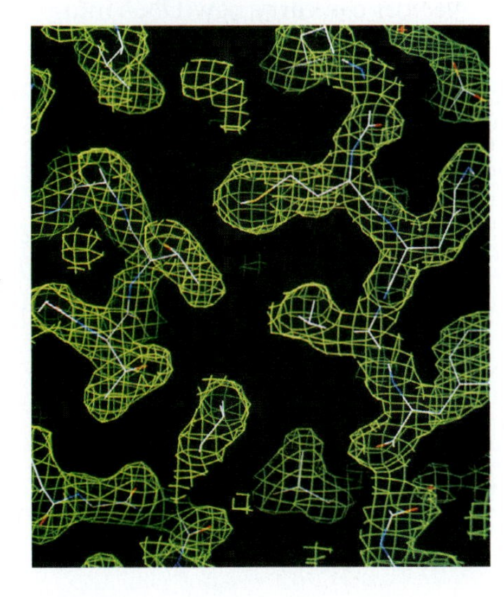

Part Three

EVOLUTIONARY PROCESSES

Part Four
THE EVOLUTION OF DIVERSITY

Part Six
THE BIOLOGY OF ANIMALS

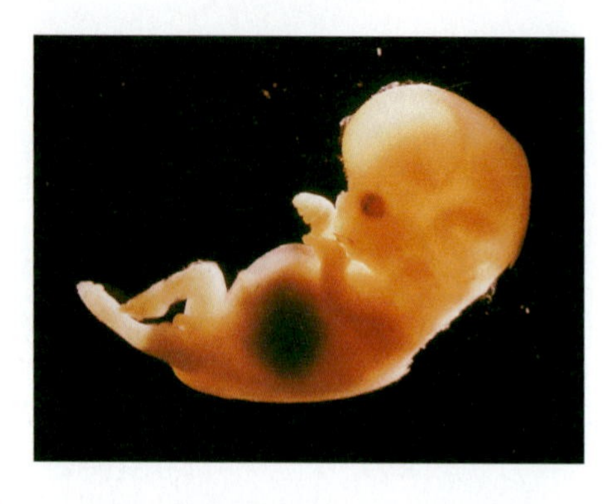

Part Seven
ECOLOGY AND BIOGEOGRAPHY

1 An Evolutionary Framework for Biology

AT MIDNIGHT ON DECEMBER 31, 1999, MASsive displays of fireworks exploded in many places on Earth as people celebrated a new millennium—the passage from one thousand-year time frame into the next—and the advent of the year 2000. One such millennial display took place above the Egyptian pyramids.

We are impressed with the size of the pyramids, how difficult it must have been to build them, and how ancient they are. The oldest of these awe-inspiring monuments to human achievement was built more than 4,000 years ago; in the human experience, this makes the Egyptian pyramids very, very old. Yet from the perspective of the age of Earth and the time over which life has been evolving, the pyramids are extremely young. Indeed, if the history of Earth is visualized as a 30-day month, recorded human history—the dawn of which coincides roughly with the construction of the earliest pyramids—is confined to the last *30 seconds* of the final day of the month (Figure 1.1).

The development of modern biology depended on the recognition that an immense length of time was available for life to arise and evolve its current richness. But for most of human history, people had no reason to suspect that Earth was so old. Until the discovery of radioactive decay at the beginning of the twentieth century, no methods existed to date prehistoric events. By the middle of the nineteenth century, however, studies of rocks and the fossils they contained had convinced geologists that Earth was much older than had generally been believed. Darwin could not have conceived his theory of evolution by natural selection had he not understood that Earth was very ancient.

In this chapter we review the events leading to the acceptance of the fact that life on Earth has evolved over several billion years. We then summarize how evolutionary mechanisms adapt organisms to their environments, and we review the major milestones in the evolution of life on Earth. Finally, we briefly describe how scientists generate new knowledge, how they develop and test hypotheses, and how that knowledge can be used to inform public policy.

A Celebration of Time
One millennial fireworks display celebrating the year 2000 took place over the ancient pyramids of Egypt, structures that represent more than 4,000 years of human history but an infinitesimal portion of Earth's geologic history.

Organisms Have Changed over Billions of Years

Long before the mechanisms of biological evolution were understood, some people realized that organisms had changed over time and that living organisms had evolved from organisms no longer alive on Earth. In the 1760s, the French naturalist Count George-Louis Leclerc de Buffon (1707–1788) wrote his *Natural History of Animals*, which contained a clear statement of the possibility of evolution. Buffon originally believed that each species had been divinely created for a particular way of life, but as he studied animal anatomy, doubts arose. He observed that the limb bones of all mammals, no matter what their way of life, were re-

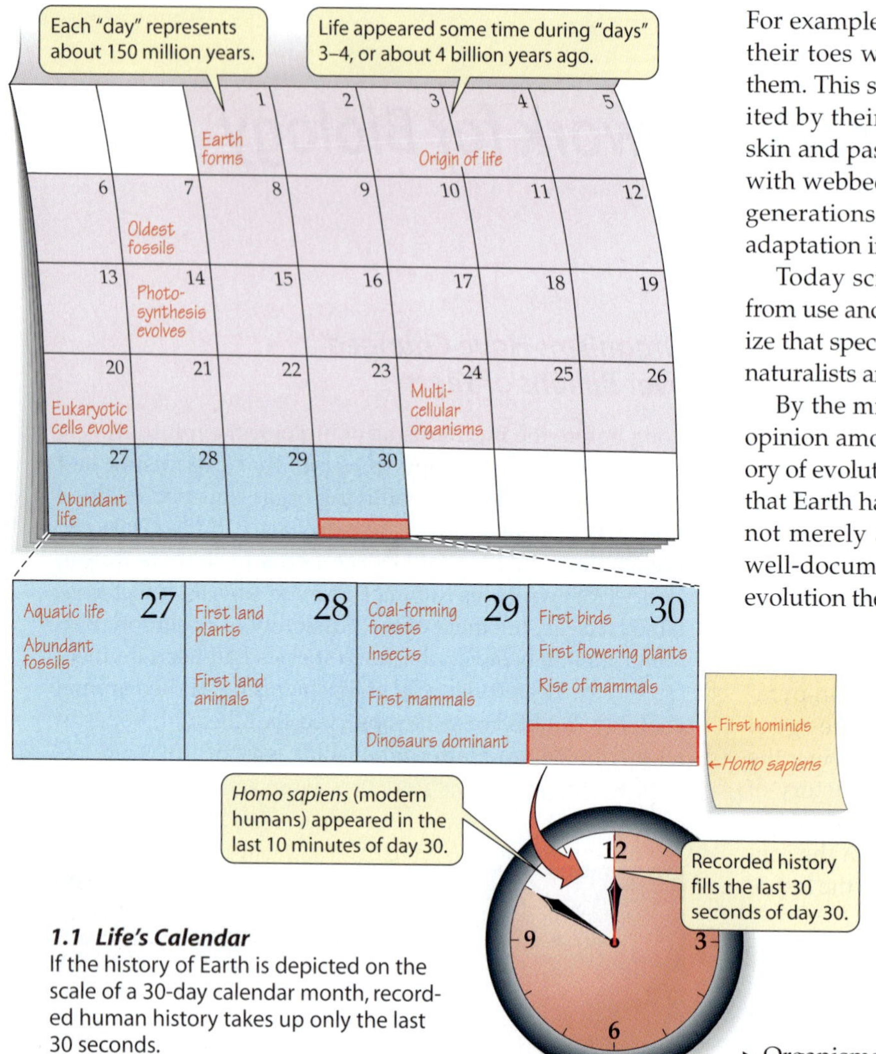

1.1 Life's Calendar
If the history of Earth is depicted on the scale of a 30-day calendar month, recorded human history takes up only the last 30 seconds.

markably similar in many details (Figure 1.2). Buffon also noticed that the legs of certain mammals, such as pigs, have toes that never touch the ground and appear to be of no use. He found it difficult to explain the presence of these seemingly useless small toes by special creation.

Both of these troubling facts could be explained if mammals had not been specially created in their present forms, but had been modified over time from an ancestor that was common to all mammals. Buffon suggested that the limb bones of mammals might all be similar, and that the functionless toes of pigs might be inherited from ancestors with fully formed and functional toes. Buffon's idea was an early statement of evolution (descent with modification), although he did not attempt to explain how such changes took place.

Buffon's student Jean Baptiste de Lamarck (1744–1829) was the first person to propose a mechanism of evolutionary change. Lamarck suggested that lineages of organisms may change gradually over many generations as offspring inherit structures that have become larger and more highly developed as a result of continued use or, conversely, have become smaller and less developed as a result of disuse. For example, Lamarck suggested that aquatic birds extend their toes while swimming, stretching the skin between them. This stretched condition, he thought, could be inherited by their offspring, which would in turn stretch their skin and pass this condition along to their offspring; birds with webbed feet would thereby evolve over a number of generations. Lamarck explained many other examples of adaptation in a similar way.

Today scientists do not believe that changes resulting from use and disuse can be inherited. But Lamarck did realize that species change with time. And after Lamarck, other naturalists and scientists speculated along similar lines.

By the middle of the nineteenth century, the climate of opinion among many scholars was receptive to a new theory of evolutionary processes. By then geologists had shown that Earth had existed and changed over millions of years, not merely a few thousand years. The presentation of a well-documented and thoroughly scientific argument for evolution then triggered a transformation of biology.

The theory of evolution by natural selection was proposed independently by Charles Darwin and Alfred Russel Wallace in 1858. We will discuss evolutionary theory in detail in Chapter 21, but its essential features are easy to understand. The theory rests on two facts and one inference drawn from them. The two facts are:

- The reproductive rates of all organisms, even slowly reproducing ones, are sufficiently high that populations would quickly become enormous if mortality rates did not balance reproductive rates.
- Organisms of all types are variable, and offspring are similar to their parents because they inherit their features from them.

The inference is:

- The differences among individuals influence how well those individuals survive and reproduce. Traits that increase the probability that their bearers will survive and reproduce are more likely to be passed on to their offspring and to their offspring's offspring.

Darwin called the differential survival and reproductive success of individuals **natural selection**. The remarkable features of all organisms have evolved under the influence of natural selection. Indeed, *the ability to evolve by means of natural selection clearly separates life from nonlife.*

Biology began a major conceptual shift a little more than a century ago with the general acceptance of long-term evolutionary change and the recognition that differential survival and reproductive success is the primary process that adapts organisms to their environments. The shift has taken a long time because it required abandoning many components of an earlier worldview. The pre-Darwinian view held that the world was young, and that organisms had been created in their current forms. In the Darwinian view,

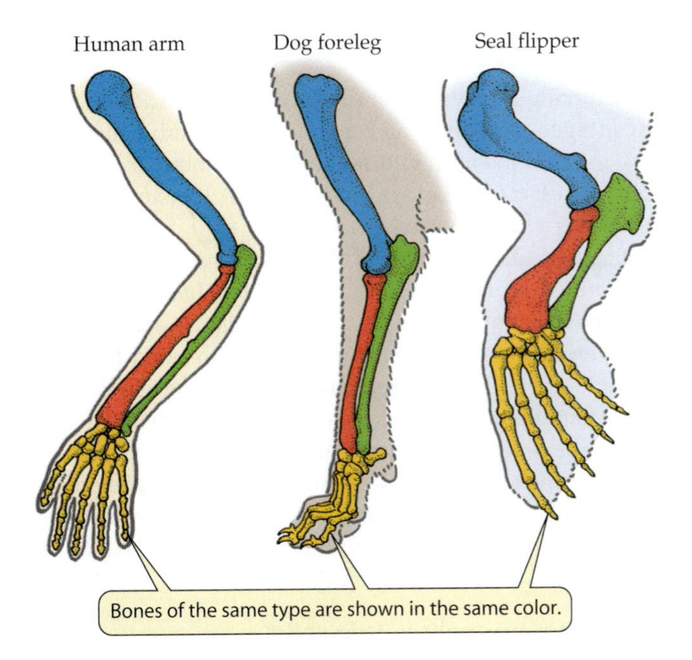

1.2 Mammals Have Similar Limbs
Mammalian forelimbs have different purposes, but the number and types of their bones are similar, indicating that they have been modified over time from a common ancestor.

the world is ancient, and both Earth and its inhabitants have been continually changing. In the Darwinian view of the world, organisms evolved their particular features because individuals with those features survived and reproduced better than individuals with different features.

Adopting this new view of the world means accepting not only the processes of evolution, but also the view that the living world is constantly evolving, and that evolutionary change occurs without any "goals." The idea that evolution is not directed toward a final goal or state has been more difficult for many people to accept than the process of evolution itself. But even though evolution has no goals, evolutionary processes have resulted in a series of profound changes—milestones—over the nearly 4 billion years life has existed on Earth.

Evolutionary Milestones

The following overview of the major milestones in the evolution of life provides both a framework for presenting the characteristics of life that will be described in this book and an overview of how those characteristics evolved during the history of life on Earth.

Life arises from nonlife

All matter, living and nonliving, is made up of chemicals. The smallest chemical units are atoms, which bond together into molecules; the properties of those molecules are the subject of Chapter 2. The processes leading to life began nearly 4 billion years ago with interactions among small molecules that stored useful information.

The information stored in these simple molecules eventually resulted in the synthesis of larger molecules with complex but relatively stable shapes. Because they were both complex and stable, these units could participate in increasing numbers and kinds of chemical reactions. Some of these large molecules—carbohydrates, lipids, proteins, and nucleic acids—are found in all living systems and perform similar functions. The properties of these complex molecules are the subject of Chapter 3.

Cells form from molecules

About 3.8 billion years ago, interacting systems of molecules came to be enclosed in compartments surrounded by membranes. Within these membrane-enclosed units, or **cells**, control was exerted over the entrance, retention, and exit of molecules, as well as the chemical reactions taking place within the cell. Cells and membranes are the subjects of Chapters 4 and 5.

Cells are so effective at capturing energy and replicating themselves—two fundamental characteristics of life—that since the time they evolved, they have been the unit on which all life has been built. Experiments by the French chemist and microbiologist Louis Pasteur and others during the nineteenth century convinced most scientists that, under present conditions on Earth, cells do not arise from noncellular material, but must come from other cells.

For 2 billion years, cells were tiny packages of molecules each enclosed in a single membrane. These **prokaryotic cells** lived autonomous lives, each separate from the other. They were confined to the oceans, where they were shielded from lethal ultraviolet sunlight. Some prokaryotes living today may be similar to these early cells (Figure 1.3).

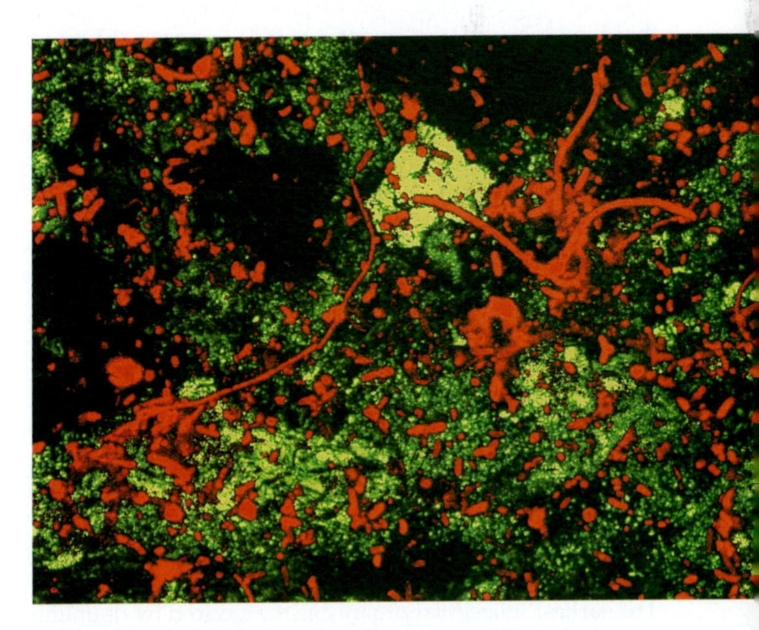

1.3 Early Life May Have Resembled These Cells
"Rock-eating" bacteria, appearing red in this artificially colored micrograph, were discovered in pools of water trapped between layers of rock more than 1,000 meters below Earth's surface. Deriving chemical nutrients from the rocks and living in an environment devoid of oxygen, they may resemble some of the earliest prokaryotic cells.

To maintain themselves, to grow, and to reproduce, these early prokaryotes, like all cells that have subsequently evolved, obtained raw materials and energy from their environment, using these as building blocks to synthesize larger, carbon-containing molecules. The energy contained in these large molecules powered the chemical reactions necessary for the life of the cell. These conversions of matter and energy are called **metabolism**.

All organisms can be viewed as devices to capture, process, and convert matter and energy from one form to another; these conversions are the subjects of Chapters 6 and 7. *A major theme in the evolution of life is the development of increasingly diverse ways of capturing external energy and using it to drive biologically useful reactions.*

Photosynthesis changes Earth's environment

About 2.5 billion years ago, some organisms evolved the ability to use the energy of sunlight to power their metabolism. Although they still took raw materials from the environment, the energy they used to metabolize these materials came directly from the sun. Early photosynthetic cells were probably similar to present-day prokaryotes called cyanobacteria (Figure 1.4). The energy-capturing process they used—**photosynthesis**—is the basis of nearly all life on Earth today; it is explained in detail in Chapter 8. It used new metabolic reactions that exploited an abundant source of energy (sunlight), and generated a new waste product (oxygen) that radically changed Earth's atmosphere.

The ability to perform photosynthetic reactions probably accumulated gradually during the first billion years or so of evolution, but once this ability had evolved, its effects were dramatic. Photosynthetic prokaryotes became so abundant that they released vast quantities of oxygen gas (O_2) into the atmosphere. The presence of oxygen opened up new avenues of evolution. Metabolic reactions that use O_2, called **aerobic metabolism**, came to be used by most organisms on Earth. The oxygen in the air we breathe today would not exist without photosynthesis.

Over a much longer time, the vast quantities of oxygen liberated by photosynthesis had another effect. Formed from O_2, ozone (O_3) began to accumulate in the upper atmosphere. The ozone slowly formed a dense layer that acted as a shield, intercepting much of the sun's deadly ultraviolet radiation. Eventually (although only within the last 800 million years of evolution), the presence of this shield allowed organisms to leave the protection of the oceans and establish new lifestyles on Earth's land surfaces.

Sex enhances adaptation

The earliest unicellular organisms reproduced by doubling their hereditary (genetic) material and then dividing it into two new cells, a process known as mitosis. The resulting progeny cells were identical to each other and to the parent. That is, they were clones. But **sexual reproduction**—the combining of genes from two cells in one cell—appeared early during the evolution of life. Sexual reproduction is advantageous because an organism that combines its genetic information with information from another individual produces offspring that are more variable. *Reproduction with variation is a major characteristic of life.*

Variation allows organisms to adapt to a changing environment. **Adaptation** to environmental change is one of life's most distinctive features. An organism is adapted to a given environment when it possesses inherited features that enhance its survival and ability to reproduce in that environment. Because environments are constantly changing, organisms that produce variable offspring have an advantage over those that produce genetically identical "clones," because they are more likely to produce some offspring better adapted to the environment in which they find themselves.

Eukaryotes are "cells within cells"

As the ages passed, some prokaryotic cells became large enough to attack, engulf, and digest smaller cells, becoming the first predators. Usually the smaller cells were destroyed within the predators' cells. But some of these smaller cells survived and became permanently integrated into the operation of their hosts' cells. In this manner, cells with complex internal compartments arose. We call these cells **eukaryotic cells**. Their appearance slightly more than 1.5 billion years ago opened more new evolutionary opportunities.

Prokaryotic cells—the Bacteria and Archaea—have no membrane-enclosed compartments. Eukaryotic cells, on the

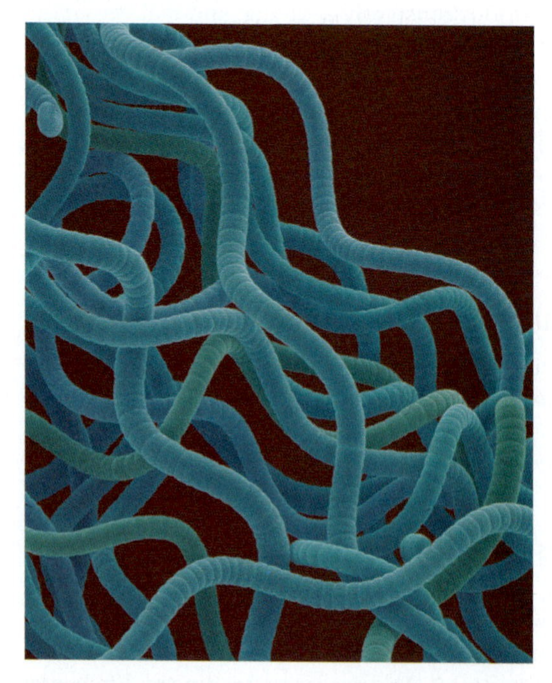

1.4 Oxygen Produced by Prokaryotes Changed Earth's Atmosphere
These modern cyanobacteria are probably very similar to early photosynthetic prokaryotes.

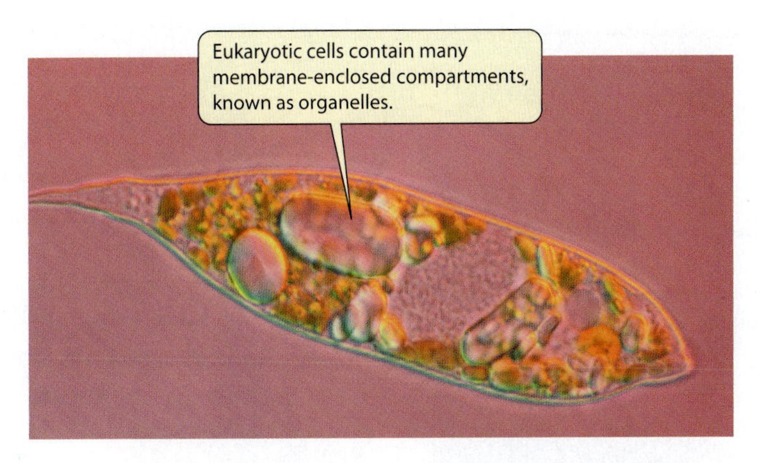

1.5 Multiple Compartments Characterize Eukaryotic Cells The nucleus and other specialized organelles probably evolved from small prokaryotes that were ingested by a larger prokaryotic cell. This is a photograph of a single-celled eukaryotic organism known as a protist.

other hand, are filled with membrane-enclosed compartments. In eukaryotic cells, genetic material—genes and chromosomes—became contained within a discrete nucleus and became increasingly complex. Other compartments became specialized for other purposes, such as photosynthesis. We refer to these specialized compartments as **organelles** (Figure 1.5).

Multicellularity permits specialization of cells

Until slightly more than 1 billion years ago, only single-celled organisms existed. Two key developments made the evolution of multicellular organisms—organisms consisting of more than one cell—possible. One was the ability of a cell to change its structure and functioning to meet the challenges of a changing environment. This was accomplished when prokaryotes evolved the ability to change from rapidly growing cells into resting cells called **spores** that could survive harsh environmental conditions. The second development allowed cells to stick together in a "clump" after they divided, forming a multicellular organism.

Once organisms could be composed of many cells, it became possible for the cells to specialize. Certain cells, for example, could be specialized to perform photosynthesis. Other cells might become specialized to transport chemical materials such as oxygen from one part of an organism to another. Very early in the evolution of multicellular life, certain cells began to be specialized for sex—the passage of new genetic information from one generation to the next.

With the presence of specialized sex cells, genetic transmission became more complicated. Simple nuclear division—mitosis—was and is sufficient for the needs of most cells. But among the sex cells, or gametes, a whole new method of nuclear division—meiosis—evolved. Meiosis allows gametes to combine and rearrange the genetic information from two distinct parent organisms into a genetic package that contains elements of both parent cells but is different from either. The recombinational possibilities generated by meiosis had great impact on variability and adaptation and on the speed at which evolution could occur.

Mitosis and meiosis are covered in detail in Chapter 9.

Controlling internal environments becomes more complicated

The pace of evolution, quickened by the emergence of sex and multicellular life, was also heightened by changes in Earth's atmosphere that allowed life to move out of the oceans and exploit environments on land. Photosynthetic green plants colonized the land, providing a rich source of energy for a vast array of organisms that consumed them. But whether it is made up of one cell or many, an organism must respond appropriately to its external environment. Life on land presented a new set of environmental challenges.

In any environment, external conditions can change rapidly and unpredictably in ways that are beyond an organism's control. An organism can remain healthy only if its internal environment remains within a given range of physical and chemical conditions. Organisms maintain relatively constant internal environments by making metabolic adjustments to changes in external and internal conditions such as temperature, the presence or absence of sunlight, the presence or absence of specific chemicals, the need for nutrients (food) and water, or the presence of foreign agents inside their bodies. Maintenance of a relatively stable internal condition—such as a constant human body temperature despite variation in the temperature of the surrounding environment—is called **homeostasis**. *A major theme in the evolution of life is the development of increasingly complicated systems for maintaining homeostasis.*

Multicellular organisms undergo regulated growth

Multicellular organisms cannot achieve their adult shapes or function effectively unless their growth is carefully regulated. Uncontrolled growth—one example of which is cancer—ultimately destroys life. *A vital characteristic of living organisms is regulated growth.* Achieving a functional multicellular organism requires a sequence of events leading from a single cell to a multicellular adult. This process is called **development**.

The adjustments that organisms make to maintain constant internal conditions are usually minor; they are not obvious, because nothing appears to change. However, at some time during their lives, many organisms respond to changing conditions not by maintaining their status, but by undergoing major cellular and molecular reorganization. An early form of such developmental reorganization was the prokaryotic spores that were generated in response to environmental stresses. A striking example that evolved much later is **metamorphosis**, seen in many modern in-

1.6 Organisms May Change Dramatically During Their Lives
The caterpillar, pupa, and adult are all stages in the life cycle of a monarch butterfly. The transition from one stage to another is triggered by internal signals.

sects, such as butterflies. In response to internal chemical signals, a caterpillar changes into a pupa and then into an adult butterfly (Figure 1.6).

The activation of gene-based information within cells and the exchange of signal information among cells produce the well-timed events that are required for the transition to the adult form. Genes control the metabolic processes necessary for life. The nature of the genetic material that controls these lifelong events has been understood only within the twentieth century; it is the story to which much of Part Two of this book is devoted.

Altering the timing of development can produce striking changes. Just a few genes can control processes that result in dramatically different adult organisms. Chimpanzees and humans share more than 98 percent of their genes, but the differences between the two in form and in behavioral abilities—most notably speech—are dramatic (Figure 1.7). When we realize how little information it sometimes takes to create major transformations, the still mysterious process of **speciation** becomes a little less of a mystery.

Speciation produces the diversity of life

All organisms on Earth today are the descendants of a kind of unicellular organism that lived almost 4 billion years ago. The preceding pages described the major evolutionary events that have led to more complex living organisms. The course of this evolution has been accompanied by the storage of larger and larger quantities of information and increasingly complex mechanisms for using it. But if that were the entire story, only one kind of organism might exist on Earth today. Instead, Earth is populated by many millions of kinds of organisms that do not interbreed with one another. We call these genetically independent groups of organisms **species**.

As long as individuals within a population mate at random and reproduce, structural and functional changes may occur, but only one species will exist. However, if a population becomes divided into two or more groups, and individuals can mate only with individuals in their own group, differences may accumulate with time, and the groups may evolve into different species.

The splitting of groups of organisms into separate species has resulted in the great variety of life found on Earth today, as described in Chapter 20. How species form is explained in Chapter 22. From a single ancestor, many species may arise as a result of the repeated splitting of populations. How biologists determine which species have descended from a particular ancestor is discussed in Chapter 23.

1.7 Genetically Similar Yet Very Different
By looking at the two, you might be surprised to learn that chimpanzees and humans share more than 98 percent of their genes.

1.8 Adaptations to the Environment
(*a*) The long, pointed wings of the peregrine falcon allow it to accelerate rapidly as it dives on its prey. (*b*) The action of a hummingbird's wings allows it to hover in front of a flower while it extracts nectar. (*c*) In a water-limited environment, this saguaro cactus stores water in its fleshy trunk. Its roots spread broadly to extract water immediately after it rains. (*d*) The aboveground root system of mangroves is an adaptation that allows these plants to thrive while inundated by salt water—an environment that would kill most terrestrial plants.

Sometimes humans refer to species as "primitive" or "advanced." These and similar terms, such as "lower" and "higher," are best avoided because they imply that some organisms function better than others. In this book, we use the terms "ancestral" and "derived" to distinguish characteristics that appeared earlier from those that appeared later in the evolution of life.

It is important to recognize that *all* living organisms are successfully adapted to their environments. The wings that allow a bird to fly and the structures that allow green plants to survive in environments where water is either scarce or overabundant are examples of the rich array of adaptations found among organisms (Figure 1.8).

The Hierarchy of Life

Biologists study life in two complementary ways:

- They study structures and processes ranging from the simple to the complex and from the small to the large.
- They study the patterns of life's evolution over billions of years to determine how evolutionary processes have resulted in lineages of organisms that can be traced back to recent and distant ancestors.

These two themes of biological investigation help us synthesize the hierarchical relationships among organisms and the role of these relationships in space and time. We first describe the hierarchy of interactions among the units of biology from the smallest to the largest—from cells to the biosphere. Then we turn to the hierarchy of evolutionary relationships among organisms.

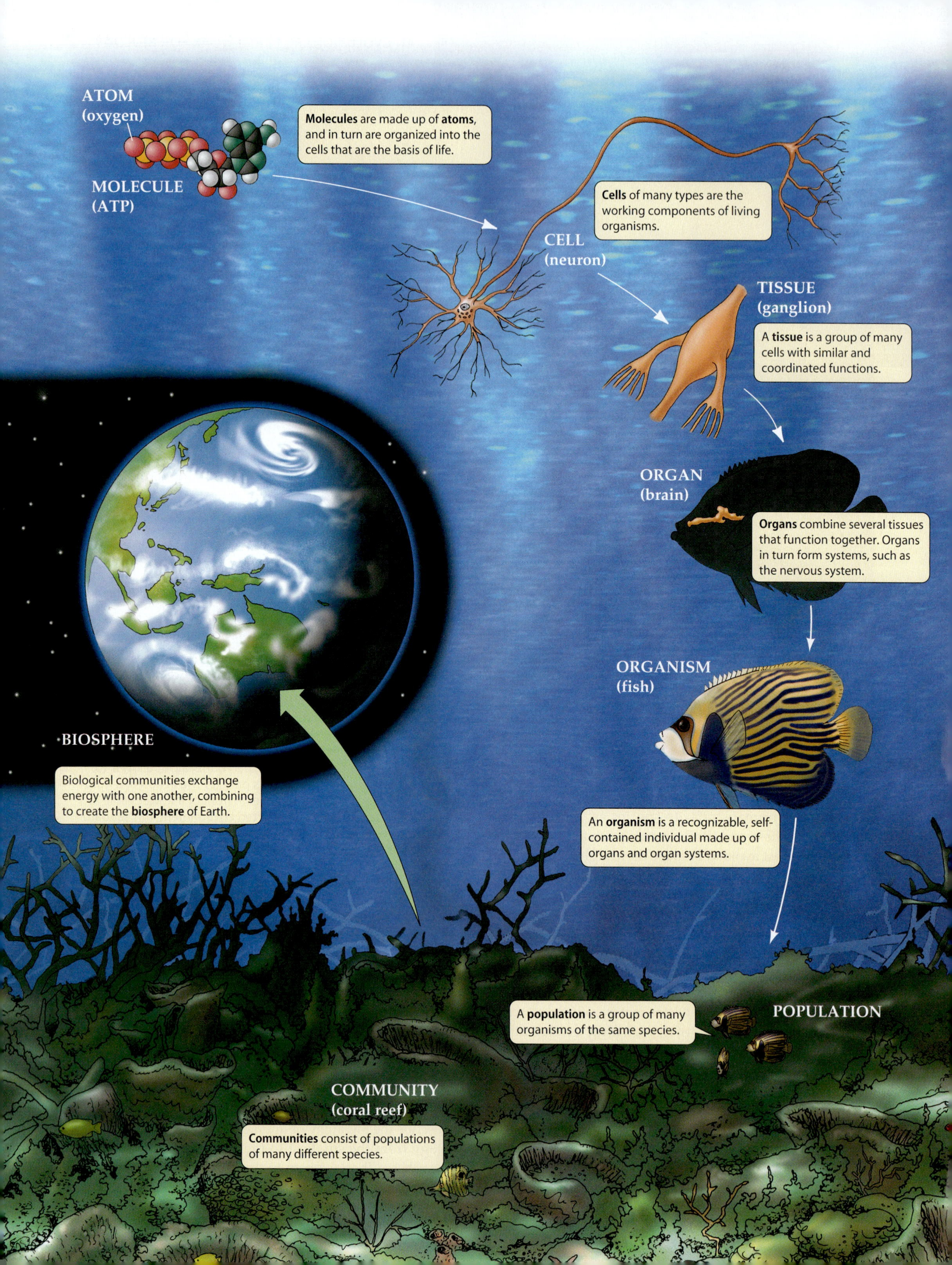

ATOM
(oxygen)
MOLECULE
(ATP)
Molecules are made up of **atoms**, and in turn are organized into the cells that are the basis of life.
Cells of many types are the working components of living organisms.
CELL
(neuron)
TISSUE
(ganglion)
A **tissue** is a group of many cells with similar and coordinated functions.
ORGAN
(brain)
Organs combine several tissues that function together. Organs in turn form systems, such as the nervous system.
ORGANISM
(fish)
An **organism** is a recognizable, self-contained individual made up of organs and organ systems.
BIOSPHERE
Biological communities exchange energy with one another, combining to create the **biosphere** of Earth.
A **population** is a group of many organisms of the same species.
POPULATION
COMMUNITY
(coral reef)
Communities consist of populations of many different species.

1.9 The Hierarchy of Life
The individual organism is the central unit of study in biology, but understanding it requires a knowledge of many levels of biological organization both above and below it. At each higher level, additional and more complex properties and functions emerge.

Biologists study life at different levels

Biology can be visualized as a hierarchy in which the units, from the smallest to the largest, include atoms, molecules, cells, tissues, organs, organisms, populations, and communities (Figure 1.9).

The organism is the central unit of study in biology. Parts Five and Six of this book discuss organismal biology in detail. But to understand organisms, biologists must study life at all its levels of organization. Biologists study molecules, chemical reactions, and cells to understand the operations of tissues and organs. They study organs and organ systems to determine how organisms function and maintain internal homeostasis. At higher levels in the hierarchy, biologists study how organisms interact with one another to form social systems, populations, ecological communities, and biomes, which are the subjects of Part Seven of this book.

Each level of biological organization has properties, called **emergent properties**, that are not found at lower levels. For example, cells and multicellular organisms have characteristics and carry out processes that are not found in the molecules of which they are composed.

Emergent properties arise in two ways. First, many *emergent properties of systems result from interactions among their parts*. For example, at the organismal level, developmental interactions of cells result in a multicellular organism whose adult features are vastly richer than those of the single cell from which it grew. Other examples of properties that emerge through complex interactions are memory and emotions. In the human brain, these properties result from interactions among the brain's 10^{12} (trillion) cells with their 10^{15} (quadrillion) connections. No single cell, or even small group of cells, possesses them.

Second, *emergent properties arise because aggregations have collective properties* that their individual units lack. For example, individuals are born and they die; they have a life span. An individual does not have a birth rate or a death rate, but a population (composed of many individuals) does. Birth and death rates are emergent properties of a population. Evolution is an emergent property of populations that depends on variation in birth and death rates, which emerges from the different life spans and reproductive success of individuals in the various populations.

Emergent properties do not violate the principles that operate at lower levels of organization. However, emergent properties usually cannot be detected, predicted, or even suspected by studying lower levels. Biologists could never discover the existence of human emotions by studying single nerve cells, even though they may eventually be able to explain it in terms of interactions among many nerve cells.

Biological diversity is organized hierarchically

As many as 30 million species of organisms inhabit Earth today. Many times that number lived in the past but are now extinct. If we go back four billion years, to the origin of life, all organisms are believed to be descended from a single *common ancestor*. The concept of a common ancestor is crucial to modern methods of classifying organisms. *Organisms are grouped in ways that attempt to define their evolutionary relationships, or how recently the different members of the group shared a common ancestor.*

To determine evolutionary relationships, biologists assemble facts from a variety of sources. Fossils tell us where and when ancestral organisms lived and what they looked like. The physical structures different organisms share—toes among mammals, for example—can be an indication of how closely related they are. But a modern "revolution" in classification has emerged because technologies developed in the past 30 years now allow us to compare the genomes of organisms: We can actually determine how many genes different species share. The more genes species have in common, the more recently they probably shared a common ancestor.

Because no fossil evidence for the earliest forms of life remains, the decision to divide all living organisms into three major **domains**—the deepest divisions in the evolutionary history of life—is based primarily on molecular evidence (Figure 1.10). Although new evidence is constantly being brought to light, it seems clear that organisms belonging to a particular domain have been evolving separately from organisms in the other two domains for more than a billion years.

Organisms in the domains **Archaea** and **Bacteria** are prokaryotes—single cells that lack a nucleus and the other internal compartments found in the Eukarya. Archaea and Bacteria differ so fundamentally from each other in the chemical reactions by which they function and in the products they produce that they are believed to have separated into distinct evolutionary lineages very early during the evolution of life. These domains are covered in Chapter 26.

Members of the third domain have eukaryotic cells containing nuclei and complex cellular compartments called organelles. The **Eukarya** are divided into four groups—the protists and the classical kingdoms Plantae, Fungi, and Animalia (see Figure 1.10). Protists, the subject of Chapter 27, are mostly single-celled organisms. The remaining three kingdoms, whose members are all multicellular, are believed to have arisen from ancestral protists.

Some bacteria, some protists, and most members of the kingdom Plantae (plants) convert light energy to chemical energy by photosynthesis. The biological molecules that they produce are the primary food for nearly all other living organisms. The Plantae are covered in Chapters 28 and 29.

The Fungi, the subject of Chapter 30, include molds, mushrooms, yeasts, and other similar organisms, all of

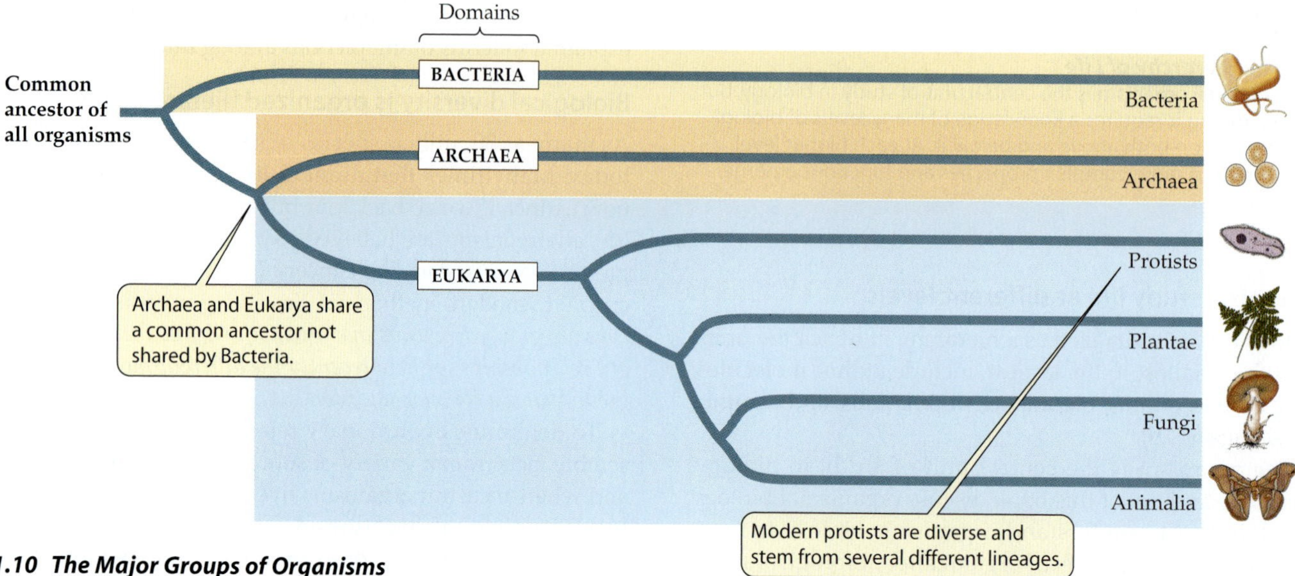

1.10 The Major Groups of Organisms
The classification system used in this book divides Earth's organisms into three domains. The domain Eukarya contains numerous groups of unicellular and multicellular organisms. This "tree" diagram gives information on evolutionary relationships among the groups, as described in Chapter 23.

which are **heterotrophs**: They require a food source of energy-rich molecules synthesized by other organisms. Fungi absorb food substances from their surroundings and break them down (digest them) within their cells. They are important as decomposers of the dead bodies of other organisms.

Members of the kingdom Animalia (animals) are also heterotrophs. These organisms ingest their food source, digest the food outside their cells, and then absorb the products. Animals get their raw materials and energy by eating other forms of life. Perhaps because we are animals ourselves, we are often drawn to study members of this kingdom, which is covered in Chapters 31, 32, and 33.

The biological classification system used today has many hierarchical levels in addition to the ones shown in Figure 1.10. We will discuss the principal levels in Chapter 23. But to understand some of the terms we will use in the intervening chapters, you need to know that each species of organism is identified by two names. The first identifies the **genus**—a group of species that share a recent common ancestor—of which the species is a member. The second name is the species name. To avoid confusion, a particular combination of two names is assigned to only a single species. For example, the scientific name of the modern human species is *Homo sapiens*.

Asking and Answering "How?" and "Why?"

Because biology is an evolutionary science, biological processes and products can be viewed from two different but complementary perspectives. Biologists ask, and try to answer, functional questions: How does it work? They also ask, and try to answer, adaptive questions: Why has it evolved to work that way?

Suppose, for example, that some marine biologists walking on mudflats in the Bay of Fundy, Nova Scotia, Canada, observe many amphipods (tiny relatives of shrimps and lobsters) crawling on the surface of the mud (Figure 1.11). Two obvious questions they might ask are

- *How* do these animals crawl?
- *Why* do they crawl?

To answer the "how" question, the scientists would investigate the molecular mechanisms underlying muscular contraction, nerve and muscle interactions, and the receipt of stimuli by the amphipods' brains. To answer the "why" question, they would attempt to determine why crawling on the mud is adaptive—that is, why it improves the survival and reproductive success of amphipods.

Is either of these two types of questions more basic or important than the other? Is any one of the answers more fundamental or more important than the other? Not really. The richness of possible answers to apparently simple questions makes biology a complex field, but also an exciting one. Whether we're talking about molecules bonding, cells dividing, blood flowing, amphipods crawling, or forests growing, we are constantly posing both how and why questions. To answer these questions, scientists generate hypotheses that can be tested.

Hypothesis testing guides scientific research

The most important motivator of most biologists is curiosity. People are fascinated by the richness and diversity of life, and they want to learn more about organisms and how they function and interact with one another. Curiosity is probably an adaptive trait. Humans who were motivated to learn about their surroundings are likely to have survived and reproduced better, on average, than their less curious relatives. We hope this book will help you share in the ex-

1.11 An Amphipod from the Mud Flats
Scientists studied this tiny crustacean (whose actual size of approximately 1 centimeter is shown by the scale bar) in an attempt to see whether its behavior changes when it is infected by a parasitic worm. The female of this amphipod species is at the top; the lower specimen is a male.

1.12 Sandpipers Feed on Amphipods
Migrating sandpipers crowd the exposed tidal flats in search of food. By consuming infected amphipods, the sandpipers also become infected, serving as hosts and allowing the parasitic worm to complete its life cycle.

citement biologists feel as they develop and test hypotheses. There are vast numbers of how and why questions for which we do not have answers, and new discoveries usually engender questions no one thought to ask before. Perhaps *your* curiosity will lead to an important new idea.

Underlying all scientific research is the **hypothetico-deductive (H-D) approach** by which scientists ask questions and test answers. The H-D approach allows scientists to modify and correct their beliefs as new observations and information become available. The method has five stages:

- Making observations.
- Asking questions.
- Forming **hypotheses,** or tentative answers to the questions.
- Making predictions based on the hypotheses.
- Testing the predictions by making additional observations or conducting experiments.

The data gained may support or contradict the predictions being tested. If the data support the hypothesis, it is subjected to still more predictions and tests. If they continue to support it, confidence in its correctness increases, and the hypothesis comes to be considered a **theory**. If the data do not support the hypothesis, it is abandoned or modified in accordance with the new information. Then new predictions are made, and more tests are conducted.

Applying the hypothetico-deductive method

The way in which marine biologists answered the question "Why do amphipods crawl on the surface of the mud rather than staying hidden within?" illustrates the H-D approach. As we saw above, the biologists observed something occurring in nature and formulated a question about it. To begin answering the question, they assembled available information on amphipods and the species that eat them.

They learned that during July and August of each year, thousands of sandpipers assemble for four to six weeks on the mudflats of the Bay of Fundy, during their southward migration from their Arctic breeding grounds to their wintering areas in South America (Figure 1.12). On these mudflats, which are exposed twice daily by the tides, they feed vigorously, putting on fat to fuel their next long flight. Amphipods living in the mud form about 85 percent of the diet of the sandpipers. Each bird may consume as many as 20,000 amphipods per day!

Previous observations had shown that a nematode (roundworm) parasitizes both the amphipods and the sandpipers. To complete its life cycle, the nematode must develop within both a sandpiper and an amphipod. The nematodes mature within the sandpipers' digestive tracts, mate, and release their eggs into the environment in the birds' feces. Small larvae hatch from the eggs and search for, find, and enter amphipods, where they grow through several larval stages. Sandpipers are reinfected when they eat parasitized amphipods.

GENERATING A HYPOTHESIS AND PREDICTIONS. Based on the available information, biologists generated the following hypothesis: *Nematodes alter the behavior of their amphipod hosts in a way that increases the chance that the worms will be*

1.13 Collecting Field Data
Amphipods are collected from the mud to be tested for infection by parasites. Some of these crustaceans will be used in laboratory experiments.

passed on to sandpiper hosts. From this general hypothesis they generated two specific predictions.

- First, they predicted that amphipods infected by nematodes would increase their activity on the surface of the mud during daylight hours, when the sandpipers hunted by sight, but not at night, when the sandpipers fed less and captured prey by probing into the mud.
- Second, they predicted that only amphipods with late-stage nematode larvae—the only stage that can infect sandpipers—would have their behavior manipulated by the nematodes.

For each hypothesis proposing an effect, there is a corresponding **null hypothesis**, which asserts that the proposed effect is absent. For the hypothesis we have just stated, the null hypothesis is that nematodes have no influence on the behavior of their amphipod hosts. The alternative predictions that would support the null hypothesis are (1) that infected amphipods show no increase their activity either during the day or at night and (2) that all larval stages affect their hosts in the same manner. It is important in hypothesis testing to generate and test as many alternate hypotheses and predictions as possible.

TESTING PREDICTIONS. Investigators collected amphipods in the field, taking them from the surface and from within the mud, during the day and at night (Figure 1.13). They found that during the day, amphipods crawling on the surface were much more likely to be infected with nematodes than were amphipods collected from within the mud. At night, however, there was no difference between the proportion of infected amphipods on the surface and those burrowing within the mud. This evidence supported the first prediction.

The field collections also showed that a higher proportion of the amphipods collected on the surface than of those collected from within the mud were parasitized by late-stage nematode larvae. However, amphipods crawling on the surface were no more likely to be infected by early-stage nematode larvae than were amphipods collected from the mud. These findings supported the *second* prediction.

To test the prediction that nematode larvae are more likely to affect amphipod behavior once they become infective, biologists performed laboratory experiments. They artificially infected amphipods with nematode eggs they obtained from sandpipers collected in the field. The infected amphipods established themselves in mud in laboratory containers.

By examining infected amphipods, investigators determined that it took about 13 days for the nematode larvae to reach the late, infective stage. By monitoring the behavior of the amphipods in the test tubes, the researchers determined that the amphipods were more likely to expose themselves on the surface of the mud once the parasites had reached the infective stage (Figure 1.14). This finding supported the second prediction.

Thus a combination of field and laboratory experiments, observation, and prior knowledge all supported the hypothesis that nematodes manipulate the behavior of their amphipod hosts in a way that decreases the survival of the amphipods, but increases the survival of the nematodes.

As is common practice in all the sciences, the researchers gathered all their data and collected them in a report, which they submitted to a scientific journal. Once such a report is published,* other scientists can evaluate the data, make their own observations, and formulate new ideas and experiments.

Experiments are powerful tools

The key feature of **experimentation** is the control of most factors so that the influence of a single factor can be seen clearly. In the laboratory experiments with amphipods, all individuals were raised under the same conditions. As a result, the nematodes reached the infective stage at about the same time in all of the infected amphipods.

Both laboratory and field experiments have their strengths and weaknesses. The advantage of working in a laboratory is that control of environmental factors is more

*In the case illustrated here, the data on amphipod behavior were published in the journal *Behavioral Ecology*, Volume 10, Number 4 (1998). D. McCurdy et al., "Evidence that the parasitic nematode *Skrjabinoclava* manipulates host *Corephium* behavior to increase transmission to the sandpiper, *Calidris pusilla*."

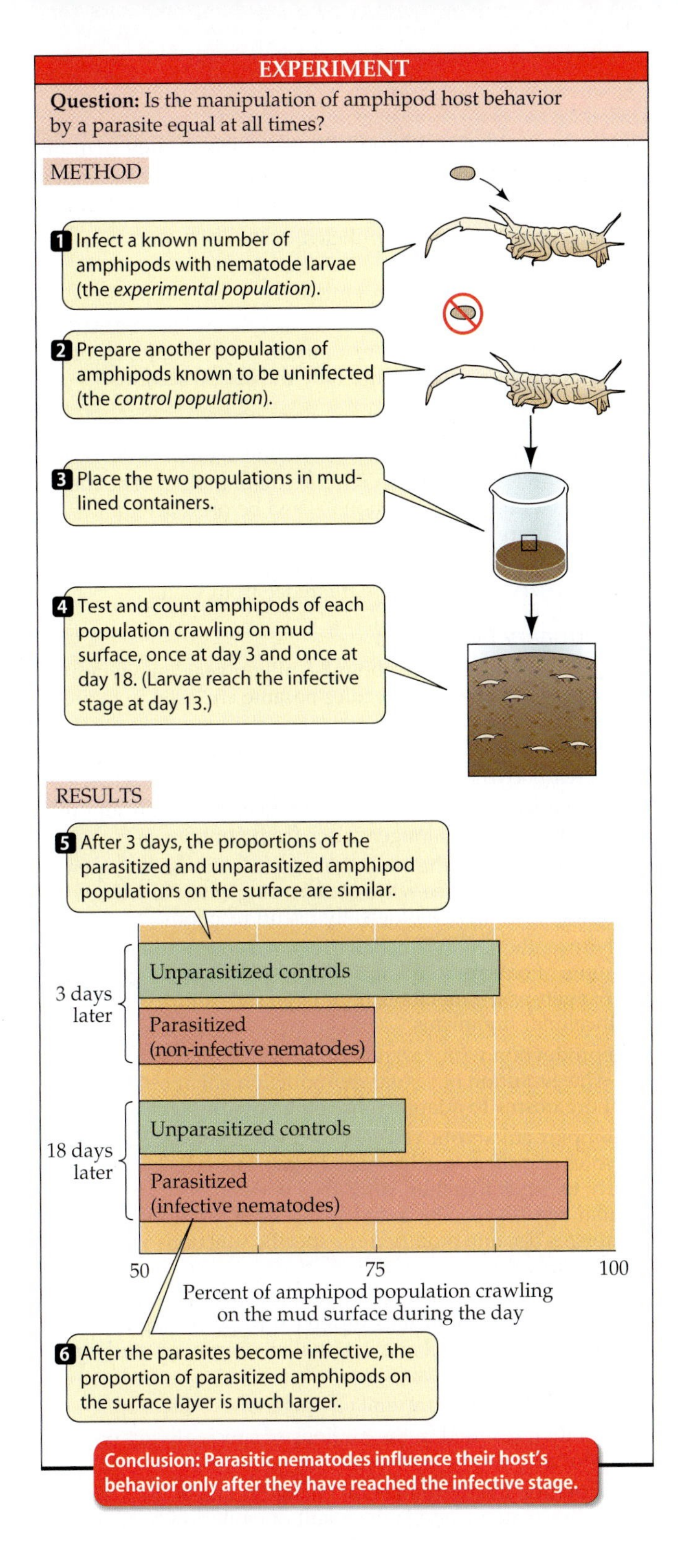

1.14 An Experiment Demonstrates that Parasites Influence Amphipod Behavior
Amphipods are more likely to crawl on the surface of the mud, exposing themselves to being captured by sandpipers, when their parasitic nematodes have reached the stage at which they can infect a sandpiper.

complete. Field experiments are more difficult because it is usually impossible to control more than a small number of environmental factors. But field experiments have one important advantage: Their results are more readily applicable to what happens where the organisms actually live and evolve. Just because an organism does something in the laboratory does not mean that it behaves the same way in nature. Because biologists usually wish to explain nature, not processes in the laboratory, combinations of laboratory and field experiments are needed to test most hypotheses about what organisms do.

A single piece of supporting evidence rarely leads to widespread acceptance of a hypothesis. Similarly, a single contrary result rarely leads to abandonment of a hypothesis. Results that do not support the hypothesis being tested can be obtained for many reasons, only one of which is that the hypothesis is wrong. Incorrect predictions may have been made from a correct hypothesis. A negative finding can also result from poor experimental design, or because an inappropriate organism was chosen for the test. For example, a species of sandpiper that fed only by probing in the mud for its prey would have been an unsuitable subject for testing the hypothesis that nematodes alter their hosts in a way to make them more visible to predators.

Accepted scientific theories are based on many kinds of evidence

A general textbook like this one presents hypotheses and theories that have been extensively tested, using a variety of methods, and are generally accepted. When possible, we illustrate hypotheses and theories with observations and experiments that support them, but we cannot, because of space constraints, detail all the evidence. Remember as you read that statements of biological "fact" are mixtures of observations, predictions, and interpretations.

No amount of observation could possibly substitute for experimentation. However, this does not mean that scientists are insensitive to the welfare of the organisms with which they work. Most scientists who work with animals are continually alert to finding ways of getting answers that use the smallest number of experimental subjects and that cause the subjects the least pain and suffering.

Not all forms of inquiry are scientific

If you understand the methods of science, you can distinguish science from non-science. Recently some people have claimed that "creation science," sometimes called "scientific creationism," is a legitimate science that deserves to be taught in schools together with the evolutionary view of the world presented in this book. In spite of these claims, creation science is not science.

Science begins with observations and the formulation of hypotheses that can be tested and that will be rejected if significant contrary evidence is found. Creation science begins with the assertions, derived from religious texts, that Earth is only a few thousand years old and that all species of organisms were created in approximately their present forms. These assertions are not presented as a hypothesis from which testable predictions can be derived. Advocates of creation science assume their assertions to be true and that no tests are needed, nor are they willing to accept any evidence that refutes them.

In this chapter we have outlined the hypotheses that Earth is about 4 billion years old, that today's living organisms evolved from single-celled ancestors, and that many organisms dramatically different from those we see today lived on Earth in the remote past. The rest of this book will provide evidence supporting this scenario. To reject this view of Earth's history, a person must reject not only evolutionary biology, but also modern geology, astronomy, chemistry, and physics. All of this extensive scientific evidence is rejected or misinterpreted by proponents of "creation science" in favor of their particular religious beliefs.

Evidence gathered by scientific procedures does not diminish the value of religious accounts of creation. Religious beliefs are based on faith—not on falsifiable hypotheses, as science is. They serve different purposes, giving meaning and spiritual guidance to human lives. They form the basis for establishing values—something science cannot do. The legitimacy and value of both religion and science is undermined when a religious belief is presented as scientific evidence.

Biology and Public Policy

During the Second World War and immediately thereafter, the physical sciences were highly influential in shaping public policy in the industrialized world. Since then, the biological sciences have assumed increasing importance. One reason is the discovery of the genetic code and the ability to manipulate the genetic constitution of organisms. These developments have opened vast new possibilities for improvements in the control of human diseases and agricultural productivity. At the same time, these capabilities have raised important ethical and policy issues. How much, and in what ways, should we tinker with the genetics of people and other species? Does it matter whether organisms are changed by traditional breeding experiments or by gene transfers? How safe are genetically modified organisms in the environment and in human foods?

Another reason for the importance of the biological sciences is the vastly increased human population. Our use of renewable and nonrenewable natural resources is stressing the ability of the environment to produce the goods and services upon which society depends. Human activities are causing the extinction of a large number of species and are resulting in the spread of new human diseases and the resurgence of old ones. Biological knowledge is vital for determining the causes of these changes and for devising wise policies to deal with them.

Therefore, biologists are increasingly called upon to advise governmental agencies concerning the laws, rules, and regulations by which society deals with the increasing number of problems and challenges that have at least a partial biological basis. We will discuss these issues in many chapters of this book. You will see how the use of biological information can contribute to the establishment and implementation of wise public policies.

Chapter Summary

▶ If the history of Earth were a month with 30 days, recorded human history would occupy only the last 30 seconds. **Review Figure 1.1**

Organisms Have Changed over Billions of Years

▶ Evolution is the theme that unites all of biology. The idea of, and evidence for, evolution existed before Darwin. **Review Figure 1.2**

▶ The theory of evolution by natural selection rests on two simple observations and one inference from them.

Evolutionary Milestones

▶ Life arose from nonlife about 3.8 billion years ago when interacting systems of molecules became enclosed in membranes to form cells.

▶ All living organisms contain the same types of large molecules—carbohydrates, lipids, proteins, and nucleic acids.

▶ All organisms consist of cells, and all cells come from preexisting cells. Life no longer arises from nonlife.

▶ A major theme in the evolution of life is the development of increasingly diverse ways of capturing external energy and using it to drive biologically useful reactions.

▶ Photosynthetic single-celled organisms released large amounts of oxygen into Earth's atmosphere, making possible the oxygen-based metabolism of large cells and, eventually, multicellular organisms.

▶ Reproduction with variation is a major characteristic of life. The evolution of sexual reproduction enhanced the ability of organisms to adapt to changing environments.

▶ Complex eukaryotic cells evolved when some large prokaryotes engulfed smaller ones. Eukaryotic cells evolved the ability to "stick together" after they divided, forming multicellular organisms. The individual cells of multicellular organisms became modified for specific functions within the organism.

▶ A major theme in the evolution of life is the development of increasingly complicated systems for responding to changes in the internal and external environments and for maintaining homeostasis.

▶ Regulated growth is a vital characteristic of life.

▶ Speciation resulted in the millions of species living on Earth today.

▶ Adaptation to environmental change is one of life's most distinctive features and is the result of evolution by natural selection.

The Hierarchy of Life

▶ Biology is organized into a hierarchy of levels from molecules to the biosphere. Each level has emergent properties that are not found at lower levels. **Review Figure 1.9**

▶ Species are classified into three domains: Archaea, Bacteria, and Eukarya. The domains Archaea and Bacteria consist of prokaryotic cells. The domain Eukarya contains the protists and the kingdoms Plantae, Fungi, and Animalia, all of which have eukaryotic cells. **Review Figure 1.10**

Asking and Answering "How?" and "Why?"

▶ Biologists ask two kinds of questions. "How" questions ask how organisms work. "Why" questions ask why they evolved to work that way.

▶ Both how and why questions are usually answered using a hypothetico-deductive (H-D) approach. Hypotheses are tentative answers to questions. Predictions are made on the basis of a hypothesis. The predictions are tested by observations and experiments, the results of which may support or refute the hypothesis. **Review Figure 1.14**

▶ Science is based on the formulation of testable hypotheses that can be rejected in light of contrary evidence. The acceptance on faith of already refuted, untested, or untestable assumptions is not science.

Biology and Public Policy

▶ Biologists are often called upon to advise governmental agencies on the solution of important problems that have a biological component.

For Discussion

1. According to the theory of evolution by natural selection, a species evolves certain features because they improve the chances that its members will survive and reproduce. There is no evidence, however, that evolutionary mechanisms have foresight or that organisms can anticipate future conditions. What, then, do biologists mean when they say, for example, that wings are "for flying"?
2. Why is it so important in science that we design and perform tests capable of rejecting a hypothesis?
3. One hypothesis about the manipulation of a host's behavior by a parasite was discussed in this chapter, and some tests of that hypothesis were described. Suggest some other hypotheses about the ways in which parasites might change the behavior and physiology of their hosts. Develop some critical tests for one of these alternatives. What are the appropriate associated null hypotheses?
4. Some philosophers and scientists believe that it is impossible to prove any scientific hypothesis—that we can only fail to find a reason to reject it. Evaluate this view. Can you think of reasons why we can be more certain about rejecting a hypothesis than about accepting it?
5. Discuss one current environmental problem whose solution requires the use of biological knowledge. How well is biology being used? What factors prevent scientific data from playing a more important role in finding a solution to the problem?

Part Five

The Biology of Flowering Plants

34 The Plant Body

THE OLDEST KNOWN PLANT IS A BRISTLECONE pine that has been living for more than 4,900 years—almost 50 centuries. In contrast, it is doubtful that any animal has ever lived as long as 2 centuries. The extreme ages achieved by some trees prove that plants can cope very successfully with their environments.

Plants cannot move, but they have mechanisms for coping with environmental changes that they can't escape. They create and maintain an internal environment that differs from the external environment. They also regulate their own metabolism, which enables them to perform their necessary functions.

Motion is not a characteristic of plants; instead, we may think of plants as "growing machines." By growing, plants accomplish some of the same things that animals achieve through motion. Growing roots, for example, can reach into new supplies of water and nutrients.

Although they have simpler nutritional needs than animals do, plants must nevertheless obtain nutrients—not only the raw materials of photosynthesis (carbon dioxide and water), but also mineral elements such as nitrogen, potassium, and calcium. Seed plants—even the tallest trees—transport water from the soil to their tops, and they transport the products of photosynthesis from the leaves to their roots and other parts.

Plants also interact with their living and nonliving environments. They respond to environmental cues as they grow and develop. Their responses are mediated by chemical signals that move within cells and throughout the plant body. Among the resulting changes are ones that lead to growth, development, and reproduction.

Because we can understand the function of these growing machines only in terms of their underlying structure, this chapter focuses on the structure of the plant body, with primary emphasis on flowering plants. We'll examine plant structure at the level of the organs, cells, tissues, and tissue systems. Then we'll see how meristems—organized groups of dividing cells—serve the growth of the plant body, both in length and, in woody plants, in width. The chapter concludes with a consideration of how leaf structure supports photosynthesis.

An Ancient Individual
Bristlecone pines (*Pinus aristata*) can live for centuries. The oldest known living organism is a bristlecone pine that has been alive for almost 5,000 years—long enough to have witnessed all of recorded human history.

Vegetative Organs of the Flowering Plant Body

You will recall from Chapter 29 that flowering plants (angiosperms) are tracheophytes that are characterized by double fertilization, a triploid endosperm, and seeds en-

34.1 Monocots versus Eudicots The possession of a single cotyledon clearly distinguishes the monocots from the other angiosperms. Several other anatomical characteristics also differ between the monocots and the eudicots. Most angiosperms that do not belong to either lineage resemble eudicots in the characteristics shown here.

	Cotyledons	Veins in leaves	Flower parts	Arrangement of primary vascular bundles in stem
Monocots	One	Usually parallel	Usually in multiples of three	Scattered
Eudicots	Two	Usually netlike	Usually in fours or fives	In a ring

closed in modified leaves called carpels. Their xylem contains cells called vessel elements and fibers, and their phloem contains companion cells.

Flowering plants possess three kinds of *vegetative* (nonreproductive) organs: roots, stems, and leaves. Flowers, which are the plant's devices for sexual reproduction, consist of other types of organs that will be considered in a later chapter.

Most flowering plants belong to one of two major lineages. **Monocots** (Monocotyledones) are generally narrow-leaved flowering plants such as grasses, lilies, orchids, and palms. **Eudicots** (Eudicotyledones) are broad-leaved flowering plants such as soybeans, roses, sunflowers, and maples. These two monophyletic classes account for 97 percent of the species of flowering plants (Figure 34.1). Most of the remaining species (including water lilies and the lineage that includes magnolias) are structurally similar to the eudicots.*

The basic body plans of a generalized monocot and a generalized eudicot are shown in Figure 34.2. In both lineages, the vegetative plant body consists of two systems: the shoot system and the root system.

The **shoot system** of a plant consists of the stems and the leaves, as well as flowers (which contain leaflike parts). Broadly speaking, the **leaves** are the chief organs of photosynthesis. The **stems** hold and display the leaves to the sun and provide connections for the transport of materials between roots and leaves. The points where leaves attach to a stem are called **nodes**, and the stem regions between successive nodes are **internodes**.

*Traditionally, botanists have referred to all flowering plants other than monocots as *dicots*. However, the dicots do not constitute a monophyletic lineage (see Figure 29.13). Because we wish to emphasize lineages, we do not use that term here.

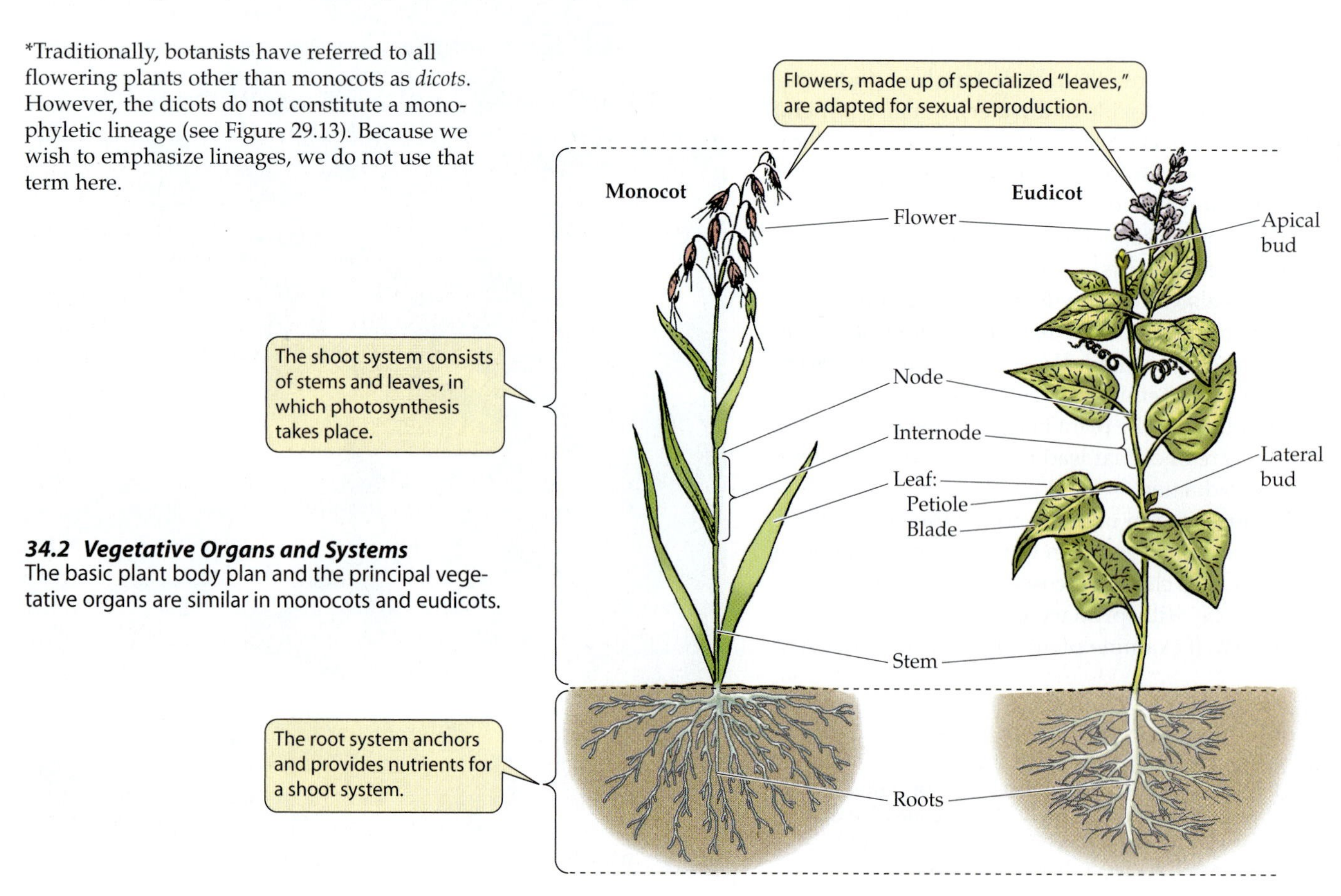

34.2 Vegetative Organs and Systems The basic plant body plan and the principal vegetative organs are similar in monocots and eudicots.

(a)

(b)

34.3 Root Systems
The taproot system of a beet (*a*) contrasts with the fibrous root system of a grass (*b*).

The **root system** provides support and nutrition. Roots anchor the plant in place, and their extreme branching and high surface area-to-volume ratio adapt them to absorb water and mineral nutrients from the soil.

Each of the vegetative organs can be understood in terms of its structure. By *structure* we mean both its overall form and the anatomy of its component cells and tissues, as well as their arrangement. Let's first consider the overall forms of roots, stems, and leaves.

Roots anchor the plant and take up water and minerals

Water and minerals usually enter the plant through the root system, which usually lies in the soil, mostly in complete darkness. There are two principal types of root systems. Many eudicots have a *taproot system:* a single, large, deep-growing primary root accompanied by less prominent lateral roots (Figure 34.3*a*). The taproot itself often functions as a food storage organ, as in carrots and radishes.

By contrast, monocots and some eudicots have a *fibrous root system*, which is composed of numerous thin roots roughly equal in diameter (Figure 34.3*b*). Many fibrous root systems have a large surface area for the absorption of water and minerals. A fibrous root system holds soil very well. Grasses with fibrous root systems, for example, may protect steep hillsides where runoff from rain would otherwise cause erosion.

Some plants have what are called *adventitious roots*. These roots arise from points along the stem where roots would not usually occur; some even arise from the leaves. In many species, adventitious roots also form when a piece of shoot is cut from the plant and placed in water or soil. Adventitious rooting enables the cutting to establish itself in the soil. Some plants—corn, banyan trees, and some palms, for example—use adventitious roots as props to help support the shoot.

Stems bear buds, leaves, and flowers

Unlike roots, stems bear buds of various types. A **bud** is an embryonic shoot. A stem bears leaves at its nodes, and where each leaf meets the stem there is a **lateral bud** (see Figure 34.2), which can develop into a new *branch*, or extension of the stem, if it becomes active. The branching patterns of plants are highly variable, depending on the species, environmental conditions, and a gardener's pruning activities.

At the tip of each stem or branch is an **apical bud**, which produces the cells for the growth and development of that stem or branch. At times that vary depending on the species, buds form that develop into flowers.

Some stems are highly modified. The *tuber* of a potato, for example—the part of the plant eaten by humans—is a portion of the stem rather than a root. Its "eyes" contain lateral buds; thus, a sprouting potato is just a branching stem (Figure 34.4*a*). The *runners* of strawberry plants and Bermuda grass are horizontal stems from which roots grow at frequent intervals (Figure 34.4*b*). If the links between the rooted portions are broken, independent plants can develop on each side of the break. This is a form of vegetative reproduction, which we will discuss in a later chapter.

Unlike most roots, stems may be green and capable of photosynthesis. But stems are not the principal sites of photosynthesis.

Leaves are the primary sites of photosynthesis

In gymnosperms and most flowering plants, the leaves are responsible for most of the plant's photosynthesis, producing energy-rich organic molecules and releasing oxygen gas. Most leaves also carry out metabolic reactions that make nitrogen available to the plant for the synthesis of proteins and nucleic acids. In certain plants, leaves are highly modified for more specialized functions, as we will see below.

As photosynthetic organs, leaves are marvelously adapted for gathering light. Typically, the **blade** of a leaf is a thin, flat structure attached to the stem by a stalk called a **petiole**. During the daytime the leaf blade is held by its petiole at an angle almost perpendicular to the rays of the sun. This placement, with the leaf surface facing the sun, maximizes the amount of light available for photosynthesis. Some leaves track the sun, moving so that they constantly face it. If leaves were thicker than they are, the outer layers of cells would absorb so much of the light that the interior layers would be too dark and would be unable to photosynthesize.

The leaves at different sites on a single plant may have quite different shapes. These shapes result from a combination of genetic, environmental, and developmental influences. Most species, however, bear leaves of a particular broadly defined type. A leaf may be **simple**, consisting of a

(a)

(b)

(c)

34.4 Modified Stems
(*a*) A potato is a modified stem called a tuber; the sprouts that grow from its eyes are branches, not roots. (*b*) Runners produce roots at intervals, providing a local water supply and allowing rooted portions to live independently if the runner is cut. (*c*) The thorny spines are leaves on the enlarged stem of a cactus plant.

single blade, or **compound**, with blades, or *leaflets*, arranged along an axis or radiating from a central point (Figure 34.5). In a simple leaf, or in a leaflet of a compound leaf, the veins may be parallel to one another, as in monocots, or in a netlike arrangement, as in eudicots.

The general development of a specific leaf pattern is programmed in the plant's genes and is expressed by differential growth of the leaf veins and of the tissue between the veins. As a result, plant taxonomists have often found leaf forms (outlines, margins, tips, bases, and patterns of arrangement) to be reliable characters for classification and identification. At least some of the forms in Figure 34.5 probably look familiar to you.

In some plant species, leaves are highly modified for special functions. Some leaves serve as storage depots for energy-rich molecules, as in the bulbs of onions. In other species—some of the succulents—the leaves store water. The spines of cacti are modified leaves. Certain leaves of poinsettias, dogwoods, and some other plants are brightly colored and help attract pollinating animals to the often less striking flowers. Many plants, such as peas, have tendrils—modified leaves that support the plant by wrapping around other plants.

Leaves and other plant organs are composed of cells, tissues, and tissue systems. Let's now consider plant cells—the basic building blocks of plant organs.

34.5 The Diversity of Leaves
Simple leaves are those with a single blade. Compound leaves consist of leaflets arranged along a central axis. Further division of leaflets results in a doubly compound leaf. Other characters of leaves can also be used to identify plants.

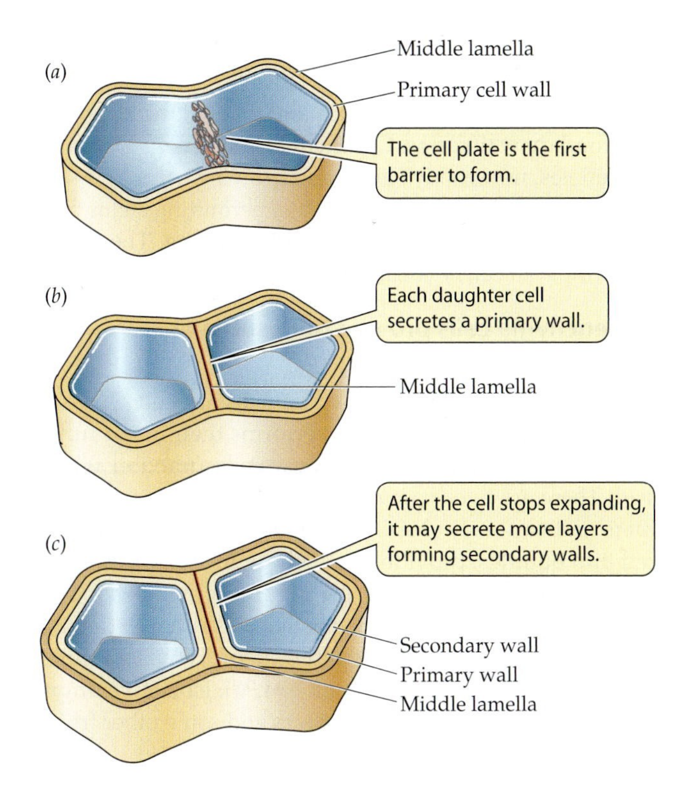

34.6 Cell Wall Formation
Cell walls form as the final step in plant cell division.

Plant Cells

Living plant cells have all the essential organelles common to eukaryotes (see Figure 4.7). In addition, they have some distinguishing structures and organelles.

- Some plant cells contain chloroplasts or other plastids.
- Many plant cells contain vacuoles.
- Every plant cell is surrounded by a cellulose-containing cell wall.

Although most kinds of plant cells are alive when they perform their functions, certain others function only after their living parts have died and disintegrated. Other plant cells develop special chemical capabilities; for example, some can perform photosynthesis, and others produce and secrete waterproofing materials. Several plant cell types differ dramatically in the structure of their cell walls.

Cell walls may be complex in structure

The cytokinesis of a plant cell is completed when cell walls form, separating the two daughter cells. The daughter cells secrete a glue that constitutes the **middle lamella**, which forms a layer between them. Then each daughter cell secretes cellulose and other polysaccharides to form a **primary wall**, which continues to grow as the cell grows to its final size (Figure 34.6).

Once cell expansion stops, a plant cell may deposit more polysaccharides, sometimes impregnated with further materials—such as **lignin**, characteristic of wood, or **suberin**, a complex lipid characteristic of cork—in one or more layers internal to the primary wall. These layers collectively form the **secondary wall**, which often serves supporting or waterproofing roles.

Although the cell wall lies outside the plasma membrane of the cell, it is not a chemically inactive region. Chemical reactions in the wall play important roles in cell expansion and defense. Cell walls may thicken or be sculpted or perforated as part of differentiation into specialized cell types. Except where the secondary wall is made waterproof by added substances, the wall is porous to water and to most small molecules.

Localized modifications in the walls of adjacent cells allow water and dissolved materials to move easily from cell to cell. In cells that have not developed a secondary wall, the primary wall usually has thin regions. Strands of cytoplasm called **plasmodesmata** (singular plasmodesma) pass through the primary wall in these regions, allowing substances to move freely from cell to cell without having to cross a plasma membrane (Figure 34.7*a*). The plasmodes-

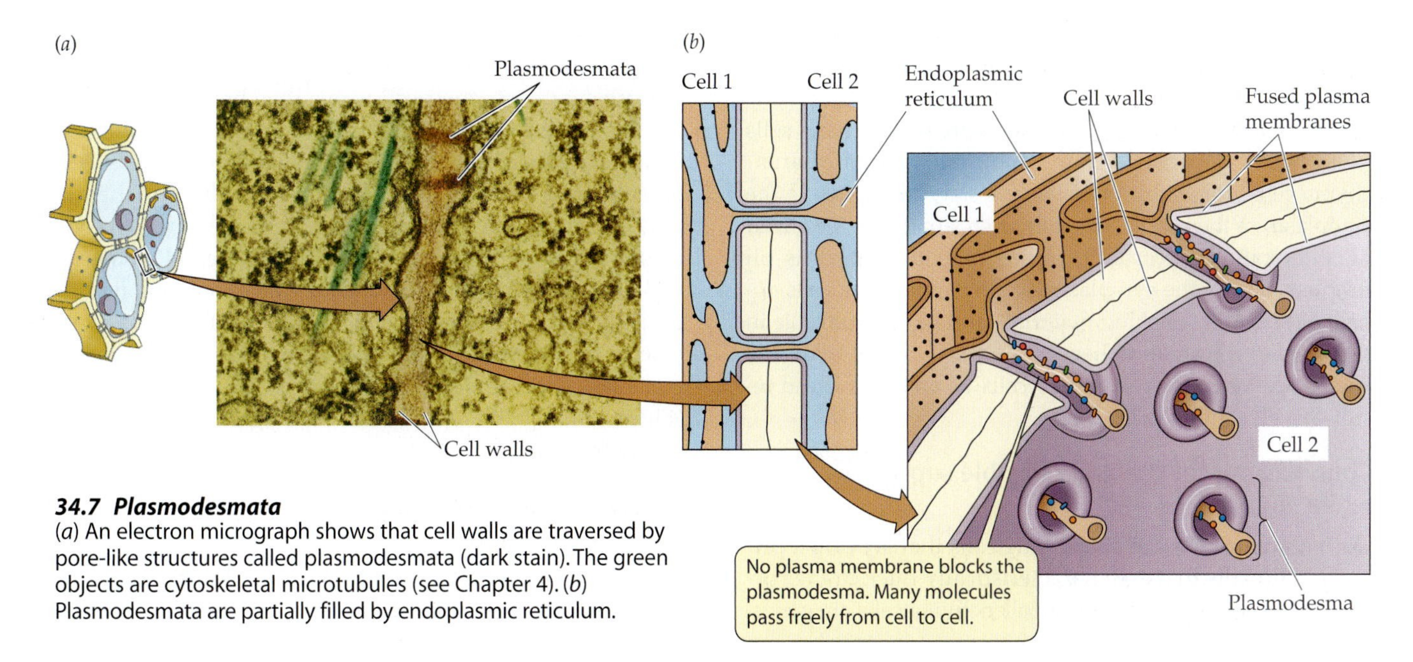

34.7 Plasmodesmata
(*a*) An electron micrograph shows that cell walls are traversed by pore-like structures called plasmodesmata (dark stain). The green objects are cytoskeletal microtubules (see Chapter 4). (*b*) Plasmodesmata are partially filled by endoplasmic reticulum.

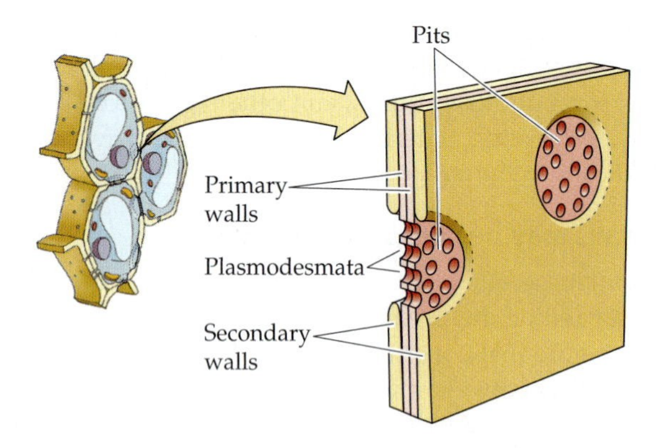

34.8 Pits Allow Materials to Move between Cells with Secondary Walls
Secondary cell walls may be interrupted by pits that allow the passage of water and other materials between cells.

mata allow direct communication between plant cells, as discussed in Chapter 15.

A plasmodesma is wide enough so that portions of the endoplasmic reticulum extend between cells (Figure 34.7*b*). Under certain circumstances, a plasmodesma can enlarge dramatically, allowing even macromolecules and viruses to pass directly between cells. Even in cells with a waterproofed secondary wall, water and dissolved materials can pass from cell to cell by way of structures called **pits** (Figure 34.8). Pits are interruptions in the secondary wall that leave the thin regions of the primary wall, and thus the plasmodesmata, unobstructed.

Parenchyma cells are alive when they perform their functions

The most numerous cell type in young plants is the **parenchyma** cell (Figure 34.9*a*). Parenchyma cells usually have thin walls, consisting only of a primary wall and the shared middle lamella. Many parenchyma cells have shapes similar to those of soap bubbles crowded into a limited space—shapes with 14 faces. They are usually not elongated or otherwise asymmetrical. Most have large central vacuoles.

The photosynthetic cells in leaves are parenchyma cells filled with chloroplasts. Some nonphotosynthetic parenchyma cells store substances such as starch or lipids. In the cytoplasm of these cells, starch is often stored in specialized plastids called *leucoplasts*. Lipids may be stored as oil droplets, also in the cytoplasm. Some parenchyma cells appear to serve as "packing material" and play a vital role in supporting the stem. Others retain the capacity to divide and hence may give rise to new cells, as when a wound results in cell proliferation.

Collenchyma cells provide flexible support while alive

Collenchyma cells are supporting cells that lay down primary cell walls that are characteristically thick in the corners of the cells (Figure 34.9*b*). Collenchyma cells are generally elongated. In these cells the primary wall thickens, but no secondary wall forms. Collenchyma provides support to leaf petioles, nonwoody stems, and growing organs. Tissue made of collenchyma cells is flexible, permitting stems and petioles to sway in the wind without snapping. The familiar "strings" in celery consist primarily of collenchyma.

Sclerenchyma cells provide rigid support after they die

In contrast to collenchyma, **sclerenchyma** cells have a thickened secondary wall that performs their major function: support. Many sclerenchyma cells function when dead. There are two types of sclerenchyma cells: elongated **fibers** and variously shaped **sclereids**. Fibers, often organized into bundles, provide relatively rigid support both in wood and in other parts of the plant (Figure 34.9*c*). The bark of trees owes much of its mechanical strength to long fibers. Sclereids may pack together densely, as in a nut's shell or in some seed coats (Figure 34.9*d*). Isolated clumps of sclereids, called *stone cells*, in pears and some other fruits give them their characteristic gritty texture.

Xylem transports water from roots to stems and leaves

The **xylem** of tracheophytes conducts water from roots to aboveground plant parts. It contains conducting cells called **tracheary elements**, which undergo programmed cell death before they assume their function of transporting water and dissolved minerals. The evolutionarily more ancient tracheary elements, found in gymnosperms and other tracheophytes, are **tracheids**—spindle-shaped cells interconnected by numerous pits in their cell walls (Figure 34.9*e*). When the cell contents—nucleus and cytoplasm—disintegrate upon cell death, water can move with little resistance from one tracheid to its neighbors by way of pits.

Flowering plants evolved a water-conducting system made up of *vessels*. The individual cells that form vessels, called **vessel elements**, also die and become empty before they can transport water. These cells secrete a waterproofing substance into their cell walls, then break down their end walls, and finally die and disintegrate. The result is a hollow tube through which water can flow freely. Vessel elements are generally larger in diameter than tracheids; they are laid down end-to-end, so that each vessel is a continuous hollow tube consisting of many vessel elements and providing an open pipeline for water conduction (Figure 34.9*f*). In the course of angiosperm evolution, vessel elements have become shorter, and their end walls have become less and less obliquely oriented and less obstructed (Figure 34.10). The xylem of many angiosperms also includes tracheids.

Phloem translocates carbohydrates and other nutrients

The transport cells of the **phloem**, unlike those of the xylem, are living cells. In flowering plants the characteristic

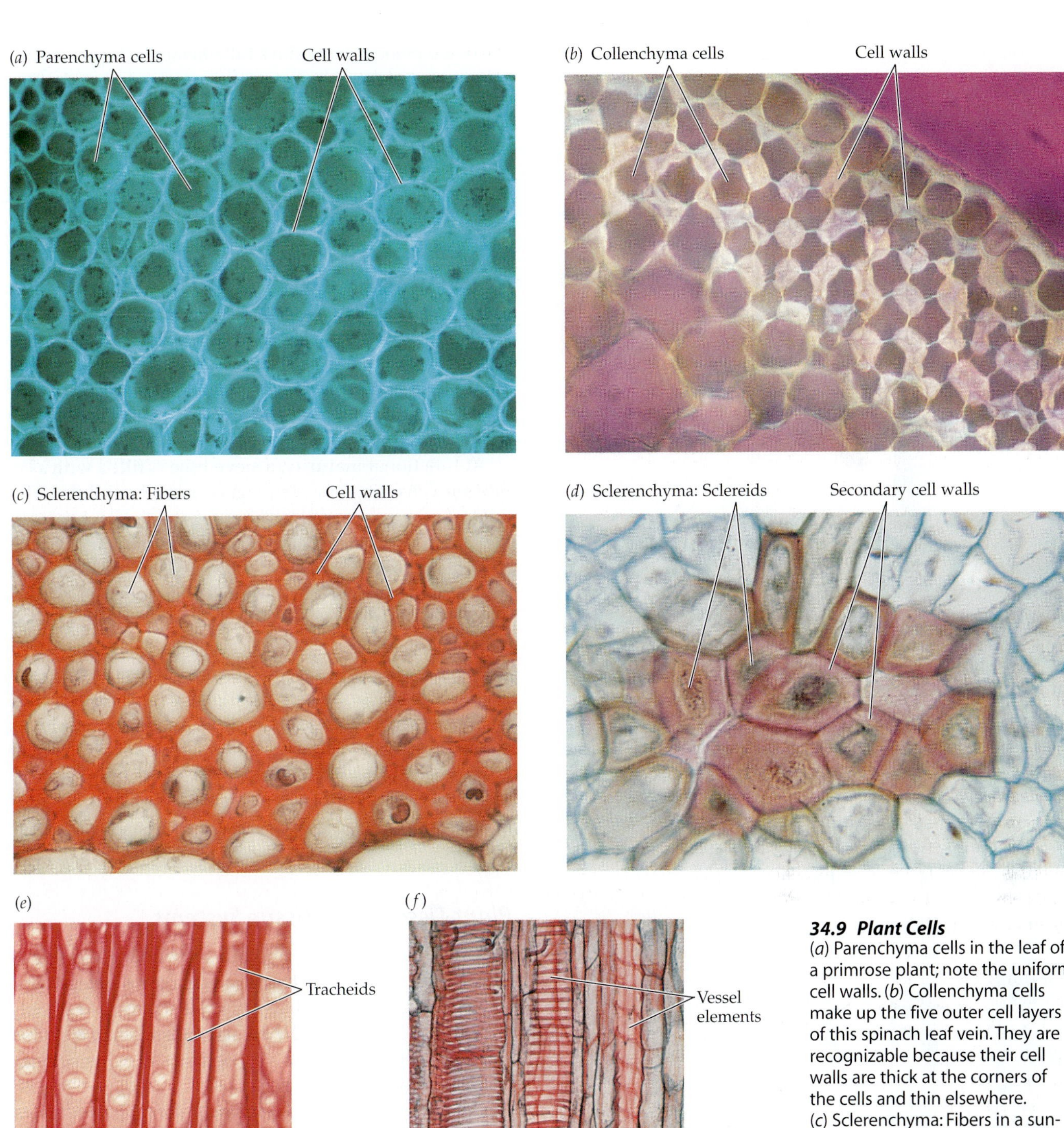

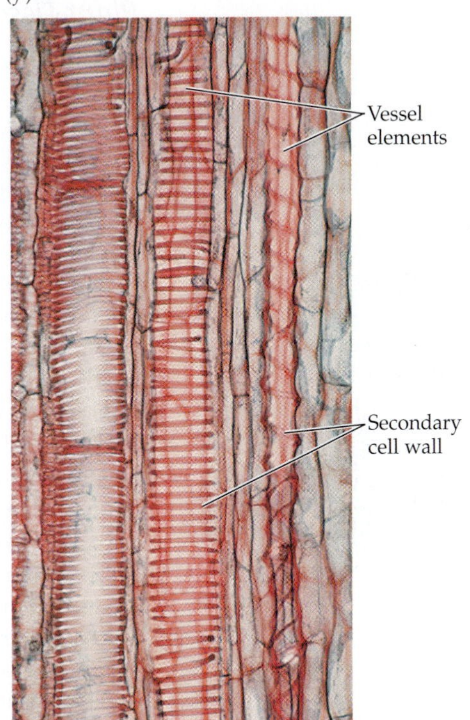

34.9 Plant Cells
(*a*) Parenchyma cells in the leaf of a primrose plant; note the uniform cell walls. (*b*) Collenchyma cells make up the five outer cell layers of this spinach leaf vein. They are recognizable because their cell walls are thick at the corners of the cells and thin elsewhere. (*c*) Sclerenchyma: Fibers in a sunflower plant (*Helianthus*). The thickened walls are stained red. (*d*) Sclerenchyma: These extremely thick secondary cell walls of sclereids are laid down in layers. They provide support and a hard texture to structures such as nuts and seeds. (*e*) Water-conducting tracheids in pine wood. The thick cell walls are stained dark red. (*f*) Vessel elements in the stem of a squash. The secondary walls are stained red; note the different patterns of thickening, including rings and spirals.

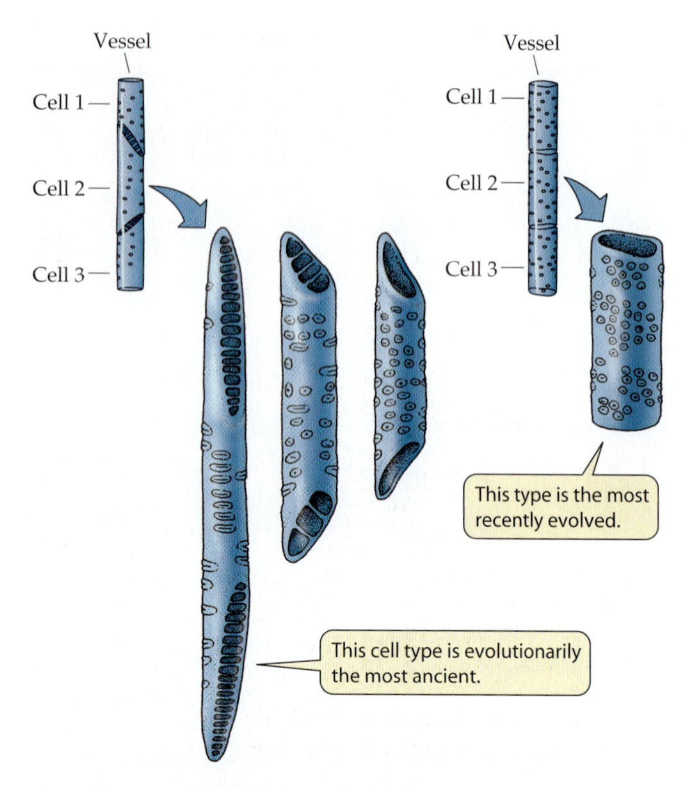

34.10 Evolution of the Conducting Cells of Vascular Systems
The xylem of tracheophytes has changed over time. The cells that conduct water and mineral nutrients have become shorter and the end walls have become more perpendicular to the side walls.

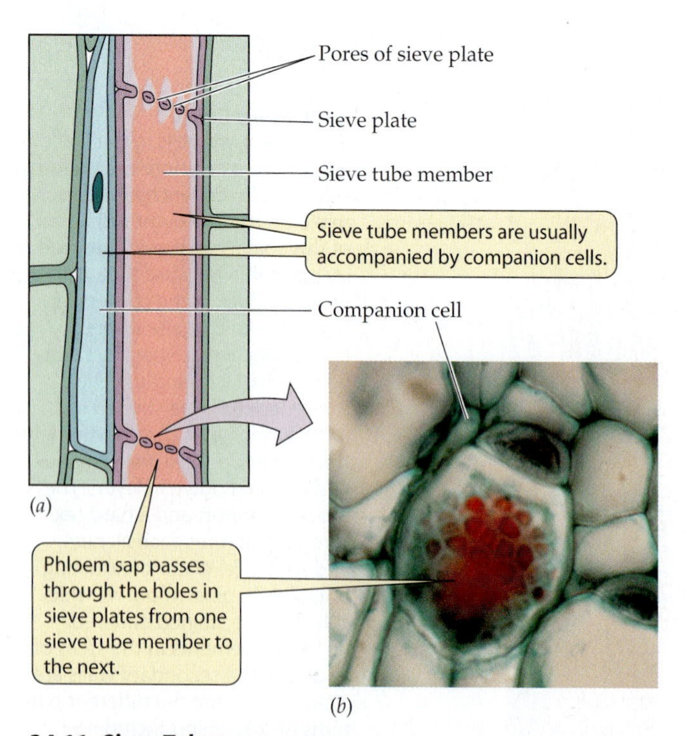

34.11 Sieve Tubes
(*a*) Individual sieve tube members join to form long tubes that transport carbohydrates and other nutrient molecules. (*b*) Sieve plates form at the ends of each sieve tube member.

cell of the phloem is the **sieve tube member** (Figure 34.11*a*). Like vessel elements, these cells meet end-to-end. They form long *sieve tubes*, which transport carbohydrates and many other materials from their sources to tissues that consume or store them. In plants with mature leaves, for example, excess products of photosynthesis move from leaves to root tissues.

As sieve tube members mature, plasmodesmata in the end walls enlarge, enhancing the connection between the contents of neighboring cells. The result is end walls that look like sieves, and are called *sieve plates* (Figure 34.11*b*). As the holes in the sieve plates expand, the *tonoplast* (the membrane around the central vacuole) disappears. The nucleus and some of the other organelles in the sieve tube member also break down and thus do not clog the holes of the sieve.

At functional maturity, a sieve tube is filled with *sieve tube sap*, consisting of water, sugars, and other solutes. This mixture moves from cell to cell along the sieve tube, carrying its dissolved sugars and other important materials with it. At the periphery of a sieve tube member, next to the cell wall and distinct from the sieve tube sap, is a layer of cytoplasm. This stationary layer of cytoplasm confines the remaining organelles.

The sieve tube members have adjacent **companion cells** (see Figure 34.11*a*), produced along with the sieve tube member when a parent cell divides. Companion cells retain all their organelles and may, through the activities of their nuclei, regulate the performance of the sieve tube members.

All these kinds of plant cells play important roles. Next let's see how they are organized into tissues and tissue systems.

Plant Tissues and Tissue Systems

A *tissue* is an organized group of cells, working together as a functional unit. Parenchyma cells make up parenchyma tissue, a *simple tissue*—that is, a tissue composed of only one type of cell. Sclerenchyma and collenchyma are other simple tissues, composed, respectively, of sclerenchyma and collenchyma cells.

Different cell types also combine to form *complex tissues*. Xylem and phloem are complex tissues, composed of more than one type of cell. As a result of its cellular complexity, xylem can perform a variety of functions, including transport, support, and storage. The xylem of angiosperms contains vessel elements and tracheids as conducting cells, thick-walled sclerenchyma fibers for support, and parenchyma cells that store food. The phloem of angiosperms includes sieve tube members, companion cells, fibers, sclereids, and parenchyma cells.

Tissues, in turn, are grouped into *tissue systems* that extend throughout the body of the plant, from organ to organ. There are three tissue systems: vascular, dermal, and ground (Figure 34.12).

The **vascular tissue system**, which includes the xylem and phloem, is the conductive, or "plumbing," system of

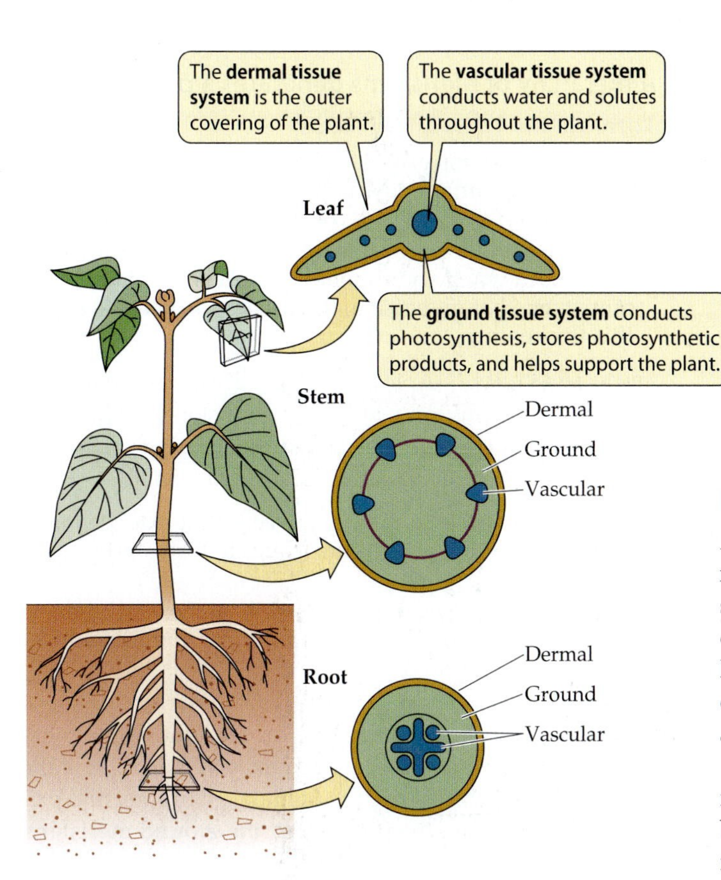

34.12 Three Tissue Systems Extend throughout the Plant Body
The arrangement shown here is typical of eudicots.

the plant. All the living cells of the plant body require a source of energy and chemical building blocks. The phloem transports carbohydrates from sites of production—called *sources* (commonly the leaves)—to sites of utilization or storage—called *sinks*—elsewhere in the plant. The xylem distributes water and mineral ions taken up by the roots to the stem and leaves.

The **dermal tissue system** is the outer covering of the plant. All parts of the young plant body are covered by an **epidermis**, which may be a single layer of cells or several layers. The epidermis contains *epidermal cells* and may also include specialized cell types, such as the guard cells that form stomata (pores) in leaves. The shoot epidermis secretes a layer of wax-covered cutin, the **cuticle**, that helps retard water loss from stems and leaves. The stems and roots of woody plants have an additional protective covering called the periderm, which will be discussed later in this chapter.

The **ground tissue system** makes up the rest of a plant and consists primarily of parenchyma tissue, often supplemented by collenchyma or sclerenchyma. Ground tissue functions primarily in storage, support, photosynthesis, and the production of defensive and attractive substances.

In the discussions that follow, we'll examine how the tissue systems are organized in the different organs of a flowering plant. Let's begin by seeing how this organization develops as the plant grows.

Forming the Plant Body

In its early embryonic stages, a plant establishes the basic body plan for its mature form (Figure 34.13). Two patterns contribute to the plant body plan:

- The *apical–basal pattern* is the arrangement of cells and tissues along the main axis from root to shoot.
- The *radial pattern* is the concentric arrangement of tissue systems.

Both patterns arise through orderly development and are best understood in developmental terms.

Plants and animals develop differently

As the plant body grows, it may lose parts, and it forms new parts that may grow at different rates. The growing stem consists of *modules* or *units*, laid down one after another. Each unit consists of a node with its attached leaf or leaves, the internode below that node, and the lateral bud or buds at the base of that internode. New units are formed as long as the stem continues to grow.

Each branch of a plant may be thought of as a unit that is in some ways independent of the other branches. A branch of a plant does not bear the same relationship to the remainder of the plant body as an arm does to the remainder of the human body. Among other things, branches form one after another, unlike arms, which form simultaneously during embryonic development. Also, branches often differ from one another in number of leaves and in the degree to which they themselves branch.

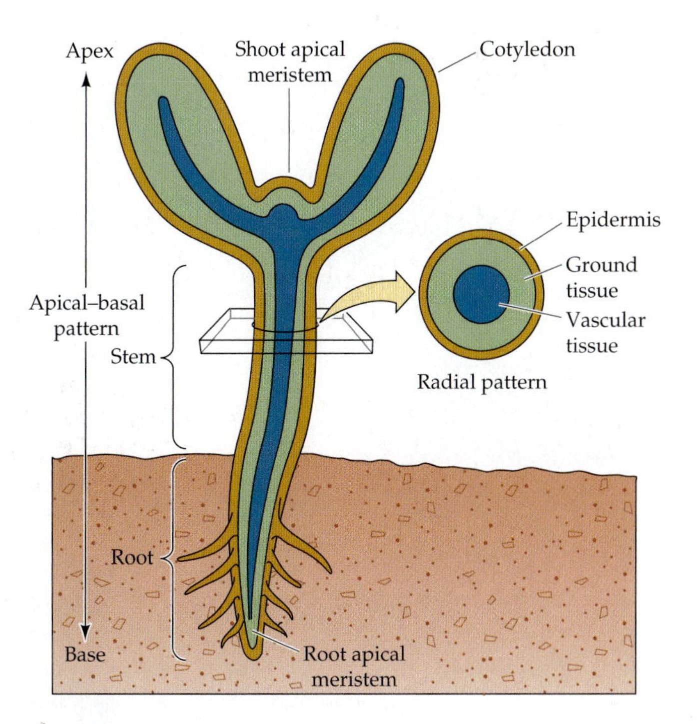

34.13 Patterns in the Plant Body Plan
The embryonic plant establishes the basic body plan for its mature form. Two patterns are evident in this body plan: the apical–basal pattern and the radial pattern.

Leaves are units of another sort, produced in fresh batches to take over the daily function of gathering energy for the plant. Leaves are usually short-lived, lasting weeks to a few years. Branches and stems are longer-lived, lasting from years to centuries.

Root systems are also branching structures, and lateral roots are semi-independent units. As the root system grows, penetrating and exploring the soil environment, many roots die and are replaced by new ones.

All parts of the animal body grow as an individual develops from embryo to adult, but this growth is *determinate*. That is, the growth of the individual and its parts ceases when the adult state is reached. The growth of stems and roots, by contrast, is *indeterminate* and is generated from specific regions of active cell division and cell expansion.

The localized regions of cell division in plants are called **meristems**. Meristematic tissues are forever young, retaining the ability to produce new cells indefinitely. They are comparable to the stem cells found in animals. When a meristem cell divides, the two resulting cells initially take up no more volume than did the single cell prior to division. One daughter cell develops into another meristem cell the size of its parent, while the other daughter cell develops into a more specialized cell.

A hierarchy of meristems generates a plant's body

There are two types of meristems:

- **Apical meristems** give rise to the *primary plant body*, which is the entire body of many plants.
- **Lateral meristems** give rise to the *secondary plant body*. The stems and roots of some plants (most obviously trees) form wood and become thick; it is the lateral meristems that give rise to the tissues responsible for this thickening.

APICAL MERISTEMS. Apical meristems are located at the tips of roots and stems, and in buds. They elongate the plant body by producing the cells that subsequently expand and differentiate to form all plant organs (Figure 34.14).

- *Shoot apical meristems* supply the cells that extend stems and branches, allowing more leaves to form and photosynthesize.
- *Root apical meristems* supply the cells that extend roots, enabling the plant to "forage" for water and minerals.

Both root and shoot apical meristems give rise to a set of cylindrical *primary meristems* that produce the primary tissues of the plant body. From the outside to the inside of the root or shoot, which are both cylindrical organs, the primary meristems are the **protoderm**, the **ground meristem**,

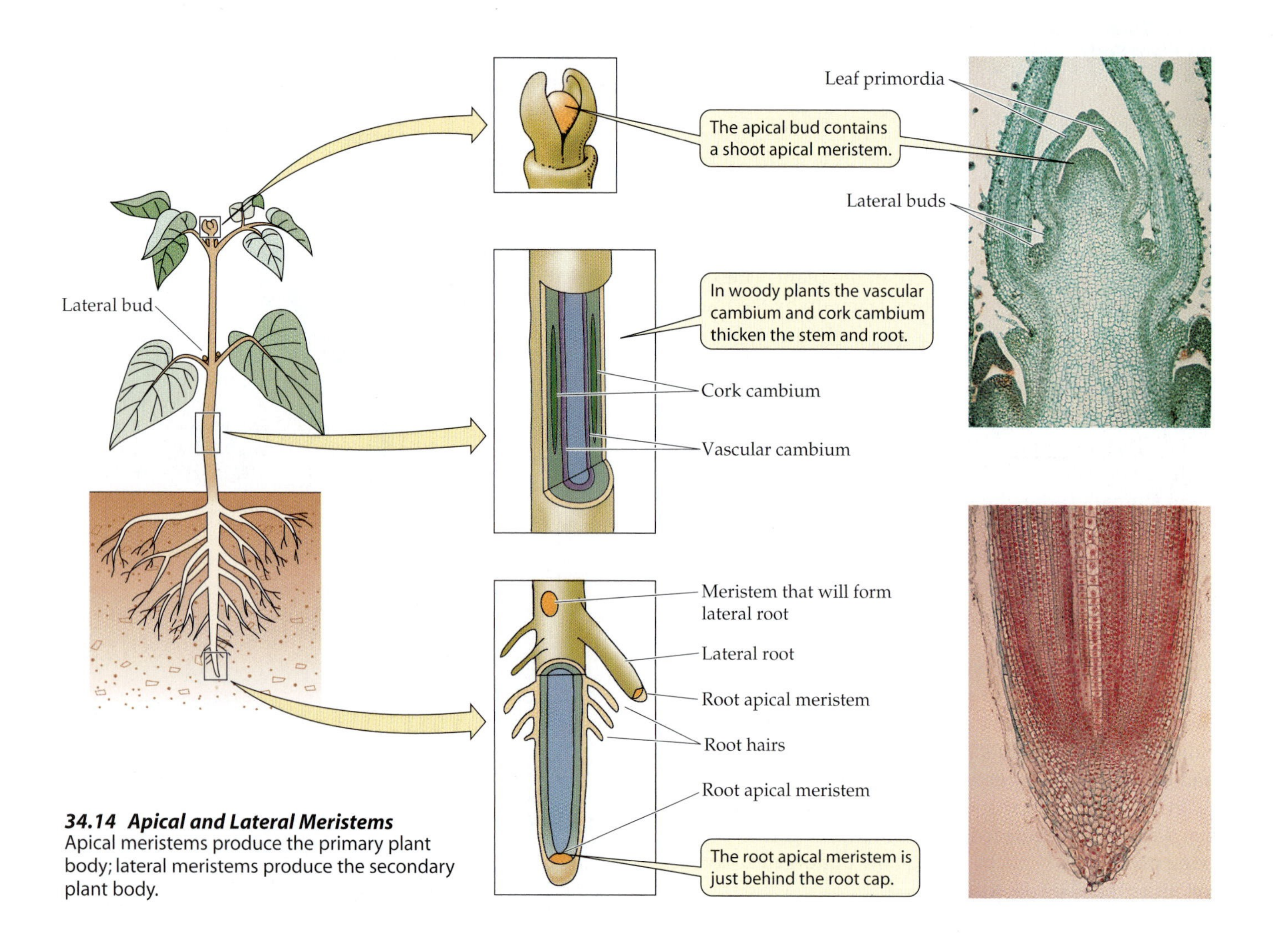

34.14 Apical and Lateral Meristems
Apical meristems produce the primary plant body; lateral meristems produce the secondary plant body.

and the **procambium**. They give rise to the three tissue systems as follows:

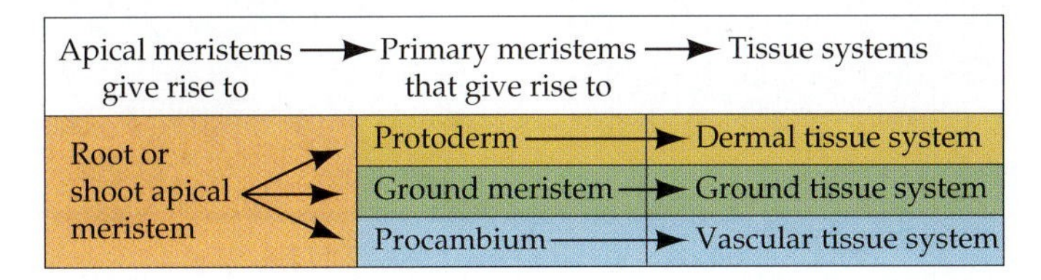

Apical meristems give rise to	Primary meristems that give rise to	Tissue systems
Root or shoot apical meristem	Protoderm	Dermal tissue system
	Ground meristem	Ground tissue system
	Procambium	Vascular tissue system

Apical meristems are responsible for **primary growth**, which leads to elongation and organ formation. All plant organs arise ultimately from cell divisions in the apical meristems, followed by cell expansion and differentiation. Primary growth gives rise to the entire body of many plants.

Because meristems can continue to produce new organs, the plant body is much more variable in form than the animal body, whose organs are laid down only once.

LATERAL MERISTEMS. Some roots and stems develop a secondary body—what we commonly refer to as wood and bark. In an oak tree, for example, the secondary body constitutes almost the entire stem and root system. These complex tissues are derived from two lateral meristems: the vascular cambium and the cork cambium (see Figure 34.14).

The **vascular cambium** is a cylindrical tissue consisting of vertically elongated cells that divide frequently. Toward the inside of the stem or root the dividing cells form new xylem, and toward the outside they form new phloem.

As a tree grows in diameter, the outermost layers of the stem crack and fall off. Without the activity of the **cork cambium**, this sloughing off of the dermal tissues would expose the tree to potential damage, including excessive water loss or invasion by microorganisms. The cork cambium produces new protective cells, primarily in the outward direction. The walls of these cells become impregnated with suberin, which makes them waterproof. The layer of growth produced by the cork cambium is called the **periderm**.

Growth in the diameter of stems and roots, produced by the vascular and cork cambia, is called **secondary growth**. It is the source of wood and bark. **Wood** is secondary xylem. **Bark** (periderm plus secondary phloem) is everything external to the vascular cambium.

In some plants, meristems may remain active for years—even centuries. The bristlecone pine mentioned at the beginning of this chapter provides a dramatic example. Such plants grow in size, or at least in diameter, throughout their lives. Recall that this pattern of continuous growth is known as indeterminate growth. Determinate growth, which stops at some point, is characteristic of some plant parts, such as leaves, flowers, and fruits, as well as most animals.

In the sections that follow, we'll examine how the various meristems give rise to the plant body.

The root apical meristem gives rise to the root cap and the primary meristems

The root apical meristem produces all the cells that contribute to growth in the length of the root. Some of the daughter cells from the apical end of the root apical meristem contribute to a **root cap**, which protects the delicate growing region of the root as it pushes through the soil. Cells of the root cap are often damaged or scraped away and must therefore be replaced constantly. The root cap is also the structure that detects the pull of gravity and thus controls the downward growth of roots.

The daughter cells that are produced at the other end of the meristem elongate and lengthen the root. Following elongation, these cells differentiate, giving rise to the various tissues of the mature root. The growing region above the apical meristem—away from the root cap—comprises the three cylindrical primary meristems that give rise to the three tissue systems of the root: the protoderm, the ground meristem, and the procambium (Figure 34.15).

The apical and primary meristems constitute the **zone of cell division**, the source of all the cells of the root's primary tissues. Just above this zone is the **zone of cell elongation**, where the newly formed cells are elongating and thus causing the root to reach farther into the soil. Above this is the **zone of cell differentiation**, where the cells are taking on specialized forms and functions. These three zones grade imperceptibly into one another; there is some cell division even as far up as the zone of cell differentiation, and some cells differentiate even in the zone of cell division.

The products of the root's primary meristems become root tissues

What are the products of the three primary meristems? The protoderm gives rise to the outer layer of cells—the epidermis—which is adapted for protection of the root and for the absorption of mineral ions and water (Figure 34.16). Epi-

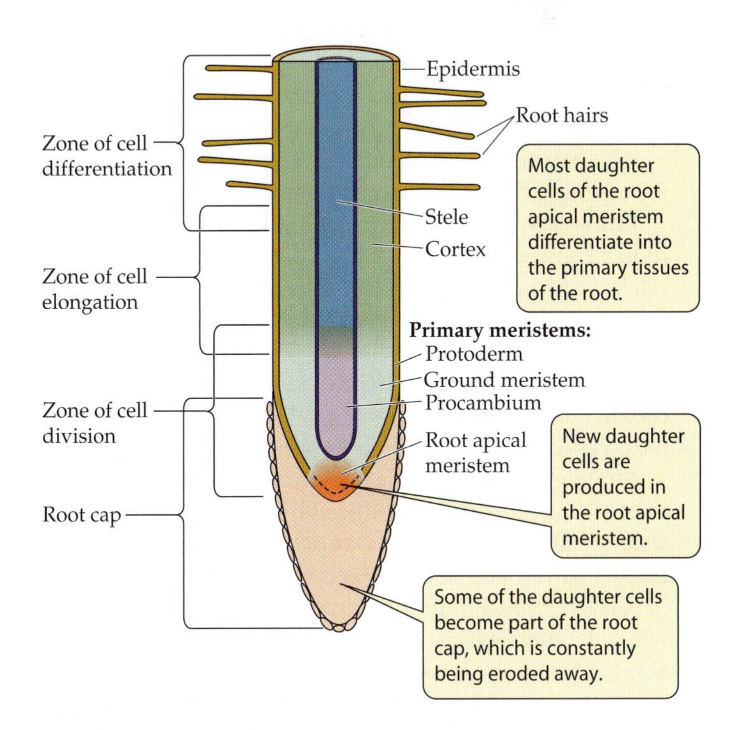

34.15 Tissues and Regions of the Root Tip
Extensive cell division creates the complex structure of the root.

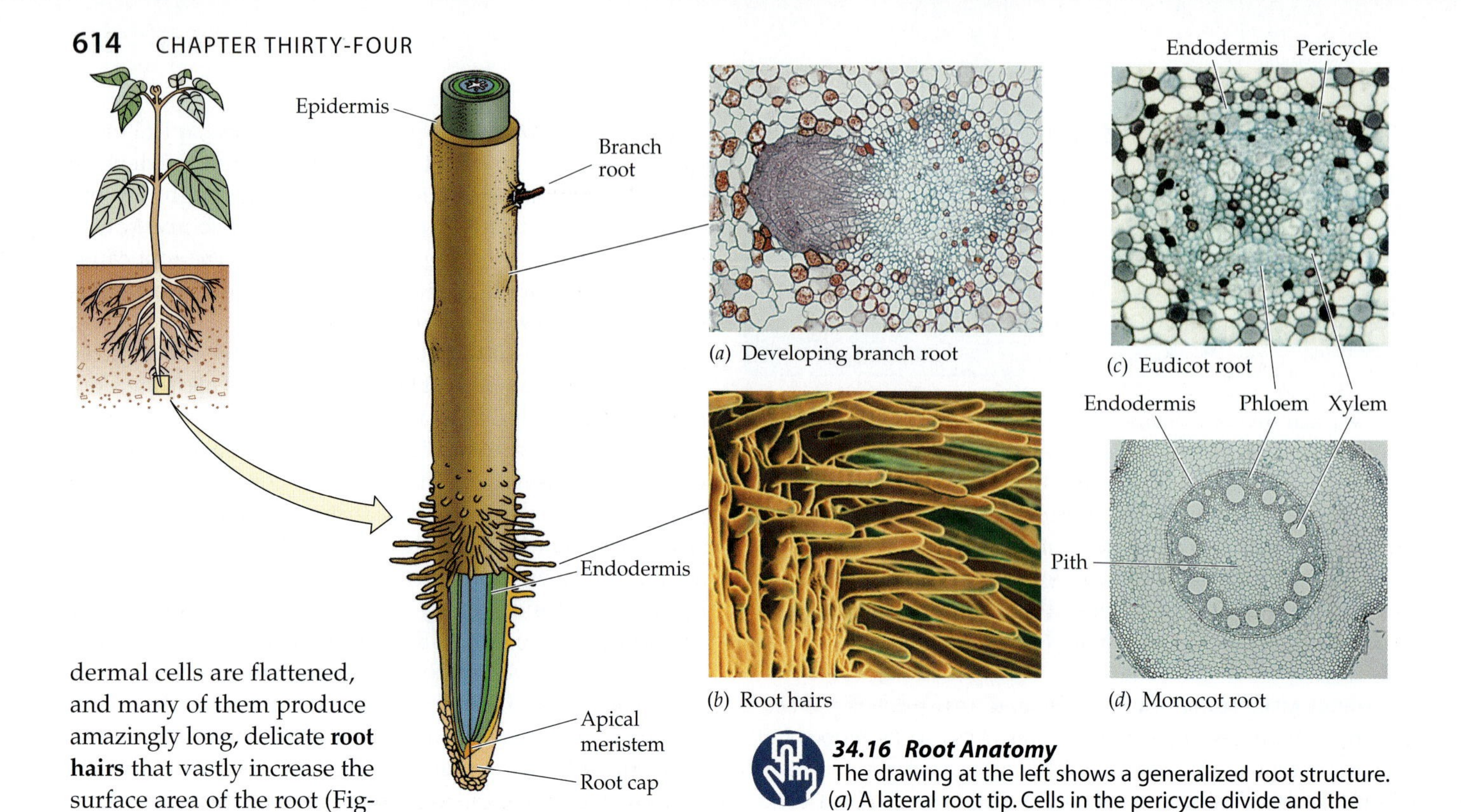

(*a*) Developing branch root

(*b*) Root hairs

(*c*) Eudicot root

(*d*) Monocot root

34.16 Root Anatomy
The drawing at the left shows a generalized root structure. (*a*) A lateral root tip. Cells in the pericycle divide and the products differentiate, forming the tissues of a lateral root. (*b*) Root hairs, seen with a scanning electron micrograph. (*c,d*) The primary root tissues of a eudicot and a monocot. The monocot has a central pith region; the eudicot does not.

dermal cells are flattened, and many of them produce amazingly long, delicate **root hairs** that vastly increase the surface area of the root (Figure 34.16*b*). It has been estimated that the root system of a mature rye plant has a total absorptive surface of more than 600 square meters. Root hairs grow out among the soil particles, probing nooks and crannies and taking up water and minerals.

Internal to the epidermis, the ground meristem gives rise to a region of ground tissue that is many cells thick, called the **cortex**. The cells of the cortex are relatively unspecialized and often function in food storage.

In many plants, but especially in trees, epidermal and sometimes cortical cells form an association with a fungus. This association, called a *mycorrhiza*, increases the absorption of minerals and water by the plant (see Figure 30.16). Some plant species have poorly developed root hairs or no root hairs. These plants cannot survive unless they develop mycorrhizae that help them with mineral absorption.

Proceeding inward, we come to the **endodermis** of the root, a single cylindrical layer of cells that is the innermost cell layer of the cortex. The cell walls of the endodermal cells differ markedly from those of the other cortical cells. Endodermal cell walls contain suberin, which forms a waterproof seal wherever it is present. The placement of the seal in just certain parts of the wall enables endodermal cells to control the access of water and dissolved ions to the vascular tissues.

Moving inward past the endodermis, we enter the vascular cylinder, or **stele**, produced by the procambium. The stele consists of three tissues: pericycle, xylem, and phloem (Figure 34.17). The **pericycle** consists of one or more layers of relatively undifferentiated cells. It is the tissue within which lateral roots arise (see Figure 34.16*a*). The pericycle also contributes to secondary growth.

At the very center of the root of a eudicot lies the xylem—seen in cross section as a star with a variable number of points. Between the points are bundles of phloem. In monocots, a region of parenchyma cells, called the **pith**, lies in the center of the root. The pith often stores carbohydrate reserves. It is useful to try picturing these structures in three dimensions (as in Figure 34.17), rather than attempting to understand their functions solely on the basis of two-dimensional cross sections (as in Figure 34.16*c,d*).

The products of the stem's primary meristems become stem tissues

The shoot apical meristem, like the root apical meristem, forms three primary meristems, which in turn give rise to the three tissue systems. Leaves arise from bulges called *leaf*

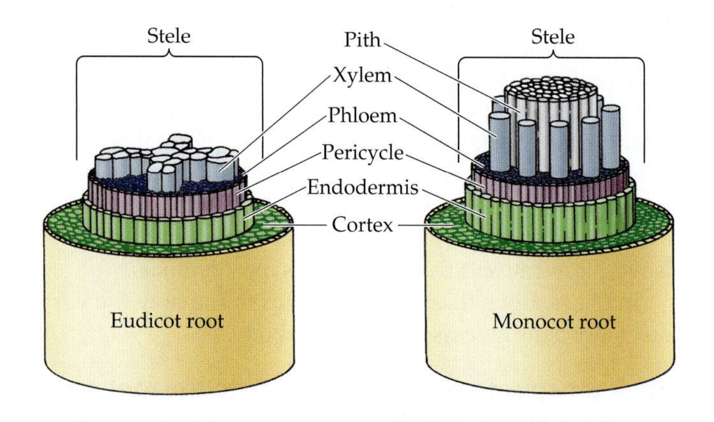

34.17 The Stele
The distribution of tissues in the stele—the region internal to the endodermis—differs in the roots of eudicots and monocots.

primordia, which form as cells divide on the sides of shoot apical meristems (see Figure 34.14). The growing stem has no cap analogous to the root cap, but the leaf primordia can act as a protective covering.

The plumbing of angiosperm stems differs from that of roots. In a root, the vascular tissue lies deep in the interior, with the xylem at or near the very center. The vascular tissue of a young stem, however, is divided into discrete **vascular bundles**, which in eudicots generally form a cylinder but in monocots are seemingly scattered throughout a cross section of the stem (Figure 34.18). Each vascular bundle contains both xylem and phloem.

The stem contains other important tissues in addition to the vascular tissues. Internal to the vascular bundles of eudicots is a storage tissue, the pith, and to the outside lies a similar storage tissue, the cortex. The cortex may contain strengthening collenchyma cells with thickened walls. The pith, the cortex, and the regions between the vascular bundles in eudicots—called *pith rays*—constitute the ground tissue system of the stem. The outermost cell layer of the young stem is the epidermis, the primary function of which is to minimize the loss of water from the cells within.

Many stems and roots undergo secondary growth

Some stems and roots remain slender and show little or no growth in diameter (secondary growth). However, in many eudicots, stems and roots thicken considerably. This thickening is of great importance and interest because it gives rise to wood and bark, and it makes the support of tall trees possible.

Secondary growth results from the activity of the two lateral meristems: vascular cambium and cork cambium (see Figure 34.14). Vascular cambium consists of cells that divide to produce new (secondary) xylem and phloem cells, while cork cambium produces mainly waxy-walled cork cells.

Initially, the vascular cambium is a single layer of cells lying between the primary xylem and the primary phloem. The root or stem increases in diameter when the cells of the vascular cambium divide, producing secondary xylem cells toward the inside of the root or stem and producing secondary phloem cells toward the outside (Figure 34.19). In the stems of woody plants, cells of the pith rays between the vascular bundles also divide, forming a continuous cylinder of vascular cambium running the length of the stem. This cylinder in turn gives rise to complete cylinders of secondary xylem (wood) and secondary phloem from which bark will develop.

As the vascular cambium produces secondary xylem and phloem, its principal cell products are vessel elements, supportive fibers, and parenchyma cells in the xylem, and sieve tube members, companion cells, fibers, and parenchyma cells in the phloem. The parenchyma cells in the xylem and phloem store carbohydrate reserves in the stem and root.

Living tissues such as this storage parenchyma must be connected to the sieve tubes of the phloem, or they will starve to death. The connections are provided by **vascular rays**, which are composed of cells derived from the vascular cambium. The rays, laid down progressively as the cam-

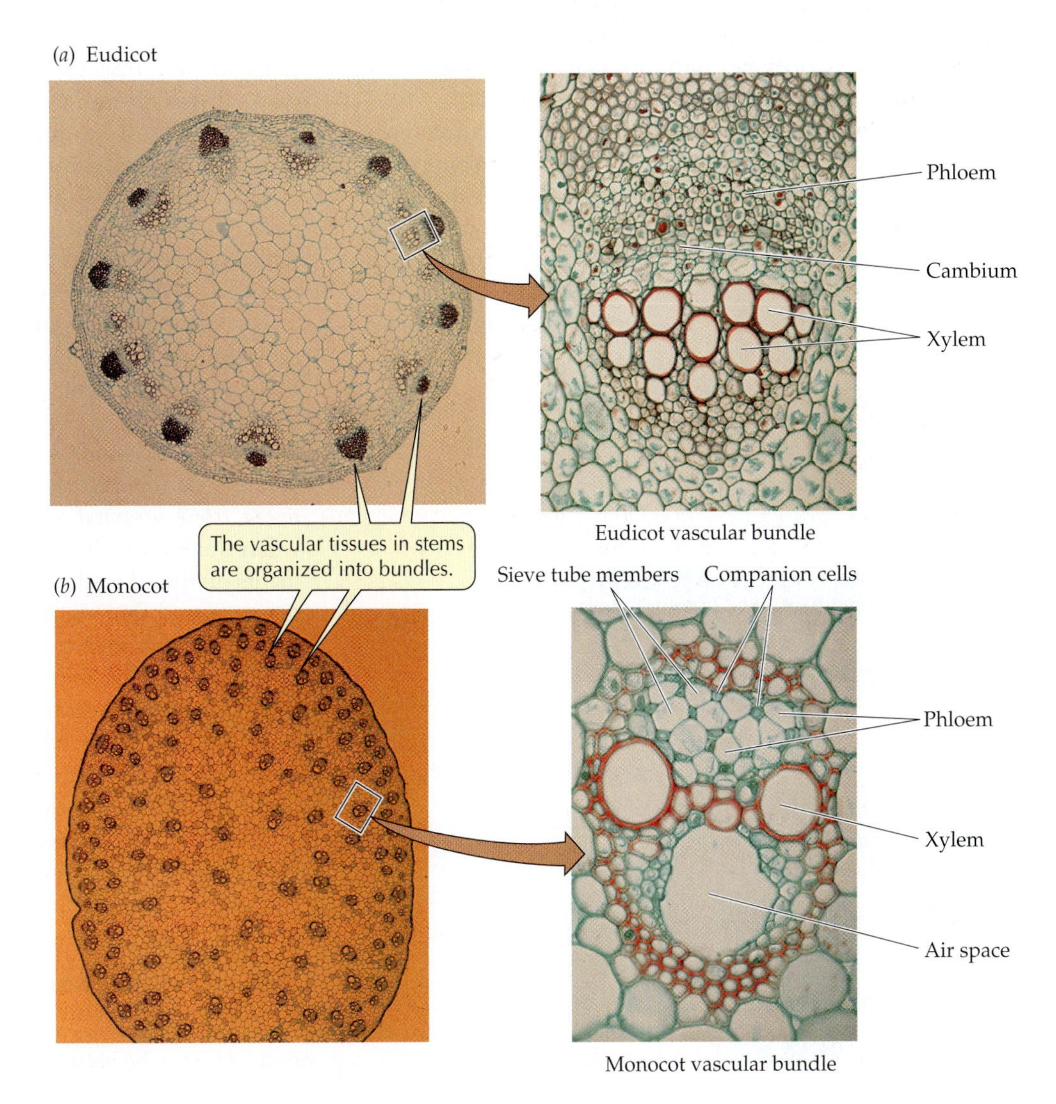

34.18 Vascular Bundles in Stems (*a*) In eudicots the vascular bundles are arranged in a cylinder, with pith in the center and cortex outside the ring. (*b*) This scattered arrangement of bundles is typical of monocot stems.

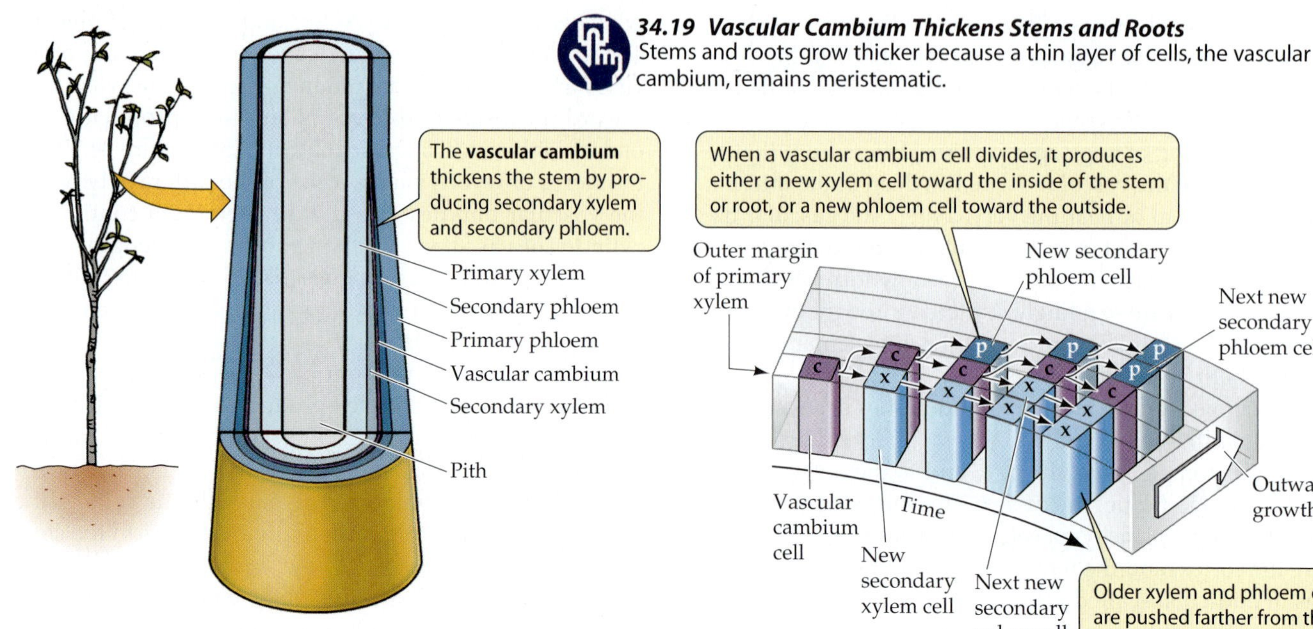

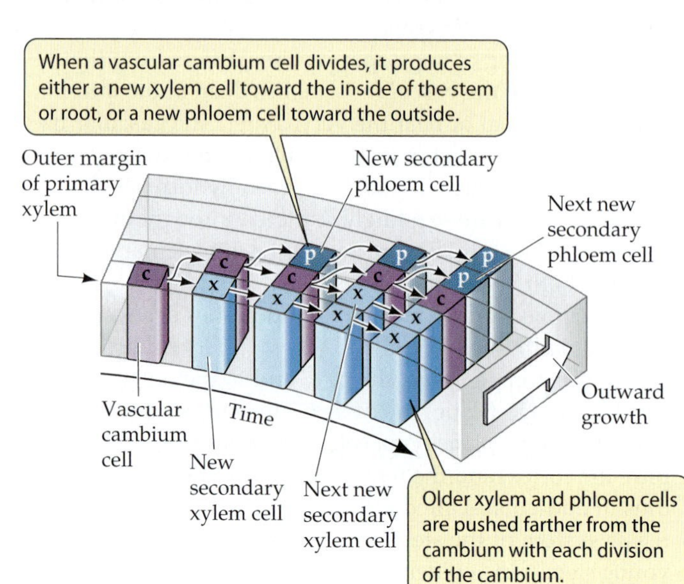

34.19 Vascular Cambium Thickens Stems and Roots
Stems and roots grow thicker because a thin layer of cells, the vascular cambium, remains meristematic.

bium divides, are rows of living parenchyma cells that run perpendicular to the xylem vessels and phloem sieve tubes (Figure 34.20). As the root or stem continues to increase in diameter, new vascular rays are initiated so that this storage and transport tissue continues to meet the needs both of the bark and of the living cells in the xylem.

The vascular cambium itself increases in circumference with the growth of the root or stem. To do this, some of its cells divide in a plane at right angles to the plane that gives rise to secondary xylem and phloem. The products of each of these divisions lie within the vascular cambium itself and increase its circumference.

Only eudicots have a vascular cambium and a cork cambium and thus undergo secondary growth. In the rare cases in which monocots form thickened stems—palm trees, for example—they do so without using vascular cambium or cork cambium. Palm trees have a very wide apical meristem that produces a wide stem, and dead leaf bases also add to the diameter of the stem. Basically, monocots grow in the same way as do other angiosperms that lack secondary growth.

Wood and bark are unique to plants showing secondary growth. These tissues have their own patterns of organization and development.

WOOD. Cross sections of most tree trunks (stems) in temperate-zone forests show *annual rings* (Figure 34.21), which result from seasonal environmental conditions. In spring, when water is relatively plentiful, the tracheids or vessel elements produced by the vascular cambium tend to be large in diameter and thin-walled. As water becomes less available during the summer, narrower cells with thicker walls are produced, making this summer wood darker and perhaps more dense than the wood formed in spring. Thus each year is usually recorded in a tree trunk by a clearly visible annual ring consisting of one light and one dark layer. Trees in the wet tropics do not lay down such obvious regular rings.

The difference between old and new regions also contributes to the appearance of wood. As a tree grows in di-

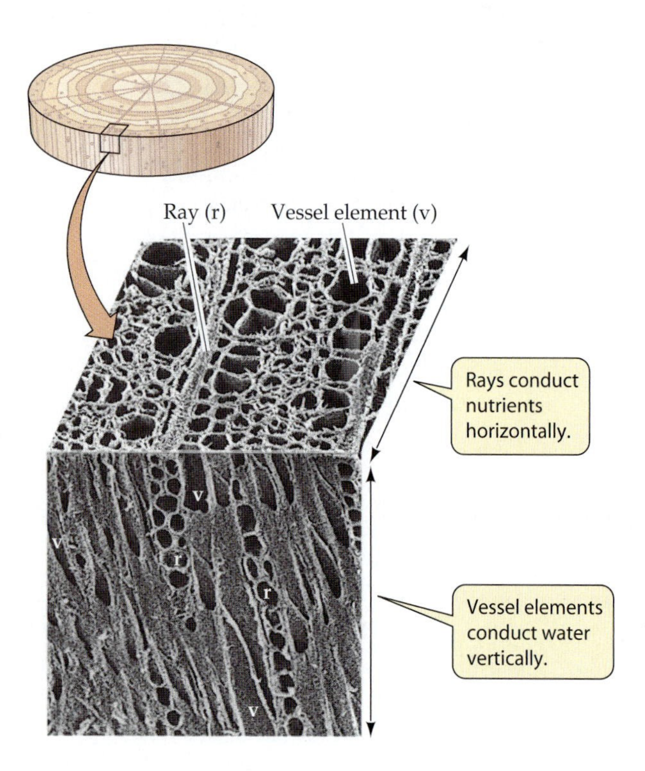

34.20 Vascular Rays and Vessel Elements
Wood of the tulip poplar, showing that the orientation of xylem vessels is perpendicular to that of vascular rays. The rays transport sieve tube sap horizontally from the phloem to storage parenchyma cells.

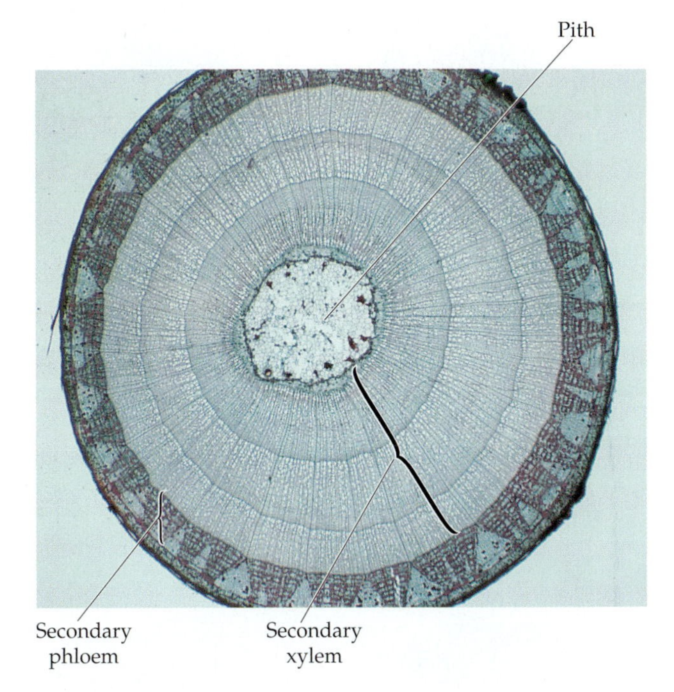

34.21 Annual Rings
Rings of secondary xylem are the most noticeable feature of this cross section from a 3-year-old basswood stem.

ameter, the xylem toward the center becomes clogged with water-insoluble substances and ceases to conduct water and minerals; this is *heartwood* and appears darker in color. The portion of the xylem that is actively conducting all water and minerals in the tree is called *sapwood* and is lighter in color and more porous than heartwood.

The knots that we find attractive in knotty pine but regard as a defect in structural timbers are cross sections of branches: As a trunk grows, the bases of branches become buried in the trunk's new wood and appear as knots when the trunk is cut lengthwise.

BARK. As secondary growth of stems or roots continues, the expanding vascular tissue stretches and breaks the epidermis and cortex, which ultimately flake away and are lost. Tissue derived from the secondary phloem then becomes the outermost part of the stem. Before the dermal tissues are broken away, cells lying near the surface of the secondary phloem begin to divide and produce layers of **cork**, a tissue composed of cells with thick walls, waterproofed with suberin. The cork soon becomes the outermost tissue of the stem or root. The dividing cells, derived from the secondary phloem, form a cork cambium. Sometimes the cork cambium produces cells to the inside as well as to the outside; these cells constitute what is known as the **phelloderm**.

Cork, cork cambium, and phelloderm make up the periderm of the secondary plant body. As the vascular cambium continues to produce secondary vascular tissue, the corky layers are in turn lost, but the continuous formation of cork cambium in the underlying phloem gives rise to new corky layers.

When bark forms on stems and roots, the underlying tissues still need to release carbon dioxide and take up oxygen. **Lenticels** are spongy regions in the cork of stems and roots that allow such gas exchange (Figure 34.22).

Leaf Anatomy Supports Photosynthesis

We can think of roots and stems as important supporting actors that sustain the activities of the real stars of the plant body, the leaves—the organs of photosynthesis. Leaf anatomy is beautifully adapted to carry out photosynthesis and to support photosynthesis by exchanging the gases O_2 and CO_2 with the environment, limiting evaporative water loss, and transporting the products of photosynthesis to the rest of the plant. Figure 34.23*a* shows a typical eudicot leaf in cross section.

(*a*)

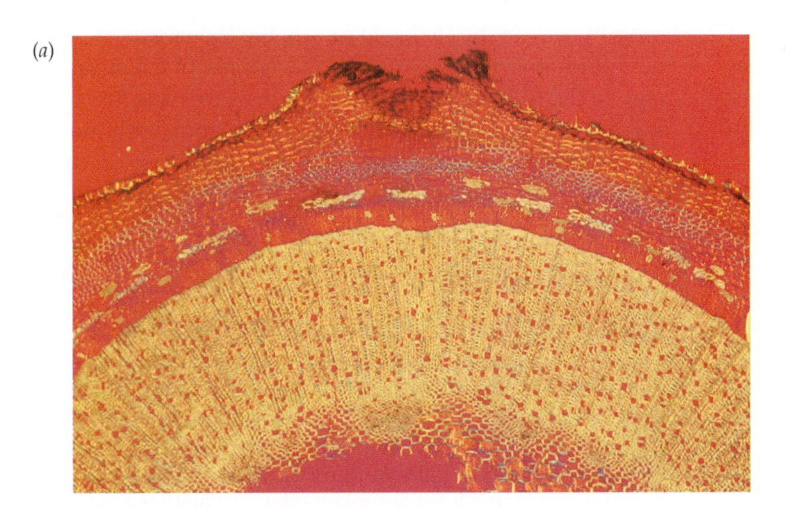

(*b*)

34.22 Lenticels Allow Gas Exchange through the Periderm
(*a*) The region of periderm that appears broken open is a lenticel in a year-old elder twig; note the spongy tissue that constitutes the lenticel. (*b*) The rough areas on the trunk of this Chinese plum tree are lenticels. Most tree species have lenticels much smaller than these.

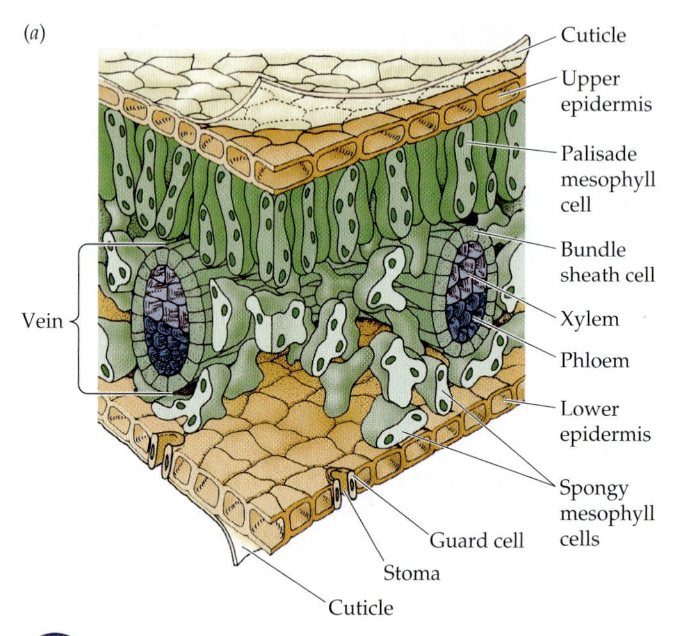

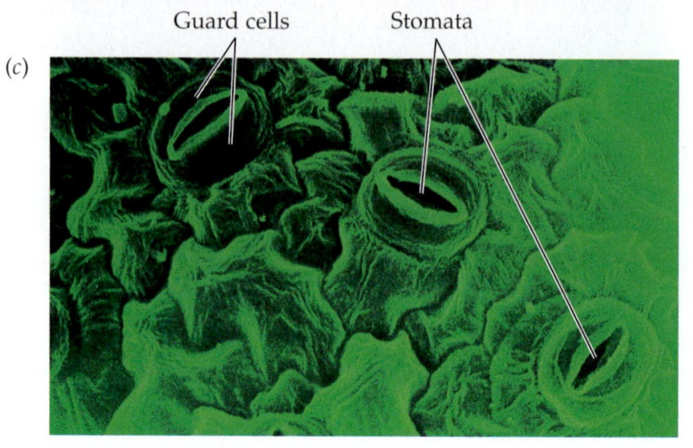

34.23 The Eudicot Leaf
(*a*) Cross section of a eudicot leaf. (*b*) The network of fine veins in this maple leaf carries water to the mesophyll cells and carries photosynthetic products away from them. (*c*) These paired cells on the lower epidermis of a eudicot leaf are guard cells; the gaps between them are stomata, through which carbon dioxide enters the leaf.

Most eudicot leaves have two zones of photosynthesizing parenchyma tissue referred to as **mesophyll**, which means "middle of the leaf." The upper layer or layers of mesophyll consist of roughly cylindrical cells. This zone is referred to as *palisade mesophyll*. The lower layer or layers consist of irregularly shaped cells; this zone is called *spongy mesophyll*. Within the mesophyll is a great deal of air space through which carbon dioxide can diffuse to surround all photosynthesizing cells.

Vascular tissue branches extensively in the leaf, forming a network of **veins** (Figure 34.23*b*). Veins extend to within a few cell diameters of all the cells of the leaf, ensuring that the mesophyll cells are well supplied with water and minerals. The products of photosynthesis are loaded into the phloem of the veins for export to the rest of the plant.

Covering the entire leaf is a layer of nonphotosynthetic cells constituting the epidermis. The epidermal cells have an overlying waxy cuticle that is highly impermeable to water. But this impermeability poses a problem: While keeping water in the leaf, the epidermis also keeps carbon dioxide, the other raw material of photosynthesis, out.

The problem of balancing water retention and carbon dioxide availability is solved by an elegant regulatory system that will be discussed in more detail in the next chapter. *Guard cells* are modified epidermal cells that change their shape, thereby opening or closing pores called *stomata*, which serve as passageways between the environment and the leaf's interior (Figure 34.23*c*). When the stomata are open, carbon dioxide can enter and oxygen can leave, but some water can also be lost.

In Chapter 8 we described C_4 plants, which can fix carbon dioxide efficiently even when the carbon dioxide supply in the leaf decreases to a level at which the photosynthesis of C_3 plants is inefficient. One adaptation that helps C_4 plants do this is their modified leaf anatomy, as shown in Figure 8.19. Notice that the photosynthetic cells in the C_4 leaf are grouped around the veins in concentric layers, forming an outer mesophyll layer and an inner *bundle sheath*. These layers each contain different types of chloroplasts, leading to the biochemical division of labor described in Chapter 8.

Leaves receive water and mineral nutrients from the roots by way of the stems. In return, the leaves export products of photosynthesis, providing a supply of chemical energy to the rest of the plant body. And, as we have just seen, leaves exchange gases, including water vapor, with the environment by way of the stomata. All three of these processes will be considered in detail in the next chapter.

Chapter Summary

Vegetative Organs of the Flowering Plant Body

- Monocots typically have a single cotyledon, narrow leaves with parallel veins, flower parts in threes or multiples of three, and stems with scattered vascular bundles.
- Eudicots typically have two cotyledons, broad leaves with netlike veins, flower parts in fours or fives, and vascular bundles in a ring. Flowering plants that are neither monocots nor eudicots are generally similar in structure to eudicots. **Review Figure 34.1**
- The vegetative organs of flowering plants are roots, which form a root system, and stems and leaves, which form a shoot system. **Review Figure 34.2**

▶ Roots anchor the plant and take up water and minerals.

▶ Stems bear leaves and buds. Lateral buds form branches. Apical buds produce cells that contribute to the elongation of the stem.

▶ Leaves are responsible for most photosynthesis, for which their flat blades, oriented perpendicular to the sun's rays, are well adapted. **Review Figure 34.5**

Plant Cells

▶ The walls of plant cells have a structure that often corresponds to the special functions of the cell. The walls of individual cells are separated by a middle lamella common to two neighboring cells; each cell also has its own primary wall. **Review Figure 34.6**

▶ Some cells produce a thick secondary wall. Adjacent cells are connected by plasmodesmata that extend through both cell walls. **Review Figures 34.7, 34.8**

▶ Parenchyma cells have thin walls. Many parenchyma cells store starch or lipids; some others carry out photosynthesis. **Review Figure 34.9*a***

▶ Collenchyma cells provide flexible support. **Review Figure 34.9*b***

▶ Sclerenchyma cells provide strength and function when dead. **Review Figure 34.9*c, d***

▶ Tracheids and vessel elements are xylem cells that conduct water and minerals after the cells die. **Review Figures 34.9*e, f*, 34.10**

▶ Sieve tube members are the conducting cells of the phloem. Their activities are often controlled by companion cells. **Review Figure 34.11**

Plant Tissues and Tissue Systems

▶ Three tissue systems extend throughout the plant body. The vascular tissue system, consisting of xylem and phloem, conducts water, minerals, and the products of photosynthesis. The dermal tissue system protects the body surface. The ground tissue system produces and stores food materials and performs other functions. **Review Figure 34.12**

Forming the Plant Body

▶ The apical–basal pattern and the radial pattern are parts of the plant body plan; they arise through orderly development. **Review Figure 34.13**

▶ The plant body is modular, and the growth of stems and roots is indeterminate. Leaves, flowers, and fruits show determinate growth.

▶ Meristems are localized regions of cell division. A hierarchy of meristems generates the plant body.

▶ Apical meristems at the tips of stems and roots produce the primary growth of those organs. **Review Figure 34.14**

▶ Shoot apical meristems and root apical meristems give rise to primary meristems: the protoderm, the ground meristem, and the procambium. **Review Figure 34.15**

▶ In some plants, the products of primary growth constitute the entire plant body. Many other plants show secondary growth. Two lateral meristems, the vascular cambium and cork cambium, are responsible for secondary growth. **Review Figure 34.14**

▶ The young root has an apical meristem that gives rise to the root cap and to the three primary meristems, which in turn produce the three tissue systems. The protoderm produces the dermal tissue system, the ground meristem produces the ground tissue system, and the procambium produces the vascular tissue system. **Review Figure 34.15**

▶ Root tips have three overlapping zones: the zone of cell division, the zone of cell elongation, and the zone of cell differentiation. **Review Figure 34.15**

▶ The dermal tissue system consists of the epidermis, part of which forms the root hairs that are responsible for absorbing water and minerals. **Review Figure 34.16**

▶ The ground tissue system of a young root is the cortex, whose innermost cell layer, the endodermis, controls access to the stele.

▶ The stele, consisting of the pericycle, xylem, and phloem, is the root's vascular tissue system. Lateral roots arise in the pericycle. Monocot roots have a central pith region. **Review Figure 34.17**

▶ The shoot apical meristem also gives rise to three primary meristems, with roles similar to their counterparts in the root. Leaf primordia on the sides of the apical meristem develop into leaves.

▶ The vascular tissue in young stems is divided into vascular bundles, each containing both xylem and phloem. Pith occupies the center of the eudicot stem. Cortex lies to the outside of the vascular bundles in eudicots, with pith rays lying between the vascular bundles. **Review Figure 34.18**

▶ Many eudicot stems and roots show secondary growth, in which vascular and cork cambia give rise to secondary xylem and secondary phloem. As secondary growth continues, the products are wood and bark. **Review Figure 34.19**

▶ The vascular cambium lays down layers of secondary xylem and phloem. Living cells within these tissues are nourished by vascular rays. **Review Figure 34.20**

▶ The periderm consists of cork, cork cambium, and phelloderm, all pierced at intervals by lenticels that allow gas exchange.

Leaf Anatomy Supports Photosynthesis

▶ The photosynthetic tissue of a leaf is called mesophyll. Veins bring water and minerals to the mesophyll and carry the products of photosynthesis to other parts of the plant body.

▶ A waxy cuticle prevents water loss from the leaf, but is impermeable to carbon dioxide. Guard cells control the opening of stomata, openings in the leaf that allow CO_2 to enter but also allow some water to escape. **Review Figure 34.23**

For Discussion

1. When a young oak was 5 m tall, a thoughtless person carved his initials in its trunk at a height of 1.5 m above the ground. Today that tree is 10 m tall. How high above the ground are those initials? Explain your answer in terms of the manner of plant growth.
2. Consider a newly formed sieve tube member in the secondary phloem of an oak tree. What kind of cell divided to produce the sieve tube member? What kind of cell divided to produce that parent cell? Keep tracing back until you arrive at a cell in the apical meristem.
3. Distinguish between sclerenchyma cells and collenchyma cells in terms of structure and function.
4. Distinguish between primary and secondary growth. Do all angiosperms undergo secondary growth? Explain.
5. What anatomical features make it possible for a plant to retain water as it grows? Describe the plant tissues and how and when they form.

35 Transport in Plants

ABOUT 40 YEARS AGO THE BIOLOGIST PER Scholander was studying water movement to the top of an 80-meter Douglas fir. To collect samples rapidly from the treetop, he hired a sharpshooter, who aimed a high-powered rifle at a twig high in the tree and fired. From high above, a twig fluttered to the ground, and Scholander quickly inserted it into an instrument for measuring tension in the xylem sap. As we will soon see, Scholander's measurements increased our understanding of how water and minerals reach the tops of tall trees.

The water and minerals in a plant's xylem must be transported to the entire shoot system, all the way to the highest leaves and apical buds. Carbohydrates produced in all the leaves, including the highest, must be translocated to all the living nonphotosynthetic parts of the plant. Before we consider the mechanisms underlying these processes, we should consider two questions: How much water is transported? And how high can water be transported?

In answer to the first question, consider the following example: A single maple tree 15 meters tall was estimated to have some 177,000 leaves, with a total leaf surface area of 675 square meters—half again the area of a basketball court. During a summer day, that tree lost 220 liters of water *per hour* to the atmosphere by evaporation from the leaves. To prevent wilting, the xylem needed to transport 220 liters of water from the roots to the leaves every hour. (By comparison, a 50-gallon drum holds 189 liters.)

The second question can be rephrased: How tall are the tallest trees? The tallest gymnosperms, the coast redwoods—*Sequoia sempervirens*—exceed 110 meters in height, as do the tallest angiosperms, the Australian *Eucalyptus regnans*. Any successful explanation of water transport in the xylem must account for transport to these great heights.

In this chapter we consider the uptake and transport of water and minerals by plants, the control of evaporative water loss through the stomata, and the translocation of substances in the phloem.

A Long Way to the Top
Water and minerals must defy gravity and climb over 80 meters to reach the top branches of these Douglas firs (*Pseudotsuga menziesii*).

Uptake and Transport of Water and Minerals

Terrestrial plants obtain both water and mineral nutrients from the soil, usually by way of their roots. You already know that water is one of the ingredients required for carbohydrate production by photosynthesis in the leaves. Water is also essential for transporting solutes, for cooling the plant, and for developing the internal pressure that supports the plant body.

How do leaves high in a tree obtain water from the soil? What are the mechanisms by which water and mineral ions enter the plant body through the roots and ascend as sap in the xylem? Because neither water nor minerals can move through the plant into the xylem without crossing at least one plasma membrane, we focus first on osmosis. Then we examine the uptake of mineral ions, and follow the pathway by which both water and minerals move through the root to gain entry to the xylem.

Water moves through a membrane by osmosis

Osmosis, the movement of water through a membrane in accordance with the laws of diffusion, was described in Chapter 5. The **solute potential** (osmotic potential) of a solution is a measure of the effect of dissolved solutes on the osmotic behavior of the solution. The greater the solute concentration of a solution, the more negative its solute potential, and the greater the tendency of water to move into it from another solution of lower solute concentration (and less negative solute potential). For osmosis to occur, the two solutions must be separated by a membrane permeable to water but relatively impermeable to the solute. Recall, too, that osmosis is a passive process—energy is not required.

Unlike animal cells, plant cells are surrounded by a relatively rigid cell wall. As water enters a plant cell, the entry of more water is increasingly resisted by an opposing **pressure potential** (turgor pressure), owing to the rigidity of the wall. As more and more water enters, the pressure potential becomes greater and greater.

Pressure potential is a hydraulic pressure analogous to the air pressure in an automobile tire; it is a real pressure that can be measured with a pressure gauge. Cells with walls do not burst when placed in pure water; instead, water enters by osmosis until the pressure potential exactly balances the solute potential. At this point, the cell is *turgid*; that is, it has a high pressure potential.

The overall tendency of a solution to take up water from pure water, across a membrane, is called its **water potential**, represented as ψ, the Greek letter psi (Figure 35.1). The water potential is simply the sum of the (negative) solute potential (ψ_s) and the (usually positive) pressure potential (ψ_p):

$$\psi = \psi_s + \psi_p$$

For pure water under no applied pressure, all three of these parameters are zero.

We can measure solute potential, pressure potential, and water potential in *megapascals* (MPa), a unit of pressure. (Atmospheric pressure is about 0.1 MPa, or 14.7 pounds per square inch; typical pressure in an automobile tire is about 0.2 MPa.)

In all cases in which water moves between two solutions separated by a membrane, the following rule of osmosis applies: *Water always moves across a differentially permeable membrane toward the region of more negative water potential.*

Osmotic phenomena are of great importance to plants. The structure of many plants is maintained by the pressure potential of their cells; if the pressure potential is lost, a plant *wilts*. Within living tissues, the movement of water from cell to cell by osmosis follows a gradient of *water potential*. Over longer distances, in open tubes such as xylem vessels and phloem sieve tubes, the flow of water and dissolved solutes is driven by a gradient in *pressure potential*. The movement of a solution due to a difference in pressure potential between two parts of a plant is called **bulk flow**.

Uptake of mineral ions requires transport proteins

Mineral ions, which carry electric charges, cannot move across a plasma membrane unless they are aided by trans-

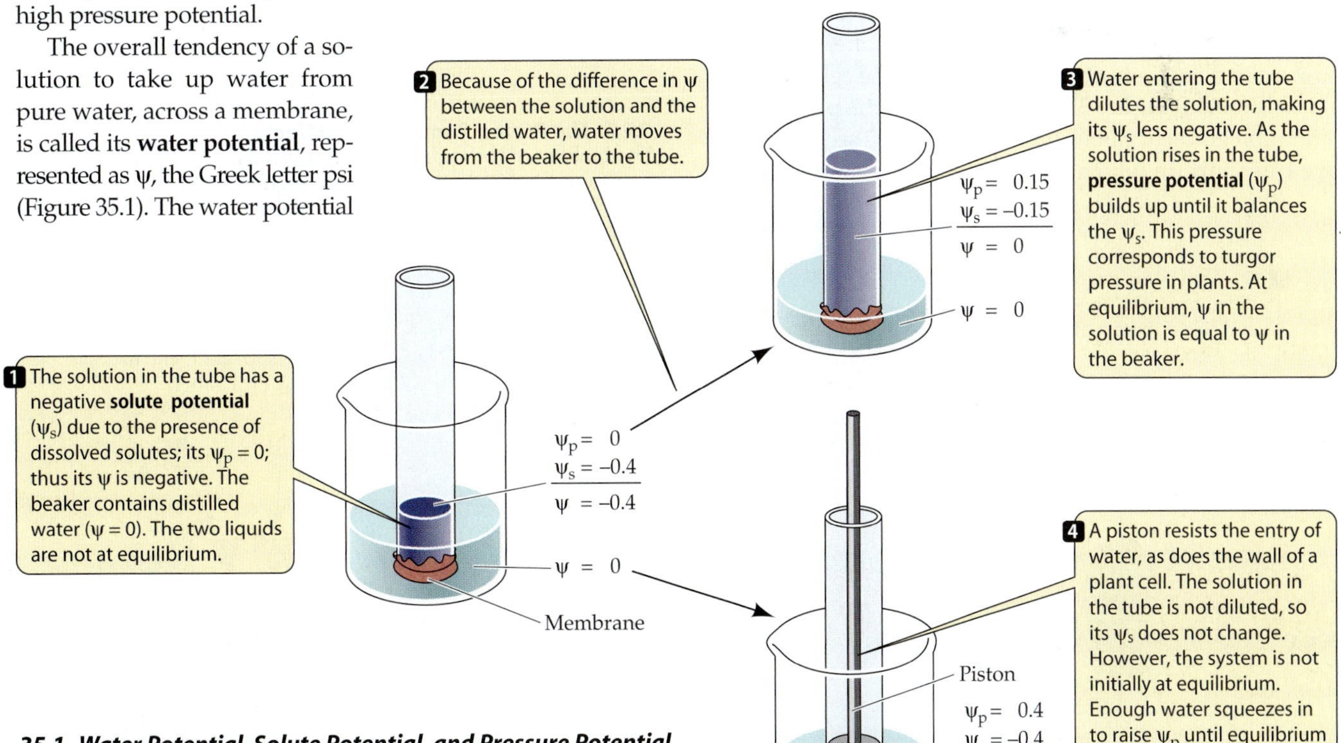

35.1 Water Potential, Solute Potential, and Pressure Potential Water potential (ψ) is the tendency of a solution to take up water from pure water. The water potential is the sum of the solute potential (ψ_s) and the pressure potential (ψ_p). For pure water under no applied pressure, all three of these parameters are equal to zero.

port proteins. (You may wish to review the description of transport proteins in Chapter 5.) When the concentration of these charged ions in the soil is greater than that in the plant, ion channels and carrier proteins can move them into the plant by facilitated diffusion.

The concentrations of some ions in the soil solution, however, are lower than those required inside the plant. Thus the plant must take up these ions against a concentration gradient. Electric potential also plays a role in this process: To move a negatively charged ion into a negatively charged region is to move it against an electrical gradient. The combination of concentration and electrical gradients is called an *electrochemical gradient*. Uptake against an electrochemical gradient is active transport, an energy-requiring process, depending on cellular respiration for ATP. Active transport, of course, requires specific carrier proteins.

Unlike animals, plants do not have a sodium–potassium pump for active transport. Rather, plants have a **proton pump**, which uses energy obtained from ATP to move protons out of the cell against a proton concentration gradient (Figure 35.2*a*). Because protons (H^+) are positively charged, their accumulation on one side of a membrane has two results:

- The region outside the membrane becomes positively charged with respect to the region inside.
- A proton concentration gradient develops.

Each of these results has consequences for the movement of other ions. Because of the charge difference across the membrane, the movement of positively charged ions, such as potassium (K^+), into the cell through their membrane channels is enhanced. These positive ions move into the now more negatively charged interior of the cell by facilitated diffusion (Figure 35.2*b*). In addition, the proton concentration gradient can be harnessed to drive secondary active transport, in which negatively charged ions such as chloride (Cl^-)are moved into the cell against an electrochemical gradient by a symport protein that couples their movement with that of H^+ (Figure 35.2*c*). In sum, there is vigorous traffic of ions across plant membranes.

The proton pump and the coordinated activities of other membrane transport proteins cause the interior of a plant cell to be strongly negative with respect to the exterior. Such a difference in charge across a membrane is called a *membrane potential*. Biologists can measure the membrane potential of a plant cell with microelectrodes, just as they can measure similar charge differences in neurons (nerve cells) and other animal cells. Most plant cells have a membrane potential of at least –120 millivolts, and they maintain it at this level. The membrane potential difference affects the movements of mineral ions into and out of cells.

Water and ions pass to the xylem by way of the apoplast and symplast

Mineral ions enter and move through plants in various ways. Where bulk flow of water is occurring, dissolved minerals are carried along in the stream. Where water is moving more slowly, minerals move by diffusion. At certain points, where plasma membranes are being crossed, some mineral ions are moved by active transport. One such point is the surface of a root hair, where mineral ions first enter the cells of the plant. Later, within the stele, the ions must cross a plasma membrane before entering the lifeless cells of the xylem.

The movement of ions across membranes can also result in the movement of water. Water moves into a root because the root has a more negative water potential than does the

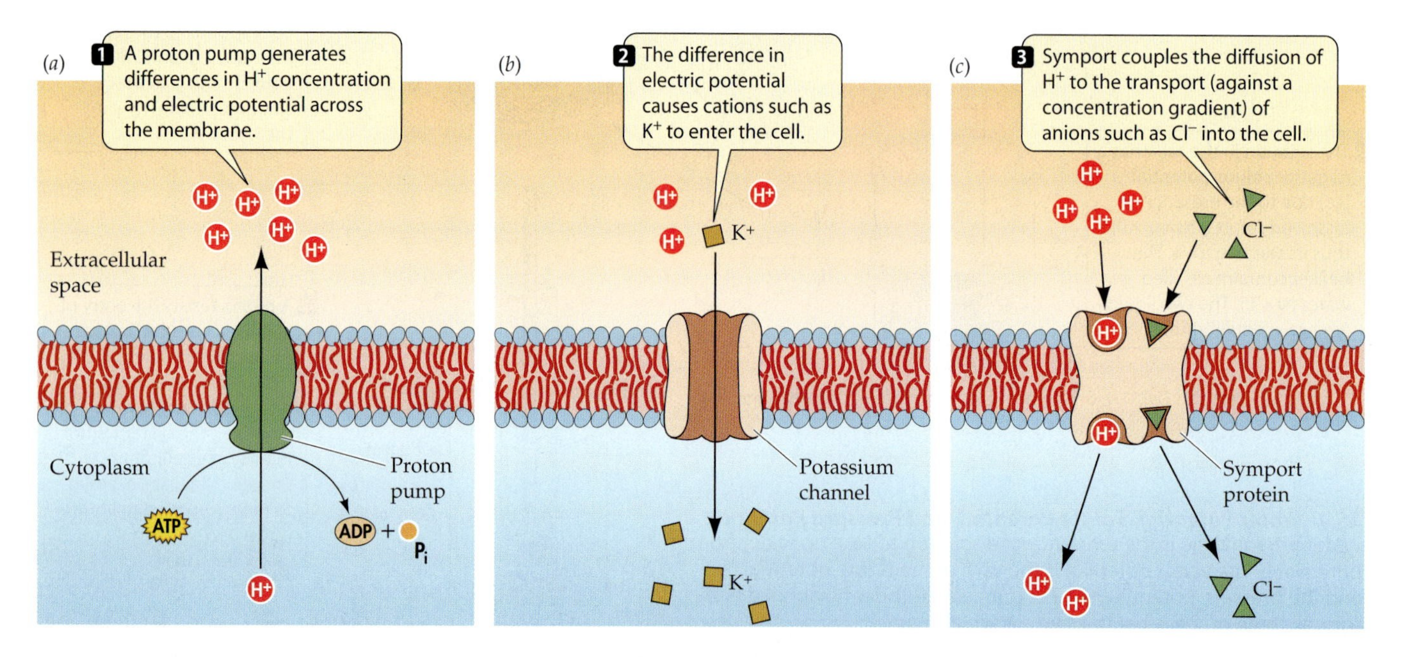

35.2 The Proton Pump and Its Effects
The buildup of hydrogen ions transported across the plasma membrane by the proton pump (*a*) triggers the movement of both cations (*b*) and anions (*c*) into the cell.

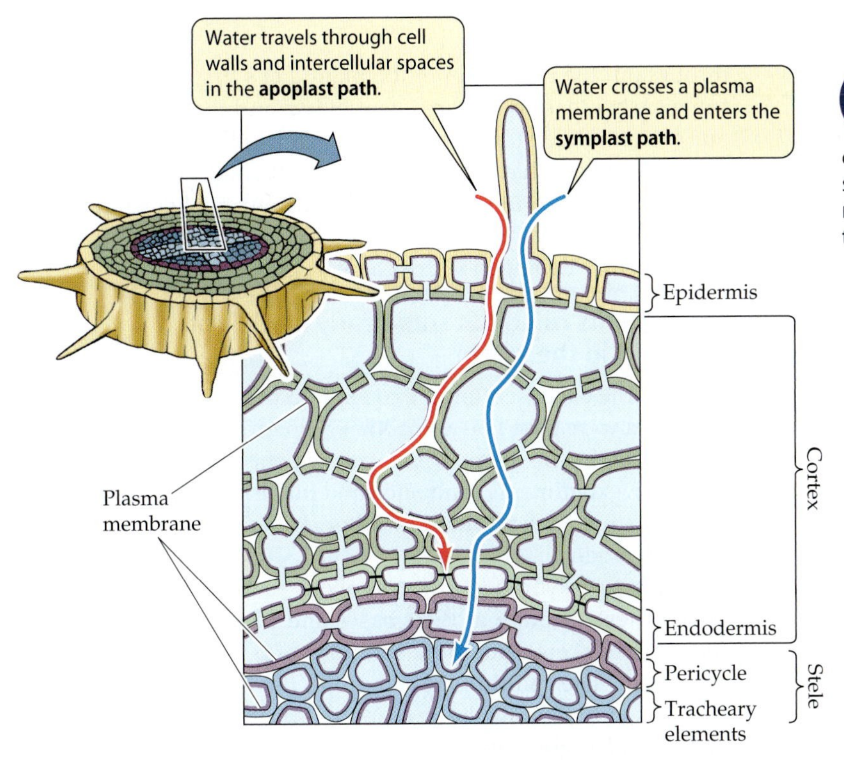

35.3 Apoplast and Symplast The plant cell walls and intercellular spaces constitute the apoplast. The symplast comprises the living cells, which are connected by plasmodesmata. To enter the symplast, water and solutes must pass through a plasma membrane. No such selective barrier limits movement through the apoplast.

soil solution. Water moves from the cortex of the root into the stele (which is where the vascular tissues are located) because the stele has a more negative water potential than does the cortex.

Water and minerals from the soil may pass through the dermal and ground tissues to the stele via two pathways: the apoplast and the symplast. Plant cells are surrounded by cell walls that lie outside the plasma membrane, and intercellular spaces (spaces between cells) are common in many tissues. The walls and intercellular spaces together constitute the **apoplast** (from the Greek for "away from living material"). The apoplast is a continuous meshwork through which water and dissolved substances can flow or diffuse without ever having to cross a membrane (Figure 35.3). Movement of materials through the apoplast is thus unregulated.

The remainder of the plant body is the **symplast** (from the Greek for "together with living material"). The symplast is the portion of the plant body enclosed by membranes—the continuous cytoplasm of the living cells, connected by plasmodesmata (see Figure 35.3). The selectively permeable plasma membranes of the cells control access to the symplast, so movement of water and dissolved substances into the symplast is tightly regulated.

Water and minerals can pass from the soil solution through the apoplast as far as the endodermis, the innermost layer of the cortex. The endodermis is distinguished from the rest of the ground tissue by the presence of **Casparian strips**. These waxy, suberin-containing structures impregnate the endodermal cell wall and form a belt surrounding the endodermal cells. The Casparian strips act as a gasket that prevents water and ions from moving between the cells (Figure 35.4).

The Casparian strips of the endodermis thus completely separate the apoplast of the cortex from the apoplast of the stele. They do not obstruct the outer or inner faces of the endodermal cells. Accordingly, water and ions can enter the stele only by way of the symplast—that is, by entering and passing through the cytoplasm of the endodermal cells. Thus transport proteins in the membranes of these cells determine which mineral ions pass into the stele, and at what rates. This is one of several ways in which plants regulate their chemical composition and ensure an appropriate balance of their constituents. This balance is essential to plant life.

Once they have passed the endodermal barrier, water and minerals leave the symplast. Parenchyma cells in the pericycle or xylem help mineral ions move back into the apoplast. Some of these parenchyma cells, called **transfer cells**, are structurally modified for transporting mineral ions from their cytoplasm (part of the symplast) into their cell walls (part of the apoplast). The wall that receives the transported ions has many knobby extensions projecting into the transfer cell, increasing the surface area of the plasma membrane, the

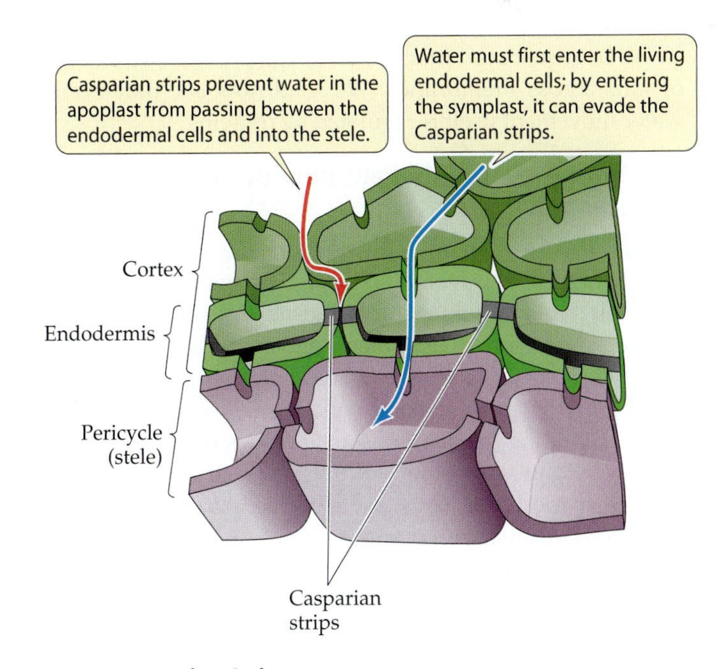

35.4 Casparian Strips
Suberin-impregnated Casparian strips prevent water and ions from moving between the endodermal cells.

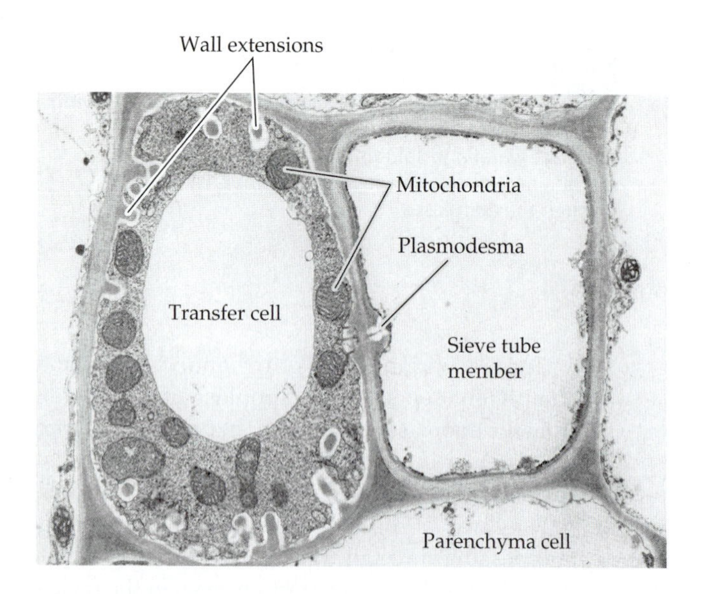

35.5 A Transfer Cell
Three walls of this transfer cell in a pea leaf have knobby extensions that face the cells from which the transfer cell imports solutes. This transfer cell exports the solutes to the neighboring sieve tube member.

number of transport proteins, and thus the rate of transport (Figure 35.5).

Transfer cells also have many mitochondria that produce the ATP needed to power the active transport of mineral ions. As mineral ions move into the solution in the walls, the water potential of the wall solution (apoplast) becomes more negative; thus water moves out of the cells and into the apoplast by osmosis. Active transport of ions moves the ions directly, and water follows passively. The end result is that water and minerals end up in the xylem, where they constitute the *xylem sap*.

We have just seen that proteins regulate the movement of ions across membranes. We shall now see that even water movement itself is regulated by proteins.

Aquaporins control the rate, but not the direction, of water movement

Aquaporins are membrane channel proteins through which water can traverse a membrane without interacting with the hydrophobic environment of its phospholipid bilayer. These proteins, important in both plants and animals, allow water to move rapidly from environment to cell and from cell to cell. The permeability of some aquaporins is subject to regulation, changing the *rate* of osmosis across the membrane. However, water movement through aquaporins is always passive, so the *direction* of water movement is unchanged by alterations in aquaporin permeability.

Transport of Water and Minerals in the Xylem

So far in this chapter we've described the movement of water and minerals into plant roots and their entry into the root xylem. Now we will consider how xylem sap moves throughout the remainder of the plant. Let's first consider some early ideas about the ascent of sap and then turn to our current understanding of how it works. We'll describe the experiments that ruled out some early models as well as some evidence in support of the current model—and we'll find out what Per Scholander's sharpshooter was up to in the story that opened this chapter.

Experiments ruled out some early models of transport in the xylem

Some of the earliest attempts to explain the rise of sap in the xylem were based on a hypothetical pumping action by living cells in the stem, which pushed the sap upward. However, experiments conducted and published in 1893 by the German botanist Eduard Strasburger definitively ruled out such models.

Strasburger worked with trees about 20 meters tall. He sawed them through at their bases and plunged the cut ends into buckets containing solutions of poisons such as picric acid. The solutions rose through the trunks, as was readily evident from the progressive death of the bark higher and higher up. When the solutions reached the leaves, the leaves died, too, at which point the solutions stopped being transported (as shown by the liquid levels in the buckets, which stopped dropping).

This simple experiment established three important points:

- Living, "pumping" cells were not responsible for the upward movement of the solutions, because the solutions themselves killed all living cells with which they came in contact.
- The leaves play a crucial role in transport. As long as they were alive, solutions continued to be transported upward; when the leaves died, transport ceased.
- Transport was not caused by the roots, because the trunks had been completely separated from the roots.

Root pressure does not account for xylem transport

In spite of Strasburger's observations, some plant physiologists turned to a model of transport based on **root pressure**—pressure exerted by the root tissues that would force liquid up the xylem. The basis for root pressure is a higher solute concentration, and accordingly a more negative water potential, in the xylem sap than in the soil solution. This negative potential draws water into the stele; once there, the water has nowhere to go but up.

There is good evidence for root pressure—for example, the phenomenon of *guttation*, in which liquid water is forced out through openings in the leaves (Figure 35.6). Guttation occurs only under conditions of high atmospheric humidity and plentiful water in the soil, which occur most commonly at night. Root pressure is also the source of the sap that oozes from the cut stumps of some plants, such as *Coleus*, when their tops are removed. Root pressure, however, cannot account for the ascent of sap in trees.

Root pressure seldom exceeds 0.1–0.2 MPa (one or two times atmospheric pressure). If root pressure were driving

35.6 Guttation
Root pressure is responsible for forcing water through openings in the tips of this strawberry leaf.

sap up the xylem, we would observe a positive pressure potential in the xylem at all times. In fact, as we are about to see, the xylem sap is under tension—a negative pressure potential—when it is ascending. Furthermore, as Strasburger had already shown, materials can be transported upward in the xylem even when the roots have been removed. If the roots aren't pushing the xylem sap upward, what does cause it to rise?

The transpiration–cohesion–tension mechanism accounts for xylem transport

The obvious alternative to pushing is pulling: The leaves pull the xylem sap upward. **Transpiration**, the evaporative loss of water from the leaves, generates a pulling force (tension) on the water in the apoplast of the leaves. Hydrogen bonding between water molecules makes the sap in the xylem cohesive enough to withstand the tension and rise by bulk flow. Let's see how this **transpiration–cohesion–tension** mechanism works.

We'll start with transpiration. Water vapor diffuses from the intercellular spaces of the leaf, by way of the stomata, to the outside air because the water vapor concentration is greater inside the leaf than outside. Where did this water vapor come from?

As water vapor diffuses out of the leaf, more water evaporates from the moist walls of the mesophyll cells (Figure 35.7). Evaporation of water from the thin film surrounding the cell wall causes the film to shrink into the cellulose meshwork of the wall. The surface of the film curves where the water retreats into microscopic pores. The surface tension of the curved surfaces generates a **tension**—a negative pressure potential, a pull—in the film. The tension increases as more water leaves the film. This tension is what causes the bulk flow of water all the way from the roots.

The tension in the mesophyll draws water from the vessels or tracheids in the xylem of the nearest vein. The water, with its dissolved solutes, moves by bulk flow through the apoplast. The removal of water from the xylem of the veins establishes tension on the entire column of water contained within the xylem, so the column is drawn upward all the way from the roots.

The ability of water to be pulled upward through tiny tubes results from the remark-

START
Leaf
Stoma
Mesophyll cell
Vein
Stem
Xylem
Root
Root hair
H_2O
Xylem
1 Water vapor diffuses out of the stomata.
2 Water evaporates from mesophyll cell walls.
3 Tension pulls water from the veins into the apoplast about the mesophyll cells.
4 Tension pulls the water column upward and outward in the xylem of veins in the leaves.
5 Tension pulls the water column upward in the xylem of the stem.
6 Tension pulls the water column upward in the xylem of the root.
7 Water molecules form a cohesive column.
8 Water moves into the stele by osmosis.

35.7 Water Transport in Plants
Evaporation from surface cells, tension generated by the curvature of the shrinking surface film, and the cohesive nature of water molecules all account for the bulk flow of water from the soil to the atmosphere.

able cohesiveness of water—the tendency of water molecules to cohere to one another through hydrogen bonding. The narrower the tube, the greater the tension the water column can withstand without breaking. The integrity of the column is also maintained by the adhesion of water to the cell walls. In the tallest trees, such as a 100-meter redwood, the difference in pressure potential between the top and the bottom of the column may be as great as 3 MPa. The cohesiveness of water in the xylem is great enough to withstand even that great a tension.

In summary, the key elements of water transport in the xylem are:

- *Transpiration*, followed by evaporation from the moist cell walls in the leaves, resulting in…
- *tension* in the remainder of the xylem's water owing to the…
- *cohesion* of water, which pulls up more water to replace water that has been lost.

These elements require no work—no expenditure of energy—on the part of the plant. At each step between soil and atmosphere, water moves passively to a region with a more strongly negative water potential. Dry air has the most negative water potential (–95 MPa at 50% relative humidity), and the soil solution has the least negative water potential (between –0.01 and –3 MPa). Xylem sap has a water potential more negative than that of cells in the root, but less negative than that of mesophyll cells in the leaf.

Mineral ions contained in the xylem sap rise passively with the solution as it ascends from root to leaf. In this way the nutritional needs of the shoot are met. Some of the mineral elements brought to the leaves are subsequently redistributed to other parts of the plant by way of the phloem, but the initial delivery from the roots is through the xylem.

In addition to promoting the transport of minerals, transpiration contributes to temperature regulation. As water evaporates from mesophyll cells, heat is taken up from the cells, and the leaf temperature drops. This cooling effect is important in enabling plants to live in hot environments. A farmer can hold a leaf between thumb and forefinger to estimate its temperature; if the leaf doesn't feel cool, that means that transpiration is not occurring, so it must be time to water.

A pressure bomb measures tension in the xylem sap

The transpiration–cohesion–tension model can be true only if the column of sap in the xylem is under tension (negative pressure potential). The most elegant demonstrations of this tension, and of its adequacy to account for the ascent of sap in tall trees, were performed by Per Scholander. He measured tension in stems with an instrument called a **pressure bomb**.

The principle of the pressure bomb is as follows: Consider a stem in which the xylem sap is under tension. If the stem is cut, the sap pulls away from the cut, into the stem. Now the stem is placed in a device called a pressure bomb,

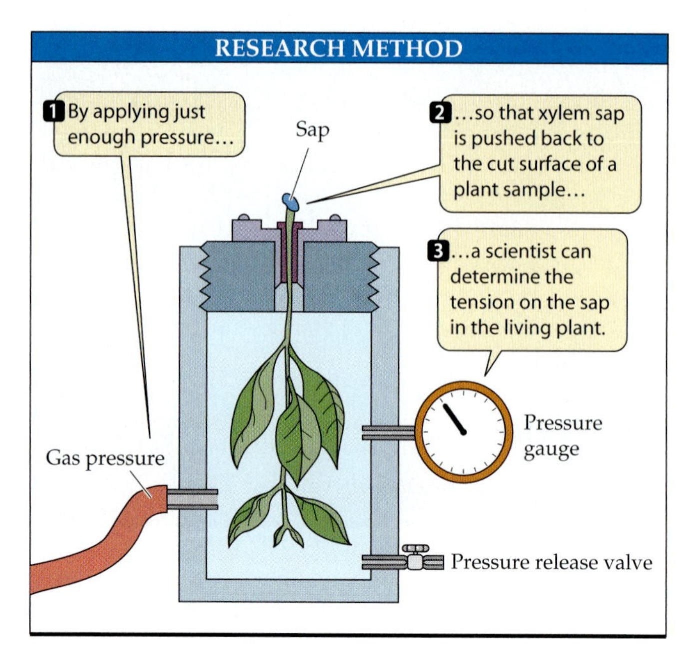

35.8 A Pressure Bomb
The amount of tension on the sap in different types of plants can be measured with this laboratory device.

in which the pressure may be raised. The cut surface remains outside the bomb. As pressure is applied to the plant parts within the bomb, the xylem sap is forced back to the cut surface. When the sap first becomes visible again at the cut surface, the pressure in the bomb is recorded. This pressure is equal in magnitude but opposite in sign to the tension (negative pressure potential) originally present in the xylem (Figure 35.8).

Scholander used the pressure bomb to study dozens of plant species, from diverse habitats, growing under a variety of conditions. In all cases in which xylem sap was ascending, it was found to be under tension. The tension disappeared in some of the plants at night, when transpiration ceased. In developing vines, the xylem sap was under no tension until leaves formed. Once leaves developed, transport in the xylem began, and tensions were recorded.

Suppose you wanted to measure tensions in the xylem at various heights in a large tree, to confirm that the tensions are sufficient to account for the rate at which sap is moving up the trunk. How would you obtain stem samples for measurement? Scholander used surveying instruments to determine the heights of particular twigs and then had a sharpshooter shoot the twigs from the tree with a high-powered rifle. As quickly as the twigs fell to the ground, Scholander inserted them in the pressure bomb and recorded their xylem tension. In every case, the differences in tensions at different heights were great enough to keep the xylem sap ascending.

Although transpiration provides the impetus for transport of water and minerals in the xylem, it also results in the loss of tremendous quantities of water from the plant. How do plants control this loss?

Transpiration and the Stomata

The epidermis of leaves and stems minimizes transpirational water loss by secreting a waxy cuticle, which is impermeable to water. However, the cuticle is also impermeable to carbon dioxide. This poses a problem: How can the leaf balance its need to retain water with its need to obtain carbon dioxide for photosynthesis?

Plants have evolved an elegant compromise in the form of **stomata** (singular stoma), or gaps, in the epidermis. A pair of specialized epidermal cells called **guard cells** controls the opening and closing of each stoma (Figure 35.9*a*). When the stomata are open, carbon dioxide can enter the leaf by diffusion, but water vapor is also lost in the same way. Closed stomata prevent water loss, but also exclude carbon dioxide from the leaf.

Most plants open their stomata only when the light intensity is sufficient to maintain a moderate rate of photosynthesis. At night, when darkness precludes photosynthesis, the stomata remain closed; no carbon dioxide is needed at this time, and water is conserved. Even during the day, the stomata close if water is being lost at too great a rate.

The stoma and guard cells in Figure 35.9*a* are typical of eudicots. Monocots typically have specialized epidermal cells associated with their guard cells. The principle of operation, however, is the same for both monocot and eudicot stomata. In what follows, we describe the regulation and mechanism of stomatal opening, the normal cycle of opening and closing, and the modified cycle used by some plants that live in dry or saline environments.

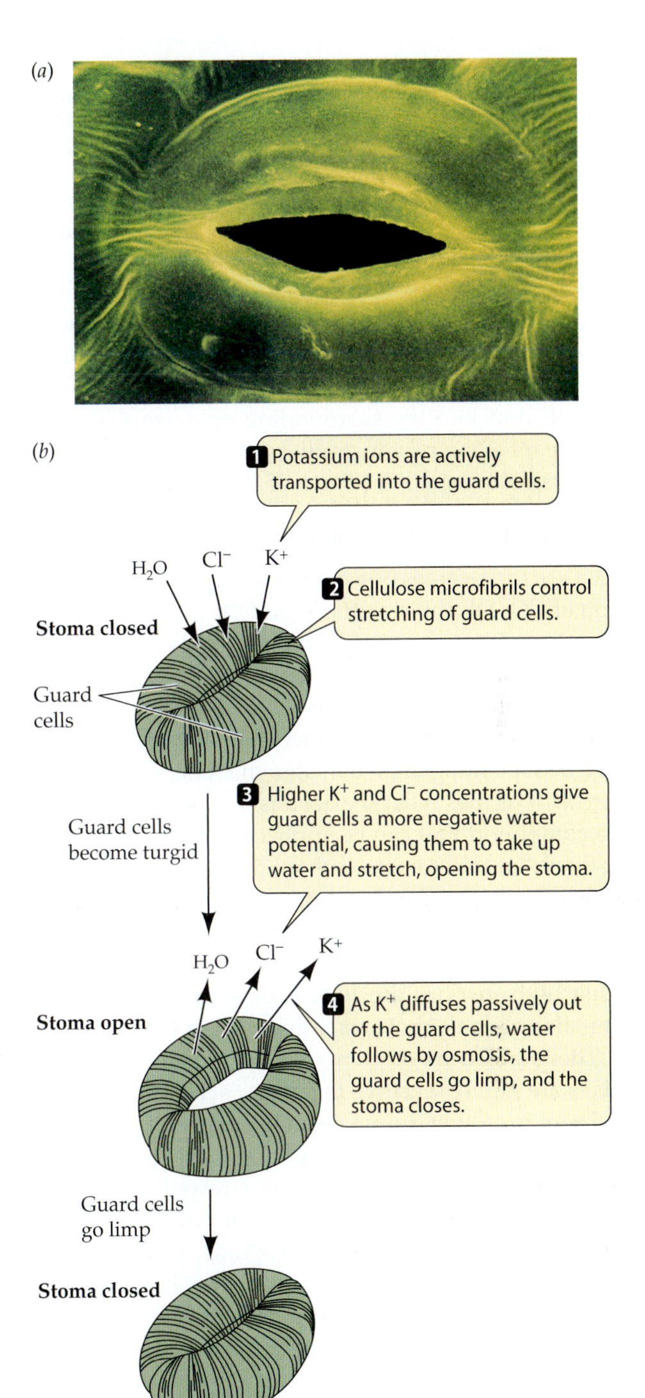

35.9 Stomata
(*a*) A scanning electron micrograph of a gaping stoma between two sausage-shaped guard cells. (*b*) Potassium ion (K^+) concentrations and water potential control the opening and closing of stomata. Negatively charged ions traveling with K^+ maintain electrical balance and contribute to the changes in osmotic potential that open and close the stomata.

The guard cells control the size of the stomatal opening

Light causes the stomata of most plants to open, admitting carbon dioxide for photosynthesis. Another cue for stomatal opening is the level of carbon dioxide in the spaces inside the leaf. A low level favors opening of the stomata, thus allowing the uptake of more carbon dioxide.

Water stress is a common problem for plants, especially on hot, sunny, windy days. Plants have a protective response to these conditions, using the water potential of the mesophyll cells as a cue. Even when the carbon dioxide level is low and the sun is shining, if the mesophyll is too dehydrated—that is, if the water potential of the mesophyll is too negative—the mesophyll cells release a plant hormone called *abscisic acid*. Abscisic acid acts on the guard cells, causing them to close the stomata and prevent further drying of the leaf. This response reduces the rate of photosynthesis, but it protects the plant.

The increasing internal concentration of potassium ions makes the water potential of the guard cells more negative. Water enters the guard cells by osmosis, increasing their pressure potential. Their cell walls contain cellulose microfibrils that cause the cells to respond to this increase by changing their shapes so that a gap—the stoma—appears between them.

The stoma closes by the reverse process when the proton pump is no longer active. Potassium ions diffuse passively out of the guard cells, water follows by osmosis, the pressure potential decreases, and the guard cells sag together and seal off the stoma. Negatively charged chloride ions and organic ions also move out of the guard cells with the potassium ions, maintaining electrical balance and contributing to the change in the solute potential of the guard cells.

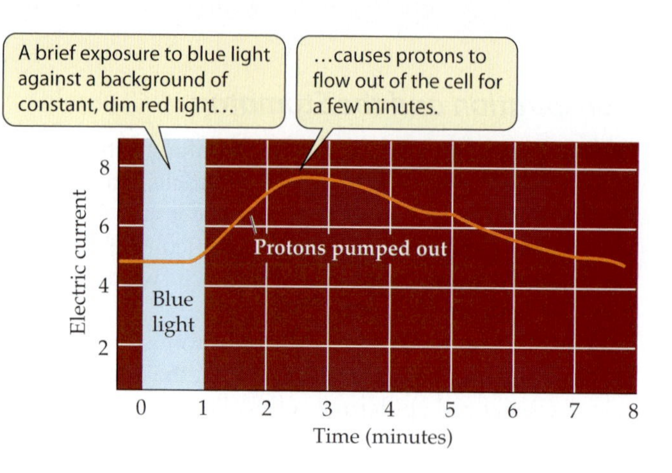

35.10 Light-Induced Proton Pumping in a Guard Cell Membrane This graph shows a trace of the tiny electric current that results from the flow of protons across the plasma membrane of a guard cell when it is exposed briefly to blue light.

What drives the opening and closing of the stomata? Certain wavelengths of blue light, absorbed by a pigment in the guard cell plasma membrane, activate a proton pump, which actively transports protons (H^+) out of the guard cells and into the surrounding epidermis (Figure 35.10). The resulting proton gradient drives the accumulation of potassium ions (K^+) (Figure 35.9*b*) in the guard cell.

Antitranspirants decrease water loss

Stomata are the referees of a compromise between the admission of CO_2 for photosynthesis and the loss of water by transpiration. Farmers would like their crops to transpire less, thus reducing the need for irrigation. Similarly, nurseries and gardeners would like to be able to reduce the amount of water lost by plants that are to be transplanted, because transplanting often damages the roots, causing the plant to wilt or die. What we need is a good **antitranspirant**: a compound that can be applied to plants, reducing water loss from the stomata without producing disastrous side effects by excessively limiting carbon dioxide uptake.

Abscisic acid and its commercial chemical analogs have been found to work as antitranspirants in small-scale tests, but their high cost has precluded commercial use. What about making plants more sensitive to their own abscisic acid? The guard cells of plants with a genetic mutation called *era* are highly sensitive to abscisic acid. These plants are resistant to wilting during drought stress.

A totally different type of antitranspirant seals off the leaves from the atmosphere for a time. Growers use a variety of compounds, most of which form polymeric films around leaves, to form a barrier to evaporation. These compounds cause undesirable side effects, however, and can be used only for relatively short periods of time. Their most common use is in the transplanting of nursery stock.

Crassulacean acid metabolism correlates with an inverted stomatal cycle

Most plants open and close their stomata on a schedule like that shown by the blue curve in Figure 35.11. The stomata are typically open for much of the day and closed at night. (They may also close during very hot days to reduce water loss.) But not all plants follow this pattern.

Many plants that live in dry areas or near the ocean have some unusual biochemical and behavioral features. One particularly surprising feature is their "backward" stomatal cycle: Their stomata are open at night and closed by day (as shown by the red curve in Figure 35.11). This behavior is part of the phenomenon of **crassulacean acid metabolism** (**CAM**), which was described in Chapter 8 (see Figure 8.21).

At night, while the stomata are open, carbon dioxide diffuses freely into the leaves of CAM plants and reacts in the mesophyll cells with phosphoenolpyruvic acid to produce organic acids. These acids accumulate to high concentrations. At daybreak the stomata close. Throughout the day, the organic acids are broken down to release the carbon dioxide they contain—behind closed stomata. Because the carbon dioxide cannot diffuse out of the plant, it is available for photosynthesis.

CAM is well adapted to environments where water is scarce: A leaf with its stomata open only at night—when the environment is cooler—loses much less water than does a leaf with its stomata open by day.

In both CAM and non-CAM plants, carbon dioxide is fixed and converted to the products of photosynthesis. How are these products delivered to other parts of the plant?

Translocation of Substances in the Phloem

Substances in the phloem move from sources to sinks. A **source** is an organ (such as a mature leaf or a storage root) that produces (by photosynthesis or by digestion of stored

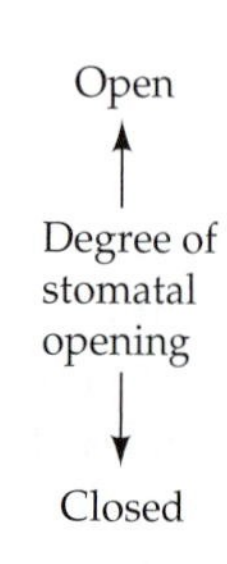

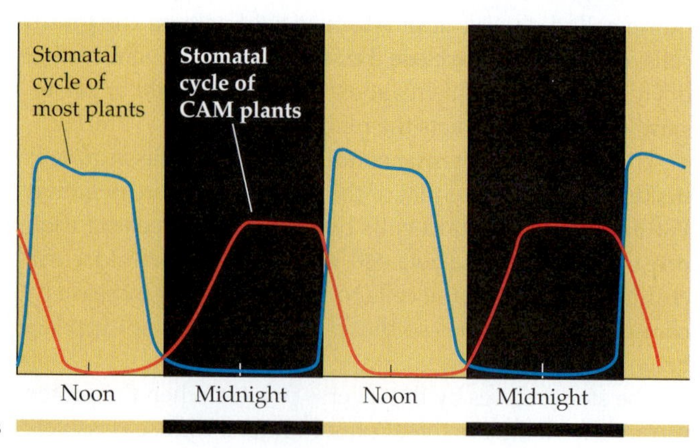

35.11 Stomatal Cycles Most plants open their stomata during the day. CAM plants reverse this stomatal cycle: Their stomata open during the night.

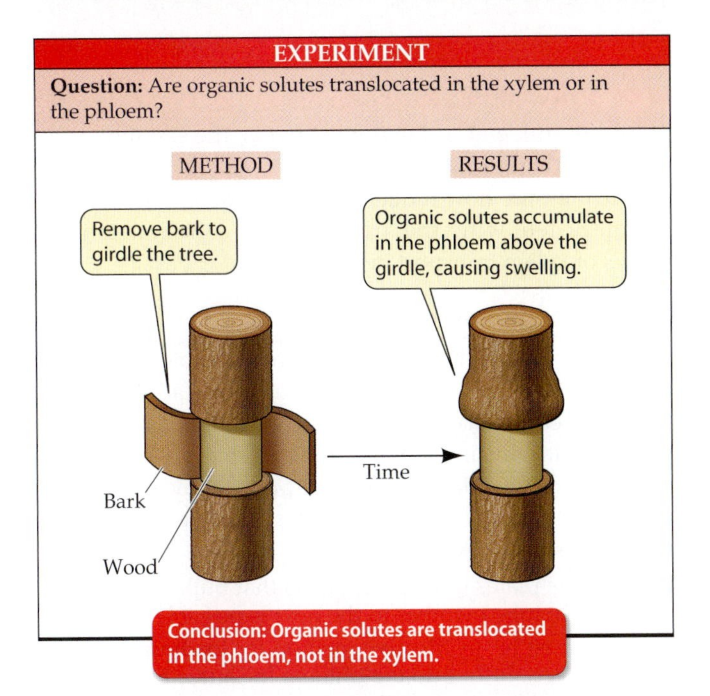

35.12 Girdling Blocks Translocation in the Phloem
By removing a ring of bark (containing the phloem), Malpighi blocked the translocation of organic solutes in a tree.

reserves) more sugars than it requires. A **sink** is an organ (such as a root, a flower, a developing tuber, or an immature leaf) that does not make enough sugar for its own growth and storage needs. Sugars (primarily sucrose), amino acids, some minerals, and a variety of other substances are translocated between sources and sinks in the phloem.

How do we know that such organic solutes are translocated in the phloem, rather than in the xylem? Just over 300 years ago, the Italian scientist Marcello Malpighi performed a classic experiment in which he removed a ring of bark (containing the phloem) from the trunk of a tree—that is, he *girdled* the tree (Figure 35.12). The bark in the region above the girdle swelled over time. We now know that the swelling resulted from the accumulation of organic solutes that came from higher up the tree and could no longer continue downward because of the disruption of the phloem. Later, the bark below the girdle died because it no longer received sugars from the leaves.

Any model to explain translocation of organic solutes must account for a few important facts:

- Translocation stops if the phloem tissue is killed by heating or other methods; thus the mechanism must be different from that of transport in the xylem.
- Translocation often proceeds in both directions—up and down the stem—simultaneously.
- Translocation is inhibited by compounds that inhibit respiration and thus limit the ATP supply in the source.

To investigate translocation, plant physiologists needed to obtain samples of pure sieve tube sap from individual sieve tube members. This difficult task was simplified when scientists discovered that a common garden pest, the aphid, feeds by drilling into a sieve tube. An aphid inserts its stylet, or feeding organ, into a stem until the stylet enters a sieve tube (Figure 35.13*a*). Within the sieve tube, the pressure is much greater than in the surrounding plant tissues, so nutritious sieve tube sap is forced up the stylet and into the aphid's digestive tract. So great is the pressure that sugary liquid is forced through the insect's body and out the anus (Figure 35.13*b*).

Plant physiologists use aphids to collect sieve tube sap. When liquid appears on the aphid's abdomen, indicating that the insect has connected with a sieve tube, the physiologist quickly freezes the aphid and cuts its body away from the stylet, which remains in the sieve tube member. For hours, sieve tube sap continues to exude from the cut stylet, where it may be collected for analysis. Chemical analysis of sieve tube sap collected in this manner reveals the contents of a single sieve tube member over time. We can also infer the rates at which different substances are translocated by measuring how long it takes for radioactive tracers administered to a leaf to appear at stylets at different distances from the leaf.

These methods have allowed us to understand how, at times, different substances might move in opposite directions in the phloem of a stem. Experiments with aphid stylets have shown that all the contents of any given sieve tube member move in the same direction. Thus, bidirectional translocation can be understood in terms of different sieve tubes conducting sap in opposite directions. Data obtained by these and other means led to the general adoption of the pressure flow model as an explanation for translocation in the phloem.

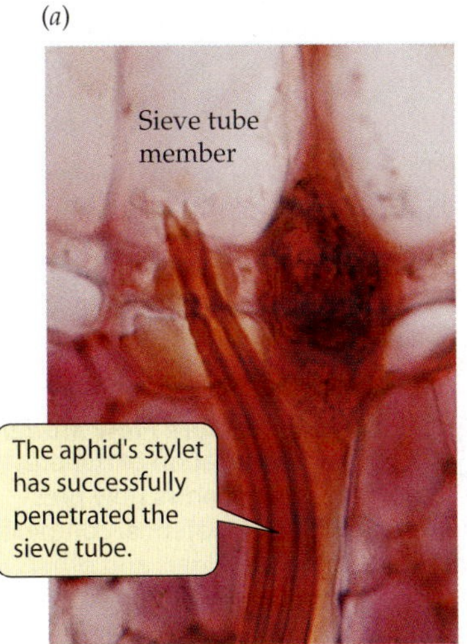

35.13 Aphids Collect Sieve Tube Sap
(*a*) Aphids feed on phloem sap drawn from the sieve tube, which they penetrate with a modified feeding organ, the stylet. (*b*) Pressure inside the sieve tube forces sap through the aphid's digestive tract, from which it can be harvested.

The pressure flow model appears to account for phloem translocation

The tonoplast breaks down during sieve tube member development, allowing the contents of the central vacuole to combine with much of the cytosol to form the sieve tube sap (see Chapter 34). The sap flows under pressure through the sieve tubes. It moves from one sieve tube member to the next by bulk flow through the sieve plates, without crossing a membrane.

Two steps in sieve tube sap flow require metabolic energy:

- Transport of sucrose and other solutes into the sieve tubes (**loading**) at sources
- Removal (**unloading**) of the solutes where the sieve tubes enter sinks

According to the **pressure flow model** of translocation in the phloem, sucrose is actively transported into sieve tube members at sources, giving these cells a much greater sucrose concentration than surrounding cells. Water therefore enters the sieve tube members by osmosis. The entry of this water causes a greater pressure potential at the source end, so the entire fluid content of the sieve tube is pushed toward the sink end of the tube—that is, the sap moves by bulk flow (Figure 35.14).

The pressure flow model of translocation in the phloem is contrasted with the transpiration–cohesion–tension model of xylem transport in Table 35.1.

Testing the pressure flow model

The pressure flow model was first proposed more than half a century ago, but some of its features are still debated. Other mechanisms have been proposed to account for translocation in sieve tubes. Some have been disproved, and none of the rest have been supported by a weight of evidence comparable to that for the pressure flow model, which must meet two requirements:

- The sieve plates must be open, so that bulk flow from one sieve tube member to the next is possible.
- There must be an effective method for loading sucrose and other solutes into the phloem in source tissues and removing them in sink tissues.

Let us see whether these requirements are met.

ARE THE SIEVE PLATES CLOGGED OR OPEN? Early electron microscopic studies of phloem samples cut from plants produced results that seemed to contradict the pressure flow

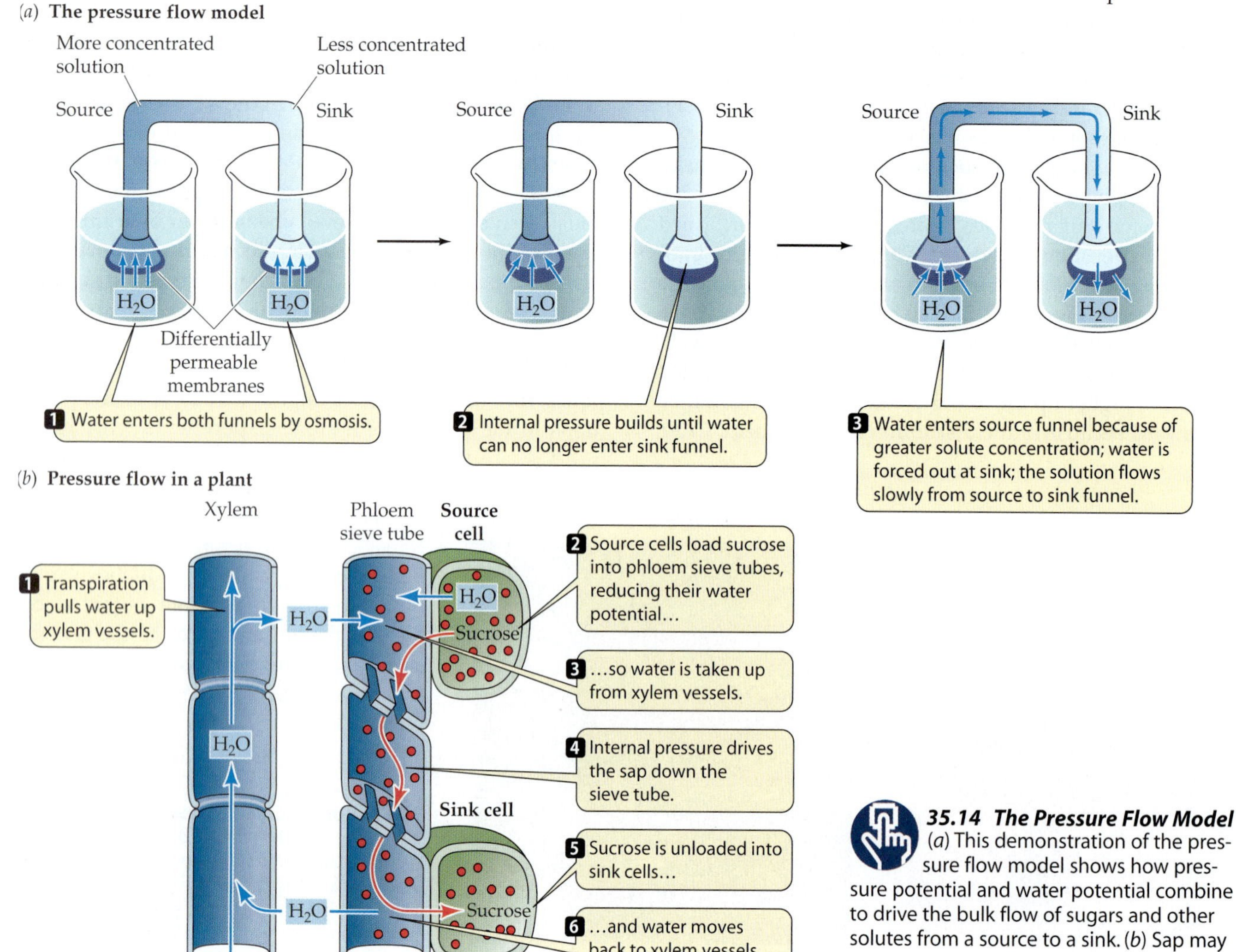

35.14 The Pressure Flow Model (*a*) This demonstration of the pressure flow model shows how pressure potential and water potential combine to drive the bulk flow of sugars and other solutes from a source to a sink. (*b*) Sap may flow through sieve tubes in this manner.

35.1 Mechanisms of Bulk Flow in Plant Vascular Tissues

	XYLEM	PHLOEM
Source of bulk flow	Transpiration from leaves	Active transport of sucrose at source
Site of bulk flow	Dead vessel elements and tracheids	Living sieve tube members
Pressure potential in sap	Negative (pull from top)	Positive (push from source)

model. The pores in the sieve plates always appeared to be plugged with masses of a fibrous protein, suggesting that sieve tube sap could not flow freely. But what is the function of that fibrous protein?

One possibility is that this protein is usually distributed more or less at random throughout the sieve tube members until the sieve tube is damaged; then the sudden surge of sap toward the cut surface carries the protein into the pores, blocking them and preventing the loss of valuable nutrients. In other words, perhaps the protein does *not* block the pores unless the phloem is damaged. How might this possibility be tested? Could we obtain phloem for microscopic observation without causing the sap to surge to the cut surface?

One way to prevent the surge of the sap is to freeze plant tissue before cutting it. Another way is to let the tissue wilt so that there is no pressure in the phloem before cutting. When these methods were used, the sieve plates were not clogged by the protein. Thus, the first condition of the pressure flow model is met.

NEIGHBORING CELLS LOAD AND UNLOAD THE SIEVE TUBE MEMBERS. If the pressure flow model is correct, there must be mechanisms for loading sugars and other solutes into the phloem in source regions and for unloading them in sink regions. One pathway of phloem loading has been demonstrated in some plant species.

Sugars and other solutes pass from cell to cell through the symplast in the mesophyll. When these substances reach cells adjacent to the ends of leaf veins, they leave the mesophyll cells and enter the apoplast, sometimes with the help of transfer cells. Then specific sugars and amino acids are actively transported into cells of the phloem, thus reentering the symplast (Figure 35.15).

Passage through the apoplast and back into the symplast selects substances to be accumulated for translocation because substances can enter the phloem only after passing through a differentially permeable membrane. In many plants, solutes reenter the symplast at the companion cells (see Chapter 34), which then transfer the solutes to the adjacent sieve tube members. As Figure 35.15 shows, in other plant species, sucrose or other sugars move from the mesophyll to the sieve tube members entirely within the symplast; that is, transfer of solutes from symplast to apoplast and back again is not a universal feature of phloem loading.

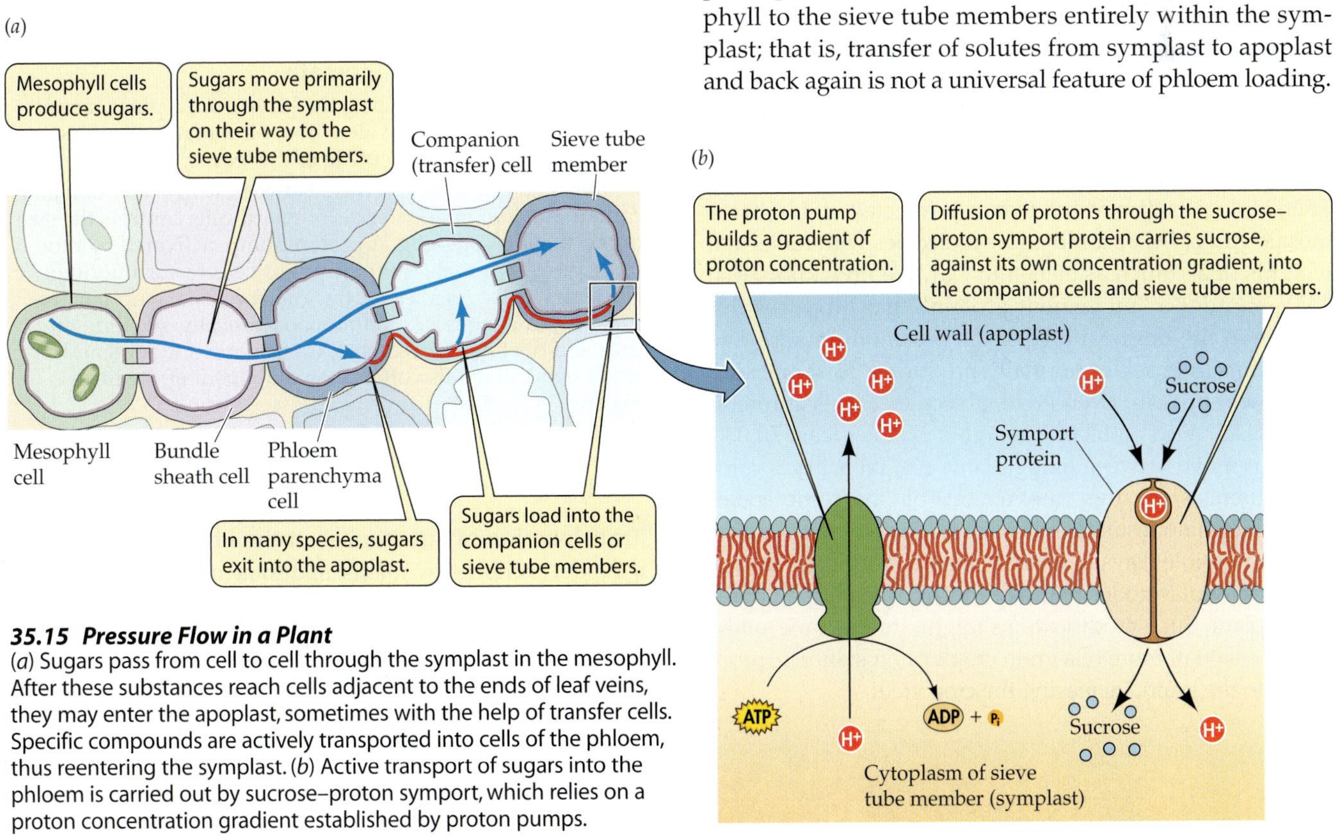

35.15 Pressure Flow in a Plant
(*a*) Sugars pass from cell to cell through the symplast in the mesophyll. After these substances reach cells adjacent to the ends of leaf veins, they may enter the apoplast, sometimes with the help of transfer cells. Specific compounds are actively transported into cells of the phloem, thus reentering the symplast. (*b*) Active transport of sugars into the phloem is carried out by sucrose–proton symport, which relies on a proton concentration gradient established by proton pumps.

A form of secondary active transport (see Chapter 5, pages 88–90) loads sucrose into the companion cells and sieve tube members. Sucrose is carried through the plasma membrane from apoplast to symplast by sucrose–proton symport; thus the entry of sucrose and of protons is strictly coupled. For this symport to work, the apoplast must have a high concentration of protons; the protons are supplied by a primary active transport system, the proton pump. The protons then diffuse back into the cell through the symport protein, bringing sucrose with them.

In sink regions, the solutes are actively transported out of the sieve tube members and into the surrounding tissues. This unloading serves two purposes: It helps maintain the gradient of solute potential and hence of pressure potential in the sieve tubes, and it promotes the buildup of sugars and starch to high concentrations in storage regions, such as developing fruits and seeds.

Plasmodesmata and material transfer between cells

Many substances move from cell to cell within the symplast by way of plasmodesmata (see Figure 34.7). Among their other roles, plasmodesmata participate in the loading and unloading of sieve tube members. Mechanisms vary among plant species, but the story in tobacco plants is a common one. In tobacco, sugars and other compounds in source tissues enter companion cells by active transport from the apoplast and move on to the sieve tube members through plasmodesmata. In sink tissues, plasmodesmata connect sieve tube members, companion cells, and the cells that will receive and use the transported compounds.

Plasmodesmata undergo developmental changes as an immature sink leaf matures into a mature source leaf. Plasmodesmata in sink tissues favor rapid unloading: They are more abundant, and they allow the passage of larger molecules. Plasmodesmata in source tissues are few in number.

It was long thought that only substances with molecular weights less than 1,000 could fit through a plasmodesma. Then biologists discovered that cells infected with tobacco mosaic virus (TMV) could allow molecules with molecular weights of as much as 20,000 to exit. We now know that TMV encodes a "movement protein" that produces this change in the permeability of the plasmodesmata—and that plants themselves normally produce at least one such movement protein. Even large molecules such as proteins and RNAs, with molecular weights up to at least 50,000, can thus move between living plant cells. We will see some consequences of this movement of macromolecules through plasmodesmata in later chapters. Biologists are exploring possible ways to regulate the permeability, number, and form of plasmodesmata as a means of modifying traffic in the plant. Such modifications might, for example, allow the diversion of more of a grain crop's photosynthetic products into the grain, increasing the crop yield.

Chapter Summary

Uptake and Transport of Water and Minerals

▶ Plant roots take up water and minerals from the soil.

▶ Water moves through biological membranes by osmosis, always moving toward cells with a more negative water potential. The water potential of a cell or solution is the sum of the solute potential and the pressure potential. All three parameters are expressed in megapascals (MPa). **Review Figure 35.1**

▶ Mineral uptake requires transport proteins. Some minerals enter the plant by facilitated diffusion; others enter by active transport. A proton pump facilitates the active transport of many solutes across membranes in plants. **Review Figure 35.2**

▶ Water and minerals pass from the soil to the xylem by way of the apoplast and symplast. In the root, water and minerals may pass from the cortex into the stele only by way of the symplast because Casparian strips in the endodermis block water and solute movement in the apoplast. **Review Figures 35.3, 35.4**

Transport of Water and Minerals in the Xylem

▶ Early experiments established that sap does not move via the pumping action of living cells.

▶ Root pressure is responsible for guttation and for the oozing of sap from cut stumps, but it cannot account for the ascent of sap in trees.

▶ Xylem transport is the result of the combined effects of transpiration, cohesion, and tension. Evaporation in the leaf produces tension in the surface film of water on the moist-walled mesophyll cells, and thus pulls water—held together by its cohesiveness—up through the xylem from the root. Dissolved minerals go along for the ride. **Review Figure 35.7**

▶ Support for the transpiration–cohesion–tension model of xylem transport came from studies using a pressure bomb. **Review Figure 35.8**

Transpiration and the Stomata

▶ Evaporation of water cools the leaves, but a plant cannot afford to lose too much water. Transpirational water loss is minimized by the waxy cuticle of the leaves.

▶ Stomata allow a compromise between water retention and carbon dioxide uptake. A pair of guard cells controls the size of the stomatal opening. A proton pump, activated by blue light, pumps protons from the guard cells to surrounding epidermal cells. As a result, the guard cells take up potassium ions, causing water to follow osmotically, swelling the cells and opening the stomata. Carbon dioxide level and water availability also affect stomatal opening. **Review Figures 35.9, 35.10**

▶ In most plants the stomata are open during the day and closed at night. CAM plants have an inverted stomatal cycle, enabling them to conserve water. **Review Figure 35.11**

Translocation of Substances in the Phloem

▶ Products of photosynthesis, and some minerals, are translocated through sieve tubes in the phloem by way of living sieve tube members. Translocation proceeds in both directions in the stem, although in a single sieve tube it goes only one way. Translocation requires a supply of ATP.

▶ Translocation in the phloem proceeds in accordance with the pressure flow model: The difference in solute concentration between sources and sinks allows a difference in pressure potential along the sieve tubes, resulting in bulk flow. **Review Figure 35.14, Table 35.1**

▶ The pressure flow model succeeds because the sieve plates are normally open, allowing bulk flow, and because neighboring cells load organic solutes into the sieve tube members in source regions and unload them in sink regions. **Review Figure 35.15**

▶ The distribution and properties of plasmodesmata differ between source and sink tissues. It may become possible to regulate plasmodesmata in crop plants.

For Discussion

1. Epidermal cells protect against excess water loss. How do they perform this function?
2. Phloem transports material from sources to sinks. What is meant by "source" and "sink"? Give examples of each.
3. What is the minimum number of plasma membranes a water molecule would have to cross in order to get from the soil solution to the atmosphere by way of the stele? To get from the soil solution to a mesophyll cell in a leaf.
4. Transpiration exerts a powerful pulling force on the water column in the xylem. When would you expect transpiration to proceed most rapidly? Why? Describe the source of the pulling force.
5. Plants that perform crassulacean acid metabolism (CAM plants) are adapted to environments in which water supply is limited; these plants open their stomata only at night. Could a non-CAM plant, such as a pea plant, enjoy an advantage if it opened its stomata only at night? Explain.

36 Plant Nutrition

AN INSECT HAS STEPPED ON A TRIGGER HAIR on the leaf of a Venus flytrap—a big mistake for the insect. The trigger hair sends an electrical signal that springs a mechanical trap. The two halves of the leaves close, and spiny outgrowths at the margins of the leaves interlock to imprison the insect. The leaf secretes enzymes that will digest its prey. The leaf then absorbs the products of digestion, especially amino acids, and uses them as a nutritional supplement.

Why does the Venus flytrap go to all this trouble? Few other plants are carnivorous—your petunia plant is not stalking you, after all. But the Venus flytrap (*Dionaea muscipula*) lives on soils in which nitrogen is scarce. Its carnivorous adaptation gives it another way to obtain needed nitrogen.

Why do plants need nitrogen? The answer is simple if we recall the chemical structures of proteins and nucleic acids that we looked at in Chapter 3. These vital components of all living things contain nitrogen, as do chlorophyll and many other important biochemical compounds. If a plant cannot get enough nitrogen, it cannot synthesize these compounds at a rate adequate to keep itself healthy.

In addition to nitrogen, plants need other materials from their environment. In this chapter we explore the differences between the basic strategies of plants and animals for obtaining nutrition. Then we look at what nutrients plants require, and how they acquire them. Because most nutrients come from the soil, we discuss the formation of soils and the effects of plants on soils. As any farmer can tell you, nitrogen is the nutrient that most often limits plant growth, so we devote a section specifically to nitrogen metabolism in plants. The chapter concludes with a look at plants that use means other than photosynthesis to supplement their nutrition.

The Acquisition of Nutrients

Every living thing must obtain raw materials from its environment. These **nutrients** include the major ingredients of macromolecules: carbon, hydrogen, oxygen, and nitrogen. Carbon and oxygen enter the living world through the carbon-fixing reactions of photosynthesis, in which photosynthetic organisms obtain them from atmospheric carbon dioxide. Hydrogen enters living systems through the light reactions of photosynthesis, which split water. For carbon, oxygen, and hydrogen, photosynthesis is the gateway to the living world.

The movement of nitrogen into organisms begins with processing by some highly specialized bacteria living in the soil. Some of these bacteria act on nitrogen gas, converting it into a form usable by plants. The plants in turn provide organic nitrogen and carbon to animals, fungi, and many microorganisms.

In addition to carbon, oxygen, hydrogen, and nitrogen, other **mineral nutrients** are essential to living systems. The proteins of organisms contain sulfur (S), and their nucleic acids contain phosphorus (P). There is magnesium (Mg) in chlorophyll, and iron (Fe) in many important compounds, such as the cytochromes. Within the soil, these and other minerals dissolve in water, forming a solution—called the **soil solution**—that contacts the roots of plants. Plants take up most of these mineral nutrients from the soil solution in ionic form.

A Meat-Eating Plant
Dionaea muscipula, the Venus flytrap, has adapted to a nitrogen-poor environment by becoming carnivorous. It obtains this necessary mineral from the bodies of insects trapped inside the plant when its hinges snap shut.

Autotrophs make their own organic compounds

The plants provide carbon, oxygen, hydrogen, nitrogen, and sulfur to most of the rest of the living world. Plants, some protists, and some bacteria are *autotrophs;* that is, they make their own organic compounds from simple inorganic nutrients—carbon dioxide, water, nitrate or ammonium ions containing nitrogen, and a few other soluble mineral nutrients (Figure 36.1). *Heterotrophs* are organisms that require preformed *organic* compounds (compounds that contain carbon) as food. Herbivores and carnivores are heterotrophs that depend directly or indirectly on autotrophs as their source of nutrition.

Most autotrophs are *photosynthesizers*—that is, they use light as the source of energy for synthesizing organic compounds from inorganic raw materials. Some autotrophs, however, are *chemosynthesizers*, deriving their energy not from light, but from reduced inorganic substances, such as hydrogen sulfide (H_2S), in their environment. All chemosynthesizers are bacteria. Some chemosynthetic bacteria in the soil contribute to the nutrition of plants by increasing the availability of nitrogen and sulfur. But how does a plant obtain its nutrients, whether they come with or without bacterial action?

How does a stationary organism find nutrients?

An organism that cannot move must exploit energy that is somehow brought to it. Most sessile animals depend primarily on the movement of water to bring energy, in the form of food, to them, but a plant's supply of energy arrives at the speed of light from the sun. A plant's supply of nutrients, however, is strictly local, and the plant may use up the water and mineral nutrients in its local environment as it develops. How does a plant cope with such a problem?

One answer is to extend itself by growing into new resources. *Growth is a plant's version of locomotion.* Root systems mine the soil; by growing, they reach new sources of mineral nutrients and water. Growth of leaves helps a plant secure light and carbon dioxide. A plant may compete with other plants for light by outgrowing them, both capturing more light for itself and preventing the growth of its neighbors by shading them.

As it grows, a plant—or even a single root—must deal with environmental heterogeneity. Animal droppings create high local concentrations of nitrogen. A particle of calcium carbonate in the soil may make a tiny area alkaline, while dead organic matter may make a nearby area acidic.

Mineral Nutrients Essential to Plants

What important mineral nutrients do plants take up from their environment, and what are their roles? Table 36.1 lists the mineral nutrients that have been proved to be essential for plants. Except for nitrogen, they all come from the soil solution and derive ultimately from rock.

There are three criteria for calling something an **essential element**:

- The element must be *necessary* for normal growth and reproduction.
- The element cannot be *replaceable* by another element.
- The requirement must be *direct*—that is, not the result of an indirect effect, such as the need to relieve toxicity caused by another substance.

In this section, we'll consider the symptoms of particular mineral deficiencies, the roles of some of the mineral nutrients, and the technique by which the essential elements for plants were identified.

The essential elements in Table 36.1 are divided into two categories: the macronutrients and the micronutrients. Plant tissues need **macronutrients** in concentrations of at least 1 gram per kilogram of their dry matter, and they need **micronutrients** in concentrations of less than 100 milligrams per kilogram of their dry matter. (Dry matter, or dry weight, is what remains after all the water has been removed from a tissue sample.) Both the macronutrients and the micronutrients are essential for the plant to complete its life cycle from seed to seed.

Deficiency symptoms reveal inadequate nutrition

Before a plant that is deficient in an essential element dies, it usually displays characteristic deficiency symptoms. Table 36.2 describes the symptoms of some common mineral deficiencies. Such symptoms help horticulturists diagnose mineral nutrient deficiencies in plants.

36.1 What Do Plants Need?
To survive, plants require only light plus carbon dioxide, water, and several essential mineral elements. These plants are growing on nothing more than a solution that contains water and mineral elements. This technique is known as hydroponic culture.

36.1 Mineral Elements Required by Plants

ELEMENT	ABSORBED FORM	MAJOR FUNCTIONS
Macronutrients		
Nitrogen (N)	NO_3^- and NH_4^+	In proteins, nucleic acids, etc.
Phosphorus (P)	$H_2PO_4^-$ and HPO_4^{2-}	In nucleic acids, ATP, phospholipids, etc.
Potassium (K)	K^+	Enzyme activation; water balance; ion balance; stomatal opening
Sulfur (S)	SO_4^{2-}	In proteins and coenzymes
Calcium (Ca)	Ca^{2+}	Affects the cytoskeleton, membranes, and many enzymes; second messenger
Magnesium (Mg)	Mg^{2+}	In chlorophyll; required by many enzymes; stabilizes ribosomes
Micronutrients		
Iron (Fe)	Fe^{2+}	In active site of many redox enzymes and electron carriers; chlorophyll synthesis
Chlorine (Cl)	Cl^-	Photosynthesis; ion balance
Manganese (Mn)	Mn^{2+}	Activation of many enzymes
Boron (B)	$B(OH)_3$	Possibly carbohydrate transport (poorly understood)
Zinc (Zn)	Zn^{2+}	Enzyme activation; auxin synthesis
Copper (Cu)	Cu^{2+}	In active site of many redox enzymes and electron carriers
Nickel (Ni)	Ni^{2+}	Activation of one enzyme
Molybdenum (Mo)	MoO_4^{2-}	Nitrogen fixation; nitrate reduction

Nitrogen deficiency is the most common mineral deficiency in plants. Plants in natural environments are almost always deficient in nitrogen, but they seldom display deficiency symptoms. Instead, their growth slows to match the available supply of nitrogen. Crop plants, on the other hand, show deficiency symptoms if a formerly abundant supply of nitrogen runs out. The visible symptoms of nitrogen deficiency include uniform yellowing, or *chlorosis*, of older leaves. Chlorophyll, which is responsible for the green color of leaves, contains nitrogen. Without nitrogen there is no chlorophyll, and without chlorophyll, the yellow pigments become visible.

36.2 Some Mineral Deficiencies in Plants

DEFICIENCY	SYMPTOMS
Calcium	Growing points die back; young leaves are yellow and crinkly
Iron	Young leaves are white or yellow with green veins
Magnesium	Older leaves have yellow in stripes between veins
Manganese	Younger leaves are pale with stripes of dead patches
Nitrogen	Oldest leaves turn yellow and die prematurely; plant is stunted
Phosphorus	Plant is dark green with purple veins and is stunted
Potassium	Older leaves have dead edges
Sulfur	Young leaves are yellow to white with yellow veins
Zinc	Young leaves are abnormally small; older leaves have many dead spots

Inadequate available iron in the soil can also cause chlorosis because, although it is not contained in the chlorophyll molecule, iron is required for chlorophyll synthesis. However, iron deficiency commonly causes chlorosis of the *youngest* leaves, with their veins sometimes remaining green. The reason for this difference is that nitrogen is readily translocated in the plant and can be redistributed from older tissues to younger tissues to favor their growth. Iron, on the other hand, cannot be readily redistributed. Younger tissues that are actively growing and synthesizing compounds needed for their growth show iron deficiency before older leaves, which have already completed their growth.

Several essential elements fulfill multiple roles

Essential elements can play several different roles—some structural, others catalytic. Magnesium, as we have mentioned, is a constituent of the chlorophyll molecule and hence is essential to photosynthesis. It is also required as a cofactor by numerous enzymes in cellular respiration and other metabolic pathways.

Phosphorus, usually in phosphate groups, is found in many organic compounds, particularly in nucleic acids and in the intermediates of the energy pathways of photosynthesis and glycolysis. The transfer of phosphate groups occurs in many energy-storing and energy-releasing reactions, notably those that use or produce ATP. Other roles of phosphate groups include the activation and inactivation of enzymes.

Calcium plays many roles in plants. Its function in the processing of hormonal and environmental cues is the subject of great biological interest, as we'll see in the next chapter. Calcium also affects membranes and cytoskeleton activity, participates in spindle formation for mitosis and meiosis, and is a constituent of the middle lamella of cell

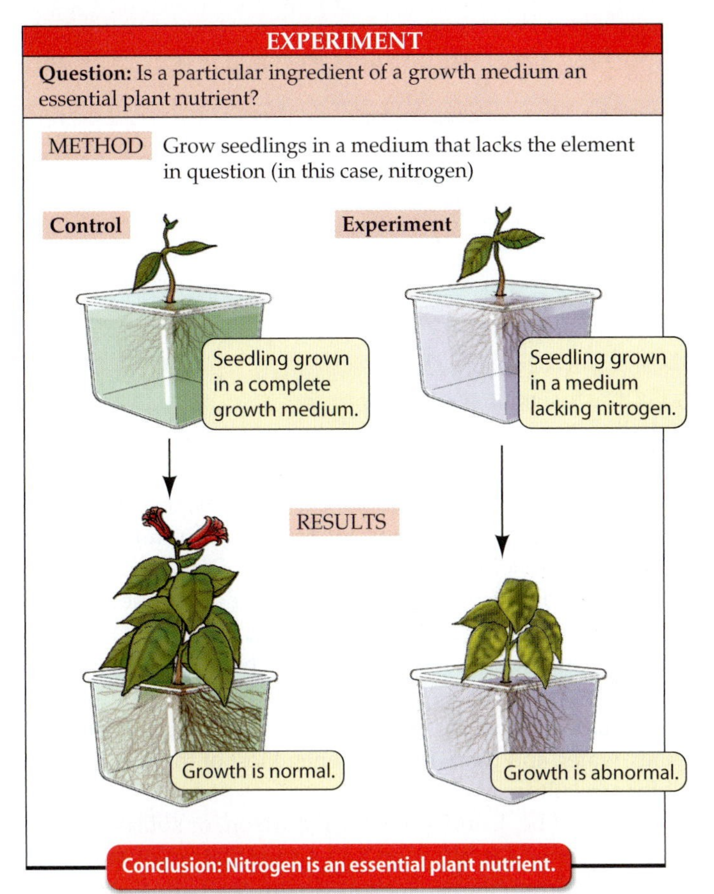

36.2 Identifying Essential Elements for Plants The diagram shows the procedure for identifying nutrients essential to plants, using nitrogen as an example. The environment in such experiments must be rigorously controlled because some essential elements are needed in only tiny amounts, and may be present in sufficient quantities as contaminants.

walls. Other elements, such as iron and potassium, also play multiple roles.

All of these elements are essential to the life of all plants. How did biologists discover which elements are essential?

The identification of essential elements

An element is considered essential if a plant fails to complete its life cycle, or grows abnormally, when that element is not available, or is not available in sufficient quantities. Plant physiologists identified most of the essential elements for plants by the technique outlined in Figure 36.2. This technique is limited, however, by the possibility that some elements thought to be absent from the test solutions are actually present. Impurities and contamination are always possible.

In early experiments on plant nutrition, some of the chemicals used were so impure that they provided micronutrients that the investigators thought they had excluded. Some mineral nutrients are required in such tiny amounts that there may be enough in a seed to supply the embryo and the resultant plant throughout its lifetime and leave enough in the next seed to get the next generation well started. Simply touching a plant may give it a significant dose of chlorine in the form of chloride ions from sweat. Only rarely are new essential elements reported now. Either the list is nearly complete, or more likely, we will need more sophisticated techniques to add to it.

Where does the plant find its essential mineral nutrients? How does it absorb them?

Soils and Plants

Soils are very important to plants, and plant interactions with the soil are complex. Plants obtain their mineral nutrients from the soil solution or the water in which they grow. Water for terrestrial plants also comes from the soil, as does the supply of oxygen for the roots. Soil also provides mechanical support for plants on land, and it harbors bacteria that perform chemical reactions leading to products required for plant growth. On the other hand, soil may also contain organisms harmful to plants.

In the pages that follow, we'll examine the composition and structure of soils, their formation, their role in plant nutrition, their care and supplementation in agriculture, and their modification by the plants that grow in them.

Soils are complex in structure

Soils are complex systems of living and nonliving components. The living components include plant roots, as well as populations of bacteria, fungi, and animals such as earthworms and insects (Figure 36.3). The nonliving portion of the soil includes rock fragments ranging in size from large

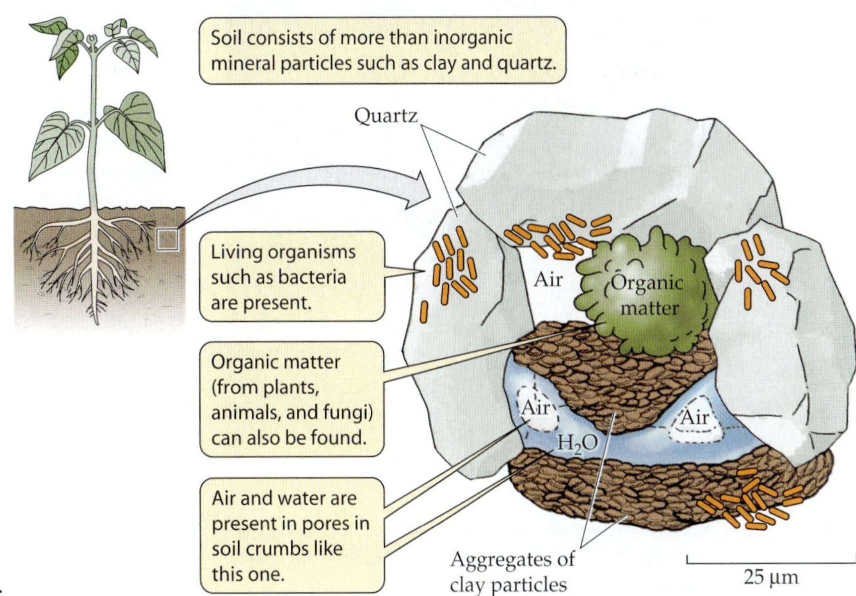

36.3 The Complexity of Soil Soil has both organic and inorganic components.

36.3 Two Systems for Classifying Soil Particles

UNITED STATES DEPARTMENT OF AGRICULTURE		INTERNATIONAL SOCIETY FOR SOIL SCIENCE	
SOIL TYPE	PARTICLE SIZE (MM)	SOIL TYPE	PARTICLE SIZE (MM)
Sand	0.05–2.0	Coarse sand	0.2–2.0
		Fine sand	0.02–0.2
Silt	0.002–0.05	Silt	0.002–0.02
Clay	<0.002	Clay	<0.002

boulders to tiny particles called **clay** that are 2 μm or less in diameter (Table 36.3). Soils also contain water and dissolved mineral nutrients, air spaces, and dead organic matter. The air spaces are crucial sources of oxygen for plant roots. Soils change constantly through natural causes—such as rain, temperature extremes, and the activities of plants and animals—as well as human activities—farming in particular.

The structure of many soils changes with depth, revealing a **soil profile**. Although soils differ greatly, almost all soils consist of two or more **horizons**—recognizable horizontal layers—lying on top of one another. Mineral nutrients tend to be **leached**—dissolved in rain or irrigation water and carried to deeper horizons.

Soil scientists recognize three major zones (A, B, and C) in the profile of a typical soil (Figure 36.4). **Topsoil** is the A horizon, from which mineral nutrients may be depleted by leaching. Most of the organic matter in the soil is in the A horizon, as are most roots, earthworms, insects, nematodes, and microorganisms. Successful agriculture depends on the presence of a suitable A horizon. Pure sand contains plenty of air spaces, but is low in water and mineral nutrients. Clay contains lots of nutrients and more water than sand does, but it is low in air. A little bit of clay goes a long way in affecting soil properties. A **loam** has significant amounts of sand, silt, and clay, and thus has good levels of air, water, and nutrients for plants. Most of the best topsoils for agriculture are loams.

Below the A horizon is the B horizon, or subsoil, which is the zone of infiltration and accumulation of materials leached from above. Farther down, the C horizon is the original parent rock from which the soil is derived. Some deep-growing roots extend into the B horizon, but roots rarely enter the C horizon.

36.4 A Soil's Profile
The A, B, and C horizons can sometimes be seen in road cuts such as this one in Australia. The dark upper layer (A horizon) is home to most of the living organisms in the soil.

Soils form through the weathering of rock

The type of soil in a given area depends on the type of rock from which it formed, the climate, the landscape features, the organisms living there, and the length of time that soil-forming processes have been acting (sometimes millions of years). Rocks are broken down in part by **mechanical weathering**, which is the physical breakdown—without any accompanying chemical changes—of materials by wetting, drying, and freezing. The most important parts of soil formation, however, include **chemical weathering**, the chemical alteration of at least some of the materials in the rocks.

The key process is the formation of clay. Both the physical and the chemical properties of soils depend on the amount and kind of clay particles they contain. Just grinding up rocks does not produce a clay that binds mineral nutrients and aggregates into particles. Such a clay results only from chemical weathering.

Soils are the source of plant nutrition

The supply of mineral nutrients for plants depends on the presence of clay particles in the soil. Many of the minerals that are important for plant nutrition, such as potassium (K^+), magnesium (Mg^{2+}), and calcium (Ca^{2+}), exist in soil as positively charged ions, or *cations*. Clay particles have a net negative charge, which they get from negatively charged ions that are permanently attached to them. Cations in solution are attracted to these negative ions. To become available to plants, the cations must be detached from the clay particles.

This task is accomplished by reactions with protons (hydrogen ions, H^+). Roots release protons into the soil, and they are also released by the ionization of carbonic acid (H_2CO_3), which is formed whenever CO_2 from respiring roots or from the atmosphere dissolves in water. Protons bond more strongly to the clay particles than do the mineral cations, so they trade places with the cations, thus putting the nutrients back into the soil solution. This trading of places is called **ion exchange** (Figure 36.5). The fertility of a soil is determined in part by its ability to provide nutrients in this manner.

Clay particles effectively hold and exchange positively charged ions, but there is no comparable mechanism for exchanging negatively charged ions. As a result, important negative ions such as nitrate (NO_3^-) and sulfate (SO_4^{2-})—the primary and direct sources of nitrogen and sulfur—leach rapidly from soil, whereas positive ions tend to be retained in the A horizon. The reservoir of soil nitrogen is the organic matter in the soil, which slowly decomposes to release nitrogen in a form available to plants.

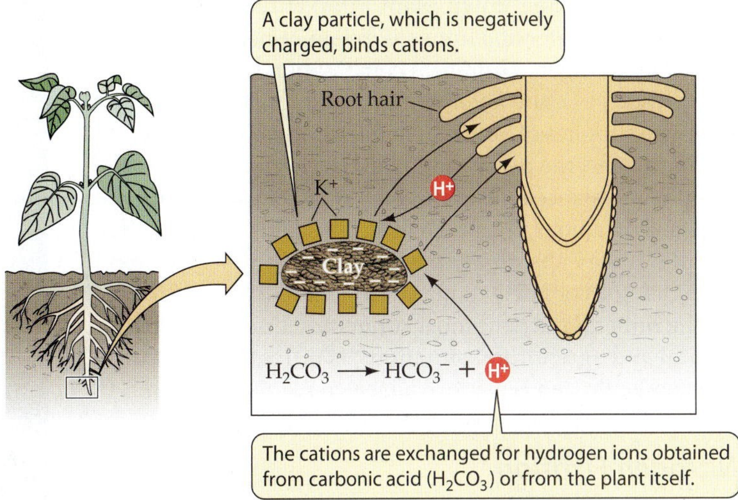

36.5 Ion Exchange
Plants obtain mineral nutrients from the soil primarily in the form of positive ions; potassium is the example shown here.

Fertilizers and lime are used in agriculture

Agricultural soils often require fertilizers because irrigation and rainwater leach mineral nutrients from the soil, and the harvesting of crops removes the nutrients that the plants took up from the soil during their growth. Crop yields decrease if any essential element is depleted. Mineral nutrients may be replaced by organic fertilizers, such as rotted manure, or inorganic fertilizers of various types. The three elements most commonly added to agricultural soils are nitrogen (N), phosphorus (P), and potassium (K). Commercial fertilizers are characterized by their "N-P-K" percentages. A 5-10-10 fertilizer, for example, contains 5 percent nitrogen, 10 percent phosphate (P_2O_5), and 10 percent potash (K_2O) by weight.* Sulfur, in the form of a sulfate, is also occasionally added to soils.

Either organic or inorganic fertilizers can provide the necessary mineral nutrients for plants. Organic fertilizers release nutrients slowly, which results in less leaching than a one-time application of inorganic fertilizer. Organic fertilizers also contain materials that improve the physical properties of the soil, providing spaces for gas movement, root growth, and drainage. Inorganic fertilizers, on the other hand, provide an almost instantaneous supply of soil nutrients and can be formulated to meet the requirements of a particular soil and a particular crop.

The availability of nutrient ions, whether they are naturally present in the soil or added as fertilizer, is altered by changes in soil pH. The optimal soil pH for most crops is about 6.5, but so-called acid-loving crops such as blueberries prefer a pH closer to 4. Rainfall and the decomposition of organic substances in the soil lower its pH. Such acidification of the soil can be reversed by **liming**—the application of compounds commonly known as lime, such as calcium carbonate, calcium hydroxide, or magnesium carbonate. The addition of these compounds leads to the removal of H^+ ions from the soil. Liming also increases the availability of calcium to plants, which require it as a macronutrient.

It is easy to guess how humans learned to use fertilizer: It didn't take much insight to notice improved plant growth around animal feces. Perhaps a similar observation of limestone, or chalk, or oyster shells—all sources of calcium carbonate—led to the practice of liming. Sometimes, on the other hand, a soil is not acidic enough. In this case, a farmer can add sulfur, and soil bacteria will convert it to sulfuric acid. Iron and some other elements are more available to plants at a slightly acidic pH. Soil pH testing is useful for home gardens and lawns as well as for agriculture.

Spraying leaves with a nutrient solution is another effective way to deliver some essential elements to growing plants. Plants take up more copper, iron, and manganese when these elements are applied as *foliar* (leaf) sprays than when they are added to the soil as fertilizer. Adjusting the concentrations of nutrient ions and pH in order to optimize uptake and to minimize toxicity can yield excellent results. Such foliar application of mineral nutrients is increasingly used in wheat production, but fertilizer is still delivered most commonly by way of the soil.

Plants affect soils

The soil that forms in a particular place also depends on the types of plants growing there. Plant litter, such as dead fallen leaves, is the major source of carbon-rich materials that break down to form **humus**—dark-colored organic material, each particle of which is too small to be recognizable with the naked eye. Soil bacteria and fungi produce

*The analysis is by weight and is not reported as elemental N, P, and K. A 5-10-10 fertilizer actually *does* contain 5 percent nitrogen, but only 4.3 percent phosphorus and 8.3 percent potassium on an elemental basis.

humus by breaking down plant litter, animal feces, and other organic material. Humus is rich in mineral nutrients, especially nitrogen. Humus in combination with clay promotes a soil structure favorable to plant growth, promoting adequate supplies of both water and oxygen to the roots.

Plants affect the pH of the soil in which they grow. Roots maintain a balance of electric charges. If they absorb more cations than anions, they excrete H^+ ions, thus lowering the soil pH. If they absorb more anions than cations, they excrete OH^- ions or HCO_3^- ions, raising the soil pH.

The mineral nutrient most commonly limiting, in both natural and agricultural situations, is nitrogen. Let's consider how nitrogen is made available to plants.

Nitrogen Fixation

Earth's atmosphere is a vast reservoir of nitrogen in the form of nitrogen gas (N_2). N_2 constitutes almost four-fifths of the atmosphere. However, plants cannot use N_2 directly as a nutrient. It is a highly unreactive substance—the triple bond linking the two nitrogen atoms is extremely stable, and a great deal of energy is required to break it.

A few species of bacteria have an enzyme that enables them to convert N_2 into a more reactive form by a process called **nitrogen fixation**. These prokaryotic organisms—*nitrogen fixers*—convert N_2 to ammonia (NH_3). There are relatively few species of nitrogen fixers, and their biomass is small relative to the mass of other organisms that depend on them for survival on Earth. This talented group of prokaryotes is just as essential to the biosphere as are the photosynthetic autotrophs.

Nitrogen fixers make all other life possible

By far the greatest share of total world nitrogen fixation is performed biologically by nitrogen-fixing organisms, which fix approximately 170 million Mg (megagrams, metric tons) of nitrogen per year. About 80 million Mg is fixed industrially by humans. A smaller amount of nitrogen is fixed in the atmosphere by nonbiological means such as lightning, volcanic eruption, and forest fires. Rain brings these atmospherically formed products to the ground.

Several groups of bacteria fix nitrogen. In the oceans, various photosynthetic bacteria, including cyanobacteria, fix nitrogen. In fresh water, cyanobacteria are the principal nitrogen fixers. On land, free-living soil bacteria make some contribution to nitrogen fixation, but they fix only what they need for their own use and release the fixed nitrogen only when they die. Other nitrogen-fixing bacteria live in close association with plant roots. They release up to 90 percent of the nitrogen they fix to the plant and excrete some amino acids into the soil, making nitrogen immediately available to other organisms.

Bacteria of the genus *Rhizobium* fix nitrogen only in close association with the roots of plants in the legume family (Figure 36.6). The legumes include peas, soybeans, clover, alfalfa, and many tropical shrubs and trees. The bacteria infect the plant's roots, and the roots develop nodules in response to their presence. The various species of *Rhizobium* show a fairly high specificity for the species of legume they infect. Farmers and gardeners coat legume seeds with *Rhizobium* to make sure the bacteria are present. Some farmers alternate their crops, planting clover or alfalfa occasionally to increase the available nitrogen content of the soil.

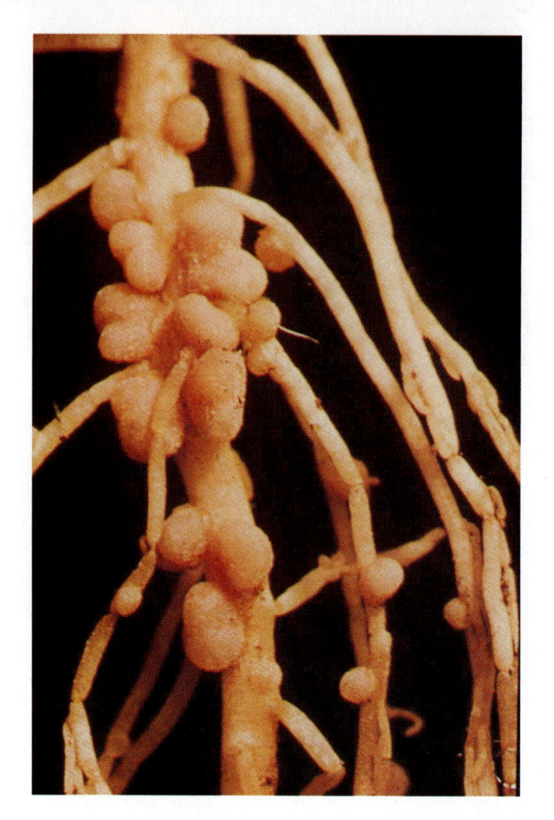

36.6 Root Nodules
Large, round, tumorlike nodules are visible in the root system of a broad bean. These nodules house nitrogen-fixing bacteria.

The legume–*Rhizobium* association is not the only bacterial association that fixes nitrogen. Some cyanobacteria fix nitrogen in association with fungi in lichens or with ferns, cycads, or nontracheophytes. Rice farmers can increase crop yields by growing the water fern *Azolla*, with its symbiotic nitrogen-fixing cyanobacterium, in the flooded fields where rice is grown. Another group of bacteria, the filamentous actinomycetes, fix nitrogen in association with root nodules on woody species such as alder and mountain lilacs.

How does biological nitrogen fixation work? In the four sections that follow, we'll consider the role of the enzyme nitrogenase, the mutualistic collaboration of plant and bacterial cells in root nodules, the need to supplement biological nitrogen fixation in agriculture, and the contributions of plants and bacteria to the global nitrogen cycle.

Nitrogenase catalyzes nitrogen fixation

Nitrogen fixation is the reduction of nitrogen gas. It proceeds by the stepwise addition of three pairs of hydrogen atoms (Figure 36.7). In addition to N_2, these reactions require a strong reducing agent to transfer hydrogen atoms to nitrogen and the intermediate products, as well as a great deal of energy, which is supplied by ATP. Depending on the species of nitrogen fixer, either respiration or photosynthe-

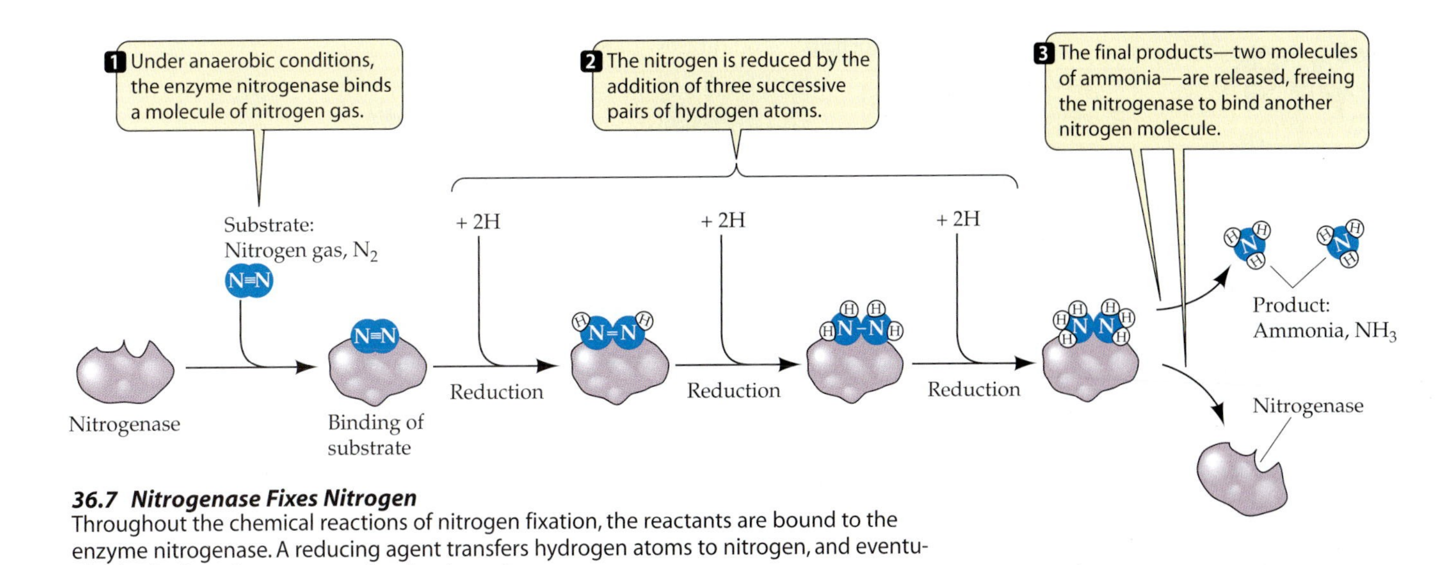

36.7 Nitrogenase Fixes Nitrogen
Throughout the chemical reactions of nitrogen fixation, the reactants are bound to the enzyme nitrogenase. A reducing agent transfers hydrogen atoms to nitrogen, and eventually the final product—ammonia—is released.

sis may provide both the necessary reducing agent and ATP. The reactants are firmly bound to the surface of a single enzyme, called **nitrogenase**.

Nitrogenase is so strongly inhibited by oxygen (O_2) that its discovery was delayed because investigators had not thought to seek it under anaerobic conditions. Because nitrogenase cannot function in the presence of oxygen, it is not surprising that many nitrogen fixers are anaerobes. Legumes respire aerobically, as do *Rhizobium*. Within a root nodule, oxygen is maintained at a level sufficient to support respiration but not so high as to inactivate nitrogenase.

Some plants and bacteria work together to fix nitrogen

The legume nodule provides an excellent example of *symbiosis*, in which two different organisms live in physical contact. In the form of symbiosis called *mutualism*, both organisms benefit from their relationship. The legume obtains fixed nitrogen from the bacterium, and the bacterium obtains the products of photosynthesis from the plant. Neither free-living *Rhizobium* species nor uninfected legumes can fix nitrogen. Only when the two are closely associated in root nodules does the reaction take place.

The establishment of this symbiosis between *Rhizobium* and a legume requires a complex series of steps, with active contributions by both the bacteria and the plant root (Figure 36.8). First the root releases flavonoids and other chemical signals that attract the *Rhizobium* to the vicinity of the root. Flavonoids trigger the transcription of bacterial *nod* genes, which encode Nod (nodulation) factors. These factors, secreted by the bacteria, cause cell divisions in the root cortex, leading to the formation of a primary nodule meristem. Within the nodules, the bacteria take the form of **bacteroids** within membranous vesicles. Bacteroids are swollen, deformed bacteria that can fix nitrogen.

Before the bacteroids can begin to fix nitrogen, the plant must produce the protein **leghemoglobin**, which surrounds the bacteroids. Leghemoglobin is a close relative of hemoglobin, the oxygen-carrying pigment of animals. Some plant nodules contain enough of it to be bright pink when viewed in cross section. Leghemoglobin, with its iron-containing heme, transports oxygen to the bacteroids to support their respiration.

The partnership between bacterium and plant in nitrogen-fixing nodules is not the only case in which plants depend on other organisms for assistance with their nutrition. Another example that we considered earlier is that of *mycorrhizae*, root–fungus associations in which the fungus greatly increases the absorption of water and minerals (especially phosphorus) by the plant (see Figure 30.16). A growing body of evidence suggests that nodule formation depends on some of the same genes and mechanisms that allow mycorrhizae to develop.

Biological nitrogen fixation does not always meet agricultural needs

Bacterial nitrogen fixation is not sufficient to support the needs of agriculture. Traditional farmers used to plant dead fish along with corn so that the decaying fish would release fixed nitrogen that the developing corn could use. Industrial nitrogen fixation is becoming ever more important to world agriculture because of the degradation of soils and the need to feed a rapidly expanding population. Research on biological nitrogen fixation is being vigorously pursued, with commercial applications very much in mind.

Most industrial nitrogen fixation is done by a chemical process called the Haber process, which requires a great deal of energy. An alternative is urgently needed because of the cost of energy. At present, the manufacture of nitrogen-containing fertilizer takes more energy than does any other aspect of crop production in the United States.

One line of investigation centers on recombinant DNA technology as a means of engineering new plants that produce their own nitrogenase. Workers in many industrial

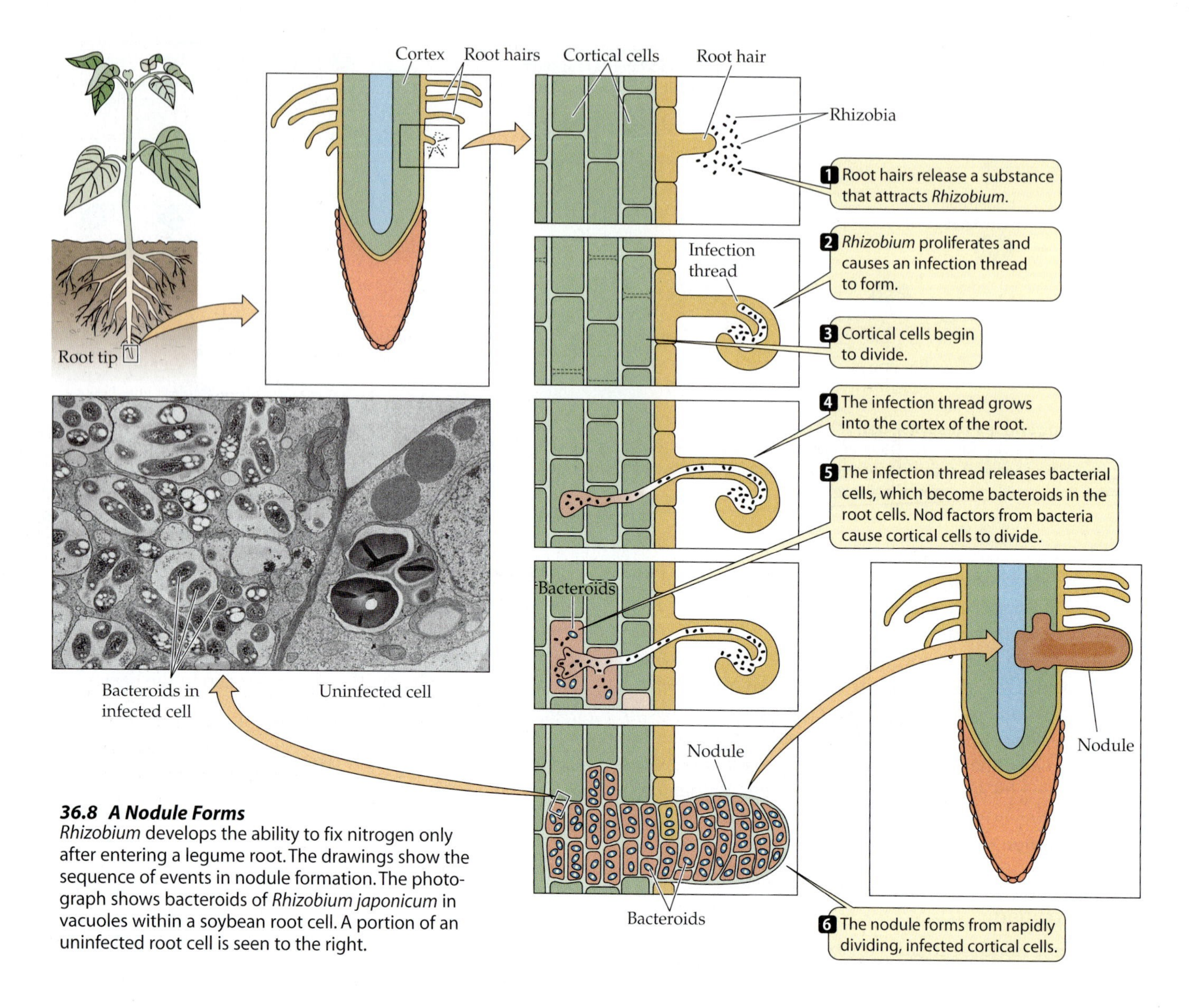

36.8 A Nodule Forms
Rhizobium develops the ability to fix nitrogen only after entering a legume root. The drawings show the sequence of events in nodule formation. The photograph shows bacteroids of *Rhizobium japonicum* in vacuoles within a soybean root cell. A portion of an uninfected root cell is seen to the right.

and academic laboratories are attempting to insert bacterial genes coding for nitrogenase into the cells of angiosperms, particularly crop plants. Developing crops that can fix their own nitrogen, however, will take more than just the insertion of genes for nitrogenase. Biotechnology must also find ways to exclude O_2 and obtain strong reducing agents and an energy source. Ultimately, the need for ATP represents a greater technical challenge than the insertion of nitrogenase genes. The stakes, however—especially the financial ones—are high, and a huge amount of effort is being invested in research along these lines.

Plants and bacteria participate in the global nitrogen cycle

The reduced nitrogen released into the soil by nitrogen fixers is primarily in the form of ammonia (NH_3) and ammonium ions (NH_4^+). Although ammonia is toxic to plants, ammonium ions can be taken up safely at low concentrations. Soil bacteria called *nitrifiers,* which we described in Chapter 26, oxidize ammonia to nitrate ions (NO_3^-)—another form that plants can take up—by the process of **nitrification**. Soil pH affects uptake: Nitrate ions are taken up preferentially under more acidic conditions, ammonium ions under more basic ones.

The steps that we have followed so far are carried out by bacteria: N_2 is *reduced* to ammonia in nitrogen fixation and ammonia is *oxidized* to nitrate in nitrification. The next steps are carried out by plants, which *reduce* the nitrate they have taken up all the way back to ammonia (Figure 36.9). All the reactions of **nitrate reduction** are carried on by the plant's own enzymes. The later steps, from nitrite (NO_2^-) to ammonia, take place in the chloroplasts, but this conversion is not part of photosynthesis. The plant uses the ammonia thus formed to manufacture amino acids, from which the plant's proteins and all its other nitrogen-containing compounds are formed. Animals cannot reduce nitrogen, and they depend on plants to supply them with reduced nitrogenous compounds.

Bacteria called *denitrifiers* return nitrogen from animal wastes and dead organisms to the atmosphere as N_2. This process, described in Chapter 26, is called **denitrification**.

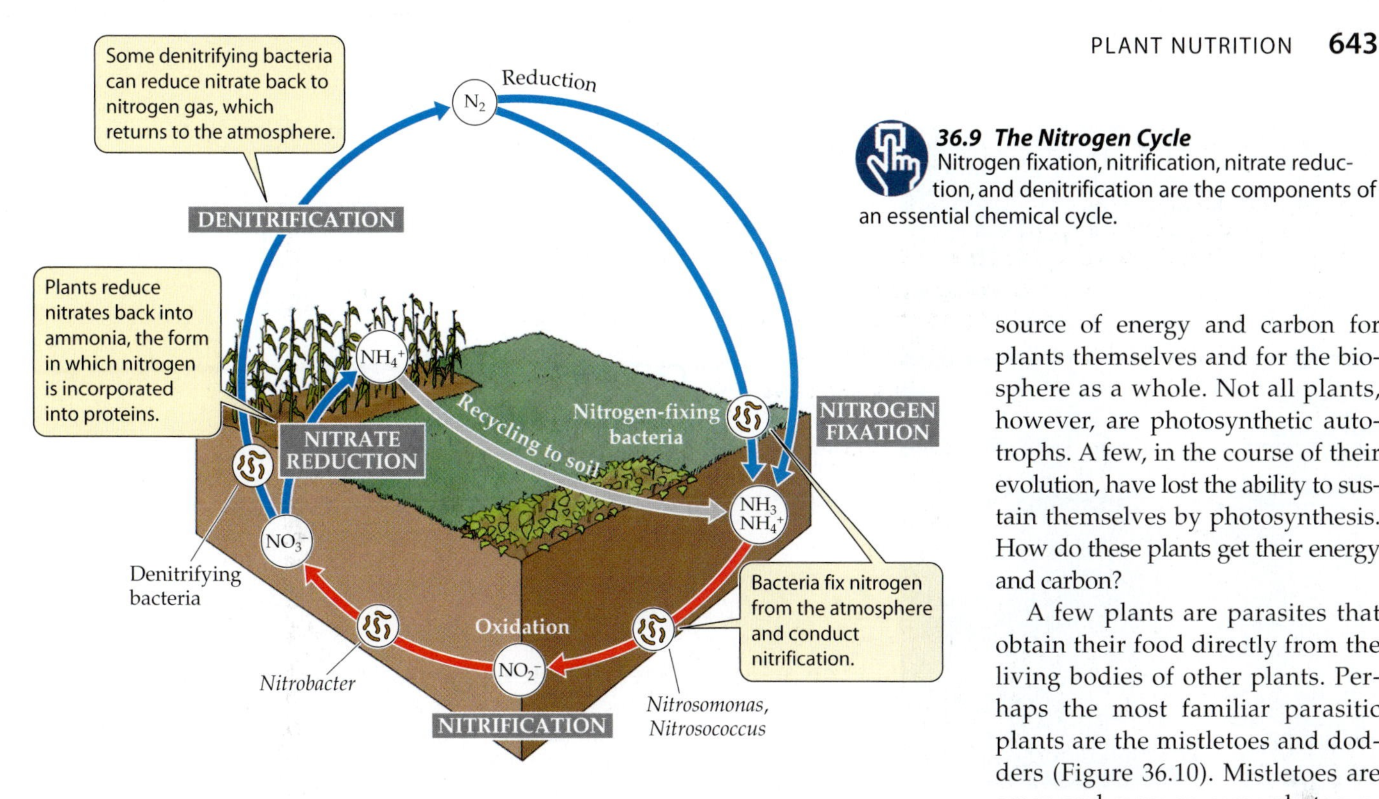

36.9 The Nitrogen Cycle
Nitrogen fixation, nitrification, nitrate reduction, and denitrification are the components of an essential chemical cycle.

In combination with leaching and the removal of crops, it keeps the level of available nitrogen in soils low.

Nitrogen metabolism, in bacteria and in plants, is complex. It is also of great importance: Nitrogen-containing compounds constitute 5 to 30 percent of a plant's total dry weight. The nitrogen content of animals is even higher, and all the nitrogen in the animal world arrives there by way of the plant kingdom.

Sulfur Metabolism

All living things require sulfur, which is a constituent of two amino acids, cysteine and methionine, and hence of almost all proteins. Sulfur is also a component of other biologically crucial compounds, such as coenzyme A. Animals must obtain their cysteine and methionine from plants, but plants can make their own, using sulfate ions obtained from the soil or from a liquid environment.

Except for oxygen, all of the most abundant elements in plants are taken up from the environment in their most oxidized forms—sulfur as sulfate, carbon as carbon dioxide, nitrogen as nitrate, phosphorus as phosphate, and hydrogen as water. In plants, sulfate is reduced and incorporated into cysteine. From this amino acid all the other sulfur-containing compounds in the plant are made. Sulfate reduction and the utilization of cysteine are analogous to the reduction of nitrate to ammonia and the subsequent utilization of ammonia by plants.

Heterotrophic and Carnivorous Seed Plants

Thus far in this chapter we have considered the mineral nutrition of plants. As you already know, another crucial aspect of plant nutrition is photosynthesis—the principal source of energy and carbon for plants themselves and for the biosphere as a whole. Not all plants, however, are photosynthetic autotrophs. A few, in the course of their evolution, have lost the ability to sustain themselves by photosynthesis. How do these plants get their energy and carbon?

A few plants are parasites that obtain their food directly from the living bodies of other plants. Perhaps the most familiar parasitic plants are the mistletoes and dodders (Figure 36.10). Mistletoes are green and carry on some photosynthesis, but they parasitize other plants for water and mineral nutrients and may derive photosynthetic products from them as well. Mistletoes and dodders extract nutrients from the vascular tissues of their hosts by forming absorptive organs called *haustoria*, which invade the host plant's tissue. Another parasitic plant, the Indian pipe, once was thought to obtain its food from dead organic matter. It is

36.10 A Parasitic Plant
Tendrils of dodder wrap around other plants. This parasitic plant obtains water, sugars, and other nutrients from its host through tiny, rootlike protuberances that penetrate the surface of the host.

36.11 A Carnivorous Sundew
Sundews trap insects on their sticky hairs. Secreted enzymes will digest the carcasss externally.

now known to get its nutrients, with the help of fungi, from nearby actively photosynthesizing plants. Hence it too is a parasite.

Some other plants that do not live by photosynthesis alone are the 450 or so carnivorous species—those that augment their nitrogen and phosphorus supply by capturing and digesting flies and other insects (Figure 36.11; also shown at the start of this chapter). The best-known carnivorous plants are Venus flytraps (genus *Dionaea*), sundews (genus *Drosera*), and pitcher plants (genus *Sarracenia*).

Carnivorous plants are normally found in boggy regions where the soil is acidic. Most decay-causing organisms require a less acidic pH to break down the bodies of dead organisms, so relatively little nitrogen is recycled into these acidic soils. Accordingly, the carnivorous plants have adaptations that allow them to augment their supply of nitrogen by capturing animals and digesting their proteins.

Sarracenia produces pitcher-shaped leaves that collect small amounts of rainwater. Insects are attracted into the pitchers either by bright colors or by scent and are prevented from getting out again by stiff, downward-pointing hairs. The insects eventually die and are digested by a combination of enzymes and bacteria in the water. Even rats have been found in large pitcher plants.

Sundews have leaves covered with hairs that secrete a clear, sticky, sugary liquid. An insect touching one of these hairs becomes stuck, and more hairs curve over the insect and stick to it as well. The plant secretes enzymes to digest the insect and later absorbs the carbon- and nitrogen-containing products of digestion.

None of the carnivorous plants must feed on insects. They grow adequately without insects, but in their natural habitats they grow faster and are a darker green when they succeed in catching insects. The additional nitrogen from the insects is used to make more proteins, chlorophyll, and other nitrogen-containing compounds.

Chapter Summary

The Acquisition of Nutrients

▶ Plants are photosynthetic autotrophs that can produce all the compounds they need from carbon dioxide, water, and minerals, including a nitrogen source. They obtain energy from sunlight, carbon dioxide from the atmosphere, and nitrogen-containing ions and mineral nutrients from the soil.

▶ Plants explore their surroundings by growing rather than by locomotion.

Mineral Nutrients Essential to Plants

▶ Plants require 14 essential mineral elements, all of which come from the soil. Several essential elements fulfill multiple roles. **Review Table 36.1**

▶ The six mineral nutrients required in substantial amounts are called macronutrients; the eight required in much smaller amounts are called micronutrients. **Review Table 36.1**

▶ Deficiency symptoms suggest what essential element a plant lacks. **Review Table 36.2**

▶ Biologists discovered the essentiality of each mineral nutrient by growing plants on nutrient solutions lacking the test element. **Review Figure 36.2**

Soils and Plants

▶ Soils are complex in structure, with living and nonliving components. They contain water, gases, and inorganic and organic substances. They typically consist of two or three horizontal zones called horizons. **Review Figures 36.3, 36.4, and Table 36.3**

▶ Soils form by mechanical and chemical weathering of rock.

▶ Plants obtain some mineral nutrients by ion exchange between the soil solution and the surface of clay particles. **Review Figure 36.5**

▶ Farmers use fertilizer to make up for deficiencies in soil mineral nutrient content, and they apply lime to raise low soil pH.

▶ Plants affect soils in various ways, helping them form, adding material such as humus, and removing nutrients (especially in agriculture).

Nitrogen Fixation

▶ A few species of soil bacteria are responsible for almost all nitrogen fixation. Some nitrogen-fixing bacteria live free in the soil; others live symbiotically as bacteroids within the roots of plants.

▶ In nitrogen fixation, nitrogen gas (N_2) is reduced to ammonia (NH_3) or ammonium ions (NH_4^+) in a reaction catalyzed by nitrogenase. **Review Figure 36.7**

▶ Nitrogenase requires anaerobic conditions, but the bacteroids in root nodules require oxygen for their respiration. Leghemoglobin helps maintain the oxygen supply to the bacteroids.

▶ The formation of a nodule requires an interaction between the root system of a legume and *Rhizobium* bacteria. **Review Figure 36.8**

▶ Nitrogen-fixing bacteria reduce atmospheric N_2 to ammonia, but most plants take up both ammonium ions and nitrate ions. Nitrifying bacteria oxidize ammonia to nitrate. Plants take up nitrate and reduce it back to ammonia, a feat of which animals are incapable. **Review Figure 36.9**

▶ Denitrifying bacteria return N_2 to the atmosphere, completing the biological nitrogen cycle. **Review Figure 36.9**

Sulfur Metabolism

▶ Plants take up sulfate ions and reduce them, forming the amino acids cysteine and methionine. Cysteine is the major precursor for other sulfur-containing compounds in plants and in animals, which must obtain their organic sulfur from plants.

Heterotrophic and Carnivorous Seed Plants

▶ A few heterotrophic plants are parasitic on other plants.

▶ Carnivorous plant species are autotrophs that supplement their nitrogen supply by feeding on insects.

For Discussion

1. Methods for determining whether a particular element is essential have been known for more than a century. Since these methods are so well established, why was the essentiality of some elements discovered only recently?
2. If a Venus flytrap were deprived of soil sulfates and hence made unable to synthesize the amino acids cysteine and methionine, would it die from lack of protein?
3. Soils are dynamic systems. What changes might result when land is subjected to heavy irrigation for agriculture after being relatively dry for many years? What changes in the soil might result when a virgin deciduous forest is cut down and replaced by crops that are harvested each year?
4. We mentioned that important positively charged ions are held in the soil by clay particles, but other, equally important, negatively charged ions are leached deeper into the soil's B horizon. Why doesn't leaching cause an electrical imbalance in the soil? (*Hint*: Think of the ionization of water.)
5. The biosphere of Earth as we know it depends on the existence of a few species of nitrogen-fixing prokaryotes. What do you think might happen if one of these species were to become extinct? If all of them were to disappear?

37 Plant Growth Regulation

MORE THAN A CENTURY AGO, CHARLES Darwin and his son studied the growth of plant shoots toward the light. Their findings, which we will detail in this chapter, pointed the way to the eventual discovery of the photoreceptor molecules that capture light signals and the hormones that transmit those signals to other parts of the plant. Light and hormones affect processes in plants as diverse as stem growth, flowering, bud dormancy, and the dropping of leaves in autumn. Several of the hormones now find important commercial applications, including the regulation of fruit ripening and enhanced germination of barley for the brewing industry.

Recent advances in understanding plant development have come largely from work with *Arabidopsis thaliana*, a little mustard-like weed. This plant is useful to researchers because its body and seeds are tiny, and its genome is unusually small for a flowering plant. It also flowers and forms seeds in a relatively short time after growth begins. *Arabidopsis* mutants with altered developmental patterns provide evidence for the existence of hormones and for the mechanisms of hormone and photoreceptor action.

In this chapter we first give a brief overview of the life of a flowering plant and its developmental stages. We explore the environmental cues, photoreceptors, and hormones that regulate plant development, and consider the multiple roles that each plays in normal development.

Interacting Factors in Plant Development

The *development* of a plant—the series of progressive changes that take place throughout its life—is regulated in complex ways. Four factors take part in this regulation:

- The plant senses and responds to *environmental cues*.
- The plant's *genome* encodes enzymes that catalyze the biochemical reactions of development, including the ones that make hormones and receptors, produce chemical building blocks, and participate in protein synthesis and energy metabolism.
- In order to sense environmental cues, the plant uses *receptors*, such as photoreceptors that absorb light.
- Chemical messages, or *hormones*, mediate the effects of the environmental cues sensed by the receptors.

Several hormones and photoreceptors regulate plant growth

Hormones are regulatory compounds that act at very low concentrations at sites distant from where they are produced. They mediate many developmental phenomena in plants, such as stem growth and autumn leaf fall. Unlike

Catching Some Rays
Most of us have observed the manner in which plants turn toward sunlight. Light signals caught by photoreceptor proteins are transmitted by hormones to other parts of the plant in a finely tuned developmental dance.

37.1 Plant Hormones

HORMONE	TYPICAL ACTIVITIES
Abscisic acid	Maintains seed dormancy and winter dormancy; closes stomata
Auxin	Promotes stem elongation, adventitious root initiation, and fruit growth; inhibits lateral bud outgrowth and leaf abscission
Brassinosteroids	Promote elongation of stems and pollen tubes; promote vascular tissue differentiation
Cytokinins	Inhibit leaf senescence; promote cell division and lateral bud outgrowth; affect root growth
Ethylene	Promotes fruit ripening and leaf abscission; inhibits stem elongation and gravitropism
Gibberellins	Promote seed germination, stem growth, and fruit development; break winter dormancy; mobilize nutrient reserves in grass seeds
Jasmonates	Trigger defenses against pathogens and herbivores
Oligosaccharins	Trigger defenses against pathogens; limit effects of high auxin concentrations; regulate cell differentiation
Salicylic acid	Triggers resistance to pathogens
Systemin	Causes jasmonate production in response to tissue damage

animals, which produce each hormone in a specific part of the body, plants produce hormones in many of their cells. Each plant hormone plays multiple regulatory roles, affecting several different aspects of development (Table 37.1). Interactions among the hormones are often complex.

Like hormones, **photoreceptors** regulate many developmental processes in plants. Unlike the hormones, which are small molecules, plant photoreceptors are proteins. Light (an environmental cue) acts directly on photoreceptors, which in turn regulate processes such as the many changes accompanying the growth of a young plant out of the soil and into the light.

No matter what cues direct development, ultimately the plant's genome determines the limits within which the plant and its parts will develop. The genome encodes the master plan, but its interpretation depends on conditions in the environment. It is also the target for some hormone actions. For several decades hormones and photoreceptors were the focus of most work on plant development, but recent advances in molecular genetics allow us to focus on underlying processes such as signal transduction pathways.

Signal transduction pathways mediate hormone and photoreceptor action

We introduced the topic of signal transduction pathways in Chapter 15. Plants, like other organisms, make extensive use of these pathways. Cell signaling in plant development involves three steps: a receptor (for a hormone or for light), a signal transduction pathway, and the ultimate cellular response (see Figure 15.3). Protein kinase cascades amplify responses to receptor binding in plants, as they do in other organisms (see Figure 15.11). The signal transduction pathways of plants differ from those of animals only in the details; for example, their protein kinases phosphorylate the amino acid residues serine or threonine but not tyrosine.

Before concerning ourselves with molecular details, let's set a broader context. What is the general pattern of plant development?

From Seed to Death: An Overview of Plant Development

Let's review the life history of a flowering plant, from seed to death, focusing on how the developmental events are regulated. As plants develop, environmental cues, photoreceptors, and hormones affect three fundamental processes: cell division, cell expansion, and cell differentiation.

The seed germinates and forms a growing seedling

All developmental activity may be suspended in a seed, even when conditions appear to be suitable for its growth. In other words, a seed may be **dormant**. Typically, only 5 to 20 percent of a seed's weight is water, whereas most plant parts contain far more water.

Cells in dormant seeds do not divide, expand, or differentiate. For the embryo to begin developing, seed dormancy must be broken by one of several physical mechanisms, such as exposure to light, mechanical abrasion, fire, or leaching of inhibitors by water.

As the seed **germinates** (begins to develop), it first imbibes (takes up) water. The growing embryo must then obtain building blocks—monomers—for its development by digesting the polysaccharides, fats, and proteins stored in the cotyledons or in the endosperm. The embryos of some plant species secrete hormones that direct the mobilization of these reserves.

If the seed germinates underground, the new seedling must elongate rapidly and cope with life in darkness or dim

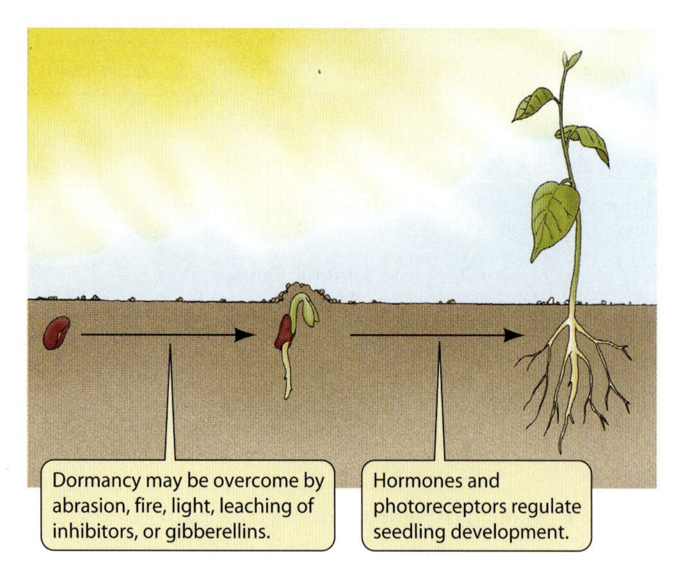

37.1 From Seed to Seedling
Environmental factors, hormones, and photoreceptors regulate the first stages of plant growth.

light. A photoreceptor controls this stage, and ends it when the shoot is exposed to sufficient light to begin photosynthesis (Figure 37.1).

Early shoot development varies among the flowering plants. Figure 37.2 presents the distinctive shoot development patterns of monocots and eudicots. Plant growth from seedling to adult, both in darkness and in light, also involves several hormones.

The plant flowers and sets fruit

Flowering—the formation of reproductive organs—may be initiated when the plant reaches an appropriate age or size. Some plant species, however, flower at particular times of the year, meaning that the plant must sense the appropriate time. In these plants, the leaves measure the length of the night (shorter in summer, longer in winter) with great precision. Light absorbed by photoreceptors affects this time-measuring process.

Once a leaf has determined that it is time for the plant to flower, that information must be transported as a signal to the places where flowers will form. The means by which this signal is transmitted remains a mystery, but it is likely that a "flowering hormone" travels from the leaf to the point of flower formation.

After flowers form, hormones play further roles. Hormones and other substances control the growth of a pollen tube down the style of a pistil. Following fertilization, a fruit develops and ripens under hormonal control (Figure 37.3).

The plant senesces and dies

Some plants, known as **perennials**, continue to grow year after year. Many perennials have buds that enter a state of winter dormancy during the cold season. A hormone called abscisic acid helps maintain this dormancy.

In many species, leaves **senesce** (deteriorate because of aging) and fall at the end of the growing season, shortly before the onset of the severe conditions of winter. Leaf fall (abscission) is regulated by an interplay of the hormones ethylene and auxin. Finally, the entire plant senesces and dies.

37.2 Patterns of Early Shoot Development
(*a*) In grasses and some other monocots, growing shoots are protected by a coleoptile until they reach the surface. (*b*) In most eudicots, the growing point of the shoot is protected by the cotyledons. (*c*) In some other eudicots, the cotyledons remain in the soil, and the growing point is protected by the first true leaves.

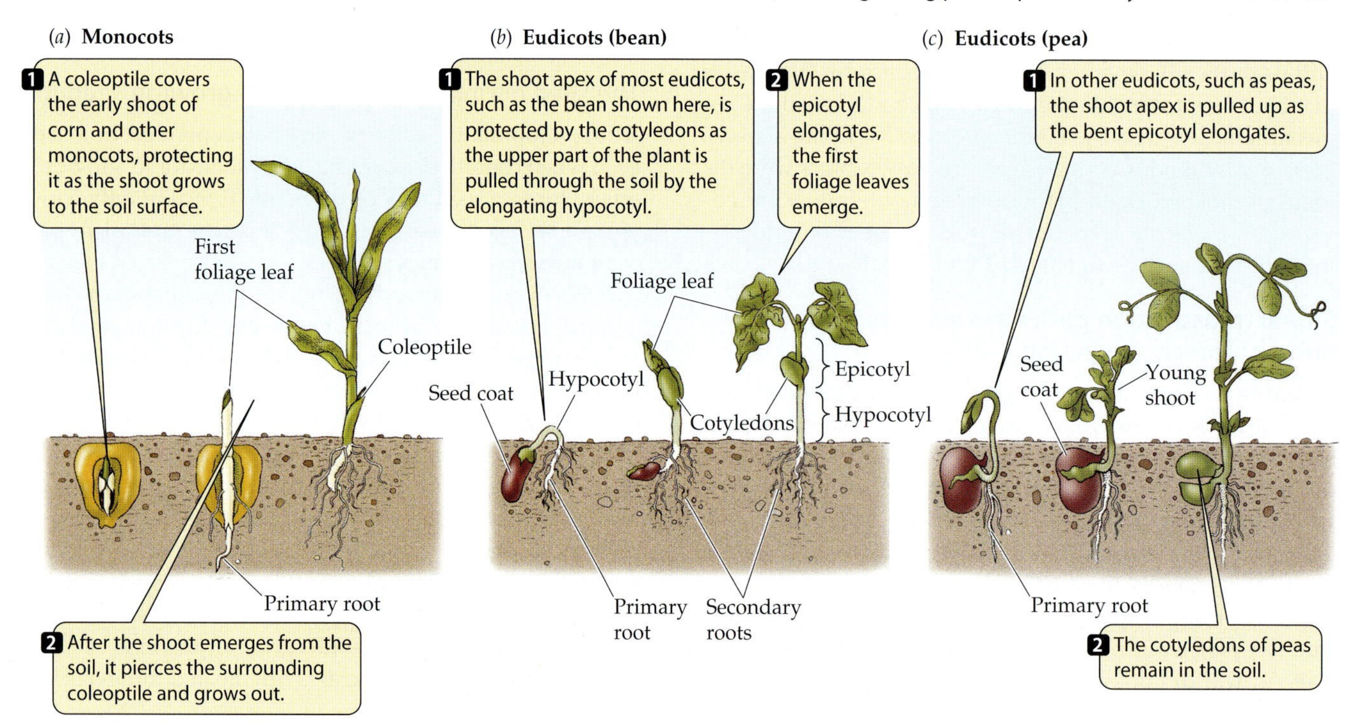

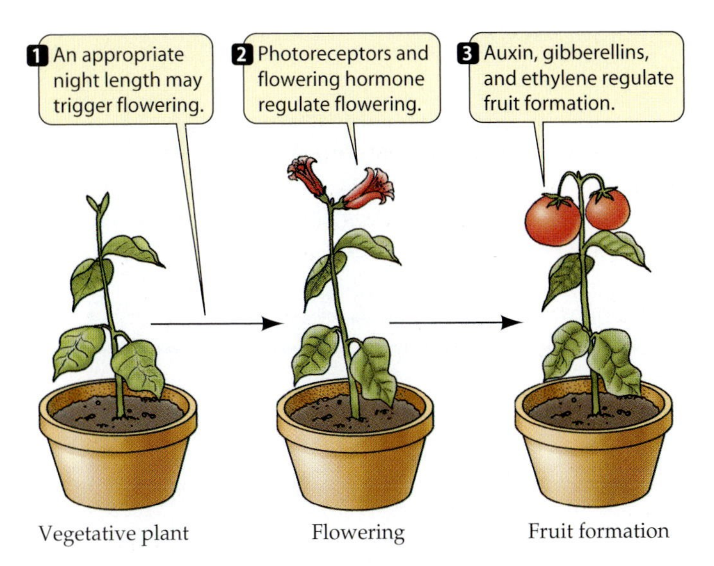

37.3 Flowering and Fruit Formation
Environmental cues, photoreceptors, and hormones regulate plant reproduction.

Death, which may be initiated by signals from the environment, follows senescent changes that are controlled by hormones such as ethylene. This life history pattern appears to be an adaptation for producing more offspring by pumping energy (food) and nutrients into the seeds; in so doing, the parent plant essentially starves itself to death.

We have reached the end of the plant's life history. Now let's examine how the various steps are regulated. We'll begin with regulation at the start of the life history—the seed and its germination.

Ending Seed Dormancy and Beginning Germination

The seeds of some species are, in effect, instant plants: All they need for germination is water. But many other species have seeds whose germination is regulated in more complex ways.

Seed dormancy may last for weeks, months, years, or even centuries. The mechanisms of dormancy are numerous and diverse, but three principal strategies dominate:

- Exclusion of water or oxygen from the embryo by means of an impermeable seed coat
- Mechanical restraint of the embryo by means of a tough seed coat
- Chemical inhibition of embryo development

The dormancy of seeds with impermeable coats can be broken if the seed coat is abraded as the seed tumbles across the ground or through creek beds, or passes through the digestive tract of an animal. Soil microorganisms probably play a major role in softening seed coats. Fire can release mechanical restraint. It can also melt wax in seed coats, removing the waterproofing and allowing water to reach the embryo (Figure 37.4). *Leaching*—prolonged exposure to water—is one way to reduce the level of a water-soluble chemical inhibitor and end dormancy. Scorching of seeds by fire can also break down some inhibitors.

Seed dormancy affords adaptive advantages

What are the potential advantages of seed dormancy? For many species, dormancy assures survival through unfavorable conditions and results in germination when conditions are more favorable. To avoid germination in the dry days of late summer, for example, some seeds must be exposed to a long cold period before they will germinate. Other seeds will not germinate until a certain amount of time has passed, regardless of how they are treated. This strategy prevents germination while the seed of a cereal grain, for example, is still attached to the parent plant.

Seeds that must be scorched by fire in order to germinate avoid competition by germinating only when an area has been cleared by fire. Light-requiring seeds, which germinate only at or near the surface of the soil, are generally tiny seeds with few food reserves. Conversely, germination of some seeds is inhibited by light; these seeds germinate only when buried and thus kept in darkness. Light-inhibited seeds are usually large and well stocked with nutrients.

Seed dormancy helps annual plants counter the effects of year-to-year variation in the environment. The seeds of some annuals remain dormant throughout an unfavorable year. The seeds of other plants germinate at different times

37.4 Fire and Seed Germination
This fireweed germinated and flourished after a great fire along the Alaska Highway.

37.5 Leaching of Germination Inhibitors
The seeds of the cypress, a swamp-adapted tree, germinate only after being leached by water, which increases the chances that they will germinate in a situation suitable for their growth.

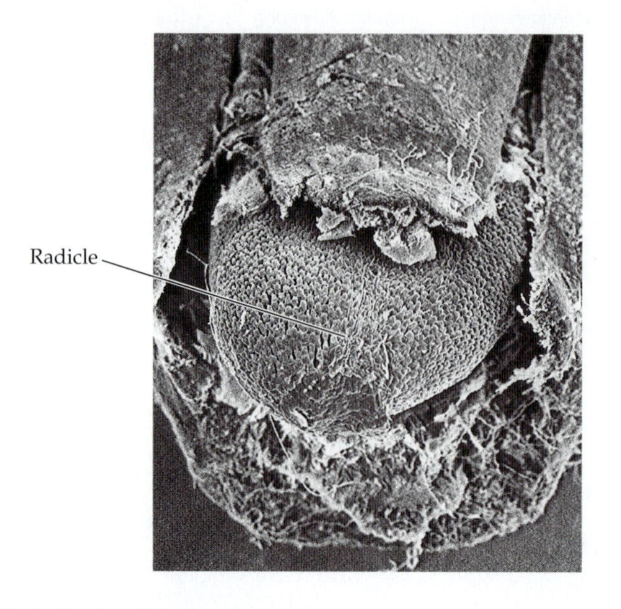

37.6 The Radicle Emerges
The tip of this barley seed's radicle has just broken through its protective sheath. The appearance of the radicle—the embryonic root—is one of the first externally visible events in seed germination.

during the year, increasing the likelihood that at least some of the seedlings will encounter favorable conditions.

Dormancy may also increase the likelihood of a seed's germinating in the right place. Some cypress trees, for example, grow in standing water, and their seeds germinate only if inhibitors are leached by water (Figure 37.5).

Seed germination begins with the uptake of water

The first step in seed germination is the uptake of water, called **imbibition**. A seed's water potential (see Chapter 35) is very negative, and water can be taken up readily if the seed coat allows it. The magnitude of this water potential is demonstrated by the force exerted by seeds expanding in water. Cocklebur seeds that are imbibing can exert a pressure of up to 1,000 atmospheres (about 100 megapascals) against a restraining force.

As a seed takes up water, it undergoes metabolic changes: Certain existing enzymes become activated, RNA and then proteins are synthesized, the rate of cellular respiration increases, and other metabolic pathways become activated. In many seeds there is no DNA synthesis and no cell division during these early stages of germination. Initially, growth results solely from the expansion of small, preformed cells. DNA is synthesized only after the embryonic root, called the **radicle**, begins to grow and poke out beyond the seed coat (Figure 37.6).

The embryo must mobilize its reserves

Until the young plant (the **seedling**) becomes able to photosynthesize, it depends on reserves stored in the endosperm or cotyledons. The principal reserve of energy and carbon in many seeds is starch. Other seeds store fats or oils. Usually, the endosperm of the seed holds amino acid reserves in the form of proteins, rather than as free amino acids.

The giant molecules of starch, lipids, and proteins must be digested by enzymes into monomers that can enter the cells of the embryo. The polymer starch yields glucose for energy metabolism. The digestion of reserve proteins provides the amino acids the embryo needs to synthesize its own proteins. The digestion of lipids releases glycerol and fatty acids, both of which can be metabolized for energy. Glycerol and fatty acids can also be converted to glucose, which permits fat-storing plants to make all the building blocks they need for growth.

In germinating barley and other cereal seeds, the embryo secretes **gibberellins**, one of several classes of plant growth hormones. Gibberellins diffuse through the endosperm to a surrounding tissue called the **aleurone layer**, which lies inside the seed coat. The gibberellins trigger a crucial series of events in the aleurone layer, culminating in the release of enzymes that digest proteins and starch stored in the endosperm (Figure 37.7). Commercially, gibberellins are used in the brewing industry to enhance the "malting" (germination) of barley and the breakdown of its endosperm, producing sugar that is fermented to alcohol.

Gibberellins: Regulators from Germination to Fruit Growth

Gibberellins produce a wide variety of effects on plant development in addition to triggering digestive enzyme synthesis. We begin our discussion of the different plant growth hormones by discussing the discovery of the gibberellins, as well as their many effects.

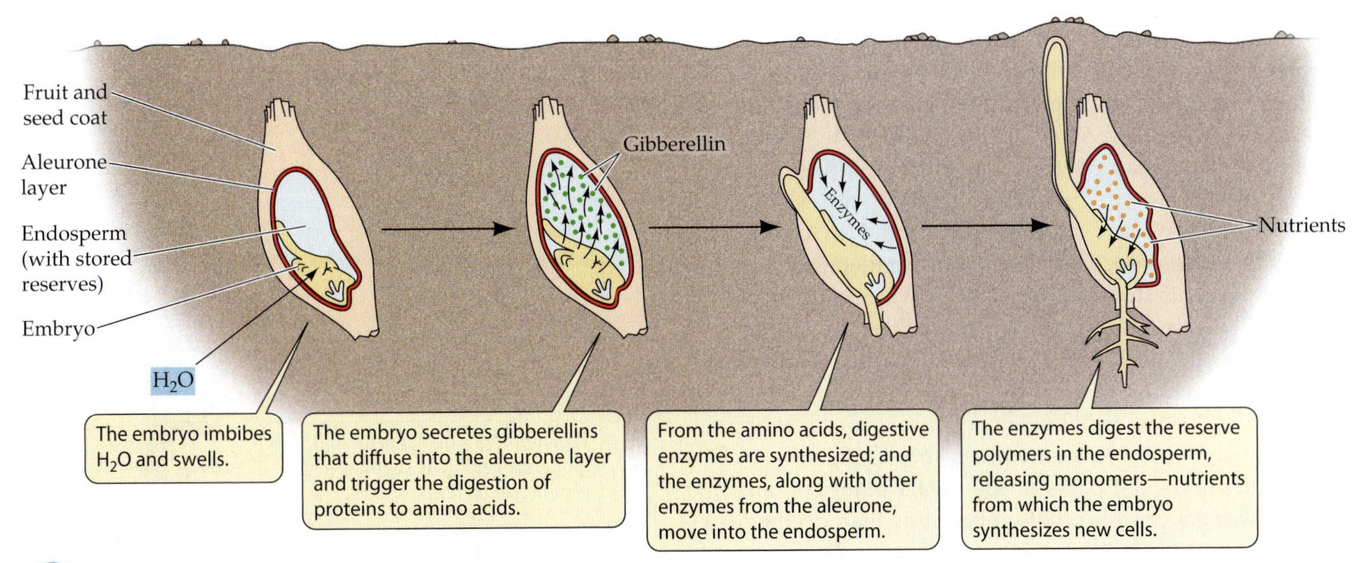

37.7 Embryos Mobilize Polymer Reserves
Seed germination in cereal grasses consists of a cascade of processes. Gibberellins trigger the conversion of reserve polymers into monomers that can be used by the developing embryo.

Foolish seedlings led to the discovery of the gibberellins

The gibberellins are a large family of closely related compounds. Some are found in plants and others in a pathogenic (disease-causing) fungus, where they were first discovered.

Gibberellin A_1
(important in stem growth)

Gibberellin A_3
(commercially available)

In 1809, the study of the gibberellins began indirectly with observations of the *bakanae*, or "foolish seedling," disease of rice. Seedlings affected by this disease grow tall more rapidly than their healthy neighbors, but this rapid growth gives rise to spindly plants that die before producing seed (the rice grain used for food). The disease has had considerable economic impacts in several parts of the world. It is caused by the ascomycete fungus *Gibberella fujikuroi*.

In 1925, the Japanese biologist Eiichi Kurosawa grew *G. fujikuroi* on a liquid medium, then separated the fungus from the medium by filtering. He heated the filtered medium to kill any remaining fungus, but the resulting heat-treated filtrate still caused rapid growth in rice seedlings. Medium that had never contained the fungus did not stimulate seedling growth. This experiment established that *G. fujikuroi* produces a growth-promoting chemical substance, which Kurosawa called a gibberellin.

Were the gibberellins simply exotic products of an obscure fungus, or did they play a more general role in the growth of plants? Bernard O. Phinney of the University of California, Los Angeles, answered this question in part in 1956, when he reported the spectacular growth-promoting effect of gibberellins on dwarf corn seedlings. He used plants that were known to be genetic dwarfs; each phenotype was produced when a particular recessive allele (say, *d1*) was present in the homozygous condition (*d1*/*d1*). Gibberellins applied to nondwarf—normal—corn seedlings had almost no effect, but gibberellins applied to the dwarfs caused them to grow as tall as their normal relatives. (A comparable effect of gibberellins applied to a dwarf tomato plant is shown in Figure 37.8.)

37.8 The Effect of Gibberellins on Dwarf Plants
In this experiment, the effect of gibberellins was tested on two dwarf tomato plants. Both plants were the same size when the one on the right was treated with gibberellins.

This result suggested to Phinney (1) that gibberellins are normal constituents of corn, and perhaps of all plants, and (2) that dwarf plants are short because they cannot produce their own gibberellins. According to Phinney's hypothesis, nondwarf plants manufacture enough gibberellins to promote their full growth, but dwarf plants do not. Extracts from numerous plant species were found to promote growth in dwarf corn. These findings provided direct evidence that plants that are not genetic dwarfs contain gibberellin-like substances. Phinney's work set the stage for today's use of mutant plants to investigate the control of plant development.

The roots, leaves, and flowers of a dwarf corn plant appear normal, but the stems are much shorter than those of wild-type plants. All parts of the dwarf plant contain a much lower concentration of gibberellins than do the organs of a wild-type plant. We may infer, then, that stem elongation *requires* gibberellins or the products of gibberellin action. We can further conclude that gibberellins play a less essential role in the development of roots, leaves, and flowers.

Although more than 80 gibberellins have been identified, only one, *gibberellin* A_1, actually controls stem elongation in most plants. The other gibberellins found in stems are simply intermediates in the production of gibberellin A_1. As we will see in the next section, gibberellins affect processes other than stem elongation, but we do not yet know which gibberellin has any other particular effect.

37.9 Bolting
Spraying with gibberellins causes cabbage and some other plants to bolt.

The gibberellins have many effects

Gibberellins and other hormones regulate the growth of fruits. It has long been known that seedless grapes (an inbred strain) form smaller fruit than their seeded relatives. Experimental removal of seeds from very young seeded grapes prevented normal fruit growth, suggesting that the seeds are sources of a fruit growth regulator. It was then shown that spraying young seedless grapes with a gibberellin solution caused them to grow as large as seeded ones. It is now a standard commercial procedure to spray seedless grapes with gibberellins. Subsequent biochemical studies showed that the developing seeds produce gibberellins, which diffuse out into the immature fruit tissue.

Some biennial plants respond dramatically to an increased level of gibberellins. **Biennials** grow vegetatively in their first year and flower and die in their second year. In the second year, the apical meristems of biennials respond to environmental cues by producing elongated shoots that eventually bear flowers. This elongation is called **bolting**. When the plant senses the appropriate environmental cue—longer days or a sufficient winter chilling—it produces more gibberellins, raising the gibberellin concentration to a level that causes the shoot to bolt. Plants of some biennial species will bolt when sprayed with a gibberellin solution without the environmental cue (Figure 37.9).

Gibberellins also cause fruit to grow from unfertilized flowers, promote seed germination in lettuce and some other species, and help bring spring buds out of winter dormancy. Most hormones have multiple effects within the plant, and they often interact with one another in regulating developmental processes. In controlling stem elongation, for example, gibberellins interact with another hormone, auxin.

Auxin Affects Plant Growth and Form

If you pinch off the apical bud at the top of a bean plant, inactive lateral buds become active, developing into branches. Similarly, pruning a shrub causes an increase in branching. If you cut off the blade of a leaf but leave its petiole (stalk) attached to the plant, the petiole drops off sooner than it would have if the leaf were intact. If a plant is kept indoors, its shoot system grows toward a window. These diverse responses of shoot systems are all mediated by a plant hormone called **auxin**, or *indoleacetic acid* (IAA).

Auxin (indoleacetic acid)

$H_2C{-}COOH$ ... N–H

In the discussions that follow, we will look at the discovery of auxin, its transport within the plant, and its role as mediator of the effects of light and gravity on plant growth. We'll discover its many effects on vegetative growth and on fruit development. Then we'll examine its mechanism of action.

Plant movements led to the discovery of auxin

The discovery of auxin and its numerous physiological effects can be traced back to work done in the 1880s by Charles Darwin and his son Francis. The Darwins were interested in plant movements. One type of movement they studied was **phototropism**, the growth of plant structures toward light (as in most shoots) or away from it (as in roots). They asked, What part of the plant senses the light?

To answer this question, the Darwins worked with canary grass (*Phalaris canariensis*) seedlings grown in the dark. A young grass seedling has a **coleoptile**—a cylindrical sheath a few cells thick that protects the delicate shoot as it pushes through the soil (see Figure 37.2*a*). When the coleoptile breaks through the surface of the soil, it soon stops growing, and the first leaves emerge unharmed. The coleoptiles of grasses are *phototropic*—they grow toward the light.

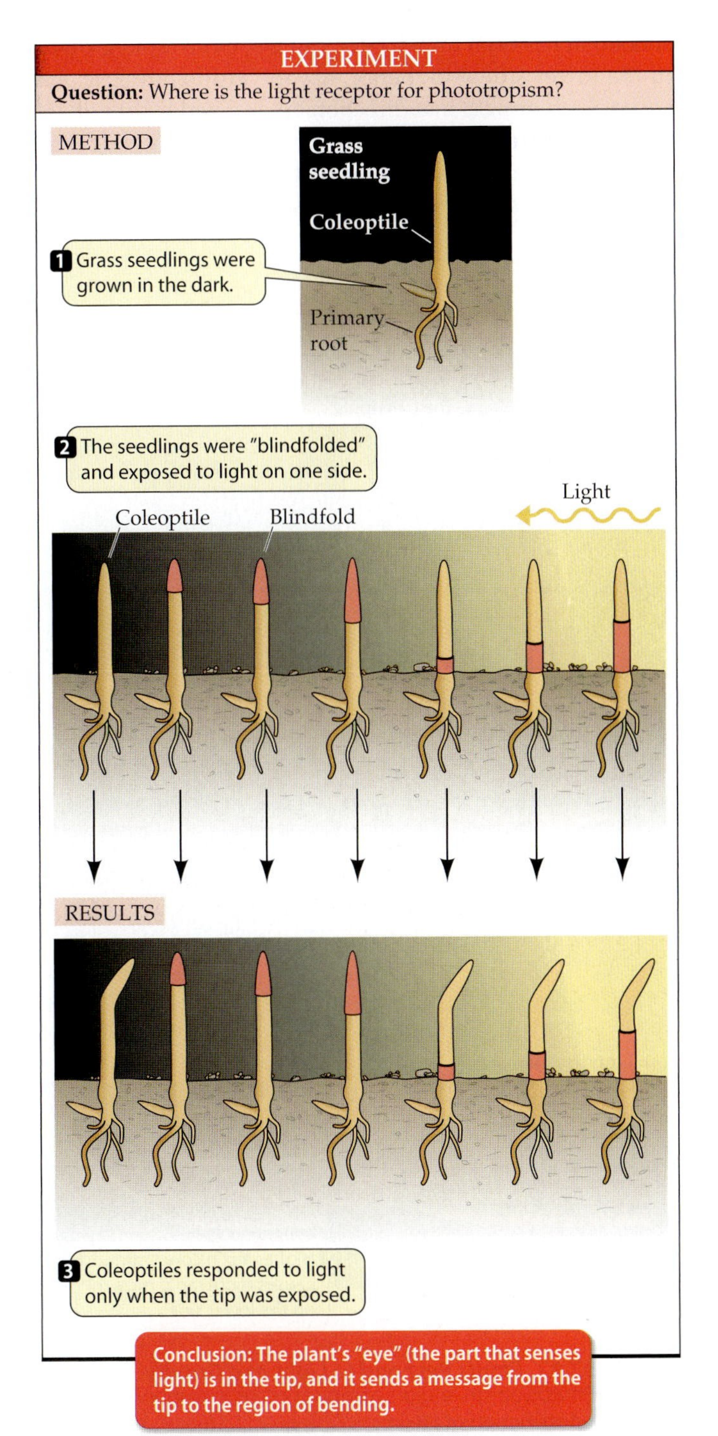

To find the light-receptive region of the coleoptile, the Darwins tried "blindfolding" the coleoptiles of dark-grown canary grass seedlings in various places, then illuminating them from one side (Figure 37.10). The coleoptile grew toward the light whenever its tip was exposed. If the top millimeter or more of the coleoptile was covered, however, there was no phototropic response. Thus the tip contains the photoreceptor that responds to light. The actual bending toward the light, however, takes place in a growing region a few millimeters below the tip. Therefore, the Darwins reasoned, some type of message must travel within the coleoptile from the tip to the growing region. Others later demonstrated that the message is a chemical substance by showing that it can move through certain nonliving materials, such as gelatin, but not through others, such as a metal barrier.

Further experiments showed that the tip of the coleoptile produces a hormone that moves down the coleoptile to the growing region. If the tip is removed, the growth of the coleoptile is sharply inhibited. If the tip is carefully replaced, growth resumes, even if the tip and base are separated by a thin layer of gelatin. The hormone moves down from the tip, but it does not move from one side of the coleoptile to the other. If the tip is cut off and replaced so that it covers only one side of the cut end of the coleoptile, the coleoptile curves as the cells on the side below the replaced tip grow more rapidly than those on the other side.

The Dutch botanist Frits W. Went removed coleoptile tips and placed their cut surfaces on a block of gelatin. Then he placed pieces of the gelatin block on decapitated coleoptiles—positioned to cover only one side, just as coleoptile tips had been placed in earlier experiments (Figure 37.11). As they grew, the coleoptiles curved toward the side away from the gelatin. This curvature demonstrated that a hormone had indeed diffused into the gelatin block from the isolated coleoptile tips. Went had at last isolated a hormone from a plant. Later chemical analysis showed that this hormone, named auxin, was indoleacetic acid.

Auxin transport is polar

Since being isolated, auxin has been intensively studied. Early experiments showed that its movement through certain plant tissues is strictly *polar*—that is, unidirectional along a line from apex to base. By inverting some plants or plant parts, scientists determined that the apex-to-base di-

37.10 The Darwins' Phototropism Experiment The top drawings show some of the ways in which seedlings grown in the dark were "blindfolded"; the lower drawings show what the Darwins observed in each case. Their observations led them to hypothesize the existence of a growth-promoting "messenger" substance produced by the coleoptile.

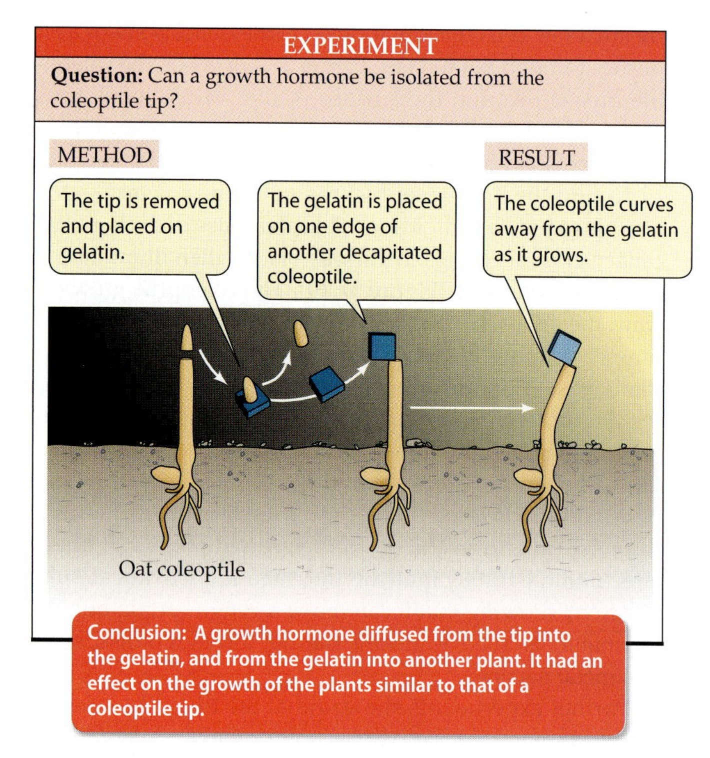

37.11 Went's Experiment
Went succeeded in isolating the growth-promoting chemical substance whose existence the Darwins had hypothesized by placing coleoptile tips on a block of gelatin.

rection of auxin movement has nothing to do with gravity; the polarity of this movement is a totally biological matter. Many plant parts show complete or partial polarity of auxin transport. For example, in most leaf petioles auxin moves only from the blade end toward the stem end.

Auxin carrier proteins move auxin into and out of cells

In one of the most intense areas of current research on auxin, biologists are using *Arabidopsis* plants that have mutations affecting the transport of auxin. By cloning genes from these plants and characterizing their products, they are finding a growing number of auxin carrier proteins. In polar transport in the stem, a carrier protein imports auxin at the end of the cell toward the shoot apex, and it or another carrier exports auxin at the other end of the cell. Auxin carrier proteins contribute to the establishment of auxin gradients in the plant. As a result, auxin acts as a morphogen, telling cells where they lie within the plant and determining how they differentiate. There are probably auxin carrier proteins specific to different tissues and different cells, participating in different auxin responses.

Light and gravity affect the direction of plant growth

While polar auxin transport establishes the orientation of growth, *lateral* (side-to-side) redistribution of auxin appears to be the mechanism that explains both phototropism and another response depending on differential growth, gravitropism. This redistribution may be carried out by other auxin carrier proteins.

When light strikes a coleoptile from one side, auxin at the tip moves laterally toward the shaded side. The imbalance thus established is maintained down the coleoptile, so that in the growing region below, there is more auxin on the shaded side, causing the unequal growth that results in a coleoptile bent toward the light. This bending toward light is phototropism (Figure 37.12*a*). If you have noticed a house plant bending and pointing toward a window, you have seen phototropism.

Even in the dark, auxin moves to the lower side of a shoot that has been tipped over, causing more rapid growth in the lower side and, hence, an upward bending of the shoot. Such growth in a direction determined by gravity is called **gravitropism** (Figure 37.12*b*). The upward gravitropic response of shoots is defined as negative; the gravitropism of roots, which bend downward, is positive.

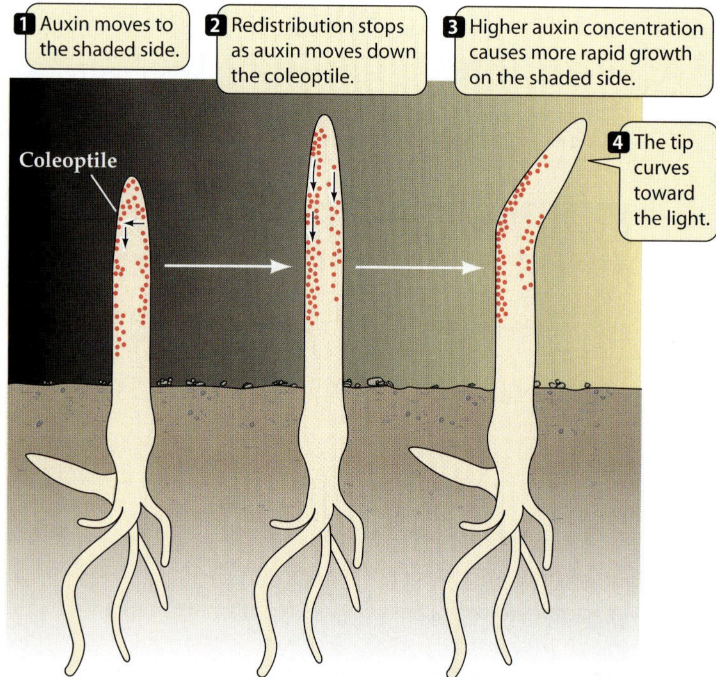

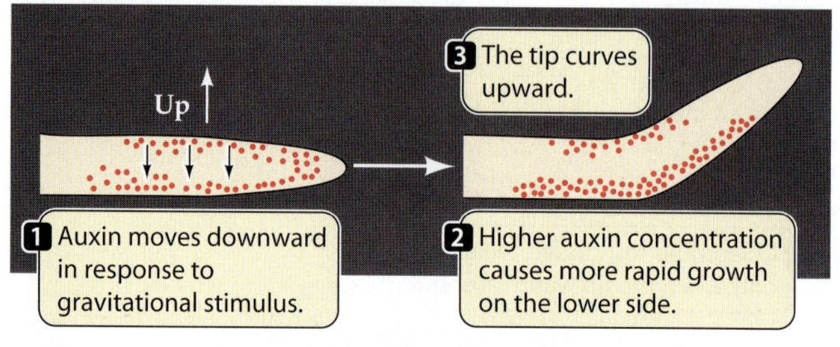

37.12 Plants Respond to Light and Gravity
Phototropism and gravitropism occur in response to a redistribution of auxin.

Auxin affects vegetative growth in several ways

Like the gibberellins, auxin has many roles in plant development. It affects the vegetative growth of plants in several ways, including:

- Initiating root growth in cuttings
- Stimulating the detachment of old leaves from their stems (abscission)
- Maintaining apical dominance
- Promoting stem elongation and inhibiting root elongation

Let's examine each of these aspects in turn.

Cuttings from the shoots of some plants can produce roots and grow into entire new plants. For this to happen, certain undifferentiated cells in the *interior* of the shoot, originally destined to function only in food storage, must set off on an new mission: They must differentiate and become organized into the apical meristem of a new root.

These changes are similar to those in the pericycle of a root when a lateral root forms (see Chapter 34). Shoot cuttings of many species can be stimulated to grow profuse roots by dipping the cut surfaces into an auxin solution; this observation suggests that the plant's own auxin plays a role in the initiation of lateral roots. Commercial preparations that enhance the rooting of plant cuttings typically contain mostly synthetic auxins.

The effect of auxin on the detachment of old leaves from stems is quite different from root initiation. This process, called **abscission**, is the cause of autumn leaf fall. Leaves consist of a blade and a petiole that attaches the blade to the stem. Abscission results from the breakdown of a specific part of the petiole, the *abscission zone* (Figure 37.13). If the blade of a leaf is cut off, the petiole falls from the plant more rapidly than if the leaf had remained intact. If the cut surface is treated with an auxin solution, however, the petiole remains attached to the plant, often longer than an intact leaf would have (Figure 37.14). The time of abscission of leaves in nature appears to be determined in part by a decrease in the movement of auxin, produced in the blade, through the petiole.

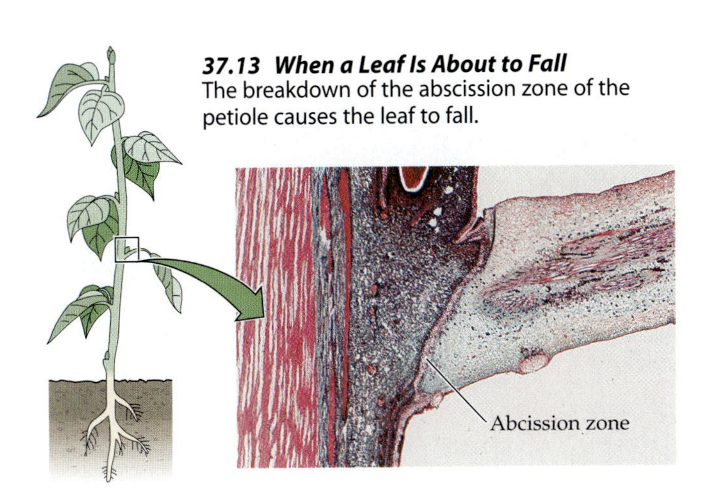

37.13 When a Leaf Is About to Fall
The breakdown of the abscission zone of the petiole causes the leaf to fall.

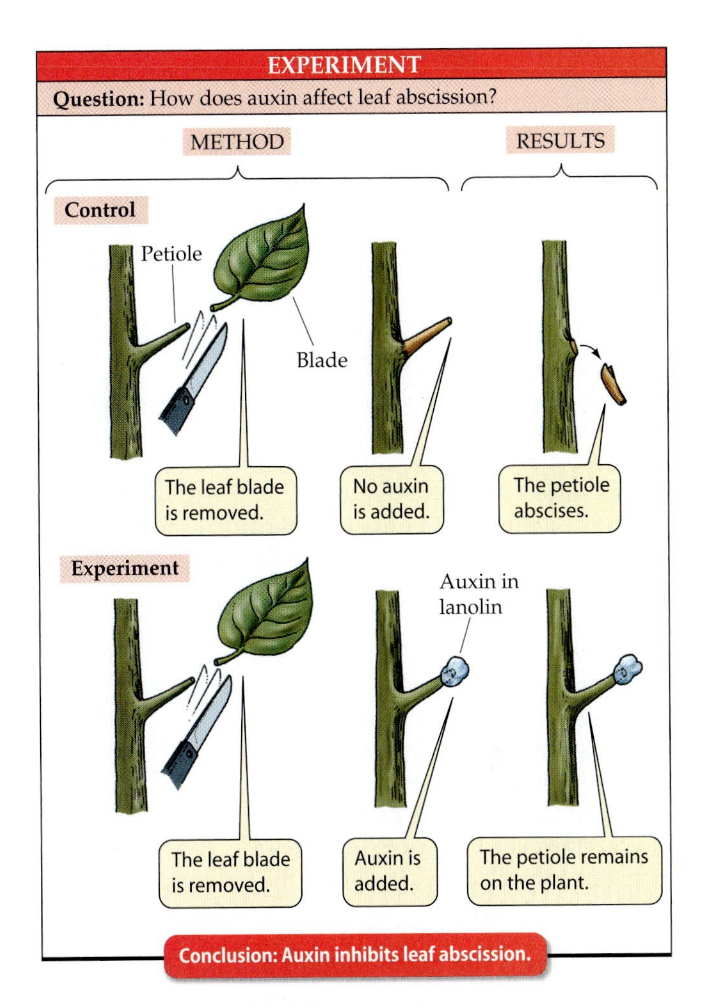

37.14 Auxin and Leaf Abscission
The leaf blade is a source of auxin throughout the growing season; without auxin, the petiole falls from the plant.

Auxin maintains **apical dominance**, a phenomenon in which apical buds inhibit the growth of lateral buds. This phenomenon can be demonstrated by an experiment with young seedlings. If the plant remains intact, the stem elongates, and the lateral buds remain inactive. Removal of the apical bud—the major site of auxin production—permits the lateral buds to grow out vigorously. If the cut surface of the stem is treated with an auxin solution, however, the lateral buds do not grow (Figure 37.15). Apical buds of branches also exert apical dominance: The lateral buds on the branch are inactive unless the apex of the branch is removed.

In the two experiments on leaves and stems that we have just discussed, removal of a particular part of the plant produces an effect—abscission or loss of apical dominance—and that effect is prevented by treatment with auxin. These results are consistent with other data showing that the excised part of the leaf or stem is an auxin source and that auxin in the intact plant helps maintain apical dominance and delays the abscission of leaves. As we will discover later, other hormones can modify the effects of auxin. *Plant*

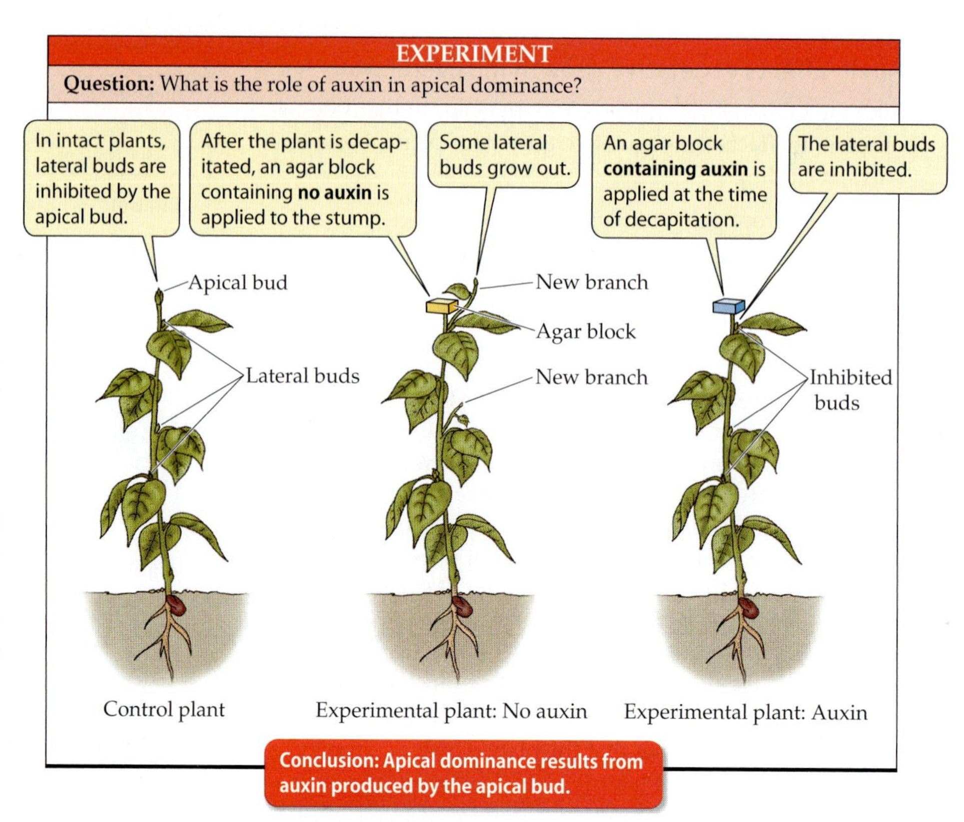

37.15 Auxin and Apical Dominance Auxin produced by the apical bud maintains apical dominance—the growth of a single main stem with minimal branching.

growth is regulated more by hormone interactions than by a single hormone.

Auxin promotes stem elongation, but it inhibits the elongation of roots. The question of why different organs should respond in opposite ways to the same chemical signal remains unanswered, but is a subject of current research.

Many synthetic auxins—chemical analogs of indoleacetic acid—have been produced and studied. One of them, 2,4-dichlorophenoxyacetic acid (2,4-D), has the striking property of being lethal to eudicots at concentrations that are harmless to monocots. This property made 2,4-D an effective *selective herbicide* that could be sprayed on a lawn or a cereal crop to kill those weeds that are eudicots. However, because 2,4-D takes a long time to break down, it pollutes the environment, so scientists are seeking new approaches to selective weed killing.

Auxin controls the development of some fruits

Although fruit development normally depends on prior fertilization of the egg, in many species treatment of an unfertilized ovary with auxin or gibberellins causes **parthenocarpy**—fruit formation without fertilization of the egg. Parthenocarpic fruits form spontaneously in some plants, including dandelions, seedless grapes, and cultivated bananas.

All of these activities illustrate the great diversity of important roles that auxin plays. Now let's see *how* auxin plays one of its roles—promoting stem elongation through effects on the cell wall.

Auxin promotes growth by acting on cell walls

CELL WALLS ARE A KEY TO PLANT GROWTH. The principal strengthening component of the plant cell wall is *cellulose*, a large polymer of glucose. In the wall, cellulose molecules tend to associate in parallel with one another. Bundles of approximately 250 cellulose molecules make up *microfibrils*

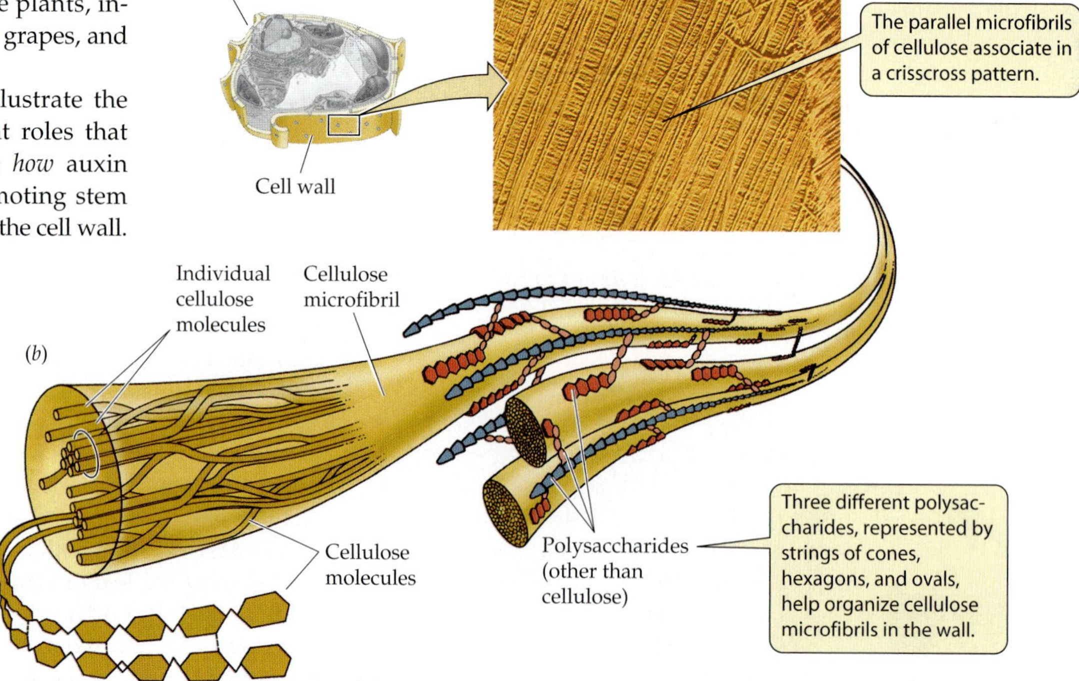

37.16 Cellulose in the Cell Wall The plant cell wall is a network of cellulose microfibrils linked by other polysaccharides.

that are visible with an electron microscope (Figure 37.16). What makes the cell wall rigid is a network of cellulose microfibrils connected by bridges of other, smaller polysaccharides (Figure 37.16*b*). The orientation of the cellulose microfibrils determines the direction of cell expansion (Figure 37.17).

The growth of a plant cell is driven primarily by the uptake of water, which enters the cytoplasm of the cell and accumulates in its central vacuole. As the vacuole expands, the cell grows rapidly, with the vacuole often making up more than 90 percent of the volume of a mature cell. As the vacuole expands, it presses the cytoplasm against the cell wall, and the wall resists this force.

For the cell to grow, its wall must loosen and be stretched. If the wall simply stretched, it would become thinner. However, new polysaccharides are deposited throughout the wall and new cellulose microfibrils are deposited at the inner surface of the wall, maintaining its thickness. Thus the cellulose microfibrils in the outermost part of the wall are the oldest, and those in the innermost part the youngest.

The cell wall plays key roles in controlling the rate and direction of growth of a plant cell. How does the plant determine the behavior of its cell walls?

AUXIN LOOSENS THE CELL WALL. Experiments with segments of oat coleoptiles showed that plant cell walls recover incompletely from being stretched (Figure 37.18). Reversible stretching is called *elasticity*, and irreversible stretching is called *plasticity*. Pretreating the coleoptile segments with auxin significantly increased their plasticity; in other words, it loosened the cell walls. This result suggested that auxin-induced cell expansion might result from just such a loosening effect.

Auxin acts by causing the release of a "wall-loosening factor" from the cytoplasm. Studies in the 1970s indicated that the wall-loosening factor was sometimes simply hydrogen ions (protons, H^+). Acidifying the growth medium (that is, adding H^+) causes segments of stems or coleoptiles to grow as rapidly as segments treated with auxin. Furthermore, treating coleoptile segments with auxin causes acidification of the growth medium. Treatments that block acidification by auxin also block auxin-induced growth. It was suggested that hydrogen ions secreted into the cell wall as a result of auxin action might activate one or more proteins in the wall.

Proteins called *expansins*, isolated from plant cell walls in the 1990s, were found to cause the extension of isolated cell walls of several species. Expansins are widespread among land plants. Expansin action is pH-dependent, and the expansins appear to be activated by hydrogen ions. These proteins apparently modify hydrogen bonding between polysaccharides in the plant cell wall. The changed hydrogen bonding pattern may allow the polysaccharide macromolecules to slip past each other, so that the wall stretches and the cell expands.

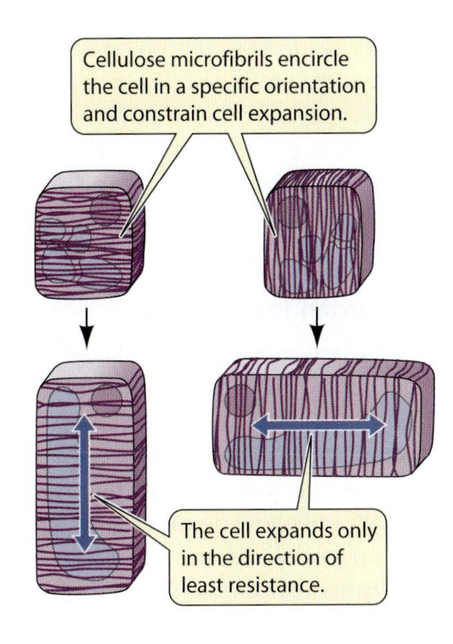

37.17 Plant Cells Expand The orientation of cellulose microfibrils in the plant's cell walls determines the direction of cell expansion.

Plants contain specific auxin receptor proteins

The initial step in the action of any plant hormone is the binding of the hormone to specific receptor proteins. Several proteins can bind various plant hormones, but some of this binding may be nonspecific. It must be shown that auxin-binding proteins actually mediate the effects of auxin.

Plant molecular biologists showed that the protein ABP1 (*A*uxin-*B*inding *P*rotein *1*) functions as an auxin receptor. They inserted the *ABP1* gene of *Arabidopsis* into other species and then induced the expression of the gene in cells that normally show a limited response to auxin. Upon expression of the inserted *ABP1* gene, the cells showed

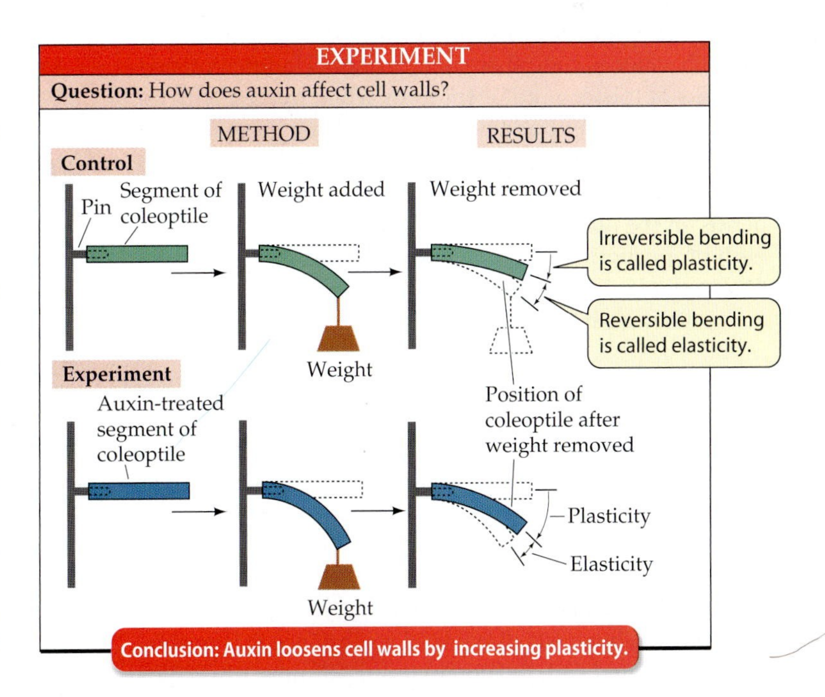

37.18 Auxin Affects Cell Walls Auxin increases the plasticity, but not the elasticity, of cell walls.

greater responses to both endogenous and applied auxin. Subsequent work has conclusively shown the existence and importance of other auxin receptor proteins. Given the number of processes regulated by auxin, it is hardly surprising that there appear to be multiple receptors and signal transduction pathways for this hormone.

Auxin and other hormones evoke differentiation and organ formation

What plant substance signals the different types of cells and organs to form? Much of the research on such questions has been done with plant tissues grown in culture outside the plant body. One easily grown tissue is pith—the spongy, innermost tissue of a stem. Pith tissue cultures proliferate rapidly, but show no differentiation. All the cells are similar and unspecialized; they grow into a lump on the surface of a culture medium.

Cutting a notch in the cultured pith tissue and inserting a stem tip into the notch causes the pith cells below the inserted tip to differentiate. Some of them differentiate to form water-conducting xylem cells. Differentiation of pith cells can also be initiated by adding to the notch a mixture of auxin and coconut milk (a rich source of plant hormones).

A similar effect can be observed in intact plants. If notches are cut in the stems of *Coleus blumei* plants, interrupting some of the strands of vascular tissue, the strands gradually regenerate from the upper side of the cut to the lower (recall that auxin moves from the tip to the base of a stem). If the leaves above the cut are removed, regeneration is slowed. However, when the missing leaves are replaced with an auxin solution, vascular tissue regenerates. Auxin and other plant hormones signal the formation of specific cell types.

Experiments with cultured tissues have helped clarify which hormones control organ formation. Undifferentiated cultures of tobacco pith form roots when treated with an appropriate concentration of auxin. Another group of hormones—the **cytokinins**—causes buds and then shoots to form in such cultures. The pattern of organ formation depends on the ratio of auxin to cytokinin in the medium. A high proportion of auxin favors roots, and a high proportion of cytokinins favors buds, but both processes are most active when both hormones are present.

Cytokinins Are Active from Seed to Senescence

Besides stimulating bud formation, the cytokinins promote cell division in cultured plant tissues, an activity that led to their discovery. In addition, cytokinins aid germination, inhibit stem elongation, stimulate lateral bud growth, and delay leaf senescence.

Cytokinins are derivatives of adenine. In studies of plant cell division, botanists discovered a substance that powerfully stimulated cell division in tissue cultures. This compound, *kinetin*, consists of adenine with an attached group. We now know that kinetin is just one of a family of compounds, which are now called cytokinins. Kinetin may be considered a synthetic cytokinin, because it has never been isolated from plant tissue. However, two closely related compounds, called *zeatin* and *isopentenyl adenine*, occur naturally in plants.

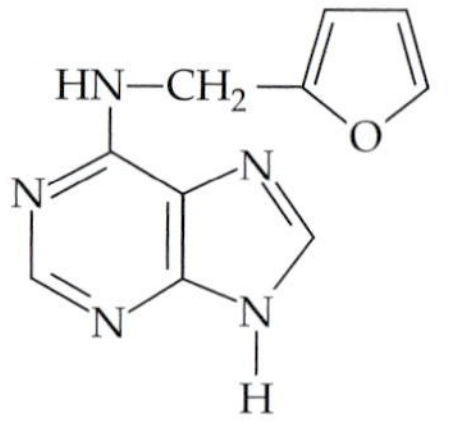

Kinetin
(a cytokinin discovered in aged DNA)

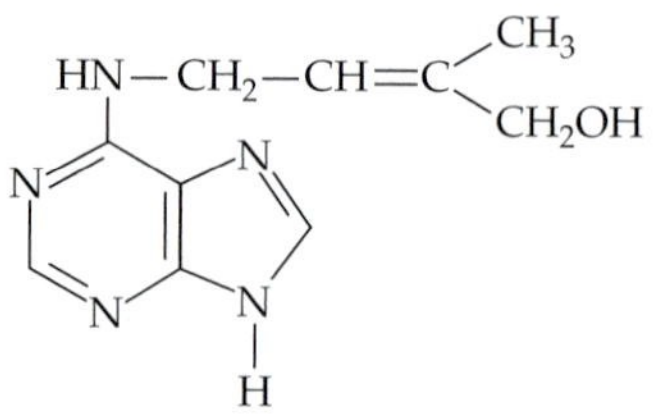

Zeatin
(a naturally occurring cytokinin in plants)

Cytokinins form primarily in the roots and move to other parts of the plant. They have several effects:

- Adding an appropriate combination of auxin and cytokinins to a growth medium yields rapid growth of plant tissues.
- Cytokinins can cause certain light-requiring seeds to germinate when the seeds are kept in constant darkness.
- Cytokinins usually inhibit the elongation of stems, but they cause lateral swelling of stems and roots (the fleshy roots of radishes are an extreme example).
- Cytokinins stimulate lateral buds to grow into branches; thus the balance between auxin and cytokinin levels controls the bushiness of a plant.
- Cytokinins increase the expansion of cut pieces of leaf tissue, and may regulate normal leaf expansion.
- Cytokinins delay the senescence of leaves. If leaf blades are detached from a plant and floated on water or a nutrient solution, they quickly turn yellow and show other signs of senescence. If instead they are floated on a solution containing a cytokinin, they remain green and senesce much more slowly.

Ethylene: A Gaseous Hormone That Promotes Senescence

Whereas the cytokinins oppose or delay senescence, another plant hormone promotes it. This hormone is the gas **ethylene**, which is sometimes called the senescence hormone. Ethylene can be produced by all parts of the plant, and like all plant hormones, it has several effects.

Ethylene
(the "senescence hormone")

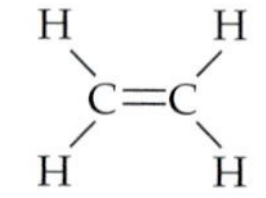

Back when streets were lit by gas rather than by electricity, leaves on trees near street lamps abscised earlier than those on trees farther from the lamps. We now know that ethylene, a combustion product of the illuminating gas, is what caused the abscission. Auxin delays leaf abscission, but ethylene strongly promotes it; thus a balance of auxin and ethylene controls abscission.

Ethylene hastens the ripening of fruit

By promoting senescence, ethylene speeds the ripening of fruit. The old saying "one rotten apple spoils the barrel" is true. That rotten apple is a rich source of ethylene, which speeds the ripening and subsequent rotting of the others in the barrel. As the fruit ripens, it loses chlorophyll and its cell walls break down. Ethylene produced in the fruit tissue promotes both processes. Ethylene also causes an increase in its own production. Thus, once ripening begins, more and more ethylene forms, and because it is a gas, it diffuses readily throughout the fruit and even to neighboring fruits on the same or other plants.

Farmers in ancient times used to slash developing figs to hasten their ripening. We now know that wounding causes an increase in ethylene production by the fruit, and that the raised ethylene level promotes ripening. Today commercial shippers and storers of fruit hasten ripening by adding ethylene to storage chambers. This use of ethylene is the single most important use of a plant hormone in agriculture and commerce. Ripening can also be delayed by the use of "scrubbers" and adsorbents to remove ethylene from the atmosphere in fruit storage chambers.

As flowers senesce, their petals may abscise, to the detriment of the cut-flower industry. Florists or their suppliers often spray their flowers with dilute solutions of silver thiosulfate. Silver salts inhibit ethylene action, probably by interacting directly with the ethylene receptor, and thus delay senescence—enabling florists to keep their wares salable longer.

Ethylene affects stems in several ways

Although associated primarily with senescence, ethylene is active at other stages of plant development as well. The stems of many eudicot seedlings form an **apical hook** that protects the delicate shoot apex while the stem grows through the soil (Figure 37.19). The apical hook is maintained through an asymmetrical production of ethylene gas, which inhibits the elongation of cells on the inner surface of the hook. Once the seedling breaks through the soil surface and is exposed to light, ethylene synthesis stops, and the cells of the inner surface are no longer inhibited. These cells now elongate, and the hook opens, raising the shoot apex and expanding leaves into the sun.

37.19 The Apical Hook of a Eudicot
Asymmetrical production of ethylene is responsible for the apical hook of this seedling, which was grown in the dark.

Ethylene also inhibits stem elongation in general, promotes lateral swelling of stems (as do the cytokinins), and causes stems to lose their sensitivity to gravitropic stimulation.

The ethylene signal transduction pathway is well understood

Analysis of *Arabidopsis* mutants has revealed the steps in the mechanism of ethylene action. Some of these mutants do not respond to applied ethylene, and others act as if they have been exposed to ethylene even though they haven't. Studies of genes from these mutants and their protein products, coupled with comparisons of their amino acid sequences with those of other known proteins, have revealed some of the details of the signal transduction pathway through which ethylene produces its effects (Figure 37.20).

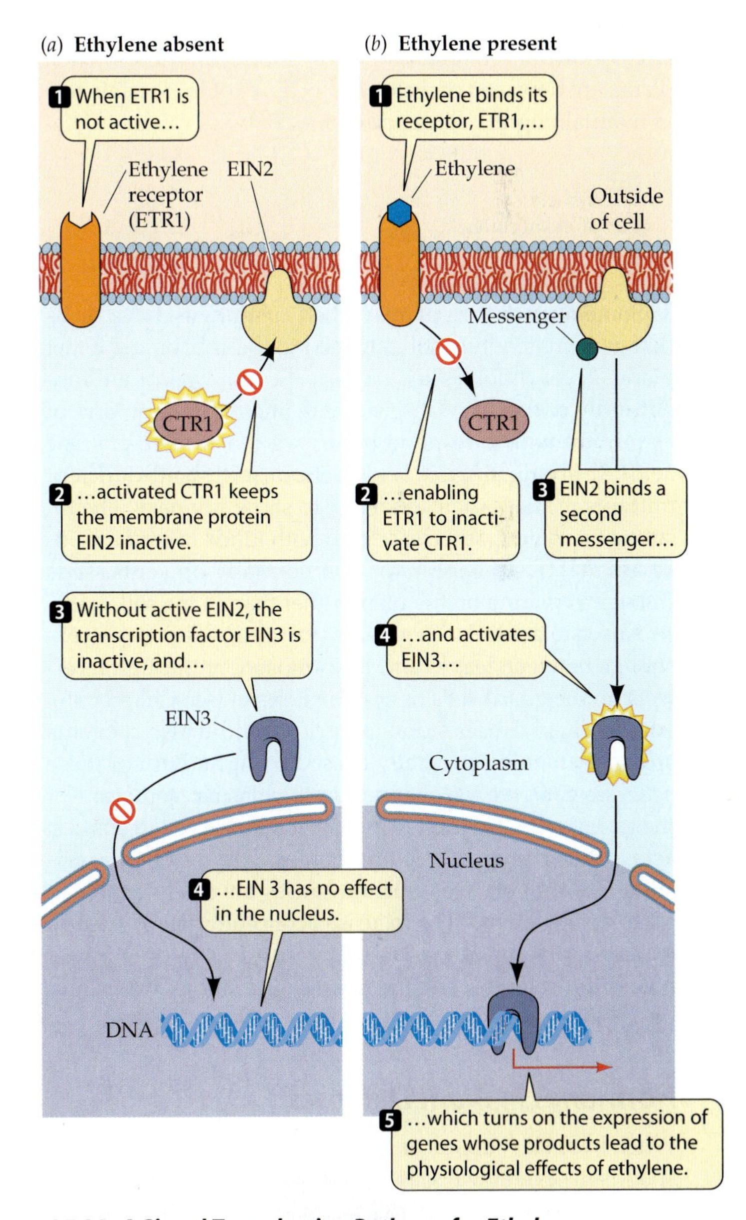

37.20 A Signal Transduction Pathway for Ethylene
This slightly simplified scheme shows the roles of four gene products (ETR1, CTR1, EIN2, and EIN3) in the signal transduction pathway through which ethylene exerts its effects.

The pathway includes two membrane proteins: The first is an ethylene receptor (ETR1), and the second is a channel (EIN2) that acts through a second messenger to activate a transcription factor that turns on genes. The resolution of this pathway has been one of the high points of plant biological research over the last 30 years.

Abscisic Acid: The Stress Hormone

Abscisic acid is another hormone that has multiple effects in the living plant. During embryo formation, abscisic acid promotes the accumulation of storage proteins in seeds by allowing the expression of the genes that encode those proteins. It is generally present in high concentrations in dormant buds and some dormant seeds, and it is the most common inhibitor of seed germination. Abscisic acid also inhibits stem elongation. It is sometimes referred to as the stress hormone of plants because it accumulates when plants are deprived of water and because of its possible role in maintaining the winter dormancy of buds.

Abscisic acid
(the "stress hormone")

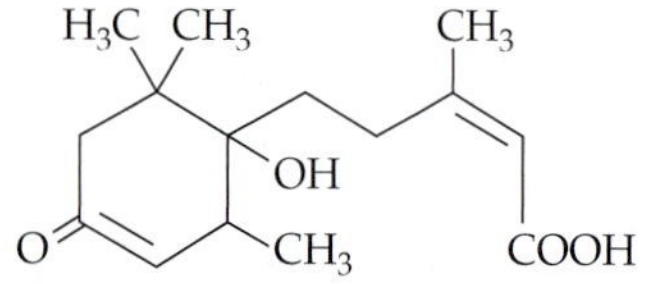

Some mutant corn plants, called *vp* mutants, have seeds that germinate while still attached to the cob, on the intact plant—a condition called *vivipary*. Several *vp* mutants are naturally deficient in abscisic acid, and a different kind of *vp* mutant fails to respond in any way to applied abscisic acid. Applying abscisic acid to the abscisic acid-deficient mutants reduces their tendency to show vivipary. The results of applying abscisic acid to both kinds of mutants indicate that it is the inhibitor that normally prevents seeds from germinating on the plant rather than in the soil.

Abscisic acid also regulates gas and water vapor exchange between leaves and the atmosphere through its effects on the guard cells of the leaf stomata (see Chapter 35). Abscisic acid causes stomata to close, and it also prevents stomatal opening normally caused by light. Both of these processes involve ion channels in the plasma membrane of the guard cells. The first response of a guard cell to abscisic acid is the opening of calcium channels and the entry of calcium into the cell. This calcium causes the cell's vacuole to release calcium, too. The increased concentration of calcium leads to a chain of events that result in the opening of potassium channels and the release of K^+ and of water, and the closing of the stoma as the guard cells sag together.

Hormones in Plant Defenses

When bacteria, viruses, or fungi attack a plant, the plant responds in several ways, as we will see in Chapter 39. One of its first responses is to release hormones called **oligosaccharins**. These hormones, as their name implies, are oligosaccharides—compounds consisting of a few sugar or derivative sugar units. They are actually fragments of the cell wall, which are released when enzymes from an attacker degrade it. They act as signals that trigger the plant's defenses.

Because auxin modifies the cell wall, it is not surprising that auxin, too, causes the release of oligosaccharins. The interactions between auxin and oligosaccharins may be complex. One oligosaccharin, at an extremely low concentration, has been found to inhibit auxin-induced growth promotion. Other oligosaccharins may regulate aspects of cell differentiation.

Three other hormones—*jasmonates*, *salicylic acid*, and *systemin*—serve as important signals in plant defenses. Their activities will be discussed in Chapter 39.

Brassinosteroids: "New" Hormones with Multiple Effects

More than 20 years ago, biologists isolated an interesting steroid from the pollen of rape, a member of the Brassicaceae, or mustard family. When applied to various plant tissues, this **brassinosteroid** stimulated cell elongation, pollen tube elongation, and vascular tissue differentiation, and it inhibited root elongation. Since then, dozens of chemically related and growth-affecting brassinosteroids have been found in plants. Treatment with as little as a few nanograms of brassinosteroid per plant is enough to promote growth. However, the brassinosteroids were not at first regarded as plant hormones, in part because of similarities between their effects and those of auxin.

Brassinolide
(a brassinosteroid)

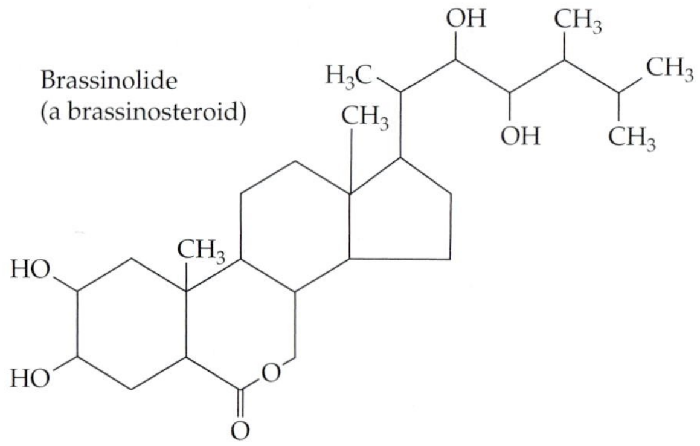

The properties of an *Arabidopsis* mutant called *det2* made it clear that brassinosteroids are naturally occurring plant hormones. When grown in darkness, seedlings homozygous for the *det2* allele differ dramatically from wild-type seedlings: In many respects, they look like wild-type seedlings grown in the light. Treatment of dark-grown *det2* mutant seedlings with brassinosteroids causes them to grow normally—that is, like wild-type plants grown in the dark. The *det2* plants are unable to synthesize their own brassinosteroids, and the lack of the hormone results in abnormal growth.

Brassinosteroids will probably be important in agriculture. They have increased the yields of some crops in field tests. Could they also be useful for keeping some plants small? What about limiting the growth rate of lawns and the height of trees and hedges? Joanne Chory and her colleagues at the University of California, San Diego and the

Salk Institute found a way to do this. They showed that a mutation of a gene called *bas-1* in *Arabidopsis* results in a dwarfed plant because the gene's product inactivates brassinosteroids in the stem. By introducing the *bas-1* mutation into selected plants, agriculturists could produce slow-growing plants and then adjust their growth rate by treatment with brassinosteroids.

Chory and others have shown that some of the effects of light on plant development result from effects on the signal transduction pathway for brassinosteroids. Let's now look more closely at the effects of environmental cues such as light.

Light and Photoreceptors

The length of the night determines the onset of winter dormancy. As summer wears on, the days become shorter (that is, the nights become longer). Leaves have a mechanism for measuring the length of the night, as we will see in the next chapter. Measuring night length is an accurate way to determine the season of the year. If a plant determined the season by the temperature, it could be fooled by a winter warm spell or by unseasonably cold weather in the summer. The length of the night, on the other hand, is determined by Earth's rotation around the sun and does not vary. Plants use the environmental cue of night length to time several aspects of their growth and development.

Length of the night is one of several environmental cues detected by plants, or by individual parts such as leaves. Light—its presence or absence, its intensity, its color, and its duration—provides cues to various conditions. Temperature, too, provides important environmental cues, both by its value at any particular time and by the distribution of warmer and colder stretches over a period of time. The plant "reads" an environmental cue and then "interprets" it, often by stepping up or decreasing its production of hormones.

We'll discuss an example of a temperature cue in the next chapter. Here, we'll see how certain photoreceptors interpret light, its duration, and its wavelength distribution.

Light regulates many aspects of plant development in addition to phototropism. The affected processes range from seed germination to shoot elongation to the initiation of flowering. Several photoreceptors take part in these and other processes. Five **phytochromes** mediate the effects of red and dim blue light. Three or more **blue-light receptors**, discovered more recently, mediate the effects of higher-intensity blue light.

Phytochromes mediate the effects of red and far-red light

Some seeds will not germinate in darkness, but do so readily after even a brief exposure to light. Blue and red light are highly effective in promoting germination, whereas green light is not.

Of particular importance to plants is the fact that far-red light *reverses* the effect of a prior exposure to red light. Far-red light is a very deep red, bordering on the limit of human vision and centered on a wavelength of 730 nm; red wavelengths are around 660 nm. If exposed to brief, alternating periods of red and far-red light in close succession, lettuce seeds respond only to the final exposure: If it is red, they germinate; if it is far-red, they remain dormant (Figure 37.21). This reversibility of the effects of red and far-red light regulates many other aspects of plant development, including flowering and seedling growth.

The basis for the red and far-red effects resides in certain bluish photoreceptor proteins called **phytochromes**. They are blue because they absorb red and far-red light and transmit other light. In the cytosol of plants are two interconvertible forms of phytochromes. Light drives the interconversion of the two forms. The form that absorbs principally red light is called P_r. Upon absorption of a photon of red light, a molecule of P_r is converted into P_{fr}. The P_{fr} form absorbs far-red light; when it does so, it is converted to P_r.

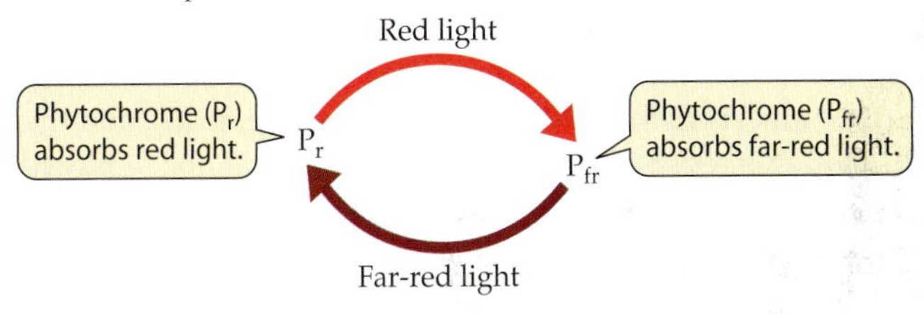

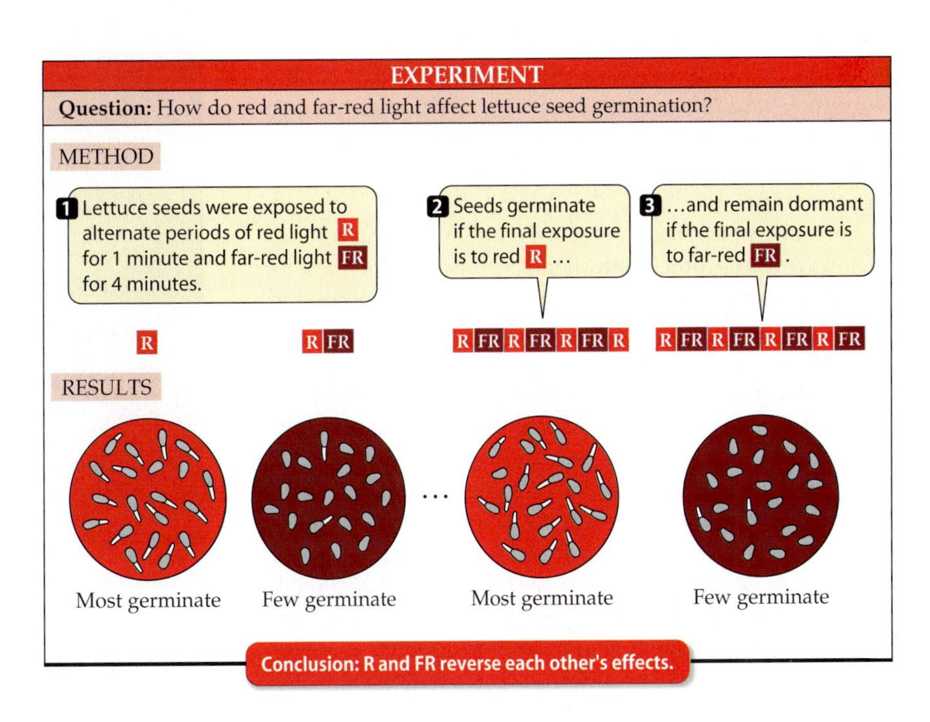

37.21 *Sensitivity of Seeds to Light*
In each case, the final exposure reverses the preceding exposure; seeds respond only to the wavelength of the final light exposure.

P_{fr} has some important biological effects. As we have just seen, one of them is to initiate germination in certain seeds, such as lettuce.

Phytochromes have many effects

Phytochromes help to regulate a seedling's early growth. The radicle, or embryonic root, is the first portion of the seedling to escape the seed coat (see Figure 37.6); the shoot emerges later. When seeds germinate in the dark below the soil surface, a pale and spindly seedling forms, with undeveloped leaves. Such an **etiolated** plant cannot carry out photosynthesis. The seedling shoot must reach the soil surface and begin photosynthesis before its nutrient reserves are expended and it starves.

Plants have evolved a variety of ways to cope with the problem of germinating underground. Etiolated flowering plants, for example, do not form chlorophyll. They turn green only when exposed to light, thereby conserving the resources needed to make chlorophyll, which would be useless in the dark. An etiolated shoot uses stored resources to elongate rapidly and hasten its arrival at the soil surface, where photosynthesis quickly begins. To break through soil yet protect delicate, underdeveloped leaves, the shoot of an etiolated eudicot seedling forms an apical hook (see Figure 37.19).

All of these etiolation phenomena (lack of chlorophyll, rapid shoot elongation, production of an apical hook, delayed leaf expansion) are regulated by the phytochromes. In a seedling that has never been exposed to light, all the phytochrome is in the red-absorbing (P_r) form. Exposure to light converts P_r to P_{fr} (the far-red-absorbing form), and the P_{fr} initiates reversal of the etiolation phenomena: Chlorophyll synthesis begins, shoot elongation slows, the apical hook straightens out, and the leaves start to expand.

There are multiple phytochromes

For years, plant biologists had difficulty accounting for some aspects of phytochrome action. A solution to these problems may lie in the discovery of multiple forms of phytochromes and other photoreceptors. *Arabidopsis* has five genes that encode different phytochromes, and this diversity has been found throughout the plant kingdom and in algae as well.

The several phytochromes may play differing roles in various phytochrome-controlled responses. Some of them may even play off each other to fine-tune plant growth during the day. Consider, for example, the light spectrum available to a seedling that is growing in the shade of other plants. Because chlorophyll in the leaves above it absorbs the light first, the shaded seedling "sees" a spectrum relatively rich in far-red (and poor in red); the ratio of far-red to red is increased as much as 10-fold to 20-fold in the shade. The interplay among signal transduction pathways initiated by the different phytochromes may lead to an increased rate of stem elongation that tends to bring the leaf into full sunlight.

We do not yet know how the various phytochromes produce their many effects, although it is evident that phytochromes act through the plant's genome. Phytochromes appear to activate one or more G proteins. G proteins are membrane proteins that must bind to guanosine triphosphate (GTP) to exert their effects (see Chapter 15). The phytochrome-activated G proteins may convert GTP into the second messenger cGMP (cyclic guanosine monophosphate) and open channels that admit calcium ions into the cell, where they bind to the protein calmodulin. Both cGMP and the calcium–calmodulin complex can trigger changes leading eventually to the activation of specific genes.

Cryptochromes and phototropin are blue-light receptors

Cryptochromes are yellow photoreceptor pigments that absorb blue and ultraviolet light. They affect some of the same developmental processes, including seedling development and flowering, as do phytochromes. Unlike phytochromes, cryptochromes are present and play important roles in animals as well as plants.

In contrast to phytochromes, cryptochromes are located primarily in the plant nucleus. The exact mechanism of cryptochrome action is not yet known. It may be significant that phytochromes behave like protein kinases, and that cryptochromes can be substrates of such enzymes. It is likely that both classes of photoreceptors participate in protein kinase-based signaling pathways (see Chapter 15).

We began this chapter with a photo of a plant's phototropic response. Later we saw that the study of phototro-

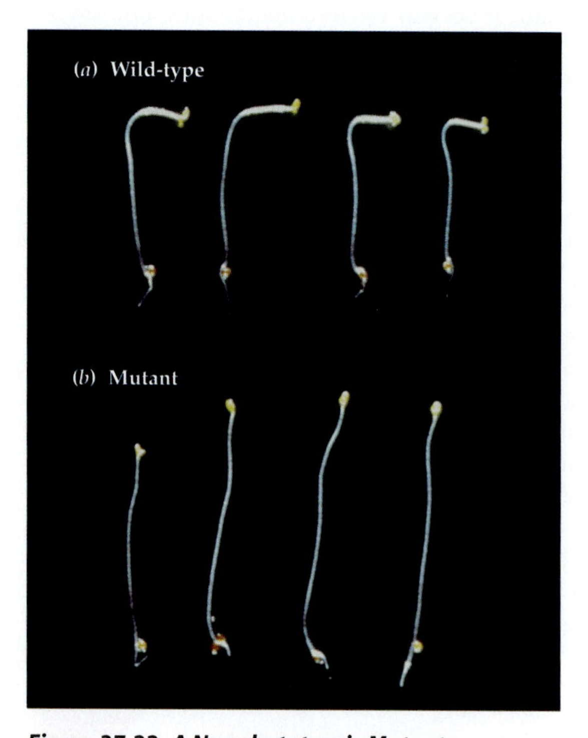

Figure 37.22 A Nonphototropic Mutant
(*a*) The four etiolated wild-type *Arabidopsis* seedlings in the top row are demonstrating normal phototropism. (*b*) These mutant seedlings cannot produce phototropin, the photoreceptor that signals the plant to curve toward light.

pism led to the discovery of auxin. But a question remained: What is the photoreceptor for phototropism? Plant scientists working with phototropic mutants of *Arabidopsis* have recently showed convincing evidence that it is a yellow protein, which they named **phototropin**. Upon absorbing blue light, phototropin initiates a signal transduction pathway leading to phototropic curvature (Figure 37.22). Still another type of blue-light receptor may be responsible for the light-induced closure of stomata.

Plants respond to light in many ways, and their responses are mediated by interactions of several photoreceptors, as we have seen.

Chapter Summary

Interacting Factors in Plant Development

▶ The environment, photoreceptors, hormones, and the plant's genome all play roles in the regulation of plant development.

▶ Hormones mediate many developmental phenomena in plants. Each plant hormone plays multiple regulatory roles, affecting several different aspects of development. Interactions among the hormones are often complex. **Review Table 37.1**

▶ Hormones and photoreceptors act through signal transduction pathways.

From Seed to Death: An Overview of Plant Development

▶ Cell division, cell expansion, and cell differentiation all contribute to plant development.

▶ The dormant seed eventually germinates and forms a growing seedling. Photoreceptors and hormones regulate seedling development, including growth. **Review Figures 37.1, 37.2**

▶ Eventually the plant flowers and forms fruit. Flowering in some plants is controlled by the length of the night. Hormones, probably including a flowering hormone, play roles in plant reproduction. **Review Figure 37.3**

▶ Some plant buds demonstrate winter dormancy. Eventually, all plants senesce and die. Dormancy and senescence are triggered by environmental cues, mediated by photoreceptors and hormones.

Ending Seed Dormancy and Beginning Germination

▶ Seed dormancy may be caused by exclusion of water or oxygen from the embryo, mechanical restraint of the embryo, or chemical inhibition of embryo development. In nature, seed dormancy is broken in various ways, including scarification, fire, leaching, and low temperatures.

▶ Seed dormancy offers adaptive advantages, such as an increased likelihood of germination in a place and at a time favorable for seedling growth.

▶ Seed germination begins with the imbibition of water. Then the embryo mobilizes its reserves to obtain building blocks and energy. The embryos of cereal seeds secrete gibberellins, which cause the aleurone layer to synthesize and secrete digestive enzymes that break down large molecules stored in the endosperm. **Review Figure 37.7**

Gibberellins: Regulators from Germination to Fruit Growth

▶ There are dozens of gibberellins. One, gibberellin A_1, regulates stem growth in most plants.

▶ Mutant plants that cannot produce normal amounts of gibberellins are dwarfs: Their stems are shorter than wild-type stems.

▶ Gibberellins regulate the growth of some fruits and cause bolting in some biennial plants. **Review Figure 37.9**

Auxin Affects Plant Growth and Form

▶ Studies of phototropism led to the discovery and isolation of auxin (indoleacetic acid). In grass seedlings, the photoreceptor for phototropism is in the tip of the coleoptile, and auxin is a messenger from the photoreceptor to the growing region of the coleoptile. **Review Figures 37.10, 37.11**

▶ Auxin transport is polar. Lateral movement of auxin establishes shoot and root responses to light and gravity: phototropism and gravitropism, respectively. Auxin carrier proteins move auxin into and out of cells. **Review Figure 37.12**

▶ Auxin plays roles in root formation, leaf abscission, apical dominance, and parthenocarpic fruit development. Certain synthetic auxins are used as selective herbicides. **Review Figures 37.13, 37.14**

▶ The arrangement of cellulose microfibrils in the plant cell wall limits the rate and direction of cell growth. Auxin increases the plasticity of the cell wall, promoting cell expansion. Part of the auxin response results from the pumping of protons from the cytoplasm into the cell wall, where the lowered pH activates proteins called expansins. **Review Figures 37.15, 37. 16, 37.17**

▶ Like all plant hormones, auxin is bound by receptor proteins.

▶ Auxin and other plant hormones signal cell differentiation and organ formation.

Cytokinins Are Active from Seed to Senescence

▶ Cytokinins are adenine derivatives. Zeatin and isopentenyl adenine are naturally occurring cytokinins, and kinetin is a synthetic cytokinin.

▶ First studied as promoters of plant cell division, cytokinins also promote seed germination in some species, inhibit stem elongation, promote lateral swelling of stems and roots, stimulate the growth of lateral buds, promote the expansion of leaf tissue, and delay leaf senescence.

Ethylene: A Gaseous Hormone That Promotes Senescence

▶ A balance between auxin and ethylene controls leaf abscission.

▶ Ethylene promotes senescence and fruit ripening.

▶ Ethylene causes the formation of a protective apical hook in eudicot seedlings that have not been exposed to light. In stems, it inhibits elongation, promotes lateral swelling, and causes a loss of gravitropic sensitivity.

▶ Ethylene acts through a signal transduction pathway that includes two proteins in the plasma membrane and that leads to the expression of genes. **Review Figure 37.20**

Abscisic Acid: The Stress Hormone

▶ Abscisic acid appears to maintain winter dormancy in buds. It prevents seeds from germinating while still attached to the parent plant, and it inhibits stem elongation. Through its effects on stomatal opening, it also regulates gas and water exchange between leaves and the atmosphere.

Hormones in Plant Defenses

▶ Oligosaccharins are hormones released by the cell wall in response to an attack by a pathogen. They participate in plant defenses against pathogens, and they interact in complex ways with auxin.

Brassinosteroids: "New" Hormones with Multiple Effects

▶ There are dozens of brassinosteroids. They affect cell elongation, pollen tube elongation, vascular tissue differentiation, and root elongation.

Light and Photoreceptors

▶ Phytochromes are bluish proteins found in the cytosol. Each phytochrome exists in two forms, P_r and P_{fr}, that are interconvertible by light. P_r absorbs red light (with a maximum at 660 nm), and P_{fr} absorbs far-red light (730 nm). **Review Figure 37.21**

▶ Phytochromes have many effects, including the various manifestations of etiolation.

▶ There are five phytochromes. They may play different roles in development, and their signal transduction pathways may interact to mediate the effects of light environments of differing spectral distribution. They mediate the effects of red and low-energy blue light.

▶ Cryptochromes, yellow photoreceptor proteins that absorb blue and ultraviolet light, interact with phytochromes in controlling seedling development and floral initiation. Cryptochromes mediate high-energy blue light effects.

▶ The signaling pathways for phytochromes and cryptochromes are based on protein kinases.

▶ Phototropin, another yellow protein, is the photoreceptor for phototropism.

For Discussion

1. How may it be advantageous for some species to have seeds whose dormancy is broken by fire?
2. Cocklebur fruits contain two seeds each, and the two seeds are kept dormant by two different mechanisms. How may this use of two mechanisms of dormancy be advantageous to cockleburs?
3. Corn stunt virus causes a great reduction in the growth rate of infected corn plants, so the diseased plants take on a dwarfed form. Since their appearance is reminiscent of the genetically dwarfed corn studied by Phinney, you suspect that the virus may inhibit the synthesis of gibberellins by the corn plants. Describe two experiments you might conduct to test this hypothesis, only one of which should require chemical measurement.
4. Whereas relatively low concentrations of auxin promote the elongation of segments cut from young plant stems, higher concentrations generally inhibit growth, as shown in the figure.

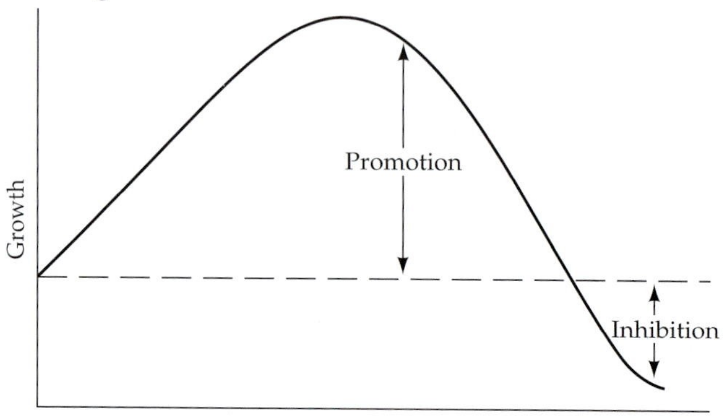

In some plants, the inhibitory effects of high auxin concentrations appear to be secondary: High auxin concentrations cause the synthesis of ethylene, which is what causes the growth inhibition. Silver thiosulfate inhibits ethylene action. How do you think the addition of silver thiosulfate to the solutions in which the stem segments grew would affect the appearance of the above graph?

5. Some etiolated seedlings develop hairs on their epidermis when exposed to dim light. Describe an experiment to test the hypothesis that a phytochrome is the photoreceptor for this effect.

38 Reproduction in Flowering Plants

Biologists have known for more than 60 years that the leaves of some plants, such as the cocklebur, contain built-in timers that measure the length of every night. When the night is of the appropriate length, the leaves—even a single leaf on a plant stripped of all other leaves—send a signal to other parts of the plant, telling them to form flowers. The evidence for this signal is substantial and convincing, yet nobody has been able to isolate and identify it.

After years of frustration, we may soon solve this mystery. The probable key lies in recent discoveries, described earlier in this book, about the functioning of plasmodesmata, the minute passageways connecting adjacent plant cells. Studies using mutant plants may allow scientists to identify the signal and learn how it triggers flowering. This knowledge will be a major advance in our understanding of reproduction in plants.

Why do plants expend energy and resources to produce flowers? The answer is simple: Flowers are sexual reproductive structures, and reproduction is one of the most important events in a plant's—or any organism's—life.

In this chapter we look at several aspects of plant reproduction, including some that are still not well understood. We contrast sexual and asexual reproduction, and we consider sexual reproduction in detail. In doing so, we look at angiosperm gametophytes, pollination, double fertilization, embryonic development, and the roles of fruits in seed dispersal. The transition to the flowering state is a key event in plant development, and we'll see how changing seasons trigger flowering in some plants—and speculate on the existence of a flowering hormone. We conclude the chapter with an examination of asexual reproduction in nature and in agriculture.

Many Ways to Reproduce

Plants have many ways of reproducing themselves—and with humans helping, there are even more ways. Flowers contain the sex organs of plants; it is thus no surprise that almost all flowering plants reproduce sexually. But many reproduce asexually as well; some even reproduce asexually most of the time. What are the advantages and disadvantages of these two kinds of reproduction? The answers to this question involve genetic recombination. As we have seen, sexual reproduction produces new genetic combinations and diverse phenotypes. Asexual reproduction, in contrast, produces a clone of genetically identical individuals.

Both sexual and asexual reproduction are important in agriculture. Many important annual crops are grown from seeds, which are the products of sexual reproduction. Seed-grown crops include wheat, rice, millet, and corn—the great grain crops, all of which are grasses—as well as plants in other families, such as soybeans and safflower. Other crops are produced asexually from grafts, or by other asexual means.

Orange trees, which have been under cultivation for centuries, can be grown from seed—except for one type, the navel orange. This plant apparently arose only once in history. Early in the nineteenth century, on a plantation on the Brazilian coast, one seed gave rise to one tree that had aberrant flowers. Parts of the flowers aborted, and seedless

Where It All Began
Some of the best early evidence for a flowering hormone came from studies of cockleburs (*Xanthium* sp.).

fruits formed. Every navel orange in the world comes from a navel orange tree derived asexually from that original Brazilian tree. Asexual reproduction is the only way of propagating this plant.

Unlike navel oranges, strawberries need not be propagated asexually, because they are capable of forming seeds. Nonetheless, asexual propagation of strawberries is common because vast numbers of plants that are genetically identical to a particularly desirable plant can be produced in this way.

We will treat asexual reproduction in greater detail at the end of this chapter. We begin, however, by considering sexual reproduction.

Sexual Reproduction

Sexual reproduction provides genetic diversity through recombination (see Chapter 9). Meiosis and mating shuffle genes into new combinations, giving a population a variety of genotypes in each generation. This genetic diversity may serve the population well as the environment changes or as the population expands into new environments. The adaptability resulting from genetic diversity is the major advantage of sexual reproduction over asexual reproduction.

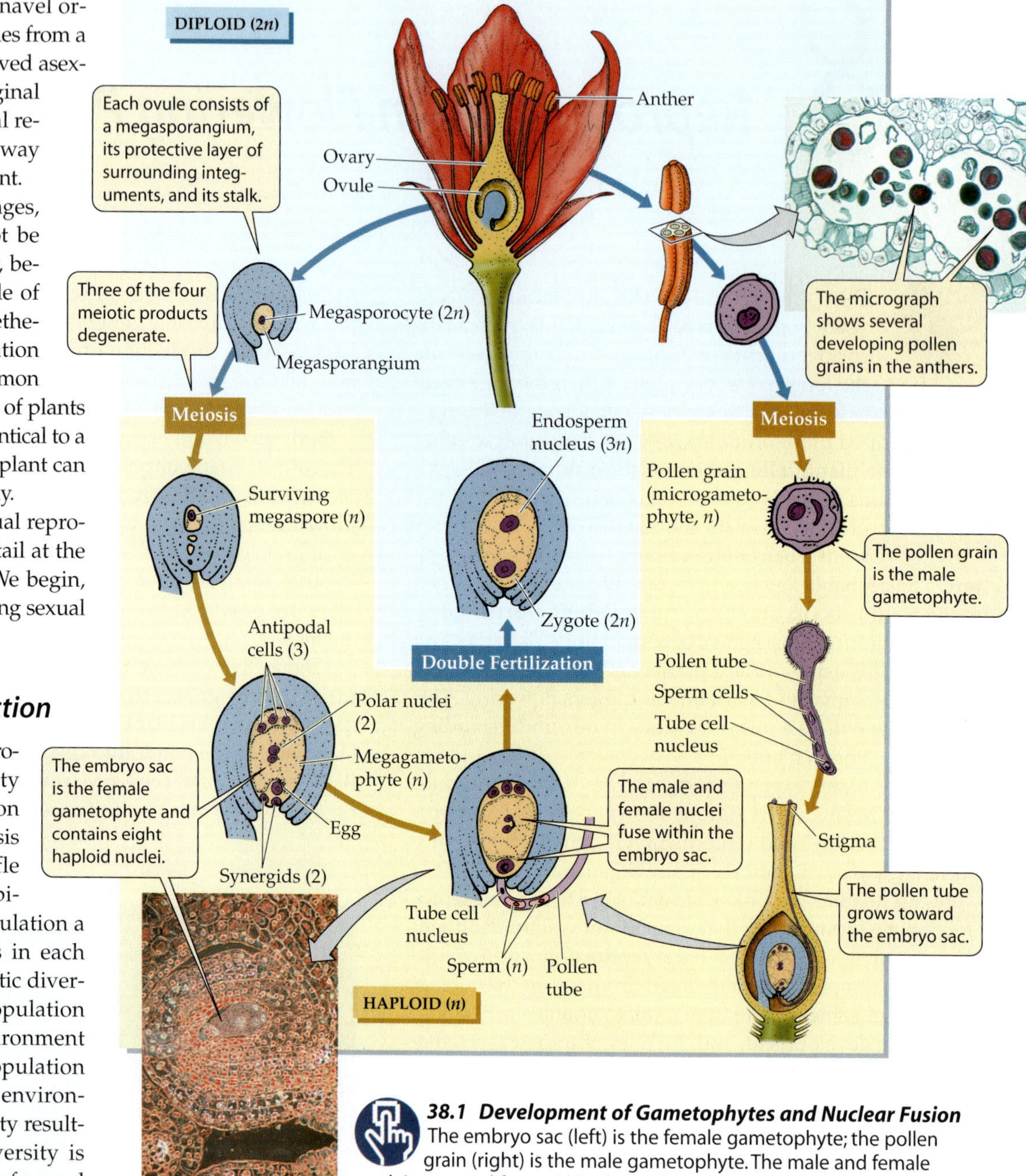

38.1 Development of Gametophytes and Nuclear Fusion The embryo sac (left) is the female gametophyte; the pollen grain (right) is the male gametophyte. The male and female nuclei meet and fuse within the embryo sac.

The flower is an angiosperm's device for sexual reproduction

A complete flower consists of four groups of organs that are modified leaves: the carpels, stamens, petals, and sepals (see Figure 29.6 for review). The *carpels* and *stamens* are, respectively, the female and male sex organs. A *pistil* is a structure composed of one or more carpels. The base of the pistil, called the *ovary*, contains one or more *ovules*, each of which contains a megasporangium. The stalk of the pistil is the *style*, and the end of that stalk is the *stigma*. Each stamen is composed of a *filament* bearing a two-lobed *anther*, which consists of four microsporangia fused together.

The *petals* and *sepals* of many flowers are arranged in whorls (circles) around the carpels and stamens. Together, the petals constitute the *corolla*. Below them, the sepals constitute the *calyx*. The petals are often colored; the sepals are often green and photosynthetic. All the parts of the flower are borne on a stem tip, the *receptacle*.

Flowering plants have microscopic gametophytes

Before reading this section, you may wish to review the section in Chapter 28 entitled "Life cycles of plants feature alternation of generations." The concept of alternation of generations is central to an understanding of plant reproduction.

In plants, the sporophyte generation produces flowers. The flowers produce spores, which develop into tiny gametophytes. The flower is more than just a place where the egg and sperm are eventually found—it is also the place where the alternate generation resides.

The gametophytes—the gamete-producing generation—of flowering plants develop from haploid spores in sporangia within the flower (Figure 38.1).

- Female gametophytes (megagametophytes), which are called **embryo sacs**, develop in megasporangia.
- Male gametophytes (microgametophytes), which are called **pollen grains**, develop in microsporangia.

Within the ovule, a megasporocyte—a cell within the megasporangium—divides meiotically to produce four haploid megaspores. All but one of these megaspores then degenerate. The surviving megaspore undergoes mitotic divisions, usually producing eight haploid nuclei, all initially contained within a single cell—three nuclei at one end, three at the other, and two in the middle. Subsequent cell wall formation leads to an elliptical, seven-celled megagametophyte with a total of eight nuclei.

- At one end of the elliptical megagametophyte are three tiny cells: the egg and two cells called **synergids**. The egg is the female gamete, and the synergids participate indirectly in fertilization.
- At the opposite end of the megagametophyte are three **antipodal cells**, which eventually degenerate.
- In the large central cell are two **polar nuclei**.

The embryo sac is the entire seven-celled, eight-nucleus structure. (Follow the arrows down the left-hand side of Figure 38.1 to review the development of the embryo sac.)

The male gametophyte, or pollen grain, consists of fewer cells than the female gametophyte. The development of a pollen grain begins when a microsporocyte within the anther divides meiotically. Each resulting haploid microspore normally undergoes one mitotic division within the spore wall before the anthers open and release these two-celled pollen grains. Further development of the pollen grain, which we will describe shortly, is delayed until the pollen arrives at a stigma. In angiosperms, the transfer of pollen from the anther to the stigma is referred to as **pollination**.

Pollination enables fertilization in the absence of liquid water

Gymnosperms and angiosperms evolved independence from liquid water as a medium for gamete travel and fertilization—a freedom not shared by other plant groups. The male gametes of gymnosperms and angiosperms travel within pollen grains (Figure 38.2). But how do angiosperm pollen grains travel from an anther to a stigma?

Many different mechanisms have evolved for pollen transport. In some plants, such as peas and their relatives, pollination is accomplished before the flower bud opens. Pollen is transferred by the direct contact of anther and stigma within the same flower, resulting in *self-fertilization*.

Wind is the vehicle for pollen transport in many species. Wind-pollinated flowers have sticky or featherlike stigmas, and they produce pollen grains in great numbers (Figure 38.3). Some aquatic angiosperms are pollinated by water action, with water carrying pollen grains from plant to

38.2 A Pollen Grain Sampler Each species' pollen has a characteristic size, shape, and cell wall formation.

38.3 Wind Pollination The numerous anthers on these inflorescences (groups of flowers) of a hazelnut tree all point away from the stalk and stand free of the plant, promoting dispersal of the pollen by wind.

plant. Animals, including insects, birds, and bats, carry pollen among the flowers of many plants.

Some plants practice "mate selection"

In our discussion of Mendel's work (see Chapter 10), we saw that some plants can reproduce sexually either by cross-pollinating or by self-pollinating. Many plants demonstrate **self-incompatibility**; that is, they reject pollen from their own flowers. This rejection promotes genetic variation and limits inbreeding. A single gene, the *S* gene, is responsible for self-incompatibility. The *S* gene has dozens of alleles. A pollen grain is haploid and possesses a single *S* allele; the recipient stigma is diploid. In self-incompatible plants, pollen fails to germinate, or develops abnormally, on a stigma that possesses the same *S* allele (Figure 38.4).

The stigma plays an important role in "mate selection" by flowering plants. The stigmas of wind-pollinated plants are exposed to the pollen of many other species as well as their own, and even the flowers of plants with coevolved, specific animal pollinators may receive pollen from other plant species. Pollen from the same species binds strongly to the stigma due to cell–cell signaling by the cell wall of pollen grains of the same species. In contrast, foreign pollen falls off readily, without germinating.

A pollen tube delivers male cells to the embryo sac

When a pollen grain lands on the stigma of a compatible pistil, a **pollen tube** develops from the pollen grain (Figure 38.5). The pollen tube either digests its way through the spongy tissue of the style or, if the style is hollow, grows downward on the inner surface of this female organ. The pollen tube grows millimeters or even centimeters in the process .

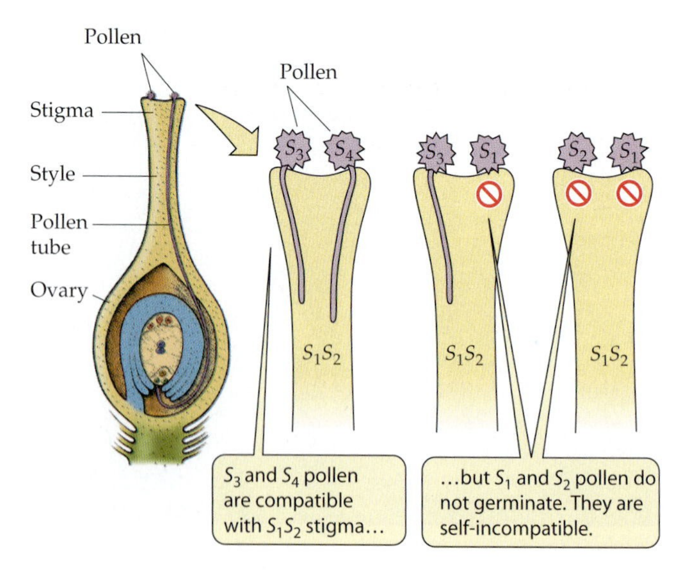

38.4 Self-Incompatibility
Pollen grains do not germinate normally if their *S* allele matches one of the *S* alleles of the stigma. Thus, the egg cannot be fertilized by a sperm from the same plant.

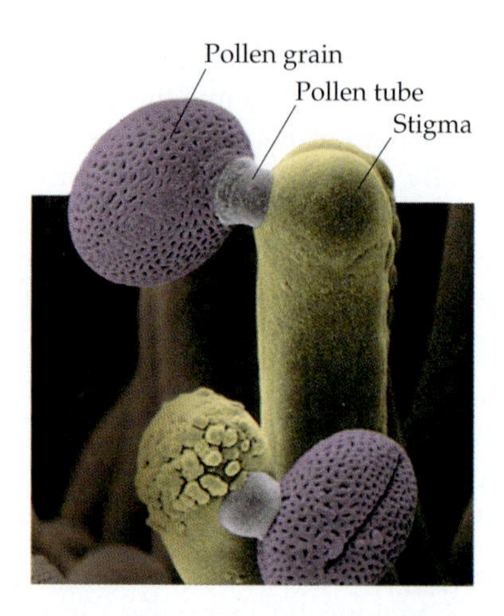

38.5 Pollen Tubes Begin to Grow
Pollen grains have landed on hairlike structures on the stigma of an *Arabidopsis* flower. Pollen tubes have started to form.

The rapid growth of the pollen tube requires calcium ions, taken up at the growing tip of the tube, as well as cell adhesion proteins. The downward growth of the pollen tube is guided by a long-distance signal from the ovule.

Angiosperms perform double fertilization

The pollen grain consists of two cells. The larger **tube cell** encloses the much smaller **generative cell** (Figure 38.6). Guided by the tube cell nucleus, the pollen tube eventually grows through megasporangial tissue and reaches the embryo sac. The generative cell meanwhile has undergone one mitotic division and cytokinesis to produce two **sperm cells**.

Both of the sperm cells enter the embryo sac, where they are released into the cytoplasm of one of the synergids. This synergid degenerates, releasing the sperm cells. Each sperm cell then fuses with a different cell of the embryo sac. One sperm cell fuses with the egg cell, producing the diploid zygote. The other fuses with the central cell, and that sperm cell nucleus and the two polar nuclei unite to form a triploid ($3n$) nucleus. While the zygote nucleus begins division to form the new sporophyte embryo, the triploid nucleus undergoes rapid mitosis to form a specialized nutritive tissue, the **endosperm**. The antipodal cells and the remaining synergid eventually degenerate, as does the pollen tube nucleus.

The fusion of a sperm cell nucleus with polar nuclei to form endosperm takes place only in angiosperms. This and the possession of flowers are the two most definitive characteristics shared by all angiosperms.

Embryos develop within seeds

Shortly after fertilization, highly coordinated growth and development of embryo, endosperm, integuments, and carpel ensues. The integuments develop into a double-layered seed coat, and the carpel ultimately becomes the wall of the fruit that encloses the seed.

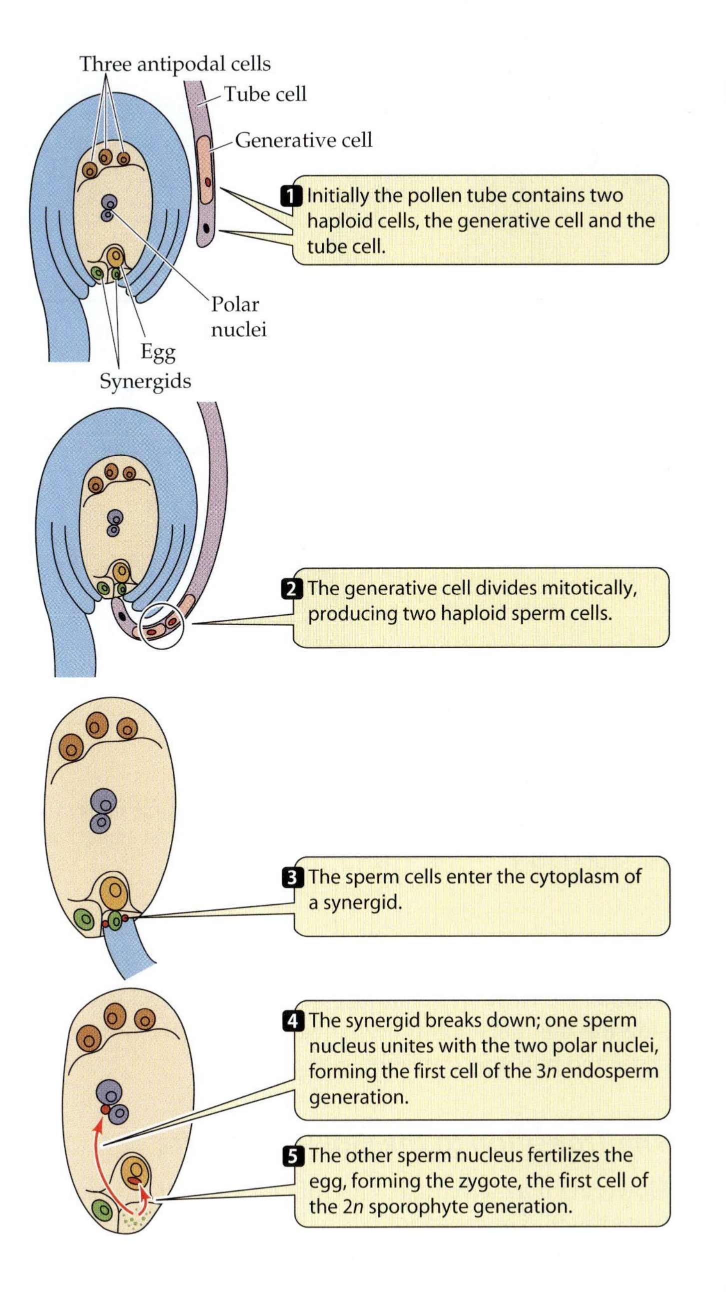

38.6 Pollen Nuclei and Double Fertilization The sperm nuclei contribute to the formation of the diploid zygote and the triploid endosperm. Double fertilization is a characteristic feature of angiosperm reproduction.

The first step in the normal formation of the embryo is a mitotic division of the zygote—the fertilized egg—giving rise to two daughter cells. Even at this stage, the two cells face different fates. An asymmetrical (uneven) distribution of cytoplasm within the zygote causes one end to produce the embryo proper and the other end to produce a supporting structure, the **suspensor** (Figure 38.7). The suspensor pushes the embryo against or into the endosperm, and provides one route by which nutrients enter the embryo.

With the asymmetrical division of the zygote, polarity has been established, as has the longitudinal axis of the new plant. A filamentous suspensor and a globular embryo are distinguishable after just four mitotic divisions. The suspensor soon ceases to elongate. In the embryo, the primary meristems form. As development continues, the first organs take form within the embryo.

In eudicots (monocots are somewhat different), the initially globular embryo takes on a characteristic *heart-stage* form as the cotyledons start to grow. Further elongation of the cotyledons and of the main axis of the embryo gives rise to what is called the *torpedo* stage (see Figure 38.7), during which some of the internal tissues begin to differentiate. The elongating region below the cotyledons is the *hypocotyl*. At the top of the hypocotyl, between the cotyledons, is the shoot apex; at the other end is the root apex. Each of these apical regions contains an apical meristem whose dividing cells will give rise to the organs of the mature plant.

Large amounts of nutrients are moved in from other parts of the plant, and the endosperm accumulates starch,

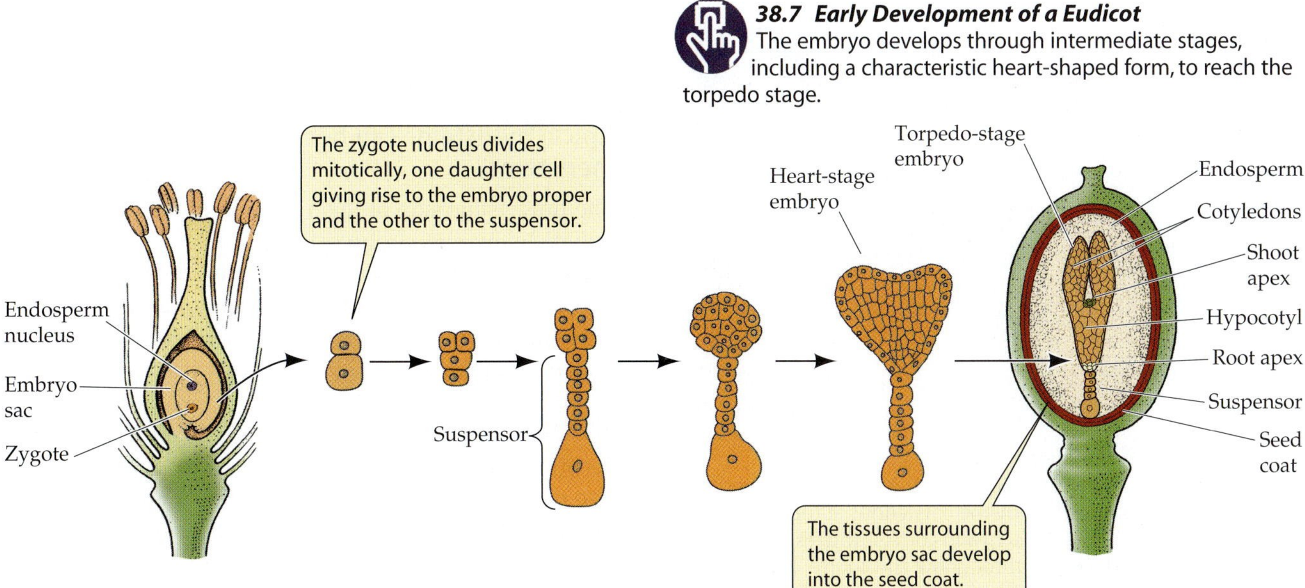

38.7 Early Development of a Eudicot The embryo develops through intermediate stages, including a characteristic heart-shaped form, to reach the torpedo stage.

lipids, and proteins. In many species, the cotyledons absorb the nutrient reserves from the surrounding endosperm and grow very large in relation to the rest of the embryo (Figure 38.8*a*). In others, the cotyledons remain thin (Figure 38.8*b*); they draw on the reserves in the endosperm as needed when the seed germinates.

In the late stages of embryonic development, the seed loses water—sometimes as much as 95 percent of its original water content. In its dried state, the embryo is incapable of further development. It remains in this quiescent state until the conditions are right for germination. (Recall from Chapter 37 that a necessary first step in seed germination is the massive imbibition of water.)

Some fruits assist in seed dispersal

After fertilization, the ovary wall of a flowering plant—together with its seeds—develops into a fruit. A **fruit** may consist of only the mature ovary and its seeds, or it may include other parts of the flower or structures that are closely related to it. Some major variations on this theme are illustrated in Figure 29.11, which shows only fleshy, edible fruits. Many other fruits are dry or inedible.

Some fruits help disperse seeds over substantial distances. Various trees, including ash, elm, maple, and tree of heaven, produce a dry, winged fruit that may be blown some distance from the parent tree by the wind (Figure 38.9*a*). Water disperses some fruits; coconuts have been spread in this way from island to island in the Pacific (Figure 38.9*b*). Still other fruits travel by hitching rides with animals—either inside or outside them. Fleshy fruits such as berries provide food for mammals or birds; their seeds travel safely through the animal's digestive tract and are deposited some distance from the parent plant.

We have traced the sexual life cycle from the flower to the fruit to the dispersal of seeds. We discussed seed germination and vegetative development of the seedling in Chapter 37. Now let's complete the sexual life cycle by considering the transition from the vegetative to the flowering state, and how this transition is regulated.

38.9 Dispersal of Fruits
(*a*) A samara is a winged fruit characteristic of the maple family. (*b*) A coconut seed germinates where it washed ashore on a beach in the South Pacific.

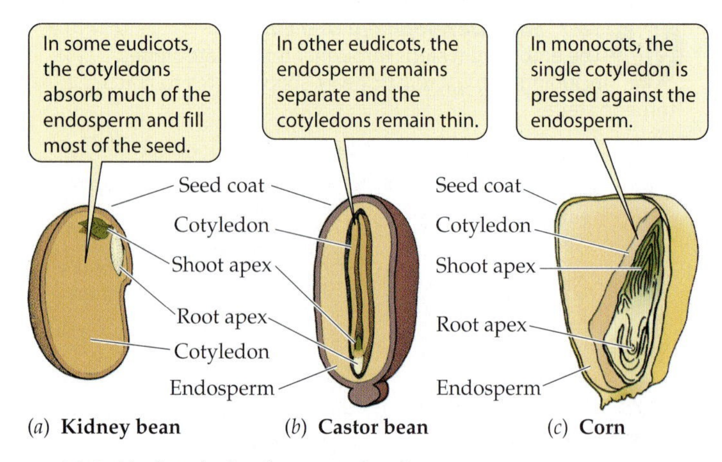

38.8 Variety in Angiosperm Seeds
In some seeds, such as kidney beans (*a*), the nutrient reserves of the endosperm are absorbed by the cotyledons at the seed stage. In others, such as castor beans (*b*) and corn (*c*), the reserves in the endosperm will be drawn on throughout the course of development.

The Transition to the Flowering State

Flowering may terminate, interrupt, or accompany vegetative growth. The transition to the flowering state marks the end of vegetative growth for some plants. If we view a plant as something produced by a seed for the purpose of bearing more seeds, then the act of flowering is one of the supreme events in a plant's life.

Apical meristems can become inflorescence meristems

The first visible sign of the transition to the flowering state may be a change in one or more apical meristems in the shoot system. During vegetative growth, an apical meristem continually produces leaves, lateral buds, and internodes (regions of stem between the nodes where leaves and buds form: Figure 38.10*a*). This unrestricted growth is *indeterminate* (see Chapter 34).

Flowers may appear singly or in an orderly cluster that constitutes a structure called an **inflorescence**. If a vegetative meristem becomes an **inflorescence meristem**, it generally produces several other structures: smaller leafy structures called **bracts**, as well as new meristems in the angles between the bracts and the internodes (Figure 38.10*b*).

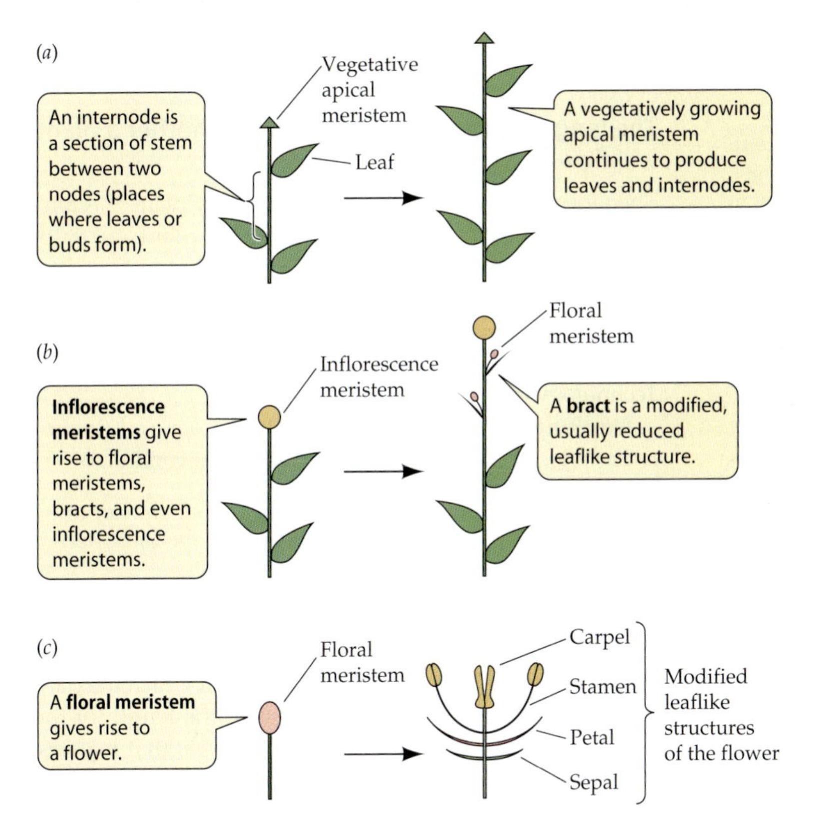

38.10 Flowering and the Apical Meristem
A vegetative apical meristem (*a*) grows without producing flowers. Once the transition to the flowering state is made, inflorescence meristems (*b*) give rise to bracts and to floral meristems (*c*), which become the flowers.

These new meristems may also be inflorescence meristems, or they may be **floral meristems**, which give rise to the flowers themselves.

Each floral meristem typically produces four consecutive whorls of organs—the sepals, petals, stamens, and carpels—separated by very short internodes, keeping the flower compact (Figure 38.10*c*). In contrast to vegetative meristems and some inflorescence meristems, floral meristems are responsible for *determinate* growth—the limited growth of the flower to a particular size and form.

A cascade of gene expression leads to flowering

How do apical meristems become inflorescence meristems, and how do inflorescence meristems give rise to floral meristems? How does a floral meristem give rise, in short order, to four different organs? How does each flower come to have the correct number of each of the floral organs? Numerous genes collaborate to produce these results. We'll refer here to some of the genes whose actions have been most thoroughly understood in *Arabidopsis* and snapdragons.

In order for an inflorescence meristem to give rise to a floral meristem, a group of *floral meristem identity genes* must be expressed. Expression of these genes initiates a cascade of further gene expression. Another set of genes, part of this cascade, participates in *pattern formation*—the spatial organization of the whorls of organs, which are still to be determined. These genes, in turn, trigger the expression of a group of *organ identity genes* that work in concert to specify the successive whorls (see Figure 16.11). They are homeotic genes, and their products are transcription factors that mediate the expression of still further genes.

Now that we have seen how flowering occurs, we will consider how the transition from the vegetative to the flowering state is initiated.

Photoperiodic Control of Flowering

The life cycles of flowering plants fall into three categories: annual, biennial, and perennial. **Annuals**, such as many food crops, complete their life cycle (seed to flower) in less than a year. **Biennials**, such as carrots and cabbage, grow for all or part of one year and live on into a second year, during which they flower, form seeds, and die. **Perennials**, such as oak trees, live for a few to many years, during which both growth and flowering occur. What control systems give rise to these and other differences in flowering behavior?

In 1920, W. W. Garner and H. A. Allard of the U.S. Department of Agriculture studied the behavior of a newly discovered mutant tobacco plant. The mutant, named 'Maryland Mammoth,' had large leaves and exceptional height. When the other plants in the field flowered, the 'Maryland Mammoth' continued to grow. Garner and Allard took cuttings of the 'Maryland Mammoth' into their greenhouse, and the plants that grew from the cuttings finally flowered in December.

Garner and Allard guessed that this pattern had something to do with the seasons. They tested several likely seasonal variables, such as temperature, but the key variable proved to be the length of the day (as they saw it). By moving plants between light and dark rooms at different times to vary the day length artificially, they were able to establish a direct link between flowering and day length. (We now know that the key variable is the length of the *night*, rather than the day, but Garner and Allard did not make that distinction.)

The 'Maryland Mammoth' plants did not flower if the light period was longer than 14 hours each day, but flowering commenced after the days became shorter than 14 hours. Thus, the **critical day length** for 'Maryland Mammoth' tobacco is 14 hours (Figure 38.11). This phenomenon of control by the length of day or night is called **photoperiodism**.

There are short-day, long-day, and day-neutral plants

Poinsettias, chrysanthemums, and 'Maryland Mammoth' tobacco are **short-day plants** (SDP's), which flower only when the day is *shorter* than a critical *maximum*. Spinach and clover are examples of **long-day plants** (LDP's), which flower only when the day is *longer* than a critical *minimum*.

38.11 Day Length and Flowering
By artificially varying the length of the day, Garner and Allard showed that the flowering of 'Maryland Mammoth' tobacco is initiated when the days become shorter than a critical length. 'Maryland Mammoth' tobacco is thus called a short-day plant. Henbane, a long-day plant, shows an inverse pattern of flowering.

Generally, LDP's are triggered to flower in midsummer and SDP's in late summer, or sometimes in the spring.

Some plants require photoperiodic signals that are more complex than just short or long days in order to flower. One group, the *short–long-day plants*, must first experience short days and then long ones. Accordingly, white clover and other short–long-day plants flower during the long days before midsummer. Another group, the *long–short-day plants*, cannot flower until the long days of summer have been followed by shorter ones, so they bloom only in the fall. *Kalanchoe*, seen in Figure 38.17*b*, is a long–short-day plant.

Other effects besides flowering are also under photoperiodic control. We have learned, for example, that short days trigger the onset of winter dormancy in plants. (Animals, too, show a variety of photoperiodic behaviors.)

The flowering of some angiosperms, such as corn and tomatoes, is not photoperiodic. In fact, there are more of these **day-neutral plants** than there are short-day and long-day plants. Some plants are photoperiodically sensitive only when young and become day-neutral as they grow older. Others require specific combinations of day length and other factors—especially temperature—to flower.

The length of the night determines whether a plant will flower

The terms "short-day plant" and "long-day plant" became entrenched before scientists learned that plants actually measure the length of the *night*, or of a period of darkness, rather than the length the of day. This fact was demonstrated by Karl Hamner of the University of California at Los Angeles and James Bonner of the California Institute of Technology (Figure 38.12).

Working with cocklebur, an SDP, Hamner and Bonner ran a series of experiments using two sets of conditions:

- The light period was kept constant—either shorter or longer than the critical day length—and the dark period was varied.
- The dark period was kept constant and the light period was varied.

The plants flowered under all treatments in which the dark period exceeded 9 hours, regardless of the length of the light period. Thus it is the length of the *night* that matters; for cocklebur, the *critical night length* is about 9 hours. Thus, it would be more accurate to call cocklebur a "long-night plant" than a short-day plant.

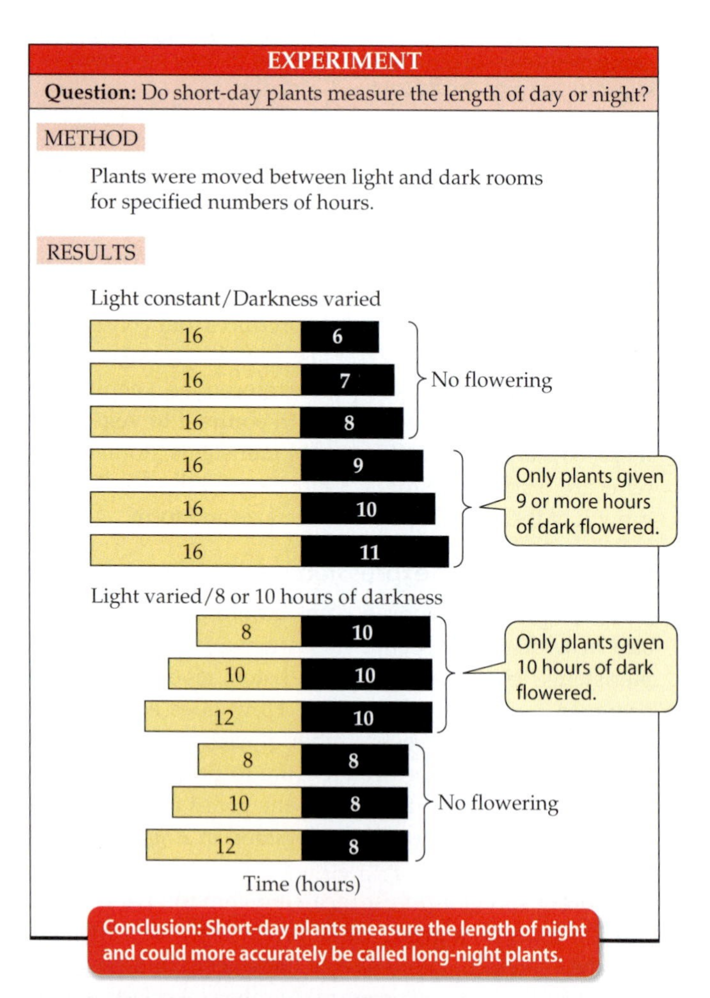

38.12 Night Length and Flowering
The length of the dark period, not the length of the light period, determines flowering.

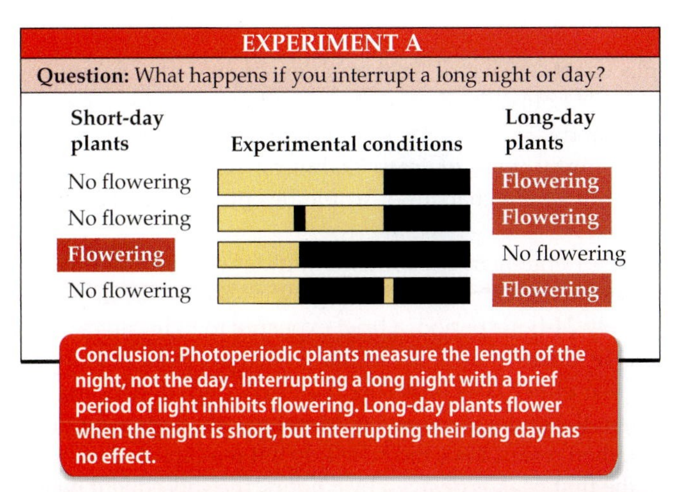

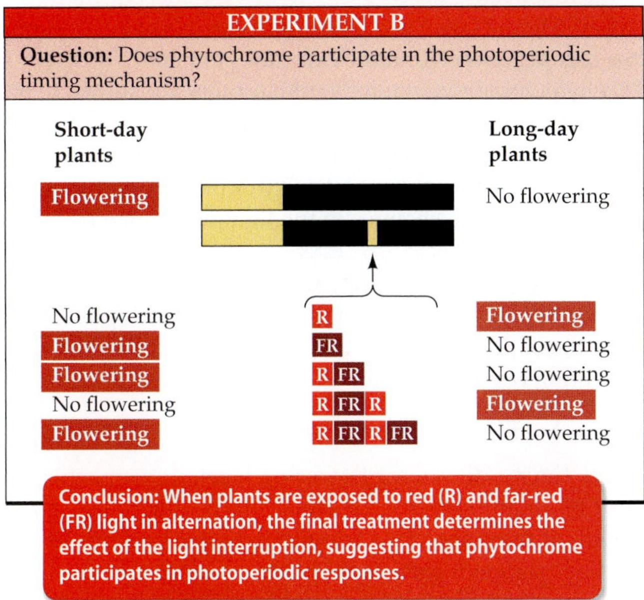

38.13 The Effect of Interrupted Days and Nights
(*a*) Experiments suggest that plants are able to measure the length of a continuous dark period and use this information to trigger flowering. (*b*) Phytochromes seem to be involved in the photoperiodic timing mechanism.

In cocklebur, a single long night is enough of a photoperiodic stimulus to trigger full flowering some days later, even if the intervening nights are short ones. Most plants are less sensitive than cocklebur, requiring from two to many nights of appropriate length to induce flowering. For some plants, a single shorter night in a series of long ones, even one day before flowering would have commenced, inhibits flowering.

Hamner and Bonner showed that plants measure the length of the night using another method as well. They grew SDP's and LDP's under a variety of light conditions. Under some conditions, the dark period was interrupted by a brief exposure to light; in others, the light period was interrupted briefly by darkness. Interruptions of the light period by darkness had no effect on the flowering of either short-day or long-day plants. Even a brief interruption of the dark period by light, however, completely nullified the effect of a long night (Figure 38.13*a*). An SDP flowered only if the long nights were uninterrupted. An LDP experiencing long nights flowered if those nights were broken by exposure to light. Thus a plant must have a timing mechanism that measures the length of a continuous dark period. Despite much study, the nature of this timing mechanism is still unknown.

Phytochromes and blue-light receptors, which affect several aspects of plant development (see Chapter 37), also participate in the photoperiodic timing mechanism. In the interrupted-night experiments, the most effective wavelengths of light were in the red range (Figure 38.13*b*), and the effect of a red-light interruption of the night could be fully reversed by a subsequent exposure to far-red light. It was once thought that the timing mechanism might simply be the slow conversion of phytochrome during the night from the P_{fr} form—produced during the light hours—to the P_r form. But this suggestion is inconsistent with most of the experimental observations and must be wrong. Phytochrome must be only a photoreceptor. The timekeeping role must be played by a biological clock.

Circadian rhythms are maintained by a biological clock

It is abundantly clear that organisms have some way of measuring time, and that they are well adapted to the 24-hour day–night cycle of our planet. Some sort of biological clock resides within the cells of all eukaryotes. The major outward manifestations of this clock are known as **circadian rhythms** (from the Latin *circa*, "about," and *dies*, "day").

We can characterize circadian rhythms, as well as other regular biological cycles, in two ways: The **period** is the length of one cycle, and the **amplitude** is the magnitude of the change over the course of a cycle (Figure 38.14).

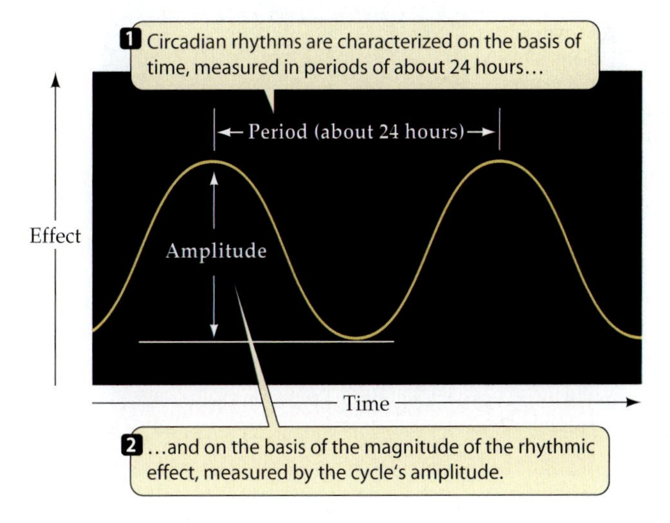

38.14 Features of Circadian Rhythms
The circadian rhythms of plants, like those of other organisms, can be characterized in two ways.

Circadian rhythms of protists, animals, fungi, and plants have been found to share some important characteristics:

- The period is remarkably *insensitive to temperature,* although lowering the temperature may drastically reduce the amplitude of the fluctuation.
- Circadian rhythms are *highly persistent;* they continue even in an environment in which there is no alternation of light and dark.
- Circadian rhythms can be *entrained,* within limits, by light–dark cycles that differ from 24 hours. That is, the period an organism expresses can be made to coincide with that of the light–dark regime.
- A brief exposure to light can shift the rhythm—it can cause a *phase shift.*

Plants provide innumerable examples of approximately 24-hour cycles. The leaflets of a plant such as clover or the tropical tree *Albizia* normally hang down and fold at night and rise and expand during the day. Flowers of many plants show similar "sleep movements," closing at night and opening during the day. They continue to open and close on an approximately 24-hour cycle even when the light and dark periods are experimentally modified (Figure 38.15).

The period of circadian rhythms in nature is approximately 24 hours. If an *Albizia* tree, for example, were to be placed under electric light on a day–night cycle totaling exactly 24 hours, the rhythm expressed would show a period of exactly 24 hours. However, if an experimenter used a day–night cycle of, say, 22 hours, then over time the rhythm would change—it would be **entrained** to a 22-hour period.

If an organism is maintained under constant darkness, with its circadian rhythm being expressed on the approximately 24-hour period, a brief exposure to light can cause a **phase shift**—that is, it can make the next peak of activity appear either later or earlier than expected, depending on when the exposure is given. Moreover, the organism does not then return to its old schedule if it remains in darkness. If the first peak is delayed by 6 hours, the subsequent peaks are all 6 hours late. Such phase shifts are permanent—until the organism receives more exposures to light.

Phytochromes and blue-light receptors are known to affect the period of the biological clock, with the different pigments reporting on different wavelengths and intensities of light. Perhaps this diversity of photoreceptors is an adaptation to the changes in the light environment that a plant experiences.

There is now ample evidence that the photoperiodic behavior of plants is based on the interaction of night length with the biological clock. But how the clock is coupled with flowering remains unclear.

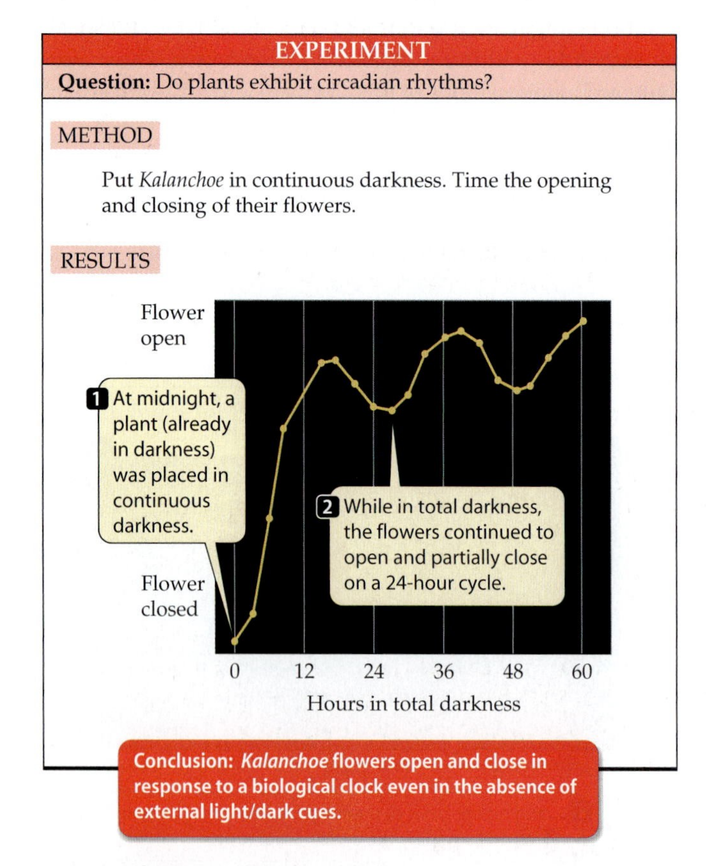

38.15 Plants Can Measure Time
Even when *Kalanchoe* is placed in continuous darkness, the opening and closing of its flowers continues to exhibit a circadian rhythm.

Is there a flowering hormone?

Is the timing device for flowering located in a particular part of an angiosperm, or are all parts able to sense the length of the night? This question was resolved by "blindfolding" different parts of the plant.

It quickly became apparent that each leaf is capable of timing the night. If a short-day plant is kept under a regime of short nights and long days, but a leaf is covered so as to give it the needed long nights, the plant will flower (Experiment A in Figure 38.16). This type of experiment works best if only one leaf is left on the plant. If one leaf is given a photoperiodic treatment conducive to flowering—an *inductive* treatment—other leaves kept under noninductive conditions will tend to inhibit flowering.

Although it is the leaves that sense an inductive dark period, the flowers form elsewhere on the plant. Thus a message must be sent from the leaf to the site of flower formation. Three lines of evidence suggest that this message is a chemical substance—a flowering hormone.

- If a photoperiodically induced leaf is removed from the plant shortly after the inductive dark period, the plant does not flower. If, however, the induced leaf remains attached to the plant for several hours, the plant flowers. This result suggests that something must be synthesized in the leaf in response to the inductive dark period, then move out of the leaf to induce flowering.
- If two cocklebur plants are grafted together, and if one plant is given inductive long nights and its graft partner is given noninductive short nights, both plants flower (Experiment B in Figure 38.16).

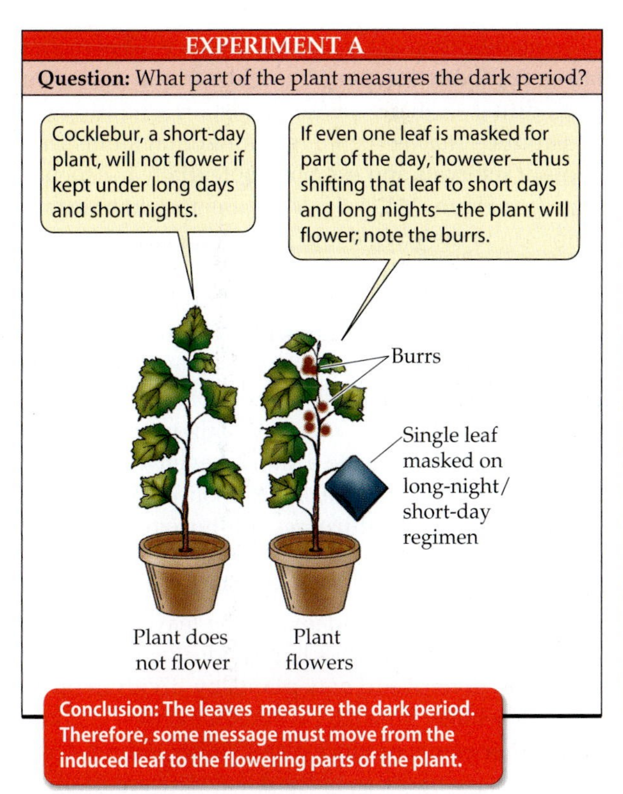

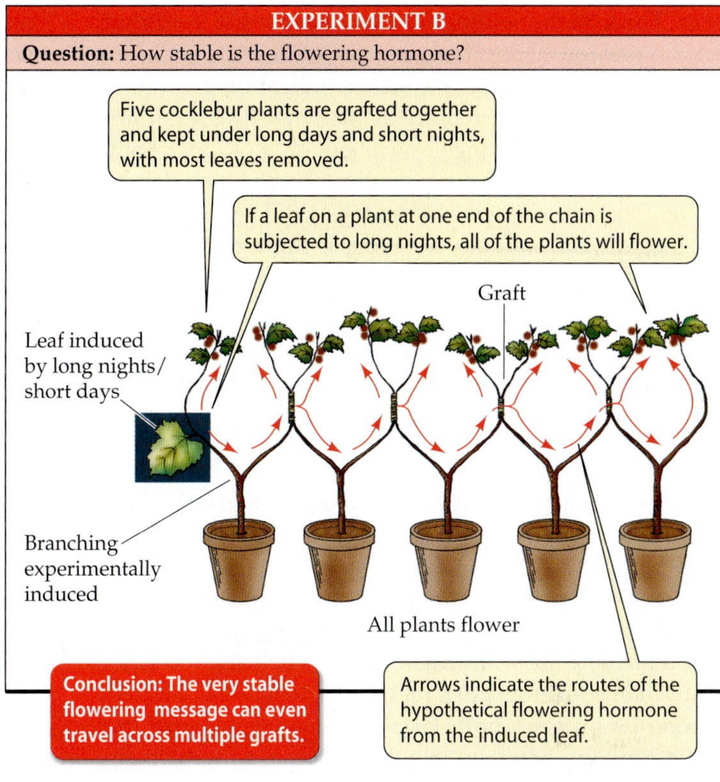

38.16 Evidence for a Flowering Hormone
If even a single leaf is exposed to inductive conditions, a "message" travels to the entire plant (and even to other plants, in grafting experiments), inducing it to flower.

- In at least one species, if an induced leaf from one plant is grafted onto another, noninduced plant, the host plant flowers.

Jan A. D. Zeevaart, a plant physiologist now at Michigan State University, performed this last experiment. He exposed a single leaf of the SDP *Perilla* to a short-day/long-night regime, inducing the plant to flower. Then he detached this leaf and grafted it onto another, noninduced, *Perilla* plant—which responded by flowering. The same leaf grafted onto successive hosts caused each of them to flower in turn. As long as 3 months after the leaf was exposed to the short-day/long-night regime, it could still cause plants to flower.

Experiments such as Zeevaart's suggest that the photoperiodic induction of a leaf causes a more or less permanent change in it, inducing it to start and continue producing a flowering hormone that is transported to other parts of the plant, switching those target parts to the reproductive state. So reasonable is this idea that biologists have named this hormone **florigen**, even though, after decades of active searching, it has not been isolated and characterized.

An elegant experiment suggested that the florigen of short-day plants is identical to that of long-day plants, even though SDP's produce it only under long nights and LDP's only under short nights. An SDP and an LDP were grafted together, and both flowered, as long as the photoperiodic conditions were inductive for one of the partners. Either the SDP or the LDP could be the one induced, but both would always flower. These results suggest that a flowering hormone—the elusive florigen—was being transferred from one plant to the other.

The direct demonstration of florigen activity remains a cherished goal of plant physiologists. For a long time it was thought that florigen could be neither a protein nor an RNA because these molecules were too large to pass from one living plant cell to another. However, we now know that such macromolecules can be transferred by way of plasmodesmata, and biologists are reexamining the possibility that an RNA or a protein is the long-sought florigen.

Vernalization and Flowering

In both wheat and rye, we distinguish two categories of flowering behavior. Spring wheat, for example, is sown in the spring and flowers in the same year. It is an annual plant. Winter wheat is biennial and must be sown in the fall; it flowers in the following summer. If winter wheat is not exposed to cold after its first year, it will not flower normally the next year. The implications of this finding were of great agricultural interest in Russia because winter wheat is a better producer than spring wheat, but it cannot be grown in some parts of Russia because the winters there are too cold for its survival.

Several studies performed in Russia during the early 1900s demonstrated that if seeds of winter wheat were premoistened and prechilled, they could be sown in the spring, and would develop and flower normally the same year. Thus, high-yielding winter wheat could be grown even in previously hostile regions. This induction of flowering by low temperatures is called **vernalization**.

Vernalization may require as many as 50 days of low temperatures (in the range from about –2° to +12°C). Some plant species require both vernalization and long days to flower. There is a long wait from the cold days of winter to the long days of summer, but because the vernalized state easily lasts at least 200 days, these plants do flower when they experience the appropriate night length.

Asexual Reproduction

Although sexual reproduction takes up most of the space in this chapter, asexual reproduction is responsible for many of the new plant individuals appearing on Earth. This fact suggests that in some circumstances, asexual reproduction must be advantageous.

Consider genetic recombination. When a plant self-fertilizes, there are fewer opportunities for genetic recombination than there are with cross-fertilization. A self-fertilizing plant that is heterozygous for a certain locus can produce among its progeny both kinds of homozygotes for that locus plus the heterozygote, but it cannot produce any progeny that carry alleles that it does not itself possess. Yet many plants continue to be self-compatible.

Asexual reproduction goes farther than self-fertilization: It eliminates genetic recombination altogether. When a plant reproduces asexually, it produces a *clone* of progeny with genotypes identical to its own. If a plant is well adapted to its environment, asexual reproduction may spread its genotype throughout that environment. This ability to exploit a particular environment is an advantage of asexual reproduction.

There are many forms of asexual reproduction

We call stems, leaves, and roots *vegetative organs*, distinguishing them from flowers, the reproductive parts of the plant. The modification of a vegetative organ is what makes **vegetative reproduction** possible. The stem is the organ that is modified in many cases. Strawberries and some grasses produce *stolons* (runners), horizontal stems that form roots at intervals and establish potentially independent plants (see Figure 34.4*b*). *Tip layers* are upright branches whose tips sag to the ground and put out roots, as in blackberry and forsythia.

Some plants, such as potatoes, form *tubers*, enlarged fleshy tips of underground stems (see Figure 34.4*a*). *Rhizomes* are horizontal underground stems that can give rise to new shoots. Bamboo is a striking example of a plant that reproduces vegetatively by means of rhizomes. A single bamboo plant can give rise to a stand—even a forest—of plants constituting a single, physically connected entity.

Whereas stolons and rhizomes are horizontal stems, bulbs and corms are short, vertical, underground stems. Lilies and onions form *bulbs* (Figure 38.17*a*), short stems with many fleshy, modified leaves. The leaves make up most of the bulb. Bulbs are thus large buds that store nutrients. They can give rise to new plants by dividing or by producing new bulbs from lateral buds. Crocuses, gladioli, and many other plants produce *corms*, underground stems that function very much as bulbs do. Corms are disclike and consist primarily of stem tissue; they lack the fleshy modified leaves that are characteristic of bulbs.

Not all vegetative organs modified for reproduction are stems. Leaves may also be the source of new plantlets, as in the succulent plants of the genus *Kalanchoe* (Figure 38.17*b*). Many kinds of angiosperms, ranging from grasses to trees such as aspens and poplars, form interconnected, genetically homogeneous populations by means of *suckers*—shoots produced by roots. What appears to be a whole stand of aspen trees, for example, may be a clone derived from a single tree by suckers (see Figure 54.1*b*).

Plants that reproduce vegetatively often grow in physically unstable environments, such as eroding hillsides. Plants with stolons or rhizomes, such as beach grasses,

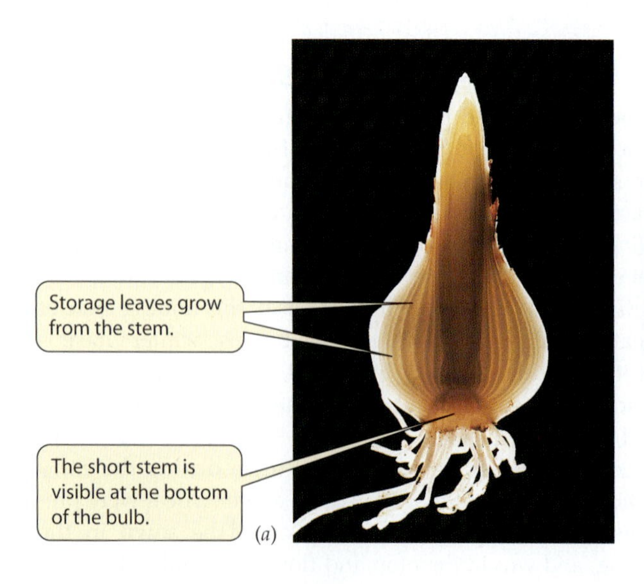

38.17 Vegetative Organs Modified for Reproduction
(*a*) Bulbs are short stems with large buds that store food and can give rise to new plants. (*b*) In *Kalanchoe*, new plantlets can form on leaves.

rushes, and sand verbena, are common pioneers on coastal sand dunes. Rapid vegetative reproduction enables these plants, once introduced, not only to multiply but also to survive burial by the shifting sand; in addition, the dunes are stabilized by the extensive network of rhizomes or stolons.

Dandelions, citrus trees, and some other plants reproduce by **apomixis**, the asexual production of seeds. As we have seen, meiosis reduces the number of chromosomes in gametes, and fertilization restores the sporophytic number of chromosomes in the zygote. Some plants can skip over *both* meiosis and fertilization and still produce seeds. Apomixis produces seeds within the female gametophyte without the mingling and segregation of chromosomes and without the union of gametes. The ovule simply develops into a seed, and the ovary wall develops into a fruit. An apomictic embryo has the sporophytic number ($2n$) of chromosomes. The result of apomixis is a fruit with seeds that are genetically identical to the parent plant.

Interestingly, apomixis sometimes requires pollination. In some apomictic species, a sperm nucleus must combine with the polar nuclei in order for the endosperm to form. In other apomictic species, the pollen provides the signals for embryo and endosperm formation, although neither sperm nucleus participates in fertilization. Pollination and fertilization are not the same thing!

Asexual reproduction is important in agriculture

Farmers take advantage of some natural forms of vegetative reproduction. Farmers and scientists have also developed new types of asexual reproduction by manipulating plants. One of the oldest methods of vegetative reproduction used in agriculture consists simply of making cuttings of stems, inserting them in soil, and waiting for them to form roots and thus become autonomous plants. The cuttings are usually encouraged to root by treatment with a plant hormone, auxin, as described in Chapter 37.

Horticulturists reproduce many woody plants by **grafting**—attaching a bud or a piece of stem from one plant to the root-bearing stem of another plant. The part of the resulting plant that comes from the root-bearing "host" is called the *stock*; the part grafted on is called the *scion* (Figure 38.18). In order for a graft to succeed, the cambium of the scion must become associated with the cambium of the stock. By cell division, both cambia form masses of wound tissue. If the two masses meet and fuse, the resulting continuous cambium can produce xylem and phloem, allowing transport of water and minerals to the scion and of photosynthate to the stock. Grafts are most often successful when the stock and scion belong to the same or closely related species.

Most fruit grown for market in the United States is produced on trees grown from grafts. There are many reasons for grafting plants for fruit production. The most common is the desire to combine a hardy root system with a shoot system that produces the best-tasting fruit. This motive is

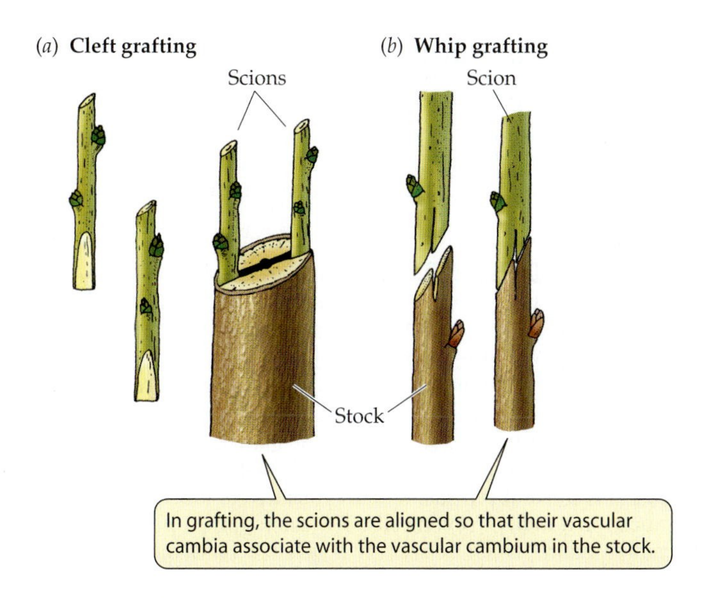

38.18 Grafting
Grafting—attaching a piece of a plant to the stem or root of another plant—is common in agriculture. The "host" stem or root is the stock; the grafted piece is the scion.

illustrated by the story of the wine grape *Vitis vinifera*. In 1863, plant lice of the genus *Phylloxera* inflicted great damage on the root systems of grapevines in French vineyards. More than 2.5 million acres of vines were destroyed. The problem was overcome by importing *V. vinifera* plants, which have *Phylloxera*-resistant root systems, from California. These plants were used as stocks to which French vines were grafted as scions. Thus the fine French grapes could be grown using roots resistant to the lice. (But the battle continues; in recent years, a new strain of *Phylloxera* has been damaging grapevines in California.)

Scientists in universities and industrial laboratories have been developing new ways to produce valuable plant materials via **tissue culture**. Because many plant cells are totipotent (see Figure 16.3), cultures of undifferentiated tissue can give rise to entire plants, as can small pieces of tissue cut directly from a parent plant. Tissue cultures are used commercially to produce orchids, rhododendrons, and many crops without resorting to seeds.

Culturing tiny bits of apical meristem can produce plants free of viruses. Because apical meristems lack developed vascular tissues, viruses tend not to enter them. Such meristem cultures have been used to increase the yields of potatoes and other crops.

Recombinant DNA techniques applied to tissue cultures can provide plants with capabilities they previously lacked such as resistance to pests, or increased nutritive value to humans. There is also interest in making certain valuable, sexually reproducing plants capable of apomikis. By causing cells of different types to fuse, one can obtain plants with exciting new combinations of properties.

Chapter Summary

Many Ways to Reproduce

▶ Almost all flowering plants reproduce sexually, and many also reproduce asexually.

▶ Both sexual and asexual reproduction are important in agriculture.

Sexual Reproduction

▶ Sexual reproduction promotes genetic diversity in a population, which may give the population an advantage under changing environmental conditions.

▶ The flower is an angiosperm's device for sexual reproduction.

▶ Flowering plants have microscopic gametophytes that develop in the flowers of the sporophytes. The megagametophyte is the embryo sac, which typically contains eight nuclei in a total of seven cells. The microgametophyte is the pollen grain, which delivers two sperm cells to the megagametophyte by means of a long pollen tube. **Review Figure 38.1**

▶ Pollination enables fertilization in the absence of liquid water.

▶ In self-incompatible species, the stigma rejects pollen from the same plant. **Review Figure 38.4**

▶ Angiosperms perform double fertilization: One sperm nucleus fertilizes the egg, forming a zygote, and the other sperm nucleus unites with the two polar nuclei to form a triploid endosperm nucleus. **Review Figure 38.6**

▶ The zygote develops into an embryo (with an attached suspensor), which remains quiescent in the seed until conditions are right for germination. The endosperm is the nutritive reserve upon which the embryo depends at germination. **Review Figures 38.7, 38.8**

▶ Flowers develop into seed-containing fruits, which often play important roles in the dispersal of the species.

The Transition to the Flowering State

▶ For a vegetatively growing plant to flower, an apical meristem in the shoot system must become an inflorescence meristem, which gives rise to bracts and more meristems. The meristems it produces may become floral meristems or additional inflorescence meristems. **Review Figure 38.10**

▶ Flowering results from a cascade of gene expression. Organ identity genes are expressed in floral meristems that give rise to sepals, petals, stamens, and carpels.

Photoperiodic Control of Flowering

▶ Photoperiodic plants regulate their flowering by measuring the length of light and dark periods.

▶ Short-day plants flower when the days are shorter than a species-specific critical day length; long-day plants flower when the days are longer than a critical day length. **Review Figure 38.11**

▶ Some angiosperms have more complex photoperiodic requirements than short-day or long-day plants have, but most are day-neutral.

▶ The length of the *night* is what actually determines whether a photoperiodic plant will flower. **Review Figure 38.12**

▶ Interruption of the nightly dark period by a brief exposure to light undoes the effect of a long night. **Review Figure 38.13**

▶ The mechanism of photoperiodic control involves a biological clock and phytochromes. **Review Figures 38.14, 38.15**

▶ Evidence suggests that there is a flowering hormone, called florigen, but the substance has yet to be isolated from any plant. **Review Figure 38.16**

Vernalization and Flowering

▶ In some plant species, exposure to low temperatures—vernalization—is required for flowering.

Asexual Reproduction

▶ Asexual reproduction allows rapid multiplication of organisms well suited to their environment.

▶ Vegetative reproduction involves the modification of a vegetative organ—usually the stem—for reproduction. Stolons, tip layers, tubers, rhizomes, bulbs, corms, and suckers are means by which plants may reproduce vegetatively.

▶ Some plant species produce seeds asexually by apomixis.

▶ Agriculturalists use natural and artificial techniques of asexual reproduction to reproduce particularly desirable plants.

▶ Horticulturists often graft different plants together to take advantage of favorable properties of both stock and scion. **Review Figure 38.18**

▶ Tissue culture techniques, based on the totipotency of many plant cells, are used to propagate plants asexually, to produce virus-free clones of crop plants, and to manipulate plants by recombinant DNA technology.

For Discussion

1. For a crop plant that reproduces both sexually and asexually, which method of reproduction might the farmer prefer?
2. Thompson seedless grapes are produced by vines that are triploid. Think about the consequences of this chromosomal condition for meiosis in the flowers. Why are these grapes seedless? Describe the role played by the flower in fruit formation when no seeds are being formed. How do you suppose Thompson seedless grapes are propagated?
3. Poinsettias are popular ornamental plants that typically bloom just before Christmas. Their flowering is photoperiodically controlled. Are they long-day or short-day plants? Explain.
4. You plan to induce the flowering of a crop of long-day plants in the field by using artificial light. Is it necessary to keep the lights on continuously from sundown until the point at which the critical day length is reached?

39 Plant Responses to Environmental Challenges

IF YOU ARE ATTACKED, IT MAKES SENSE TO call for help. Plants do this, too. When caterpillars begin to chew on the leaves of corn, cotton, or some other plant species, the plants synthesize and release chemical signals into the atmosphere. These substances attract other insects that feed on the caterpillars.

Herbivores aren't the only challenges plants face, however. The environment teems with plant pathogens. We know of more than a hundred diseases that can kill a tomato plant, each of them caused by a different pathogen (including various bacteria, fungi, protists, and viruses). Like animals, plants have a variety of defenses against pathogens. And, like the defenses of our own bodies, these mechanisms are not perfect, but they keep the plant world in competitive balance with its pathogens.

Environmental challenges to plants aren't limited to herbivores and pathogens. Some physical conditions pose substantial problems for plants and thus limit the places where different kinds of plants can live. The most challenging physical environments include ones that are very dry (deserts), that are water-saturated, that are dangerously salty, that contain high concentrations of toxic substances such as heavy metals, and that are very hot or very cold.

This chapter focuses on how plants meet the myriad challenges presented by their biological and physical environments. We begin by examining interactions between plants and pathogens and go on to consider interactions between plants and herbivores. Then we discuss the adaptations of some types of plants to their physical environments.

Plant–Pathogen Interactions

Plants and pathogens have evolved together in a continuing "arms race." Pathogens have evolved mechanisms by which to attack plants, and plants have evolved defenses against them. Each set of mechanisms uses information from the other. The pathogen's enzymes break down the plant's cell walls, for example, and the breakdown products signal to the plant that it is under attack. In turn, the plant's defenses alert the pathogen that it is under attack.

What determines the outcome of a battle between a plant and a pathogen? The key to success for the plant is to respond to the information about the pathogen quickly and massively. Plants use both mechanical and chemical defenses in this effort.

Plants seal off infected parts to limit damage

Tissues such as epidermis or cork protect the outer surfaces of plants, and these tissues are generally covered by cutin, suberin, or waxes. This protection is comparable to the nonspecific defenses of animals. When pathogens pass these barriers, other nonspecific plant defenses are activated.

The defense systems of plants and animals differ. Animals generally repair tissues that have been damaged by pathogens, but plants do not. Instead, they seal off and sacrifice the damaged tissue so that the rest of the plant does not become infected. This approach works because most plants, unlike most animals, are modular and can replace damaged parts by growing new ones.

One of a plant cell's first defensive responses is the rapid deposition of additional polysaccharides to the cell wall, reinforcing this barrier to invasion by the pathogen (Figure 39.1). These polysaccharides block the plasmodesmata, limiting the ability of viral pathogens to move from cell to cell. They also serve as a base upon which lignin may be laid down. Lignin enhances the mechanical barrier, and the toxicity of lignin building blocks makes the cell inhospitable to some pathogens.

Calling In an Air Strike
As this caterpillar of a corn earworm moth (*Helicoverpa zea*) munches on a cotton boll, it is triggering a series of reactions in the plant that may end in the attraction of other insects that will attack the caterpillar.

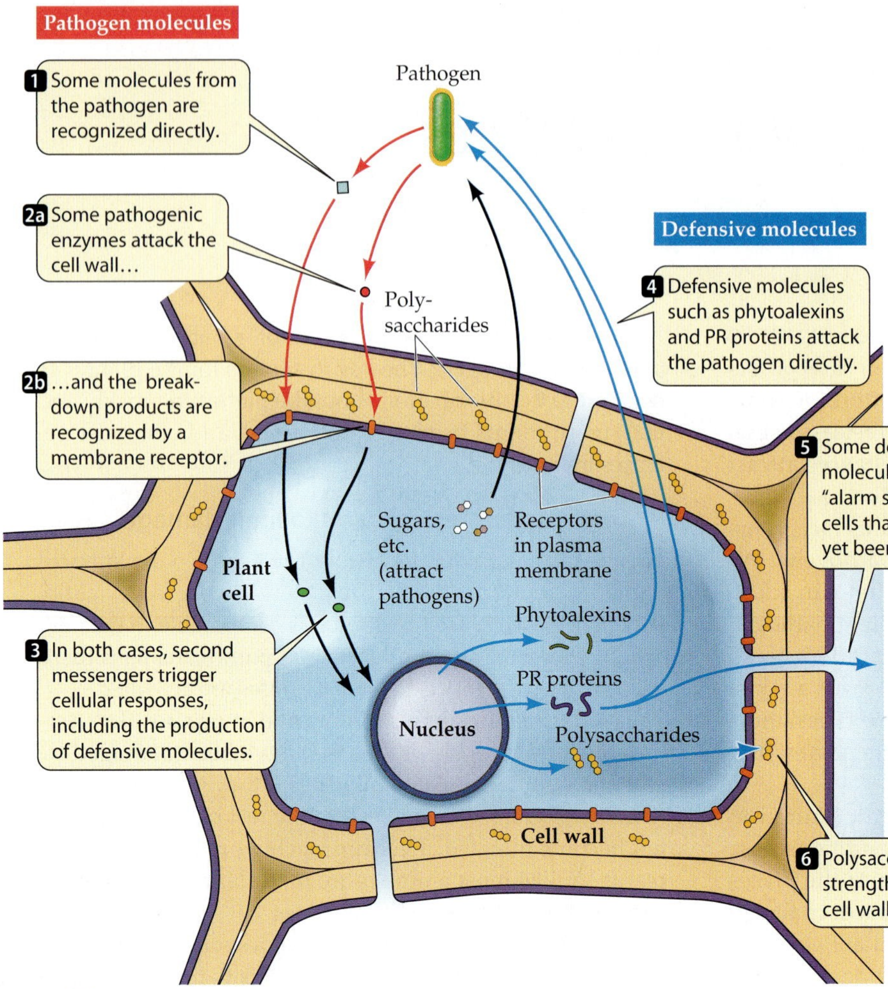

39.1 Signaling between Plants and Pathogens
Chemical interactions between plants and pathogens are highly coevolved. Plants produce molecules such as sugars that attract pathogens. But the presence of a pathogen stimulates the plant to produce defensive molecules that can work in many different ways.

Plants have potent chemical defenses against pathogens

When infected by certain fungi and bacteria, plants produce a variety of defensive compounds, among which are small molecules called phytoalexins and larger proteins called pathogenesis-related proteins (see Figure 39.1).

Phytoalexins are toxic to many fungi and bacteria. (Most are phenolics or terpenes, compounds that are also used to protect plants against herbivores; see Table 39.1.) They are produced by infected cells and their immediate neighbors within hours of the onset of infection. Enzymes from a pathogenic fungus can cause plant cell walls to release hormones called oligosaccharins (see Chapter 37), which trigger phytoalexin production. Because their antimicrobial activity is nonspecific, phytoalexins can destroy many species of fungi and bacteria in addition to the one that originally triggered their production. Physical injuries, viral infections, and chemical compounds produced in response to damage by herbivores can also induce the production of phytoalexins.

Plants also produce several types of **pathogenesis-related proteins**, or **PR proteins**. Some are enzymes that break down the cell walls of pathogens. These enzymes destroy some of the invading cells, and in some cases the breakdown products of the pathogen's cell walls serve as chemical signals that trigger further defensive responses. Other PR proteins may serve as alarm signals to plant cells that have not yet been attacked. In general, PR proteins appear not to be rapid-response weapons; rather, they act more slowly, perhaps after other mechanisms have blunted the pathogen's attack.

The hypersensitive response is a localized containment strategy

Plants that are resistant to fungal, bacterial, or viral diseases generally owe this resistance to what is known as the **hypersensitive response**. Cells around the site of microbial infection die, preventing the spread of the pathogen by depriving it of nutrients. Some of the cells produce phytoalexins and other chemicals before they die. The dead tissue, called a *necrotic lesion*, contains and isolates what is left of the microbial invasion (Figure 39.2). The rest of the plant remains free of the infecting microbe.

One of the chemicals produced during the hypersensitive response is a close relative of aspirin. Since ancient times, people in Asia, Europe, and the Americas have used willow (*Salix*) leaves and bark to relieve pain and fever. The active ingredient in willow is **salicylic acid**, the same substance from which aspirin is derived:

Salicylic acid

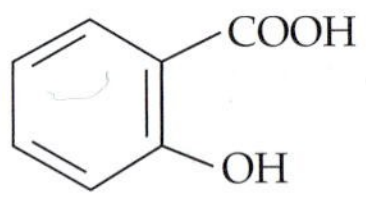

It now appears that all plants contain at least some salicylic acid. This compound plays a hormonal role in the plants' own defenses, often leading to a long-lasting effect that makes them resistant to later attacks by pathogens.

Systemic acquired resistance is a form of long-term "immunity"

Systemic acquired resistance is a general increase in the resistance of the entire plant to a wide range of pathogen species. It is not limited to the pathogen that originally triggered it or to the site of the original infection.

39.2 The Aftermath of a Hypersensitive Response
The necrotic spots on these leaves are a response to the fungus that causes strawberry blight.

The systemic acquired resistance that sometimes follows the hypersensitive response is accompanied by the synthesis of PR proteins. Treatment of plants with salicylic acid or aspirin leads to the production of PR proteins and to a resistance to pathogens. Salicylic acid treatment provides substantial protection against tobacco mosaic virus (a well-studied plant pathogen) and some other viruses.

Salicylic acid also serves as a hormone for disease resistance. In some cases, microbial infection in one part of a plant leads to the export of salicylic acid to other parts of the plant, where it causes the production of PR proteins before the infection can spread. The PR proteins then limit the extent of the infection. Infected plant parts also produce the closely related *methyl salicylate* (also known as oil of wintergreen). This volatile substance travels to other plant parts through the air. It may be that methyl salicylate can also trigger the production of PR proteins in neighboring plants that have not yet been infected.

Some plant genes match up with pathogen genes

Many plants use the hypersensitive response and systemic acquired resistance as nonspecific defenses against various pathogens. However, the triggering of these responses resides in a highly specific mechanism, called **gene-for-gene resistance**. In gene-for-gene resistance, the ability of a plant to defend itself against a specific strain of a pathogen depends on the plant's having a particular allele of a gene that corresponds to a particular allele of a gene in the pathogen (Figure 39.3). Let's see how this matching works.

Plants have a large number of ***R* genes** (resistance genes), and many pathogens have sets of ***Avr* genes** (avirulence genes). Dominant *R* alleles favor resistance, and dominant *Avr* alleles make a pathogen less effective. If a particular plant has the dominant allele of an *R* gene and a pathogen strain infecting it has the dominant allele of the corresponding *Avr* gene, the plant will be resistant to that strain. This is true even when none of the other *R–Avr* pairs features corresponding dominant alleles. (This effect, one *R–Avr* pair overruling the others, is an example of epistasis, which was discussed in Chapter 10.)

The mechanism of gene-for-gene resistance is not completely understood. There are thousands of specific *R* genes among the plants, and their products have different functions. The *Avr* genes in pathogens are simply genes that cause the pathogen to produce a substance that elicits a defensive response in the plant. Most gene-for-gene interactions trigger the hypersensitive response.

Not all biological threats to plants come from microorganisms and viruses that cause diseases. Many animals, from inchworms to elephants, *eat* plants.

Plants and Herbivores: Benefits and Losses

Herbivores—animals that eat plants—depend on plants for energy and nutrients. Plants have many defense mechanisms that protect them against herbivores, as we will see. First, let's consider how herbivores can have a *positive* effect on the plants they eat.

Grazing increases the productivity of some plants

In **grazing**, a predator eats part of a plant, such as the leaves, without killing its prey, which then has the potential

Key
Pathogen
Plant cell cytoplasm
Signal encoded by dominant allele of pathogen.
Receptor, encoded by dominant allele, in plant cell plasma membrane.

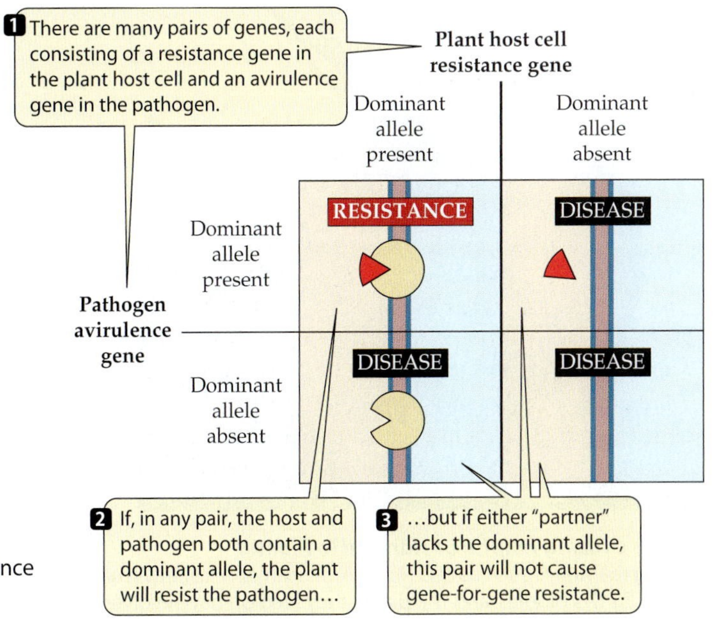

39.3 Gene-for-Gene Resistance
A single pair of corresponding dominant alleles promotes resistance even if all the other pairs are mismatches.

to grow back (Figure 39.4). What are the consequences of grazing? Is it always detrimental to plants, or are they somehow adapted to their place in the food chain? Certain plants and their predators evolved together, each acting as the agent of natural selection on the other (see Chapter 29). Because of this coevolution, grazing actually increases photosynthetic production in some plant species.

Removing some leaves from a plant may increase the rate of photosynthesis of the remaining leaves. This phenomenon probably is the result of several factors. First, nitrogen obtained from the soil by the roots no longer needs to be divided among so many leaves. Second, the export of sugars and other photosynthetic products from the leaves may be enhanced because the demand for those products in the roots is undiminished, while the sources for such products—leaves—have been decreased. The remaining leaves may compensate by photosynthesizing more rapidly.

A third and particularly significant factor increasing photosynthesis, especially in grasses, is an increase in the availability of light to the younger, more active leaves or leaf parts. The removal of older or dead leaves by a grazer decreases the shading of younger leaves. Unlike most other plants, which grow from their shoot and leaf tips, grasses grow from the base of the shoot and leaf, so their growth is not cut short by grazing.

Mule deer and elk graze many plants, including one called scarlet gilia. Their grazing removes about 95 percent of the aboveground part of each plant (Figure 39.5), but each plant quickly regrows not one, but four, replacement stems. Grazed plants produce three times as many fruits by the end of the growing season as do ungrazed plants.

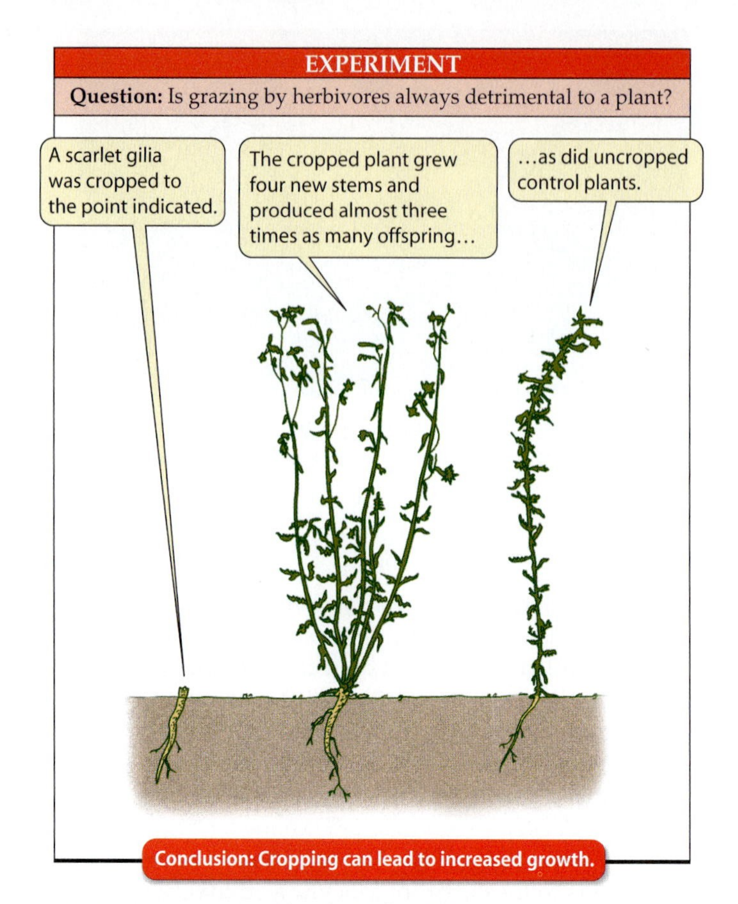

39.5 Overcompensation for Being Eaten
Experiments confirm that some plants benefit from the effects of grazing.

39.4 Is Grazing Helpful or Harmful to Plants?
Grazing mammals such as this North American elk exist in virtually all of Earth's biomes, and the plants they feed on have evolved along with them.

Some grazed trees and shrubs continue to grow until much later in the season than do ungrazed but otherwise similar plants. This longer growing season results in part because the removal of apical buds by the grazers stimulates lateral buds to become active, producing a more heavily branched plant. Leaves on ungrazed plants may also die earlier in the growing season than leaves on grazed plants.

A plant may benefit from moderate herbivory by attracting animals that spread its pollen or that eat its fruit and thus disperse its seeds. Nevertheless, resisting attack by herbivores is often to the advantage of a plant.

Some plants produce chemical defenses

Although a plant cannot flee its herbivorous enemies, it may be able to defend itself chemically. Many plants attract, resist, and inhibit other organisms by producing special chemicals known as **secondary products**. *Primary products* are substances, such as proteins, nucleic acids, carbohydrates, and lipids, that are produced and used by all living things. Although all organisms use the same kinds of primary products, plants can differ as radically in their secondary products as they do in their external appearance.

The more than 10,000 known secondary plant products range in molecular weight from about 70 to more than 400,000, but most are of low molecular weight. Some are produced by only a single species, while others are characteristic of an entire genus or even family. These compounds help plants compensate for being unable to move.

39.1 Secondary Plant Products in Defense

CLASS	TYPE	ROLE	EXAMPLE
Nitrogen-containing	Alkaloids	Affect herbivore nervous system	Nicotine in tobacco
	Glycosides	Release cyanide or sulfur compounds	Dhurrin in sorghum
	Nonprotein amino acids	Disrupt herbivore protein structure	Canavanine in jack bean
Phenolics	Flavonoids	Phytoalexins	Capsidol in peppers
	Quinones	Inhibit competing plants	Juglone in walnut
	Tannins	Herbivore and microbe deterrents	Many woods, such as oak
Terpenes	Monoterpenes	Insecticides	Pyrethroids in chrysanthemum
	Sesquiterpenes	Antiherbivores	Gossypol in cotton
	Steroids	Mimic insect hormones and disrupt insect life cycle	
	Polyterpenes	Feeding deterrent?	Rubber in rubber tree

The effects of defensive secondary products on animals are diverse. Some secondary products act on the nervous systems of herbivorous insects, mollusks, or mammals. Others mimic the natural hormones of insects, causing some larvae to fail to develop into adults. Still others damage the digestive tracts of herbivores. Some secondary products are toxic to fungal pests. Humans make commercial use of many secondary plant products as fungicides, insecticides, rodenticides, and pharmaceuticals.

While many secondary products have protective functions, others are essential as attractants for pollinators and seed dispersers. Table 39.1 lists the major classes of defensive secondary plant products and their biological roles.

Let's look at a specific example of an insecticidal secondary product, canavanine.

Some secondary products play multiple roles

Canavanine is an amino acid that is not found in proteins, but is closely similar to the amino acid arginine, which is found in almost all proteins. Canavanine has two important roles in those plants that produce it in significant quantities. The first role is as a nitrogen-storing compound in seeds. The second, defensive role is based on the similarity of canavanine to arginine:

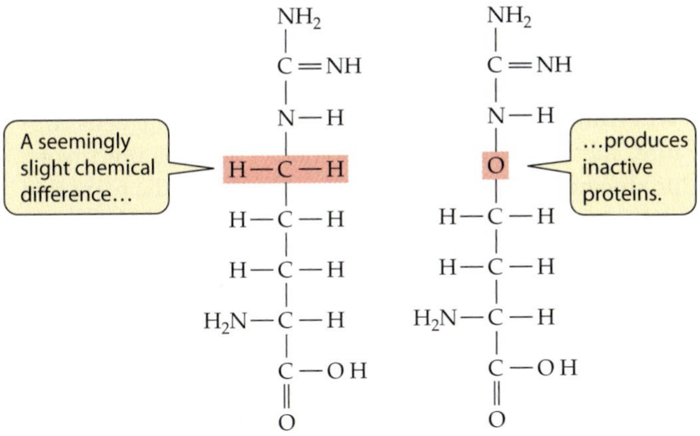

A seemingly slight chemical difference...

...produces inactive proteins.

Arginine Canavanine

Many insect larvae that consume canavanine-containing plant tissue are poisoned. The canavanine is incorporated into the insect's proteins in some of the places where the DNA has coded for arginine because the enzyme that charges the tRNA specific for arginine fails to discriminate accurately between the two amino acids. The structure of canavanine is different enough from that of arginine that some of the resulting proteins end up with a modified tertiary structure and hence reduced biological activity. These defects in protein structure and function lead to developmental abnormalities that kill the insect.

A few insect larvae are able to eat canavanine-containing plant tissue and still develop normally. How can this be? In these larvae, the enzyme that charges the arginine tRNA discriminates correctly between arginine and canavanine. The canavanine they ingest is thus not incorporated into the proteins they form, and the larvae are not harmed.

Many defenses depend on extensive signaling

Plant defenses result from a series of signals. Insects feeding on tomato leaves damage the cells, leading to a chain of events including the formation of hormones and ending with the production of an insecticide. The signaling steps in the production of one defensive compound, shown in Figure 39.6, involve two hormones. **Systemin** is a polypeptide hormone—the first polypeptide hormone to be discovered in plants. **Jasmonates** are formed from the unsaturated fatty acid linolenic acid. The final step in this series is the production of a protease inhibitor. The inhibitor, once in an insect's gut, interferes with the digestion of proteins and thus stunts the insect's growth.

Jasmonic acid (a jasmonate)

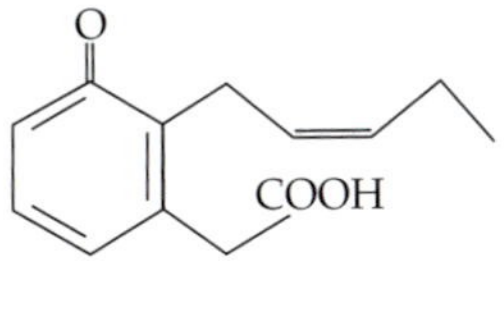

Jasmonates also take part in the "call for help" described at the beginning of this chapter. In that case, a substance re-

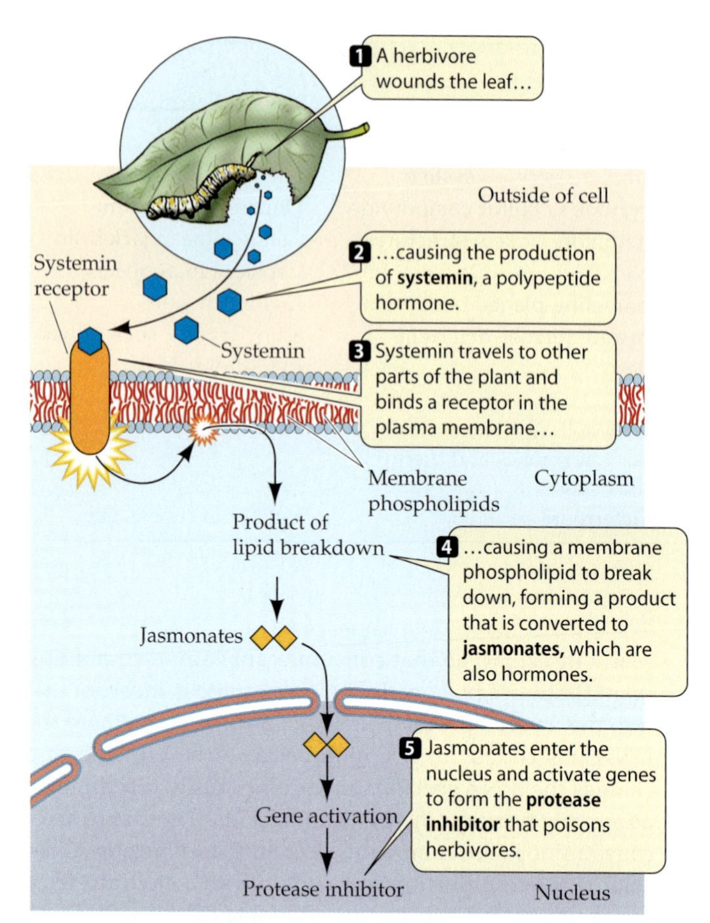

39.6 A Signaling Pathway for Synthesis of an Insecticide The chain of events initiated by an insect's attack and leading to the production of a defensive chemical can consist of many steps. These steps may include the synthesis of one or two hormones, binding of receptors, gene activation, and, finally, synthesis of a poison.

leased by chewing caterpillars is the first signal, leading to the formation of jasmonates by the plant. The jasmonates, in turn, trigger the formation of the volatile compounds that attract the insects that prey on the caterpillars.

Gene splicing may confer resistance to insects

Wild and domesticated common beans (*Phaseolus vulgaris*) differ in their resistance to attack by two species of bean weevils. Some wild bean seeds are highly resistant to these insects, but no cultivated bean seeds show such resistance. Scientists discovered that all weevil-resistant bean seeds contain a specific seed protein, *arcelin*. This protein has never been found in cultivated bean seeds. Therefore, the scientists hypothesized that arcelin is responsible for the resistance of some seeds to predation by the weevils.

To rule out other differences between wild and cultivated beans as being responsible for the resistance, the scientists performed two series of experiments. In one series, they crossed cultivated and wild bean plants. All of the progeny seeds of such crosses that contained arcelin showed resistance to weevils. In the other series of experiments, the scientists worked with artificial bean seeds made by removing the seed coats of cultivated beans and grinding the remainder of the seeds into flour. They added different concentrations of arcelin to different batches and molded the flour into artificial seeds. They then let weevils attack the artificial seeds. The more arcelin the artificial seeds contained, the more resistant they were to weevils.

In preliminary tests, arcelin in cooked beans was shown not to be harmful to rats—a first step toward determining whether arcelin is safe in food for humans. Agricultural scientists must sometimes choose between crop protection and appeal to humans. A plant with sturdy chemical defenses may taste bad, make us sick, or even kill us.

The development of crop plants that produce their own pesticides is an active area of research in agricultural biotechnology. Scientists are seeking to introduce genes for arcelin and other resistance-conferring proteins into agriculturally important crops such as beans. One of the most widely applied approaches is the engineering of several crops, such as tomato, corn, and cotton, to express the toxin genes from *Bacillus thuringiensis* discussed in Chapter 17.

Why don't plants poison themselves?

Why don't the chemicals that are so toxic to herbivores and microbes kill the plants that produce them? Plants that produce toxic secondary products generally use one of the following measures to protect themselves:

- The toxic material is isolated in a special compartment, such as the central vacuole.
- The toxic substance is produced only after the plant's cells have already been damaged.
- The plant uses modified enzymes or modified receptors that do not recognize the toxic substance.

The first method is the most common. Plants using this method store their poisons in vacuoles if they are water-soluble. If hydrophobic, the poisons are stored in **laticifers** (tubes containing a white, rubbery latex) or dissolved in waxes on the epidermal surface. This compartmentalized storage keeps the toxic substance away from the mitochondria, chloroplasts, and other parts of the plant's own metabolic machinery.

Some plants store the precursors of toxic substances in one compartment, such as the epidermis, and store the enzymes that convert the precursors to the active poison in another compartment, such as the mesophyll. These plants produce the toxic substance only after being damaged. When an herbivore chews part of the plant, the cells rupture, and the enzymes come in contact with the precursors, producing the toxic product. The only part of the plant that is damaged by the toxic material is that which was already damaged by the herbivore. Plants that respond to attack by producing cyanide—a strong inhibitor of cellular respiration in all organisms that respire—are among those that use this protective measure.

The third protective measure is used by the canavanine-producing plants described earlier. These plants produce a

39.7 Disarming a Plant's Defenses
This beetle is inactivating a milkweed's defense system by cutting its laticifer supply lines.

tRNA-charging enzyme for arginine that does not bind canavanine. However, as we have seen, some herbivores can evade being poisoned by canavanine in a similar manner, demonstrating that no plant defense is perfect.

The plant doesn't always win

Milkweeds such as *Asclepias syriaca* are latex-producing (laticiferous) plants. When damaged, a milkweed releases copious amounts of toxic latex from its laticifers. Latex has long been suspected to deter insects from eating the plant, because insects that feed on neighboring plants of other species do not attack laticiferous plants. This observed behavior is consistent with, but does not prove, the hypothesis that the latex keeps the insects at bay.

Stronger support for the hypothesis was obtained by studying field populations of *Labidomera clivicollis*, a beetle that is one of the few insects that feed on *A. syriaca*. These beetles show a remarkable prefeeding behavior: They cut a few veins in the leaves before settling down to dine (Figure 39.7). Cutting the veins, with their adjacent laticifers, causes massive latex leakage and interrupts the latex supply to a downstream portion of the leaf. The beetles then move to the relatively latex-free portion and eat their fill.

Does this behavior of the beetles negate the adaptive value of latex protection? Not entirely. There are still great numbers of potential insect pests that are effectively deterred by the latex. And evolution proceeds. Over time, milkweed plants producing higher concentrations of toxins may be selected by virtue of their ability to kill beetles that cut their laticifers.

Having discussed how plants defend themselves against other organisms, we now turn our attention to how plants adapt to environments where water is a problem.

Water Extremes: Dry Soils and Saturated Soils

Water is often in short supply in the terrestrial environment. Some terrestrial habitats, such as deserts, intensify this challenge, and many plants that inhabit particularly dry areas have one or more adaptations that allow them to conserve water. Plants adapted to dry environments are called **xerophytes**.

Some plants evade drought

Some desert plants have no special *structural* adaptations for water conservation other than those found in almost all flowering plants. Instead, they have an alternative *strategy*. These desert annuals simply evade the periods of drought. They carry out their entire life cycle—from seed to seed—during a brief period in which rainfall has made the surrounding desert soil sufficiently moist (Figure 39.8).

Some leaves have special adaptations to dry environments

Plants that remain active during dry periods must have structural adaptations that enable them to survive. The secretion of a heavier cuticle over the leaf epidermis to retard water loss is a common adaptation to dry environments. An even more common adaptation is a dense covering of epidermal hairs. Some species have stomata only in sunken

39.8 Desert Annuals Evade Drought
Seeds of desert plants often lie dormant for long periods awaiting conditions appropriate for germination. When they do germinate, they grow and reproduce rapidly before the short wet season passes. During the long dry spells, only seeds remain alive.

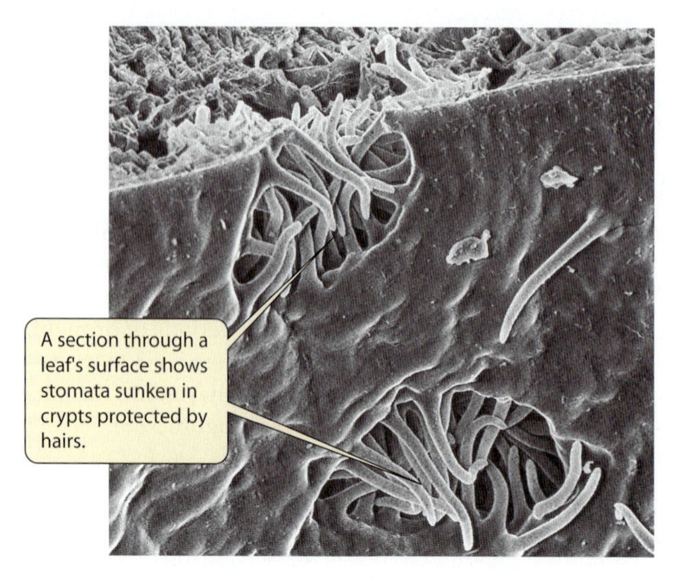

39.9 Stomatal Crypts
Stomata in the leaves of some xerophytes are in sunken pits called stomatal crypts. The hairs covering these crypts trap moist air. (A section of the leaf's interior can be seen at the top of the photo.)

39.10 Opportune Leaf Production
The ocotillo, a xerophyte that lives in the lower deserts of the southwestern United States and northern Mexico, produces leaves only when there is sufficient water for photosynthesis.

cavities below the leaf surface, which reduces the drying effects of air currents; often these stomatal cavities contain hairs as well (Figure 39.9).

Succulence—the possession of fleshy, water-storing leaves—is an adaptation to dry environments. Ice plants and their relatives have fleshy leaves in which water may be stored. Others, such as ocotillo, produce leaves only when water is abundant, shedding them as the soil dries out (Figure 39.10). Cacti and similar plants have spines rather than typical leaves, and photosynthesis is confined to the fleshy stems. The spines may reflect incident radiation, or they may dissipate heat. Corn and some related grasses have leaves that roll up during dry periods, thus reducing the leaf surface area through which water is lost. Some trees, such as eucalyptuses, that grow in arid regions have leaves that hang vertically at all times, thus evading the midday sun (Figure 39.11).

Xerophytic adaptations of leaves minimize water loss by the plant. However, such adaptations simultaneously minimize the uptake of carbon dioxide and thus limit photosynthesis. In consequence, most xerophytes grow slowly, but they utilize water more efficiently than do other plants; that is, they fix more grams of carbon by photosynthesis per gram of water lost to transpiration than other plants do.

Plants have other adaptations to a limited water supply

Roots may also be adapted to dry environments. The Atacama Desert in northern Chile often goes several years without measurable rainfall. The landscape there is almost barren of plant life, save for many surprisingly large

39.11 Shade at Midday
Because eucalyptus leaves hang vertically, their flat surfaces are not presented directly to the midday sun. This adaptation minimizes heating as well as water loss.

mesquite trees (genus *Prosopis*; Figure 39.12). These trees obtain water through taproots that grow to great depths, reaching water supplies far underground, as well as from condensation on their leaves.

A more common adaptation of desert plants is a root system that grows rapidly during rainy seasons but dies back during dry periods. Cacti have shallow but extensive fibrous root systems that effectively intercept water at the surface of the soil following even light rains.

Xerophytes and other plants that receive inadequate water may accumulate the amino acid proline to substantial concentrations in their vacuoles. As a consequence, the osmotic potential and water potential of their cells become more negative; thus these plants tend to extract more water from the soil than do plants that lack this adaptation. Plants living in salty environments share this and several other adaptations with xerophytes, as we will see.

As we have seen, there are many ways in which some plants eke out an existence in terrestrial environments with very little water. What happens if there is too much water?

In water-saturated soils, oxygen is scarce

Some plants live in environments so wet that the diffusion of oxygen to their roots is severely limited. Since most plant roots require oxygen to support respiration and ATP production, most plants cannot tolerate this situation for long.

39.13 Coming Up for Air
The roots of the mangroves in this tidal swamp obtain oxygen through pneumatophores.

39.12 Mining Water with Deep Taproots
Death Valley, California, is not as arid as the Chilean Atacama, but the mesquite must reach far down into the sand dunes for its water supply.

Some species, however, are adapted to life in a water-saturated habitat. Their roots grow slowly and hence do not penetrate deeply. Because the oxygen level is too low to support aerobic respiration, the roots carry on alcoholic fermentation (see Chapter 7), which provides ATP for the activities of the root system but explains why growth is slow.

The root systems of some plants adapted to swampy environments have **pneumatophores**, which are extensions that grow out of the water and up into the air (Figure 39.13). Pneumatophores have lenticels and contain spongy tissues that allow oxygen to diffuse through them, aerating the submerged parts of the root system. Cypresses and some mangroves are examples of plants with pneumatophores.

Submerged or partly submerged aquatic plants often have large air spaces in the leaf parenchyma and in the petioles. Tissue containing such air spaces is called **aerenchyma** (Figure 39.14). Aerenchyma stores oxygen produced by photosynthesis and permits its ready diffusion to parts of the plant where it is needed for cellular respiration.

Aerenchyma also imparts buoyancy. Furthermore, because it contains far fewer cells than most other plant tissue, respiratory metabolism in aerenchyma proceeds at a lower rate, and the need for oxygen is much reduced.

Thus far we have considered water supply—either too little or too much—as a factor limiting plant growth. Other substances also can make an environment inhospitable to plant growth. One of these is salt.

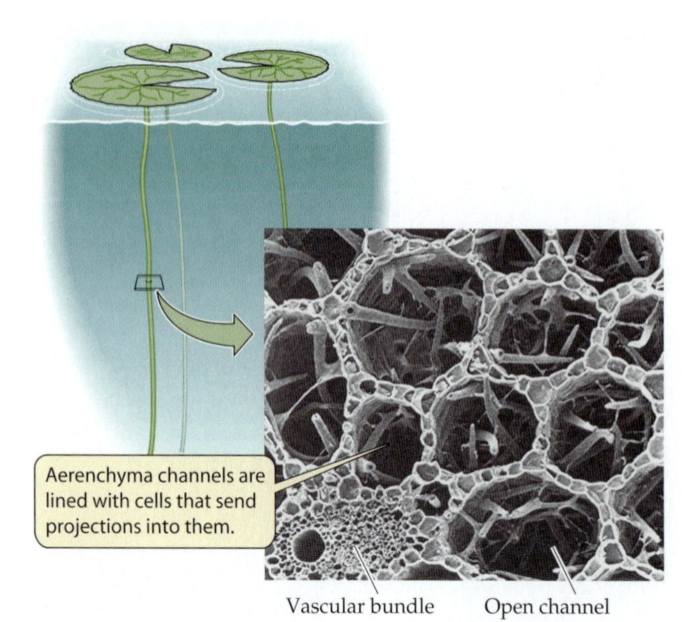

39.14 Aerenchyma Lets Oxygen Reach Submerged Tissues
The scanning electron micrograph, a cross section of a petiole of the yellow water lily, shows a vascular bundle and aerenchyma.

Too Much Salt: Saline Environments

Worldwide, no toxic substance restricts flowering plant growth more than salt (sodium chloride) does. *Saline*—salty—habitats support, at best, sparse vegetation. Saline habitats themselves are diverse, ranging from hot, dry, salty deserts to moist, cool, salty marshes. Along the seashore are saline environments created by ocean spray. The ocean itself is a saline environment, as are river estuaries, where fresh and salt water meet and mingle. The salinization of agricultural land is an increasing global problem. Even where crops are irrigated with fresh water, sodium ions from the water accumulate in the soil to ever greater concentrations as the water evaporates.

Saline environments pose an osmotic problem for plants. Because of its high salt concentration, a saline environment has an unusually negative water potential. To obtain water from such an environment, a plant must have an even more negative water potential than that of a plant in a nonsaline environment; otherwise, it will lose water, wilt, and die. A second problem is the potential toxicity of high concentrations of certain ions, notably sodium and chloride.

The **halophytes**—plants adapted to saline habitats—belong to a wide variety of flowering plant groups. How can these plants cope with a highly saline environment?

Most halophytes accumulate salt

Most halophytes share one adaptation: They accumulate sodium and, usually, chloride ions, and transport these ions to the leaves. The accumulated ions are stored in the central vacuoles of leaf cells, away from more sensitive parts of the cells. Nonhalophytes accumulate relatively little sodium, even when placed in a saline environment; of the sodium that is absorbed by their roots, very little is transported to the shoot. The increased salt concentration in the tissues of halophytes makes their water potential more negative, so they can take up water more easily from the saline environment.

In 1999, scientists reported the first success in causing overexpression of a gene in *Arabidopsis* that enables sodium uptake. This gene encodes a Na^+/H^+ antiport protein in the tonoplast (the membrane surrounding the central vacuole). By making the gene produce a greater than normal number of these antiport proteins, the scientists increased sodium transport in *Arabidopsis*, converting this nonhalophyte into a halophyte. Further research along this line may result in a great boost to agriculture in saline environments. Biologists in Israel and elsewhere have had some success in breeding crops that can be watered with seawater or diluted seawater.

Some halophytes have other adaptations to life in saline environments. Some, for example, have **salt glands** in their leaves. These glands excrete salt, which collects on the leaf surface until it is removed by rain or wind (Figure 39.15). This adaptation, which reduces the danger of poisoning by accumulated salt, is found both in some desert plants, such as tamarisk, and in some mangroves growing in seawater in the Tropics.

Salt glands can play multiple roles, as in the desert shrub *Atriplex halimus*. This shrub has glands that secrete salt into small bladders on the leaves, where, by increasing the gradient in water potential, the salt helps the leaves obtain water from the roots. At the same time, by making the water potential of the leaves more negative, the salt reduces the transpirational loss of water to the atmosphere.

The adaptations we have just discussed are specific to halophytes. Several other adaptations are shared by halophytes and xerophytes.

Halophytes and xerophytes have some similar adaptations

Many halophytes, like some xerophytes, accumulate the amino acid proline in their cell vacuoles, making the water

39.15 Secreting Salt
This salty mangrove has special salt glands that secrete salt, which appears here as crystals on the leaves.

potential of their tissues more negative. Unlike sodium, proline is relatively nontoxic.

Succulence is another adaptation that halophytes and xerophytes have in common, as might be expected, since saline environments, like dry ones, make water uptake difficult. Succulence characterizes many halophytes that occupy salt marshes. There the salt concentration in the soil solution may change throughout the day; while the tide is out, for instance, evaporation increases the salt concentration. Succulence may offer a reserve of water for the plant during the period of maximum salinity; when the salinity drops as the tide comes in, the leaf's store of water is replenished. Many succulents—both xerophytes and halophytes—use crassulacean acid metabolism (CAM) and have reversed stomatal cycles that enable them to conserve water by closing their stomata in the daytime (see Figure 35.11). Other general adaptations to a saline environment include high root-to-shoot ratios, sunken stomata, reduced leaf areas, and thick cuticles.

Salt is not the only toxic solute found in soils. Some heavy metal ions are more toxic than sodium at equivalent concentrations.

Habitats Laden with Heavy Metals

High concentrations of some heavy metal ions, such as aluminum, mercury, lead, and cadmium, poison most plants. Some geographic sites are naturally rich in heavy metals as a result of normal geological processes. Acid rain leads to the release of toxic aluminum ions in the soil. Other human activities, notably the mining of metallic ores, leave localized areas—known as *tailings*—with substantial concentrations of heavy metals and low concentrations of nutrients. Such sites are hostile to most plants, and seeds falling on them generally do not produce adult plants.

Mine tailings rich in heavy metals, however, generally are not completely barren (Figure 39.16). They may support healthy plant populations that differ genetically from populations of the same species on the surrounding normal soils. How can these plants survive?

39.16 Life after Strip Mining
Although high concentrations of heavy metals kill most plants, grass is colonizing this eroded strip mine in North Park, Colorado.

Initially, some plants were thought to tolerate heavy metals by excluding them: By not taking up the metal ions, it was believed, the plant avoided being poisoned. However, measurements have shown that tolerant plants growing on mine tailings do take up heavy metals, accumulating them to concentrations that would kill most plants. Thus the tolerant plants must have a mechanism for dealing with the heavy metals they take up. Such tolerant plants may be found to be useful agents for *bioremediation*, a decontamination process by which the heavy metal content of some contaminated soils is decreased by living organisms.

We know the mechanism of at least one case of tolerance to a heavy metal. The roots of a buckwheat grown in China secrete oxalic acid soon after they are exposed to aluminum concentrations high enough to inhibit root growth in other plants. Oxalic acid combines with aluminum ions, forming a complex that does not inhibit growth.

From mine to mine, the heavy metals in the soil differ. In Wales and Scotland, bent grass (*Agrostis*) grows near many mines. Samples of bent grass from several such sites were tested for their ability to grow in various solutions, each containing only one heavy metal. In general, the plants tolerated a particular heavy metal—the one most abundant in their habitat—but were sensitive to others. That is, they tolerated only one or two heavy metals, rather than heavy metals as a group.

Tolerant plant populations can evolve and colonize an area surprisingly rapidly. The bent grass population around a particular copper mine in Wales is resistant to copper and is relatively abundant, even though the copper-rich soil dates from mining done only a century ago.

Hot and Cold Environments

Temperatures that are too high or too low can stress plants and even kill them. Plants differ in their sensitivity to heat and cold, but all plants have their limits.

Any temperature extreme can damage cellular membranes:

- High temperatures destabilize membranes and denature many proteins, especially some of the enzymes of photosynthesis.
- Low temperatures cause membranes to lose their fluidity and alter their permeabilities to solutes.
- Freezing temperatures may cause ice crystals to form, damaging cellular membranes.

Plants have ways of coping with high temperatures

Transpiration, the evaporative loss of water, can cool a plant, but it also increases the plant's need for water. Therefore, it is not surprising that many plants living in hot environments have adaptations similar to those of xerophytes. These adaptations include epidermal hairs and spines that radiate heat, modified leaf displays that intercept less direct sunlight, and others.

Plants respond within minutes to high temperatures by producing several kinds of **heat shock proteins**. Among these are chaperonins (see Chapter 3), which help other proteins maintain their structures and avoid denaturation. Threshold temperatures for the production of heat shock proteins vary, but 40°C is sufficient to induce them in most plants. We have much to learn about the dozens of heat shock proteins, but we do know that some other types of stress also induce their formation. Among these are chilling and freezing.

Some plants are adapted to survival at low temperatures

Low temperatures above freezing injure many plants, including important crops such as rice, corn, and cotton. Many plant species can be modified to resist the effects of cold spells by a process called **cold-hardening**, which involves repeated exposure to cool, but not injurious, temperatures. The hardening process is a slow one, requiring many days. A key change that occurs during the hardening process is an increase in the relative fraction of unsaturated fatty acids in membranes. Unsaturated fatty acids solidify at lower temperatures than do saturated ones. Thus, the membranes retain their fluidity and function normally at cooler temperatures.

Low temperatures induce the formation of certain heat shock proteins that protect against chilling damage. There are also cases of "cross-protection" by heat shock proteins that are induced by one type of stress and that protect against other stresses. Tomatoes shocked by 2 days of high temperatures, for example, formed heat shock proteins and became resistant to chilling damage for the next 3 weeks.

If ice crystals form within cells, they can kill the cells by puncturing organelles and plasma membranes. Even outside cells, the growth of ice crystals can draw water from the cells and dehydrate them. Freezing-tolerant plants have a variety of adaptations to cope with these problems. A common one is the production of *antifreeze proteins* that inhibit the growth of ice crystals.

Plants have many effective mechanisms for coping with environmental challenges of many kinds. Their success is obvious—just look around you.

Chapter Summary

Plant–Pathogen Interactions

- Plants and pathogens evolve together. **Review Figure 39.1**
- Plants can strengthen their cell walls when attacked.
- Plant chemical defenses include PR proteins and phytoalexins.
- In the hypersensitive response, cells produce phytoalexins and then die, trapping the pathogens in dead tissue.
- The hypersensitive response is often followed by systemic acquired resistance, in which the hormone salicylic acid activates further synthesis of PR proteins and triggers responses in other parts of the plant.
- The hypersensitive response is nonspecific. A more specific response, called gene-for-gene resistance, matches up alleles in a plant's resistance genes and a pathogen's avirulence genes. **Review Figure 39.3**

Plants and Herbivores: Benefits and Losses

- Grazing by herbivores increases the productivity of some plants. **Review Figure 39.5**
- Some plants produce secondary products that function as chemical defenses against herbivores. **Review Table 39.1**
- Various hormones, including systemin and jasmonates, participate in the pathways leading to the production of defensive chemicals. **Review Figure 39.6**
- To avoid poisoning themselves, plants may confine the toxic substances they produce to special compartments, or they may produce the substances only after cells have been damaged, or they may form enzymes and receptors that are not affected by the substances.

Water Extremes: Dry Soils and Saturated Soils

- Desert annuals evade drought by living only long enough to take advantage of the brief period during which the soil has enough moisture to support them.
- Some leaves have special adaptations to dry environments: a thickened cuticle, epidermal hairs, sunken stomata, fleshy leaves and stems, spines, and altered leaf display angles.
- Other adaptations to dry environments include long taproots and root systems that die back seasonally.
- The submerged roots of some plants form pneumatophores to allow oxygen uptake from the air. Aerenchyma in submerged plant parts stores and permits the diffusion of oxygen. **Review Figure 39.14**

Too Much Salt: Saline Environments

- A saline environment restricts the availability of water to plants. Halophytes are plants that are adapted to such environments.
- Most halophytes accumulate salt, and some have salt glands that excrete the salt to the leaf surface.
- Halophytes and xerophytes have some adaptations in common.

Habitats Laden with Heavy Metals

- Aluminum, mercury, lead, and cadmium are among the heavy metals that are toxic to plants at high concentrations.
- Rather than excluding heavy metals, tolerant plants deal with them after taking them up. A given plant's tolerance is limited to only one or two heavy metals.

Hot and Cold Environments

- High temperatures destabilize cell membranes and some proteins.
- Adaptations to elevated temperatures include the production of heat shock proteins.
- Low temperatures cause membranes to lose their fluidity.

▶ Ice crystals can puncture organelles and plasma membranes.

▶ Adaptations to low temperatures and freezing include a change in membrane fatty acid composition and the production of antifreeze proteins.

For Discussion

1. We mentioned the possibility of designing crop plants that produce their own pesticides. Now chemical companies are designing crop plants capable of detoxifying weed killers, so that crops grow after farmers have destroyed competing vegetation. Discuss the likely usefulness and possible drawbacks of such applications of recombinant DNA technology.
2. How might plant adaptations affect the evolution of herbivores? How might adaptions of herbivores affect plant evolution?
3. The stomata of the common oleander, *Nerium oleander*, are sunk in crypts in its leaves. Whether or not you know what an oleander is, you should be able to descibe an important feature of its natural habitat; what is this feature?
4. Explain why halophytes often use the same mechanisms for coping with their challenging environments as xerophytes do for coping with theirs.
5. In ancient times, people used less sophisticated methods for mining than we use today. Thus ancient mines often yield substantial profits to modern-day miners who find and work them. On the basis of material in this chapter, how might you try to locate the site of an ancient mine?

Part Six

THE BIOLOGY OF ANIMALS

40 Physiology, Homeostasis, and Temperature Regulation

THE CAMEL IS CALLED THE "SHIP OF THE desert" because it can carry a large load across a hot desert without having to drink for days. Under similar conditions, a human sweats profusely to keep body temperature from rising, and can lose from 1 to 4 liters of water an hour. Without water, a human can become dangerously dehydrated in an hour. The dehydration causes the circulatory system and the thermoregulatory systems to fail, and body temperature begins to rise. As body temperature rises above 40°C, a person becomes dizzy and disoriented and gradually becomes delirious, loses consciousness, suffers brain damage, and dies. A person without water can die in the desert in only a few hours.

Why are camels better able to deal with desert conditions than we are? Several adaptations are important, but one of the most significant is that the camel's body temperature rises and falls more than ours does. Whereas humans try to keep their body temperatures close to 37°C by sweating, the camel allows its body temperature to rise to about 41°C over the course of the day. The camel's insulating coat of fur helps to slow its uptake of heat, but by tolerating the rise in its body temperature, it does not squander its body water for evaporative cooling.

When the air temperature falls at night, the camel passively unloads its accumulated heat to the environment. But now the situation reverses, and the camel allows its body temperature to fall. A human starts to produce heat by shivering when body temperature falls below about 36°C, but the camel's body temperature can fall below 34°C without stimulating shivering. The heat a camel stores in its body by allowing its body temperature to rise 7°C between sunup and sundown would require the evaporation of at least 5 liters of water to dissipate if the camel tried to maintain a constant body temperature. The camel has other adaptations, too. When it does reach water, it can drink about a third of its body weight to replace its losses!

The camel is an example of an animal that can live in an extreme environment. **Physiology** is the study of how organisms work—the study of the functions of all of the parts and processes of living systems. By studying the physiology of animals, we can understand how many species manage to live in extreme environments. By studying the special adaptations of such species, we frequently gain new information about how the human body works.

Animal physiology is the focus of the next twelve chapters of this book. Each chapter will present a system of structures that provides essential functions for the animal body. A structure might be a single cell, or it might be a population of similar cells that make up a tissue, or it might be a collection of different kinds of tissues that make up an organ. Two or more interacting organs constitute an organ system that serves one or more physiological functions. Organs and organ systems function to maintain the various physical and chemical aspects of an animal's internal environment at optimal levels.

Ships of the Desert
Camels conserve body water by allowing their body temperature to rise during the day rather than using evaporative cooling such as sweating and panting.

In this chapter we set the stage for our study of animal physiology by presenting an overview of how cells are organized into tissues, tissues into organs, and organs into organ systems with different physiological functions. We discuss general principles of how organ systems are controlled and regulated to achieve constancy in the internal environment. Most of this chapter deals with one feature of the internal environment: temperature. We will see how temperature influences living systems, what adaptations animals have for dealing with temperature challenges, and finally, how mammals regulate body temperature.

Homeostasis: Maintaining the Internal Environment

Single-celled organisms meet all their needs by direct exchanges with the external environment. Even the cells of some small, simple multicellular animals meet their needs in this way. Such animals are common in the sea. Seawater contains nutrients and salts, and provides a relatively unchanging physical environment. Most cells of a sponge or a jellyfish are in direct contact with seawater, or are close enough that they can receive nutrients and eliminate wastes without specialized organs to transport nutrients and wastes around their bodies. This lifestyle is quite limiting, however. No part of the animal's body can be more than a few cell layers thick, every cell must be able to take care of all its own needs, and the animal is limited to environments that provide for all of its cellular needs.

The evolution of an internal environment, distinct from the external environment, made complex multicellular animals possible. The internal environment consists of extracellular fluids that bathe every cell of the body, supplying nutrients and receiving wastes. Its physical and chemical conditions can be kept at levels favorable to the cells. The cells are thereby protected from the external environment, making it possible for an animal to occupy habitats that would not support its cells if they were exposed to it directly.

In complex multicellular organisms, it became possible for cells to become specialized for tasks that could contribute to maintaining specific aspects of the internal environment. Some cells became organized into tissues specialized to maintain the salt and water balance of the internal environment, others became specialized to provide nutrients, and still others to maintain appropriate levels of oxygen and carbon dioxide. Specialized tissues and organs form systems within the the internal environment, each providing something all the cells of an animal need (Figure 40.1).

The composition of the internal environment is constantly being perturbed by the external environment and by the activities of cells themselves. The internal environment of a person in a desert, for example, will either increase in temperature or decrease in volume and change in composition because of water loss. Simultaneously, the activities of the person's cells will be taking nutrients from and contributing wastes to the internal environment. The activities of the specialized tissues and organs must continuously correct the physical and chemical composition of the internal environment so that it remains conducive to life.

The maintenance of constant conditions in the internal environment is called **homeostasis**. Homeostasis is an essential feature of complex animals. If an organ fails to function properly, homeostasis is compromised, and as a result, cells become damaged and die. The damaged cells are not just those of the organ that functions improperly, but the cells of other organs as well. Loss of homeostasis is a serious problem that makes itself worse. To avoid loss of homeostasis, the activities of organs must be controlled and regulated in response to changes in both the external and the internal environments.

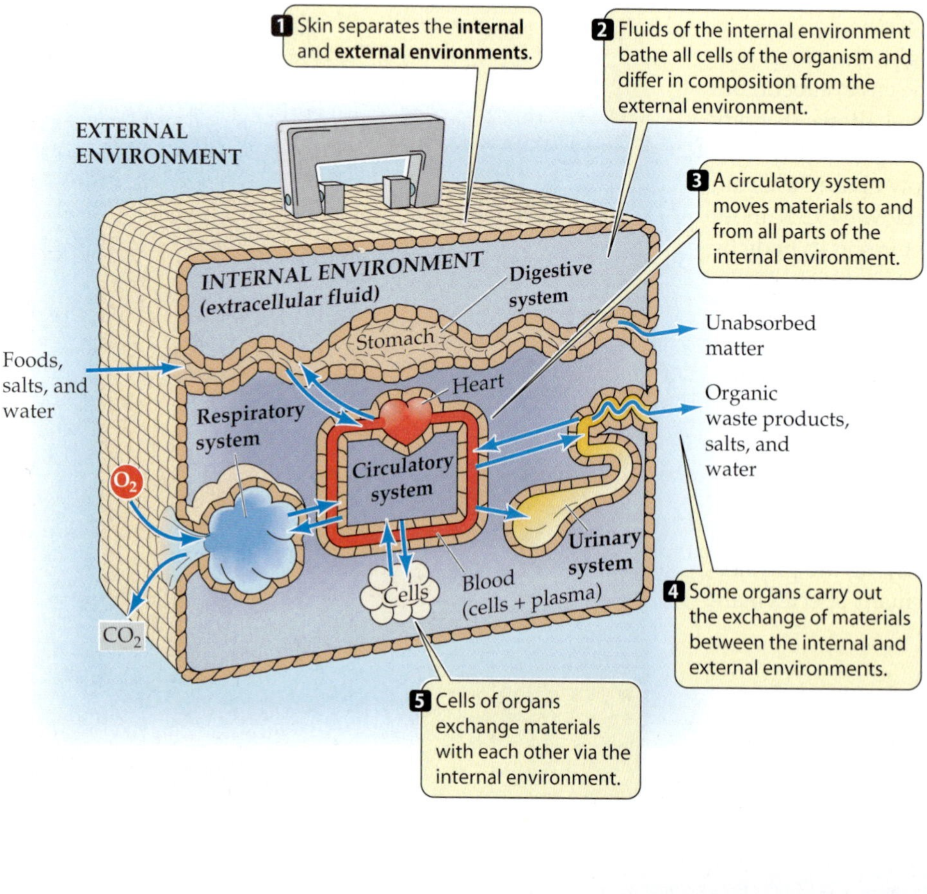

40.1 Maintaining Internal Stability While On the Go
Organ systems maintain a constant internal environment that provides for the needs of all cells of the body, making it possible for animals to travel among different and often highly variable external environments.

Control and regulation require information; hence the organ systems of information—the endocrine and nervous systems—must be included in our discussions of every physiological function. For that reason, we treat the endocrine and nervous systems early in this part of the book. Subsequent chapters deal with the systems responsible for controlling various aspects of the internal environment. Although each chapter will focus on different organs, those organs are all made of the same tissue types. What are these tissue types, and what are their general features?

Tissues, Organs, and Organ Systems

Cells are the basic building blocks of multicellular animals. When cells with the same characteristics or specializations are grouped together, they form a **tissue**. There are four basic types of tissues—epithelial, connective, muscle, and nervous—but there are variations on each basic type. An organ is usually made up of several different tissue types (Figure 40.2).

Epithelial tissues cover the body and line organs

Epithelial tissues are sheets of densely packed, tightly connected cells that cover inner and outer body surfaces. They line hollow organs of the body such as the gut, the lungs, the bladder, and the blood vessels. Some epithelial cells have secretory functions—for example, those that secrete milk, mucus, digestive enzymes, or sweat. Others have cilia and help substances move over surfaces or through tubes. Since epithelial cells create boundaries between the inside and the outside of the body and between body compartments, they frequently have absorptive and transport functions. Epithelial cells can also be receptors that provide information to the nervous system. Smell and taste receptors, for example, are epithelial cells that detect specific chemicals.

An epithelial tissue can be classified according to its structure and the appearance of its cells:

- A *simple epithelium* is a single layer of cells, such as that forming the tubules of the kidney (Figure 40.3*a*).
- A *stratified epithelium* consists of multiple layers of cells, as is the case with the skin (Figure 40.3*b*).
- A *pseudostratified epithelium* really consists of a single layer of cells, but because the cells are of different lengths, they give the appearance of multiple layers (Figure 40.3*c*).

The shapes of the cells making up an epithelium can be *squamous* (flattened), *cuboidal*, or *columnar*. Most cuboidal and columnar epithelia are involved in transport or secretory functions and have an abundance of organelles such as mitochondria and Golgi apparatus. Squamous epithelia, such as the outer layer of the skin, have fewer organelles; they frequently serve structural functions and as permeability boundaries.

Epithelial tissues have distinct inner and outer surfaces. The outer surface faces the air, as in the case of the skin and lungs, or a fluid-filled organ cavity, such as the lumen of the gut. These outer surfaces are made up of the apical ends of the epithelial cells, which may have cilia or may be highly folded to increase their surface area. The inner surface of an epithelium consists of the basal ends of the epithelial cells, which rest on an extracellular matrix called a *basal lamina* (see Figure 4.30).

The skin and the lining of the gut are examples of epithelial tissues that receive much wear and tear. Accordingly, cells in these tissues have a high rate of cell division to replace cells that die and are shed. Dandruff consists of discarded skin cells, and the well-known Pap smear test for cancer of the female reproductive tract is based on examination of shed epithelial cells.

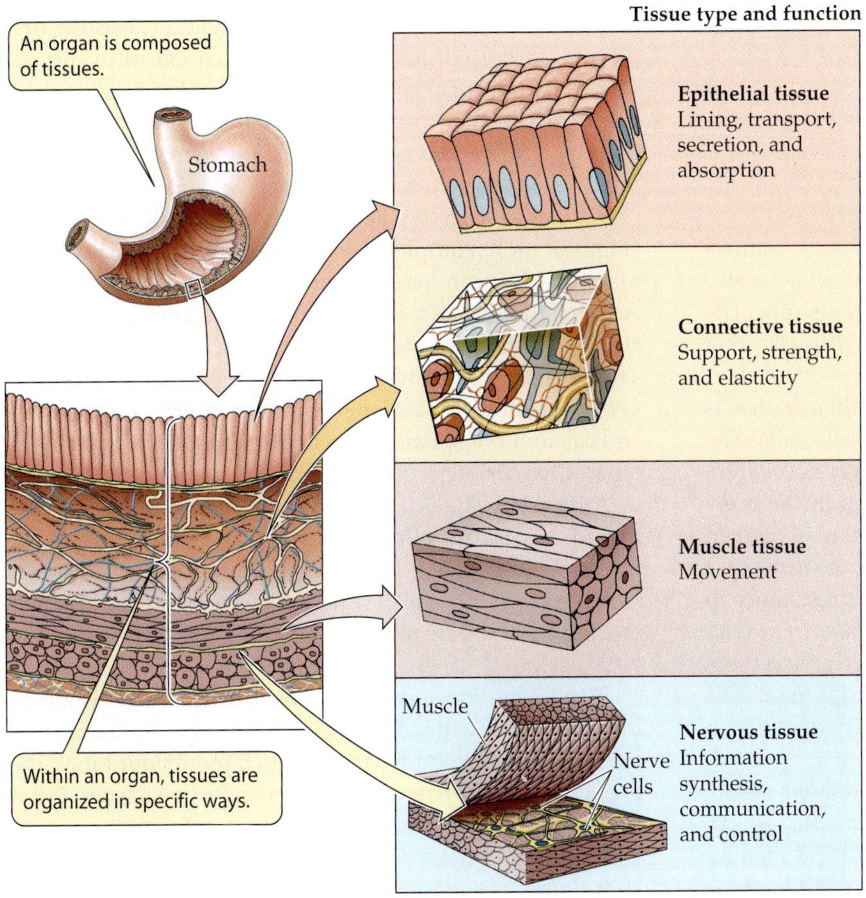

40.2 Four Types of Tissue
All cells can be classified into one of four tissue types. The cells of a given type have a similar structure and function.

(a)

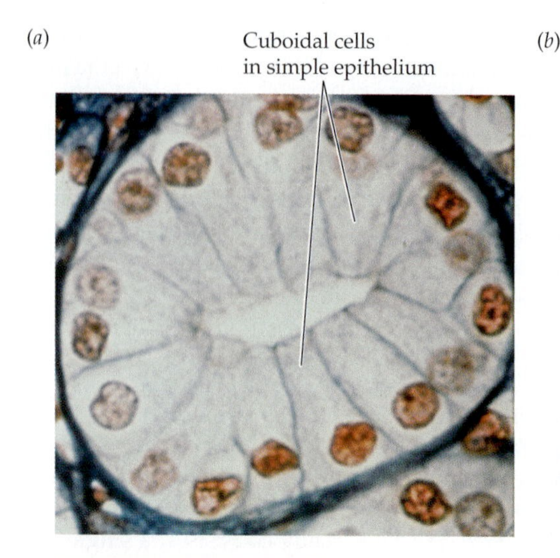

(b)

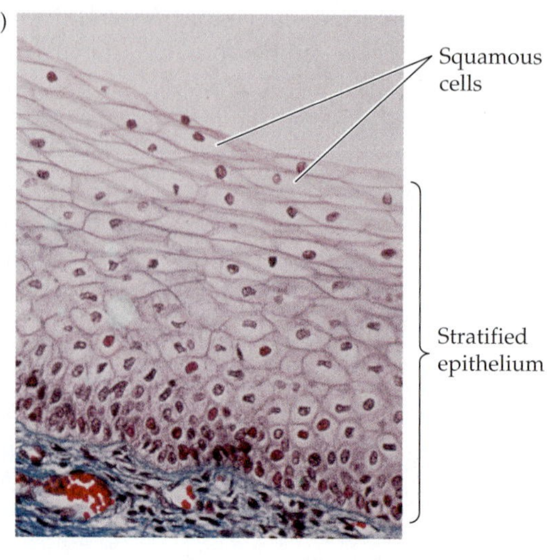

40.3 *Epithelial Tissue* (*a*) A single layer of cuboidal cells forms a simple epithelium lining the collecting ducts of a human kidney. (*b*) Multiple layers of squamous cells form a stratified epithelium. (*c*) The columnar cells of this pseudostratified epithelium lining the respiratory tract give the appearance of multiple layers.

(c)

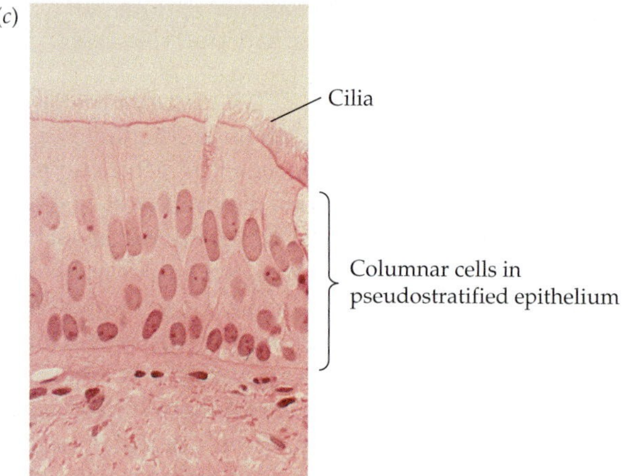

Connective tissues support and reinforce other tissues

In contrast to the densely packed epithelial tissues, **connective tissues** consist of dispersed populations of cells embedded in an extracellular matrix that they secrete. The composition and properties of the matrix differ among types of connective tissues.

An important component of the extracellular matrix is protein fibers secreted by the connective tissue cells. The dominant protein in the extracellular matrix is *collagen*, which is, in fact, the most abundant protein in the body (representing 25 percent of total body protein). Collagen fibers have high tensile strength. They give the *dense connective tissue* of skin, tendons, and ligaments resistance to stretch. Similarly, the collagen fibers of *reticular connective tissue* provide a netlike framework for organs, giving them shape and structural strength. *Loose connective tissue* fills spaces between organs and has a low density of collagen fibers.

Another type of protein fiber in the extracellular matrix of connective tissues is the stretchable protein *elastin*. It can be stretched to several times its resting length and then recoil. Fibers composed of elastin are most abundant in *elastic connective tissue*, such as that in the walls of the lungs and the large arteries. Elastin fibers in the skin are responsible for its ability to snap back when stretched, and gradual loss of these fibers with age causes gradual loss of the resiliency of the skin.

Proteoglycans are extracellular proteins that give connective tissues resistance to compression. Proteoglycans are abundant in the extracellular matrix of the connective tissues lining joints.

Cartilage and bone are connective tissues that provide rigid structural support. In **cartilage**, a network of collagen fibers is embedded in a rather flexible matrix consisting of a protein–carbohydrate complex called chondroitin sulfate. The cells that form cartilage are called *chondrocytes*, and they exist in small cavities in the cartilage (Figure 40.4*a*). Cartilage forms the entire skeletal system of sharks and rays, which are therefore called cartilaginous fishes.

Cartilage forms the skeletons of the early developmental stages of more complex vertebrates, but it is gradually replaced by **bone**, a harder connective tissue (Figure 40.4*b*). Adult vertebrates retain cartilage as the support for flexible structures such as external ears, noses, and the windpipe. The extracellular matrix in bone also contains many collagen fibers, but it is hardened by the deposition of the mineral calcium phosphate. We will discuss bone in greater detail in Chapter 44.

Adipose tissue is a form of loose connective tissue that includes adipose cells, which form and store droplets of lipids (Figure 40.4*c*). Adipose tissue is a major source of stored energy, but it also serves to cushion organs, and layers of adipose tissue under the skin can provide a barrier to heat loss.

Blood is a connective tissue consisting of cells dispersed in an extensive extracellular matrix: the blood plasma (Figure 40.4*d*). The blood plasma is much more liquid than the extracellular matrices of the other connective tissues, but it too contains an abundance of proteins. One of those proteins, *fibrinogen*, serves a structural function when it is stimulated to polymerize and form a blood clot. Many of

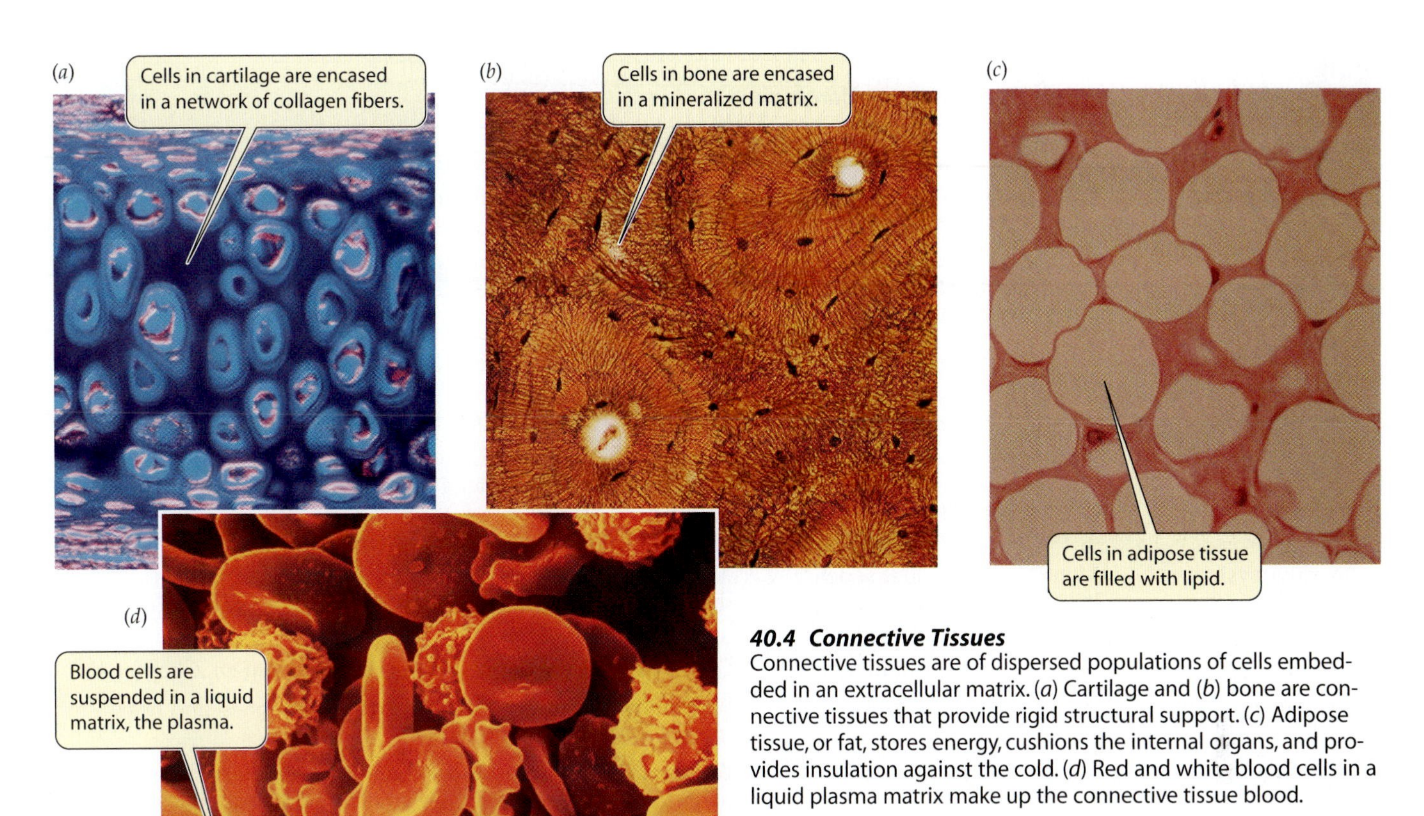

40.4 Connective Tissues
Connective tissues are of dispersed populations of cells embedded in an extracellular matrix. (*a*) Cartilage and (*b*) bone are connective tissues that provide rigid structural support. (*c*) Adipose tissue, or fat, stores energy, cushions the internal organs, and provides insulation against the cold. (*d*) Red and white blood cells in a liquid plasma matrix make up the connective tissue blood.

the proteins and cellular elements of the blood were presented in Chapter 19, and blood will be discussed again in Chapter 49.

Muscle tissues contract

Muscle tissues consist of elongated cells that can contract and cause movement. Muscle tissues are the most abundant tissues in the body, and they use most of the energy produced in the body. The contraction of muscle cells depends on intracellular protein filaments that can slide past each other. In Chapter 47, we will encounter three types of muscle tissues:

- **Skeletal muscle** connects bones to bones and is responsible for the body movements that constitute behavior.
- **Smooth muscle** is found in internal organs and is not under voluntary control; it performs functions such as moving food through the gut and constricting blood vessels.
- **Cardiac muscle** makes up the mass of the heart and pumps the blood.

Nervous tissues process information

There are two basic cell types in nervous tissues: neurons and glial cells. **Neurons**, which are extremely diverse in size and form, generate electrochemical signals. Responding to specific types of stimuli, such as light, sound, pressure, or certain molecules, neurons generate sudden voltage changes across their plasma membranes. These nerve impulses can be conducted via long extensions of the neurons to other parts of the body, where they are communicated to other neurons, muscle cells, or secretory cells. Neurons are involved in controlling the activities of most organ systems to achieve homeostasis.

Glial cells do not generate or conduct electric signals, but they provide a variety of supporting functions for neurons. There are more glial cells than neurons in our nervous systems. We will detail and illustrate the properties of nervous tissues in Chapters 44, 45, and 46.

Organs consist of multiple tissues

A discrete structure that carries out a specific function in the body is called an **organ**. Examples are the stomach, the heart, the liver, or the kidney. Most organs include all four tissue types. The wall of the stomach is a good example (see Figure 40.2). The inner surface of the stomach that contacts food is lined with a simple epithelium. Some of the epithelial cells secrete mucus, enzymes, or stomach acid.

Beneath the epithelial lining is connective tissue. Within this connective tissue are nerves, glands (secretory epithelial cells), and blood vessels. Concentric layers of muscle tissue enable the stomach to contract to mix food with the digestive juices. A network of neurons between the muscle layers controls these movements and also partially controls the secretions of the stomach. Surrounding the stomach is a layer of connective tissue called the serosa.

40.1 The Major Organ Systems of Mammals

SYSTEM	ORGANS	FUNCTIONS
Nervous system	Brain, spinal cord, sensory organs, peripheral nerves	Receives, integrates, stores information and controls muscles and glands. Chapters 44, 45, 46
Endocrine system	Glands: pituitary, thyroid, parathyroid, pineal, adrenal, testes, ovaries, pancreas	A system of glands releases chemical messages (hormones) that control and regulate other tissues and organs. Chapter 41
Muscle system	Skeletal muscle, smooth muscle, cardiac muscle	Produces forces and motion. Chapter 47
Skeletal system	Bones	Provides structural support for the body. Chapter 47
Reproductive system	Female: ovaries, oviducts, uterus, vagina, mammary glands Male: testes, sperm ducts, accessory glands, penis	Produces sex cells and hormones necessary to procreate and nurture offspring. Chapter 42
Digestive system	Mouth, esophagus, stomach, intestines, liver, pancreas, rectum, anus	Acquires and digests food, absorbs and stores nutrients, then makes them available to the cells of the body. Chapter 50
Gas exchange system	Airways, lungs, diaphragm	Exchanges respiratory gases with the environment. Chapter 48
Circulatory system	Heart and blood vessels	Transports respiratory gases, nutrients, hormones, and heat around the body. Chapter 49
Lymphatic system	Lymph and lymph vessels, lymph nodes, spleen	Brings extracellular fluids back into the circulatory system; helps the immune system fight invading organisms. Chapters 40 and 19
Immune system	Many types of white blood cells	Fights invading organisms and infections. Chapter 19
Skin system	Skin, sweat glands, hair	Protects the body from invading organisms and harsh physical conditions, helps regulate body temperature. Chapter 40
Excretory system	Kidneys, bladder, ureter, urethra	Regulates the composition of the extracellular fluids; excretes waste products. Chapter 51

An individual organ is usually part of an **organ system**—a group of organs that function together. The stomach is part of the digestive system, which also includes the food tube (esophagus), the small and large intestines, the pancreas, which secretes digestive enzymes, and the liver, which secretes bile. The major organ systems of mammals are outlined in Table 40.1.

Physiological Regulation and Homeostasis

Homeostasis depends on the ability to regulate the functions of the organs and organ systems to counteract influences that would change the physical or chemical composition of the internal environment. In this section we discuss the general properties of physiological regulatory systems, and then consider temperature regulation as a specific example.

Set points and feedback information are required for regulation

In addition to control mechanisms, regulation requires information. You can regulate the speed of a car only if you know the speed at which you are traveling and the speed you wish to maintain. The desired speed is a **set point**, and the reading on your speedometer is **feedback information**. When the set point and the feedback are compared, any difference between them is an *error signal*. Error signals suggest corrective actions, which you make by using the accelerator or brake (Figure 40.5).

Physiological regulation requires actions of cells, tissues, and organs, which are called *effectors* because they effect changes. Effectors are also referred to as **controlled systems** because their activities are controlled by commands coming from **regulatory systems**. Regulatory systems obtain, process, and integrate information and issue commands to the controlled systems. A fundamental way to analyze a regulatory system is to identify its source of feedback information.

Negative feedback is the most common type of feedback information in regulatory systems. The word "negative" indicates that this feedback information causes the effectors to reduce or reverse the process or influence that created the signal. In our car analogy, the recognition that you are going too fast is negative feedback if it causes you to slow down.

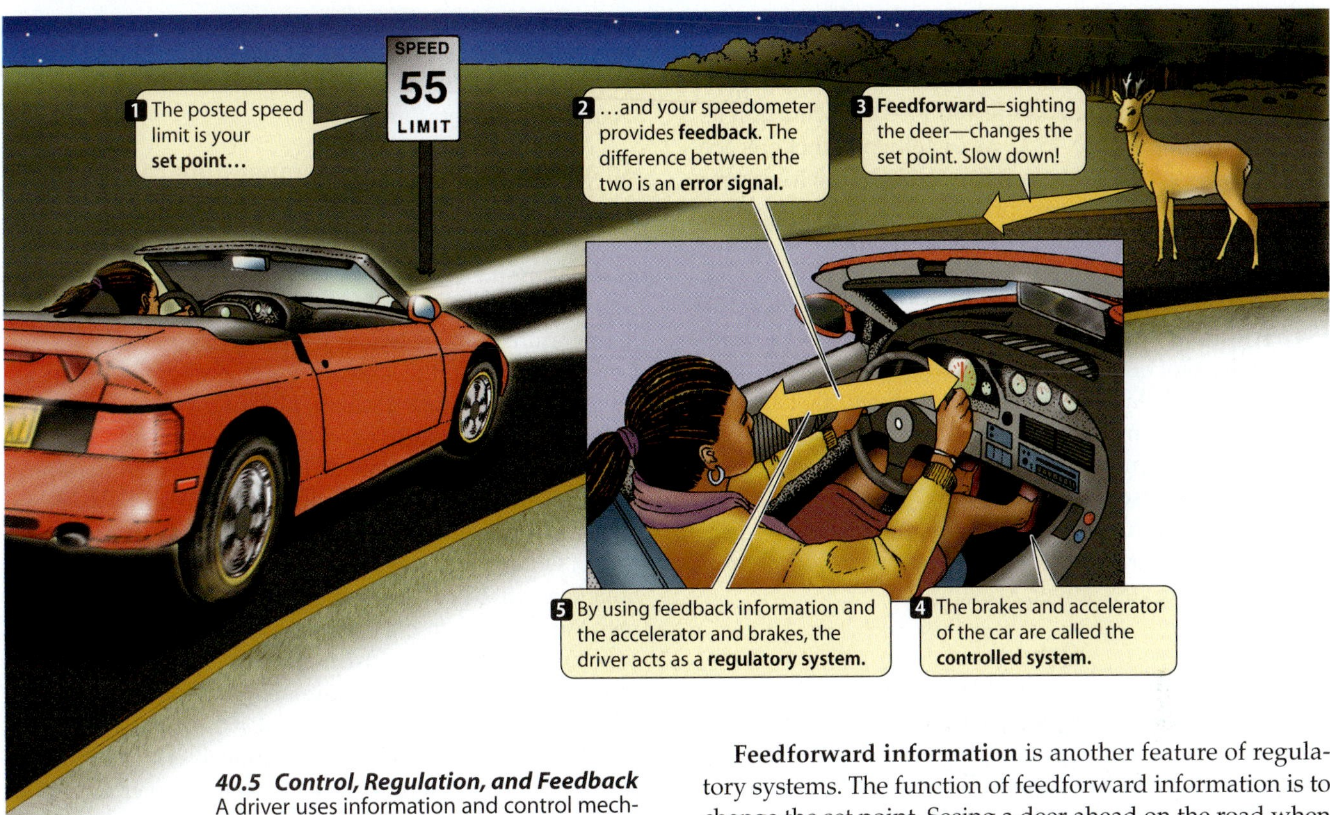

40.5 Control, Regulation, and Feedback
A driver uses information and control mechanisms to regulate the speed of the car.

Thermostats regulate temperature

The thermostat that is part of the heating–cooling system of a house is a regulatory system. It has upper and lower set points that you can adjust, and it receives feedback information from a sensor that measures room temperature. The circuitry of the thermostat converts differences between the set points and feedback information into signals that activate the controlled systems—the furnace and the air conditioner.

When room temperature rises above the upper set point, the thermostat activates the air conditioner to reduce the temperature; when room temperature falls below that upper set point, the air conditioner is turned off. If temperature falls below the lower set point, the furnace is activated, raising room temperature. The mechanism that senses room temperature provides negative feedback that is used to regulate both the air conditioner and the furnace. **Negative feedback** is a stabilizing influence in physiological regulatory systems. It contributes to homeostasis by stimulating actions that return a variable to its set point.

Is there any such thing as positive feedback in physiology? Although not as common as negative feedback, it does exist. Rather than returning a system to a set point, **positive feedback** amplifies a response. Examples of regulatory systems that use positive feedback are the responses that empty body cavities, such as urination, defecation, sneezing, and vomiting. Another example is sexual behavior, in which a little stimulation causes more behavior, which causes more stimulation, and so on.

Feedforward information is another feature of regulatory systems. The function of feedforward information is to change the set point. Seeing a deer ahead on the road when you are driving is an example of feedforward information (see Figure 40.5); this information takes precedence over the posted speed limit, and you change your set point to a slower speed. If you want the temperature of your house to be lower at night than during the day, you can add a clock to the thermostat to provide feedforward information about time of day.

These principles of control and regulation help organize our thinking about physiological systems. Once we understand how an organ or an organ system works, we can then ask how is it regulated. As an example, we will discuss in detail the system that regulates body temperature. But first, why is it necessary to regulate body temperature?

Temperature and Life

Over the face of Earth, temperatures vary enormously, from the boiling hot springs of Yellowstone National Park to the interior of Antarctica, where the temperature can fall below –80°C. Because heat always moves from a warmer to a cooler object, any change in the temperature of the environment causes a change in the temperature of an organism in that environment—unless the organism does something to regulate its temperature.

Living cells function over only a narrow range of temperatures. If cells cool to below 0°C, ice crystals damage their structures, possibly fatally. Some animals have adaptations such as antifreeze molecules in their blood that help them resist freezing; others have adaptations that enable them to survive freezing. Generally, however, cells must remain above 0°C to stay alive.

The upper temperature limit is less than 45°C for most cells. Some specialized algae can grow in hot springs at 70°C, and some archaea can live at near 100°C, but in general, proteins begin to denature and lose their function as temperatures approach 45°C. Most cellular functions are limited to the range between 0°C and 45°C, which are considered the thermal limits for life. A particular species, however, generally has much narrower limits.

Q_{10} is a measure of temperature sensitivity

Even within the range 0° to 45°C, temperature changes create problems for animals. Most physiological processes, like the biochemical reactions that constitute them, are temperature-sensitive, going faster at higher temperatures (see Figure 6.26). The temperature sensitivity of a reaction or process can be described in terms of $\mathbf{Q_{10}}$, a quotient calculated by dividing the rate of a process or reaction at a certain temperature, R_T, by the rate of that process or reaction at a temperature 10°C lower, R_{T-10}:

$$Q_{10} = \frac{R_T}{R_{T-10}}$$

Q_{10} can be measured for a simple enzymatic reaction or for a complex physiological process, such as the rate of oxygen consumption. If a reaction or process is not temperature-sensitive, it has a Q_{10} of 1. Most biological Q_{10} values are between 2 and 3, which means that reaction rates double or triple as temperature increases by 10°C (Figure 40.6).

Changes in temperature can be particularly disruptive to an animal's functioning because all the component reactions in the animal do not have the same Q_{10}. Individual reactions with different Q_{10}'s are linked together in complex networks that carry out physiological processes. Changes in temperature shift the rates of some reactions more than those of others, thus disrupting the balance and integration that the processes require. For homeostasis, organisms must be able to compensate for or prevent changes in temperature.

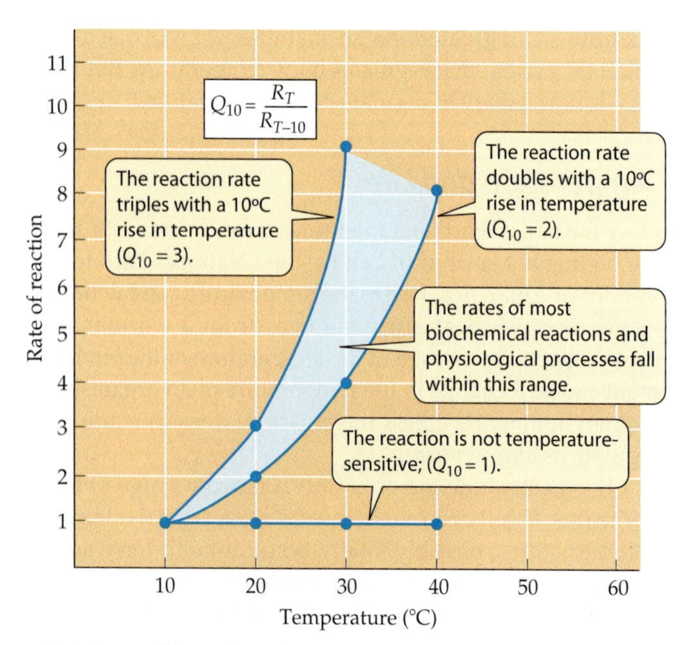

40.6 Q_{10} and Reaction Rate
The larger the Q_{10}, the faster the reaction rate rises in response to an increase in temperature.

An animal's sensitivity to temperature can change

The body temperature of some animals is tightly coupled to the environmental temperature. Think of a fish in a temperate-zone pond. As the temperature of the pond changes from 4°C in midwinter to 24°C in midsummer, the body temperature of the fish does the same (Figure 40.7). We can bring the fish into the laboratory in the summer and measure its **metabolic rate** (the sum total of the energy turnover of its cells, often measured by O_2 consumption). If we measure the metabolic rate at different water temperatures, we might plot our data as shown by the red line in Figure 40.7 and calculate a Q_{10} of 2. We predict from our graph that in winter, when the temperature is 4°C, the fish's metabolic rate will be only one-fourth of what it was in the summer. We then return the fish to its pond.

When we bring the fish back to the laboratory in the winter and repeat the measurements, we find, as the blue line shows, that its metabolic rate at 4°C is not as low as we predicted; rather, it is almost the same as it was at 24°C in the summer. If we repeat the measurement over a range of temperatures, we find that the fish's metabolic rate is always higher than the rate we predicted from the measurement we took at the same temperature in the summer. This difference is due to **acclimatization**, the process of physiological and biochemical change that an animal undergoes in response to seasonal changes in climate.

Seasonal acclimatization in the fish has produced **metabolic compensation**, which readjusts the biochemical machinery to counter the effects of temperature. What might account for such a change? Look again at Figure 6.26, which shows the different optimal temperatures of enzymes. If the fish can express similar enzymes that operate at different optimal temperatures, it can compensate metabolically by catalyzing reactions with one set of enzymes in summer and another set in winter. The end result is that metabolic functions are much less sensitive to long-term changes in temperature than they are to short-term thermal fluctuations.

Maintaining Optimal Body Temperature

Animals can be classified by how they respond to environmental temperatures.

- A **homeotherm** is an animal that maintains a constant body temperature.
- A **poikilotherm** is an animal whose body temperature changes when the temperature of its environment (the *ambient temperature*) changes.

This system of classification says something about the biology of the animals, but it presents problems. Should a fish in the deep ocean, where the temperature changes very little, be called a homeotherm? Should a hibernating mam-

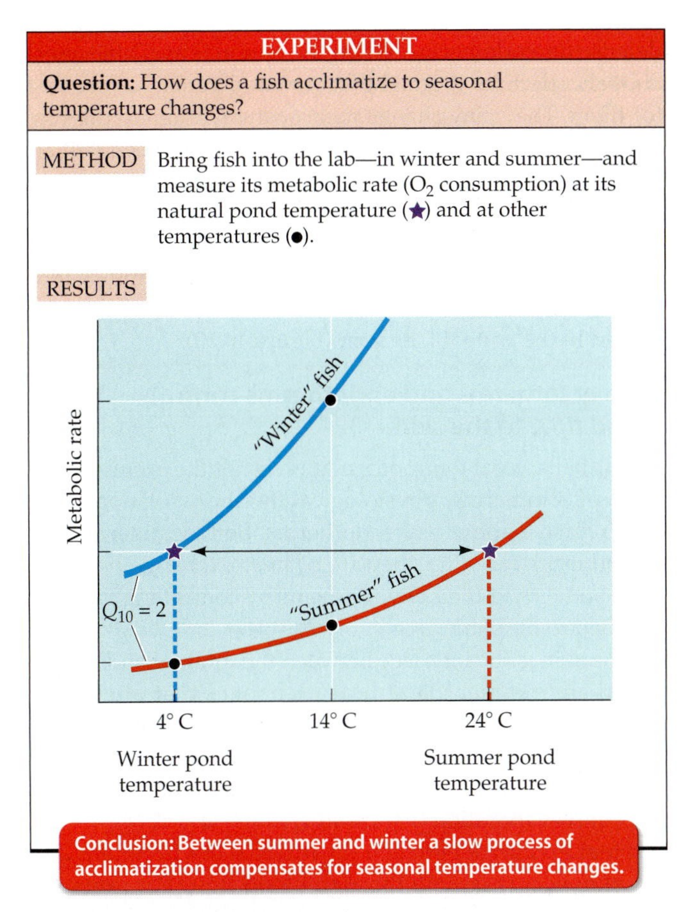

40.7 Metabolic Compensation
In its natural environment, a fish's metabolism readjusts, or acclimatizes, to compensate for seasonal changes in temperature.

mal that allows its body temperature to drop to nearly the temperature of its environment be called a poikilotherm? The problem posed by the hibernator has been solved by creating a third category: the **heterotherm**, an animal that regulates its body temperature at a constant level *some* of the time.

Another set of terms classifies animals on the basis of the sources of heat that determine their body temperatures.

- **Ectotherms** depend largely on external sources of heat, such as solar radiation, to maintain their body temperatures above the environmental temperature.
- **Endotherms** can regulate their body temperatures by producing heat metabolically or by mobilizing active mechanisms of heat loss.

Mammals and birds are endotherms; animals of all other species behave as ectotherms most of the time.

Ectotherms and endotherms respond differently in metabolic chambers

A small lizard is an example of an ectotherm. We can compare it with a mouse, which is an endotherm of the same body size. We can put each animal in a metabolic chamber and measure body temperatures and metabolic rates as we change the temperature of the chamber from 0°C to 35°C.

The results obtained from the two species differ. The body temperature of the lizard equilibrates with that of the chamber, whereas the body temperature of the mouse remains at 37°C (Figure 40.8*a*). The metabolic rate of the lizard decreases as the temperature decreases (Figure 40.8*b*). In contrast, the mouse's metabolic rate increases as chamber temperature falls below about 27°C (notice that you must read the graph right to left to see this). The lizard apparently cannot regulate its body temperature or metabolism independently of environmental temperature. The mouse, however, regulates its body temperature by in-

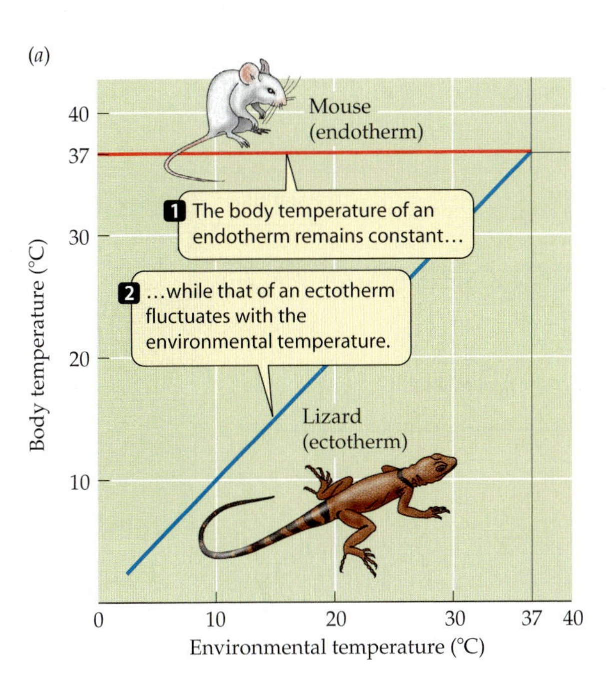

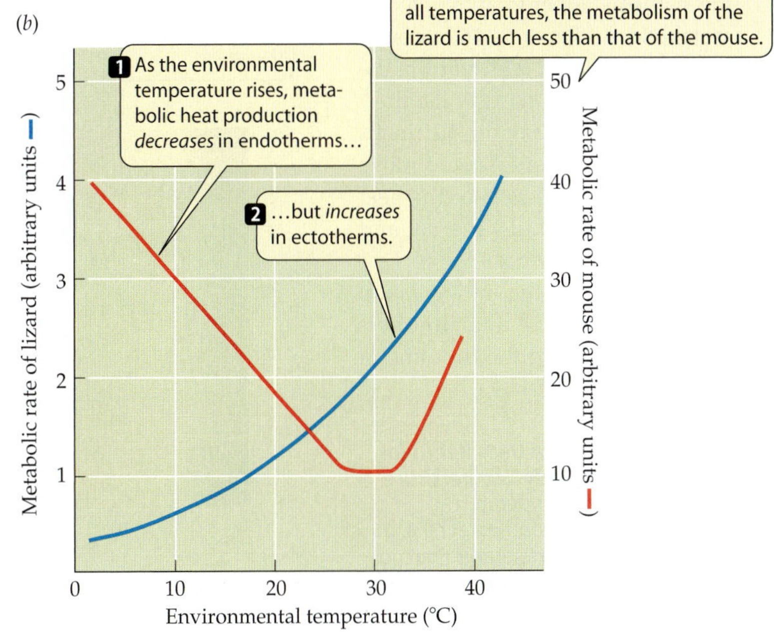

40.8 Ectotherms and Endotherms
The body temperatures of a lizard and a mouse of the same body size respond differently to changes in environmental temperature.

creasing its metabolic rate, which increases its production of body heat.

Ectotherms and endotherms use behavior to regulate body temperature

We can test our laboratory conclusion that the lizard cannot regulate its body temperature. To do this, we release the lizard in its desert habitat and measure its temperature as it goes about its normal behavior in this environment, where temperature can change 40°C in a few hours (Figure 40.9).

Unlike what we observed in the metabolic chamber, the body temperature of the lizard is at times considerably different from the environmental temperature. At night, the temperature in the desert may drop close to freezing, but the temperature of the lizard remains stable at 16°C. No mystery here: The lizard spends the night in a burrow, where the soil temperature is a constant 16°C.

Early in the morning, soon after sunrise, the lizard emerges from its burrow. The air temperature is still cool, but the lizard's body temperature rises above 30°C in less than 30 minutes. The lizard achieves this by basking on a rock with maximum exposure to the sun. As its skin absorbs solar radiation, its body temperature rises considerably above the air temperature. By altering its exposure to the sun, the lizard maintains its body temperature around 35°C all morning.

By noon, the air temperature near the surface of the desert has risen to 50°C, but the lizard maintains its body temperature around 35°C by staying mostly in the shade and frequently in the branches of bushes, where there is a cooling breeze. As afternoon progresses, the air cools, and the lizard again spends more of its time in the sun and on hot rocks to maintain its body temperature around 35°C. The lizard returns to its burrow just before sunset, and its body temperature rapidly drops to 16°C.

Our conclusion must be that the lizard can regulate its body temperature quite well by behavioral mechanisms rather than by internal metabolic mechanisms. In our laboratory experiment, the lizard in the chamber could not use its thermoregulatory behavior, but in its natural environment it could move to different places to alter the heat exchange between its internal and external environments.

But behavioral thermoregulation is not the exclusive domain of ectotherms. It is also the first line of defense for endotherms. When the option is available, most animals select the thermal microenvironments that are best for them. They may change their posture, orient to the sun, move between sun and shade, and move between still air and moving air, as demonstrated by the lizard in our field experiment. Examples of more complex thermoregulatory behavior are nest construction and social behavior such as huddling. Humans select appropriate clothing and heat or cool their buildings. Behavioral thermoregulation is widespread in the animal kingdom (Figure 40.10).

Both ectotherms and endotherms control blood flow to the skin

Just as behavioral thermoregulation is not the exclusive domain of ectotherms, physiological thermoregulation is not the exclusive domain of endotherms. Both ectotherms and endotherms can alter the rate of heat exchange between their bodies and their environments by controlling the flow of blood to the skin.

The skin is the interface between the internal and the external environment, and heat exchanges that alter body temperature occur across this interface. Heat exchange between the skin and the external environment occurs through four mechanisms: radiation, conduction, convection, and evaporation (Figure 40.11).

Heat exchange between the internal environment and the skin occurs largely through blood flow. For example, when a person's body temperature rises as a result of exercise, blood flow to the skin increases, and the skin surface becomes quite warm. The heat brought to the skin by the blood is lost to the environment, and this loss helps to bring the body temperature back to normal. In contrast, when a person is exposed to cold, the blood vessels supplying the skin constrict, decreasing blood flow and heat transport to the skin and reducing heat loss to the environment.

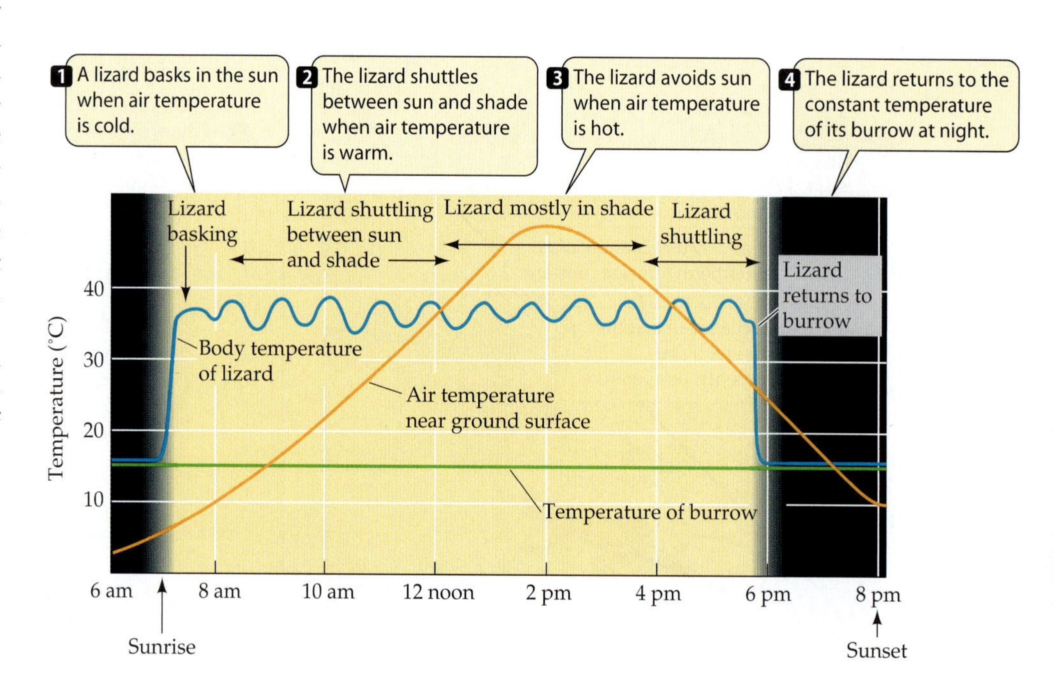

40.9 An Ectotherm Uses Behavior to Regulate Its Body Temperature
The lizard's body temperature is dependent on environmental heat, but it can regulate its temperature by moving between different environments.

40.10 Endotherms Can Use Behavior to Thermoregulate
(*a*) Humans must put on many layers of insulating clothing to help their thermoregulatory mechanisms keep pace with the extreme cold of western Siberia. (*b*) When air temperatures on the African savanna soar, an elephant may use a cool shower to thermoregulate.

The control of blood flow to the skin can be an important adaptation for an ectotherm like the marine iguana of the Galápagos archipelago. The Galápagos are volcanic islands that lie on the equator, but they are bathed by very cold oceanic currents. Marine iguanas are reptiles that bask on black lava rocks on shore and swim in the cold ocean, where they feed on algae. When the iguanas are feeding, they cool to the temperature of the sea, which makes them slower and more vulnerable to predators, and probably incapable of efficient digestion. They therefore alternate between feeding in the cold sea and basking in the sun on the hot rocks. It is advantageous for iguanas to retain body heat as long as possible while swimming and to warm up as fast as possible when basking. They adjust by changing their heart rate, and therefore their blood flow (Figure 40.12).

Some ectotherms produce heat

Some ectotherms raise their body temperatures by producing heat. For example, the powerful flight muscles of many insects must reach 35°–40°C before the insects can fly, and they must maintain these high temperatures during flight, even at air temperatures around 0°C. Such insects produce the required heat by contracting their flight muscles in a manner analogous to shivering in mammals (Figure 40.13). The heat-producing ability of these insects can be quite remarkable. Probably the most impressive case is a species of scarab beetle that lives mostly underground in mountains north of Los Angeles, California. To mate, these beetles come above ground, and males fly in search of females. They un-

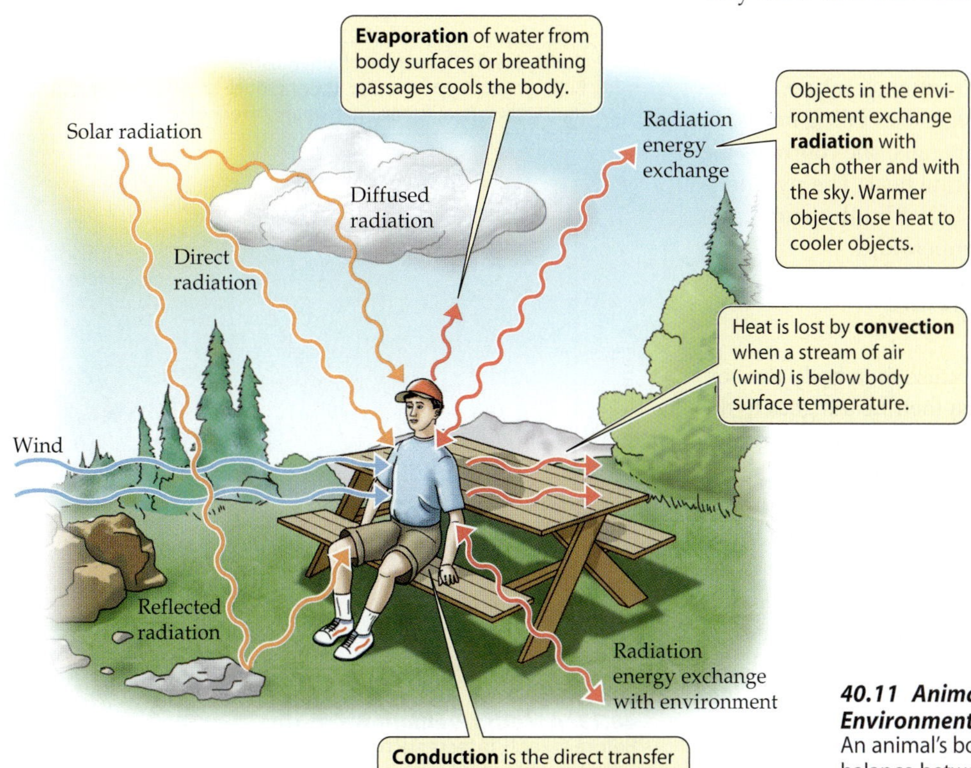

40.11 Animals Exchange Heat with the Environment
An animal's body temperature is determined by the balance between internal heat production and the avenues of heat exchange with the environment: radiation, conduction, convection, and evaporation.

40.12 Some Ectotherms Regulate Blood Flow to the Skin
Galápagos marine iguanas control blood flow to the skin to alter their heating and cooling rates.

Amblyrhynchus cristatus

1 As soon as the iguana enters the ocean, it begins to cool.

2 When the iguana leaves the ocean to bask on the shore, it begins to warm. Notice that the rate of warming is greater than the rate of cooling.

3 The iguana's heart rate drops rapidly when it enters the ocean.

4 The iguana's heart rate rises rapidly when it leaves the ocean to bask on the shore.

5 At the same body temperature, the heart rate is lower during cooling than during warming.

Body temperature
Heart rate
Body temperature (°C)
Heart rate (b/min)
Time (min)

dertake this mating ritual at night, in winter, and only during snowstorms.

Honeybees regulate temperature as a group. They live in large colonies consisting mostly of female worker bees that maintain the hive and rear the offspring of the single queen bee. During winter, honeybee workers combine their individual heat-producing abilities to regulate the temperature of the brood. They cluster around the brood and adjust their joint metabolic heat production and density of clustering so that the brood temperature remains remarkably constant, at about 34°C, even as the outside air temperature drops below freezing.

Some reptiles use metabolic heat production to raise their body temperatures above the air temperature. The female Indian python protects her eggs by coiling her body around them. If the air temperature falls, she contracts the muscles of her body wall to generate heat. The python is able to maintain the temperature of her body—and therefore that of her eggs—above air temperature.

Some fish elevate body temperature by conserving metabolic heat

It is particularly difficult for fish to raise their body temperatures because blood pumped from their hearts goes to the gills to pick up oxygen, where it comes into close contact with cool water flowing over the thin gill membranes. The blood equilibrates with the outside water temperature before it travels through the rest of the body. Any heat transferred to the blood from active muscles is lost rapidly to the environment when the blood flows through the gills. It is thus surprising to find that some large, rapidly swimming fishes, such as bluefin tuna and great white sharks, can maintain temperature differences as great as 10°–15°C between their bodies and the surrounding water. The heat comes from their powerful swimming muscles, and the ability of these "hot" fish to conserve that heat is due to remarkable arrangements of their blood vessels.

In the usual ("cold") fish circulatory system, oxygenated blood from the gills collects in a large dorsal vessel, the aorta, which travels through the center of the fish, distributing blood to all organs and muscles (Figure 40.14*a*). "Hot" fish have a smaller central dorsal aorta. Most of their oxygenated blood is transported in large vessels just under the skin (Figure 40.14*b*). Hence the cold blood from the gills is kept close to the surface of the fish. Smaller vessels trans-

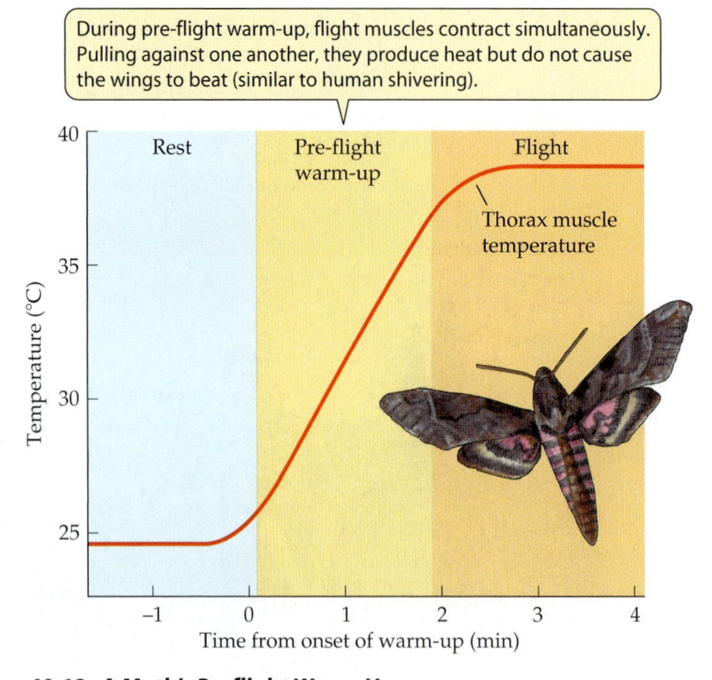

40.13 A Moth's Preflight Warm-Up
Before takeoff, insects such as the sphinx moth contract the flight muscles in the thorax to generate heat and warm the muscles up to the temperature required for flight. This mechanism enables the moth to fly even at night, when the environment is cool.

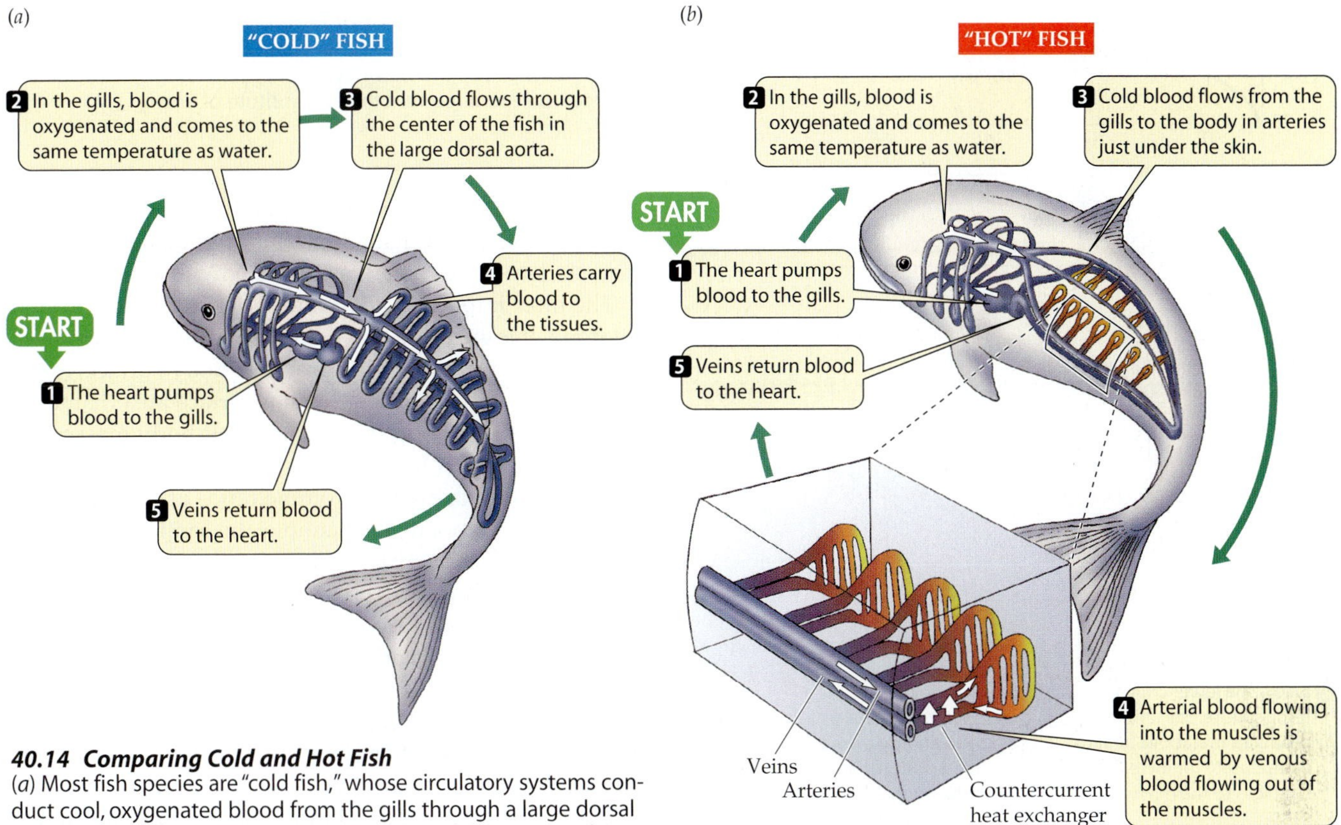

40.14 Comparing Cold and Hot Fish
(*a*) Most fish species are "cold fish," whose circulatory systems conduct cool, oxygenated blood from the gills through a large dorsal aorta to the rest of the body. (*b*) The blood vessel anatomy of "hot" fish species allows for heat exchange between the cold arterial blood entering the muscles and the departing venous blood, which has been warmed by the metabolism of the muscles.

porting this cold blood into the muscle mass run parallel to the vessels transporting warm blood from the muscle mass back toward the heart. Since the vessels carrying the cold blood into the muscle are in close contact with the vessels carrying warm blood away, heat flows from the warm to the cold blood and is therefore trapped in the muscle mass.

Because heat is exchanged between blood vessels carrying blood in opposite directions, this adaptation is called a **countercurrent heat exchanger**. It keeps the heat within the muscle mass, enabling the fish to have an internal body temperature considerably above the water temperature. Why is it advantageous for the fish to be warm? Each 10°C rise in muscle temperature increases the fish's sustainable power output almost threefold!

Thermoregulation in Endotherms

As we saw in Figure 40.8, endotherms respond to changes in environmental temperature by changing their rates of heat production, measured as metabolic rate. Within a narrow range of environmental temperatures called the **thermoneutral zone**, the metabolic rate of endotherms is low and independent of temperature. The metabolic rate of a resting animal at a temperature within the thermoneutral zone is called the **basal metabolic rate**. It is usually measured on animals that are quiet but awake, and that are not using energy for digestion, reproduction, or growth. A resting animal consumes energy at the basal metabolic rate just to carry out all of its minimal body functions.

The basal metabolic rate of an endotherm is about six times greater than the metabolic rate of an ectotherm of the same size and at the same body temperature (see Figure 40.8*b*). A gram of mouse tissue consumes energy at a much higher rate than does a gram of lizard tissue when both tissues are at 37°C. This difference results from basic changes in cell metabolism that accompanied the evolution of endotherms from their ectothermic ancestors.

Endotherms actively increase heat production or heat loss

The thermoneutral zone is bounded by a *lower critical temperature* and an *upper critical temperature*. Figure 40.15 describes the thermoregulatory responses of a mammal at temperatures ouside its thermoneutral zone.

When the environmental temperature falls below the lower critical temperature, mammals can create heat for thermoregulation through shivering and nonshivering heat production. Birds use only shivering heat production. **Shivering** uses the contractile machinery of skeletal muscles to consume ATP without causing observable behavior. The muscles pull against each other so that little movement other than a tremor results. All the energy from the conversion of ATP to ADP in this process is released as heat. Shivering heat production is perhaps too narrow a term; energy

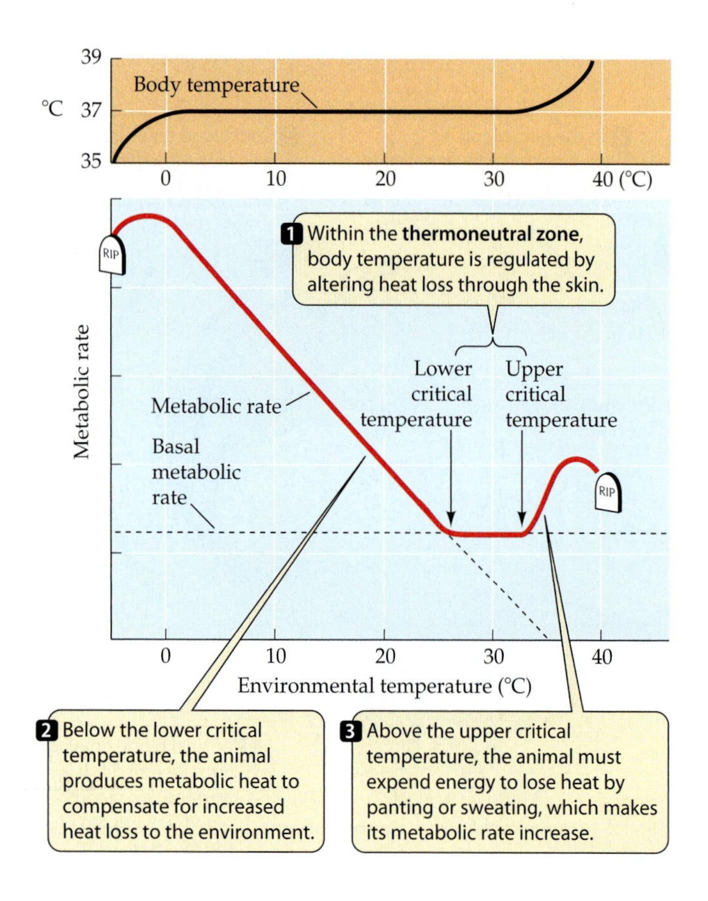

40.15 Environmental Temperature and Mammalian Metabolic Rates
Outside the thermoneutral zone, maintaining a constant body temperature requires the expenditure of energy.

expenditure due to increased muscle tone and increased body movements also contributes to increased heat production in the cold.

Most nonshivering heat production occurs in specialized adipose tissue called **brown fat** (Figure 40.16). This tissue looks brown because of its abundant mitochondria and rich blood supply. In brown fat cells, a protein called *thermogenin* uncouples proton movement from ATP production, allowing protons to leak across the inner mitochondrial membrane rather than having to pass through the ATP synthase protein and generate ATP (review the discussion of respiration in Chapter 7). As a result, metabolic fuels are consumed without producing ATP, but heat is still released. Brown fat is especially abundant in newborn infants of many mammalian species, in some adult mammals that are small and acclimatized to cold, and in mammals that hibernate.

Decreasing heat loss is important for life in the cold

The coldest habitats on Earth are in the Arctic, the Antarctic, and at the peaks of high mountains. Many birds and mammals, but almost no reptiles or amphibians, live in very cold habitats. What adaptations besides endothermy characterize species that live in the cold?

The most important adaptations of endotherms to cold environments are those that reduce heat loss to the environment. Since most heat is lost from the body surface, many cold-climate species have a smaller surface area than their warm-climate cousins, even when their body masses are the same. Rounder body shapes and shorter appendages reduce the surface area–to–volume ratios of some cold-climate species; compare, for example, the San Joaquin kit fox and the arctic fox (Figure 40.17).

Another means of decreasing heat loss is to increase thermal insulation. Animals adapted to cold have much thicker layers of fur, feathers, or fat than do their warm-climate relatives. The fur of an arctic fox or a northern sled dog provides such good thermal insulation that those animals don't even begin to shiver until the air temperature drops as low as –20°C to –30°C.

Fur and feathers are good insulators because they trap a layer of still, warm air close to the skin surface. If that air is displaced by water, insulation is drastically reduced. In many species, oil secretions spread through fur or feathers by grooming are critical for resisting wetting and maintain-

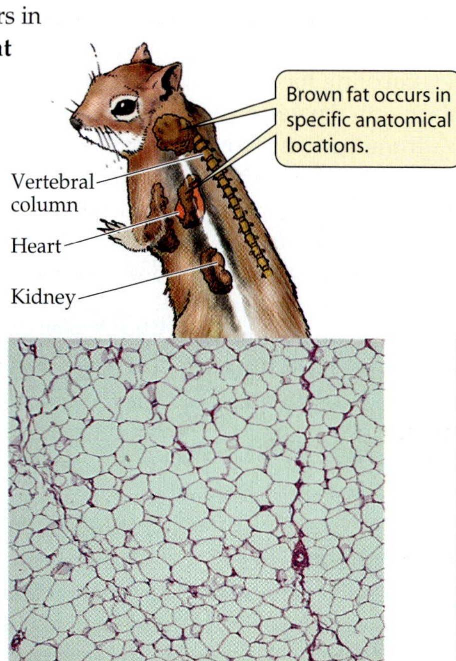

White fat viewed through a light microscope. Each cell is filled with a globule of lipid and has few organelles. The tissue has few blood vessels.

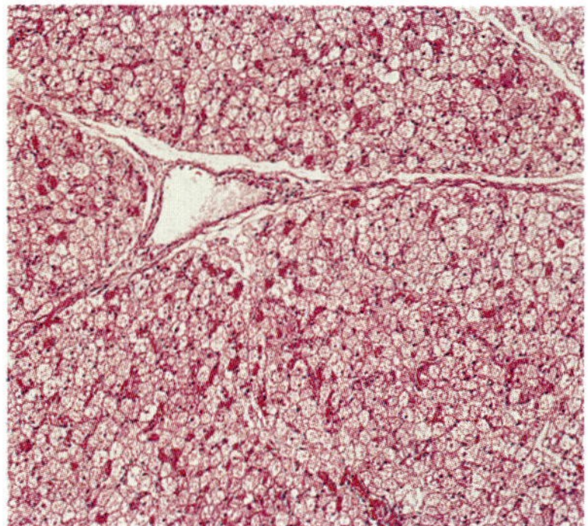

Brown fat viewed through a light microscope at the same magnification reveals cells with many intracellular structures and multiple droplets of lipid.

40.16 Brown Fat
In many mammals, specialized brown fat tissue produces heat.

(*a*) *Vulpes macrotis*

(*b*) *Alopex lagopus*

40.17 Adaptations to Hot and Cold Climates
(*a*) The San Joaquin kit fox, a desert dweller, has a large surface area for its body. The large ears serve as heat exchangers, passing heat from the fox's blood to the surrounding air. (*b*) The thick fur of the arctic fox provides insulation in the frigid winter. Its ears and extremities are relatively smaller than those of the kit fox.

ing a high level of insulation.

Humans change their thermal insulation by putting on or taking off clothes. How do other animals do it? We have already discussed one example, the ectothermic marine iguana, which controls blood flow to its skin. Increasing or decreasing blood flow to the skin is an important thermoregulatory adaptation for endotherms as well. In a hot environment, your skin feels hot because there is a high rate of blood flow through it, but when you sit in an overly airconditioned theater, your hands, feet, and other body surfaces feel cold as blood flow to those areas decreases.

Evaporation of water is an effective way to lose heat

For highly insulated arctic animals, and for many large mammals in all climates, getting rid of excess heat can be a serious problem, especially during exercise. Arctic species usually have an area on the body surface, such as the abdomen, that has only a thin layer of fur and can act as a window for heat loss. Large mammals, such as elephants, rhinoceroses, and water buffalo, have little or no fur and seek places where they can wallow in water when the air temperature is too high. Having water in contact with the skin greatly increases heat loss because water has a much greater capacity for absorbing heat than air does.

A gram of water absorbs about 580 calories of heat when it evaporates. Water is heavy, however, so animals do not carry an excess supply of it. Furthermore, hot environments tend to be arid places where water is a scarce resource. Therefore, evaporation of water by sweating or panting is usually a last resort for animals adapted to hot environments (recall the camels at the beginning of this chapter).

Sweating and panting are active processes that require the expenditure of metabolic energy. That's why the metabolic rate increases when the upper critical temperature is exceeded (see Figure 40.15). A sweating or panting animal is producing heat in the process of dissipating heat, which can be a losing battle. Animals can survive in environments that are below their lower critical temperature much better than they can in environments above their upper critical temperature.

The Vertebrate Thermostat

The thermoregulatory mechanisms and adaptations we have discussed are the controlled systems for the regulation of body temperature. These controlled systems must receive commands from a regulatory system that integrates information relevant to the regulation of body temperature. Such a regulatory system can be thought of as a *thermostat*. All animals that thermoregulate, both vertebrate and invertebrate, must have regulatory systems, but here we will focus on the vertebrate thermostat.

Where is the vertebrate thermostat? Its major integrative center is at the bottom of the brain in a structure called the **hypothalamus**. If you slide your tongue back as far as possible along the roof of your mouth, it will be just a few centimeters below your hypothalamus. The hypothalamus is a part of many regulatory systems, so we will refer to it again in the chapters to come. If the hypothalamus of a mammal's brain is damaged, the animal loses its ability to regulate its body temperature, which then rises in warm environments and falls in cold ones.

The vertebrate thermostat uses feedback information

What information does the vertebrate thermostat use? In many species, the temperature of the hypothalamus itself is the major source of feedback information to the thermostat. Cooling the hypothalamus causes fish and reptiles to seek a warmer environment, and heating the hypothalamus causes them to seek a cooler environment. In mammals, cooling the hypothalamus can stimulate constriction of the blood vessels supplying the skin and increase metabolic heat production. Because it activates these thermoregulatory responses, cooling the hypothalamus causes the body temperature to rise. Conversely, warming the hypothalamus stimulates dilation of blood vessels supplying the skin and sweating or panting, and the overall body temperature falls (Figure 40.18).

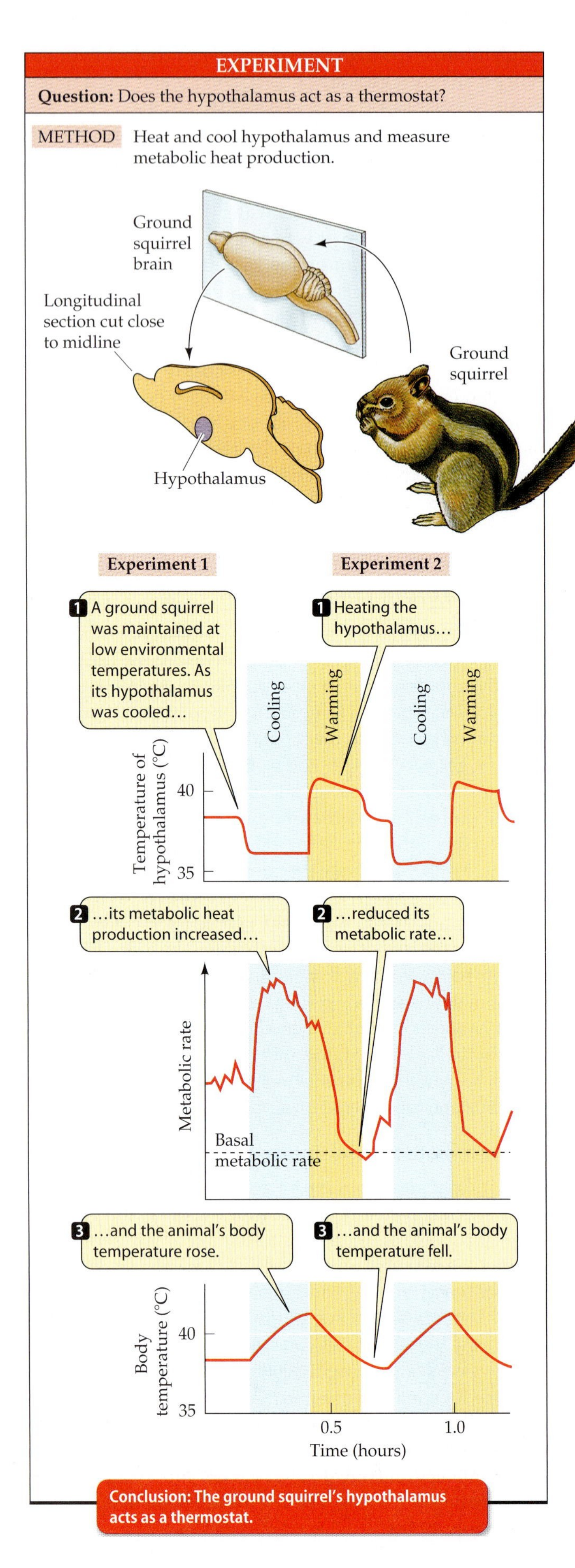

The hypothalamus appears to generate a set point like a setting on the thermostat of a house. When the temperature of the hypothalamus exceeds or drops below that set point, thermoregulatory responses (the controlled system) are activated to reverse the direction of temperature change. Hence, hypothalamic temperature is a negative feedback signal.

Heating and cooling the hypothalamus show that an animal has separate set points for activating different thermoregulatory responses. If the hypothalamus of a mammal is cooled, the vessels supplying blood to the skin constrict at a specific hypothalamic temperature. A slightly lower hypothalamic temperature initiates shivering. If the hypothalamic temperature is then raised, shivering ceases; then blood vessels supplying the skin dilate; and at still higher hypothalamic temperatures, panting starts.

We can describe the characteristics of hypothalamic control of each thermoregulatory response. For example, if we measure metabolic heat production while heating and cooling the hypothalamus (see Figure 40.18), we can describe the results graphically (Figure 40.19). Within a certain range of hypothalamic temperatures, metabolic heat production remains low and constant, but cooling the hypothalamus below a certain level—a set point—stimulates increased metabolic heat production. The increase in heat production is proportional to how much the hypothalamus is cooled below the set point. This regulatory system is much more sophisticated than a simple on–off thermostat like the one in a house.

The vertebrate thermoregulatory system integrates other sources of information in addition to hypothalamic temperature. It uses information about the temperature of the environment as registered by temperature sensors in the skin. Changes in environmental temperature shift the hypothalamic set points for thermoregulatory responses. As Figure 40.19 shows, in a warm environment you might have to cool the hypothalamus of a mammal to stimulate it to shiver, but in a cold environment you would have to warm the hypothalamus of the same animal to stop it from shivering. The set point for the metabolic heat production response is higher when the skin is cold and lower when the skin is warm.

The temperature of the skin can be considered feedforward information that adjusts the hypothalamic set point. Many other factors also shift hypothalamic set points for responses. Set points are higher during wakefulness than during sleep, and they are higher during the active part of the daily cycle than during the inactive part, even if the animal is awake at both times.

40.18 The Hypothalamus Regulates Body Temperature
The observation that damage to the hypothalamus disrupts thermoregulation led to the finding that hypothalamus acts as a thermostat in the vertebrate body.

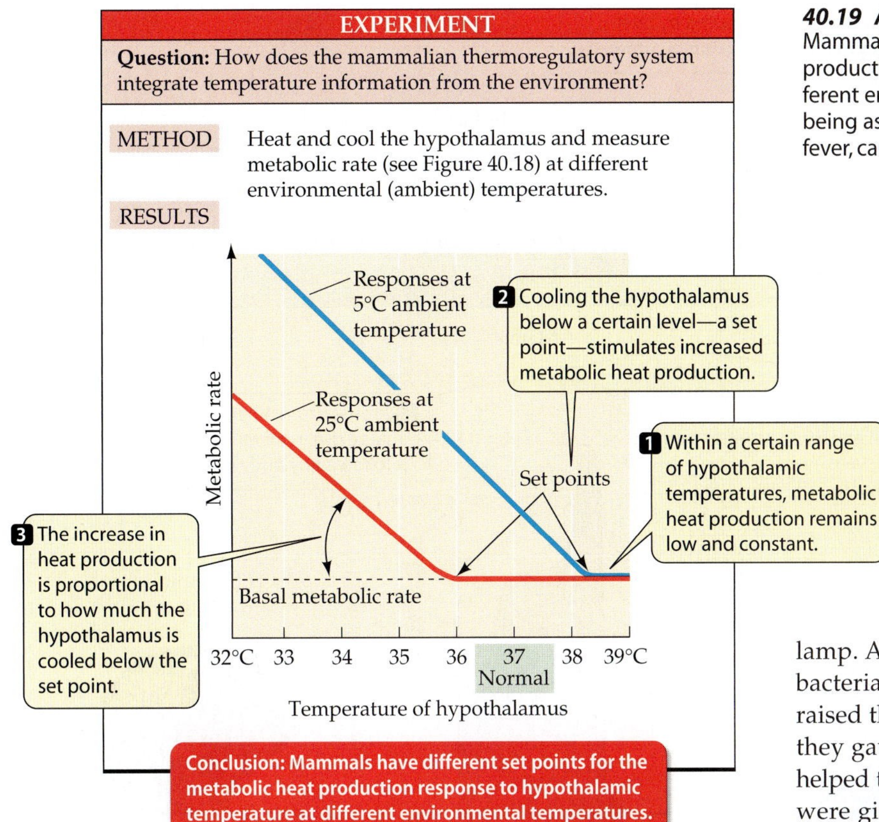

40.19 Adjustable Set Points
Mammals have different set points for the metabolic heat production response to hypothalamic temperature at different environmental temperatures. Other factors, such as being asleep or awake, the time of day, or the presence of a fever, can also affect the set point.

Fevers help the body fight infections

A **fever** is a rise in body temperature in response to substances called *pyrogens* that are derived from bacteria or viruses that invade the body. We respond to many infectious illnesses by getting a fever. Growing evidence suggests that fevers are adaptive responses that help the body fight disease-causing organisms.

The presence of a pyrogen in the body causes a rise in the hypothalamic set point for the heat production response. As a result, you shiver, put on a sweater, or crawl under a blanket, and your body temperature rises until it matches the new set point. At the higher body temperature you no longer feel cold, and you may not feel hot, but someone touching your forehead will say that you are "burning up." If you take an aspirin, it lowers your set point to normal. Now you feel hot, take off clothes, and even sweat until your elevated body temperature returns to normal.

Why do we take aspirin for fevers and "feeling crummy"? The pyrogens entering the body are attacked by cells of the immune system called macrophages (see Chapter 19). One of the things the macrophages do is to release chemicals called *interleukins*, which sound the alarm to other cells of the immune system throughout the body and trigger responses that contribute to feeling crummy. The interleukins also raise the hypothalamic set point for metabolic heat production. Among the intracellular signals triggered by interleukins are *prostaglandins*. Aspirin is a potent inhibitor of prostaglandin synthesis, thus explaining how this miracle drug reduces fever and makes us feel better.

Evidence suggests that moderate fevers help the body fight an infection. Some interesting studies were done on lizards that were given access to a heat lamp. These animals kept their body temperatures at about 38°C by adjusting their position with respect to the lamp. After they were injected with disease-causing bacteria, they spent more time close to the lamp and raised their body temperatures to 40°C and higher—they gave themselves fevers. To find out if the fever helped the lizards fight the bacteria, groups of lizards were given equal inoculations of bacteria, but were then placed in different incubators at 34°, 36°, 38°, 40°, and 42°C, respectively. All of the lizards at 34°C and 36°C died, about 25 percent at 38°C survived, and about 75 percent at 40°C and 42°C survived. Apparently fever helped the lizards fight the disease organisms.

However, extreme fevers (for example, 40°C) can be dangerous to humans and must be reduced. Even more modest fevers can be dangerous to people who have weakened hearts or those who are seriously ill. A fetus can be endangered when a pregnant woman has a fever. Fever-reducing drugs may be important in such cases.

Animals can save energy by turning down the thermostat

Hypothermia is the condition in which body temperature is below normal. It can result from a natural turning down of the thermostat, or from traumatic events such as starvation (lack of metabolic fuel), exposure, serious illness, or treatment by anesthesia. Many species of birds and mammals use regulated hypothermia as a means of surviving periods of cold and food scarcity. Some become hypothermic on a daily basis. Hummingbirds, for example, are very small endotherms and have a high metabolic rate. They could exhaust their metabolic reserves just getting through a single day without food. Hummingbirds and other small endotherms can extend the period over which they can survive without food by dropping their body temperature during the portion of day when they would normally be inactive. This adaptive hypothermia is called **daily torpor**. Body temperature can drop 10°–20°C during daily torpor, resulting in an enormous saving of metabolic energy.

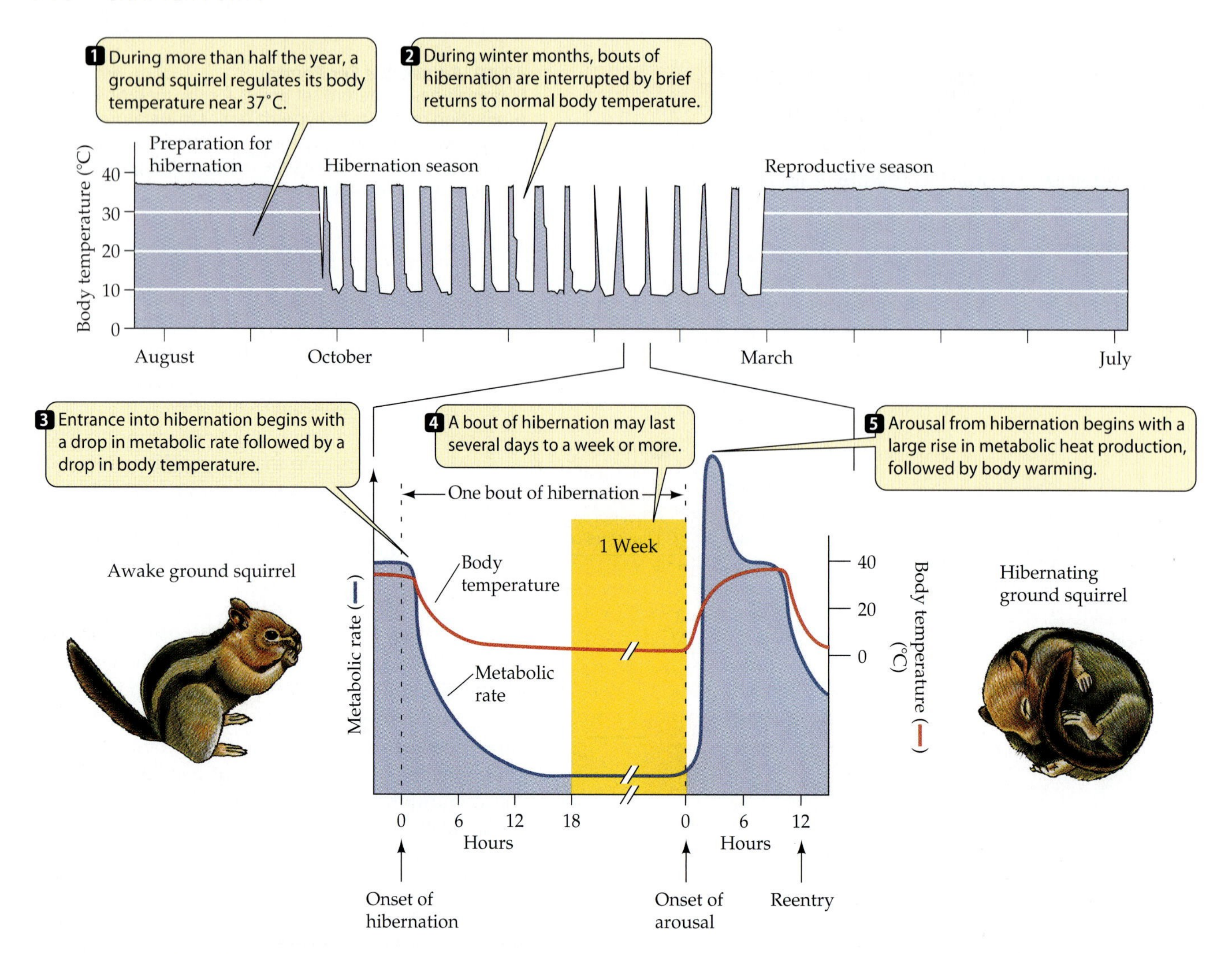

40.20 A Ground Squirrel Enters Repeated Bouts of Hibernation during Winter
At the beginning of each bout of hibernation, the ground squirrel's metabolic rate and body temperature fall. Its body temperature may come into equilibrium with the temperature of its nest and stay at that level for days. The bout is ended by a rise in metabolic heat production that returns body temperature to a normal level.

Regulated hypothermia can also last for days or even weeks, with drops to very low temperatures; this phenomenon is called **hibernation** (Figure 40.20). During the deep sleep of hibernation, the body's thermostat is turned down to an extremely low level to maximize energy conservation. Arousal from hibernation occurs when the hypothalamic set point returns to a normal level.

Many hibernators maintain body temperatures close to the freezing point during hibernation. The metabolic rate needed to sustain an animal in hibernation may be only one-fiftieth its basal metabolic rate, an enormous saving of metabolic energy. Many species of mammals, such as bats, bears, and ground squirrels, hibernate, but only one species of bird, the poorwill, has been shown to hibernate. The ability of hibernators to reduce their thermoregulatory set points so dramatically probably evolved as an extension of the set point decrease that accompanies sleep even in nonhibernating species of mammals and birds.

Chapter Summary

Homeostasis: Maintaining the Internal Environment

- The internal environment consists of the extracellular fluids. Organs and organ systems have specialized functions to keep certain aspects of the internal environment in a constant state. **Review Figure 40.1**
- Homeostasis is the maintenance of constancy in the internal environment. Homeostasis depends on the ability to control and regulate the functions of organs and organ systems.

Tissues, Organs, and Organ Systems

- Cells that have a similar structure and function make up a tissue. There are four general types of tissues: epithelial, connective, muscle, and nervous. **Review Figure 40.2**
- Epithelial tissues are sheets of tightly connected cells that cover the body surfaces and line hollow organs. **Review Figure 40.3**
- Connective tissues support and reinforce other tissues. They generally consist of dispersed cells in an extracellular matrix. Examples are cartilage, bone, blood, and adipose tissue. **Review Figure 40.4**
- Muscle tissues contract. There are three types: skeletal, cardiac, and smooth.

▶ There are two types of cells in nervous tissues: Neurons generate and transmit electrochemical signals, and glial cells provide supporting functions for neurons.

▶ Organs consist of multiple tissue types, and organs make up organ systems. **Review Table 40.1**

Physiological Regulation and Homeostasis

▶ Regulatory systems have set points and respond to feedback information. Negative feedback corrects deviations from the set point, positive feedback amplifies responses, and feedforward information changes the set point. **Review Figure 40.5**

Temperature and Life

▶ Living systems require a range of temperatures between the freezing point of water and the temperatures that denature proteins.

▶ Most biological processes and reactions are temperature-sensitive. Q_{10} is a measure of temperature sensitivity. **Review Figure 40.6**

▶ Animals that cannot avoid seasonal changes in body temperature have biochemical adaptations that compensate for those changes. These adaptations enable animals to acclimatize to seasonal changes. **Review Figure 40.7**

Maintaining Optimal Body Temperature

▶ Homeotherms maintain a fairly constant body temperature most of the time; poikilotherms do not. Endotherms produce metabolic heat to elevate body temperature; ectotherms depend mostly on environmental sources of heat. **Review Figure 40.8**

▶ Ectotherms and endotherms can regulate body temperature through behavior. **Review Figure 40.9**

▶ Heat exchange between a body and the environment is via radiation, conduction, convection, and evaporation. **Review Figure 40.11**

▶ Ectotherms and endotherms can control heat exchange with the environment by altering blood flow to the skin. **Review Figure 40.12**

▶ Some ectotherms, such as bees, nocturnal moths, and beetles, can produce metabolic heat to raise their body temperatures. **Review Figure 40.13**

▶ Some fish have circulatory systems that function as countercurrent heat exchangers to conserve heat produced by muscle metabolism. **Review Figure 40.14**

Thermoregulation in Endotherms

▶ Endotherms have high basal metabolic rates. Over a range of environmental temperatures called the thermoneutral zone, the metabolic rate of resting endotherms remains at basal levels. **Review Figure 40.15**

▶ When the environmental temperature falls below a lower critical temperature, endotherms maintain their body temperatures through shivering and nonshivering metabolic heat production. **Review Figure 40.16**

▶ When the environmental temperature rises above an upper critical temperature, metabolic rate increases as a consequence of active evaporative water loss through sweating or panting.

▶ Endotherms that live in cold climates have adaptations that minimize heat loss, including a reduced surface area-to-volume ratio and increased insulation.

▶ Endotherms may dissipate excess heat generated by exercise or the environment via evaporation. The water loss involved in this process can be dangerous to endotherms in dry environments.

The Vertebrate Thermostat

▶ The vertebrate thermostat is located in the hypothalamus. It has set points for activating thermoregulatory responses. Hypothalamic temperature provides negative feedback information.

▶ Cooling the hypothalamus induces the constriction of blood vessels and increased metabolic heat production. Heating the hypothalamus induces the dilation of blood vessels and active evaporative water loss. Thermoregulatory behaviors are also induced by changes in hypothalamic temperature. **Review Figure 40.18**

▶ Changes in set point reflect the integration of information, such as environmental temperature and time of day, that is relevant to the regulation of body temperature. **Review Figure 40.19**

▶ Fever, which results from a rise in set point, helps the body fight infections.

▶ Adaptations in which set points are reduced to conserve energy include daily torpor and hibernation. **Review Figure 40.20**

For Discussion

1. In some sheets of epithelial tissue, the cells are joined together with dense extracellular proteins that form "tight junctions," which are extremely impermeable (see Chapter 5). In other epithelial sheets the cells are joined by filamentous extracellular proteins that are strong, but not as impermeable. What do you think are the functions of tight junctions, and where would you expect to find them? Where might you expect to find epithelial sheets with the leakier connections?
2. If the major adaptation of endotherms to cold climates is their insulation, how would you compare the cold adaptations of a polar bear and a seal?
3. Why is an environment above its upper critical temperature more dangerous to an endotherm than an environment below its lower critical temperature?
4. We discussed the vertebrate thermostat by describing experiments done on mammals. Lizards also have a temperature-sensitive hypothalamus. How would you design an experiment on a lizard to see if the temperature of its hypothalamus was important feedback information for its thermoregulation? How would you modify your experiment to see if the lizard also used information from temperature sensors in its skin?
5. If the hypothalamic temperature of a mammal is the feedback information for its thermostat, why does the hypothalamic temperature scarcely change when that animal moves between environments hot enough and cold enough to stimulate the animal to pant and to shiver, respectively?

41 Animal Hormones

IN SHALLOW POOLS AROUND THE EDGE OF Lake Tanganyika in east central Africa, brightly colored male cichlid fish stake out territories and vigorously defend them against neighboring males. These dominant males constantly patrol their territories and display their colorful sexual adornments for the benefit of females, who assemble in groups at the edge of the cichlid colony. The females are hard to see because they are inactive and protectively colored. When a female is ready to spawn, and is impressed by a male's territory and display, she enters his territory and lays her eggs in a spawning pit that the male has prepared. The male then fertilizes her eggs.

At any one time, only about 10 percent of the males in the colony are displaying and holding territories. All the other males are small, nondescript, and nonaggressive like the females. If a dominant male is removed by a predator, however, the nondescript males fight over the vacated territory. The winner rapidly assumes the appearance and behavior of a dominant male: brightly colored, big, aggressive, and attractive to females.

What accounts for this dramatic change? Russell Fernald and his students at Stanford University have shown that soon after the nondescript male's victory, certain cells in his brain enlarge and secrete a chemical message. This message triggers cells in the pituitary gland, which is outside of the brain, to secrete chemical messages in turn. Although secreted in tiny quantities, these molecules enter the blood and are transported around the body. The responses of cells to these chemical messages produce the characteristics of a dominant male.

This change in the male cichlid is one example of how chemical messages, or hormones, can produce and coordinate anatomical, physiological, and behavioral changes in an animal. We explore many other examples in this chapter.

We look first at the nature of hormones and their evolution, and then examine some of their roles in the control of invertebrate life cycles. Most of this chapter is devoted to vertebrate hormones: their functions, control, and molecular mechanisms of action. We pay particular attention to the extensive interactions between the systems of neural and hormonal information. In the process, we discuss several human diseases involving hormonal dysfunction.

Hormones and Their Actions

In Chapter 40, we learned that control and regulation require information. This information is transmitted mostly as electric signals and as chemical signals. *Electric signals* are nerve impulses, a major focus of later chapters on the nervous system. Nerve impulses can be rapidly conducted over long distances to specific targets. *Chemical signals* are **hormones**, which are secreted by cells, diffuse locally in the extracellular fluid until they are picked up by the blood, and are distributed by the circulatory system.

Because the secretion, diffusion, and circulation of hormones is much slower than the transmission of nerve impulses, hormones are not useful for controlling rapid actions such as those involved in cichlid fighting. Hormone action *is* good for coordinating longer-term developmental processes, such as the transition of a nondescript cichlid into a dominant, territorial, breeding male. Hormones control many long-term physiological responses, such as the secretion of digestive enzymes by our guts and the reproductive cycles of many species.

Dominant and Non-Dominant Male Cichlids
A dominant male cichlid (*Haplochromis burtoni*) displays bright colors that attract females to his spawning pit.

A hormone is a chemical signal produced by certain cells of a multicellular organism and received by cells of the same organism. Cells that secrete hormones are called **endocrine cells**. To receive the hormonal message, a **target cell** must have appropriate *receptors* to which the hormone can bind. The binding of a hormone to its receptor activates mechanisms within the target cell that eventually lead to a response, which may be developmental, physiological, or behavioral. (In the case of the male cichlid, hormone release stimulates all three types of responses.)

Hormonal signaling systems can be distinguished according to the distance over which their messages operate. Some hormones only act on target cells close to their sites of release; others act on target cells at distant locations in the body. Some chemical messages, called *pheromones*, even exert their effects on other individuals.

Most hormones are distributed in the blood

The classic hormone is a chemical message secreted by cells in minute amounts and distributed throughout the body by the circulatory system (Figure 41.1*a*). Wherever such a hormone encounters a cell with a receptor to which it can bind, it triggers a response. The nature of the response depends on the responding cell. The same hormone can cause different responses in different types of cells.

Consider the hormone epinephrine. If you step off a curb without looking and a car screeches to a halt right next to you, you jump, your heart starts to thump, and a whole set of protective actions are set in motion. The jump and the initial heart thumping are driven by your nervous system, which can react very quickly. Simultaneously, the nervous system stimulates endocrine cells just above your kidneys (adrenal cells) to secrete epinephrine. Within seconds, epinephrine is diffusing into your blood and circulating around your body to activate the many components of the *fight-or-flight response*.

Epinephrine acts on the heart, blood vessels, liver, and fat cells. When it binds to its receptors in the heart, it causes the heart to beat faster and more strongly. It binds to receptors in the vessels that supply blood to your digestive tract, causing those vessels to constrict (digestion can wait!). Your heart is pumping more blood, and a greater percentage of that blood is going to the muscles needed for your escape. In the liver, epinephrine stimulates the breakdown of glycogen into glucose for a quick energy supply. In fatty tissue, it stimulates the breakdown of fats as another source of energy. These are just some of the many actions triggered by one hormone. They all contribute to increasing your chances of escaping a dangerous situation.

Whether or not a cell responds to a hormone depends on whether it has receptors for that hormone, and how it responds depends on what kind of a cell it is. A single hormone can stimulate many different responses.

Some hormones act locally

Some hormones are released into the extracellular fluids in such tiny quantities, or they are so rapidly inactivated by degradative enzymes, or they are taken up so efficiently by local cells, that the circulation never has the chance to distribute them to distant target cells. Thus, these hormones act only locally. When a hormone affects cells near the secreting cell, it is said to have **paracrine** function (Figure 41.1*b*).

An example of a paracrine hormone is **histamine**, one of the mediators of inflammation. Histamine is released in damaged tissues by specialized cells called *mast cells* (see Chapter 19). When the skin is cut, the area around the cut becomes inflamed—red, hot, and swollen. Histamine causes this response by dilating the local blood vessels and making them more permeable ("leaky"), which allows

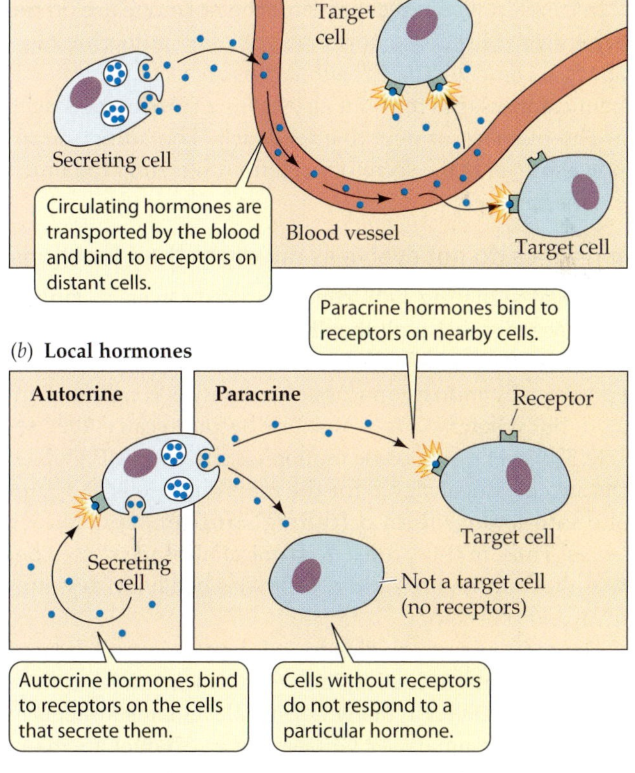

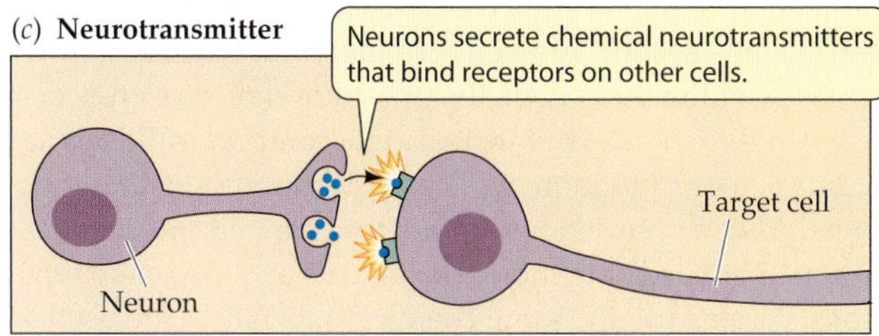

41.1 Chemical Signaling Systems
(*a*) The classic hormone is a secreted chemical message that is distributed throughout the body by the circulatory system. (*b*) An autocrine hormone influences the cell that releases it; paracrine hormones influence nearby cells. (*c*) Neurotransmitters (see Chapter 44) can be considered paracrine hormones.

blood plasma, including protective blood proteins and white blood cells, to move into the damaged tissue.

A major class of paracrine hormones consists of the various **growth factors**, which stimulate the growth and differentiation of cells. Growth factors were first discovered when scientists attempted to culture cells outside of the body. Even when given all sorts of nutrients and optimal conditions, the cells did not grow well unless blood plasma or a tissue extract was added to the medium. The components necessary for growth were found to be specific molecules present in very small quantities. At present, about 50 specific growth factors are known, along with a complex group of receptors. Some examples are:

- *Nerve growth factor* (*NGF*), which promotes the survival and growth of nerve cells.
- *Epidermal growth factor* (*EGF*), which stimulates many kinds of cells to divide
- *Vascular endothelial growth factor*, which stimulates the growth and branching of blood vessels

The nerve cells called *neurons* can also be considered paracrine cells. As we will see in Chapter 44, a neuron communicates with another cell by means of a chemical message called a *neurotransmitter*, which travels over a very small distance to the target cell (Figure 41.1*c*).

In cases in which receptors for the hormone are on the secreting cell itself, the hormone acts as an **autocrine** message (see Figure 41.1*b*). Growth factors are examples of local chemical messages that can also act as autocrine messages for the purpose of negative feedback. The autocrine response prevents the secretory cell from secreting too much of the hormone.

Hormones do not evolve as rapidly as their functions

Chemical signaling between cells exists even in single-celled organisms. Recall the life cycle of slime molds as described in Chapter 27. These protists lead solitary lives and reproduce by mitosis and fission as long as conditions are good. But when food and moisture become scarce, they secrete 3′,5′-cyclic adenosine monophosphate (cAMP), which acts as a chemical signal for the individual cells to aggregate into a slug, form a fruiting structure, and release spores. Thus, in this protist, a chemical message passed between cells influences and coordinates behavior, development, and physiology.

The molecule responsible for this very primitive form of chemical communication between slime mold cells—cAMP—is involved in many hormonal signaling systems in multicellular animals. As you learned in Chapter 15, many molecular signals cause the production of cAMP within cells. This "second messenger" mediates a variety of responses within the cell via the phosphorylation of enzymes.

With the evolution of increasingly complex multicellular animals, more and more molecules acquired signaling functions. Also, as physiological systems changed through evolution, many existing molecular signals acquired new functions. As a result, the same chemical substances are used as

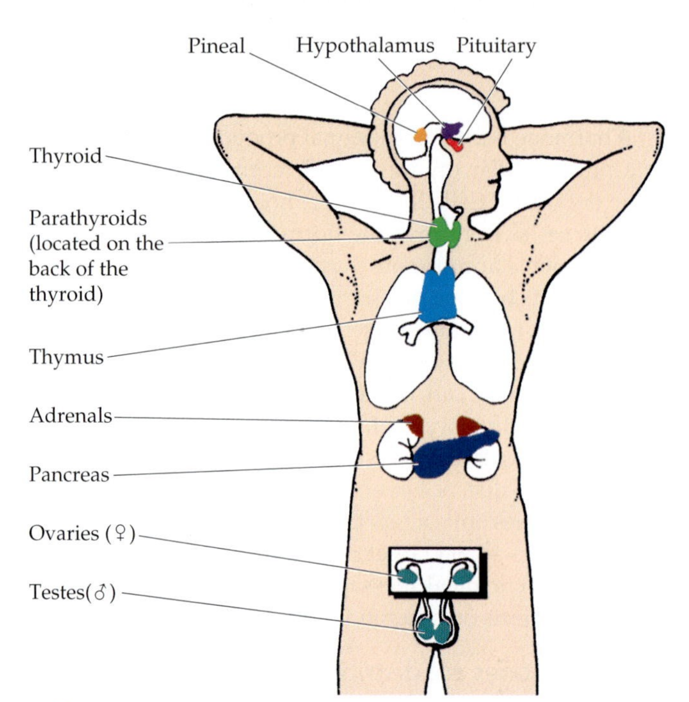

41.2 The Endocrine System of Humans
The endocrine system broadcasts chemical signals (hormones) that are received by cells with appropriate receptors. There are nine glands in the human endocrine system, but hormones are also secreted by tissues that are not part of discrete glands.

hormones in widely divergent species, but they may have completely different actions in different species.

The hormone thyroxine, for example, is found in animal species ranging from mollusks and tunicates (sea squirts) to humans. Its function is unknown in invertebrates, but it is produced in an organ that is involved in feeding. In frogs, thyroxine is essential for the metamorphosis from tadpole to adult. In mammals, thyroxine elevates cellular metabolism.

Prolactin is another example of a hormone that evolved different functions in different species. Prolactin stimulates milk production in female mammals after they give birth. In pigeons and doves, prolactin stimulates the production of crop milk, a substance secreted from the crop that is fed to the young. Prolactin causes amphibians to prepare for reproduction by seeking water. In fishes, such as salmon, that migrate between salt water and fresh water to breed, prolactin regulates the mechanisms that maintain osmotic balance with the changing environment. In all of these cases prolactin is involved in reproductive processes, but as those processes have changed through evolution, so has the information signaled by the hormone.

In summary, the structures of the molecules involved in chemical signaling are highly conserved—they have changed little throughout evolution—but their functions have changed dramatically.

Endocrine glands secrete hormones

Some endocrine cells are distributed as single cells within a tissue. Many hormones of the digestive tract, for example, are produced and secreted by isolated cells in the lining of the tract. As the contents of the digestive tract come into contact with these cells, they release their hormones, which enter the blood and, like epinephrine, circulate throughout the body and activate cells that have appropriate receptors. Many hormones, however, are secreted by aggregations of endocrine cells that form secretory organs called endocrine glands.

Animals have two types of glands. **Exocrine glands**, such as sweat glands and salivary glands, release secretions that are not hormones through ducts that lead outside the body. Sweat gland ducts, for example, open onto the surface of the skin, and salivary gland ducts open into the mouth. Glands that secrete hormones and do not have ducts are called **endocrine glands**; they secrete their products directly into the extracellular fluid. Vertebrates have nine discrete endocrine glands, which collectively make up the **endocrine system** (Figure 41.2).

Hormonal Control of Molting and Development in Insects

Many hormones of invertebrate animals have multiple functions. In this chapter we cannot do justice to the diversity of hormones in the invertebrates, but we'll discuss two important aspects of the lives of many invertebrates that are controlled by hormonal mechanisms: molting and metamorphosis.

Hormones from the head control molting in insects

Because insects have rigid exoskeletons, their growth is episodic, punctuated with *molts* (shedding) of the exoskeleton (see Chapter 32). Each growth stage between two molts is called an *instar*. The British physiologist Sir Vincent Wigglesworth was a pioneer in the study of the hormonal control of growth and development in insects.

Wigglesworth conducted experiments on the blood-sucking bug *Rhodnius*, which undergoes *incomplete metamorphosis*. Upon hatching, *Rhodnius* is nearly a miniature version of an adult, but it lacks some adult features. *Rhodnius* molts five times before developing into a mature adult; a blood meal triggers each episode of molting and growth.

Rhodnius is a hardy experimental animal; it can live a long time even after it is decapitated. If decapitated about an hour after it has a blood meal, *Rhodnius* may live for up to a year, but it does not molt. If decapitated a week after its blood meal, it does molt (Figure 41.3, Experiment 1). These observations led Wigglesworth to the hypothesis that something diffusing slowly from the head controls molting.

The proof of this hypothesis came from a clever experiment in which Wigglesworth decapitated two *Rhodnius*: one that had just had its blood meal and another that had had its blood meal a week earlier. The two decapitated bodies

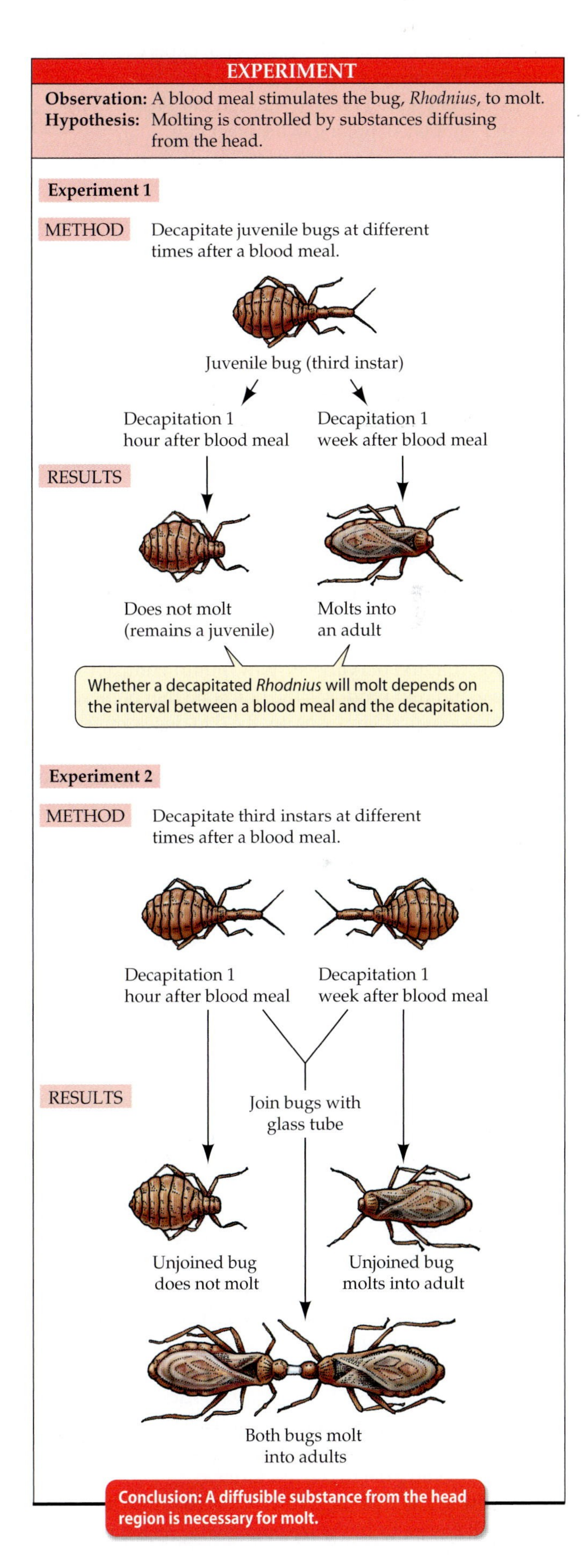

41.3 A Diffusible Substance Triggers Molting
The effect of time since the last blood meal on *Rhodnius* molting led Sir Vincent Wigglesworth to hypothesize that some substance was diffusing slowly through the insect's body. Further experiments showed that molting is indeed controlled by a substance—a hormone—diffusing from the head.

were connected with a short piece of glass tubing—and they both molted (Figure 41.3, Experiment 2). Thus one or more substances from the bug fed a week earlier crossed through the glass tube and stimulated molting in the other bug.

We now know that two hormones working in sequence regulate molting:

- Cells in the brain produce **brain hormone**.
- Brain hormone is transported to and stored in a pair of structures attached to the brain, the *corpora cardiaca* (singular corpus cardiacum).
- After appropriate stimulation (which for *Rhodnius* is a blood meal) the corpora cardiaca release brain hormone, which diffuses to an endocrine gland, the *prothoracic gland.*
- Brain hormone stimulates the prothoracic gland to release the hormone **ecdysone**.
- Ecdysone diffuses to target tissues and stimulates molting.

The control of molting by brain hormone and ecdysone is a general mechanism in insects. The nervous system receives various types of information relevant in determining the optimal timing for growth and development. It makes sense, therefore, that the nervous system should control the endocrine gland that produces the hormone that orchestrates all the physiological processes involved in development and molting. Later in this chapter we will see similar links between the nervous system and endocrine glands in vertebrates.

Juvenile hormone controls development in insects

The *Rhodnius* decapitation experiments yielded a curious result: Regardless of the instar used, the decapitated bug always molted directly into an adult form. Additional experiments by Wigglesworth demonstrated that a hormone other than those responsible for molting determines whether a bug molts into another juvenile instar or into an adult.

Because the head of *Rhodnius* is long, it was possible to remove just the front part of the head, which contains the cells that secrete and release brain hormone, while leaving intact the rear part. That rear part contains two other endocrine structures called the *corpora allata* (singular corpus allatum). When fourth-instar bugs that had been fed one week earlier were partly decapitated, leaving the corpora allata intact, they molted into fifth instars, not into adults.

This experiment, was followed up by more experiments using glass tubes to connect individual bugs, allowing body fluid transfer between them. When an unfed, completely decapitated, fifth-instar bug was connected to a fourth-instar bug that had been fed and had only the front part of its head removed, both bugs molted into juvenile forms. A substance coming from the rear part of the head of the fourth-instar bug prevented the expected result that both bugs would molt into adult forms.

We now know that the substance is **juvenile hormone** and that it comes from the corpora allata. As long as juve-

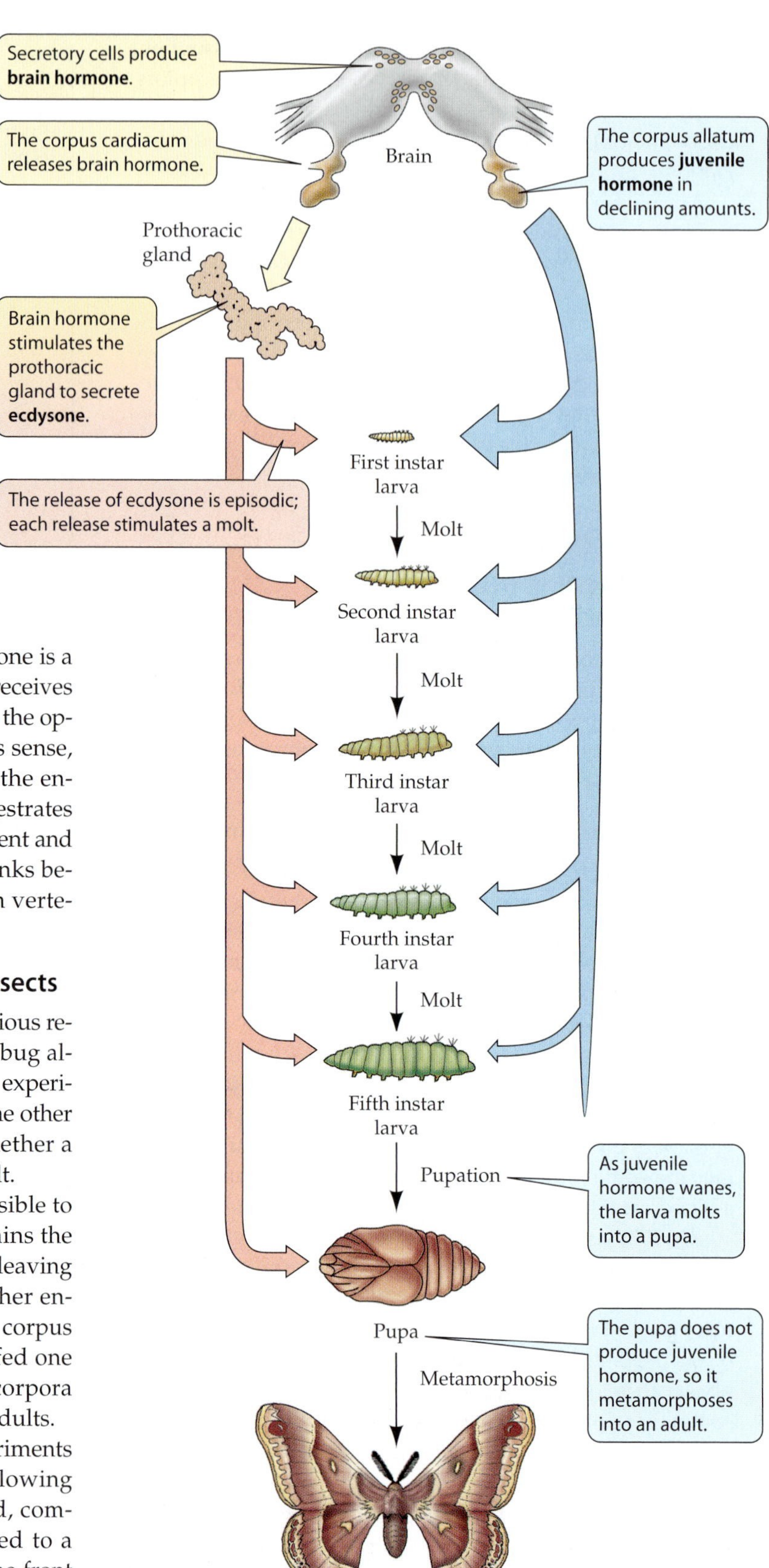

41.4 Complete Metamorphosis
Butterflies and moths undergo complete metamorphosis, in which the feeding larvae (caterpillars) bear no resemblance to the reproductive adult. Three hormones control molting and metamorphosis in the silkworm moth *Hyalophora cecropia*.

nile hormone is present, *Rhodnius* molts into another juvenile instar. The corpora allata normally stop producing juvenile hormone during the fifth instar. If juvenile hormone is absent, the bug molts into the adult form.

The control of development by juvenile hormone is more complex in insects that, like butterflies, undergo *complete metamorphosis*. These animals undergo dramatic developmental changes between instars. The fertilized egg hatches into a *larva*, which feeds and molts several times, becoming bigger and bigger. Then it enters an inactive stage called a *pupa*. It undergoes major body reorganization as a pupa, and finally emerges as an *adult*.

An excellent example of complete metamorphosis is provided by the silkworm moth, *Hyalophora cecropia* (Figure 41.4). As long as juvenile hormone is present in high concentrations, larvae molt into larvae. When the level of juvenile hormone falls, larvae molt into pupae. Because no juvenile hormone is produced in pupae, they molt into adults.

In our perpetual war against insects, juvenile hormone is a new weapon. Synthetic forms of juvenile hormone can be distributed in the environment to prevent the development of juvenile insects into adults capable of reproduction. However, as you might expect from the fact that hormone structures are highly conserved, such a weapon is not without potentially serious side effects. First, it is not selective in its effects on insects and can affect species that are beneficial as well as those that are pests. Second, the synthetic juvenile hormone could have actions in vertebrates that are not yet known.

The existence and function of insect hormones was experimentally demonstrated many years before the hormones were identified chemically. That is not surprising when you consider the tiny amounts of certain hormones that exist in an organism. In one of the earliest studies of ecdysone, biochemists produced only 250 mg of pure ecdysone (about one-fourth the weight of an apple seed) from 4 tons of silkworms!

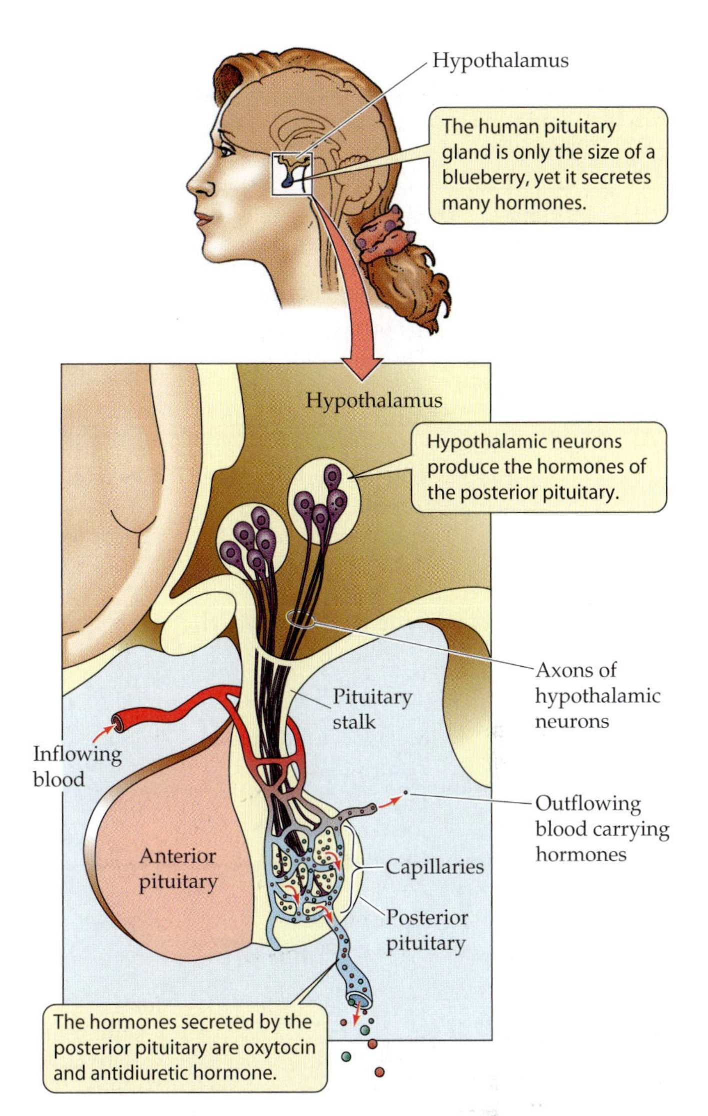

41.5 The Posterior Pituitary Releases Neurohormones
The two hormones stored and released by the posterior pituitary are neurohormones produced in the hypothalamus.

Vertebrate Endocrine Systems

The list of chemical messages in the bodies of vertebrates is long and growing longer. To make the subject manageable, we focus mostly on the hormones of humans—how they function and how they are controlled. Table 41.1 presents an overview of the hormones of humans (most of which are found in all other mammals as well). Notice that the column listing the target tissues of these hormones includes every organ system of the body.

We begin this survey with the **pituitary gland** because it plays a central role in the endocrine system. The pituitary is a link between the nervous system and many endocrine glands. It secretes some hormones that are actually produced by neurons in the brain, and under the influence of still other brain hormones, it produces a number of its own hormones, which control the activities of various endocrine glands throughout the body. For these reasons, the pituitary has been called "the master gland."

The pituitary develops from outpocketings of the mouth and brain

The pituitary gland sits in a depression at the bottom of the skull just over the back of the roof of the mouth (Figure 41.5). It is attached to the part of the brain called the *hypothalamus*, which is involved in many homeostatic regulatory systems (see Chapter 40).

The pituitary has two distinct parts that have different functions and separate origins during development. The **anterior pituitary** originates as an outpocketing of the embryonic mouth cavity, and the **posterior pituitary** originates as an outpocketing of the developing brain in the region that becomes the hypothalamus.

THE POSTERIOR PITUITARY. The posterior pituitary releases two hormones, antidiuretic hormone and oxytocin. Both are small peptides synthesized in neurons in the hypothalamus.

41.1 Principal Hormones of Humans

SECRETING TISSUE OR GLAND	HORMONE	CHEMICAL NATURE	TARGET(S)	IMPORTANT PROPERTIES OR ACTIONS
Hypothalamus	Releasing and release-inhibiting hormones (see Table 41.2)	Peptides	Anterior pituitary	Control secretion of hormones of anterior pituitary
	Oxytocin, antidiuretic hormone	Peptides	(See Posterior pituitary)	Stored and released by posterior pituitary
Anterior pituitary: Tropic hormones	Thyrotropin	Glycoprotein	Thyroid gland	Stimulates synthesis and secretion of thyroxine
	Adrenocorticotropin (ACTH)	Polypeptide	Adrenal cortex	Stimulates release of hormones from adrenal cortex
	Luteinizing hormone (LH)	Glycoprotein	Gonads	Stimulates secretion of sex hormones from ovaries and testes
	Follicle-stimulating hormone (FSH)	Glycoprotein	Gonads	Stimulates growth and maturation of eggs in females; stimulates sperm production in males
Anterior pituitary: Other hormones	Growth hormone (GH)	Protein	Bones, liver, muscles	Stimulates protein synthesis and growth
	Prolactin	Protein	Mammary glands	Stimulates milk production
	Melanocyte-stimulating hormone	Peptide	Melanocytes	Controls skin pigmentation
	Endorphins and enkephalins	Peptides	Spinal cord neurons	Decrease painful sensations
Posterior pituitary	Oxytocin	Peptide	Uterus, breasts	Induces birth by stimulating labor contractions; causes milk flow
	Antidiuretic hormone (ADH) (vasopressin)	Peptide	Kidneys	Stimulates water reabsorption and raises blood pressure
Thyroid	Thyroxine	Iodinated amino acid derivative	Many tissues	Stimulates and maintains metabolism necessary for normal development and growth
	Calcitonin	Peptide	Bones	Stimulates bone formation; lowers blood calcium
Parathyroids	Parathormone	Protein	Bones	Absorbs bone; raises blood calcium
Thymus	Thymosins	Peptides	Immune system	Activate immune responses of T cells in the lymphatic system
Pancreas	Insulin	Protein	Muscles, liver, fat, other tissues	Stimulates uptake and metabolism of glucose; increases conversion of glucose to glycogen and fat
	Glucagon	Protein	Liver	Stimulates breakdown of glycogen and raises blood sugar
	Somatostatin	Peptide	Digestive tract; other cells of the pancreas	Inhibits insulin and glucagon release; decreases secretion, motility, and absorption in the digestive tract

Hormones that are produced and released by neurons are called **neurohormones**. Antidiuretic hormone and oxytocin move down long extensions (axons) of the neurons that produce them, through the pituitary stalk into the posterior pituitary, where they are stored in the nerve endings (see Figure 41.5). How do they move down the axons? In the bodies of the neurons, these neurohormones are packaged into vesicles. Proteins called *kinesins* grab onto the vesicles and, powered by ATP, "walk" step by step down microtubules in the axons (see Figure 4.25).

The main action of **antidiuretic hormone** (**ADH**) is to increase the amount of water conserved by the kidneys. When ADH secretion is high, the kidneys resorb more water and produce only a small volume of highly concentrated urine. When ADH secretion is low, the kidneys produce a large volume of dilute urine.

The posterior pituitary increases its release of ADH whenever blood pressure falls or the blood becomes too salty. We will discuss the mechanism of ADH action in Chapter 51. ADH is also known as *vasopressin* because it

41.1 Principal Hormones of Humans (continued)

SECRETING TISSUE OR GLAND	HORMONE	CHEMICAL NATURE	TARGET(S)	IMPORTANT PROPERTIES OR ACTIONS
Adrenal medulla	Epinephrine, norepinephrine	Modified amino acids	Heart, blood vessels, liver, fat cells	Stimulate fight-or-flight reactions: increase heart rate, redistribute blood to muscles, raise blood sugar
Adrenal cortex	Glucocorticoids (cortisol)	Steroids	Muscles, immune system, other tissues	Mediate response to stress; reduce metabolism of glucose, increase metabolism of proteins and fats; reduce inflammation and immune responses
	Mineralocorticoids (aldosterone)	Steroids	Kidneys	Stimulate excretion of potassium ions and reabsorption of sodium ions
Stomach lining	Gastrin	Peptide	Stomach	Promotes digestion of food by stimulating release of digestive juices; stimulates stomach movements that mix food and digestive juices
Lining of small intestine	Secretin	Peptide	Pancreas	Stimulate secretion of bicarbonate solution by ducts of pancreas
	Cholecystokinin	Peptide	Pancreas, liver, gallbladder	Stimulates secretion of digestive enzymes by pancreas and other digestive juices from liver; stimulates contractions of gallbladder and ducts
	Enterogastrone	Polypeptide	Stomach	Inhibits digestive activities in the stomach
Pineal	Melatonin	Modified amino acid	Hypothalamus	Involved in biological rhythms
Ovaries	Estrogens	Steroids	Breasts, uterus, other tissues	Stimulate development and maintenance of female characteristics and sexual behavior
	Progesterone	Steroid	Uterus	Sustains pregnancy; helps maintain secondary female sexual characteristics
Testes	Androgens	Steroids	Various tissues	Stimulate development and maintenance of male sexual behavior and secondary male sexual characteristics; stimulate sperm production
Many cell types	Prostaglandins	Modified fatty acids	Various tissues	Have many diverse actions
Heart	Atrial natriuretic hormone	Peptide	Kidneys	Increases sodium ion excretion

also causes the constriction of peripheral blood vessels as a means of elevating blood pressure.

When a woman is about to give birth, her posterior pituitary releases **oxytocin**, which stimulates the contractions of the muscles that push the baby out of her body. Oxytocin also brings about the flow of milk from the mother's breasts. The baby's suckling stimulates nerve cells in the mother, causing the secretion of oxytocin. Even the sight and sounds of her baby can cause a nursing mother to secrete oxytocin and release milk from her breasts.

THE ANTERIOR PITUITARY. Four hormones released by the anterior pituitary (*thyrotropin*, *adrenocorticotropin*, *luteinizing hormone*, and *follicle-stimulating hormone*) control the activities of other endocrine glands and thus are called **tropic hormones** (see Figure 41.7). Each tropic hormone is produced by a different type of pituitary cell. We will say more about these tropic hormones when we describe their target glands (thyroid, adrenal cortex, testes, and ovaries) later in this chapter and in the next.

The other hormones produced by the anterior pituitary influence tissues that are not endocrine glands. These hormones are growth hormone, prolactin, melanocyte-stimulating hormone, endorphins, and enkephalins.

Growth hormone (**GH**) consists of about 200 amino acids and acts on a wide variety of tissues to promote growth directly and indirectly. One of its important direct effects is to stimulate cells to take up amino acids. Growth hormone promotes growth indirectly by stimulating the liver to produce chemical messages that stimulate the growth of bone and cartilage. Thus, in some of its actions, growth hormone can also be considered a tropic hormone.

Overproduction of growth hormone in children causes *gigantism*, and underproduction causes *dwarfism* (Figure 41.6). Beginning in the late 1950s, children diagnosed as having a serious deficiency of growth hormone were treated

(*a*)

(*b*)

41.6 Effects of Abnormal Amounts of Growth Hormone
(*a*) Overproduction of growth hormone in childhood causes gigantism. This photo from 1939 shows a young man who is more than 8 feet tall standing next to his father, who is just under 6 feet tall. (*b*) Underproduction of growth hormone during childhood results in pituitary dwarfism. The man on the left is P.T. Barnum, the circus entrepreneur. The man on the right is Charles Stratton, a dwarf, who appeared in Barnum's circus under the name General Tom Thumb.

with human growth hormone extracted from human pituitaries in cadavers. The treatment was successful in stimulating substantial growth, but it could be made available to only small numbers of patients. A year's supply of human growth hormone for one individual required up to 50 pituitaries. In the mid-1980s, scientists using genetic engineering technology isolated the gene for human growth hormone and introduced it into bacteria, which produced enough of the hormone to make it widely available.

Preventing pituitary dwarfism is now feasible and affordable, but the availability of growth hormone raises new questions. Should every child at the lower end of the height charts be treated? Should a normal child whose parents think basketball stardom is assured if she is tall be given growth hormone? These types of questions are impossible to answer with scientific data alone.

Prolactin, another hormone produced by the anterior pituitary, stimulates the production and secretion of milk in female mammals. In some mammals, prolactin also functions as an important hormone during pregnancy. In human males, prolactin plays a role along with other pituitary hormones in controlling the endocrine function of the testes.

Endorphins and **enkephalins** are the body's "natural opiates." In the brain, these molecules act as neurotransmitters in pathways that control pain. The significance of their release from the anterior pituitary is unknown. Interestingly, the production of endorphins and enkephalins in the pituitary is encoded by the same gene that encodes at least two other pituitary hormones. The gene actually encodes a large parent molecule called *pro-opiomelanocortin*. This large protein molecule is cleaved to produce several peptides, some of which have hormonal functions. Adrenocorticotropin, melanocyte-stimulating hormone, endorphins, and enkephalins all result from the cleavage of pro-opiomelanocortin.

THE ANTERIOR PITUITARY IS CONTROLLED BY HYPOTHALAMIC NEUROHORMONES Because the anterior pituitary produces tropic hormones that control other endocrine tissues, it acquired the designation "master gland." But we now

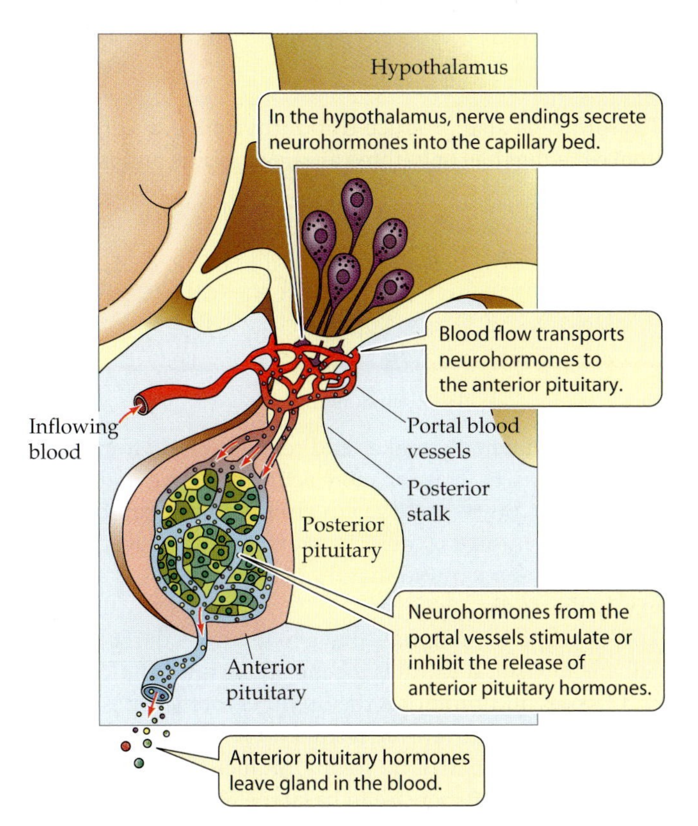

41.7 Hormones from the Hypothalamus Control the Anterior Pituitary
Neurohormones produced in tiny quantities by cells in the hypothalamus are transported to the anterior pituitary through a system of portal blood vessels. These releasing and release-inhibiting hormones control the activities of anterior pituitary endocrine cells.

know that this "master" is under still higher control by the hypothalamus, and that their interaction integrates nervous system and endocrine system functions.

The hypothalamus receives information about conditions in the body and in the external environment through the nervous system. If the connection between the hypothalamus and the pituitary is experimentally cut, pituitary hormones are no longer released in response to changes in the environment or in the body. If pituitary cells are maintained in culture, extracts of hypothalamic tissue stimulate some of those cells to release their hormones into the culture medium. Therefore, scientists hypothesized that secretions of the hypothalamic cells control the activities of anterior pituitary cells.

Although hypothalamic neurons do not extend into the anterior pituitary as they do into the posterior pituitary, a special set of **portal blood vessels** connects the hypothalamus and the anterior pituitary (Figure 41.7). It was thus proposed that secretions from nerve endings in the hypothalamus enter the blood and are conducted down the portal vessels to the anterior pituitary, where they cause the release of anterior pituitary hormones.

In the 1960s, two large teams of scientists, led by Roger Guillemin and Andrew Schally, initiated the search for the hypothalamic neurohormones. Because the amounts of such hormones in any individual mammal would be tiny, massive numbers of hypothalami from pigs and sheep were collected from slaughterhouses and shipped to laboratories in refrigerated trucks. One extraction effort began with the hypothalami from 270,000 sheep and yielded only 1 mg of purified **thyrotropin-releasing hormone**, or **TRH**, which was the first hypothalamic releasing (that is, release-stimulating) hormone isolated and characterized. Biochemical analysis of this pure sample revealed that TRH is a simple tripeptide consisting of glutamine, histidine, and proline. TRH causes certain anterior pituitary cells to release the tropic hormone *thyrotropin*, which in turn stimulates the activity of the thyroid gland.

Soon after discovering thyrotropin-releasing hormone, Guillemin's and Schally's teams identified **gonadotropin-releasing hormone**, which stimulates certain anterior pituitary cells to release the tropic hormones that control the activity of the gonads (the ovaries and the testes). For these discoveries, Guillemin and Schally received the 1972 Nobel prize in medicine. Many more hypothalamic neurohormones, including both releasing hormones and release-inhibiting hormones, are now known (Table 41.2).

Negative feedback loops control hormone secretion

As well as being controlled by hypothalamic releasing and release-inhibiting hormones, the endocrine cells of the ante-

41.2 *Releasing and Release-Inhibiting Neurohormones of the Hypothalamus*

NEUROHORMONE	ACTION
Thyrotropin-releasing hormone (TRH)	Stimulates thyrotropin release
Gonadotropin-releasing hormone (GnRH)	Stimulates release of follicle-stimulating hormone and luteinizing hormone
Prolactin release-inhibiting hormone	Inhibits prolactin release
Prolactin-releasing hormone	Stimulates prolactin release
Somatostatin (growth hormone release-inhibiting hormone)	Inhibits growth hormone release; interferes with thyrotropin release
Growth hormone–releasing hormone	Stimulates growth hormone release
Adrenocorticotropin-releasing hormone	Stimulates adrenocorticotropin release
Melanocyte-stimulating hormone release-inhibiting hormone	Inhibits release of melanocyte-stimulating hormone

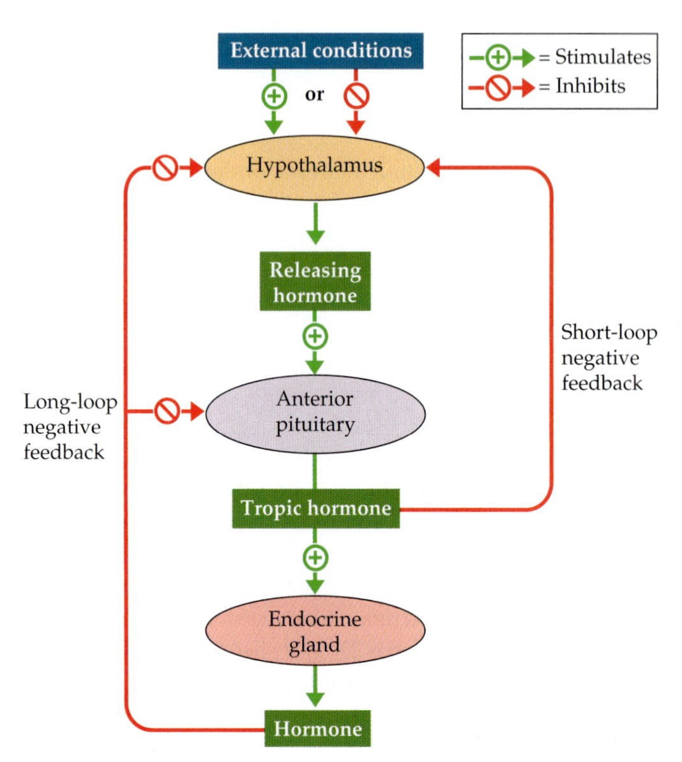

41.8 Multiple Feedback Loops Control Hormone Secretion
Multiple feedback loops regulate the chain of command from hypothalamus to anterior pituitary to endocrine glands.

rior pituitary are also under negative feedback control by the hormones of the target glands they stimulate (Figure 41.8). For example, the hormone cortisol, produced by the adrenal gland in response to adrenocorticotropin, returns to the pituitary in the circulating blood and inhibits further adrenocorticotropin release. Cortisol also acts as a negative feedback signal at the level of the hypothalamus, inhibiting the release of adrenocorticotropin-releasing hormone. In some cases a tropic hormone of the anterior pituitary also exerts negative feedback control on the hypothalamic cells producing the corresponding releasing hormone.

Thyroxine controls cell metabolism

The **thyroid gland** wraps around the front of the windpipe (*trachea*) and expands into a lobe on either side (see Figure 41.2). The thyroid gland produces the hormones thyroxine and calcitonin.

Thyroxine is synthesized in thyroid cells from two molecules of diiodotyrosine, which is the amino acid tyrosine with two atoms of iodine chemically bonded to it. Thus, a thyroxine molecule has four atoms of iodine, and is called T_4:

Thyroxine (T_4)

Thyroid cells also produce triiodothyronine, a version of thyroxine that has only three atoms of iodine and is called T_3:

Triiodothyronine (T_3)

The thyroid usually makes and releases about four times as much T_4 as T_3. T_3 is the more active hormone in the cells of the body, but when T_4 is in circulation, it can be converted to T_3 by an enzyme. Therefore, when you read about thyroxine, keep in mind that the actions discussed are primarily due to T_3.

Thyroxine in mammals plays many roles in regulating cell metabolism. It elevates the metabolic rates of most cells and tissues and promotes the use of carbohydrates rather than fats for fuel. Exposure to cold for several days leads to an increased release of thyroxine, an increased conversion of T_4 to T_3, and an increase in basal metabolic rate. Thyroxine is especially crucial during development and growth, as it promotes amino acid uptake and protein synthesis by cells. Insufficient thyroxine in a human fetus or growing child greatly retards physical and mental growth, resulting in a condition known as *cretinism*.

The tropic hormone **thyrotropin** from the anterior pituitary activates the thyroid cells that produce thyroxine (Figure 41.9). TRH (thyrotropin-releasing hormone) produced in the hypothalamus and transported to the anterior pituitary through the portal blood vessels activates the thyrotropin-producing pituitary cells. The brain uses environmental information such as temperature or day length to determine whether to increase or decrease the secretion of TRH. There is a very important negative feedback loop in this sequence of steps: Circulating thyroxine inhibits the response of the pituitary cells to TRH. Less thyrotropin is released when thyroxine levels are high, and more thyrotropin is released when thyroxine levels are low.

Thyroid dysfunction causes goiter

A *goiter* is an enlarged thyroid gland, which causes a pronounced bulge on the front and sides of the neck. Goiter can be associated with either **hyperthyroidism** (very high levels of thyroxine) or **hypothyroidism** (very low levels of thyroxine). The control diagram in Figure 41.9 helps explain how two very different conditions can result in the same symptom.

Hyperthyroid goiter results when the negative feedback mechanism fails to turn off the thyroid cells even though blood levels of thyroxine are high. The most common cause of hyperthyroidism is an autoimmune disease in which an antibody to the thyrotropin receptor is produced. This antibody can bind to the receptor and cause the thyroid cells to produce and release thyroxine. Even though blood levels of thyrotropin may be quite low because of the negative feed-

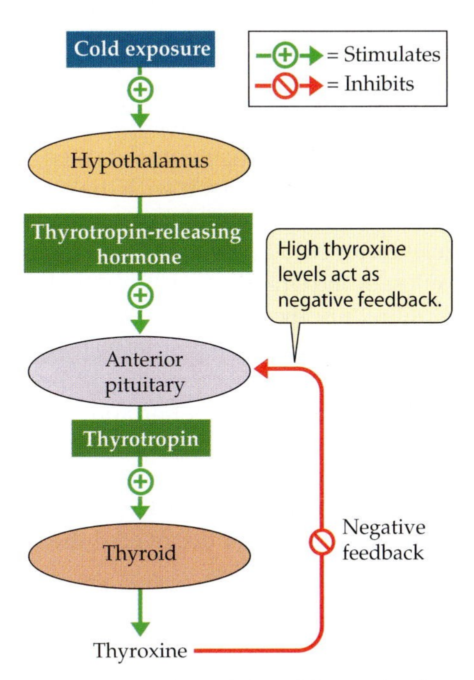

41.9 Regulation of Thyroid Function in Response to Cold
Exposure to cold temperatures stimulates the hypothalamus to produce thyrotropin-releasing hormone, which stimulates anterior pituitary cells to secrete thyrotropin, which in turn stimulates the thyroid to release thyroxine.

back from thyroxine, the thyroid remains maximally stimulated, and it grows bigger. Hyperthyroid patients have high metabolic rates, are jumpy and nervous, usually feel hot, and may have a buildup of fat behind the eyeballs, causing their eyes to bulge.

Hypothyroid goiter results when there is not enough circulating thyroxine to turn off thyrotropin production. Its most common cause is a deficiency of dietary iodide, without which the thyroid gland cannot make thyroxine. Without sufficient thyroxine, thyrotropin levels remain high, so the thyroid continues to produce large amounts of nonfunctional thyroxine and becomes very large. The symptoms of hypothyroidism are low metabolism, intolerance of cold, and general physical and mental sluggishness.

Worldwide, goiter affects about 5 percent of the population. The addition of iodide to table salt has greatly reduced the incidence of the condition in industrialized nations, but goiter is still common in the less industrialized countries of the world.

Calcitonin reduces blood calcium

Another hormone released by the thyroid gland is **calcitonin**, although it is not produced by the same cells that produce thyroxine. Calcitonin helps reduce the levels of calcium circulating in the blood (Figure 41.10).

Bone is a huge repository of calcium in the body and is continually being remodeled. Cells called **osteoclasts** break down bone and release calcium; **osteoblasts**, on the other hand, use circulating calcium to deposit new bone. Calcitonin decreases the activity of osteclasts and stimulates the activity of osteoblasts, thus shifting the

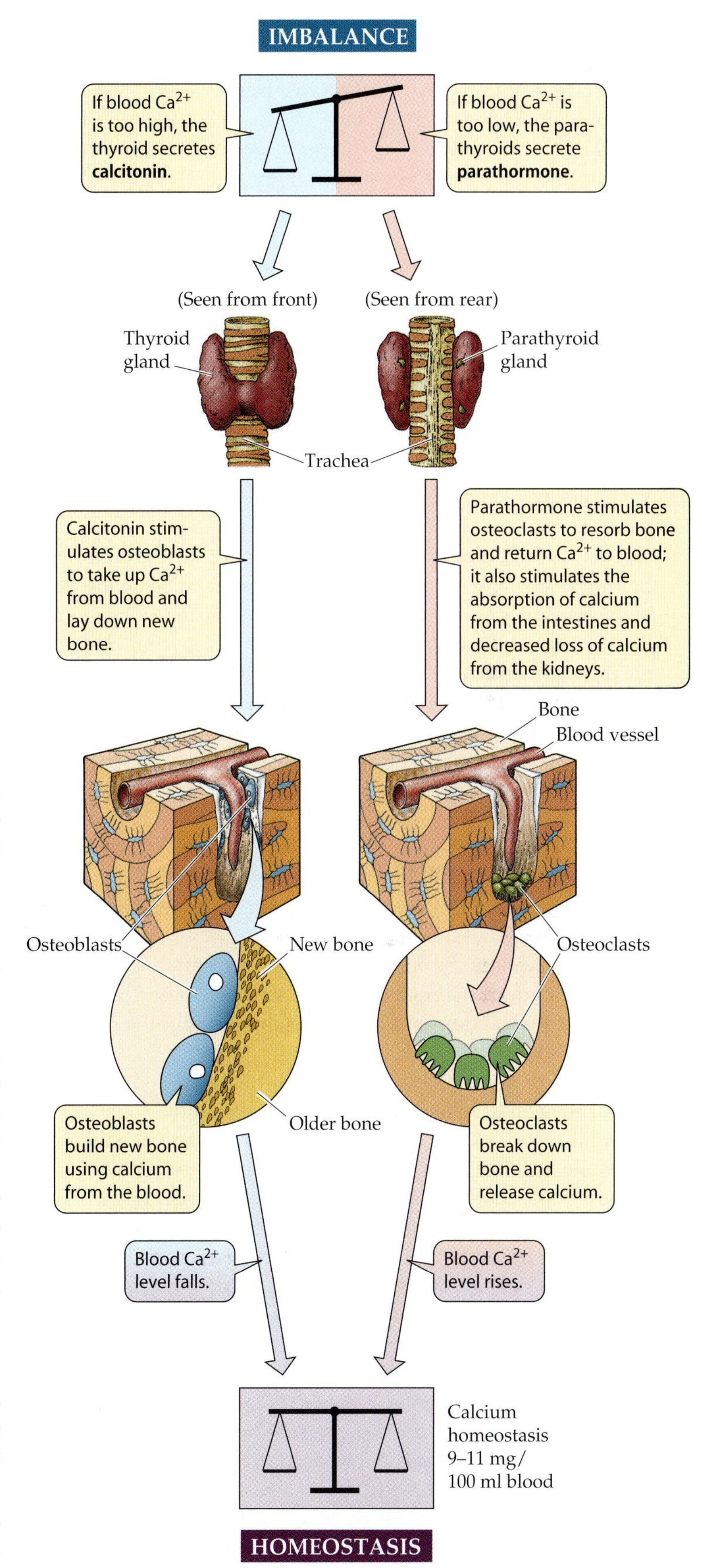

41.10 Hormonal Regulation of Calcium
Calcitonin and parathormone help regulate blood calcium levels. Bone can be a source (site of production) or a sink (site of utilization or storage) for calcium.

balance from adding calcium ions to the blood to removing them.

Calcitonin plays an important role in preventing bone loss in women during pregnancy. However, regulation of blood calcium levels is influenced more strongly by parathormone than it is by calcitonin.

Parathormone elevates blood calcium

The **parathyroid glands** are four tiny structures embedded on the surface of the thyroid gland. Their single hormone product is **parathyroid hormone**, or **parathormone**, a critical control in the regulation of blood calcium levels. Growth and remodeling of bone require calcium, and muscle contraction and nerve function are severely impaired if the blood calcium level rises or falls by as little as 30 percent of normal values.

A decrease in blood calcium triggers the release of parathormone, which stimulates osteoclasts to dissolve bone and release calcium to the blood (see Figure 41.10). Parathormone also prevents calcium in the blood from being lost in the urine by promoting calcium resorption by the kidneys. It also promotes the activation of vitamin D, which stimulates the digestive tract to absorb calcium from food.

Parathormone and calcitonin act antagonistically to regulate blood calcium levels: Parathormone elevates and calcitonin reduces. A similar antagonistic relationship is true of hormones of the pancreas, which regulate blood glucose levels.

Insulin and glucagon regulate blood glucose

Before the 1920s, *diabetes mellitus* was a fatal disease, characterized by weakness, lethargy, and body wasting. The disease was known to be connected somehow with the **pancreas**, a gland located just below the stomach, (see Figure 41.2), and with abnormal glucose metabolism, but the link was not clear.

Today we know that diabetes mellitus is caused by a lack of the hormone **insulin**, or by a lack of responsiveness of target tissues to that hormone. For patients in which the hormone is lacking, insulin replacement therapy is an extremely successful treatment. At present, more than 1.5 million people with diabetes in the United States lead almost normal lives through the use of manufactured insulin.

Insulin binds to a receptor on the plasma membranes of target cells, and this insulin–receptor complex allows glucose to enter the cell (see Figure 15.7). In the absence of insulin or insulin receptors, glucose accumulates in the blood until it is lost in the urine. High levels of blood glucose cause water to move from cells into the blood by osmosis, and the kidneys increase urine output to excrete excess fluid volume from the blood.*

Glucose uptake by most cells is impaired without insulin, so those cells must use fat and protein for fuel instead of glucose. As a result, the body of the untreated diabetic wastes away, and critical tissues and organs are damaged.

For centuries the prospects for a person with diabetes were bleak. A change in this outlook came almost overnight in 1921, when medical doctor Frederick Banting and medical student Charles Best of the University of Toronto discovered that they could reduce the symptoms of diabetes by injecting an extract they prepared from pancreatic tissue. The active component of the extract that Banting and Best prepared was found to be a small protein hormone—insulin—consisting of 51 amino acids.

Insulin is produced in clusters of endocrine cells in the pancreas. These clusters are called **islets of Langerhans** after the German medical student who discovered them. There are several types of cells in the islets:

- Beta (β) cells produce and secrete insulin.
- Alpha (α) cells produce and secrete the hormone **glucagon**, which has effects opposite those of insulin.
- Delta (δ) cells produce the hormone **somatostatin**.

The rest of the pancreas is an exocrine gland that produces enzymes and secretions that travel through ducts to the intestine, where they play roles in digestion.

After a meal, the concentration of glucose in the blood rises as glucose is absorbed from the food in the gut. This increase stimulates the pancreas to release insulin. Insulin stimulates cells to use glucose as fuel and to convert it into storage products such as glycogen and fat. When the gut contains no more food, the glucose concentration in the blood falls, and the pancreas stops releasing insulin. As a result, most cells of the body shift to using glycogen and fat rather than glucose for fuel. If the concentration of glucose in the blood falls below normal, the islet cells release glucagon, which stimulates the liver to convert glycogen back to glucose to resupply the blood. These effects will be discussed in greater detail in Chapter 50.

Somatostatin is a hormone of the brain and the gut

Somatostatin is released from the pancreas in response to rapid rises of glucose and amino acids in the blood. This hormone has paracrine functions within the islets: It inhibits the release of both insulin and glucagon. Its actions outside the pancreas slow the digestive activities of the gut. Pancreatic somatostatin extends the period of time during which nutrients are absorbed from the gut and used by the cells of the body. Somatostatin also acts as a hypothalamic neurohormone that inhibits the release of growth hormone and thyrotropin by the pituitary.

The adrenal gland is two glands in one

An **adrenal gland** sits above each kidney just below the middle of your back. Functionally and anatomically, an adrenal gland consists of a gland within a gland (Figure 41.11). The core, called the **adrenal medulla**, produces the hormone **epinephrine** (also known as *adrenaline*) and, to a lesser degree, **norepinephrine** (or *noradrenaline*). Surrounding the medulla (as an apricot surrounds its pit) is the **adre-**

*The name *diabetes* refers to the copious production of urine. *Mellitus* (Greek for "honey") reflects the fact that the urine of an untreated diabetic is sweet.

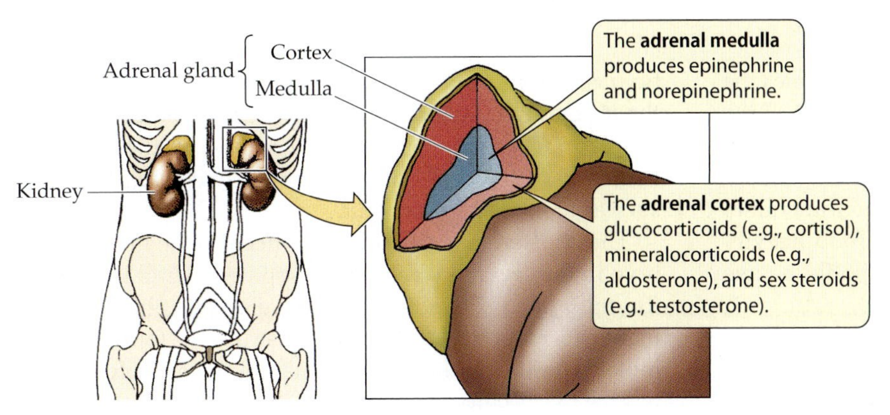

41.11 The Adrenal Gland Has an Outer and an Inner Portion
An adrenal gland, consisting of an outer cortex and an inner medulla, sits on top of each kidney. The medulla and the cortex produce different hormones.

nal cortex, which produces other hormones. The medulla develops from nervous system tissue and is under the control of the nervous system; the cortex is under hormonal control, largely by **adrenocorticotropin** (**ACTH**) from the anterior pituitary.

THE ADRENAL MEDULLA. The adrenal medulla produces epinephrine in response to stressful situations, arousing the body to action. As we saw earlier in this chapter, epinephrine increases heart rate, breathing rate, and blood pressure, and diverts blood flow to active skeletal muscles and away from the gut. These fight-or-flight reactions can be stimulated by physically threatening events, such as encountering a mugger, or by events that are mentally stressful, such as giving a public speech or taking a test. In situations of mental stress, many of the responses (such as increased heart and breathing rates) that would be useful for escaping physical danger are not useful, and can even be inconvenient or harmful.

Epinephrine and norepinephrine (a neurotransmitter involved in physiological regulation) bind to receptors on the surfaces of target cells. These receptors can be grouped into two general types, α-adrenergic and β-adrenergic receptors (each with at least two subtypes), which stimulate different actions within cells. Epinephrine acts equally on both types, but norepinephrine acts mostly on α-adrenergic receptors. Therefore, drugs called *beta blockers*, which selectively block β-adrenergic receptors, can reduce the fight-or-flight responses to epinephrine without disrupting the regulatory functions of norepinephrine.

THE ADRENAL CORTEX. The cells of the adrenal cortex use cholesterol to produce three classes of steroid hormones. (Figure 41.12). Collectively, these classes of hormones are called the **corticosteroids**.

- The **glucocorticoids** influence blood glucose concentrations as well as other aspects of fat, protein, and carbohydrate metabolism.
- The **mineralocorticoids** influence the ionic balance of extracellular fluids.
- The **sex steroids** stimulate sexual development and reproductive activity.

The sex steroids are secreted in only negligible amounts by the adrenal cortex. They will be discussed in the next section, on gonadal hormones.

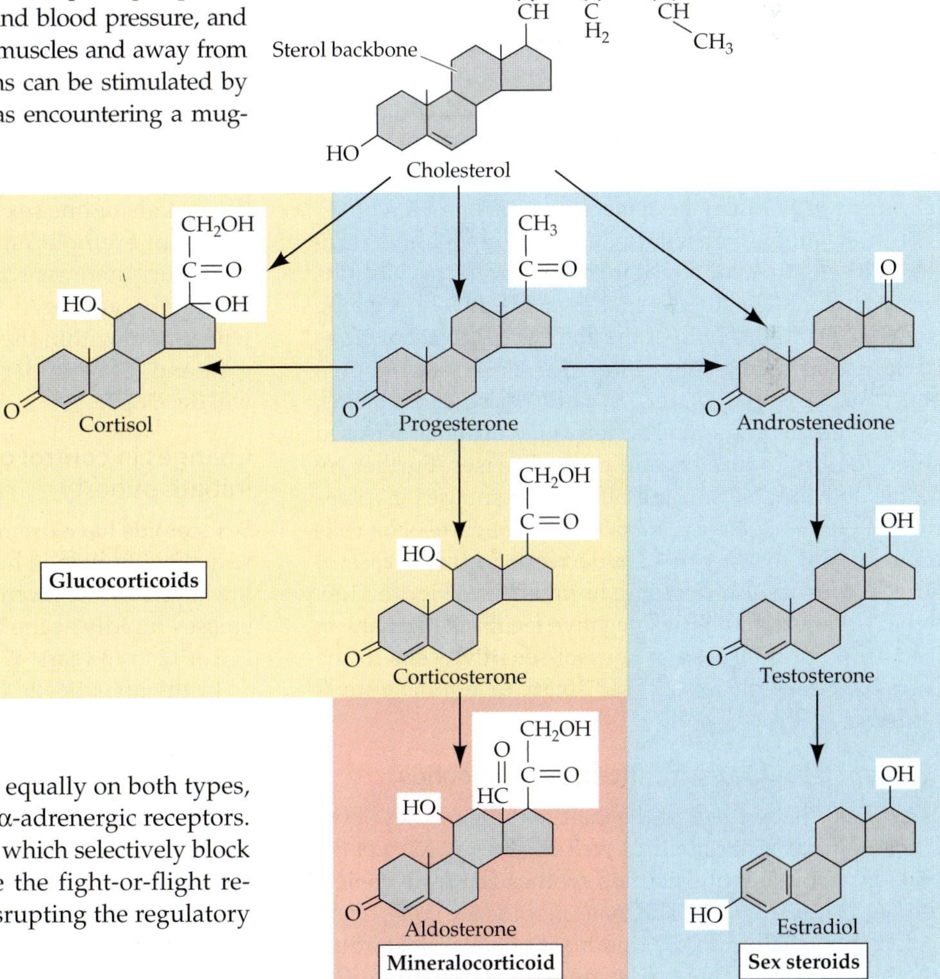

41.12 The Corticosteroids Are Built from Cholesterol
Side groups on the sterol backbone give different properties to the different corticosteroid hormones. This simplified outline of steroid biosynthesis leaves out many intermediate steps.

Aldosterone, the main mineralocorticoid, stimulates the kidney to conserve sodium and to excrete potassium. If the adrenal glands are removed from an animal, it must have sodium added to its diet, or its sodium will be depleted and it will die. One human patient with a nonfunctional adrenal gland compensated by salting her food heavily and, in addition, ate a 60-pound block of salt in the course of a year.

The main glucocorticoid, **cortisol**, is critical for mediating the body's response to stress. As we have seen, your immediate reaction to a frightening situation is stimulated by your nervous system and by the release of epinephrine. This fight-or-flight response ensures that your muscles will have enough oxygen and glucose to fuel your escape. You have a limited amount of blood glucose, however, and you need to conserve it for your muscles and your brain. Within minutes of the frightening stimulus, blood cortisol level rises. Cortisol stimulates cells not critical for your escape to decrease their use of blood glucose and shift instead to utilizing fats and proteins for energy. This is not a time to feel sick, have allergic reactions, or heal wounds, so cortisol also blocks immune system reactions. This is why cortisol is useful for reducing inflammations and allergies.

Cortisol release is controlled by ACTH from the anterior pituitary, which in turn is controlled by the hypothalamic **adrenocorticotropin-releasing hormone**. Because the cortisol response to a stressor has this chain of steps, each involving secretion, diffusion, circulation, and cell activation, it is much slower than the epinephrine response.

Turning off the cortisol response is as important as turning it on. A study of stress in rats showed that old rats could turn on their stress responses as effectively as young rats, but that they had lost the ability to turn them off as rapidly. As a result, they suffered from the well-known consequences of stress: ulcers, cardiovascular problems, strokes, impaired immune system function, and increased susceptibility to cancers and other diseases. Further research showed that turning off stress responses involves the long-loop negative feedback action of cortisol on cells in the brain, which causes a decrease in the release of adrenocorticotropin-releasing hormone (see Figure 41.8). Repeated activation of this negative feedback mechanism leads to a gradual loss of cortisol-sensitive cells in the brain, and therefore a decreased ability to terminate stress responses.

The sex steroids are produced by the gonads

The **gonads**—the testes of the male and the ovaries of the female—produce hormones as well as gametes. Most of the gonadal hormones are steroids synthesized from cholesterol (see Figure 41.12). The male steroids are collectively called **androgens**, and the dominant one is **testosterone**. The female steroids are **estrogens** and **progesterone**. The dominant estrogen is **estradiol**.

The sex steroids have important developmental effects: They determine whether a fetus develops into a female or a male. (A *fetus* is the latter stage of an embryo; a human embryo is called a fetus from the eighth week of pregnancy to the moment of birth.) After birth, the sex steroids control the maturation of the reproductive organs and the development and maintenance of secondary sexual characteristics, such as breasts and facial hair.

The sex steroids begin to exert effects in the human embryo in the seventh week of development. Until that time, the embryo has the potential to develop into either sex. In mammals and birds, the ultimate instructions for sex determination reside in the genes. In mammals, individuals that receive two X chromosomes normally become females, and individuals that receive an X and a Y chromosome normally become males (Figure 41.13). These genetic instructions are carried out through the production and action of the sex steroids, and the potential for error exists.

The presence of a Y chromosome normally causes the embryonic, undifferentiated gonads to begin producing androgens in the seventh week. In response to the androgens, the reproductive system develops into that of a male. If androgens are not produced at that time, female reproductive structures develop. In other words, androgens are required to trigger male development in humans. The opposite situation exists in birds: Male characteristics develop unless estrogens are present to trigger female development.

Occasionally the hormonal control of sexual development does not work perfectly, resulting in *intersex* individuals. The most extreme (but rare) case is a true **hermaphrodite**, who has both testes and ovaries. **Pseudohermaphrodites** have the gonads of one sex and the external sex organs of the other. For example, an XY fetus will develop testes, but if his tissues are insensitive to the androgens they produce because his androgen receptors do not function, the testes will remain within the abdomen, and the external sex organs and the secondary sexual characteristics of a female will develop.

Changes in control of sex steroid production initiate puberty

Sex steroids have dramatic effects at **puberty**—the time of sexual maturation in humans. Sex steroids are produced at low levels by the juvenile gonads, but their production increases rapidly at the beginning of puberty—around the age of 12 to 13 years. Why does this sudden increase occur?

In the juvenile, as in the adult, the production of sex steroids by the ovaries and testes is controlled by the anterior pituitary tropic hormones **luteinizing hormone** (**LH**) and **follicle-stimulating hormone** (**FSH**), which together are called the **gonadotropins**. The production of these tropic hormones is under the control of the hypothalamic gonadotropin-releasing hormone (GnRH). Prior to puberty, the gonads are capable of responding to gonadotropins, and the pituitary is capable of responding to GnRH. But prior to puberty the hypothalamus produces only very low levels of GnRH. Puberty is initiated by a reduction in the sensitivity of hypothalamic GnRH-producing cells to negative feedback from sex steroids and from gonadotropins. As a result, GnRH release increases, stimulating increased pro-

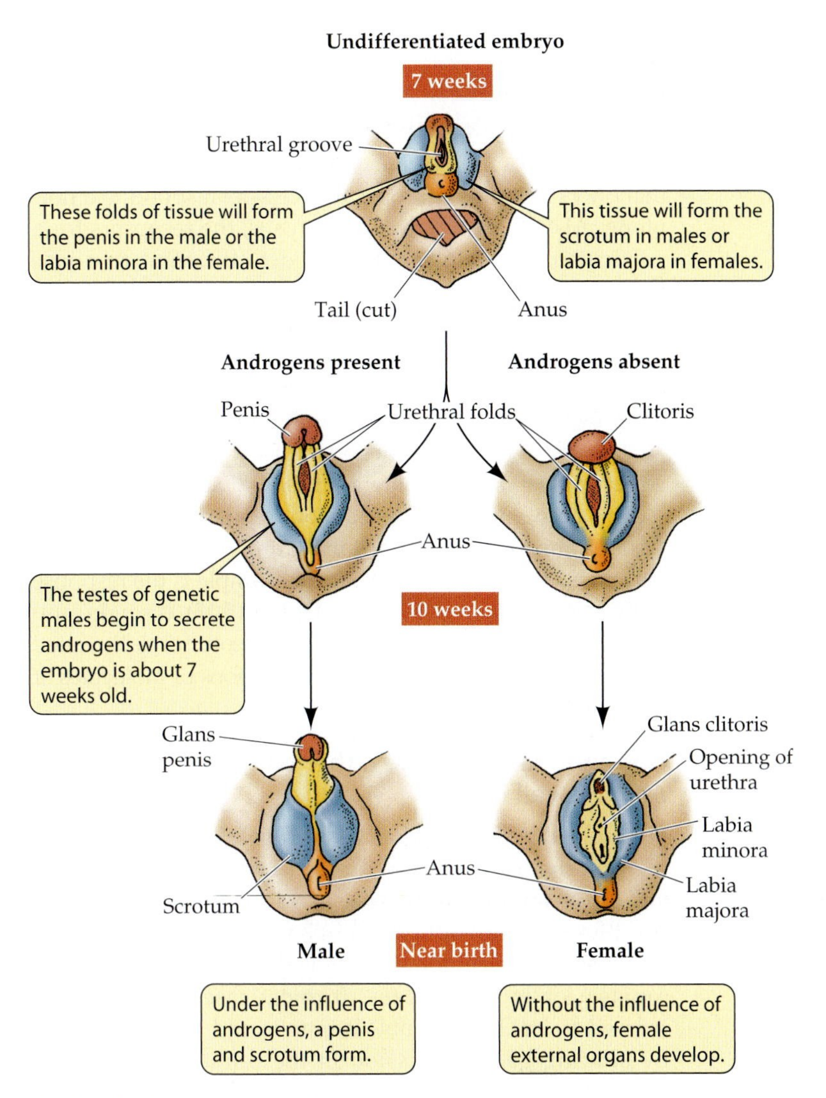

41.13 The Development of Human Sex Organs
The sex organs of early human embryos are similar. Male sex steroids (androgens) promote the development of male sex organs. Without androgen action, female sex organs form, even in genetic males.

duction of gonadotropins and hence increased production of sex steroids.

In females, increasing levels of LH and FSH at puberty stimulate the ovaries to begin producing the female sex hormones. The increased circulating levels of these hormones initiate the development of the traits of a sexually mature woman: enlarged breasts, vagina, and uterus; broad hips; increased subcutaneous fat; pubic hair; and the initiation of the menstrual cycle.

In the male, an increasing level of LH stimulates groups of cells in the testes to synthesize androgens, which in turn initiate the profound physiological, anatomical, and psychological changes associated with adolescence. The voice deepens, hair begins to grow on the face and body, and the testes and penis grow. Androgens also help skeletal muscles grow, especially when they are exercised regularly.

Natural muscle development can be exaggerated by both men and women who want to increase their maximum strength in athletic competition if they take synthetic androgens—called *anabolic steroids*. However, anabolic steroids have serious negative side effects. In women, their use causes the breasts and uterus to shrink, the clitoris to enlarge, menstruation to become irregular, facial and body hair to grow, and the voice to deepen. In men, the testes shrink, hair loss increases, the breasts enlarge, and sterility can result. You can understand the causes of some of these side effects by considering the negative feedback effects of sex steroids on the production of LH and FSH. Other side effects are even more serious. Continued use of anabolic steroids greatly increases the risk of heart disease, certain cancers, kidney damage, and personality disorders such as depression, mania, psychoses, and extreme aggression. Most official athletic organizations, including the International Olympic Committee, ban the use of anabolic steroids.

Melatonin is involved in biological rhythms and photoperiodicity

The **pineal gland** is situated between the cerebral hemispheres of the brain on a little stalk. It produces the hormone **melatonin** from the amino acid tryptophan. In various vertebrates, melatonin is involved in biological rhythms and photoperiodicity. **Photoperiodicity** is the phenomenon whereby seasonal changes in day length cause physiological changes in animals. Many species, for example, come into reproductive condition when the days begin to get longer (Figure 41.14). Humans are not photoperiodic, but melatonin in humans may play a role in entraining the daily rhythm of physiological and behavioral activities to the daily cycle of light and dark.

The release of melatonin by the pineal occurs in the dark and therefore marks the length of the night. Exposure to light inhibits the release of melatonin. The pineal of birds is directly sensitive to light, but in mammals, the light response is mediated through the eyes via a group of cells at the base of the brain, which generates a daily rhythm for many physiological functions of the body. We will learn more about this brain structure in Chapter 52.

The list of other hormones is long

We have discussed the major endocrine glands and the "classic" hormones in this chapter, but there are many hormones we have not mentioned. Examples include the hormones produced in the digestive tract that help organize the way the gut processes food (see Table 41.1 and Chapter 50). Even the heart has endocrine functions. When blood pressure rises and causes the walls of the heart to stretch, certain cells in the walls of the heart release *atrial natriuretic hormone*. This hormone increases the excretion of sodium ions and water by the kidneys, thereby lowering blood volume and blood pressure. As we discuss the organ systems

(a)

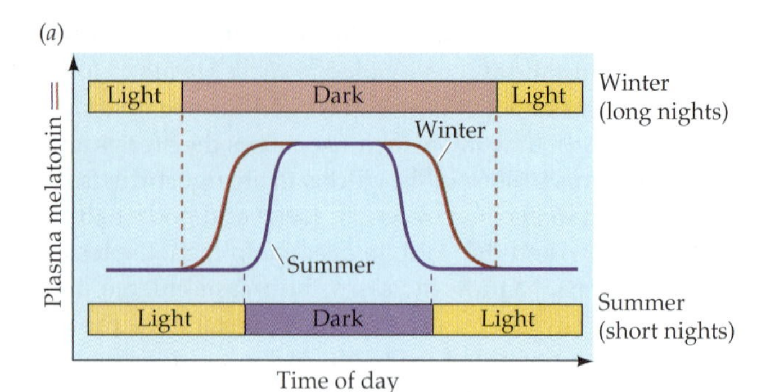

(b)

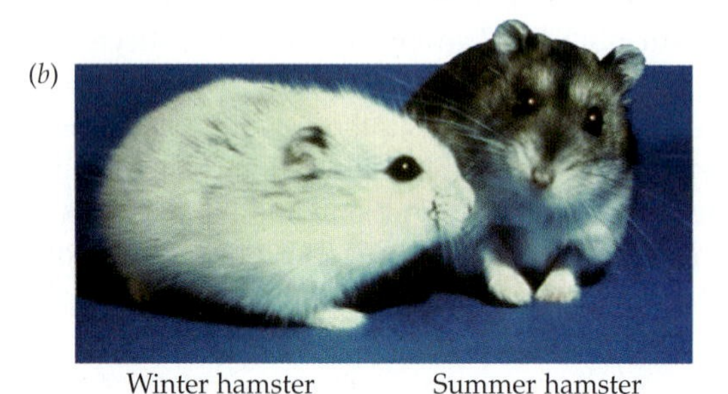

41.14 The Release of Melatonin Regulates Seasonal Changes
(*a*) Melatonin is released in the dark and inhibited by light exposure. The duration of daily melatonin release thus changes as the day length (photoperiod) changes, inducing dramatic seasonal physiological changes in some animals. (*b*) In winter these Siberian hamsters are white and are non-reproductive. In summer they are mottled brown and breed.

of the body in the chapters that follow, we will frequently mention hormones that their tissues produce, or hormones that control their functions.

Mechanisms of Hormone Action

The hormones we have discussed are released in very small quantities, yet they can cause large responses in cells or tissues all over the body, and these responses can be quite specific in different cells. For example, we have discussed the many dramatic effects of testosterone, yet its concentration in the blood of adult human males is only about 30–100 ng/ml. How can hormones in such tiny quantities have such strong and selective actions?

The actions of hormones depend on receptors and signal transduction pathways

The selective action of hormones is explained by the fact that only cells with appropriate receptors respond to a hormone. Also, in different types of cells, the receptors for a particular hormone can be linked to different response mechanisms. There are numerous signal transduction pathways in cells (see Chapter 15), and therefore the response of a cell to a hormone depends both on the receptors and on the signal transduction pathways that exist in that cell. Epinephrine, for example, acts through four different receptors and several signal transduction pathways to induce a wide variety of responses in different tissues (Table 41.3).

The strength of hormone action frequently results from signal transduction cascades that amplify the original signal, as described in Chapter 15. An example is the response of liver cells to epinephrine (see Figure 15.17). A single molecule of epinephrine binding to its receptor on a liver cell can result in that liver cell releasing millions of molecules of glucose into the blood.

Hormone receptors are either on the cell surface or in the cell interior

Hormones can be classified as lipid-soluble or water-soluble, and that classification relates to where their receptors are. **Lipid-soluble hormones**, such as the steroid hormones and thyroxine, can diffuse through plasma membranes, and therefore their receptors are inside the cell, either in the cytoplasm or in the nucleus. In most cases, the complex formed by the lipid-soluble hormone and its receptor acts by altering gene expression in the cell, as described in Chapter 15.

Water-soluble hormones cannot readily pass through plasma membranes, and their receptors are on the cell surface. Water-soluble hormones include peptides such as the hypothalamic releasing hormones, proteins such as insulin and glucagon, and some other kinds of molecules, such as epinephrine and norepinephrine.

The cell surface receptors of water-soluble hormones are large glycoprotein complexes with three domains: a *binding domain* projecting beyond the outside of the plasma membrane, a *transmembrane domain* that anchors the receptor in the membrane, and a *catalytic domain* that extends into the

41.3 *Diverse Actions of the Hormone Epinephrine*

TISSUE	RECEPTOR	SIGNAL TRANSDUCTION PATHWAY	ACTION
Arterioles in skin and gut	α_1	Stimulates IP_3	Constriction of vessels
Arterioles in leg muscles	β_2	Stimulates cAMP	Dilation of vessels
Heart muscle	β_1	Stimulates cAMP, Ca^{2+}	Increase rate and strength of contractions
Liver cells	β_2	Stimulates cAMP	Breakdown of glycogen
	α_1	Stimulates IP_3	Synthesis of glucose
Fat cells	β_1	Stimulates cAMP	Breakdown of lipids
Pancreas	β_2	Stimulates cAMP	Increased glucagon release
	α_2	Inhibits cAMP	Decreased insulin release

cytoplasm of the cell. The catalytic domain initiates cell responses by directly or indirectly activating protein kinases or protein phosphatases. In the direct pathway, the catalytic domain changes its shape and becomes capable of phosphorylating tyrosine residues on protein kinases or phosphatases, thus changing their enzymatic activities. In the indirect pathway, the catalytic domain stimulates one of the several *second messenger* pathways that were described in Chapter 15.

Regulation of hormone receptors controls sensitivity of cells to hormones

We learned above that the release of hormones can be under feedback control, usually negative feedback control. Similarly, the abundance of receptors for a hormone can be under feedback control. In some cases, continuous high levels of a hormone can decrease the number of its receptors, in a process known as **downregulation**.

An example of downregulation is type II diabetes mellitus, also called insulin-independent diabetes, or sometimes adult-onset diabetes, because it occurs more frequently in adulthood than does type I diabetes mellitus, which is due to a lack of insulin and is usually diagnosed in childhood. Type II diabetes is distinguished by elevated levels of circulating insulin, but a loss of receptors. Although genetic factors are likely to be involved, a possible immediate cause of the disease is an overstimulation of pancreatic release of insulin by excessive carbohydrate intake, which leads to downregulation of the insulin receptors.

Upregulation of receptors is a positive feedback mechanism, and is less common. One example, however, is the monthly ovarian cycle of human females, whereby an ovum matures and is released. As we will learn in the next chapter, the maturation of the ovum and its associated cells is under the control of FSH. Early in this process, FSH causes the cells associated with the ovum to increase their production of FSH receptors. This upregulation of FSH receptors accelerates the maturation process stimulated by FSH.

Responses to hormones can vary greatly

So far in this chapter we have discussed two mechanisms for regulating physiological responses to hormones: controlling the amount of hormone released, and controlling the availability of receptors. Many other factors can influence physiological responses to hormones; therefore, it is valuable to be able to characterize these responses. One way is to construct a **dose–response curve**.

To create such a curve, cells, tissues, organs, or even a whole animal are experimentally treated with different amounts of a hormone and the response is measured. The response is then plotted on the y axis of a graph, and the amount of the hormone used is plotted on the x axis (Figure 41.15). The resulting dose–response curve shows the *threshold* dose of the hormone necessary to get a response and the dose of the hormone that produces the *maximum* response. Between these two extreme values, the curve frequently has a sigmoid or S shape. The hormone dose that stimulates half the maximum response indicates the *sensitivity* of the cell, tissue, organ, or animal to the hormone.

Anything that changes the responsiveness of a system to a hormone is reflected in the dose–response curve. A change in the number of receptors in the responding cells, for example, can result in changes in threshold and in sensitivity. Changes in signaling pathways, rate-limiting enzymes, or the availability of substrates or cofactors can result in changes in the maximum response. Dose–response curves are valuable tools for studying the many factors that influence hormone-mediated processes.

Responses to hormones can also vary in their time course. Hormones are not simple on–off switches, and the time course over which it acts is an important characteristic of a hormone. This characteristic can be measured by the hormone's **half-life** in the blood. Soon after endocrine cells are stimulated to secrete their hormone, the hormone reaches its maximum concentration in the blood. By taking subsequent blood samples, researchers can determine how long it takes for the circulating hormone to drop to half of that maximum concentration.

The fight-or-flight response to epinephrine is relatively quick in its onset and termination, and the half-life of epinephrine in the blood is only 1 to 3 minutes. The actions of

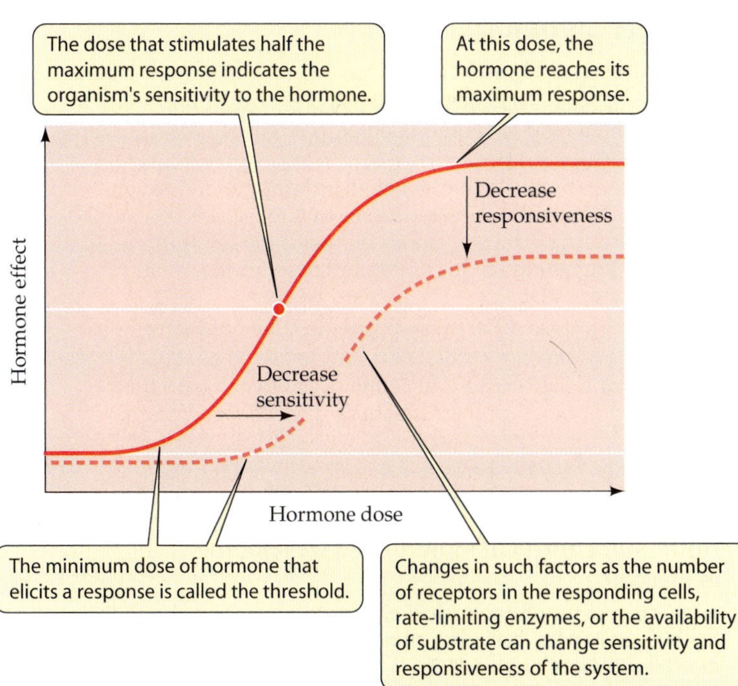

41.15 Dose–Response Curves Quantify Response to a Hormone
Between the threshold and maximum values, a dose–response curve frequently has a sigmoid or S shape. Anything that changes the responsiveness of a system—number of receptors in target cells, presence of enzymes and/or substrate, for example—affects the position of the curve.

other hormones, such as cortisol or thyroxine, are expressed over much longer periods, and their half-lives are on the order of days.

Once a hormone has been released, its half-life is partially determined by processes of degradation and elimination. Hormones are enzymatically degraded in the liver, then they are removed from the blood in the kidney and excreted in the urine. The presence of hormones or their breakdown products in the urine is the reason that urine samples can provide important information in clinical tests. In addition, most hormones are taken up and degraded by the very cells on which they act, so that they are not available to continually activate receptors.

Another factor that influences the half-life of a hormone is its ability to leave the blood. Some hormones, such as epinephrine, circulate as free molecules, but many circulate bound to carrier proteins. The extent to which hormones are bound to carrier proteins limits their ability to diffuse out of the blood to reach their target cells, to be degraded in the liver, or to be excreted by the kidney.

For example, when the mineralocorticoid aldosterone is released, about 15 percent of it binds to carrier proteins, and its half-life is 25 minutes. In contrast, when thyroxine is released, almost 100 percent of it binds to carrier proteins, and thyroxine has a half-life of 6 days. This variation in the time course of hormone responsiveness allows hormone signaling systems to have temporal characteristics that match their functions.

Of course, the nature of the target cell response to a hormone is also a factor in determining the time course of hormone action. For a hormone that stimulates a developmental effect, the time course of hormone action can be months, years, and even a lifetime. A very good example of a long-term process regulated primarily by hormones is animal reproduction, the topic of the next chapter.

Chapter Summary

Hormones and Their Actions

▶ Endocrine cells secrete chemical messages called hormones, which bind to receptors on or in target cells.

▶ Most hormones diffuse through the extracellular fluids and are picked up by the blood, which distributes them throughout the body. Some hormones diffuse to targets near the site of secretion. Autocrine hormones influence the cell that secretes them; paracrine hormones influence nearby cells. **Review Figure 41.1**

▶ Hormones cause different responses in different target cells.

▶ The chemical structures of hormones have changed little through evolution, but their functions have changed dramatically.

▶ Hormones may be secreted by single cells or by cells organized into discrete endocrine glands. **Review Figure 41.2**

Hormonal Control of Molting and Development in Insects

▶ Insects must molt their exoskeletons to grow. Two diffusible substances, brain hormone and ecdysone, control molting. **Review Figure 41.3**

▶ Juvenile hormone, another diffusible substance, prevents maturation so that juvenile instars molt into bigger juvenile instars. When the level of juvenile hormone falls low enough, the juvenile molts into the adult form.

▶ Some insects, such as butterflies, go through complete metamorphosis. When juvenile hormone drops to a low level, the larval form becomes a pupa. Because no juvenile hormone is secreted during pupation, the pupa molts into an adult. **Review Figure 41.4**

Vertebrate Endocrine Systems

▶ Vertebrates have nine endocrine glands that secrete many hormones. **Review Figure 41.2, Table 41.1**

▶ The pituitary gland is divided into two parts. The anterior pituitary develops from embryonic mouth tissue; the posterior pituitary develops from the brain.

▶ The posterior pituitary secretes the neurohormones vasopressin and oxytocin. **Review Figure 41.5**

▶ The anterior pituitary secretes tropic hormones (thyrotropin, adrenocorticotropin, and two gonadotropins), as well as growth hormone, prolactin, melanocyte-stimulating hormone, endorphins, and enkephalins.

▶ The anterior pituitary is controlled by neurohormones produced by cells in the hypothalamus and transported through portal blood vessels to the anterior pituitary. **Review Figure 41.7, Table 41.2**

▶ Hormone release in the hypothalamus/pituitary/endocrine gland axis is controlled by many feedback loops. **Review Figure 41.8**

▶ The thyroid gland is controlled by thyrotropin and secretes thyroxine, which controls cell metabolism. Goiter can be associated with too little or too much thyroxine. **Review Figure 41.9**

▶ The level of calcium in the blood is regulated by two hormones. Calcitonin, produced by the thyroid, lowers blood calcium. Parathormone, produced by the parathyroid glands, raises it. **Review Figure 41.10**

▶ The pancreas secretes three hormones. Insulin stimulates glucose uptake by cells and lowers blood glucose, glucagon raises blood glucose, and somatostatin slows the rate of nutrient absorption from the gut.

▶ The adrenal gland has two portions, one within the other. The hormones of the adrenal medulla, epinephrine and norepinephrine, stimulate the liver to supply glucose to the blood, as well as other fight-or-flight reactions. **Review Figure 41.11**

▶ The adrenal cortex produce three classes of corticosteroids: glucocorticoids, mineralocorticoids, and small amounts of sex steroids. **Review Figure 41.12**

▶ Aldosterone is a mineralocorticoid that stimulates the kidney to conserve sodium and to excrete potassium.

▶ Cortisol is a glucocorticoid that decreases glucose utilization by most cells.

▶ Sex hormones (androgens in males, estrogens and progesterone in females) are produced by the gonads in response to tropic hormones. Sex hormones control sexual development, secondary sexual characteristics, and reproductive functions. **Review Figure 41.13**

▶ The pineal hormone melatonin is involved in controlling biological rhythms and photoperiodism. **Review Figure 41.14**

Mechanisms of Hormone Action

▶ The responses of a cell to a hormone depend on what receptors it has and what signal transduction pathways those receptors activate. **Review Table 41.3**

▶ The receptors for water-soluble hormones are on the cell surface, and the receptors for lipid-soluble hormones are inside the cell.

▶ The sensitivity of a cell to hormones can be altered by up- or downregulation of the receptors in that cell.

▶ The sensitivity and time course of a response to a hormone depend on many factors, including receptor numbers, properties of signal transduction pathways, the actions of other hormones, binding of the hormone to carrier proteins, and elimination of the hormone through degradation and excretion.

▶ Important tools for characterizing hormone action are dose–response curves and measurements of half-life. **Review Figure 41.15**

For Discussion

1. Explain how both hyperthyroidism and hypothyroidism can cause goiter. Refer to the roles of the hypothalamus and the pituitary in your answer.
2. In the 1960s, women who showed signs of premature labor were sometimes treated with progestins (substances that have progesterone activity) to inhibit contractions. The female children of these women tended to be "tomboys." In recent years many of these offspring have claimed that they feel more male than female and have therefore undergone sex change operations. Can you suggest an explanation for this phenomenon? (*Hint:* Review Figure 41.12.)
3. Various side effects of anabolic steroid use were mentioned in this chapter. Some of these effects are due to the direct action of the steroid, but others are due to the negative feedback action of the steroid. Discuss an example of each and explain possible mechanisms.
4. The time course of hormone action can vary over a broad range. Compare the characteristics you would expect of a hormone signaling system that controls a short-term process, such as digestive functions, with the characteristics you would expect of a hormone signaling system that controls a long-term process, such as a developmental process.

42 Animal Reproduction

NATURAL SELECTION HAS CREATED SOME AMAZING and bizarre adaptations, but among the most unusual and diverse are the methods some animals use to reproduce. Just as "unmanned" submersibles are used in deep ocean exploration, some species of polychaete worms use "unwormed" submersibles to reproduce. The adults of these marine worms live in burrows on the ocean floor or in reefs. Predators make it dangerous for them to leave their burrows to seek a mate, and if they simply released their eggs and sperm at the mouth of the burrow, they would have a poor chance of successful fertilization. So both males and females develop specialized body segments that form at the worm's posterior end and become stuffed with sperm or eggs. These segments develop sensory organs but no mouth or gut, since they will not need to feed.

When the time is right—full moon for some species, new moon for others—these "sex-cell transporters" break loose from the main body of the worm, leave the burrow, swim up into the water column, swarm with more of their kind, and release their sperm or eggs. The sex-cell transporters die soon after they release their cargo. Union of sperm and eggs takes place in the water column, and fertilized eggs may drift a long way before they descend to the ocean floor and develop into adult worms. But in many places, the native people know when and where the sex-cell transporters will swarm, and people harvest them for food.

In this chapter you will learn how animals reproduce. We first examine *asexual* mechanisms of reproduction, in which only a single parent is involved, and then turn to *sexual* reproduction, which requires two parents. Sexually reproducing organisms produce haploid sex cells—sperm and eggs—through the process of meiosis. An egg and a sperm must unite through the process of fertilization to create a new diploid individual.

As we will see, much of the diversity in reproductive systems is in mechanisms for getting sperm and eggs together. This chapter, however, focuses the most attention on the anatomy, function, and endocrine control of the human reproductive system. This information will allow us to understand the technologies we use both to limit and to overcome infertility. We end the chapter with a discussion of sexual health and sexually transmitted diseases.

Asexual Reproduction

Sexual reproduction is a nearly universal trait of animals, although many species can reproduce asexually as well. Offspring produced asexually are genetically identical to one another and to their parents. Asexual reproduction is highly efficient because there is no mating. Mating requires energy, involves risks, and requires that resources be devoted to a large population of males, who do not produce offspring. Asexual populations can use resources efficiently because all individuals in the population can convert resources to offspring. However, asexual reproduction does not generate genetic diversity, and this can be a disadvantage in changing environments. As we learned in Chapter

Feasting on Sex Cells
During the final quarter of November's moon, the people of Samoa and Fiji harvest the reproductive segments of the palolo worm, *Eunice viridis*. The adult worms release specialized reproductive vehicles into the ocean according to a precise cycle that native people have understood for centuries. The protein-rich worm segments are prepared by roasting or frying and are eaten as a delicacy.

42.1 Asexual Reproduction in Animals
(*a*) Budding: A new individual forms as an outgrowth from an adult hydra. (*b*) Regeneration: This five-armed sea star is generating three new arms to replace amputated limbs and form an entire animal. (*c*) Parthenogenesis: When reproductive conditions are favorable, female aphids produce genetically identically offspring.

(*a*) *Hydra* sp.

(*b*) *Linickia* sp.

(*c*) *Macrosiphum rosae*

21, genetic diversity enables natural selection to shape adaptations in response to environmental change.

A variety of species, mostly invertebrates, reproduce asexually. They tend to be species that are sessile and cannot search for mates, or species that live in sparse populations and rarely encounter potential mates. Furthermore, asexually reproducing species are likely to be found in relatively constant environments, in which the potential for rapid evolutionary change is not as important as in more variable environments.

There are three common modes of asexual reproduction:

- *Budding*, in which new individuals form by mitotic cell division.
- *Regeneration*, in which a piece or section of an organism can generate an entire new individual.
- *Parthenogenesis*, in which individuals develop from unfertilized eggs.

Budding and regeneration produce new individuals by mitosis

Many simple multicellular animals produce offspring by **budding**; new individuals form as outgrowths of the bodies of older animals. These buds grow by mitotic cell division, and the cells differentiate before the buds break away from the parent (Figure 42.1*a*). The bud is genetically identical to the parent, and it may grow as large as the parent before it becomes independent.

Regeneration is usually thought of as the replacement of damaged tissues or lost limbs, but in some cases pieces of an organism can regenerate complete individuals. In a classic experiment demonstrating regeneration, a sponge is pushed through a cloth mesh, producing many little clusters of cells. Each cluster grows into a small but complete sponge. The ability of sponges to regenerate was used off the coast of Florida to restore the commercial bath-sponge fishery, which was endangered by overfishing. Echinoderms also have remarkable abilities to regenerate. If sea stars are cut into pieces, each piece that includes a portion of the central disc grows into a new animal (Figure 42.1*b*).

Regeneration frequently results when an animal is broken by an outside force. A storm, for example, can cause a heavy surf that breaks colonial cnidarians such as corals. Pieces broken off the colony can regenerate into new colonies. In some species, the breakage occurs in the absence of external forces. Some species of segmented marine worms related to the ones we discussed at the beginning of this chapter develop segments with rudimentary heads bearing sensory organs, then break apart. Each fragmented segment forms a new worm.

Parthenogenesis is the development of unfertilized eggs

Not all eggs have to be fertilized to develop. A common mode of asexual reproduction in arthropods is the development of offspring from unfertilized eggs. This phenomenon, called **parthenogenesis**, also occurs in some species of fish, amphibians, and reptiles. Most species that reproduce parthenogenetically also engage in sexual reproduction or sexual behavior.

The aphids that can rapidly populate your rosebushes in the spring and summer reproduce parthenogenetically while conditions are favorable (Figure 42.1*c*). Some of the unfertilized eggs laid in spring and summer develop into male aphids, others into females. As conditions become less favorable, the aphids mate, and the females lay fertilized eggs. These eggs do not hatch until the following spring, and they yield only females.

In some species, parthenogenesis is part of the mechanism that determines sex. For example, in many hymenopterans (ants, and most species of bees and wasps), males develop from unfertilized eggs and are haploid.

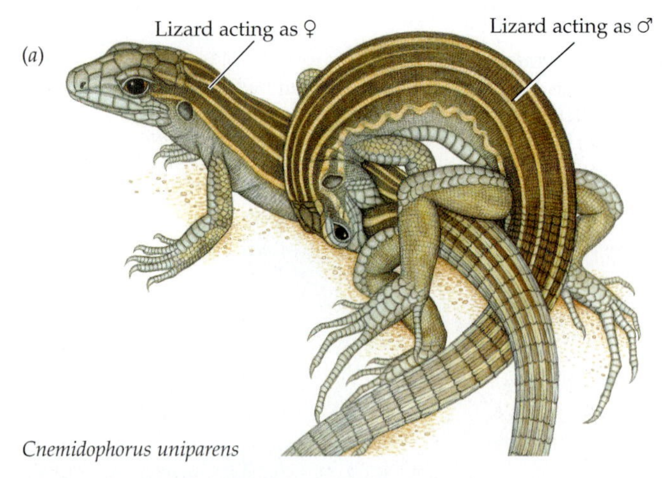

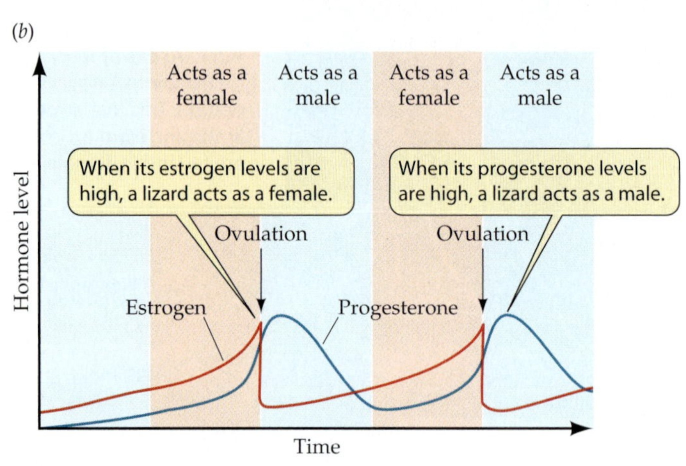

42.2 Sexual Behavior May Be Required for Asexual Reproduction
(*a*) Parthenogenetic whiptail lizards are all female, but take turns acting the male role in reproductive behavior. (*b*) The stage of the ovarian cycle determines the role an individual plays.

Females develop from fertilized eggs and are diploid. Most females are sterile workers, but a select few become fertile queens. After a queen mates, she has a supply of sperm that she controls, enabling her to produce either fertilized or unfertilized eggs. Thus the queen determines when and how much of the colony resources are expended on males.

Parthenogenetic reproduction in some species requires sexual activity even though this activity does not fertilize eggs. The eggs of parthenogenetically reproducing ticks and mites, for example, develop only after the animals have mated, even though the eggs remain unfertilized. One case that has been investigated by David Crews and his students at the University of Texas is parthenogenetic reproduction in a species of whiptail lizard. There are no males in this species, but females act as males, engaging in all aspects of courtship display and mating, even though no sperm are produced or transferred (Figure 42.2). Whether a specific female acts as a female or as a male depends on her hormonal state at the time, but sexual activity is required to stimulate ovulation.

Sexual Reproduction

A large portion of the time and energy budgets of sexually reproducing animals goes into sexual behavior, which exposes them to predation, can result in physical damage, and detracts from other useful activities such as feeding, building secure living places, and caring for existing offspring. In spite of all of these disadvantages, there is an overwhelming evolutionary advantage to sexual reproduction: It produces genetic diversity.

Sexual reproduction requires the joining of two haploid sex cells to form a diploid individual. These haploid cells, or *gametes*, are produced through **gametogenesis**, a process that involves meiotic cell divisions. Two events in meiosis contribute to genetic diversity: *crossing over* of homologous chromosomes, and the *independent assortment* of chromosomes. Both of these genetic phenomena were described in Chapter 10.

Mating behavior also contributes to genetic diversity in sexually reproducing species. The genetic variation in the gametes of a single individual and the genetic variation between any two parents produce an enormous potential for genetic variation between any two offspring of a sexually reproducing pair of individuals. This genetic diversity is the raw material for natural selection; thus evolutionary change in sexually reproducing animals can be quite rapid.

There are three fundamental phenomena of sexual reproduction in animals:

- *Gametogenesis* (making sex cells)
- *Mating* (getting sex cells together)
- *Fertilization* (getting sex cells to fuse)

There is not a great deal of diversity in gametogenesis when we compare different groups of animals. Processes of fertilization are also rather similar in widely different species. Therefore, although the discussion of gametogenesis that follows is primarily derived from information from mammals, the facts would not be terribly different if we focused on a different group of animals.

Mating, on the other hand, shows incredible evolutionary diversity. Our discussion of mating in this chapter will focus on a few specific examples as representative of the fascinating diversity that exists.

Eggs and sperm form through gametogenesis

Gametogenesis occurs in the primary sex organs, the **gonads**, which are **testes** (singular testis) in males and **ovaries** in females. The tiny gametes of males, called **sperm**, are motile and move by beating their flagella. The much larger female gametes are **eggs**, or **ova** (singular ovum), and are nonmotile (Figure 42.3).

Gametes are produced from **germ cells**, which have their origin in the earliest cell divisions of the embryo and remain distinct from the rest of the body. All the rest of the cells of the embryo are called **somatic cells**. Germ cells are sequestered in the body of the embryo until its gonads begin to form. The germ cells then migrate to the gonads, where they take up residence and proliferate by mitosis,

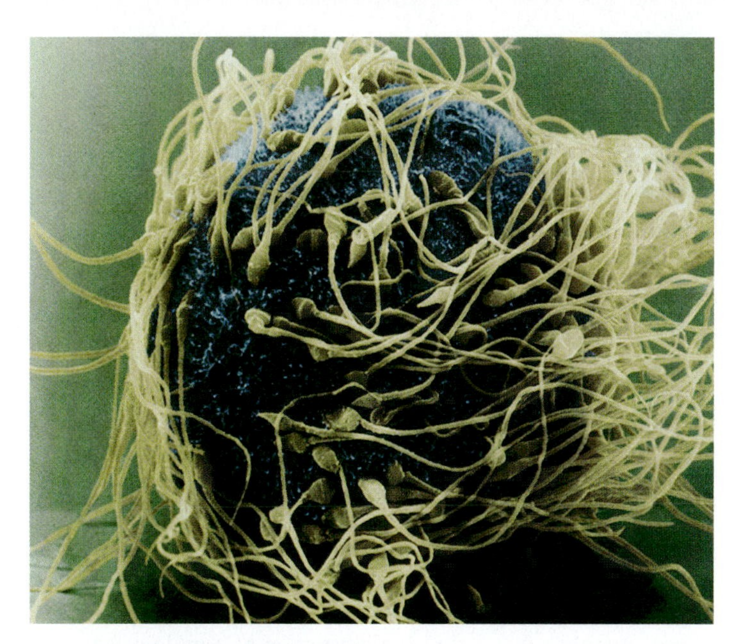

42.3 Gametes Differ in Size
Mammalian sperm (white) are much smaller than the mammalian egg (blue), as illustrated by this artificially colored micrograph of human fertilization.

producing **oogonia** (singular oogonium) in females and **spermatogonia** (singular spermatogonium) in males. Oogonia and spermatogonia, which are diploid, multiply by mitosis in turn, eventually producing **primary oocytes** and **primary spermatocytes**, which are still diploid cells.

Meiosis, the next step in gametogenesis, reduces the chromosomes to the haploid number, and these haploid cells mature into sperm and ova. (You may want to review the discussion of meiosis in Chapter 9 before reading further.) Although the steps of meiosis are very similar in males and females, there are some significant differences in gametogenesis.

SPERMATOGENESIS PRODUCES SPERM. Primary spermatocytes undergo the first meiotic division to form **secondary spermatocytes**, which are haploid. The second meiotic division produces four haploid **spermatids** for each primary spermatocyte that entered meiosis. In mammals, these cells remain connected by cross-bridges of cytoplasm after each division (Figure 42.4*a*).

The reason that mammalian spermatocytes remain in cytoplasmic contact throughout their development probably is the asymmetry of sex chromosomes in the males of

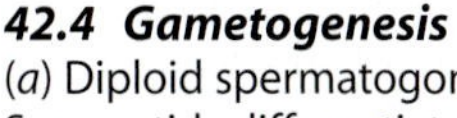

42.4 Gametogenesis
(*a*) Diploid spermatogonia develop into haploid spermatids. Spermatids differentiate into sperm. (*b*) Diploid oogonia develop into haploid secondary oocytes, which mature into ova.

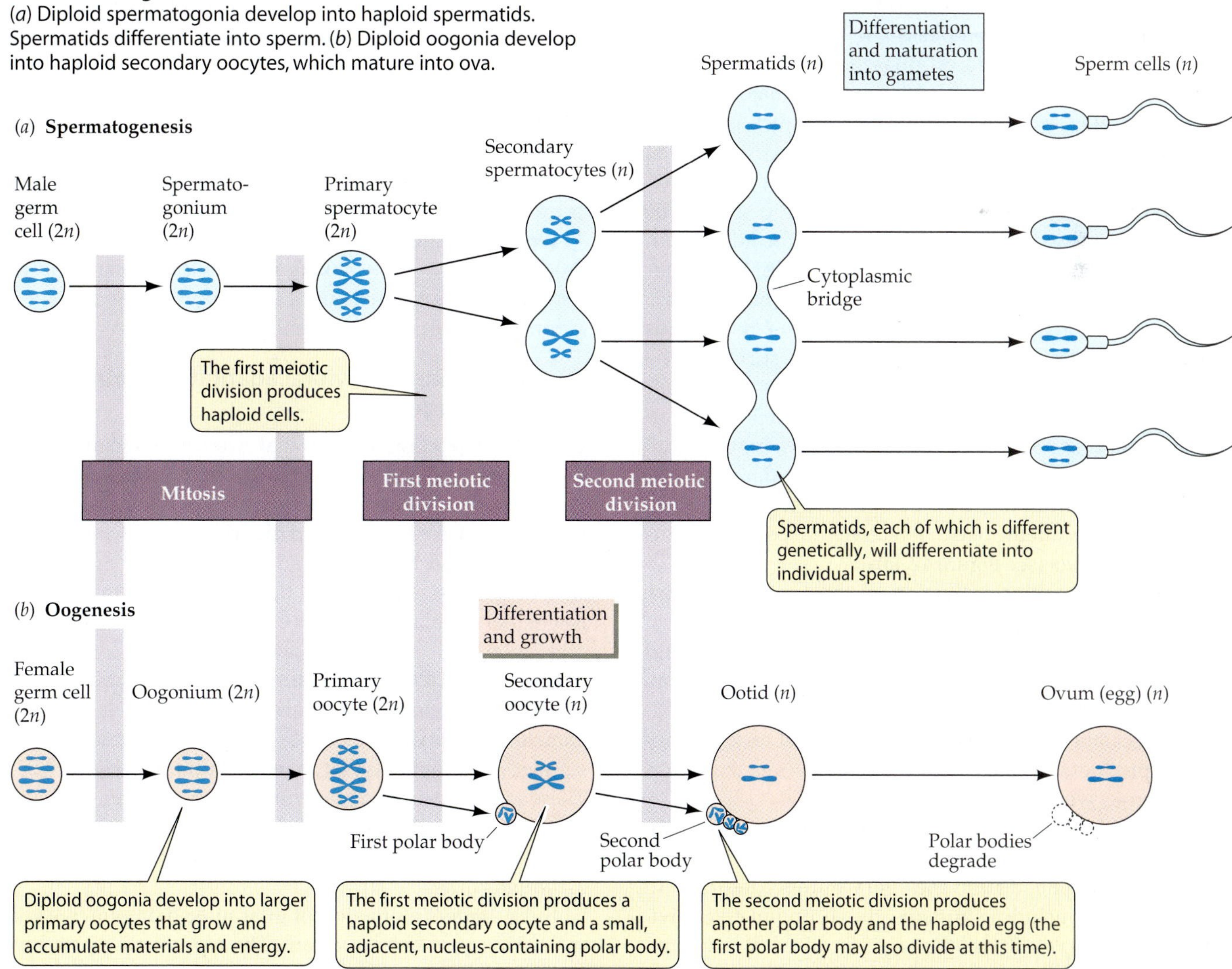

most species. Half of the secondary spermatocytes receive an X chromosome, the other half a Y chromosome. The Y chromosome contains fewer genes than the X chromosome, and apparently some of the products of genes not included in the Y chromosome are essential for spermatocyte development. By remaining in cytoplasmic contact, all spermatocytes can share the gene products of the X chromosomes, even though only half of them have an X chromosome.

Just after being produced by meiosis, a spermatid bears little resemblance to a sperm. Through further differentiation, it will become compact, streamlined, and motile. We will look at the differentiation of human sperm in more detail below.

OOGENESIS PRODUCES EGGS. Oogonia, like spermatogonia, proliferate through mitosis. The resulting egg precursor cells differentiate into primary oocytes, which immediately enter prophase of the first meiotic division. In many species, including humans, the development of the oocyte is arrested at this point, and may remain so for days, months, or years. In contrast, there is no arrest in the development of male gametes; the process goes steadily to completion once the primary spermatocyte has differentiated. In the human female, as we will see, some primary oocytes may remain in arrested prophase I for 50 years!

During this prolonged prophase I, or shortly before it ends, the primary oocyte undergoes its major growth phase. It grows larger due to increased production of ribosomes, RNA, cytoplasmic organelles, and energy stores. At this time, the primary oocyte acquires all of the energy, raw materials, and RNA that the egg will need to survive its first cell divisions after fertilization. In fact, the nutrients in the egg will have to nourish the embryo until it is either nourished by the maternal system or can feed on its own.

When a primary oocyte resumes meiosis, its nucleus completes the first meiotic division near the surface of the cell. The daughter cells of this division receive grossly unequal shares of cytoplasm. This asymmetry represents another major difference from spermatogenesis, in which cell divisions apportion cytoplasm equally. The daughter cell that receives almost all of the cytoplasm becomes the **secondary oocyte**, and the one that receives almost none forms the *first polar body* (see Figure 42.4*b*).

The second meiotic division of the large secondary oocyte is also accompanied by an asymmetrical division of the cytoplasm. One daughter cell forms the large, haploid **ootid**, which eventually differentiates into a mature ovum, and the other forms the *second polar body*. Polar bodies degenerate, so the end result of oogenesis is that each primary oocyte produces only one mature egg. However, that egg is a very large, well-provisioned cell.

A second period of arrested development occurs after the first meiotic division forms the secondary oocyte. The egg may be expelled from the ovary in this condition. In many species, including humans, the second meiotic division is not completed until the egg is fertilized by a sperm.

A single body can function as both male and female

Sexual reproduction requires both male and female haploid gametes. In most species, these gametes are produced by individuals that are either male or female. Species that have male and female members are called **dioecious** (from the Greek for "two houses"). In some species, a single individual may possess both female and male reproductive systems. Such species are called **monoecious** ("one house") or **hermaphroditic**.

Almost all invertebrate groups have hermaphroditic species. An earthworm is an example of a **simultaneous hermaphrodite**, meaning that it is both male and female at the same time. When two earthworms mate, they exchange sperm, and as a result, the eggs of each are fertilized (see Figure 31.24*b*). Some animals are **sequential hermaphrodites**, meaning that individuals function as a male or as a female at different times in their lives.

What is the selective advantage of hermaphroditism? Some simultaneous hermaphrodites have a low probability of meeting a potential mate. An example is a parasitic tapeworm. Even though it may be large and cause lots of trouble for its host, it may be the only tapeworm in the host. Tapeworms can fertilize their own eggs. Most simultaneous hermaphrodites must mate with another individual, but since each member of the population is both male and female, the probability of encountering a possible mate is double what it would be in monoecious species.

Sequential hermaphroditism can reduce the possibility of inbreeding among siblings by making them all the same sex at the same time and therefore incapable of mating with one another. In a species in which only a few males fertilize all females, sequential hermaphroditism can maximize reproductive success by making it possible for an individual to reproduce as a female until the opportunity arises for it to function successfully as a male.

Anatomical and behavioral adaptations bring eggs and sperm together

Sexual reproduction requires that two haploid gametes join together to form a diploid **zygote**. The purpose of mating behavior is to get eggs and sperm close enough together that this process—called **fertilization**—can occur. Many anatomical and behavioral adaptations have evolved to support mating. The simplest distinction in mating systems is whether fertilization occurs externally or internally.

EXTERNAL FERTILIZATION REQUIRES AN AQUATIC HABITAT. In an aquatic environment, animals can simply release their gametes into the water. *External fertilization* is common among simple aquatic animals that are not very mobile (Figure 42.5). These animals produce huge numbers of gametes. A female oyster, for example, may produce 100 million eggs in a year, and the number of sperm produced by a male oyster is astronomical.

But numbers alone do not guarantee that gametes will meet. Timing is also important. The reproductive activities

Acropora sp.

42.5 External Fertilization Is Common in Aquatic Species
External fertilization requires an aqueous environment. These staghorn corals are all releasing sperm–egg bundles into the oceans of Ningaloo Reef, Australia.

of the males and females of a population must be synchronized. Seasonal breeders may use day length, changes in temperature, or changes in weather to time their production and release of gametes. Social stimulation is also important. Sexual activity on the part of one member of a population can stimulate others to engage in mating.

Behavior can play an important role in bringing gametes together even when fertilization is external. Many species travel great distances to congregate with potential mates and release their gametes at the same time in a suitable environment. Salmon are an extreme example, traveling hundreds of miles to spawn in the stream where they hatched.

INTERNAL FERTILIZATION ENABLES TERRESTRIAL LIFE. Sperm can move only through liquid, and delicate gametes released into air would dry out and die. Terrestrial animals avoid these problems by engaging in *internal fertilization*.

Animals have evolved an incredible diversity of behavioral and anatomical adaptations toget male gametes into the female reproductive tract. As we saw above, gametogenesis occurs in the primary sex organs, the gonads. All of the additional anatomical components of an animal's reproductive system are called **accessory sex organs**. An obvious accessory sex organ in the male is a tubular structure called the **penis**, which enables the male to deposit sperm in the female's accessory sex organ, the **vagina** (or, in some species, the **cloaca**, a cavity common to the digestive, urinary, and reproductive systems). Accessory sex organs include a variety of glands, tubules, ducts, and other structures.

Copulation is the physical joining of the male and female accessory sex organs. Transfer of sperm in internal fertilization can also be indirect. Males of some species of mites and scorpions (among the arthropods) and salamanders (among the vertebrates) deposit **spermatophores**—containers filled with sperm—in the environment. When a female mite finds a spermatophore, she straddles it and opens a pair of plates in her abdomen so that the tip of the spermatophore enters her reproductive tract and allows the sperm to enter. Some female salamanders use the lips of their cloacae to scoop up the spermatophore.

Male squid and spiders play a more active role in spermatophore transfer. The male spider secretes a drop containing sperm onto a bit of web; then, with a special structure on his foreleg, he picks up the sperm-containing web and inserts it through the female's genital opening. Male squid use one special tentacle to pick up a spermatophore and insert it into the female's genital opening.

Most male insects copulate and transfer sperm to the female's vagina through a tubular penis. The **genitalia**—external sex organs—of insects often have species-specific shapes that match in a lock-and-key fashion. This mechanism ensures a tight, secure fit between the mating pair during the prolonged period of sperm transfer. The males of some insect species have elaborate structures on their penises that can scoop sperm deposited by other males out of the female's reproductive tract.

The evolution of vertebrate reproductive systems parallels the move to land

The earliest vertebrates evolved in aquatic environments. The closest living relatives of those earliest vertebrates are modern-day fishes. They remain exclusively aquatic animals, and most practice external fertilization. The most primitive of the fishes, the lampreys and hagfishes, broadcast their gametes into the environment, as do many aquatic invertebrates. In most fishes, however, fertilization is more selective: Mating behaviors bring females and males into close proximity at the time of gamete release.

In some sharks and rays, certain fins have evolved into structures that hold the male and female together and enable sperm to be transferred directly into the female reproductive tract. This internal fertilization in sharks and rays has made it possible for the females of some species to enclose fertilized eggs in protective egg cases before depositing them in the environment.

Amphibians were the first vertebrates to live in terrestrial environments. They dealt with the challenge of a dry environment by returning to water to reproduce, as most amphibians still do today. Exceptions are the terrestrial salamanders that use spermatophores to transfer sperm, as mentioned earlier. The spermatophore provides a protective, non-desiccating environment for the sperm. Other amphibians, like most fishes, rely on sexual behavior to bring eggs and sperm together. Frog mating behavior is characterized by *amplexus*, a behavior in which a male grasps a female around the middle with his forelegs and holds on until she releases her egg mass, at which time he releases his sperm (Figure 42.6).

Reptiles were the first vertebrate group to solve the problem of reproduction in the terrestrial environment.

Agalychnis calcarifer

42.6 Getting Sperm and Eggs Together
Fertilization in frogs is external, but amplexus—a behavior in which the male holds the female with his forelegs until she releases her egg mass—helps guarantee that sperm and eggs will get together.

Their solution, the shelled egg, is shared by birds (Figure 42.7). But the shelled egg created a new problem for fertilization: Sperm cannot penetrate the shell, so they have to reach the egg before the shell forms. Hence the need for internal fertilization and the evolution of the necessary accessory sex organs.

Male snakes and lizards have paired *hemipenes*, which can be filled with blood and thereby extruded from the male's cloaca to form intromittent organs. Only one hemipene is inserted in the female's cloaca at a time. It is usually rough or spiny at the end to achieve a secure hold while sperm are transferred down a groove on its surface. Retractor muscles pull the hemipene back into the male's body when mating is completed. Birds have erectile penises that channel sperm along a groove into the female cloaca.

All mammals use internal fertilization, but except for the monotremes, they have done away with the shelled egg. They keep the developing embryo in the female reproductive tract, at least through the early stages of development. Mammalian species differ enormously as to the developmental stage of the offspring at the time of birth.

Reproductive systems are distinguished by where the embryo develops

Two patterns of care and nurture of the embryo have evolved in animals: oviparity (egg bearing) and viviparity (live bearing). **Oviparous** animals lay eggs in the environment, and their embryos develop outside the mother's body. Oviparity is possible because eggs are stocked with abundant nutrients to supply the needs of the embryo.

Oviparous terrestrial animals such as insects, reptiles, and birds protect their eggs from desiccation with tough, waterproof membranes or shells. However, these egg coverings must be permeable to oxygen and carbon dioxide.

(*a*) *Cheloria mydas*

(*b*) *Aptenodytes patagonicus*

42.7 The Shelled Egg
The shelled egg was a major evolutionary step that allowed reptiles and birds to reproduce in the terrestrial environment. (*a*) A female green sea turtle deposits her eggs in the sand. (*b*) Because the terrestrial environment offers no water to bring sperm and egg together, fertilization must take place internally, as with these penguins.

Some oviparous animals engage in various forms of parental behavior to protect their eggs, but until the eggs hatch, the embryos depend entirely on the nutrients stored in the egg. The only oviparous mammalian species are the monotremes: the echidnas and the duck-billed platypus (see Figure 33.22).

Viviparous animals retain the embryo within the mother's body during its early developmental stages. Most mammals are viviparous. There are examples of viviparity in all other vertebrate groups except the crocodiles, turtles, and birds. Even some sharks retain fertilized eggs in their bodies and give birth to free-living offspring. But there is a big difference between viviparity in mammals and in other species. Mammals (except monotremes) have a specialized portion of the female reproductive tract, the **uterus**, that holds the embryo and enables it to derive nutrients from and deliver wastes to the maternal blood. In contrast, non-mammalian viviparous animals simply retain the fertilized eggs in the mother's body until they hatch. The embryos still receive their nutrition from the stores in the egg, so this reproductive adaptation is called **ovoviviparity**.

Among mammals there are various degrees of uterine adaptation. In *marsupials*, such as kangaroos and koalas, the uterus simply holds the embryo and has a limited capability for exchanging nutrients and wastes. Marsupials are born at a very early developmental stage, crawl into a pouch called a *marsupium* on the mother's belly, attach to a nipple, and complete development outside of the mother's uterus (see Figure 33.23). Mammals other than monotremes and marsupials are called *eutherians*. They are characterized by an intimate association of the blood supplies of mother and embryo in the walls of the uterus. We will now look at the reproductive system of eutherians in greater depth, using *Homo sapiens* as our model.

The Human Reproductive System

So far we have seen a small sampling of the fascinating diversity of animal reproductive systems. In this section we describe the structures and functions of the male and female sex organs in eutherian mammals, specifically in human beings, and discuss hormonal regulation of both male and female systems.

Male sex organs produce and deliver semen

Semen is the product of the male reproductive system. Besides sperm, semen contains a complex mixture of fluids and molecules that support the sperm and facilitate fertilization. Sperm make up less than five percent of the volume of the semen.

Sperm are produced in the testes, the paired male gonads. In all mammals except bats, elephants, and marine mammals, the testes are located outside the body cavity in a pouch of skin, the **scrotum** (Figure 42.8). The optimal temperature for spermatogenesis in most mammals is slightly lower than the normal body temperature. The scrotum keeps the testes at this optimal temperature. Muscles in the scrotum contract in a cold environment, bringing the testes closer to the warmth of the body; in a hot environment they relax, suspending the testes farther from the body.

A testis consists of tightly coiled **seminiferous tubules** within which spermatogenesis takes place. Each tubule is lined with a stratified epithelium. Spermatogonia reside in the outer layers of this epithelium, and moving from these outer layers toward the lumen of the tubule, we find germ cells in successive stages of spermatogenesis (Figure 42.9). These germ cells are intimately associated with **Sertoli cells**, which protect them by providing a barrier between them and any noxious substances that might be circulating in the blood. Sertoli cells also provide nutrients for the developing sperm and are involved in the hormonal control

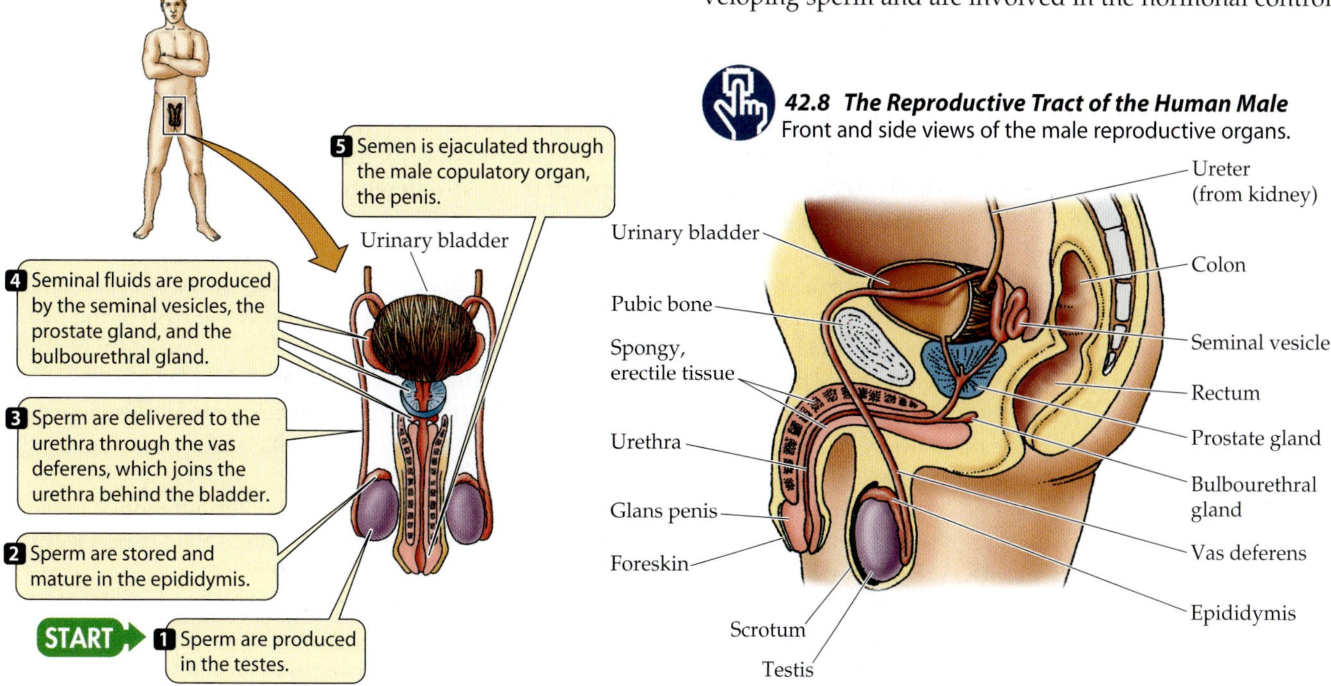

42.8 The Reproductive Tract of the Human Male Front and side views of the male reproductive organs.

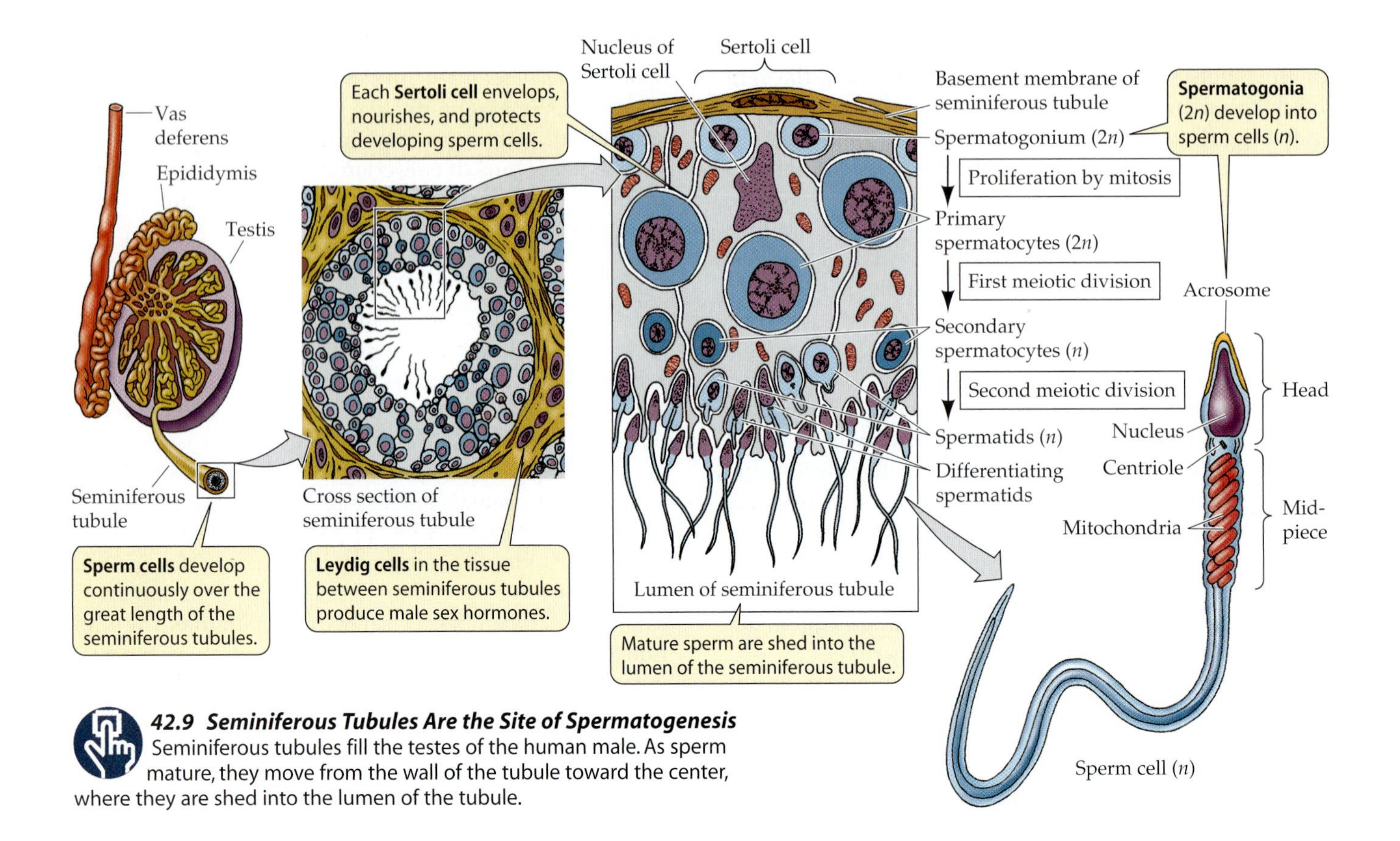

42.9 Seminiferous Tubules Are the Site of Spermatogenesis Seminiferous tubules fill the testes of the human male. As sperm mature, they move from the wall of the tubule toward the center, where they are shed into the lumen of the tubule.

of spermatogenesis. Between the seminiferous tubules are clusters of **Leydig cells**, which produce male sex hormones.

With completion of the second meiotic division, each primary spermatocyte has given rise to four spermatids (see Figure 42.4*a*), which develop into sperm as they continue to migrate toward the lumen of the seminiferous tubule. The nucleus in what will become the head of the sperm becomes compact, and the surrounding cytoplasm is lost (see Figure 42.9). A flagellum, or tail, develops. The mitochondria, which will provide energy for tail motility, become condensed into a midpiece between the head and the tail. A cap, called an *acrosome*, forms over the nucleus in the head of the sperm. The acrosome contains enzymes that enable the sperm to digest a path through protective layers surrounding the egg. Fully differentiated sperm are shed into the lumen of the seminiferous tubule.

From the tubules, sperm move into a storage structure called the **epididymis**, where they mature and become motile. The epididymis connects to the **urethra** by a tube called the **vas deferens** (plural vasa deferentia). The urethra originates in the bladder, runs through the penis, and opens to the outside of the body at the tip of the penis. It serves as the common duct for the urinary and reproductive systems (see Figure 42.8).

The penis and the scrotum are the male genitalia. The shaft of the penis is covered with normal skin, but the tip, or **glans penis**, is covered with thinner, more sensitive skin that is especially responsive to sexual stimulation. A fold of skin called the *foreskin* covers the glans of the human penis. The cultural practice of circumcision removes a portion of the foreskin.

Sexual arousal triggers responses in the the autonomic nervous system that result in the **erection** of the penis. The vessels carrying blood into the penis dilate, and this increased blood flow fills and swells shafts of spongy, erectile tissue located along the length of the penis. The enlargement of these blood-filled cavities compresses the vessels that normally carry blood out of the penis. As a result, the erectile tissue becomes more and more engorged with blood. The penis becomes hard and erect, facilitating its insertion into the female's vagina.*

The culmination of the male sex act propels semen through the vasa deferentia and the urethra in two steps, emission and ejaculation. During **emission**, rhythmic contractions of the smooth muscles of the ducts containing sperm and of the accessory glands move sperm and the various secretions into the urethra at the base of the penis. **Ejaculation**, which follows emission, is caused by contractions of other muscles at the base of the penis surrounding the urethra. The rigidity of the erect penis allows these contractions to force the gelatinous mass of semen through the urethra and out of the body.

Once a climax has been achieved, the autonomic nervous system switches signaling and causes the vessels leading

*In some species of mammals—but not humans—the penis contains a bone called the *baculum* or the *os penis*; however, even those species depend on erectile tissue for copulation.

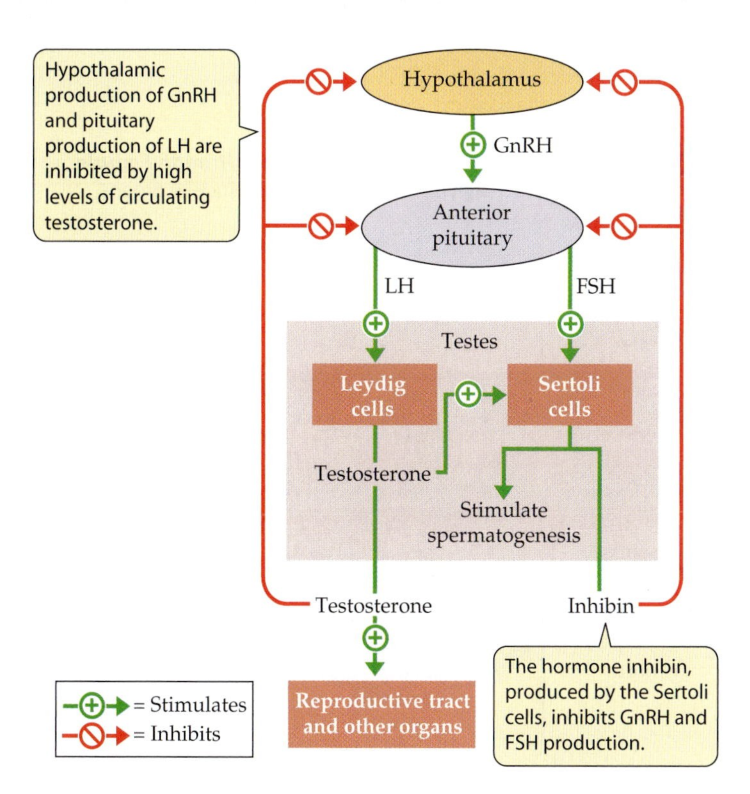

42.10 Hormones Control the Male Reproductive System
The male reproductive system is under hormonal control by the hypothalamus and the anterior pituitary.

into the penis to constrict. The resulting decrease in blood pressure in the erectile tissue relieves the compression of the blood vessels leaving the penis, and the erection declines.

The components of the semen other than sperm come from several accessory glands that contribute secretions to the urethra. A relatively small volume of fluid comes from the **bulbourethral glands**. This alkaline and mucoid secretion precedes other secretions; it neutralizes acidity in the urethra and lubricates the tip of the penis. About two-thirds of the volume of semen is seminal fluid, which comes from the **seminal vesicles**. Seminal fluid is thick because it contains mucus and protein. It also contains fructose, an energy source for the sperm, which are too small to carry much of their own fuel. Semen also carries a message for the female reproductive tract in the form of chemicals called **prostaglandins**. Prostaglandins stimulate rhythmic contractions in the female reproductive tract that help move the sperm up into the regions where fertilization can take place.

One-fourth to one-third of the volume of semen is a thin, milky fluid that comes from the **prostate gland**. Prostate fluid makes the uterine environment more hospitable to sperm. The prostate also secretes a clotting enzyme that works on the protein in seminal fluid to convert semen into a gelatinous mass. The prostate gland completely surrounds the urethra as it leaves the bladder. This gland tends to enlarge in men over 40 years of age, creating a condition known as *benign prostate hyperplasia* (*BPH*). A seriously enlarged prostate can block the urethra and make urination difficult. Unrelated to BPH, prostate cancer is the second most common cancer in men. It is relatively easy to diagnose, however, and is highly curable if detected early.

Male sexual function is controlled by hormones

Spermatogenesis and maintenance of male secondary sexual characteristics depend on testosterone, which is produced by Leydig cells in the testes. In Chapter 41 we learned that increased production of testosterone at puberty is due to an increased release of gonadotropin-releasing hormone (GnRH) by the hypothalamus, which in turn stimulates cells in the anterior pituitary to increase their secretion of luteinizing hormone (LH) and follicle-stimulating hormone (FSH) (Figure 42.10). Negative feedback loops help to regulate testis functions.

The Leydig cells are stimulated by LH to produce testosterone. The rise in the level of testosterone in the prepubertal male causes the development of secondary sexual characteristics and the pubertal growth spurt, promotes increased muscle mass, and stimulates growth and maturation of the testes. If a male is castrated (has his testes removed) before puberty, he will not develop a deep voice, typical patterns of body hair, or a muscular build, and his external genitalia will remain childlike. Continued production of testosterone after puberty is essential to maintain secondary sexual characteristics and to produce sperm. Spermatogenesis is controlled by the influence of FSH and testosterone on Sertoli cells in the seminiferous tubules.

Female sex organs produce eggs, receive sperm, and nurture the embryo

When an egg matures, it is released from the ovary directly into the body cavity. But the egg can't go far. Each ovary is enveloped by the undulating, fringed opening of an **oviduct** (also known as a *fallopian tube*), which sweeps the egg into the tube (Figure 42.11). Cilia lining the oviduct propel the egg slowly toward the uterus, or *womb*, which is a muscular, thick-walled cavity shaped like an upside-down pear. The uterus is where the embryo develops if the egg is fertilized. At the bottom of the uterus is an opening called the **cervix**, which leads into the vagina. Sperm are ejaculated into the vagina during copulation, and the fetus passes through the vagina during birth.

Two sets of skin folds surround the opening of the vagina and the opening of the urethra, through which urine passes. The inner, more delicate folds are the **labia minora** (singular labium minus); the outer, thicker folds are the **labia majora** (singular labium majus). At the anterior tip of the labia minora is the **clitoris**, a small bulb of erectile tissue that is the anatomical homolog of the penis. The clitoris is highly sensitive and plays an important role in sexual response. The labia minora and the clitoris become engorged with blood in response to sexual stimulation.

The opening of an infant female's vagina is partly covered by a thin membrane, the *hymen*. Eventually the hymen becomes ruptured by vigorous physical activity or first sexual intercourse; it can sometimes make first intercourse difficult or painful for the female.

To fertilize an egg, sperm swim and are propelled by contractions of the female reproductive tract up from the vagina, through the cervix, the uterus, and most of the

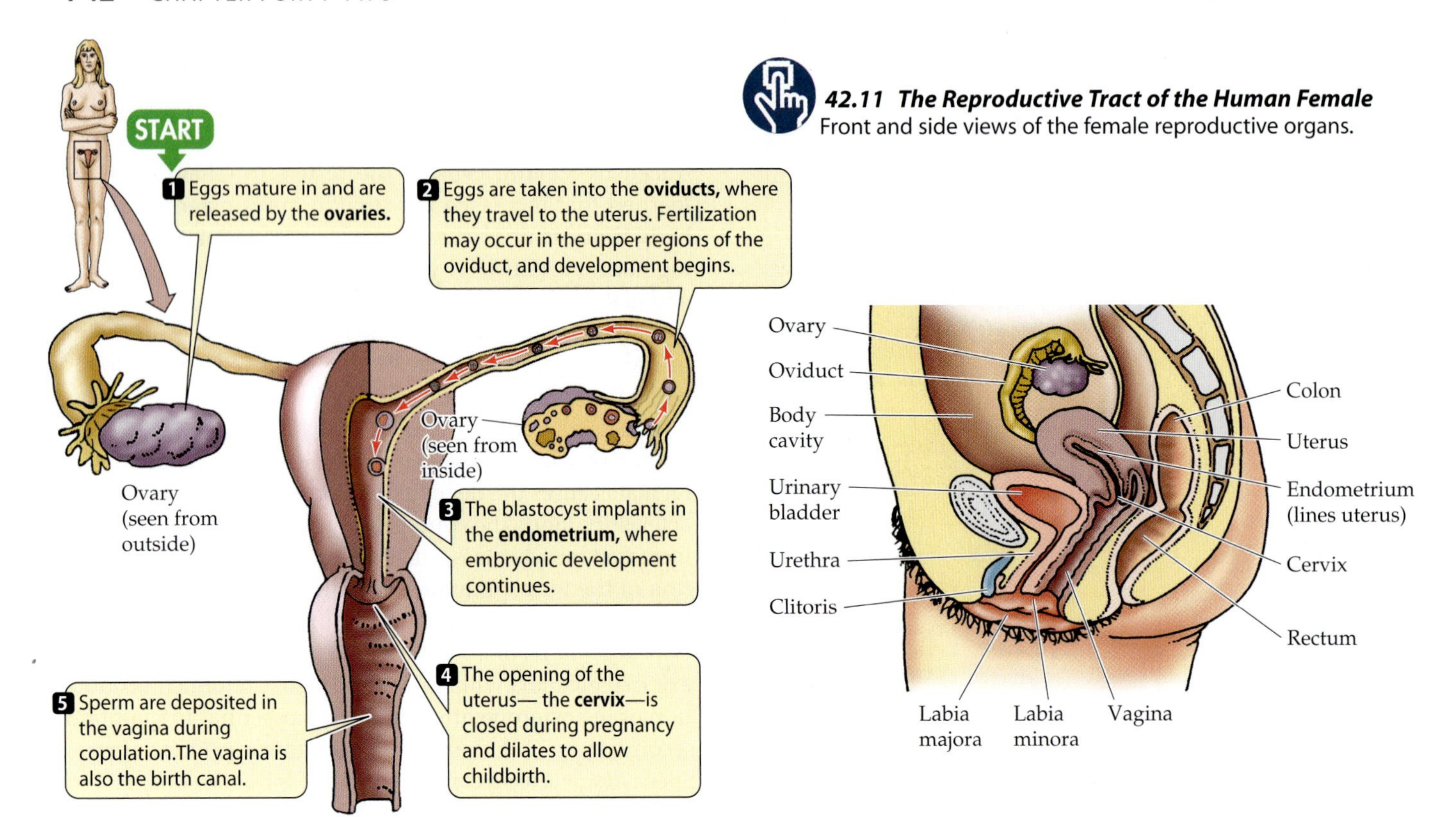

42.11 The Reproductive Tract of the Human Female Front and side views of the female reproductive organs.

oviduct. The egg (actually a secondary oocyte in humans; see Figure 42.4*b*) is fertilized in the upper region of the oviduct. Fertilization stimulates the completion of the second meiotic division, after which the haploid nuclei of the sperm and the egg can fuse to produce a diploid zygote nucleus. Still in the oviduct, the zygote undergoes its first few cell divisions to become a **blastocyst**. The blastocyst moves down the oviduct to the uterus, where it attaches itself to the epithelial lining, the **endometrium**. The endometrium and the cells of the uterine wall are stimulated by estrogen to proliferate and grow many new blood vessels in anticipation of receiving a blastocyst.

Once attached to the endometrium, the blastocyst burrows into it, a process called **implantation**, and forms a structure called the **placenta**. The placenta exchanges nutrients and waste products between the mother's blood and the baby's blood. If a blastocyst does not arrive in the uterus, the endometrium regresses or is sloughed off. Thus the female reproductive cycle actually consists of two linked cycles: an ovarian cycle that produces eggs and hormones, and a uterine cycle that creates an appropriate environment for the embryo should fertilization occur.

The ovarian cycle produces a mature egg

An **ovarian cycle** is about 28 days long in the human female,* but there is considerable variation among individuals. During the first half of the cycle, at least one primary oocyte matures into a secondary oocyte (egg) and is expelled from the ovary. During the second half of the cycle, cells in the ovary that were associated with the maturing oocyte develop endocrine functions and then regress if the egg is not fertilized. The progression of these events is shown diagrammatically in Figure 42.12.

At birth, a human female has about a million primary oocytes in each ovary. By the time she reaches sexual maturity, she has only about 200,000; the rest have degenerated. During a woman's fertile years, her ovaries will go through about 450 ovarian cycles, and during each of these cycles one oocyte will mature and be released. At about 50 years of age, she reaches **menopause**, the end of fertility, and may have only a few oocytes left in each ovary. Throughout a woman's life, oocytes are degenerating, and no new ones are produced.

Each primary oocyte in the ovary is surrounded by a layer of follicle cells. An oocyte and its follicle cells constitute the functional unit of the ovary, the **follicle**. Between puberty and menopause, six to twelve follicles begin to mature each month. In each of these follicles, the oocyte enlarges and the surrounding cells proliferate. After about a week, one of these follicles is larger than the rest and continues to grow, while the others cease to develop and shrink. In the enlarged follicle, the follicle cells nurture the growing egg, supplying it with nutrients and with macromolecules and proteins that it will use in early stages of development if it is fertilized.

After 2 weeks of follicular growth, **ovulation** occurs—the follicle ruptures, and the egg is released. Following ovulation, the follicle cells continue to proliferate and form a mass of endocrine tissue about the size of a marble. This

*Some mammals have ovarian cycles shorter than 28 days, others have longer ones. Rats and mice have ovarian cycles of about 4 days; many seasonally breeding mammals have only one ovarian cycle per year.

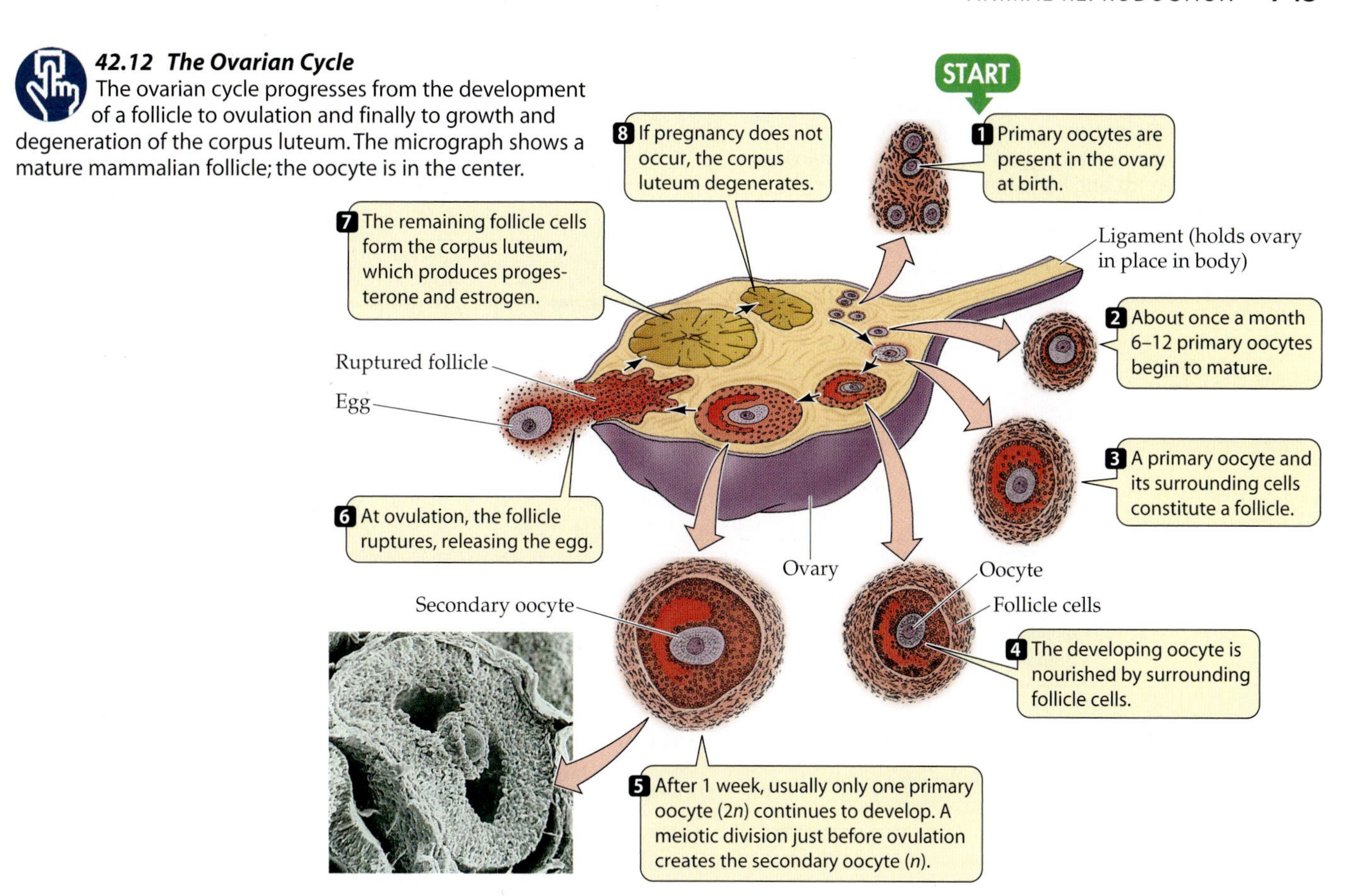

42.12 The Ovarian Cycle The ovarian cycle progresses from the development of a follicle to ovulation and finally to growth and degeneration of the corpus luteum. The micrograph shows a mature mammalian follicle; the oocyte is in the center.

structure, which remains in the ovary, is the **corpus luteum** (plural corpora lutea). It functions as an endocrine gland, producing estrogen and progesterone for about 2 weeks. It then degenerates unless the egg is fertilized.

The uterine cycle prepares an environment for the fertilized egg

The **uterine cycle** of human females parallels the ovarian cycle, and consists first of a buildup and then of a breakdown of the endometrium, or uterine lining (Figure 43.13). About 5 days into the ovarian cycle, the endometrium starts to grow in preparation for receiving a blastocyst. The uterus attains its maximum state of preparedness about 5 days after ovulation and remains in that state for another 9 days. If a blastocyst has not arrived by that time, the endometrium begins to break down, slough off, and flow from the body through the vagina—the process of **menstruation** (from *menses*, the Latin word for "months").

The uterine cycles of mammals other than humans do not include menstruation; instead, the uterine lining is resorbed. In these species the most obvious correlate of the ovarian cycle is a state of sexual receptivity called **estrus** around the time of ovulation. When the female comes into estrus, or "heat," she actively solicits male attention and may be aggressive to other females. The human female is unusual among mammals in that she is potentially sexually receptive throughout her ovarian cycle and at all seasons of the year.

Hormones control and coordinate the ovarian and uterine cycles

The ovarian and uterine cycles of human females are coordinated and timed by the same hormones that initiate sexual maturation. Gonadotropins secreted by the anterior pituitary are the central elements of this control. Before puberty (that is, before about 11 years of age), the secretion of gonadotropins is low, and the ovaries are inactive. At puberty, the hypothalamus increases its release of gonadotropin-releasing hormone (GnRH), thus stimulating the anterior pituitary to secrete follicle-stimulating hormone (FSH) and luteinizing hormone (LH).

In response to FSH and LH, ovarian tissue grows and produces estrogen. The rise in estrogen causes the development of female secondary sexual characteristics, including growth of the uterus. Between puberty and menopause, interactions of gonadotropin-releasing hormone, gonadotropins, and sex steroids control the ovarian and uterine cycles.

Menstruation marks the beginning of the uterine and ovarian cycles (see Figure 42.13). A few days before menstruation begins, the anterior pituitary begins to increase its secretion of FSH and LH. In response, some follicles begin to mature in the ovaries, and follicle cells gradually increase production of estrogen. After about a week of growth, usually all but one of these follicles wither away. Occasionally more than one follicle continues to develop, making it possible for the woman to bear fraternal (nonidentical) twins.

42.13 The Uterine and Ovarian Cycles During a woman's uterine and ovarian cycles there are coordinated changes in (*a*) gonadotropin release by the anterior pituitary, (*b*) the ovary, (*c*) the release of female sex steroids, and (*d*) the uterus. The cycles begin with the onset of menstruation; ovulation is at midcycle.

FSH and LH are under control of GnRH from the hypothalamus and the ovarian hormones estrogen and progesterone.

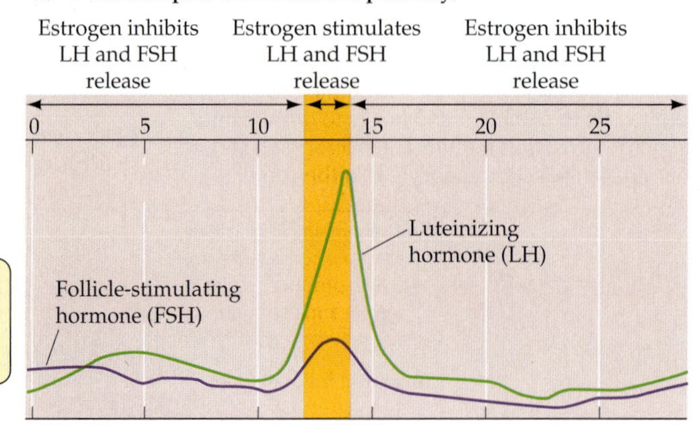

FSH stimulates the development of follicles; the LH surge causes ovulation and then the development of the corpus luteum.

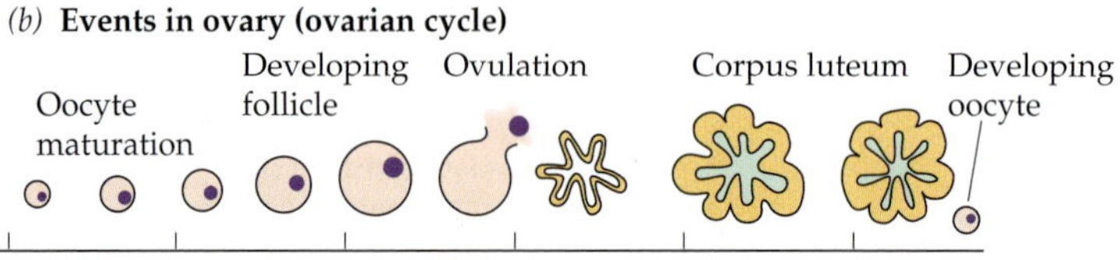

Ovarian hormones stimulate the development of the endometrium in preparation for pregnancy.

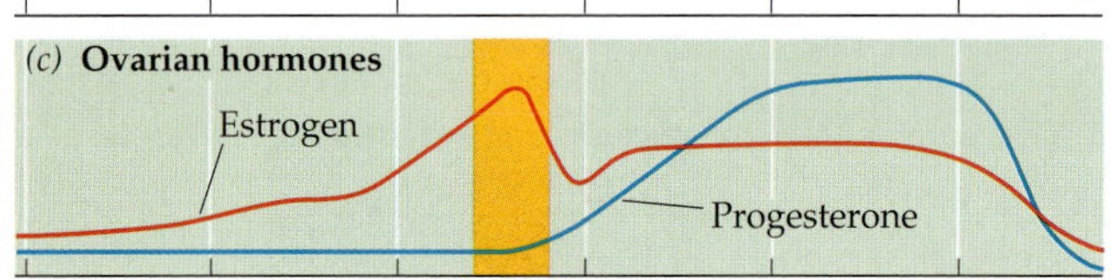

The development of the uterine lining (the endometrium) is controlled by estrogen and progesterone.

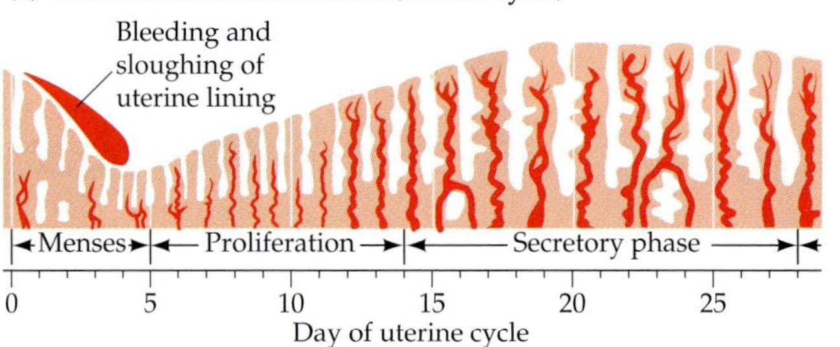

Hypothalamus

GnRH

Anterior pituitary

LH/FSH

Ovary

Estrogen and progesterone

Uterus

= Stimulates

= Inhibits

Positive feed-back occurs during days 12 through 14.

Negative feedback occurs throughout most of the cycle.

42.14 Hormones Control the Ovarian and Uterine Cycles The ovarian and uterine cycles are under a complex series of positive and negative feedback controls involving several hormones.

The follicle that is still growing secretes increasing amounts of estrogen, stimulating the endometrium to grow.

Estrogen exerts negative feedback control on gonadotropin release by the anterior pituitary during the first 12 days of the ovarian cycle. Then, on about day 12, estrogen exerts positive rather than negative feedback control on the pituitary (Figure 42.14). As a result, there is a surge of LH, and a lesser surge of FSH. The LH surge triggers the mature follicle to rupture and release its egg, and it stimulates follicle cells to develop into the corpus luteum and to secrete estrogen and progesterone.

Estrogen and especially progesterone secreted by the corpus luteum following ovulation are crucial to the continued growth and maintenance of the endometrium. In addition, these sex steroids exert negative feedback control on the pituitary, inhibiting gonadotropin release and thus preventing new follicles from beginning to mature.

If the egg is not fertilized, the corpus luteum degenerates on about day 26 of the cycle. Without the production of progesterone by the corpus luteum, the endometrium sloughs off, and menstruation occurs. The decrease in circulating steroids also releases the hypothalamus and pituitary from negative feedback control, so GnRH, FSH, and LH all increase. The increase in these hormones induces the next round of follicle development, and the ovarian cycle begins again.

If the egg is fertilized, and a blastocyst arrives in the uterus and implants itself in the endometrium, a new hormone comes into play. A layer of cells covering the blastocyst begins to secrete **human chorionic gonadotropin (hCG)**. This gonadotropin, a molecular homolog of LH, keeps the corpus luteum functional. Because hCG is present only in the blood of pregnant women, the presence of this hormone is the basis for pregnancy testing.

These tissues derived from the blastocyst also begin to produce estrogen and progesterone, eventually replacing the corpus luteum as the most important source of these sex steroids. Continued high levels of estrogen and progesterone prevent the pituitary from secreting gonadotropins; thus the ovarian cycle ceases for the duration of the pregnancy. The same mechanism is exploited by birth control pills, which contain synthetic hormones resembling estrogen and progesterone that prevent the ovarian cycle (but not the uterine cycle) by exerting negative feedback control on the hypothalamus and pituitary.

Human Sexual Behavior

The organs of the male and female reproductive systems are similar in all eutherians. However, the hormonal and emotional biology of reproductive behavior in humans is more complex. The emotional and social complexities that have evolved to such a tremendous degree in humans have affected our reproductive lives, as have our extensive technological achievements. Here we discuss human sexual responses responses and technologies for both contraception (birth control) and enhanced fertility. The chapter closes with a discussion of sexually transmitted diseases.

Human sexual responses consist of four phases

The responses of both women and men to sexual stimulation consist of four phases: excitement, plateau, orgasm, and resolution. As sexual *excitement* begins in a woman, her heart rate and blood pressure rise, muscular tension increases, her breasts swell, and her nipples become erect. Her external genitals, including the sensitive clitoris, swell as they become filled with blood, and the walls of the vagina secrete lubricating fluid that facilitates copulation.

As a woman's sexual excitement increases, she enters the *plateau* phase. Her blood pressure and heart rate rise further, her breathing becomes rapid, and the clitoris begins to retract—the greater the excitement, the greater the retraction. The sensitivity that once focused in the clitoris spreads over the external genitals, and the clitoris itself becomes even more sensitive. *Orgasm* may last as long as a few minutes, and, unlike men, some women can experience several orgasms in rapid succession. During the *resolution* phase, blood drains from the genitals, and body physiology returns to close to normal.

In the male, as in the female, the excitement phase is marked by an increase in blood pressure, heart rate, and muscle tension. The penis fills with blood and becomes hard and erect. In the plateau phase, breathing becomes rapid, the diameter of the glans increases, and a clear lubricating fluid from the bulbourethral glands oozes from the penis. Pressure and friction against the nerve endings in the glans and in the skin along the shaft of the penis eventually trigger orgasm. Massive spasms of the muscles in the genital area and contractions in the accessory reproductive organs result in ejaculation.

Within a few minutes after ejaculation, the penis shrinks to its former size, and body physiology returns to resting conditions. The male sexual response includes a *refractory period* immediately after orgasm. During this period, which may last 20 minutes or more, a man cannot achieve a full erection or another orgasm, regardless of the intensity of sexual stimulation.

Humans use a variety of technologies to control fertility

People use many methods to control the number of their children and the time between their children's births. The only absolutely sure methods of preventing fertilization and pregnancy are complete abstinence from sexual activity or surgical removal of the gonads. Since those approaches are not acceptable to most people, they turn to a variety of other methods to prevent pregnancy or conception, which therefore are called methods of **contraception**.

Some methods of contraception are used by the woman, others by the man. They vary from means of blocking gametogenesis to means of blocking development of the embryo. Contraceptive methods vary enormously in their effectiveness and in their acceptability to those who use them. Here we review some of the most common methods and their relative failure rates (Table 42.1).

NONTECHNOLOGICAL APPROACHES. An approach to contraception that does not involve physical or pharmacological technologies is to separate sperm and egg in time through the **rhythm method**. The couple avoids sex from day 10 to day 20 of the ovarian cycle, when the woman is most likely to be fertile. The cycle can be tracked by use of a calendar, supplemented by the basal body temperature method, which identifies the day of ovulation on the basis of the observation that a woman's body temperature drops on the day of ovulation and rises sharply on the day after. Changes in the stickiness of the cervical mucus also help identify the day of ovulation.

However, sperm deposited in the female reproductive tract may remain viable for up to 6 days. Similarly, the ovum remains viable for 2 to 3 days after ovulation. These facts, added to individual variation in the timing of ovulation, result in an annual failure rate of between 15 and 35 percent for the rhythm method. In other words, 15 to 35 percent of women using only the rhythm method for 1 year will become pregnant during that time.

42.1 Methods of Contraception

METHOD	MODE OF ACTION	FAILURE RATE[a]
Rhythm method	Abstinence near time of ovulation	15–35
Coitus interruptus	Prevents sperm from reaching egg	10–40
Condom	Prevents sperm from entering vagina	3–20
Diaphragm/jelly	Prevents sperm from entering uterus; kills sperm	3–25
Vaginal jelly or foam	Kills sperm; blocks sperm movement	3–30
Douche	Supposedly flushes sperm from vagina	80
Birth control pills	Prevent ovulation	0–3
Vasectomy	Prevents release of sperm	0.0–0.15
Tubal ligation	Prevents egg from entering uterus	0.0–0.05
Intrauterine device (IUD)	Prevents implantation of fertilized egg	0.5–6
RU-486	Prevents development of fertilized egg	0–15
(Unprotected)	(No form of birth control)	(85)

[a] Number of pregnancies per 100 women per year

Another approach is to try to separate sperm and egg in space through **coitus interruptus**—withdrawal of the penis before ejaculation. The annual failure rate of this method may be as high as 40 percent.

BARRIER METHODS. Techniques for placing a physical barrier between egg and sperm have been used for centuries. The **condom** is a sheath made of an impermeable material such as latex that can be fitted over the erect penis. A condom traps the semen so that sperm do not enter the vagina. Condoms also help prevent the spread of sexually transmitted diseases such as AIDS, syphilis, and gonorrhea. In theory, the use of a condom can be highly effective, with a failure rate near zero; in practice, the annual failure rate is about 15 percent, because of faulty technique. There is also a female condom, which can be used by the woman to create an impermeable lining of the vagina.

The **diaphragm** is a dome-shaped piece of rubber with a firm rim that fits over the woman's cervix and thus blocks sperm from entering the uterus. Smaller than the diaphragm is the **cervical cap**, which fits snugly just over the tip of the cervix. Both the diaphragm and the cervical cap are treated first with jelly or cream containing a *spermicide*—a chemical that kills or incapacitates sperm—and then inserted through the vagina before sexual intercourse. Annual failure rates are about the same as for condoms—about 15 percent.

Spermicidal foams, jellies, and creams can be used alone by placing them in the vagina with special applicators. Used in this way, they have an annual failure rate of 25 percent or more. *Douching* (flushing the vagina with liquid) after intercourse, in spite of popular belief, is almost useless as a method of birth control. Remember that sperm can reach the upper regions of the oviducts within 10 minutes after ejaculation.

The effectiveness of barrier methods can be greatly improved if different ones are used in combination. For example, if the man uses a condom and the woman a diaphragm, the failure rate is extremely low.

PREVENTING OVULATION. The widely used oral contraceptives, or **birth control pills**, work by preventing ovulation. Their mechanisms of action take advantage of the roles of estrogens and progesterone as negative feedback signals that work on both the hypothalamus and the pituitary to inhibit gonadotropin release. The most common pills contain low doses of synthetic estrogens and progesterones (progestins). By keeping the circulating levels of gonadotropins low, the birth control pill interferes with the maturation of follicles and ova. The ovarian cycle (but not the uterine cycle) is suspended.

The negative side effects of oral contraceptives have been the topic of extensive discussion. These side effects include increased risk of blood clot formation, heart attack, stroke, and breast cancer. However, these side effects are associated mostly with pills containing higher hormone concentrations than are used in modern pills. For pills in use today, the risk of these side effects is low, except for women over 35 years old who smoke, for whom the risk is significantly greater. Risk of death from using "the pill" is less than the risk associated with a full-term pregnancy. The pill is the most effective method of contraception other than sterilization or perhaps combined barrier methods. Oral contraceptives have an annual failure rate of less than 1 percent.

The "mini-pill" is an oral contraceptive that contains very low doses of progestins. Although it may interfere with the normal maturation and release of ova, its principal mode of action is to alter the environment of the female reproductive tract so that it is not hospitable to sperm. Cervical mucus normally becomes watery at the time of ovulation, but low levels of progestin keep the mucus thick and sticky so that it blocks the passage of sperm.

Long-lasting injectable or implantable steroids are also used to block ovulation. DepoProvera is an injectable progestin that blocks normal pituitary function for several months. Another device, called Norplant, consists of thin, flexible tubes filled with progestin. Several of these tubes are inserted under the skin, where they continue to release progestin slowly for years.

PREVENTING IMPLANTATION. A highly effective method of contraception (with a failure rate varying from 1 percent to about 7 percent) is the **intrauterine device**, or **IUD**. The IUD is a small piece of plastic or copper that is inserted in the uterus. The IUD probably works by preventing implantation of the fertilized egg.

Another way of interfering with implantation is through the use of "morning-after pills," which deliver high doses of steroids, primarily estrogens. By acting in several ways on the oviduct and the uterine lining, this treatment prevents implantation. Morning-after pills can be effective up to several days after intercourse.

A recent addition to birth control technology is a drug, **RU-486**, developed in France. RU-486 is not a contraceptive pill, but a *contragestational* pill. It is a progesterone-like molecule that blocks progesterone receptors. It therefore interferes with the normal action of progesterone produced by the corpus luteum, which is necessary for the maintenance of the uterine lining in early pregnancy. If RU-486 is administered as a "morning-after pill", it prevents implantation. However, RU-486 can be effective even if taken at the time of the first missed menses after fertilization, after implantation has been initiated. After a few days of treatment with RU-486, the endometrium regresses and sloughs off along with the embryo, which is in very early stages of development.

STERILIZATION. One virtually foolproof method of contraception is sterilization of either the man or the woman. Male sterilization by **vasectomy** is a simple operation that can be performed under a local anesthetic in a doctor's office (Figure 42.15*a*). After this minor surgery, the semen no longer contains sperm. Sperm production continues, but since the sperm cannot move out of the testes, they are destroyed by macrophages. Vasectomy does not affect a man's hormone levels or his sexual responses, and even the amount of semen he ejaculates is unchanged, because the sperm constitute so little of its volume.

In female sterilization, the aim is to prevent the egg from traveling to the uterus and to block sperm from reaching the egg. The most common method is **tubal ligation**—cutting and tying the oviducts (Figure 42.15*b*). Alternatively, the oviducts may be burned (cauterized) to seal them off, using a surgical technique called *endoscopy*. As in the male, these procedures do not alter reproductive hormones or sexual responses.

ABORTION. Once a fertilized egg has successfully implanted itself in the uterus, the termination of a pregnancy is called an **abortion**. A *spontaneous abortion* is the medical term for what is usually called a *miscarriage*. Miscarriages are common early in pregnancy; most of them occur because of an abnormality in the fetus or in the process of implantation. Abortions that are not spontaneous, but are the result of medical intervention, may be done either for therapeutic or for contraceptive purposes. A therapeutic abortion may be necessary to protect the health of the mother, or it may be performed because prenatal testing (which will be discussed in the following chapter) reveals that the fetus has a severe defect.

When performed in the first third of a pregnancy, a medical abortion carries less risk than a full-term pregnancy. The method is to dilate the cervix and then remove the

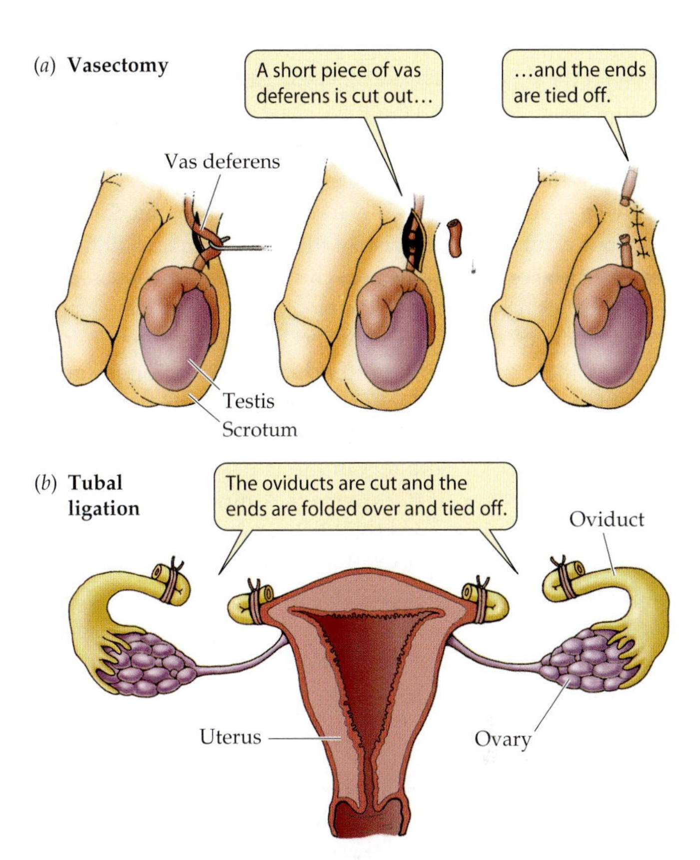

42.15 Sterilization Techniques
(*a*) Vasectomy is the technique for male sterilization. (*b*) Tubal ligation is the sterilization procedure performed on human females.

fetus and the endometrium by physical means. After the first 12 weeks of pregnancy, the risk associated with a medical abortion rises substantially.

CONTROLLING MALE FERTILITY. You may ask why all the chemical approaches to controlling fertility apply to females and none have been devised to block male fertility. The control of male fertility is a difficult problem. First, since the production of sperm is a continuous, rather than a cyclical, event, it is not possible to block a particular step in a cyclical process. The ovarian cycle is vulnerable to manipulation because certain events must happen at certain times for ovulation and implantation to occur. Second, the suppression of spermatogenesis must be total to be effective, since it takes only a single sperm to fertilize an egg, and normally millions are produced. Such suppression requires powerful and continuous chemical intervention with associated side effects.

Reproductive technologies help solve problems of infertility

There are many reasons why a man and woman may not be able to have children. The man's rate of sperm production may be low, or his sperm may lack motility. The mucus in the woman's reproductive tract may be thick and not conducive to sperm moving up to the oviducts. Structural problems may also exist, such as a woman's oviducts being blocked by scar tissue or by *endometriosis*, a proliferation of endometrial cells outside of the uterus. In some cases, treatment with powerful chemicals to cure cancer damages the ability of the gonads to produce gametes.

Even a couple who are fully fertile and who want children may not want to take the risk of the natural process of fertilization if one or both parents are carriers for a genetic disease. A number of reproductive technologies have been developed to overcome these and other barriers to childbearing.

The oldest and simplest reproductive technology is **artificial insemination**, which involves placing sperm in the appropriate place in the female's reproductive tract for fertilization to occur. This technique is useful if the male's sperm count is low, if the sperm lack motility, or if problems in the female's reproductive tract prevent the normal movement of sperm up to and through the oviducts. Artificial insemination is used widely in the production of domesticated animals such as cattle.

More recent advances, called **assisted reproductive technologies**, or **ART's**, involve procedures that remove unfertilized eggs from the ovary, combine them with sperm outside of the body, and then place fertilized eggs or egg/sperm mixtures in the appropriate location in the female's reproductive tract for development to take place.

The first successful ART was **in vitro fertilization (IVF)**. In IVF, the mother is treated with hormones that stimulate many follicles in her ovaries to mature. Eggs are harvested from these follicles, and sperm are collected from the father. Eggs and sperm are combined in a culture medium outside the body (in vitro), where fertilization takes place. The resulting embryos can then be injected into the mother's uterus in the blastocyst stage or kept frozen for implantation later. The first "test tube baby" resulting from IVF was born in 1978. Since that time, thousands of babies have been produced by this ART. IVF is useful when the woman's oviducts are blocked. It has a success rate of 20 to 25 percent.

A technique called **gamete intrafallopian transfer (GIFT)** can be used when only the entrance to the oviducts from the ovaries, or the upper segment of the oviducts, is blocked. In this procedure, harvested eggs and sperm are

42.2 *Some Sexually Transmitted Diseases*

DISEASE	INCIDENCE IN UNITED STATES	SYMPTOMS
Syphilis	80,000 new cases/yr	Primary stage (weeks): skin lesion (chancre) at site of infection Secondary stage (months): skin rash and flu-like symptoms, may be followed by a latent period Tertiary stage (years): deterioration of the cardiovascular and central nervous systems
Gonorrhea	800,000 new cases/yr	Pus-filled discharge from penis or vagina; burning urination. Infection can also start in throat or rectum
Chlamydia	>4,000,000 new cases/yr	Symptoms similar to gonorrhea, although often there are no obvious symptoms. Can result in pelvic inflammatory disease in females (see below)
Genital herpes	500,000 new cases/yr	Small blisters that can cause itching or burning sensations are accompanied by inflammation and by secondary infections
Genital warts	10% of adults infected	Small growths on genital tissues. Increases risk of cervical cancer in women
Hepatitis B	5–20% of population	Fatigue, fever, nausea, loss of appetite, jaundice, abdominal pain, muscle and joint pain. Can lead to destruction of liver or liver cancer
Pelvic inflammatory disease	1,000,000 new cases/yr (females only)	Fever and abdominal pain. Frequently results in sterility
AIDS	Approximately 900,000 cases[a]	Failure of the immune system (see Chapter 19)

[a]AIDS is widespread in other parts of the world, most notably in the southern part of the African continent, where some 9 million people are infected. The virus is also spreading rapidly in India and Southeast Asia.

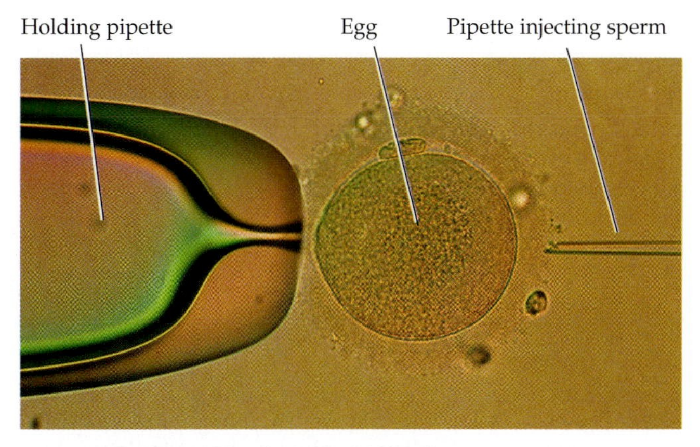

42.16 Intracytoplasmic Sperm Injection
In this procedure, sperm are injected directly into a mature egg cell. The fertilized egg can then be placed back in the female reproductive tract.

injected directly into the upper regions of the oviduct, where fertilization takes place, and the blastocyst enters the uterus via the normal route. GIFT has a success rate of about 30 percent.

A major cause of failure of IVF and GIFT is the failure of sperm to gain access to the plasma membranes of eggs. To solve this problem, methods have been developed to inject a sperm cell directly into the cytoplasm of an egg. In **intracytoplasmic sperm injection** (**ICSI**), a harvested egg is held in place by suction applied to a polished glass pipette. A slender, sharp pipette is then used to penetrate the egg and inject a sperm (Figure 42.16). This ART was successful for the first time in 1992; now thousands of these procedures are performed in U.S. clinics each year, with a success rate of about 25 percent.

IVF coupled with sensitive techniques of genetic analysis, can eliminate the risk that parents who are carriers for genetic diseases will produce affected children. With in vitro fertilization, it is possible to take a cell from the blastocyst at the 4- or 8-cell stage without damaging its developmental potential. The sampled cell can then be subjected to molecular techniques to determine whether it carries the harmful gene. This information guarantees that only embryos that will not develop the genetic disease are implanted in the mother's uterus.

CAUSE	MODE OF TRANSMISSION	CURE/TREATMENT
Spirochete bacterium (*Treponema pallidum*) that penetrates mucosal membranes and abraded skin	Intimate sexual contact (even kissing)	Antibiotics
Bacterium (*Neisseria gonorrheoeae*)	Communicated across mucous membranes	Antibiotics (but antibiotic-resistant strains have arisen)
Bacterium (*Chlamydia trachomatis*)	Communicated across mucous membranes	Antibiotics
Herpes simplex virus	Communicated by contact with infected surfaces, which can be mucous membranes or skin	No cure. Symptoms can be alleviated. Antiviral drugs may lessen subsequent outbreaks
Human papillomavirus	Communicated across mucous membranes through sexual contact	No cure for the virus. Warts can be removed surgically or by burning, freezing, or chemical treatment
Virus	Sexual contact or blood transfusions	No cure. Symptoms can be treated. A vaccine is available that can protect only if given before infection occurs
A variety of bacteria that migrate to the uterus and fallopian tubes	Sexual intercourse	Antibiotics
HIV (see Chapter 13)	The virus enters the bloodstream via cuts or abrasions, including minute ones in the genitalia. Spread primarily by intimate sexual contact, but can also be transmitted via contaminated needles	No cure. Treatments with a variety of medications can slow the course of the infection

Sexual behavior transmits many disease organisms

Disease-causing organisms are parasites, and have a very limited ability to survive outside a host organism. Therefore, getting from host to host is a major evolutionary challenge for disease-causing organisms. One of the most intimate types of contact that hosts can have is copulation. It is not surprising, then, that many parasitic organisms have evolved to depend on sexual contact between their hosts as their means of transmission. These organisms are the causes of **sexually transmitted diseases** (commonly referred to as **STD's**), and they include viruses, bacteria, yeasts, and protozoans.

STD's have been with humans since ancient times, and they are one of the most serious public health problems today. Over 10 million new cases of STD's occur each year in the United States, and about two-thirds of these cases occur in people between the ages of 15 and 30. About half of the youth in this country will contract an STD before the age of 25.

A summary of the most common STD's is presented in Table 42.2 on the preceding page. The highly prevalent bacterially transmitted diseases chlamydia and gonorrhea are generally not fatal, but when untreated are a major cause of infertility and painful inflammatory diseases. Syphilis is transmitted by a spirochete and is fatal in about half of untreated cases. It is believed that syphilis was brought to the Old World from the New World by the crew of Christopher Columbus, who himself died of advanced syphilis 16 years after his first voyage to North America. AIDS, transmitted by a virus, is an STD of recent origin and currently has a high rate of mortality. The only device that is effective against the transmission of STD's is the condom.

Chapter Summary

Asexual Reproduction

▶ Sexual reproduction is almost universal in animals, but some animals can reproduce asexually, producing offspring that are genetically identical to their parent and to one another. A disadvantage of asexual reproduction is that no genetic diversity is produced.

▶ Means of asexual reproduction include budding, regeneration, and parthenogenesis. **Review Figures 42.1, 42.2**

Sexual Reproduction

▶ Sexual reproduction consists of three basic steps: gametogenesis, mating, and fertilization. Gametogenesis and fertilization are similar in all animals, but mating includes a great variety of anatomical, physiological, and behavioral adaptations.

▶ In sexually reproducing species, genetic diversity is created by the recombination of genes during gametogenesis and by the independent assortment of chromosomes. Mating and fertilization also contribute to genetic diversity.

▶ Gametogenesis occurs in testes and ovaries. In spermatogenesis (the production of sperm) and oogenesis (the production of eggs), the primary germ cells proliferate mitotically, undergo meiosis, and mature into gametes. **Review Figure 42.4**

▶ Spermatogonia continue to proliferate by mitosis throughout the life span of the male. Each primary spermatocyte can produce four haploid sperm through the two divisions of meiosis. **Review Figure 42.4*a***

▶ Primary oocytes immediately enter prophase of the first meiotic division, and in many species, including humans, their development is arrested at this point. Each oogonium produces only one egg through meiosis. **Review Figure 42.4*b***

▶ Hermaphroditic species have both male and female reproductive systems in the same individual, either sequentially or simultaneously.

▶ Fertilization can occur externally, which is common in aquatic species, or internally, which is common in terrestrial species. Internal fertilization usually involves copulation.

▶ Internal fertilization is necessary for nonaquatic species. The shelled egg is an important adaptation to the desiccating terrestrial environment, but it must be fertilized before the shell forms. All mammals except monotremes retain the embryo internally and have done away with shelled eggs.

▶ Animals can be classified as oviparous or viviparous depending on whether the early stages of development occur outside or inside the mother's body.

The Human Reproductive System

▶ Males produce and deliver semen into the female reproductive tract. Semen consists of sperm suspended in a fluid that nourishes them and facilitates fertilization.

▶ Sperm are produced in the seminiferous tubules of the testes, mature in the epididymis, and are delivered to the urethra through the vasa deferentia. Other components of semen are produced in the seminal vesicles, prostate gland, and bulbourethral glands. All components of the semen join in the urethra at the base of the penis and are ejaculated through the erect penis by muscle contractions at the culmination of copulation. **Review Figures 42.8, 42.9**

▶ Spermatogenesis depends on testosterone secreted by Leydig cells in the testes, which are under control of luteinizing hormone from the pituitary. Spermatogenesis is also controlled by follicle-stimulating hormone from the pituitary. Hypothalamic gonadotropin-releasing hormone controls pituitary secretion of LH and FSH. The production of these hormones by the hypothalamus and pituitary is controlled by negative feedback from testosterone and another hormone produced by the testes, inhibin. **Review Figure 42.10**

▶ Eggs (ova) mature in the female's ovaries and are released into the oviducts, which deliver the eggs to the uterus. Sperm deposited in the vagina during copulation move up through the cervix into the uterus, some continuing up through the oviducts. **Review Figure 42.11**

▶ Fertilization occurs in the upper regions of the oviducts. The zygote becomes a blastocyst through repeated cell divisions as it passes down the oviduct. Upon arrival in the uterus, the blastocyst implants itself in the endometrium, where a placenta forms and the embryo develops.

▶ The maturation and release of ova constitute an ovarian cycle under the control of the anterior pituitary hormones FSH and LH. In humans, this cycle takes about 28 days. **Review Figures 42.12, 42.13**

▶ The uterus also undergoes a cycle that prepares it for receipt of a blastocyst. If no blastocyst is implanted, the lining of the uterus deteriorates and sloughs off, which is the process of menstruation. **Review Figure 42.13**

▶ Both the ovarian and the uterine cycles are under the control of hypothalamic and pituitary hormones, which in turn are under the feedback control of estrogen and progesterone. **Review Figure 42.14**

Human Sexual Behavior

- Human sexual responses consist of four phases: excitement, plateau, orgasm, and resolution. In addition, males have a refractory period during which renewed excitement is not possible.
- Methods to prevent pregnancy include abstention from copulation or the use of technologies that decrease the probability of fertilization. **Review Table 42.1**
- Barrier methods of contraception, such as condoms, diaphragms, and spermicidal substances, block the passage of sperm in the female reproductive tract or weaken and kill them.
- Methods to prevent ovulation, such as birth control pills and other hormonal treatments, interfere with the ovarian cycle so that mature, fertile ova are not produced and released.
- Males and females can be sterilized by surgical blockage of the vasa deferentia (vasectomy) or oviducts (tubal ligation). **Review Figure 42.15**
- Methods to prevent implantation of a blastocyst include intrauterine devices, excess doses of steroids, and a progesterone receptor blocker. After implantation, the termination of a pregnancy is called an abortion.
- Assisted reproductive technologies have been developed to increase fertility. ARTs include in vitro fertilization and gamete intrafallopian transfer.
- Sexually transmitted diseases result from the transmission of disease-causing organisms through sexual behavior. Many STD's are curable if treated early, but can have serious long-term consequences if not treated. **Review Table 42.2**

For Discussion

1. In the very deep ocean, there are species of fish in which the male is very much smaller than the female and actually lives attached to her body. In terms of the selective pressures that operate on sexual and asexual reproduction and in terms of the deep-sea environment, what factors do you think resulted in the evolution of this extreme sexual dimorphism?
2. What are two main differences between the immediate products of the first and second meiotic division in spermatogenesis and oogenesis? Why do these differences exist?
3. At the beginning of each ovarian cycle in humans, about six follicles begin to develop in response to rising levels of FSH, but after a week, only one follicle continues to develop, and the others wither away. Given the facts that follicles produce estrogen, estrogen stimulates follicle cells to produce FSH receptors, and estrogen exerts negative feedback on FSH production in the pituitary, how can you explain how one follicle gets "selected" to grow?
4. Compare the actions of LH and FSH in the ovaries and testes.
5. Ovarian and uterine events in the month following ovulation differ depending on whether fertilization occurs. Describe the differences and explain their hormonal controls.

43 Animal Development

THE WHALE BLOWS ITS NOSE OUT THE TOP of its head—as in "thar she blows," the infamous whalers' cry. The spout coming out of the blowhole is the whale's exhalation coming out of its nasal passages. It is convenient for a marine mammal to be able to breathe out of the top of its head. Not much of its body has to come out of the water for it to breathe, and it can continue moving forward through the water while breathing. But in most mammals, the nose is on the front of the head. How did the whale's nose happen to get to the top of its head? This is an evolutionary question, but the answer is to be found in embryological development.

The vertebrate body varies enormously among species in form and function, yet its basic structural design is highly conserved. For example, the whale flipper, the bat wing, and the human arm all have the same bones. However, during development these bones assume different shapes and dimensions to adapt the forelimbs to their various functions: swimming, flying, and tool use.

Similarly, vertebrates all have the same bones in their heads, but through development, these bones grow differentially, and therefore the skull takes on different shapes. In both the whale and the human, the nasal passages are in the nasal bone, which in both is just above the bones of the upper jaw. In the human, that places the nasal bone just above the jaw on the front of the face. During development of the whale skull, however, the bones of the upper jaw grow enormously relative to the other bones of the skull, and project far forward to form the cavernous mouth. As a result of this differential forward growth of the jaw bone, the nasal bone ends up on the top of the skull rather than on the front. The answer to why the whale's nose is on the top of its head and how forelimbs become flippers is found in the processes of development that form and shape the components of the basic vertebrate body plan.

Starting from the organization of the fertilized egg, signaling cascades unfold that control the building of the body. Typically, molecules acting as transcription factors influence the temporal and spatial expression of still other genes that control the growth and differentiation of cells. Interactions between cells set up ever more complex signaling relationships that result in the structural and functional differences between species. The chemical and genetic nature of these signals is remarkably similar over widely different species, but the exciting story of development is how these signals are used in different patterns and combinations to produce what we see as different species.

In this chapter we trace the early stages of embryonic development. We begin with the events of fertilization, which can be taken as the starting point for the development of the organism. From there we see how the zygote is converted by rapid cell divisions into a mass of cells. These cells then change position relative to each other, establishing new contacts between different groups of cells. These contacts initiate sequences of changes in cell growth, cell movements, and cell differentiation from which emerge the overall body plan, the various tissues, and the rudimentary organs of the adult.

To appreciate both the diversity and the similarity in the development of different animals, we discuss these early developmental steps in four organisms studied extensively by developmental biologists: sea urchins (invertebrates) and frogs, chicks, and humans (all vertebrates).

Thar She Blows!
The nasal passages of the whale *Orcinus orca* are on top of its head because of the extreme growth of its jaw bones during development.

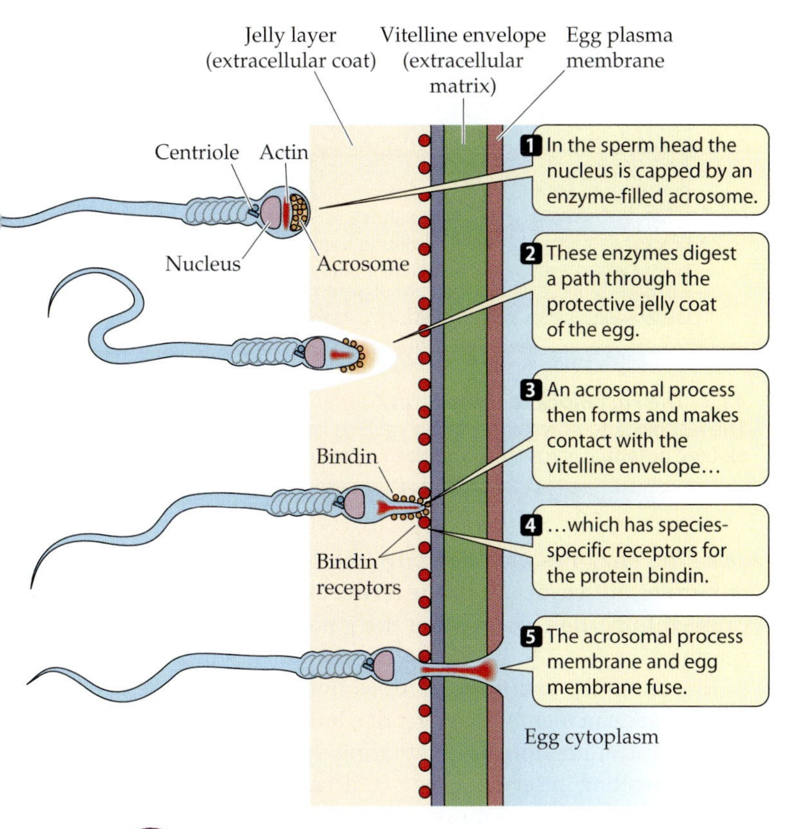

43.1 The Acrosomal Reaction
The acrosomal reaction allows a sea urchin sperm to recognize an egg of the same species and pass through its protective layers.

Fertilization: Interactions of Sperm and Egg

Development in sexually reproducing animals begins with fertilization, the union of sperm and egg to create a single cell: a diploid zygote. Fertilization requires a complex series of events:

- The sperm and the egg must recognize each other.
- The sperm must be activated so that it is capable of gaining access to the plasma membrane of the egg.
- The plasma membranes of the sperm and the egg must fuse.
- The egg must block entry by additional sperm.
- In mammals, the egg nucleus must complete its final meiotic division.
- The egg and sperm nuclei must fuse to create the diploid nucleus of the zygote.

We will look at each of these steps in turn.

Recognition molecules assure specificity in sperm–egg interactions

Specific recognition molecules mediate interactions between sperm and eggs. These molecules assure that the activities of the sperm are directed toward eggs and not other cells, and they help prevent eggs from being fertilized by sperm from the wrong species. The latter function is particularly important in species that engage in external fertilization. The sea urchin is such a species, and its mechanisms of fertilization have been well studied.

The plasma membrane of the sea urchin egg is protected by a proteinaceous **vitelline envelope**, and surrounding that is a jelly coat. The sperm must get through these two protective layers before it can fuse with the plasma membrane of the egg. To accomplish this, the sperm has a membrane-enclosed structure called an **acrosome** containing enzymes and other proteins. The acrosome is located at the front end of the sperm head and covers the nucleus.

When the sperm makes contact with an egg of its own species, substances in the jelly coat trigger an **acrosomal reaction**, which begins with the release of acrosomal enzymes that digest the jelly coat. Next, an *acrosomal process* extends out of the head of the sperm. The acrosomal process forms from globular actin behind the acrosome, which polymerizes when the acrosomal membrane breaks down (Figure 43.1).

The acrosomal process extends through the jelly coat to make contact with the vitelline envelope. Herein lies another species recognition mechanism: The acrosomal process is coated with a membrane-bound protein called **bindin**. Different species have different bindin molecules. The egg plasma membrane has species-specific bindin receptors that extend through the vitelline envelope. The reaction of acrosomal bindin with these receptors stimulates the egg membrane to form a *fertilization cone* that engulfs the sperm head, bringing it into the egg cytoplasm.

In animals that practice internal fertilization, mating behaviors help guarantee species specificity, but egg–sperm recognition mechanisms still exist. The mammalian egg is surrounded by a thick layer called the **cumulus**, which consists of follicle cells in a gelatinous matrix (Figure 43.2). Beneath the cumulus is a glycoprotein envelope called the **zona pellucida**, which is functionally similar to the vitelline envelope of sea urchin eggs. When sperm are first de-

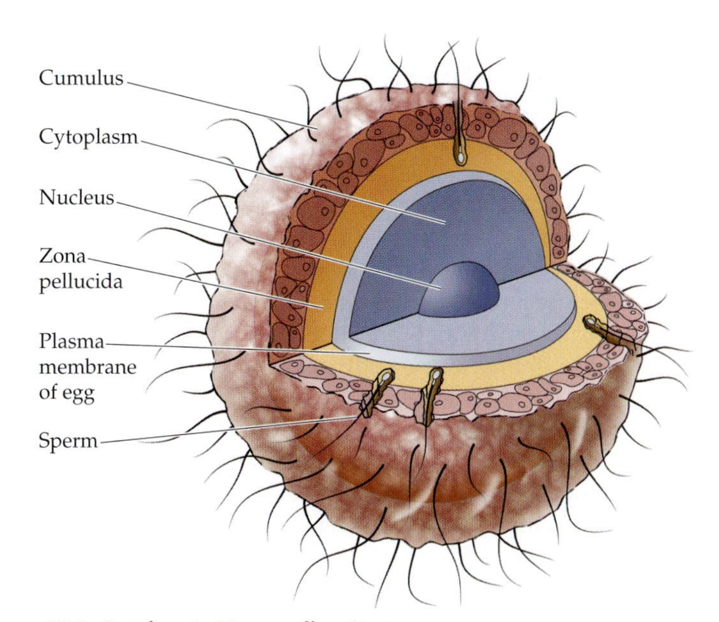

43.2 Barriers to Mammalian Sperm
This human egg is protected by the cumulus and zona pellucida, both of which a sperm must penetrate to fertilize the egg.

posited in the vagina, they are not capable of mounting an acrosomal reaction, but after being in the female reproductive tract for a time, the sperm undergo **capacitation**. A capacitated sperm is capable of interacting with an egg and its barriers, probably because certain critical proteins on its surface have been altered to make them reactive.

Mammalian sperm have enzymes on their surface that help them digest a path through the cumulus. When they make contact with the zona pellucida, a species-specific glycoprotein in the zona binds to recognition molecules on the head of the sperm. This binding triggers the acrosomal reaction, releasing acrosomal enzymes that digest a path through the zona. Other egg plasma membrane proteins then bind to adhesive proteins on the sperm, facilitating the fusion of sperm and egg plasma membranes.

The importance of the zona pellucida and its binding molecules in protection against heterospecific interactions between sperm and eggs was revealed in experiments in which the zona was stripped from eggs before they were exposed to sperm in a culture dish. In these experiments, it was possible for hamster sperm to fertilize human eggs, creating a hamster–human hybrid zygote. The hybrid zygote did not survive its first cell division because of chromosomal incompatibilities, but the experiment demonstrated that the mammalian species recognition mechanism resides in the zona.

Sperm entry triggers blocks to polyspermy and activates the egg

The fusion of the sperm and egg plasma membranes and the entry of the sperm into the egg initiate a programmed sequence of events. The first responses to sperm entry are **blocks to polyspermy**—that is, mechanisms that prevent more than one sperm from entering the egg. If more than one sperm enters the egg, the resulting embryo probably will not survive.

Blocks to polyspermy have been studied intensively in sea urchin eggs, which can be fertilized in a dish of seawater. Within a tenth of a second after a sperm enters a sea urchin egg, there is an influx of sodium ions, which changes the electric potential across the egg's plasma membrane. This *fast block to polyspermy* prevents the fusion of other sperm with the egg plasma membrane (Figure 43.3).

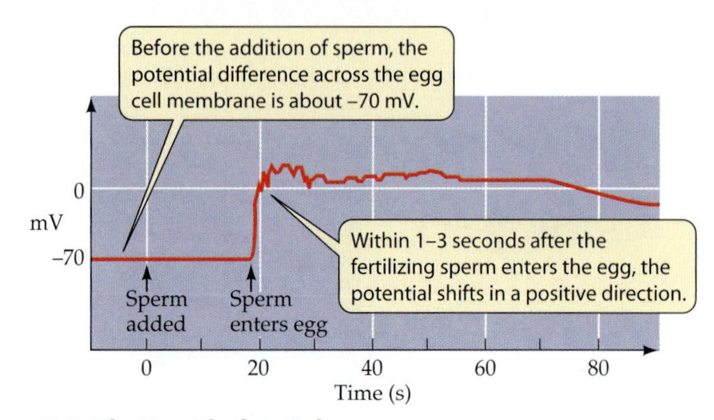

43.3 The Fast Block to Polyspermy
The fast block to polyspermy in sea urchins is a change in the electric potential across the egg plasma membrane.

The *slow block to polyspermy* takes about a minute (Figure 43.4). Before fertilization, the vitelline envelope is bonded to the plasma membrane. Just under the plasma membrane are *cortical granules* consisting of enzymes and other proteins. The sea urchin egg, like all animal cells, contains calcium that is sequestered in endoplasmic reticulum.

When a sperm enters, the sea urchin egg releases calcium from its endoplasmic reticulum into its own cytoplasm. This increase in cytoplasmic calcium causes the cortical granules to fuse with the plasma membrane. The cortical granule enzymes are released by exocytosis, breaking the bonds between the vitelline envelope and the plasma membrane. Water then flows into the space between the vitelline envelope and the plasma membrane, raising the vitelline envelope to form the *fertilization envelope*. The enzymes also degrade unused sperm-binding molecules at the surface of the fertilization envelope and cause it to harden, preventing the passage of additional sperm.

In mammals, sperm entry does not seem to cause a rapid change in membrane potential, but it does trigger the phosphatidylinositol (PTI) signaling system (see Figure 15.13), resulting in several events. Calcium is released from the endoplasmic reticulum, and the egg's metabolism is activated. The pH of the cytoplasm increases, oxygen consumption rises, and protein synthesis increases.

In mammals, the nuclei of the sperm and the egg do not fuse until about 12 hours after the sperm nucleus is taken into the egg cytoplasm because the nucleus of the egg must still complete its second meiotic division (see Figure 42.4*b*). Sea urchin eggs have already completed the second meiotic division at the time of sperm entry, and the nuclei fuse within an hour to create the zygote.

The sperm and the egg make different contributions to the zygote

Nearly all of the cytoplasm of the zygote comes from the egg. This cytoplasm is well stocked with nutrients, ribosomes, and a variety of molecules, including mRNAs. Moreover, because the sperm mitochondria degenerate, all of the mitochondria in the zygote come from the egg. In addition to its haploid nucleus, the sperm makes one other important contribution to the zygote in some species: a centriole. This centriole becomes the centrosome of the zygote, which produces the mitotic spindles for subsequent cell divisions.

For a long time, it was assumed that the one thing that sperm and egg contributed equally to the zygote was their haploid nuclei. However, we now know that even though they are equivalent in terms of genetic material, mammalian sperm and eggs are not equivalent in terms of their roles in development. In the laboratory, it is possible to construct zygotes in which both haploid nuclei come from the mother or both come from the father. In neither case does development progress normally. Apparently, in mammals at least, certain genes involved in development are active only if they come from a sperm and others are active only if they come from an egg—a phenomenon that has been termed **genomic imprinting**.

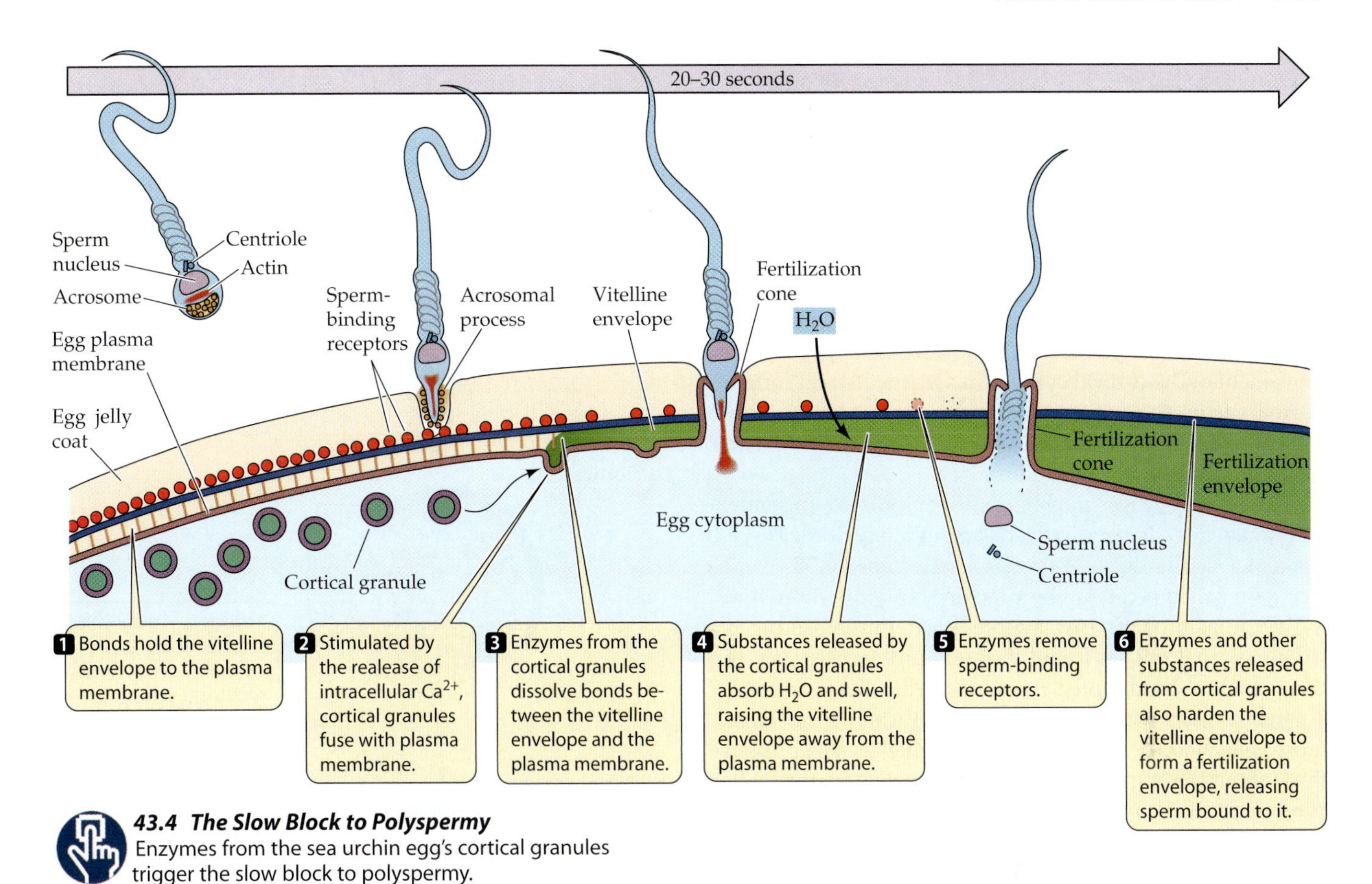

43.4 The Slow Block to Polyspermy
Enzymes from the sea urchin egg's cortical granules trigger the slow block to polyspermy.

Fertilization causes rearrangements of egg cytoplasm

The entry of the sperm into the egg stimulates changes and rearrangements of the cytoplasm that establish the symmetry and the body axes of the embryo. The nutrients and molecules in the cytoplasm of the zygote are not homogeneously distributed, and are not divided equally among all daughter cells when cell divisions begin. This unequal division of cytoplasmic factors sets the stage for the unfolding series of signals that orchestrates the sequential steps of development: determination, differentiation, and morphogenesis (see Chapter 16).

The rearrangements of egg cytoplasm in some species of frogs can be easily observed because of pigments in the egg cytoplasm. The nutrient molecules in an unfertilized frog egg are dense, and are therefore concentrated by gravity in the lower half of the egg, which is called the **vegetal hemisphere**. The haploid nucleus of the egg is located at the opposite end of the egg, in the **animal hemisphere**. The outermost (cortical) cytoplasm of the animal hemisphere is heavily pigmented, and the underlying cytoplasm has more diffuse pigmentation. The vegetal hemisphere is not pigmented.

Sperm always enter the frog egg in the animal hemisphere.When this occurs, the cortical cytoplasm rotates toward the site of sperm entry. This rotation exposes a band of diffusely pigmented cytoplasm on the side of the egg opposite the site of sperm entry. This band, called the **gray crescent**, will be the site of important developmental events (Figure 43.5).

The cytoplasmic rearrangements that create the gray crescent bring different regions of cytoplasm into contact on opposite sides of the egg. Therefore, bilateral symmetry is imposed on what was a radially symmetrical egg. Instead of just the up–down difference of the animal and vegetal hemispheres, the movement of the cytoplasm sets the stage for the creation

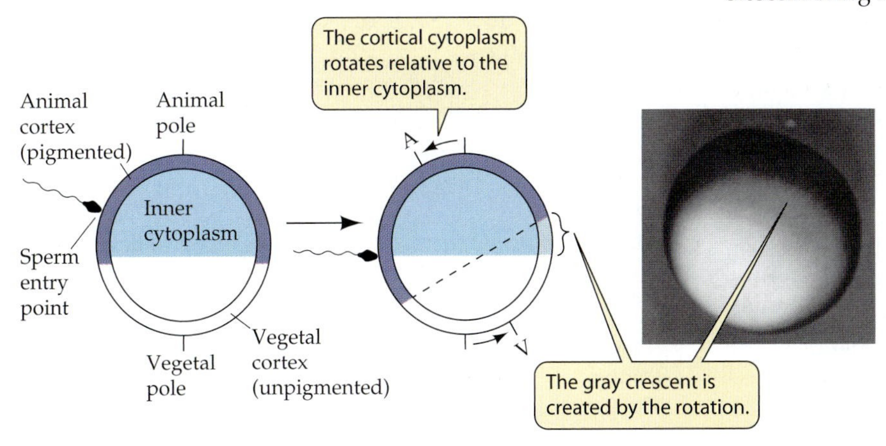

43.5 The Gray Crescent
Rearrangements of the cytoplasm of frog eggs after fertilization create the gray crescent.

of the anterior–posterior and left–right axes. In the frog, the site of sperm entry will become the *ventral* (belly) region, and the gray crescent will become the *dorsal* (back) region of the embryo. Since the gray crescent also marks the posterior end of the embryo, these relationships specify the anterior–posterior and left–right axes as well.

The molecular mechanism of this critical first step in embryo formation is beginning to be understood. The sperm centriole rearranges the microtubules in the vegetal pole cytoplasm into a parallel array that presumably guides the movement of the cortical cytoplasm. Organelles and certain proteins from the vegetal hemisphere move to the gray crescent region even faster than the cortical cytoplasm rotates.

As a result of these movements of cytoplasm, proteins, and organelles, changes in the distribution of critical developmental signals occur. A protein kinase called GSK-3 and a protein called β-catenin become unevenly distributed in the cytoplasm of the frog zygote. β-catenin is produced from maternal mRNA found throughout the cytoplasm of the egg. Once the zygote becomes a multicellular embryo, β-catenin will act as a transcription factor in the nuclei of those cells where it is present. GSK-3, which is also present throughout the egg cytoplasm, causes the degradation of β-catenin. However, inhibitors of GSK-3 migrate along the vegetal microtubules and prevent GSK-3 from degrading β-catenin (Figure 43.6). As a result, there is a higher concentration of β-catenin on what will become the dorsal side of the embryo.

Evidence supports the hypothesis that β-catenin is a key player in the cell–cell signaling cascade that begins the formation of the embryo in the region of the gray crescent. But before there can be cell–cell signaling, there must be multiple cells.

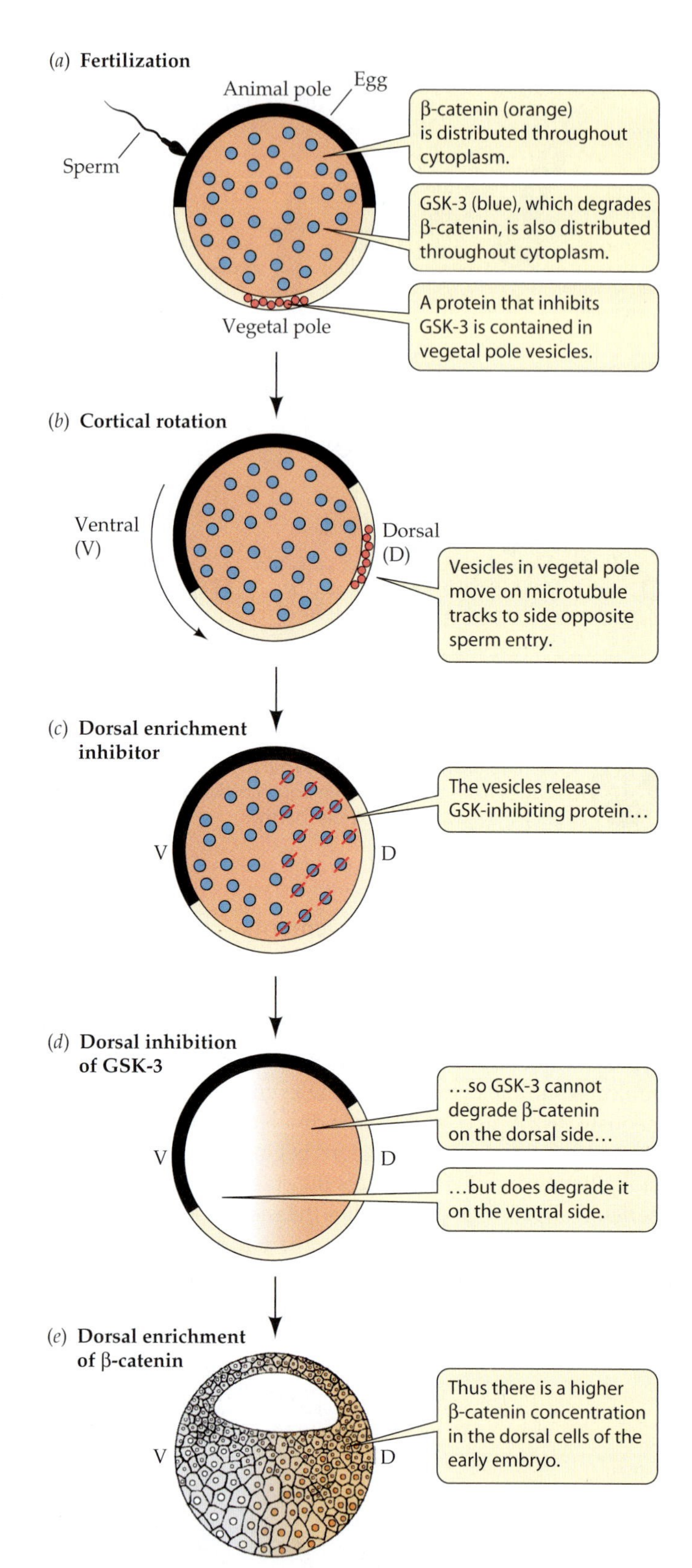

43.6 Cytoplasmic Factors Set Up Signaling Cascades Cytoplasmic movement changes the distributions of critical developmental signals. In the frog zygote, the interaction of the protein kinase GSK-3, its inhibitor, and the protein β-catenin are crucial in specifying the dorsal–ventral (back–belly) axis of the embryo.

Cleavage: Repackaging the Cytoplasm

At some point after fertilization is complete (the exact timing differs among species), a rapid series of cell divisions called **cleavage** takes place. Because the cytoplasm of the zygote is not homogeneous, these first cell divisions result in the differential distribution of nutrients and informational molecules among the cells of the early embryo. In most animals, cleavage proceeds with rapid DNA synthesis and mitosis, but no cell growth and little gene expression. The embryo becomes a ball of smaller and smaller cells. This ball forms a central cavity called a **blastocoel** (a process called *blastulation*), at which point it is called a **blastula**. The individual cells are called **blastomeres**.

Cleavage, and therefore the formation of the blastula, is influenced in different species by two major factors. First, in some species, massive amounts of nutrients, or **yolk**, are stored in the egg. The yolk influences the pattern of cell divisions by impeding the formation of *cleavage furrows* between the daughter cells. Second, proteins and mRNAs stored in the egg by the mother guide the formation of mitotic spindles and the timing of cell divisions.

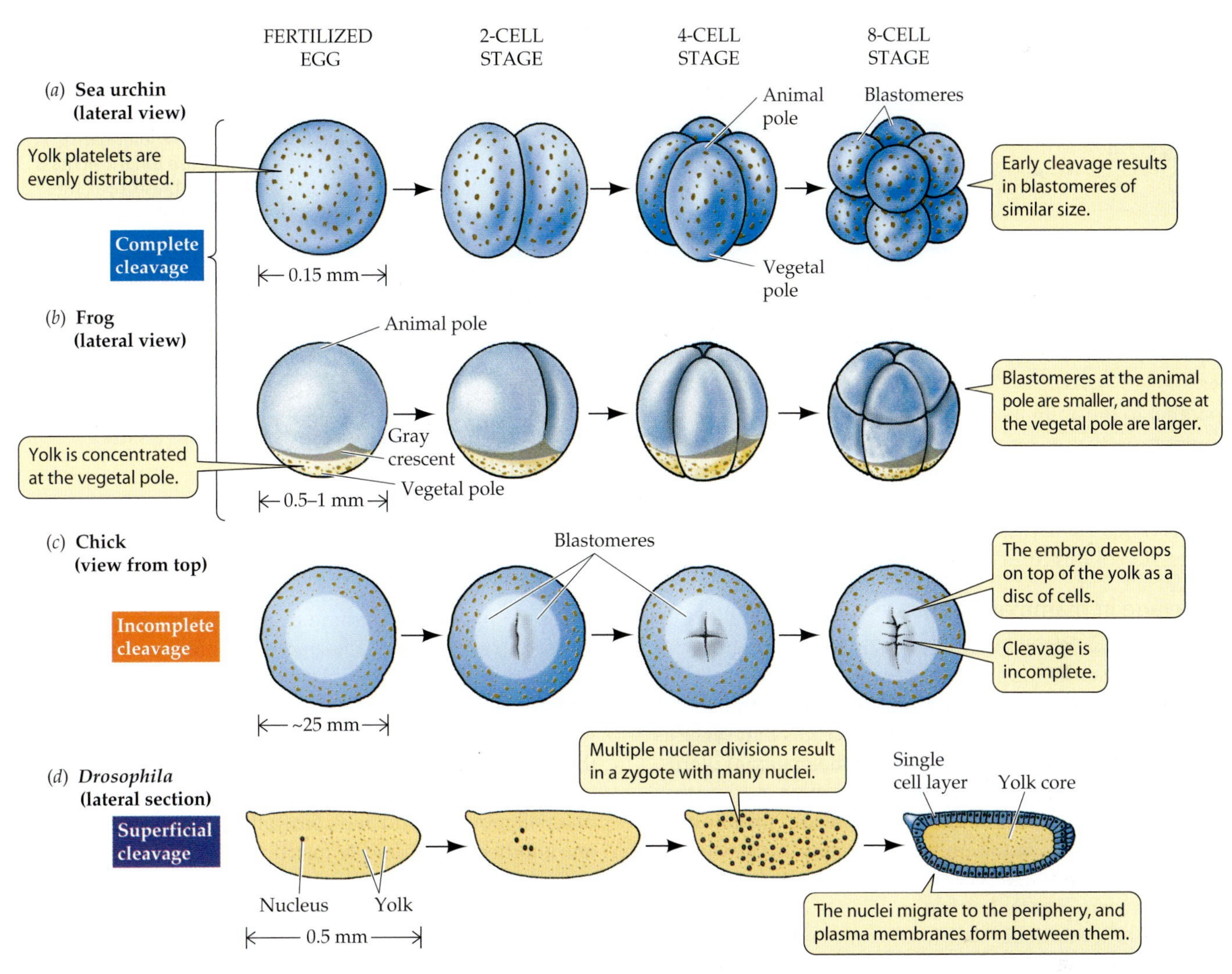

43.7 Patterns of Cleavage in Four Model Organisms
Patterns of early embryonic development reflect differences in the way the egg cytoplasm is organized.

The amount of yolk influences cleavage

When the yolk content of the egg is sparse, there is little interference with cleavage furrow formation, and all the daughter cells are of similar size; the sea urchin egg provides an example (Figure 43.7*a*). More yolk means more resistance to cleavage furrow formation; therefore, cell divisions progress more rapidly in the animal hemisphere than in the vegetal hemisphere, where the yolk is concentrated. As a result, the cells derived from the vegetal hemisphere are fewer and larger; the frog egg provides an example (Figure 43.7*b*).

In spite of this difference between sea urchin and frog eggs, the cleavage furrows completely divide the egg mass in both cases; thus these animals are said to have *complete cleavage*. In contrast, in an egg, such as the chicken egg, that contains a lot of yolk, the cleavage furrows do not penetrate the yolk. As a result, cleavage is incomplete, and the embryo forms as a disc of cells, called a **blastodisc**, on top of the yolk mass (Figure 43.7*c*). This type of cleavage, called *incomplete cleavage*, is common in fishes, reptiles, and birds.

Another type of incomplete cleavage, called *superficial cleavage*, occurs in insects such as the fruit fly (*Drosophila*). In the insect egg, the massive yolk is centrally located (Figure 43.7*d*). Early in development, cycles of mitosis occur without cytokinesis. Eventually the resulting nuclei migrate to the periphery of the egg, and after several more mitotic cycles, the plasma membrane of the egg grows inward, partitioning the nuclei off into individual cells.

The orientation of mitotic spindles influences the pattern of cleavage

The positions of the mitotic spindles during cleavage are not random; rather, they are determined by cytoplasmic factors. In turn, the orientation of the mitotic spindles determines the cleavage planes and, therefore, the arrangement of the daughter cells.

If the mitotic spindles of successive cell divisions form at right angles or parallel to the animal–vegetal axis of the zygote, the cleavage pattern is *radial*, as in the sea urchin and the frog (see Figure 43.7*a* and *b*). The orientations of successive mitotic spindles in mammals are also at right angles or

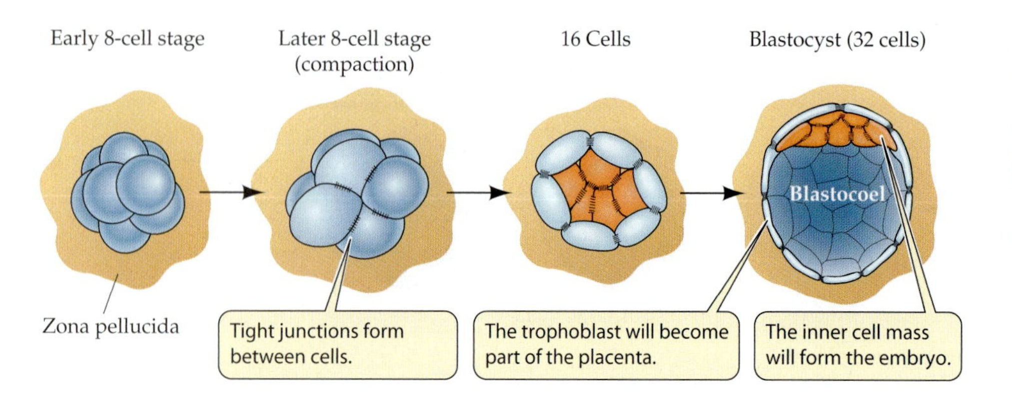

43.8 A Blastocyst Forms in Mammals
Starting at the late 8-cell stage, the mammalian embryo undergoes compaction of its cells, resulting in a blastocyst—a dense inner cell mass on top of a hollow blastocoel, surrounded by a trophoblast.

parallel to the animal–vegetal axis, but the sequence is different from that in sea urchins, resulting in a pattern called *rotational cleavage*. In mollusks, the successive mitotic spindles are not at right angles; the resulting pattern has a twist, and is called *spiral cleavage*. The coiling of snail shells is an expression of this form of cleavage.

Cleavage in mammals is unique

The early development of placental mammals is quite different from that of the other species we have discussed. The zygote of a placental mammal produces both an embryo and the elaborate extraembryonic structures that serve as the interface between the embryo and the maternal uterus. Cell divisions are slower and genes are expressed during cleavage in placental mammals. As a result, proteins encoded by the genes of the embryo play a role in cleavage. In other species, cleavage is directed almost entirely by molecules that were present in the egg.

As in other animals that have complete cleavage, the early cell divisions in a mammalian zygote produce a loosely associated ball of cells. However, at about the 8-cell stage, the behavior of the cells changes. They suddenly maximize their surface contact with each other, form tight junctions, and become a very compact mass of cells.

At the transition from the 16-cell to the 32-cell stage, the cells separate into two groups. The innermost cells form the **inner cell mass** that will become the embryo, while the outermost cells become an encompassing sac called the **trophoblast**, which will become part of the placenta. The trophoblast cells secrete fluid, thus creating a cavity with the inner cell mass at one end (Figure 43.8). At this stage, the mammalian embryo is called a **blastocyst** to distinguish it from the blastula of other animals.

When the blastocyst reaches the uterus, the trophoblast adheres to the uterine wall, beginning the process of implantation that embeds the embryo in the endometrium. In humans, implantation begins about the sixth day after fertilization. Implantation must not occur as the blastocyst moves down the oviduct to the uterus, or the result will be an *ectopic* or *tubal pregnancy*—a very dangerous condition. Early implantation is normally prevented by the zona pellucida, which remains around the cleaving ball of cells. At about the time the blastocyst reaches the uterus, it "hatches" from the zona pellucida, and implantation can occur.

Specific blastomeres generate specific tissues and organs

Cleavage in all species results in a repackaging of the cytoplasm of the egg into a large number of small cells surrounding a central cavity. Little cell differentiation has occurred during cleavage, and in most nonmammalian species none of the genome of the embryo has been expressed. Nevertheless, cells in different regions of the blastula possess different complements of the nutrients and informational molecules that were present in the egg.

The blastocoel prevents cells from different regions of the blastula from interacting, but that will soon change. During the next stage of development, the cells of the blastula will move around and come into new associations with one another, communicate instructions to one another, and begin to differentiate.

In many animals, the movements of the blastomeres are so regular and well orchestrated that it is possible to label a specific blastomere and identify the tissues and organs that form from its progeny. Such labeling experiments produce **fate maps** of the blastula (Figure 43.9).

Blastomeres become *determined*—committed to specific fates—at different times in different species. In some

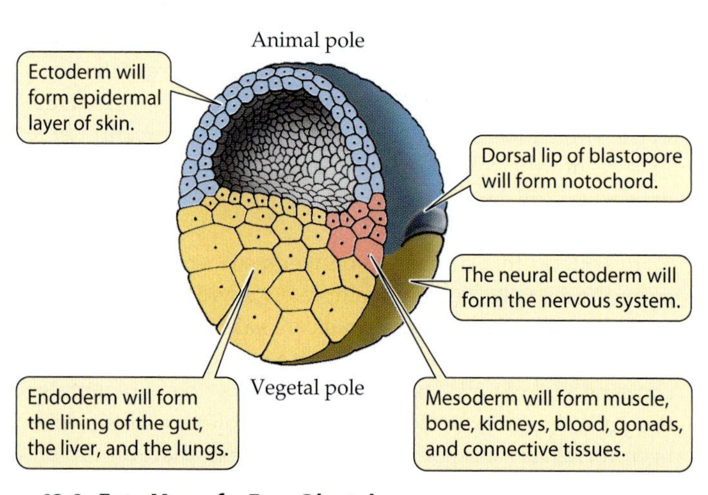

43.9 Fate Map of a Frog Blastula
The colors indicate the portions of the *Xenopus* blastula that will form the three germ layers, and subsequently the frog's tissues and organs.

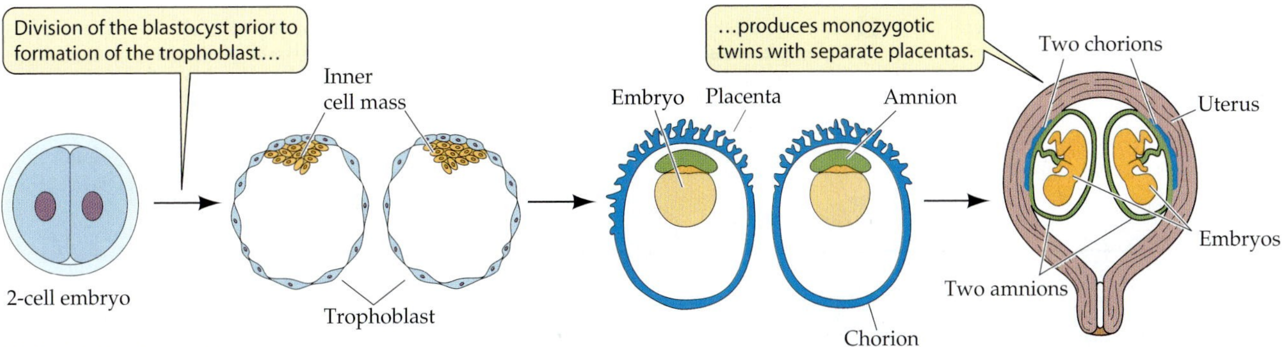

43.10 Twinning in Humans
In humans, monozygotic (identical) twins result when groups of cells in the blastula become physically separated and both groups produce embryos.

species, such as roundworms and clams, blastomeres at the 8-cell stage are already determined. If one of these blastomeres is experimentally removed, a particular portion of the embryo will not form. This type of development has been called **mosaic development** because each blastomere seems to contribute a specific set of "tiles" to the final "mosaic" that is the adult animal. In contrast, other species, such as sea urchins, frogs, and vertebrates, have **regulative development**: The loss of some cells during cleavage does not affect the developing embryo because the remaining cells compensate for the loss.

If some blastomeres can change their fate to compensate for the loss of other cells during cleavage and blastula formation, are these cells capable of forming an entire embryo? To a certain extent they are. During cleavage or early blastula formation, if the blastomeres are physically separated into two groups, both groups can produce complete embryos (Figure 43.10). Since the two embryos come from the same zygote, they will be *monozygotic twins*—genetically identical. Non-identical twins occur when two separate eggs are fertilized by two separate sperm. Thus, while identical twins are always of the same sex, non-identical twins have a 50 percent chance of being the same sex.

Gastrulation: Producing the Body Plan

The blastula is typically a fluid-filled ball of cells. How does this ball of cells become an embryo, made up of multiple tissue layers, with head and tail ends and dorsal and ventral sides? **Gastrulation** is the process by which layers of tissue, called **germ layers**, form and take specific positions relative to one another (Figure 43.9; Table 43.1). The resulting spatial relations between tissues make possible **inductive interactions**: exchanges of signals among tissues that trigger differentiation and organ formation.

During gastrulation, the animal body forms three germ layers:

- Some blastomeres move as a sheet to the inside of the embryo, creating an inner germ layer, the **endoderm**, which will give rise to gut tissues.
- The cells remaining on the outside become the outer germ layer, the **ectoderm**, and give rise to the epidermis and the nervous system.
- Other cells migrate between these two layers to become the middle germ layer, or **mesoderm**, which will contribute tissues to many organs, including bones, muscles, liver, heart, and blood vessels.

Some of the most challenging and interesting questions in animal development have concerned what directs the cell movements of gastrulation and what is responsible for the resulting patterns of cell differentiation and organ formation. In the past 20 years scientists have answered many of these questions at the molecular level. In the discussion that follows, we consider the similarities and differences of gastrulation in sea urchins, frogs, reptiles, birds, and mammals. We also review some of the exciting discoveries about the mechanisms underlying these phenomena.

Involution of the vegetal pole characterizes gastrulation in the sea urchin

The sea urchin blastula is a simple, hollow ball of cells that is only one cell layer thick. The end of the blastula stage is marked by a dramatic slowing of the rate of mitosis, and the beginning of gastrulation is marked by a flattening of the vegetal hemisphere (Figure 43.11). Some cells at the vegetal pole bulge into the blastocoel, break free, and migrate into the cavity. These cells become **primary mes-**

43.1 Fates of Embryonic Germ Layers in Vertebrates[a]

GERM LAYER	FATE
Ectoderm	Brain and nervous system; lens of eye; inner ear; lining of mouth and of nasal canal; epidermis of skin; hair and nails; sweat glands, oil glands, milk secretory glands
Mesoderm	Skeletal system (bones, cartilage, notochord); gonads; muscle; outer coverings of internal organs; dermis of skin; circulatory system (heart, blood vessels, blood cells); kidneys
Endoderm	Inner linings of gut; respiratory tract (including lungs); liver; pancreas; thyroid; urinary bladder

[a] The final structures are complex, containing cells from more than one germ layer. Interactions among tissues are usually important in determining the composition and structure of an organ.

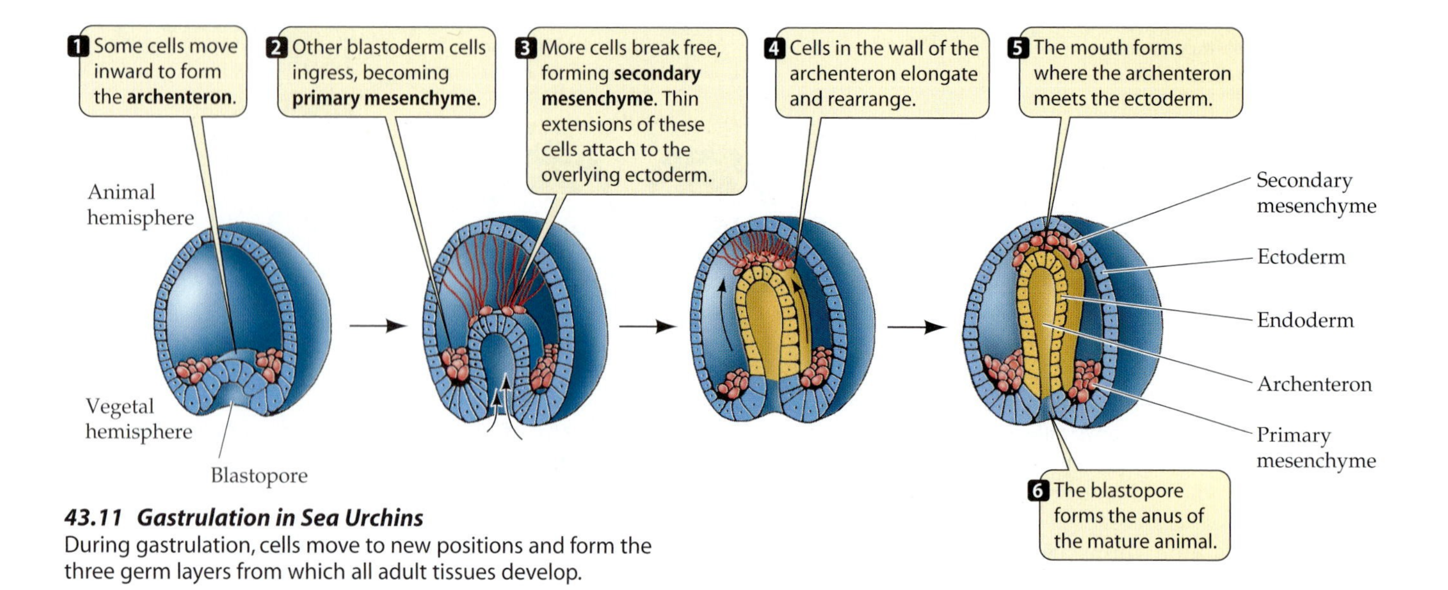

43.11 Gastrulation in Sea Urchins
During gastrulation, cells move to new positions and form the three germ layers from which all adult tissues develop.

enchyme cells—cells of the middle germ layer, the mesoderm. (The word "mesenchyme" means "a loosely organized group of cells," in contrast to cells formed into a tightly packed sheet.)

The flattening at the vegetal pole becomes an *involution*, as if someone were poking a finger into a hollow ball. The cells that involute become the endoderm and form the primitive gut, the **archenteron**. At the tip of the archenteron more cells break free, entering the blastocoel to form more mesoderm, the **secondary mesenchyme**.

The archenteron continues to move inward, partly because of changes in the shapes of its cells and partly because it is pulled by secondary mesenchyme cells. These cells, attached to the tip of the archenteron, send out extensions that adhere to the overlying ectoderm and contract. Where the archenteron eventually makes contact with the ectoderm, the mouth will form, and the **blastopore**, the opening created by the invagination of the vegetal pole, will become the anus of the animal.

What mechanisms control the various cell movements of sea urchin gastrulation? The immediate answer is that specific properties of particular cells change. For example, some vegetal cells migrate into the blastocoel to form the primary mesenchyme because they lose their attachments to neighboring cells. Once they bulge into the blastocoel, they move by extending long processes called *filopodia* along an extracellular matrix of proteins that is laid down by the ectodermal cells lining the blastocoel.

A deeper understanding of gastrulation requires that we discover the molecular mechanisms whereby certain cells

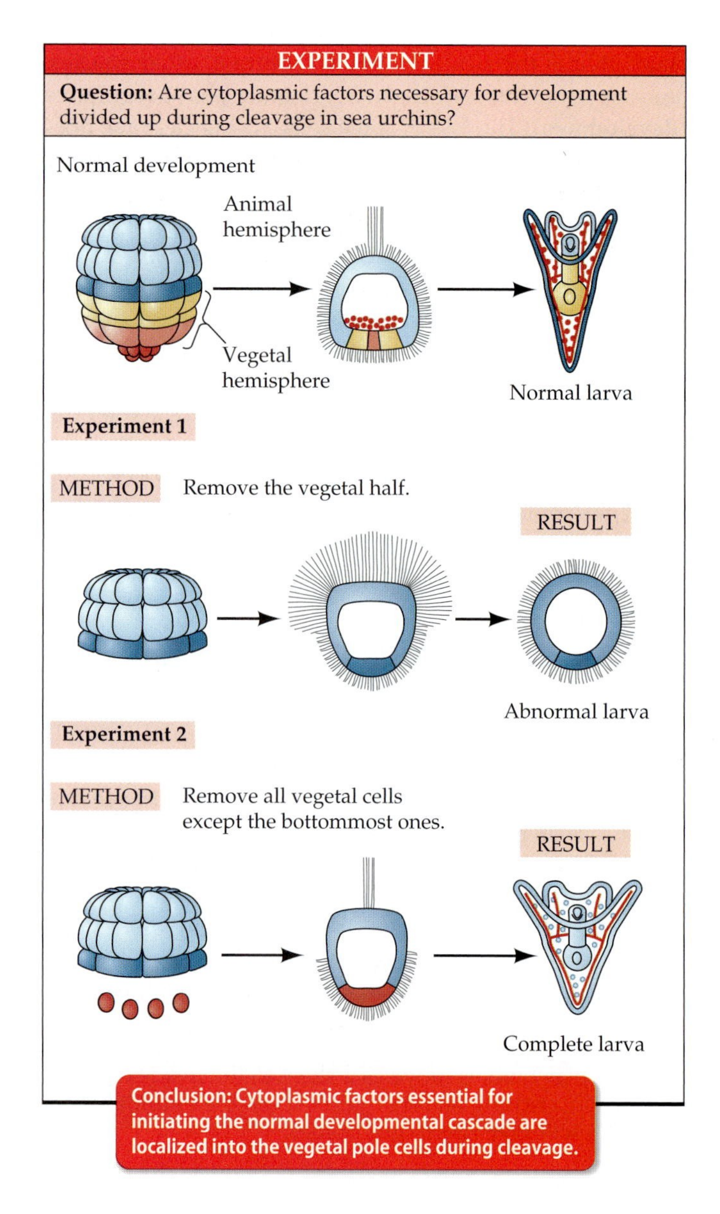

43.12 Vegetal Pole Cells Contain Essential Cytoplasmic Factors
If a sea urchin blastula is divided into animal pole and vegetal pole cells, only the vegetal pole cells have the capacity to develop normally. If a few vegetal pole cells are added to a mass of animal pole cells, normal development can occur.

of the blastula develop properties different from those of others. Cleavage divides up the cytoplasm of the egg in a very systematic way. The sea urchin blastula at the 64-cell stage can be viewed as consisting of tiers of cells. As in the frog blastula, the top is the animal pole and the bottom the vegetal pole. If different tiers of blastula cells are separated, they show different developmental potentials (Figure 43.12). Only cells from the vegetal pole are capable of initiating the development of a complete larva.

It has been proposed that transcriptional regulatory proteins are unevenly distributed in the egg cytoplasm and therefore end up in particular groups of cells as cleavage progresses. Possibly, transcription factors found in the lowest vegetal cells become active in early cleavage. Indeed, one of these factors appears to be β-catenin, the transcription factor that is differentially distributed in the frog zygote. The genes activated by β-catenin seem to produce proteins that set up a signaling cascade that initiates the processes of cell determination and differentiation. Let's turn now to gastrulation in the frog, in which a number of key signaling molecules have been identified.

Gastrulation in the frog begins at the gray crescent

Amphibian blastulas have considerable yolk and are more than one cell layer thick; therefore, gastrulation is more complex in the frog than in the sea urchin. Gastrulation begins when certain cells in the gray crescent region of the blastula change their shape and their cell adhesion properties. These are the cells that received high concentrations of β-catenin in their cytoplasm and therefore made certain proteins that cells in other parts of the blastula did not make. In some way, this pattern of gene expression causes the bodies of these cells to bulge into the blastocoel while they remain attached to the outer surface by slender necks. Because of their shape, these cells are called *bottle cells*.

As the bottle cells move into the interior of the blastula, they appear to pull other surface cells in after them (Figure 43.13). This process creates a lip over which a sheet of cells moves into the blastocoel. The first involuting cells are the prospective endoderm, and they form the primitive gut, the archenteron. These cells also bring into the blastocoel the cells that will form the mesoderm. The initial site of involution is called the **dorsal lip** of the blastopore, and it plays a central role in vertebrate development.

As gastrulation proceeds, cells from all over the surface of the blastula move toward the site of invagination. This movement of cells toward the blastopore is called *epiboly*. The dorsal lip of the amphibian blastopore widens and eventually forms a complete circle surrounding a plug of yolk-rich cells. As cells continue to move in through the blastopore, the archenteron grows and gradually displaces the blastocoel.

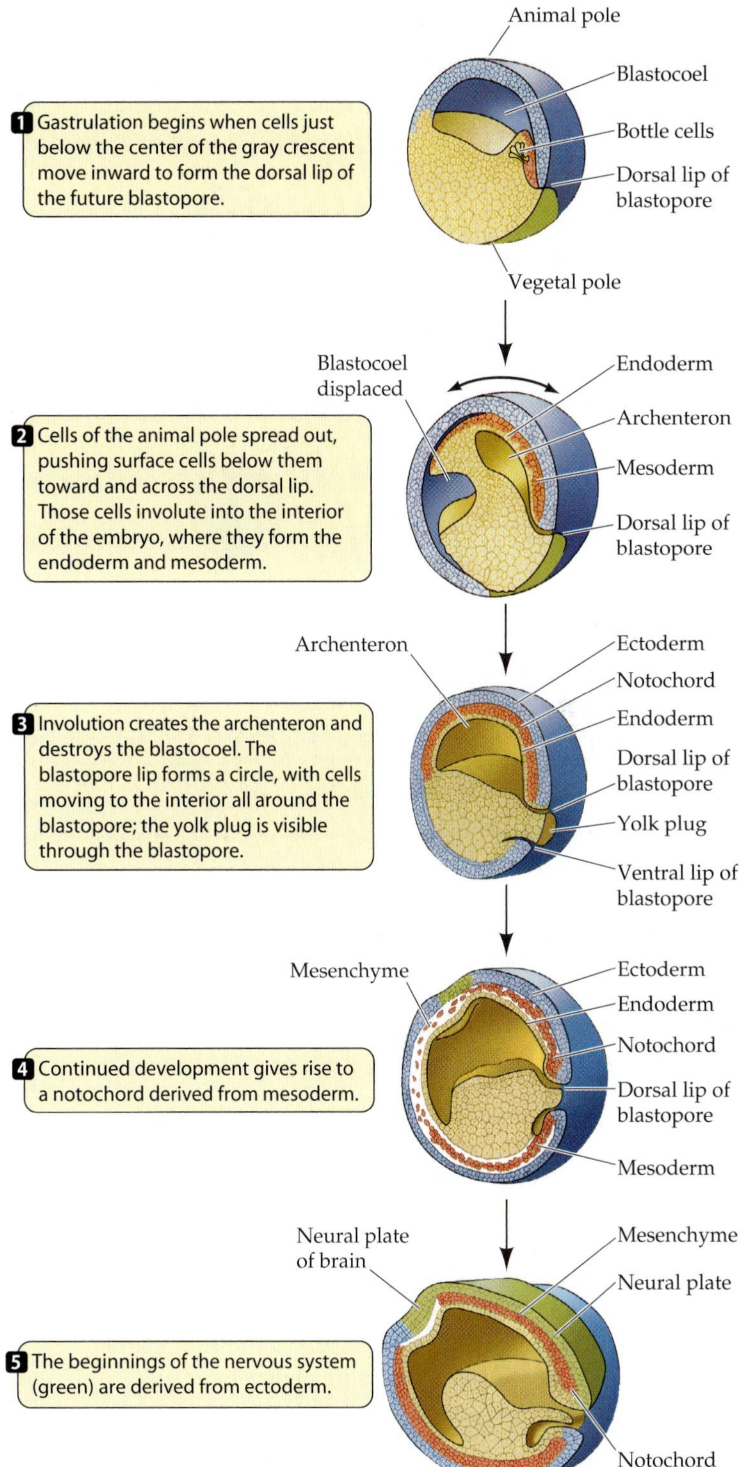

***43.13 Gastrulation in the Frog* Xenopus**
The colors in this diagram are matched to those in the frog fate map (Figure 43.9).

As gastrulation comes to an end, the embryo consists of three germ layers: ectoderm on the outside, endoderm on the inside, and mesoderm in the middle. The embryo also has a dorsal–ventral and anterior–posterior organization. Most importantly, however, the fates of specific regions of

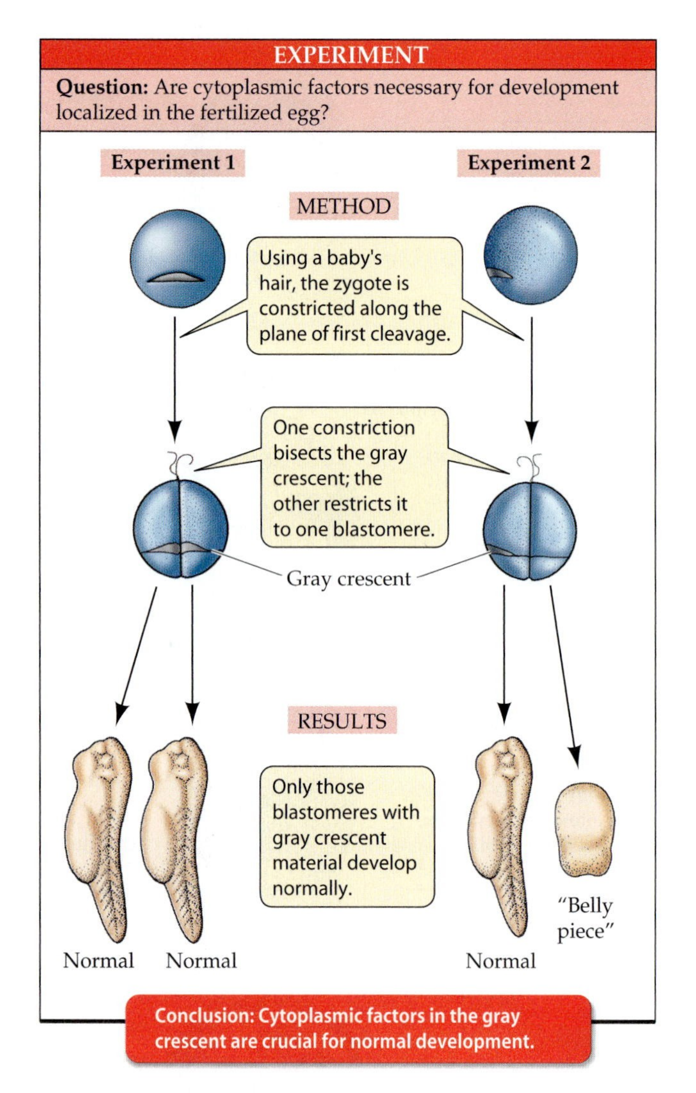

43.14 Spemann's Experiment
Spemann's research revealed that gastrulation and subsequent normal development in salamanders depended on cytoplasmic factors localized in the gray crescent.

the endoderm, mesoderm, and ectoderm have become "determined" such that they will differentiate into specific tissue types. The discovery of determination, our next topic, is one of the most exciting stories in animal development.

The dorsal lip of the blastopore organizes formation of the embryo

Early in the 1900s, German biologist Hans Spemann was studying the development of salamander eggs. He was interested in finding out whether nuclei remain *totipotent*—capable of directing the development of a complete embryo—during cleavage and blastula formation. With great patience and dexterity, he formed a loop from a single human baby hair to constrict fertilized eggs.

When Spemann's loop bisected the gray crescent, the cells on both sides of the constriction developed into complete embryos (Figure 43.14, left). But when the gray crescent was on only one side of the constriction, only that side gastrulated and developed into a complete embryo. The side lacking gray crescent material became a clump of undifferentiated cells Spemann called the "belly piece" (Figure 43.14, right). Spemann thus hypothesized that cytoplasmic factors contained in the region of the gray crescent are necessary for gastrulation and thus for the development of a normal organism.

To test his hypotheses, Spemann and his student Hilde Mangold conducted a series of delicate tissue transplantation experiments. They transplanted pieces of early gastrulas to various locations on other gastrulas. Guided by fate maps (see Figure 43.9), they were able to take a piece of ectoderm they knew would develop into epidermis and transplant it to a region that normally becomes nervous system, and vice versa (Figure 43.15).

When they did these transplants in early gastrulas, the transplanted pieces always developed into tissues that were appropriate for the location where they were placed. Donor presumptive epidermis (that is, cells destined to become skin in their original location) developed into host nervous system, and donor presumptive neural ectoderm developed into host skin. Thus, the fates of the transplanted cells had not been determined before the transplantation.

In late gastrulas, however, the same experiment yielded opposite results. Donor presumptive epidermis produced patches of skin cells in the host nervous system, and donor presumptive neural ectoderm produced neurons in the host skin. Something had occurred during gastrulation to determine the fates of the embryonic cells. In other words, as Spemann had hypothesized, the path of differentiation a cell would follow was determined during gastrulation.

Then Spemann and Mangold did an experiment that produced momentous results: they transplanted the dorsal lip of the blastopore. When this small piece of tissue was transplanted into the presumptive belly area of another gastrula, it stimulated a second site of gastrulation, and another whole embryo formed belly-to-belly with the original embryo. The dorsal lip of the blastula was apparently capable of inducing the formation of an embryo. Thus, Spemann and Mangold called the dorsal lip of the blastopore the **primary embryonic organizer**.

The primary embryonic organizer has been studied intensively, and we are beginning to understand the molecular mechanisms involved in its action. The distribution of β-catenin corresponds to the location of the primary embryonic organizer, so it is a candidate molecule for initiating organizer activity. Correlation is not enough, however. To prove that a molecule is an inductive signal, it has to be shown that it is both *necessary* and *sufficient* for the proposed effect. In other words, the effect should not occur where expected if the candidate molecule is eliminated (necessity), and the candidate molecule must be capable of inducing the effect where it would otherwise not occur (sufficiency).

The criteria of necessity and sufficiency have been satisfied for β-catenin. If β-catenin transcripts are depleted by injections of antisense RNA into the egg (see Chapter 17), gastrulation does not occur. If β-catenin is experimentally

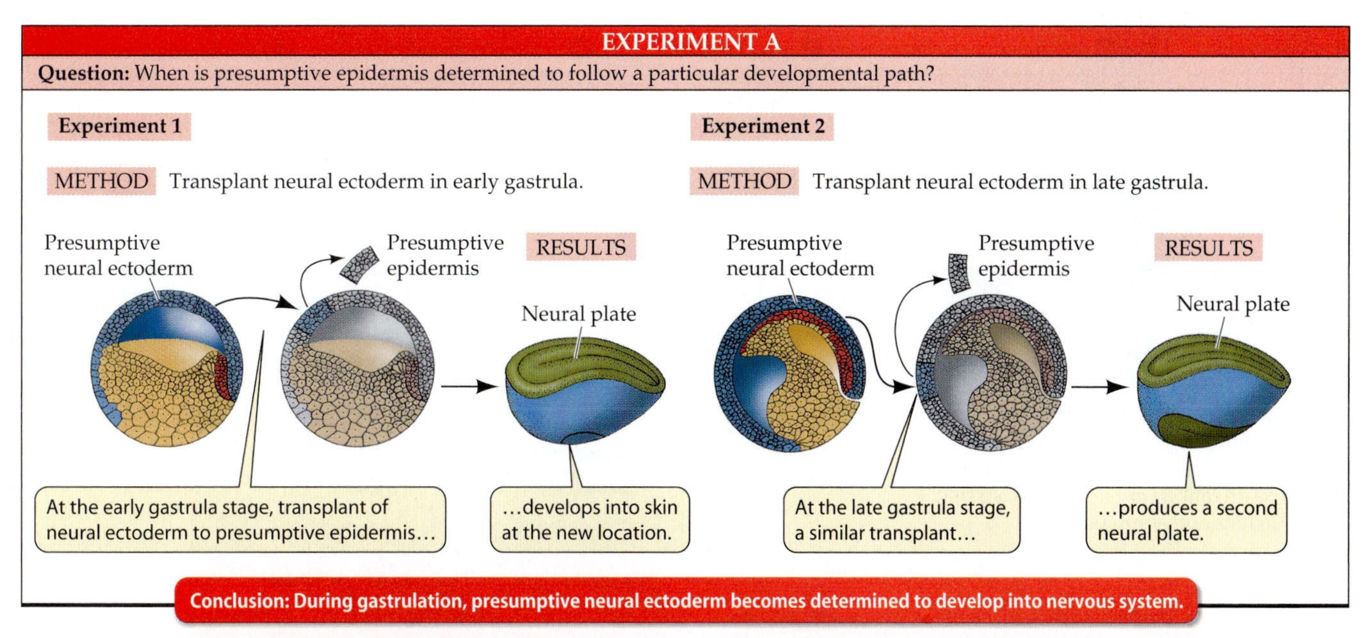

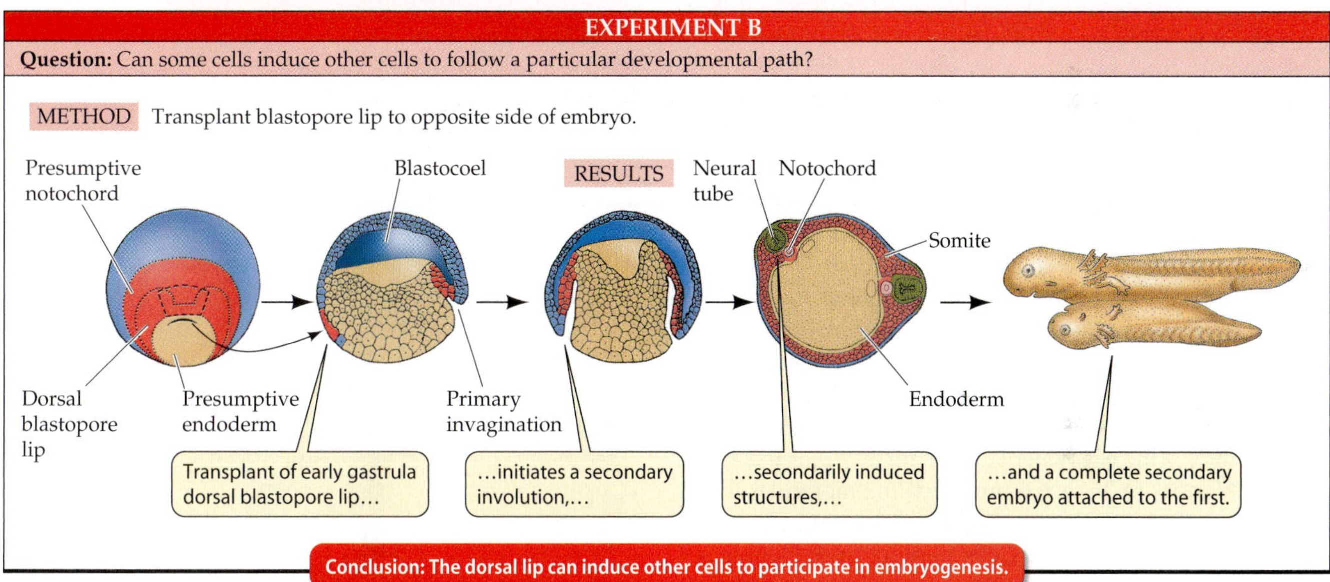

43.15 Tissue Transplants Reveal the Process of Determination
Tissue transplant experiments by Spemann and Mangold revealed that cell fate becomes increasingly determined during gastrulation. Transplanting the dorsal lip of the blastopore resulted in a second initiation of gastrulation and the formation of a second embryo.

overexpressed in another region of the blastula, it can induce a second axis of embryo formation, as the dorsal lip did in the Spemann–Mangold transplantation experiments. Thus, β-catenin appears to be both necessary and sufficient for the formation of the primary embryonic organizer, but it is only one component of a complex signaling cascade that is still the subject of intense investigation.

Reptilian and avian gastrulation is an adaptation to yolky eggs

The eggs of reptiles and birds have a massive yolk content, and therefore the blastulas of these species develop as a disc of cells on top of the yolk. We will use the chicken egg to show how gastrulation occurs in a flat disc of cells rather than in a ball of cells.

Cleavage in the chick results in a flat, circular layer of cells called the blastodisc. Between the blastodisc and the yolk mass is a fluid-filled space. Some cells from the blastodisc break free and move into this space. Other cells grow into this space from the posterior margin of the blastodisc. These cells come together to form a continuous layer called the **hypoblast**. The overlying cells are called the **epiblast**. Thus, the avian blastula is a flattened structure consisting of an upper epiblast and a lower hypoblast, which are joined at the margins of the blastodisc. The blastocoel is the fluid-filled space between the epiblast and hypoblast.

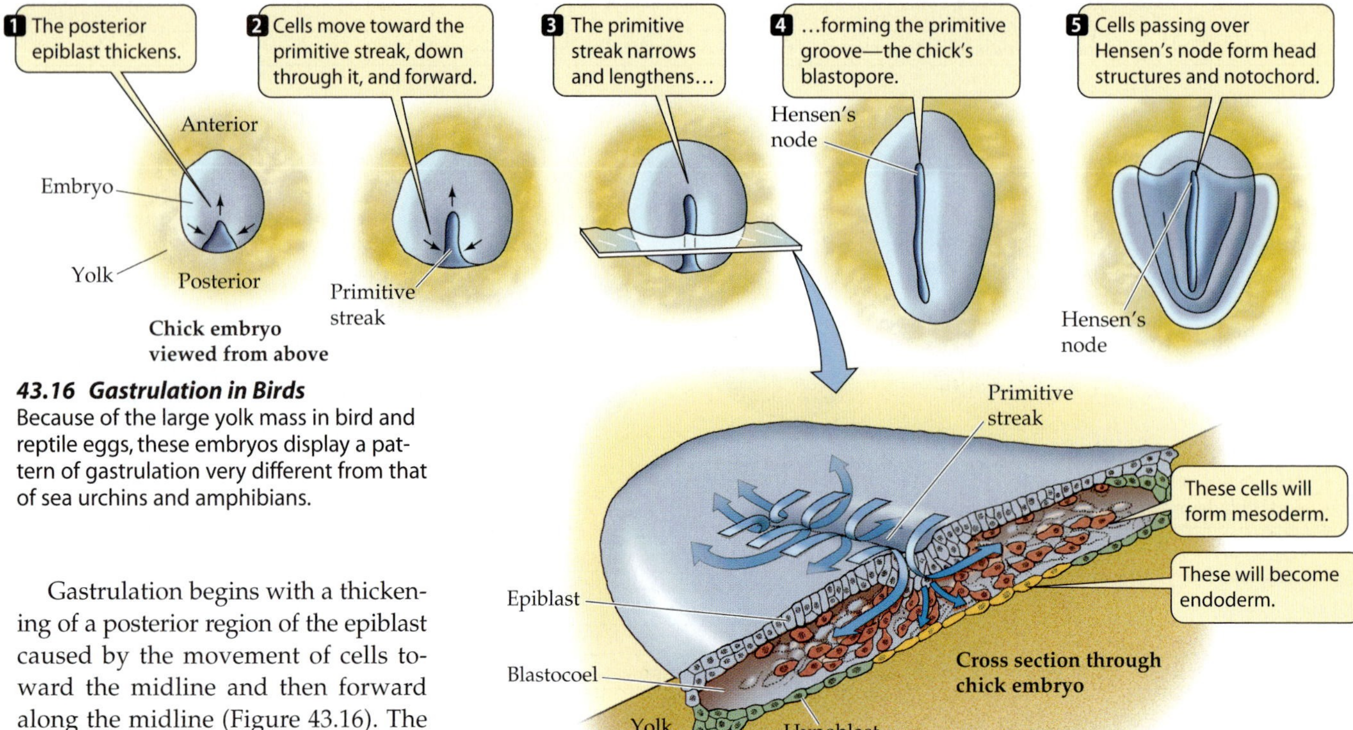

43.16 Gastrulation in Birds
Because of the large yolk mass in bird and reptile eggs, these embryos display a pattern of gastrulation very different from that of sea urchins and amphibians.

Gastrulation begins with a thickening of a posterior region of the epiblast caused by the movement of cells toward the midline and then forward along the midline (Figure 43.16). The result is a midline ridge called the **primitive streak**. A depression called the **primitive groove** forms along the length of the primitive streak. The primitive groove becomes the blastopore as cells migrate through it into the blastocoel to become endoderm and mesoderm.

In the avian embryo, no archenteron forms, but the prospective endoderm and mesoderm migrate forward to form gut and other structures. At the extreme forward end of the primitive groove is a thickening called **Hensen's node**, which is the equivalent of the dorsal lip of the amphibian blastopore. In fact, many signaling molecules that have been identified in the frog organizer are also expressed in Hensen's node. Cells passing over Hensen's node become determined to differentiate into the tissues and structures that make up the head and the dorsal midline of the embryo.

Mammals have no yolk, but retain the avian–reptilian gastrulation pattern

Mammals and birds both evolved from reptilian ancestors, so it is not surprising that they share patterns of early development, even though the eggs of placental mammals have no yolk. Earlier we described the development of the mammalian trophoblast and the inner cell mass, which is the equivalent of the avian epiblast. Keeping avian gastrulation in mind, think of the mammalian inner cell mass as sitting on top of an imaginary body of yolk (Figure 43.17).

As in avian development, the inner cell mass splits into an upper layer called the epiblast and a lower layer called the hypoblast with a fluid-filled cavity, or blastocoel, between them. The embryo will form from the epiblast, and the hypoblast will contribute to the extraembryonic membranes. The epiblast also contributes to the extraembryonic

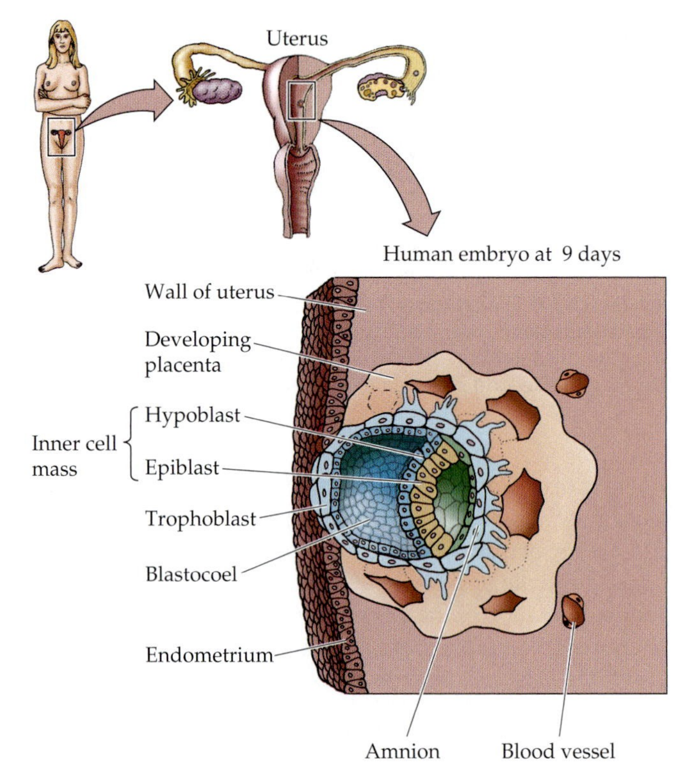

43.17 A Human Blastocyst before Gastrulation
The mammalian inner cell mass becomes the epiblast; it can be compared to the avian embryo by picturing it sitting on top of a mass of imaginary yolk.

membranes; specifically, it splits off an upper layer of cells that will form the *amnion*. The amnion will grow to surround the developing embryo as a sac filled with amniotic fluid.

Gastrulation occurs in the mammalian epiblast just as it does in the avian epiblast. A primitive groove forms, and epiblast cells migrate through the groove to become layers of endoderm and mesoderm. The cells migrating over Hensen's node become the cells that form dorsal structures such as the brain and spinal cord.

Neurulation: Initiating the Nervous System

Gastrulation produces an embryo with three germ layers that are positioned to influence one another through inductive tissue interactions. During the next phase of development, called **organogenesis**, many organs and organ systems develop simultaneously and in coordination with one another. An early process of organogenesis that is directly related to gastrulation is **neurulation**, the initiation of the nervous system. We will examine this event in the amphibian embryo, but it occurs in a similar fashion in reptiles, birds, and mammals. Many of the genes involved are highly conserved all the way from worms to humans.

The stage is set by the dorsal lip of the blastopore

The cells that pass through the dorsal lip of the blastopore and move anteriorly in the blastocoel during gastrulation are determined to become mesoderm. The dorsal mesoderm closest to the midline is further determined to become **chordomesoderm**, which forms a rod along the dorsal midline. This rod, called the **notochord**, gives structural support to the developing embryo. The notochord eventually will be replaced by the vertebral column, but after gastrulation it induces the overlying ectoderm to begin forming the nervous system. Neurulation involves the formation of an internal tube from an external sheet of cells.

The first signs of neurulation are flattening and thickening of the ectoderm overlying the notochord; this thickened area forms the **neural plate** (Figure 43.18). The edges of the neural plate that run in an anterior–posterior direction continue to thicken to form ridges or folds. Between the folds a groove forms and deepens as the folds roll over it to converge on the midline. The folds fuse, forming a cylinder, the **neural tube**, and a continuous overlying layer of epidermal

43.18 Neurulation in the Frog
Continuing the sequence from Figures 43.9 and 43.13, these drawings outline the development of the frog's neural tube. The midsagittal section (*b*) shows the development of the notochord and its position relative to the neural tube.

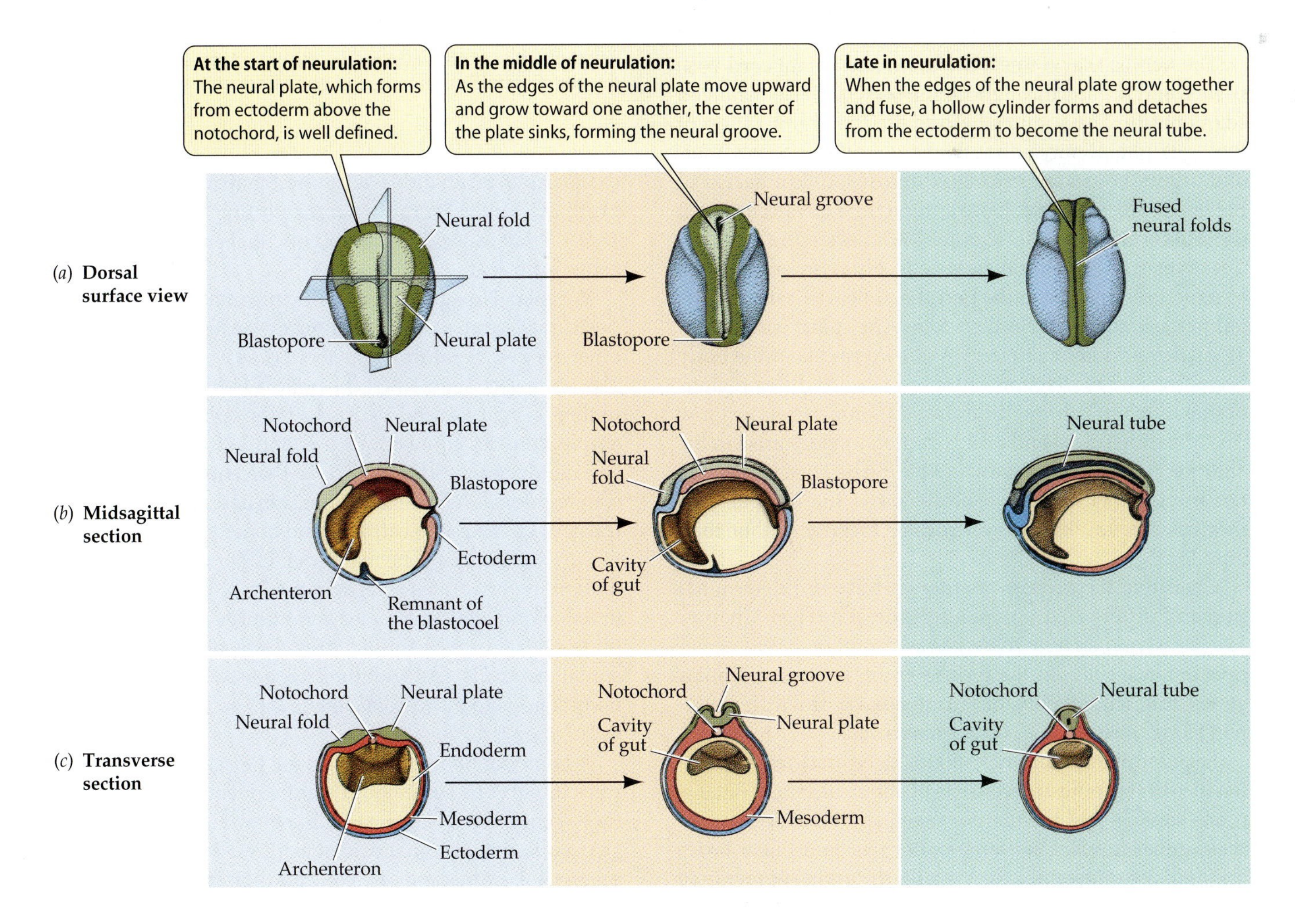

ectoderm. The neural tube develops bulges at the anterior end, which become the major divisions of the brain; the rest of the tube becomes the spinal cord.

Failure of the neural tube to develop normally can result in serious birth defects. If the neural tube fails to close in a posterior region, the result is a condition known as *spina bifida*. If it fails to close at the anterior end, an infant can develop without a forebrain—a condition called *anencephaly*. Whereas several genetic factors have been identified that can cause neural tube defects, there are also environmental factors, including dietary ones. The incidence of neural tube defects used to be about 1 in 300 live births, but we now know that this incidence can be cut in half if pregnant women have an adequate amount of folic acid (a B vitamin) in their diets.

Body segmentation develops during neurulation

Like the fruit flies whose development we traced in Chapter 16, vertebrates have a body plan consisting of repeating segments that are modified during development. These segments are most evident as the repeating patterns of vertebrae, ribs, nerves, and muscles along the anterior–posterior axis.

As the neural tube forms, mesodermal tissues gather along the sides of the notochord to form separate blocks of cells called **somites** (Figure 43.19). The somites produce cells that will become vertebrae, ribs, and muscles of the trunk and limbs.

The nerves that connect the brain and spinal cord with tissues and organs throughout the body are also arranged segmentally. The somites help guide the organization of these peripheral nerves, but the nerves are not of mesodermal origin. When the neural tube closes, cells adjacent to the line of closure break loose and migrate inward between the epidermis and the somites and under the somites. These cells, called **neural crest cells**, give rise to a number of structures, including the peripheral nerves, which grow out to the body tissues and back into the spinal cord.

As development progresses, the segments of the body become different. Regions of the spinal cord differ, regions of the vertebral column differ in that some vertebrae grow ribs of various sizes and others do not, forelegs arise in the anterior part of the embryo, and hind legs arise in the posterior region. How does a somite in the anterior part of a mouse embryo "know" to produce forelegs rather than hind legs?

Central to the process of anterior–posterior determination and differentiation are homeotic genes (see Chapter 16). We have seen how these genes control body segmentation in *Drosophila*. In the mouse, four families of similar genes, called homeobox or Hox genes, control differentiation along the anterior–posterior body axis.

Each family of mammalian Hox genes resides on a different chromosome and consists of about 10 genes. What is remarkable is that the temporal and spatial expression of these genes follows the same pattern as their linear order on their chromosomes. As a result, different segments of

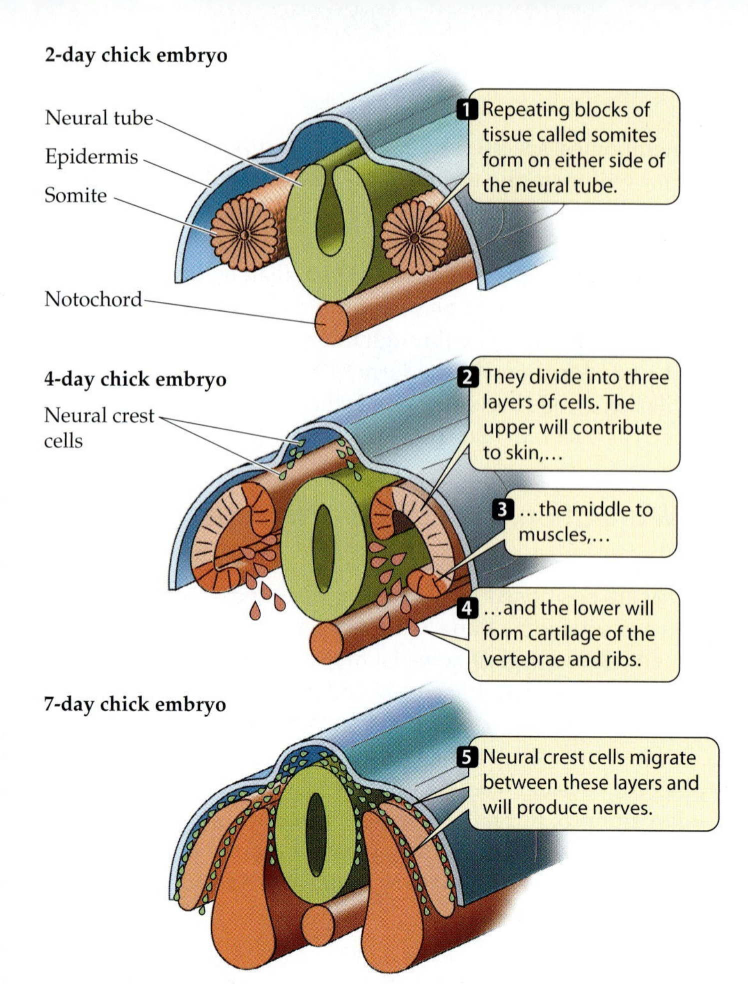

43.19 The Development of Body Segmentation
Repeating blocks of tissue called somites form on either side of the neural tube. Skin, muscle, and bone form from the somites.

the embryo receive different combinations of Hox gene products, which serve as transcription factors (Figure 43.20). What causes the linear, sequential expression of Hox genes is unclear.

Whereas Hox genes give cells information about their positional location on the anterior–posterior body axis, other genes give information about dorsal–ventral position. Tissues in each segment of the body differentiate according to their dorsal–ventral location. In the spinal cord, for example, sensory connections develop in the dorsal region and motor connections develop in the ventral region. In the somites, dorsal cells develop into skin and muscle and ventral cells develop into cartilage and bone.

An example of a gene that provides dorsal–ventral information in vertebrates is *sonic hedgehog*, which is expressed in the mammalian notochord and induces cells in the overlying neural tube to follow fates characteristic of ventral spinal cord cells. (As with the Hox genes, *sonic hedgehog* is homologous to a *Drosophila* gene, which is known simply as *hedgehog*.)

A family of homeobox genes, the Pax genes, play many roles in nervous system and somite development. One of these genes, *Pax3*, is expressed in neural tube cells that develop into dorsal spinal cord structures. *Sonic hedgehog* represses the expression of *Pax3*, and their interaction is one

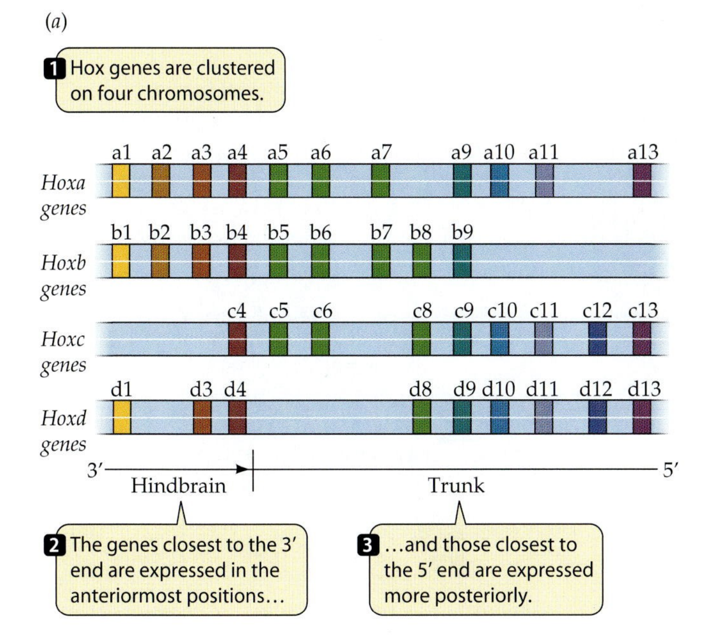

43.20 Hox Genes Control Body Segmentation
The expression of Hox genes is patterned along the anterior–posterior axis of the embryo and from the 3′ to the 5′ ends of the chromosomes.

source of dorsal–ventral information for differentiation of the spinal cord.

With the development of body segmentation, the formation of organs and organ systems progresses rapidly. The development of an organ involves extensive inductive tissue interactions, which are a current focus of study for developmental biologists. In Chapter 16, you encountered the organogenesis of the vertebrate limb. In the next chapter you will learn about the development of the brain—another example of organogenesis.

Extraembryonic Membranes

The embryos of reptiles, birds, and mammals are surrounded by several **extraembryonic membranes**, which originate from the embryo but are not part of it. The extraembryonic membranes play important roles in development, especially in mammals, in which they constitute the placenta that nourishes the embryo.

Four extraembryonic membranes form with contributions from all germ layers

We will use the chick to demonstrate how the extraembryonic membranes form from the germ layers created during gastrulation. The **yolk sac** is the first extraembryonic membrane to form, and it does so through extensions of the endodermal tissue of the hypoblast layer, which enclose the entire body of yolk in the egg (Figure 43.21, left). This yolk sac constricts at the top to create a tube that is continuous

43.21 The Extraembryonic Membranes
In birds, reptiles, and mammals, the embryo constructs four extraembryonic membranes. The yolk sac encloses the yolk, and the amnion and chorion enclose the embryo. Fluids secreted by the amnion fill the amniotic cavity, providing an aqueous environment for the embryo. The chorion, along with the allantois, mediates gas exchange between the embryo and its environment. The allantois stores the embryo's waste products.

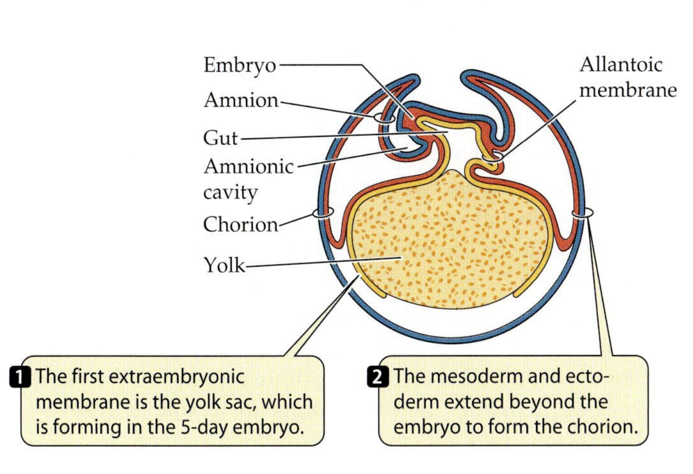

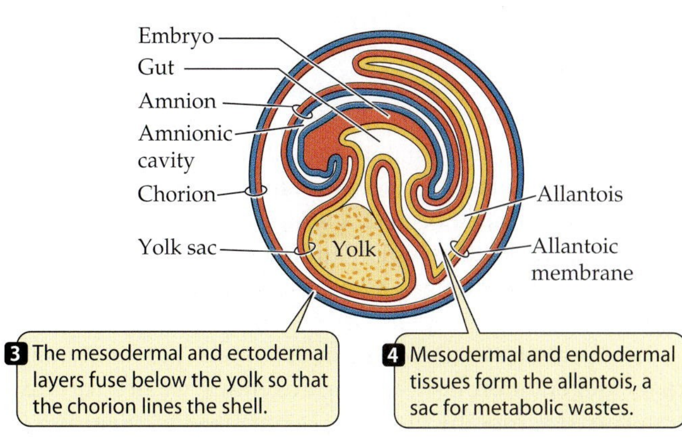

with the gut of the embryo. However, yolk does not pass through this tube. Yolk is digested by the endodermal cells of the yolk sac, and the nutrients are then transported to the embryo through blood vessels lining the outer surface of the yolk sac.

Just as the endoderm grows out from the embryo to form the yolk sac, the ectoderm and mesoderm also extend beyond the limits of the embryo. These two layers of cells extend all along the inside of the eggshell, both over the embryo and below the yolk sac. Where they meet, they fuse, forming two membranes, the **amnion** and the **chorion**. The amnion, which is the inner membrane, surrounds the embryo, forming the **amniotic cavity**. The amnion secretes fluid into the cavity, providing an aqueous environment for the embryo. The outer membrane, the chorion, forms a continuous membrane just under the eggshell (Figure 43.21, right). It limits water loss from the egg and also mediates the exchange of respiratory gases between the embryo and the outside world.

The fourth membrane to form, the allantoic membrane, is another outgrowth of the embryonic endoderm. It forms the **allantois**, a sac for storage of metabolic wastes. The allantois, in combination with the chorion, is also involved in respiratory gas exchange.

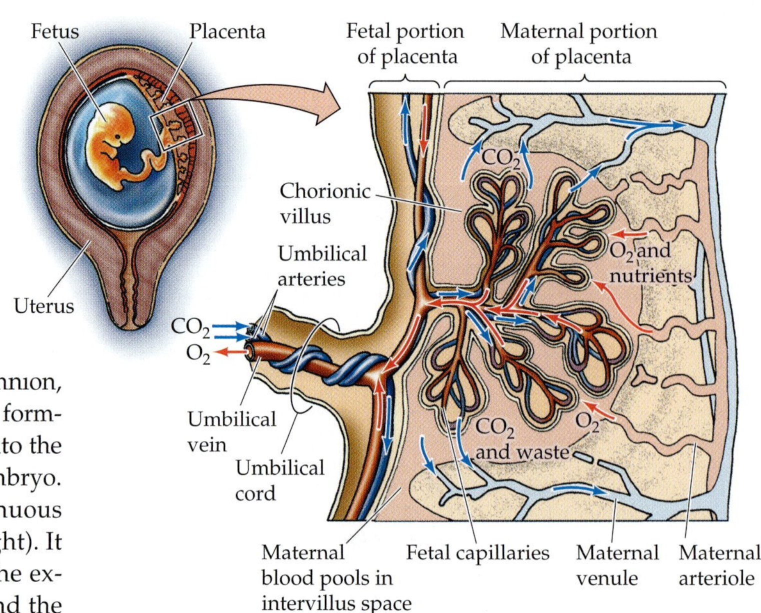

43.22 The Mammalian Placenta
In most mammals, nutrients and wastes are exchanged between maternal and fetal blood in the placenta, which forms from the chorion and tissues of the uterine wall. The embryo is attached to the placenta by the umbilical cord, and embryonic blood vessels invade the placental tissue to form fingerlike villi. Maternal blood flows into the spaces surrounding the villi.

Extraembryonic membranes in mammals form the placenta

In mammals, the first extraembryonic membrane to form is the trophoblast, whichis already apparent by the fifth cell division (see Figure 43.8). When the blastocyst reaches the uterus and "hatches" out of its encapsulating zona pellucida, the trophoblast cells interact directly with the endometrium. Adhesion molecules expressed on the surfaces of these cells attach them to the uterine wall, and by excreting proteolytic enzymes, the trophoblast burrows into the endometrium, beginning the process of implantation (see Figure 43.17). Eventually. the entire trophoblast is within the wall of the uterus. The trophoblastic cells then send out numerous projections, or villi, to increase the surface area of contact with maternal blood.

Meanwhile, the hypoblast cells extend to form what in the bird would be the yolk sac. But there is no yolk in placental mammalian eggs, so the yolk sac contributes mesodermal tissues that interact with trophoblast tissues to form the chorion. The chorion, along with tissues of the uterine wall, produces the **placenta**, the organ of nutrition for the embryo (Figure 43.22).

At the same time the yolk sac is forming from the hypoblast, the epiblast produces the amnion, which grows to enclose the entire embryo in a fluid-filled amniotic cavity. The rupturing of the amnion and the loss of the amniotic fluid ("water breaking") herald the onset of labor in humans.

An allantois also develops in mammals, but its importance depends on how well nitrogenous wastes can be transferred across the placenta. In humans the allantois is minor; in pigs it is important. In humans and other placental mammals, allantoic tissues contribute to the formation of the *umbilical cord* by which the embryo is attached to the chorionic placenta. It is through the blood vessels of the umbilical cord that nutrients and oxygen from the mother reach the developing fetus and wastes, including carbon dioxide and urea, are removed (see Figure 43.22).

The extraembryonic membranes provide means of detecting genetic diseases

Cells slough off of the embryo and float in the amniotic fluid that bathes it. Later in development, a small sample of the amniotic fluid may be withdrawn with a needle as the first step of a process called **amniocentesis**. Some of these cells can be cultured and used for biochemical and genetic analyses that can reveal the sex of the fetus, as well as genetic markers for diseases such as cystic fibrosis, Tay-Sachs disease, and Down syndrome.

If amniocentesis is performed, it is usually not until after the fourteenth week of pregnancy, and the tests require two weeks to complete. If abnormalities in the fetus are detected, termination of the pregnancy at that stage would put the mother's health at greater risk than would an earlier abortion. Therefore a newer technique, called **chorionic villus sampling**, is now in common use. In this test a small sample of the tissue from the surface of the chorion is taken (Figure 43.23). This test can be done as early as the eighth week of pregnancy, and the results are available in several days.

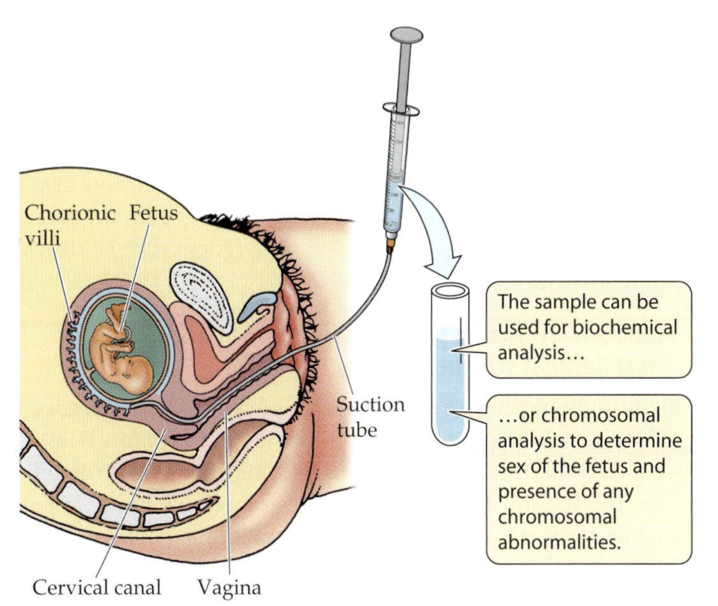

43.23 Chorionic Villus Sampling
Information about genetic defects can be obtained from chorionic tissues.

Human Pregnancy and Birth

Human pregnancy can be divided into three trimesters

Gestation, or pregnancy, in humans has a duration of about 266 days, or 9 months. In smaller mammals gestation is shorter—for example, 21 days in mice—and in larger mammals it is longer—for example, 330 days in horses and 600 days in elephants. To follow the temporal sequence of events in human pregnancy, we can divide it into three trimesters of about 3 months each.

THE FIRST TRIMESTER. Implantation of the human blastocyst begins on about the sixth day after fertilization. After implantation, the placenta develops and the tissues and organs of the embryo begin to differentiate. The human heart begins to beat in week 4, and limbs form by week 8 (Figure 43.24*a*). Most organs have started to form by the end of the first trimester. By the end of the first trimester, the embryo looks like a miniature version of the adult, and is called a **fetus**.

Because the first trimester is a time of rapid cell division and differentiation, it is the period during which the embryo is most sensitive to radiation, drugs, and chemicals that can cause birth defects. An embryo can be damaged before the mother even knows she is pregnant.

Hormonal changes cause major and noticeable responses in the mother during the first trimester, even though the fetus at the end of that time is still so small that it would fit into a teaspoon. Soon after the blastocyst implants itself, it begins to secrete human chorionic gonadotropin (hCG), the hormone that stimulates the corpus luteum to continue producing estrogen and progesterone. The high levels of these hormones have negative feedback effects on the production and secretion of hypothalamic

43.24 Stages of Human Development
(*a*) The first trimester of pregnancy is a period of rapid cell division and differentiation. The organs and body structures of this 6-week old embryo are forming rapidly. (*b*) At 4 months, the fetus moves freely within its protective amniotic cavity. The fingers and toes are fully formed.

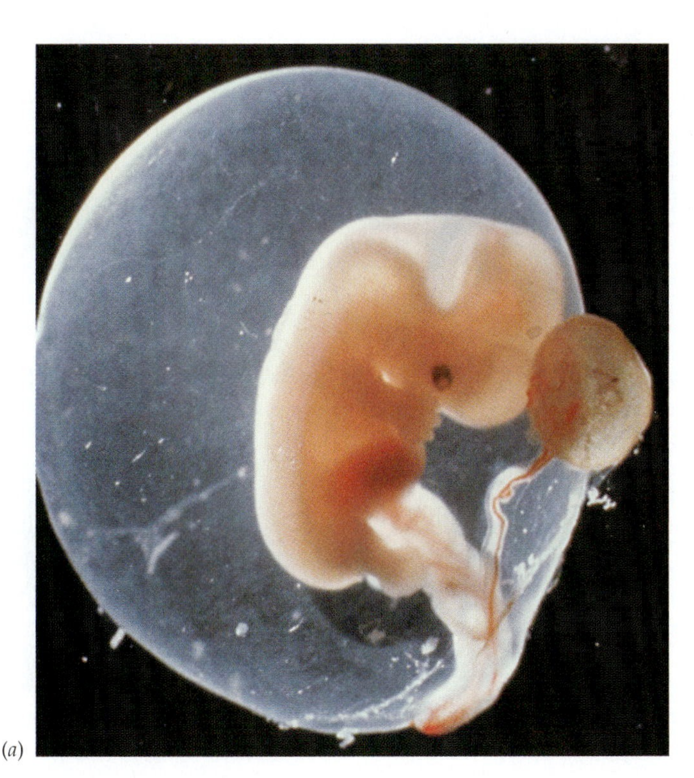

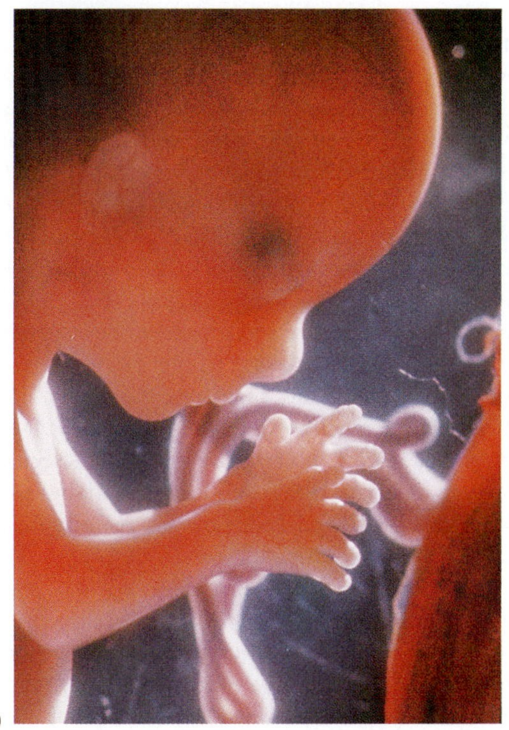

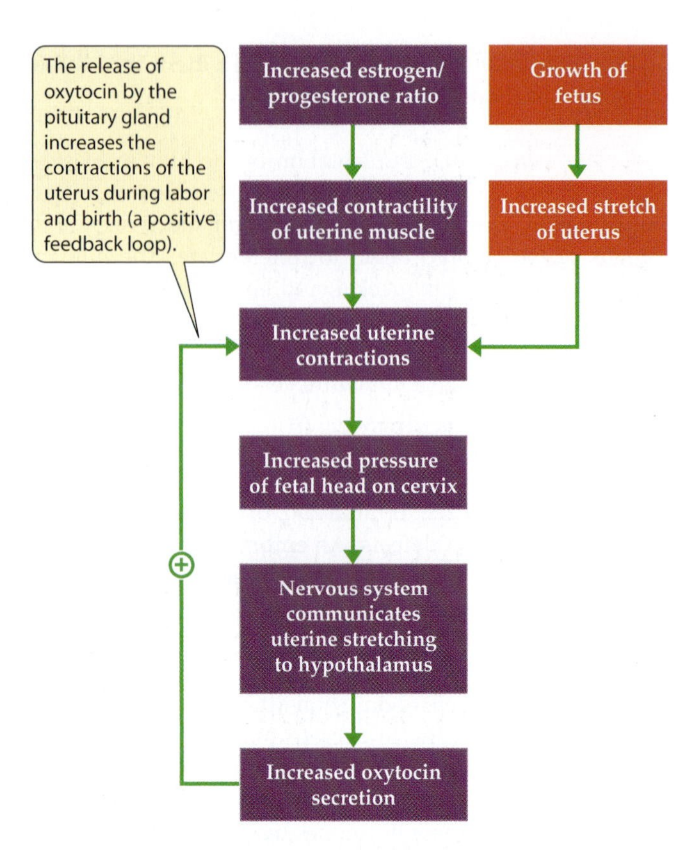

43.25 Control of Uterine Contractions and Parturition
Both mechanical and hormonal signals are involved in the onset of parturition.

and pituitary hormones (see Chapter 42). These dramatic hormonal shifts cause the well-known symptoms of pregnancy: morning sickness, mood swings, changes in the senses of taste and smell, and swelling of the breasts.

THE SECOND TRIMESTER. During the second trimester the fetus grows rapidly to a weight of about 600 g, and the mother's abdomen enlarges considerably. The limbs of the fetus elongate, and the fingers, toes, and facial features become well formed. Fetal movements are first felt by the mother early in the second trimester, and they become progressively stronger and more coordinated. By the end of the second trimester, the fetus may suck its thumb (Figure 43.24*b*).

The production of estrogen and progesterone by the placenta increases during the second trimester. As placental production of these hormones increases, the level of hCG decreases and the corpus luteum degenerates. Ovulation and menstruation are still inhibited by the steroids secreted by the placenta. Along with these hormonal changes, the unpleasant symptoms of early pregnancy usually disappear.

THE THIRD TRIMESTER. The fetus and the mother continue to grow rapidly during the third trimester. As the fetus approaches its full size, pressure on the mother's internal organs can cause indigestion, constipation, frequent urination, shortness of breath, and swelling of the legs and ankles. Since the fourth week of pregnancy the heart of the fetus has been beating, and as the third trimester approaches its end, other internal organs mature. The digestive system begins to function, the liver stores glycogen, the kidneys produce urine, and the brain undergoes cycles of sleep and waking.

Parturition is triggered by hormonal and mechanical stimuli

Throughout pregnancy, the uterus periodically undergoes slow, weak, rhythmic contractions called Braxton Hicks contractions. These contractions become gradually stronger during the third trimester and are sometimes called false labor contractions. True labor contractions usually mark the beginning of childbirth, or **parturition**.

Many factors contribute to the onset of labor. Both hormonal and mechanical stimuli increase the contractility of the uterus. Progesterone inhibits and estrogen stimulates contractions of uterine muscle. Toward the end of the third trimester, the estrogen–progesterone ratio shifts in favor of estrogen. Oxytocin stimulates uterine contraction; its secretion by the pituitaries of both mother and fetus increases at the time of labor.

Mechanical stimuli come from the stretching of the uterus by the fully grown fetus and the pressure of the fetal head on the cervix. These mechanical stimuli increase the pituitary release of oxytocin, which in turn increases the activity of the uterine muscle, which causes even more pressure on the cervix. This positive feedback loop converts the weak, slow, rhythmic contractions of the uterus into stronger labor contractions (Figure 43.25).

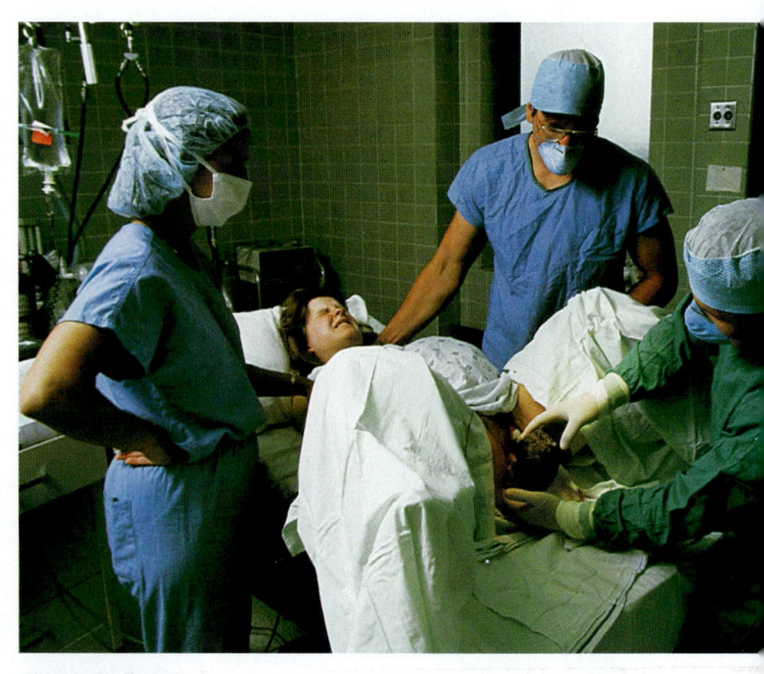

43.26 Delivery
A new person enters the world head first.

In the early stage of labor, the contractions of the uterus are 15 to 20 minutes apart, and each lasts 45 to 60 seconds. During this time, the contractions pull the cervix open until it is large enough to allow the baby to pass through. This stage of labor lasts an average of 12 to 15 hours in a first pregnancy and 8 hours or less in subsequent ones. Gradually the contractions become more frequent and more intense.

The second stage of labor, called *delivery*, begins when the cervix is fully dilated. The baby's head moves into the vagina and becomes visible from the outside (Figure 43.26). The usual head-down position of the baby at the time of delivery comes about when the fetus shifts its orientation during the seventh month. If the fetus fails to move into the head-down position, a different part of the fetus enters the vagina first, and the birth is more difficult.

Passage of the fetus through the vagina is assisted by the woman's bearing down with her abdominal and other muscles to help push the baby along. Once the head and shoulders of the baby clear the cervix, the rest of its body eases out rapidly, but it is still connected to the placenta in the mother by the umbilical cord. Delivery may take as little as a minute, or up to half an hour or more in a first pregnancy.

As soon as the baby clears the birth canal, it can start breathing and become independent of its mother's circulation. The umbilical cord may then be clamped and cut. The segment still attached to the baby dries up and sloughs off in a few days, leaving behind its distinctive signature, the belly button—more properly called the *umbilicus*. The detachment and expulsion of the placenta and fetal membranes takes from a few minutes to an hour, and may be accompanied by uterine contractions.

In humans, the completion of delivery is the start of many years of nurturing and care for the young organism. Many processes of development continue throughout childhood, and indeed, throughout life.

Chapter Summary

Fertilization: Interactions of Sperm and Egg

▶ Fertilization involves interactions between sperm and egg, including sperm activation, species-specific binding of sperm to the outer covering of the egg, the acrosomal reaction, digestion of a path through the outer covering of the egg, and fusion of sperm and egg plasma membranes. **Review Figure 43.1**

▶ The entry of the sperm into the egg triggers fast and slow blocks to polyspermy, which prevent additional sperm from entering the egg and, in mammals, signals the egg to complete meiosis and begin development. **Review Figures 43.2, 43.3, 43.4**

▶ Sperm and egg contribute differentially to the zygote. The sperm contributes a haploid nucleus and, in some species, a centriole. The egg contributes a haploid nucleus, nutrients, ribosomes, mitochondria, and informational molecules that will control the early stages of development.

▶ The cytoplasmic contents of the egg are not distributed homogeneously, and they are rearranged after fertilization to set up the major axes of the future embryo. **Review Figures 43.5, 43.6**

Cleavage: Repackaging the Cytoplasm

▶ In most animals, cleavage is a period of rapid cell division without cell growth or gene expression. During cleavage, the cytoplasm of the zygote is repackaged into smaller and smaller cells.

▶ The pattern of cleavage is influenced by the amount of yolk that impedes cleavage furrow formation and by the orientation of the mitotic spindles. The result of cleavage is a ball or mass of cells called a blastula. **Review Figure 43.7**

▶ Cleavage in mammals is unique in that cell divisions are much slower and genes are expressed early in the process. Cleavage results in an inner cell mass that becomes the embryo and an outer cell mass that becomes the trophoblast. The mammalian embryo at this stage is called a blastocyst. **Review Figure 43.8**

▶ Fate maps, which identify what tissues and organs will form from particularblastomeres, can be created for the blastula. **Review Figure 43.9**

▶ Some species undergo mosaic development, in which the fate of each cell is determined by the 8-cell stage. Other species, including vertebrates, undergo regulative development, in which cells are not determined so early and can change developmental fates. In these species, blastomeres separated at early stages can develop into complete embryos, which are then monozygotic, or identicial, twins. **Review Figure 43.10**

Gastrulation: Producing the Body Plan

▶ Gastrulation involves massive cell movements that produce three primary germ layers and place cells from various regions of the blastula into new associations with one another. **Review Table 43.1**

▶ The initial step of sea urchin and amphibian gastrulation is inward movement of certain blastomeres. The site of inward movement becomes the blastopore. Cells that move into the blastula become the endoderm and mesoderm; cells remaining on the outside become the ectoderm. Cytoplasmic factors in the vegetal pole cells are essential to initiate development. **Review Figure 43.11, 43.12**

▶ Gastrulation in frogs is initiated when cells in the gray crescent move into the blastocoel. This inward migration creates the blastopore. The dorsal lip of the blastopore is a critical site for the determination of tissues. It has been called the primary embryonic organizer. **Review Figures 43.13, 43.14, 43.15**

▶ The anterior–posterior axis of the frog blastula appears to be determined by the distribution of the protein β-catenin, which activates a signaling cascade that induces the primary embryonic organizer.

▶ Gastrulation in reptiles and birds is different from that in sea urchins and frogs because the large amount of yolk in their eggs causes the blastula to form a flattened disc of cells. **Review Figure 43.16**

▶ Mammals have a pattern of gastrulation similar to that of birds, even though they have no yolk. **Review Figure 43.17**

Neurulation: Initiating the Nervous System

▶ Neurulation follows gastrulation. Cells that migrate over the dorsal lip of the blastopore are determined to be chordomesoderm, which forms the notochord. The notochord induces the overlying ectoderm to thicken, form parallel ridges, and fold in on itself to form a neural tube below the epidermal ectoderm. The nervous system develops from this neural tube. **Review Figure 43.18**

▶ The notochord and neural crest cells participate in the segmental organization of tissues called somites along the body axis. Rudimentary organs and organ systems form during this stage. **Review Figure 43.19**

▶ Four families of Hox genes determine the pattern of anterior–posterior differentiation along the body axis in mammals. Other genes, such as *sonic hedgehog*, contribute to dorsal–ventral differentiation. **Review Figure 43.20**

Extraembryonic Membranes

▶ The embryos of reptiles, birds, and mammals are protected and nurtured by four extraembryonic membranes. In birds and reptiles the yolk sac surrounds the yolk and provides nutrients to the embryo, the chorion lines the eggshell and participates in gas exchange, the amnion surrounds the embryo and encloses it in an aqueous environment, and the allantois stores metabolic wastes. **Review Figure 43.21**

▶ In mammals, the chorion and the trophoblast cells interact with the maternal uterus to form a placenta, which provides for nutrient and gas exchange. The amnion encloses the embryo in an aqueous environment. **Review Figure 43.22**

▶ Samples of amniotic fluid or pieces of chorion can be taken during pregnancy and analyzed for evidence of genetic disease. **Review Figure 43.23**

Human Pregnancy and Birth

▶ Pregnancy in humans can be divided into three trimesters. Early embryogenesis occurs in the first trimester; during this time, the embryo is vulnerable to damage that could lead to birth defects. Hormonal changes, including high hCG, estrogen, and progesterone levels, block further ovulation and menstruation and also cause symptoms of pregnancy.

▶ During the second and third trimesters the embryo grows, the limbs elongate, and organ systems mature.

▶ The onset of labor is due to many factors, both hormonal and mechanical, that increase contractility of uterine muscles. Oxytocin plays a major role in a positive feedback loop. **Review Figure 43.25**

▶ Birth is not the end of development, which continues throughout childhood and throughout life.

For Discussion

1. Knowing what you do about sperm–egg interactions at the time of fertilization, propose a line of research that could produce a new contraceptive method. What is the rationale for your method, and what problems will you have to overcome to make it successful?
2. If you found a protein that was localized to a small group of cells in the frog blastula, how would you determine if that protein played a role in development? Address the issues of sufficiency and necessity.
3. Find out what a chimeric mouse is and explain how it is produced.
4. During gastrulation in the chick, a gene called *sonic hedgehog* is expressed only on the left side of Hensen's node. What could be the significance of this expression pattern?
5. Much of the early work of describing animal development was done on sea urchins, amphibians, and chicks. Most recent work on the molecular mechanisms of animal development have been done on roundworms, fruit flies, zebrafish, and mice. Why do you think there has been a shift in the animal models used by developmental biologists?

44 Neurons and Nervous Systems

"ON YOUR MARK." "GET SET." BANG! 10.75 seconds later Marion Jones crossed the finish line 100 meters away, winning an Olympic gold medal.

An amazing number of events took place in her nervous system during those 10.75 seconds. Cells in her ear converted the gunshot sound waves into bursts of electrical activity called nerve impulses, which traveled via nerve cells to a part of her brain that processes sound information. Cells from that part of the brain carried nerve impulses to another part of the brain that commands and coordinates the movements of the body. Command cells controlling the muscles in her legs were triggered to fire bursts of nerve impulses. These cells carried the bursts of nerve impulses down the spinal cord to about the middle of her back region. There the connections were made to other command cells, sending nerve impulses to the muscles of Marion's legs and feet. Those muscles contracted and propelled her off the blocks. Each time a cell in this chain passed information to another cell, it released small packets of chemicals that stimulated the next cell to generate bursts of nerve impulses.

All of these events took about a tenth of a second. For the next 10.65 seconds, hundreds of thousands of nerve cells fired repeated bursts of nerve impulses in just the right patterns to cause the many muscles of Marion's legs, feet, arms, and torso to repeatedly contract and relax, powering the 80 or so strides it took her to reach the finish line. Each of the millions of cell–cell interactions that occurred involved a small release of chemicals from one cell and a receptor-mediated response from another cell. What amazing complexity and precision—yet running 100 meters is a simple task for the nervous system (although very few people will ever do it as fast as Marion Jones did).

The nervous system is the most complex system of the human body. As in other organs and organ systems, its basic building blocks are cells. In the nervous system, cells are arranged in networks and circuits that process information. Within those circuits resides not only the capacity to generate immediate responses to stimuli on the many sensory systems, but also the capacity to remember those stimuli and responses, to relate stimuli and responses to other experiences, and to learn from the information that is processed. Within the cells and connections of the nervous system reside our individual personalities, our egos, and our ability to love and to hate.

Understanding the human nervous system is probably the biggest challenge in all of biology. We start with the basic building blocks: nerve cells. This chapter describes the cells of nervous systems. It then details how nerve cells generate electric signals and conduct those signals from place to place. Finally, it explains how those signals are communicated from cell to cell and how receiving cells respond.

Nervous Systems: Cells and Functions

In multicellular animals, nerve cells, called **neurons**, are specialized to receive information, encode it, and transmit it to other cells. Together with their specialized supportive cells, neurons make up **nervous systems**, whose functions can be described in terms of networks of interacting neurons.

Animals receive various kinds of information from both inside and outside their bodies. This information is received and converted, or *transduced*, by *sensory cells* (also called *receptor cells*) into electric signals that can be transmitted and processed by neurons. To cause behavioral or phys-

On the Fast Track
Marion Jones can run fast because her nervous system processes sensory information and commands her muscles so quickly.

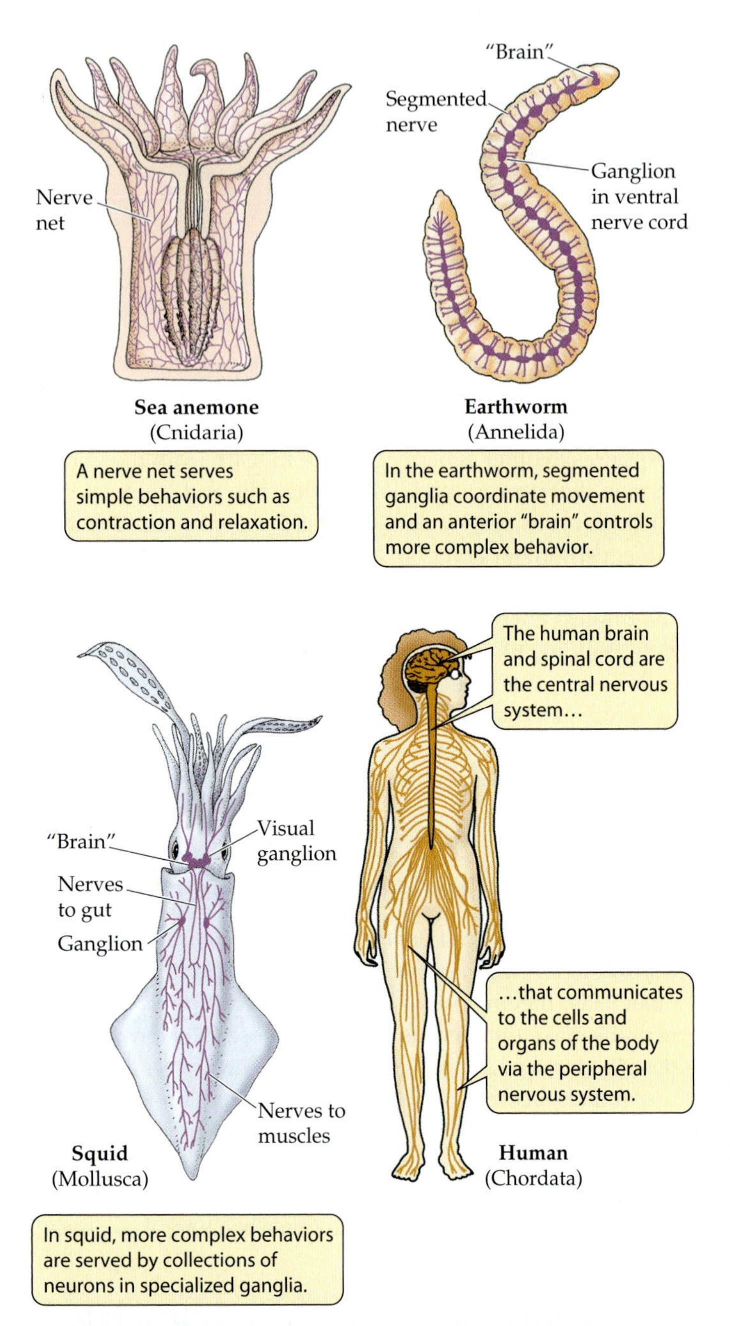

44.1 Nervous Systems Vary in Size and Complexity
As we compare animals that have increasingly complex sensory and behavioral abilities, we find information processing increasingly centralized in ganglia (collections of nerve cells) or in a brain.

iological responses, a nervous system communicates these signals to *effectors*, such as muscles and glands.

Nervous systems process information

A simple animal that remains fixed to its substrate, such as a sea anemone (a cnidarian), can process information with a simple network of neurons that does little more than provide direct lines of communication from sensory cells to effectors (Figure 44.1). The cnidarian's *nerve net* merely detects food or danger and causes its tentacles and body to extend or retract

More complex animals that move around the environment and hunt for food and mates need to process and integrate larger amounts of information. Even animals such as flatworms fit this description, and their increased need for information processing is met by clusters of neurons called **ganglia**. Ganglia serving different functions may be distributed around the body, as in the earthworm or the squid. Frequently one pair of ganglia is larger and more central than the others, and is therefore given the designation of **brain**.

In vertebrates, most of the cells of the nervous system are found in the brain and the **spinal cord**, the site of most information processing, storage, and retrieval (see Figure 44.1). Therefore, the brain and spinal cord are called the **central nervous system** (**CNS**). Information is transmitted from sensory cells to the CNS and from the CNS to effectors via neurons that extend or reside outside of the brain and the spinal cord; these neurons and their supporting cells are called the **peripheral nervous system**.

Vertebrates have highly developed central nervous systems, but they vary greatly in behavioral complexity and physiological capabilities. Figure 44.2 shows the brains of

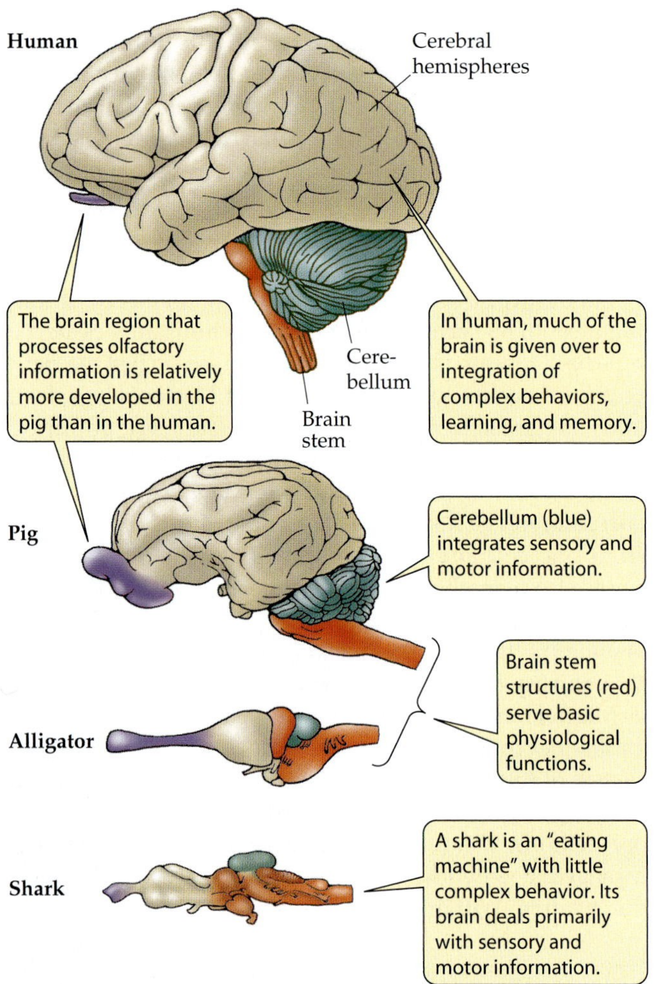

44.2 Brains Vary in Size and Complexity
The brains of four vertebrate species—all of which may have a similar body mass—show immense differences.

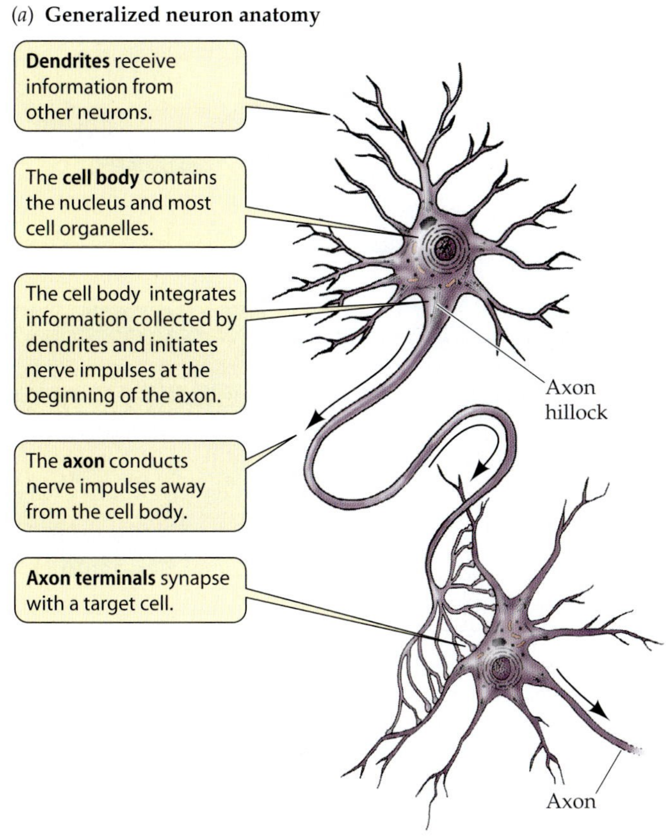

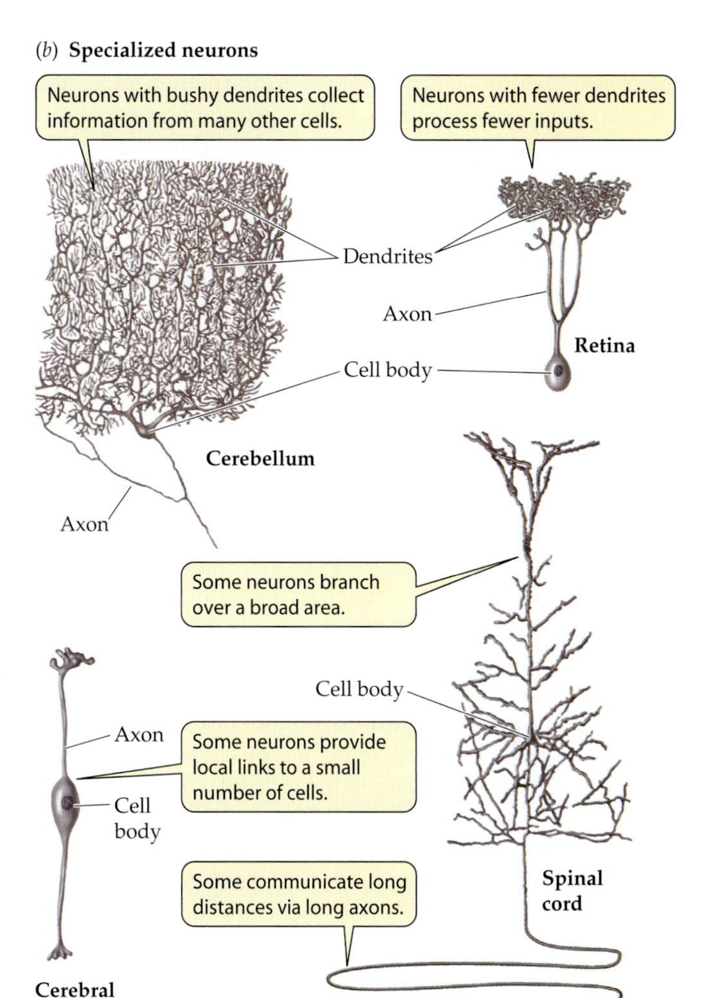

44.3 Neurons
(*a*) A generalized diagram of a neuron. (*b*) Neurons from different parts of the mammalian nervous system are specifically adapted to their functions.

four similar-sized vertebrates drawn to the same scale. But even small nervous systems are remarkably complex. Consider the nervous systems of small spiders that have programmed within them the thousands of precise movements necessary to construct a beautiful web without prior experience.

Neurons are the functional units of nervous systems

Although nervous systems vary enormously in structure and function, neurons function similarly in animals as different as squids and humans. Their plasma membranes generate electric signals—called *nerve impulses* or *action potentials*—and conduct these signals from one location on a cell to the most distant reaches of that cell—a distance that can be more than a meter.

Most neurons have four regions—a cell body, dendrites, an axon, and axon terminals (Figure 44.3*a*)—but the variation among different types of neurons is considerable (Figure 44.3*b*). The **cell body** contains the nucleus and most of the cell's organelles. Many projections may sprout from the cell body. Most of these projections are bushlike **dendrites** (from the Greek *dendron*, "tree"), which bring information from other neurons or sensory cells to the cell body. In most neurons, one projection is much longer than the others, and is called the **axon**. Axons usually carry information away from the cell body. The length of the axon differs in different types of neurons—some axons are remarkably long, such as those that run from the spinal cord to the toes. The degree of branching of the dendrites can also be quite different among types of neurons.

Axons are the "telephone lines" of the nervous system. Information received by the dendrites can cause the cell body to generate a nerve impulse, which is then conducted along the axon to the cell that is its target. At the target cell—which can be another neuron, a muscle cell, or a gland cell—the axon divides into a spray of fine nerve endings. At the tip of each of these tiny nerve endings is a swelling called an **axon terminal** that comes very close to the target cell.

Where an axon terminal comes close to another cell, the membranes of both cells are modified to form a *synapse*. In most cases, a space or cleft only about 25 nm wide separates the two membranes at the synapse. A nerve impulse arriving at an axon terminal causes molecules called *neurotransmitters* stored in the axon terminal to be released. The released neurotransmitters diffuse across the synaptic cleft and bind receptors on the plasma membrane of the target cell.

Thousands of synapses impinge on most individual neurons. A neuron generally receives information (synaptic inputs) from many sources before producing nerve impulses that travel down its single axon to target cells. We will discuss synaptic transmission in more detail later in the chapter.

Glial cells are also important components of nervous systems

Neurons are not the only type of cell in the nervous system. In fact, there are more **glial cells** than neurons in the human brain. Like neurons, glial cells come in several forms and have a diversity of functions. Some glial cells physically support and orient the neurons and help them make the right contacts during their embryonic development. Other glial cells insulate axons.

In the peripheral nervous system, **Schwann cells** wrap around the axons of neurons, covering them with concentric layers of insulating plasma membrane (Figure 44.4). Other glial cells called **oligodendrocytes** perform a similar function in the central nervous system. The covering produced by Schwann cells and oligodendrocytes, called **myelin**, gives many parts of the nervous system a glistening white appearance. Later in the chapter we will see how the electrical insulation provided by myelin increases the speed with which axons can conduct nerve impulses.

Glial cells are well known for the many housekeeping functions they perform. Some glial cells supply neurons with nutrients; others consume foreign particles and cell debris. Glial cells also help maintain the proper ionic environment around neurons. Although they have no axons and do not generate or conduct nerve impulses, some glial cells communicate with one another electrically through a special type of contact called a gap junction, a connection that enables ions to flow between cells (see Chapter 5).

Glial cells called **astrocytes** (because they look like stars) contribute to the **blood–brain barrier**, which protects the brain from toxic chemicals in the blood. Blood vessels throughout the body are very permeable to many chemicals, including toxic ones, which would reach the brain if this special barrier did not exist. Astrocytes help form this barrier by surrounding the smallest, most permeable blood vessels in the brain.

Protection of the brain is crucial because, unlike other tissues of the body, the brain has a limited capacity to recover from damage by generating new neurons and new neuronal connections. Throughout life, neurons are progressively lost. Without the blood–brain barrier, the rate of neuron loss could be much greater. However, the barrier is not perfect. Since it consists of plasma membranes, it is permeable to fat-soluble substances. Anesthetics and alcohol, both of which have well-known effects on the brain, are fat-soluble chemicals.

Neurons function in networks

As we learn more about the properties of neurons, it is important to keep in mind that nervous systems depend on neurons working together. The simplest *neuronal network* consists of three cells: a sensory neuron connected to a motor neuron connected to a muscle cell. Most of the neuronal networks that carry out the functions of the human nervous system are much more complex and consist of many more neurons. The human brain contains between 10^9 and 10^{11} neurons. Most of these neurons receive information from a thousand or more synapses. Thus there may be as many as 10^{14} synapses in the human brain. Therein lies the incredible ability of the brain to process information.

This astronomical number of neurons and synapses is divided into thousands of distinct but interacting networks that function in parallel and accomplish the many different tasks of the nervous system. But before we can understand how even one of these circuits works, we must understand the properties of individual neurons.

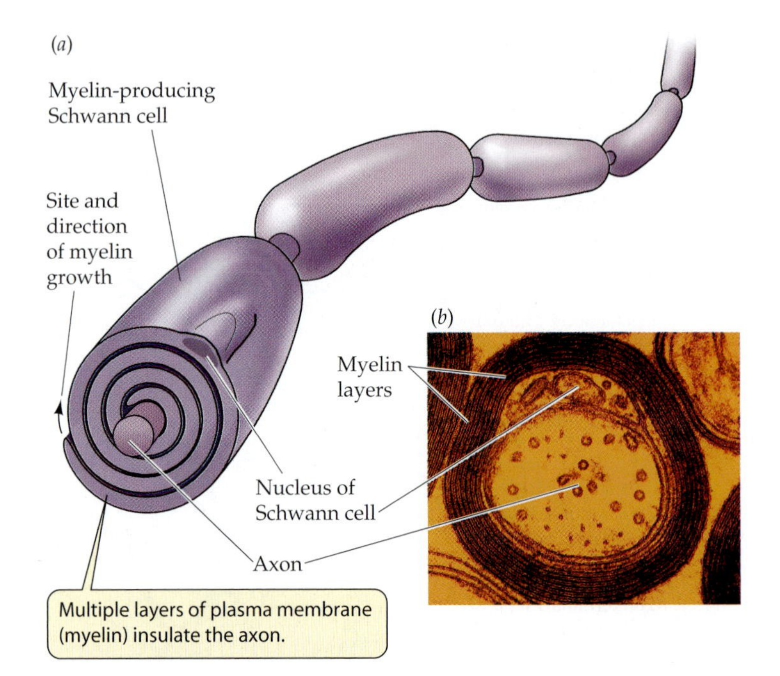

44.4 Wrapping Up an Axon
(*a*) Schwann cells wrap axons with layers of myelin, a type of plasma membrane that provides electrical insulation. (*b*) A group of myelinated axons, seen in cross section through an electron microscope.

Neurons: Generating and Conducting Nerve Impulses

In this section, we explore the electrical properties of cells. After reviewing some basic electrical concepts, we examine in detail the roles of ions, ion pumps, and ion channels in establishing and altering the electrical properties of neurons. These electrical changes generate action potentials, the language by which the nervous system processes and communicates information.

The insides of cells are electrically negative in comparison to the outsides. The difference in voltage across the plasma membrane of a neuron is called its **membrane potential**. In an *unstimulated* neuron, this voltage difference is called a **resting potential**.

Membrane potentials can be measured with *electrodes*. An electrode can be made from a glass pipette pulled to a very sharp tip and filled with a solution containing ions that conduct electric charge. Using such electrodes, we can record very tiny local electrical events that occur across plasma membranes. If a pair of electrodes is placed one on each side of the plasma membrane of a resting axon, they measure a voltage difference of about –60 millivolts (mV)—the resting potential (Figure 44.5).

The resting potential provides a means for neurons to respond to a stimulus. A neuron is sensitive to any chemical or physical factor that causes a change in the resting potential across a portion of its plasma membrane. The most extreme change in membrane potential is a nerve impulse, which is a sudden and rapid reversal in the voltage across a portion of the plasma membrane. For a brief moment—1 or 2 milliseconds—the inside of a part of the cell becomes *more positive* than the outside. Nerve impulses are also called **action potentials**, a name that conveys their contrast with the resting potential. An action potential can travel along the plasma membrane from one part of a neuron to its farthest extensions.

Simple electrical concepts underlie neuronal functions

To understand how resting potentials are created, how they are perturbed, and how action potentials are generated and conducted along plasma membranes, it is necessary to know a little about electricity, ions, and the special ion channel proteins in the plasma membranes of neurons.

Voltage (potential) is the tendency for electrically charged particles like electrons or ions to move between two points. Voltage is to the flow of electrically charged particles what pressure is to the flow of water. If the negative and the positive poles of a battery are connected by a copper wire, electrons flow from negative to positive because there is a voltage difference between them. This flow of electrons is an electric current, and it can be used to do work, just as a current of water can be used to do work such as turning a turbine.

Electric charges move across cell membranes not as electrons, but as charged ions. The major ions that carry electric charges across the plasma membranes of neurons are sodium (Na^+), chloride (Cl^-), potassium (K^+), and calcium (Ca^{2+}). It is also important to remember that ions with opposite charges attract each other. With these basics of bioelectricity in mind, we can ask how the resting potential of the neuronal plasma membrane is created, and how the flow of ions through membrane channels is turned on and off to generate action potentials.

Ion pumps and channels generate resting and action potentials

The plasma membranes of neurons, like those of all other cells, are lipid bilayers that are impermeable to ions. However, these impermeable lipid bilayers contain many protein molecules that serve as ion channels and ion pumps (see Chapter 5). Ion pumps and channels are responsible for resting and action potentials.

44.5 Measuring the Resting Potential
The difference in electric charge across the plasma membrane of a neuron can be measured using two electrodes, one inside and one outside the cell. In an unstimulated neuron, this difference is constant (about –60 mV), and is known as the resting potential.

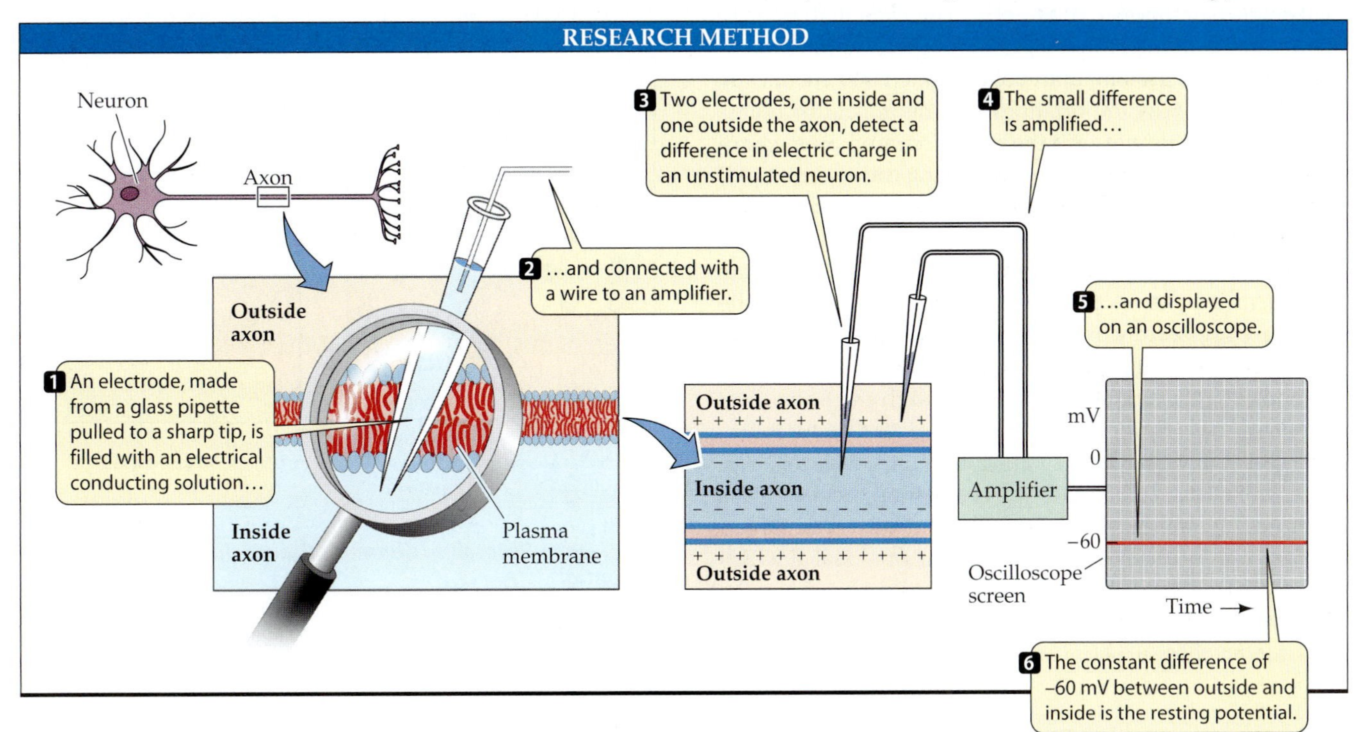

Ion pumps use energy to move ions or other molecules against their concentration gradients. The major ion pump in neuronal membranes is the **sodium–potassium pump**. The action of this pump expels Na^+ ions from the cell, exchanging them for K^+ ions from outside the cell (Figure 44.6*a*). The sodium–potassium pump keeps the concentration of K^+ inside the cell greater than that of the external medium, and the concentration of Na^+ inside the cell less than that of the external medium. The concentration differences established by the pump mean that K^+ would diffuse out of the cell and Na^+ would diffuse in if the ions could cross the lipid bilayer.

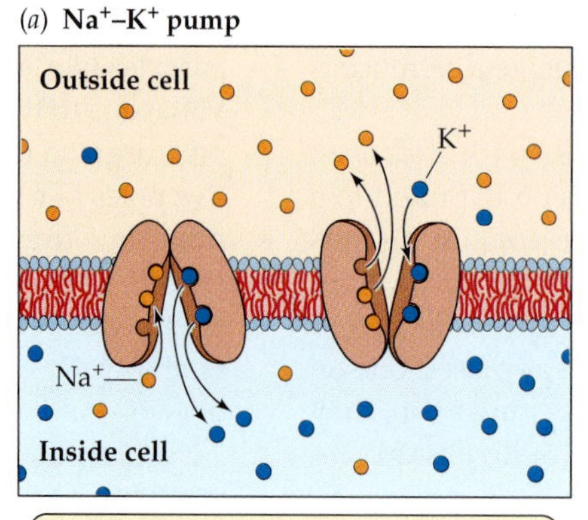

44.6 Ion Pumps and Channels
(*a*) The sodium–potassium pump actively moves K^+ ions to the inside of a neuron and Na^+ ions to the outside. (*b*) Channels allow ions to diffuse down their concentration gradient when they are open; thus K^+ ions tend to leave neurons and Na^+ ions tend to enter neurons through ion channels. The sodium channels shown here are gated, and one is closed.

Ion channels are pores formed by proteins in the lipid bilayer. These water-filled pores allow ions to pass through, but they cannot actively transport ions as ion pumps do (Figure 44.6*b*). They are generally *selective;* that is, some types of ions pass through them more easily than others. Thus there are potassium channels, sodium channels, chloride channels, and calcium channels, and there are different kinds of each. Ions move through channels by diffusion, and therefore they can move in either direction. The direction and magnitude of net movement of ions through a channel depends on the concentration gradient of that ion type across the plasma membrane.

Many ion channels in the plasma membranes of neurons behave as if they contain a "gate" that opens to allow ions to pass under some conditions, but closes under other conditions. **Voltage-gated channels** open or close in response to a change in the voltage across the plasma membrane. **Chemically gated channels** open or close depending on the presence or absence of a specific chemical that binds to the channel protein, or to a separate receptor that in turn alters the channel protein. Both voltage-gated and chemically gated channels play important roles in neuronal function.

Potassium channels are the most common open channels in the plasma membranes of resting (non-stimulated) neurons. As a consequence, resting neurons are more permeable to K^+ than to any other ion. As Figure 44.7 shows, this characteristic explains the resting potential. Because the plasma membrane is permeable to K^+, and because the sodium–potassium pump keeps the concentration of K^+ inside the cell much higher than that outside the cell, K^+ tends to diffuse out of the cell through the potassium channels.

As positively charged K^+ ions diffuse out of the cell, they leave unbalanced negative charges behind, generating a membrane potential that tends to pull positively charged K^+ ions back into the cell. The membrane potential at which the tendency of the K^+ ions to diffuse out is balanced by the negative electric potential pulling them back in is called the *potassium equilibrium potential.*

The value of the potassium equilibrium potential can be calculated from the concentrations of K^+ on the two sides of the membrane using an equation called the Nernst Equation, which is derived from the laws of physical chemistry (Figure 44.8). In general, the resting potential is a bit less negative than this potential predicts, bccause resting neurons are also slightly permeable to other ions, for example Na^+ and Cl^-.

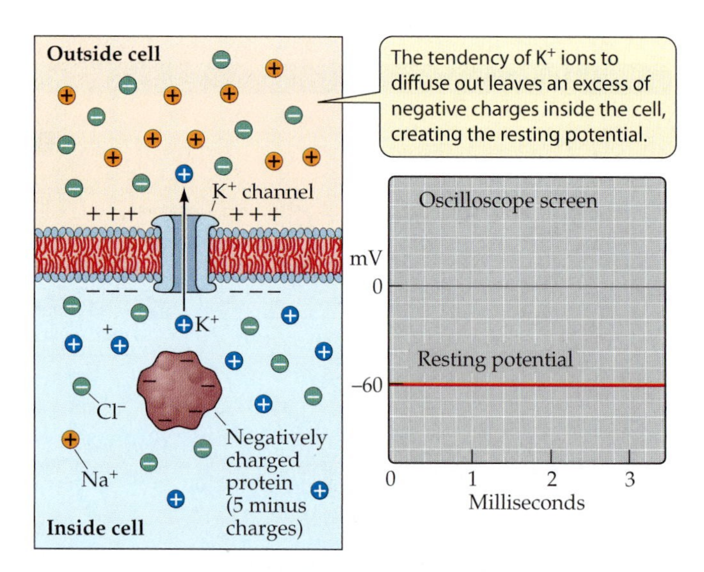

44.7 Open Potassium Channels Create the Resting Potential
Open potassium channels allow K^+ ions to diffuse out of the cell, leaving unbalanced negative charges behind (mostly on chloride ions and protein molecules).

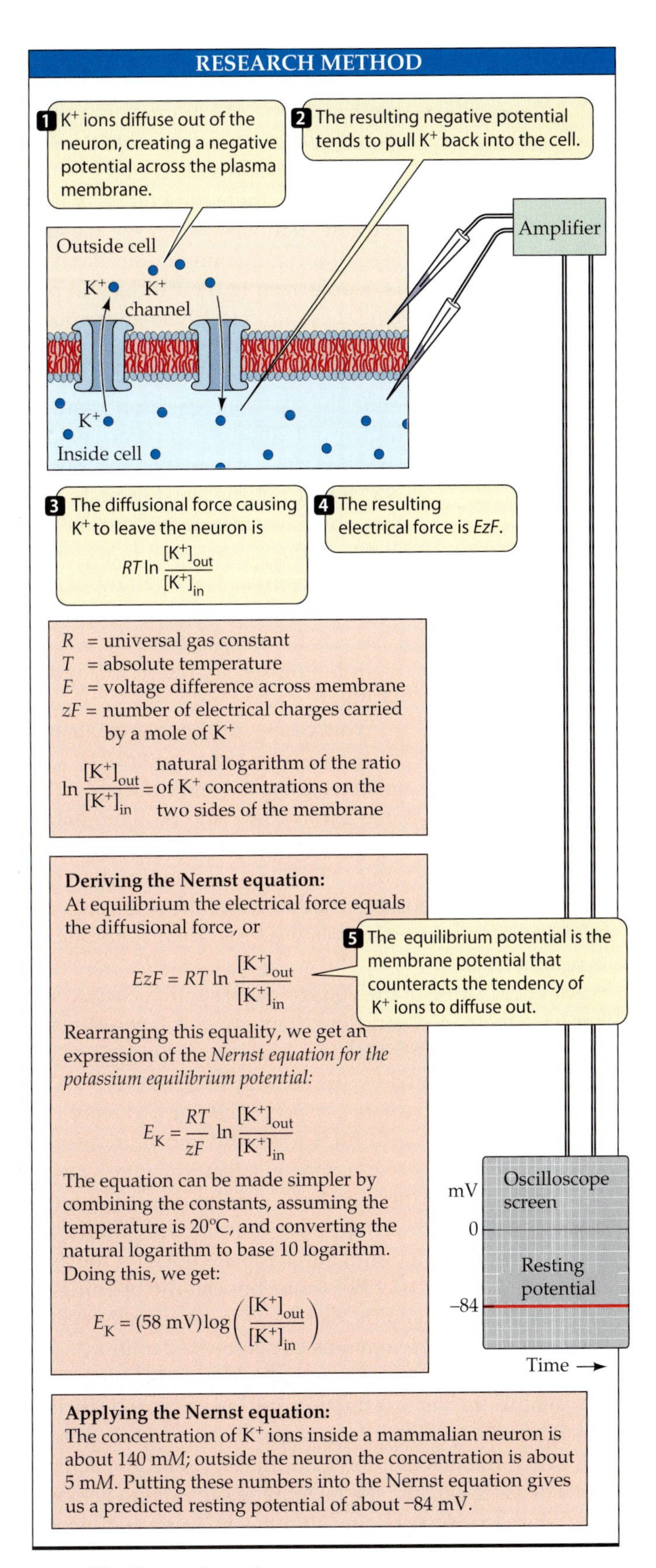

44.8 The Nernst Equation
The Nernst equation calculates membrane potential when only one type of ion can cross a membrane that separates solutions with different concentrations of that ion. The resting neuron comes close to that situation, since its permeability to K^+ ions is high and its permeability to all other ions is low.

Ion channels can alter membrane potential

Changes in the gated channels may perturb the resting potential. Imagine what would happen, for example, if sodium channels in the plasma membrane opened. Na^+ ions would diffuse into the cell because of their higher concentration on the outside, and they would also be attracted into the cell by the negative membrane potential. As a result of the entry of Na^+ ions, the inside of the cell would tend to become less negative. When the inside of a neuron becomes *less* negative in comparison to its resting condition, its plasma membrane is said to be **depolarized** (Figure 44.9*a*).

An opposite change in the resting potential would occur if gated Cl^- channels opened. The concentration of Cl^- ions is normally greater in the extracellular fluid than inside the neuron. This difference is large enough so that, in many neurons, the opening of Cl^- channels causes Cl^- to enter the cell, even though the membrane potential is negative. The entry of negative charges causes the membrane potential to become even more negative. When the inside of a neuron becomes *more* negative in comparison to its resting condition, its plasma membrane is said to be **hyperpolarized** (Figure 44.9*b*).

The opening and closing of ion channels, which result in changes in the polarity of the plasma membrane, are the basic mechanisms by which neurons respond to electrical, chemical, or other stimuli, such as touch, sound, and light. How does a neuron use a change in its resting membrane potential to process and transmit information?

A change in resting potential may result from input at a synapse. This input, however, is a very local event that affects only a small patch of plasma membrane. How can that information be passed to other parts of the cell? A local perturbation of membrane potential causes a flow of electrically charged ions, which tends to spread the change in membrane potential to adjacent regions of the membrane. This flow of electrically charged ions is an electric current. For example, if positively charged Na^+ ions enter the cell through sodium channels at one location, that positively charged area on the inside of the membrane attracts negative charges from surrounding areas, and thus there is a flow of electric current. However, this local flow of electric current does not spread very far before it diminishes and disappears.

The reason why electric currents do not travel very far in cell membranes is that these membranes are permeable to ions. An electric current traveling along a membrane is like water flowing through a leaky hose. Communication of a stimulus by the flow of electric current along plasma membranes is useful over only very short distances. Therefore, axons (the long processes of neurons) do not transmit information as a continuous flow of electric current (as telephone wires do). However, the local flow of electric current is an important part of the mechanism that generates the signals that axons do transmit over long distances: action potentials.

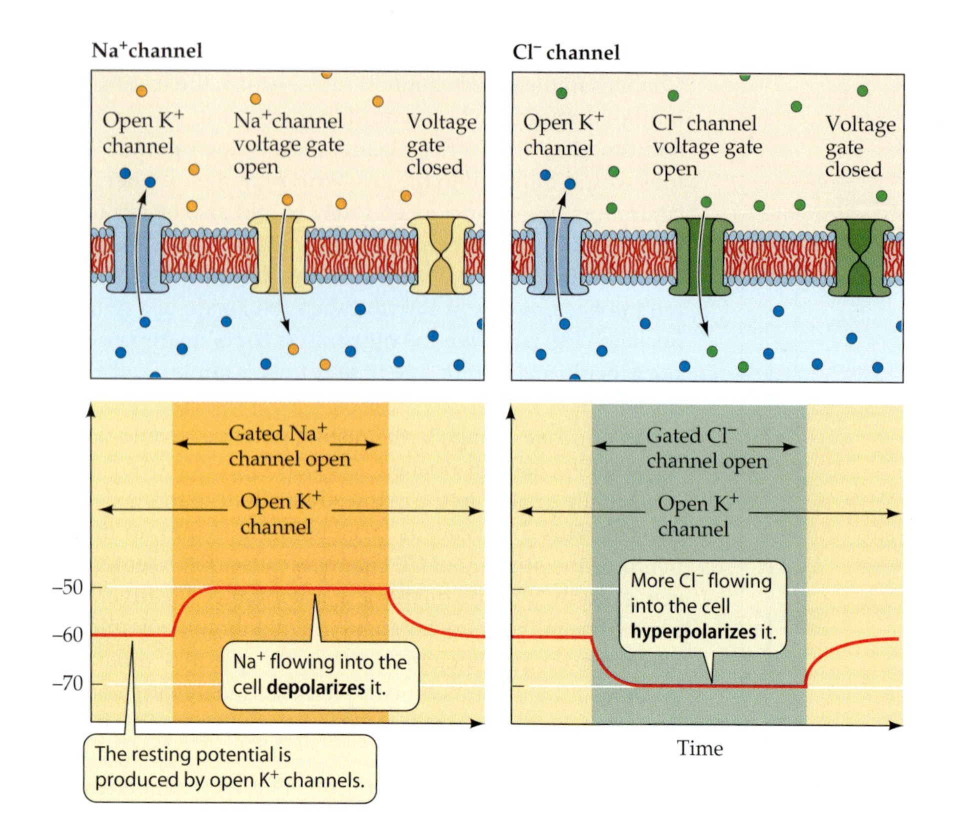

44.9 Membranes Can Be Depolarized or Hyperpolarized
The resting potential created by the diffusion of K$^+$ out of the cell can change (see Figure 44.7). A shift to a less negative membrane potential, as occurs when a gated sodium channel opens, is called depolarization. Hyperpolarization occurs when the membrane potential becomes more negative, as when more Cl$^-$ enters the cell through a gated channel.

Sudden changes in ion channels generate action potentials

An action potential is a sudden and major change in membrane potential that lasts for only 1 or 2 milliseconds. Action potentials are conducted along the axon of a neuron at speeds of up to 100 meters per second, which is equivalent to running the length of a football field in 1 second. How is an action potential generated, and how does it move down an axon?

If we place the tips of a pair of electrodes on either side of the plasma membrane of a resting axon and measure the voltage difference, the reading is about –60 mV, as we saw in Figure 44.5. If these electrodes are in place when an action potential travels down the axon, they register a rapid change in membrane potential, from –60 mV to about +50 mV. The membrane potential rapidly returns to its resting level of –60 mV as the action potential passes.

Voltage-gated sodium channels in the plasma membrane of the axon are primarily responsible for action potentials. At the normal membrane resting potential, most of these channels are closed. They are called voltage-gated channels because there is a particular membrane potential, or **threshold potential**, that causes them to open. For example, if synaptic input to some part of a neuron causes the plasma membrane of its cell body to depolarize, that depolarization can spread to the base of the axon, where there are voltage-gated sodium channels. When the plasma membrane containing these channels is depolarized to their threshold potential (about 5 to 10 mV more positive than the resting potential), they open briefly—for less than a millisecond. The Na$^+$ concentration is much higher outside the axon than inside, so when the sodium channels open, Na$^+$ ions from the outside rush into the axon. The entering Na$^+$ makes the inside of the plasma membrane electrically positive. The opening of the sodium channels causes the *rising phase* of the action potential, which neurobiologists call the *spike* (Figure 44.10).

What causes the depolarized axon to return to resting potential? The main reason is the rapid closing of the sodium channels. Some axons also have voltage-gated potassium channels. These channels open more slowly than the sodium channels and they stay open longer, allowing K$^+$ to carry excess positive charges out of the axon. As a result, they help return the voltage across the membrane to its resting level.

Another feature of the voltage-gated sodium channels is that once they open and close, they can be triggered again only after a short delay of 1 to 2 milliseconds. This property can be explained by the assumption that they have two voltage-sensitive gates, an *activation gate* and an *inactivation gate* (see Figure 44.10). Under resting conditions, the activation gate is closed and the inactivation gate is open. Depolarization of the membrane to the threshold level causes both gates to change state, but the activation gate responds faster. As a result, the channel is open for the passage of Na$^+$ ions for a brief time between the opening of the activation gate and the closing of the inactivation gate. The inactivation gate remains closed for 1–2 milliseconds before it spontaneously opens again, thus explaining why the membrane has a **refractory period** (a period during which it cannot act) before it can fire another action potential. When the inactivation gate finally opens, the activation gate is closed, and the membrane is poised to respond once again to a depolarizing stimulus by firing another action potential.

The difference in concentration of Na$^+$ across the plasma membrane of neurons is the "battery" that drives the action potential. How rapidly does the battery run down? It might seem that a substantial number of Na$^+$ and K$^+$ ions would have to cross the membrane for the membrane potential to change from –60 mV to +50 mV and back to –60 mV again.

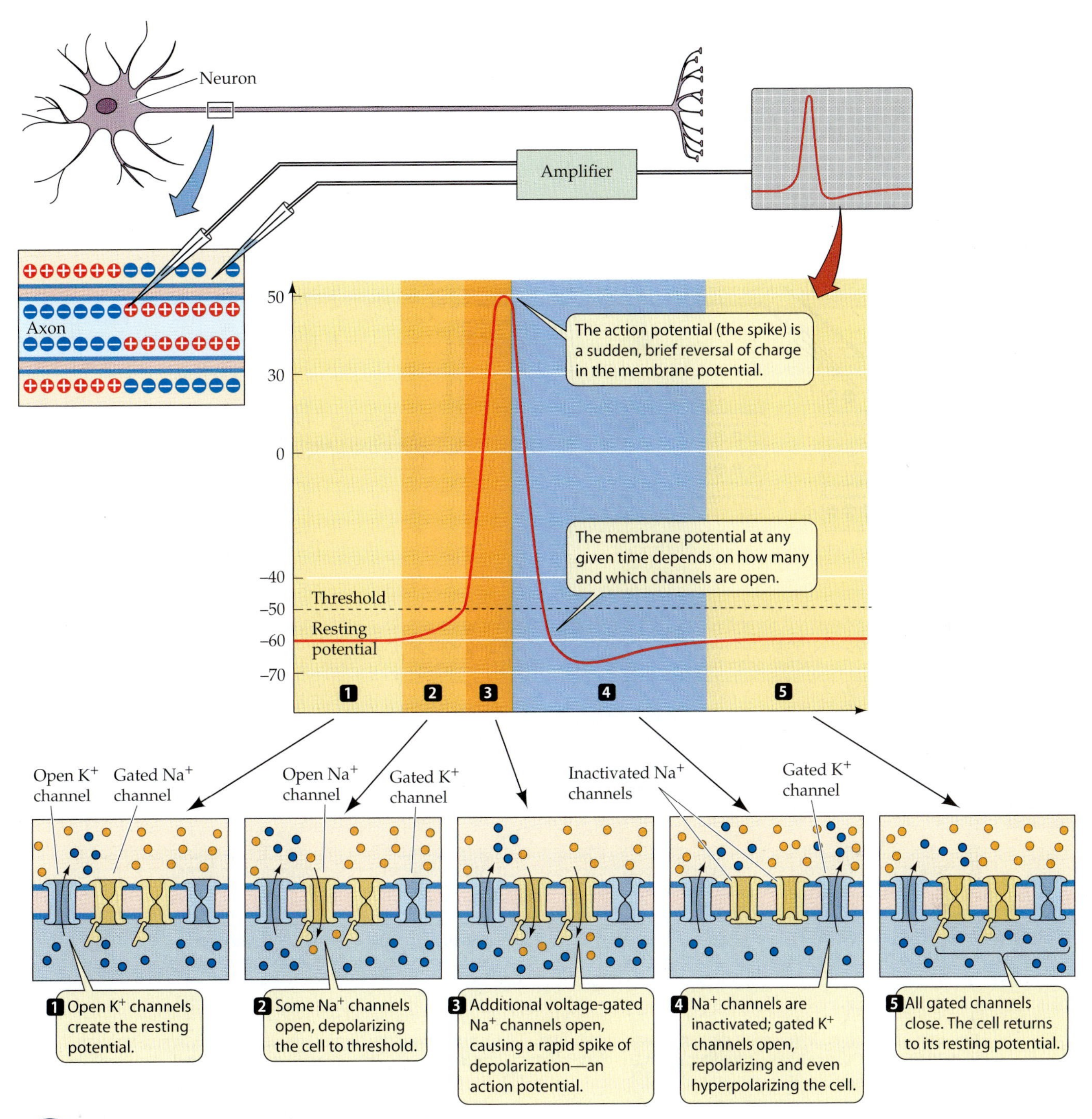

44.10 The Course of an Action Potential
Action potentials result from rapid changes in voltage-gated sodium and potassium channels. Like the resting potential, they can be measured using two electrodes (see Figure 44.5).

In fact, only about one Na^+ (or K^+) ion for every 10 million present actually moves through the channels during the passage of an action potential. Thus the effect of a single action potential on the concentration ratios of Na^+ or K^+ is very small, and it is not difficult for the sodium–potassium pump to keep the "battery" charged, even when the cell is generating many action potentials every second.

Action potentials are conducted down axons without reduction in the signal

Action potentials can travel over long distances with no loss of the signal. If we place two pairs of electrodes at two different locations along an axon, we can record an action potential at those two locations as it travels down the axon (Figure 44.11*a*). The magnitude of the action potential does not change between the two recordings. The action potential is an all-or-nothing, self-regenerating event.

The action potential is *all-or-nothing* because of the interaction between the voltage-gated sodium channels and the membrane potential. If the membrane is depolarized

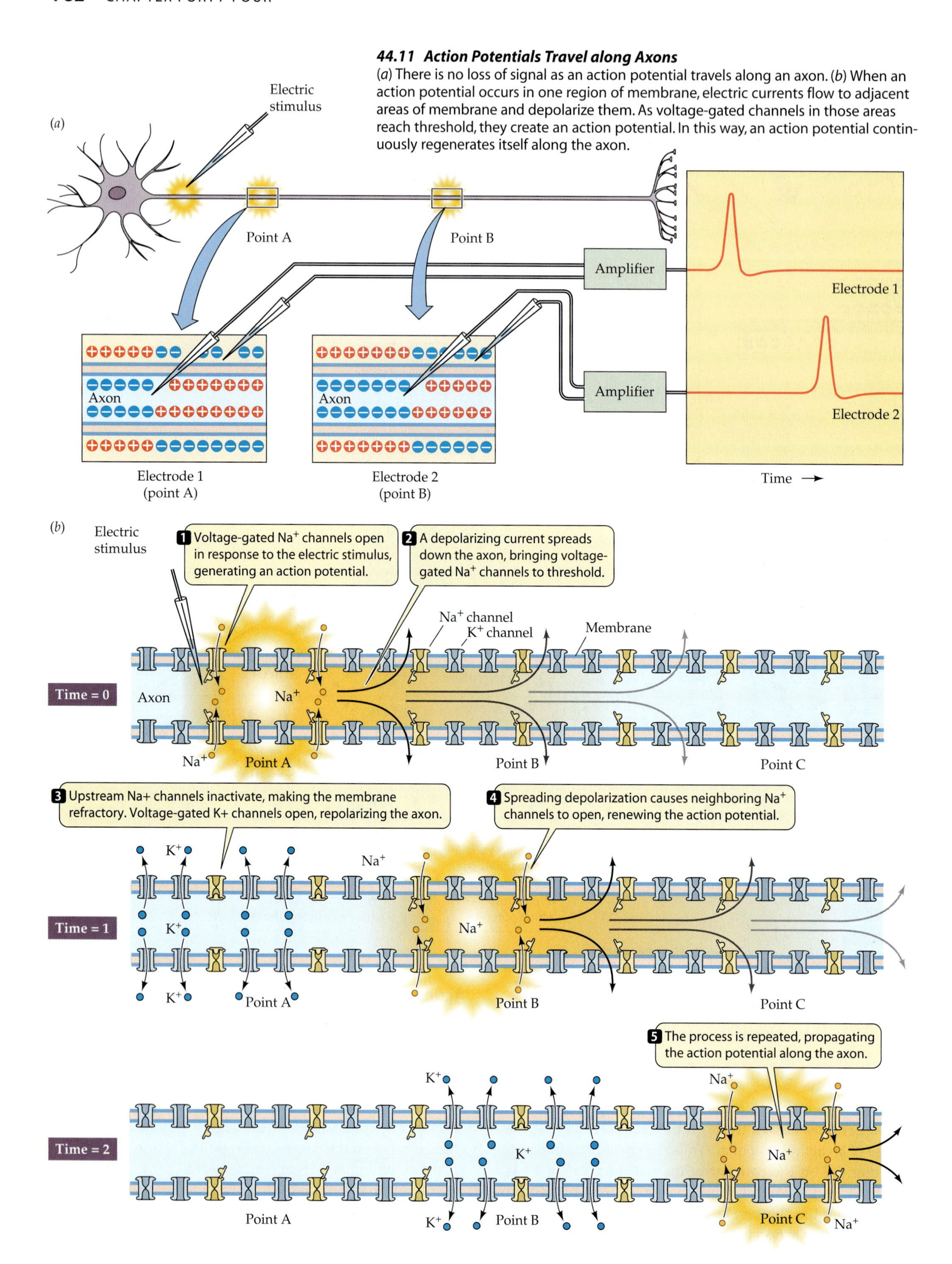

44.11 Action Potentials Travel along Axons
(*a*) There is no loss of signal as an action potential travels along an axon. (*b*) When an action potential occurs in one region of membrane, electric currents flow to adjacent areas of membrane and depolarize them. As voltage-gated channels in those areas reach threshold, they create an action potential. In this way, an action potential continuously regenerates itself along the axon.

slightly and approaches the threshold potential for voltage-gated sodium channels, some of them open. Some Na^+ ions cross the plasma membrane and depolarize it even more, bringing more voltage-gated sodium channels to threshold, and so on. This positive feedback mechanism ensures that action potentials always rise to their maximum value.

The action potential is *self-regenerating* because it spreads by current flow to adjacent regions of the membrane.

The resulting depolarization brings those areas of plasma membrane to the threshold potential for the voltage-gated Na^+ channels. So when an action potential occurs at one location on an axon, it stimulates the adjacent region of axon to generate an axon potential, and so on down the length of the axon (Figure 44.11*b*).

We can initiate an action potential by shocking an axon with an electric current delivered through a stimulating electrode that depolarizes the membrane enough to reach threshold. Now we can observe the changes in membrane potential associated with the passage of that action potential past the recording electrodes (Figure 44.11*b*).

Positive ions flood into the axon at the site of the action potential. Once inside, positive ions spread by current flow to adjacent regions of the axon, making those regions less negative. As this depolarization of the adjacent plasma membrane brings it to the threshold potential, an action potential is generated there. Because an action potential always brings the adjacent area of membrane to threshold, the action potential propagates itself along the axon. The action potential is self-regenerating and propagates itself in only one direction, however; it cannot reverse itself, because the part of the membrane it came from is in its refractory period.

Action potentials do not travel along all axons at the same speed. They travel faster in large-diameter axons than in small-diameter axons. In invertebrates, the axon diameter determines the rate of conduction, and axons that transmit messages involved in escape behavior are very thick. The giant axons that enable squid to escape predators are almost 1 mm in diameter.

Ion channels and their properties can be studied directly

The size of the squid giant axon made it possible for the British neurophysiologists A. L. Hodgkin and A. F. Huxley to study the electrical properties of axonal membranes 60 years ago. They used thin electrodes to measure voltage differences across the plasma membrane of the squid giant axon and to pass electric current into the axon to change its resting potential. They also changed the concentrations of Na^+ and K^+ ions both inside and outside the axon and measured the changes in resting and action potentials.

On the basis of their many careful experiments, Hodgkin and Huxley developed the story we have discussed so far. However, they were working long before technology enabled the actual demonstration of ion channels. Hodgkin and Huxley could only hypothesize the existence of ion channels and their properties.

With current techniques, neurobiologists can record currents caused by the openings and closings of single ion channels directly. **Patch clamping**, developed in the 1980s by Bert Sakmann and Erwin Neher, made possible the study of single ion channels in plasma membranes. In patch clamping, a glass pipette with a tip only 1 or 2 microns in diameter is placed in contact with the plasma membrane (Figure 44.12). Slight suction makes a seal between the pipette and the patch of membrane under its tip. Retracting the pipette can detach a bit of plasma membrane. Movements of ions, and therefore electric charges, through channels in the patch of membrane can be recorded through the pipette. Frequently, a patch will contain only one or a few ion channels; thus the electrical recording from that patch can show individual channels opening and closing.

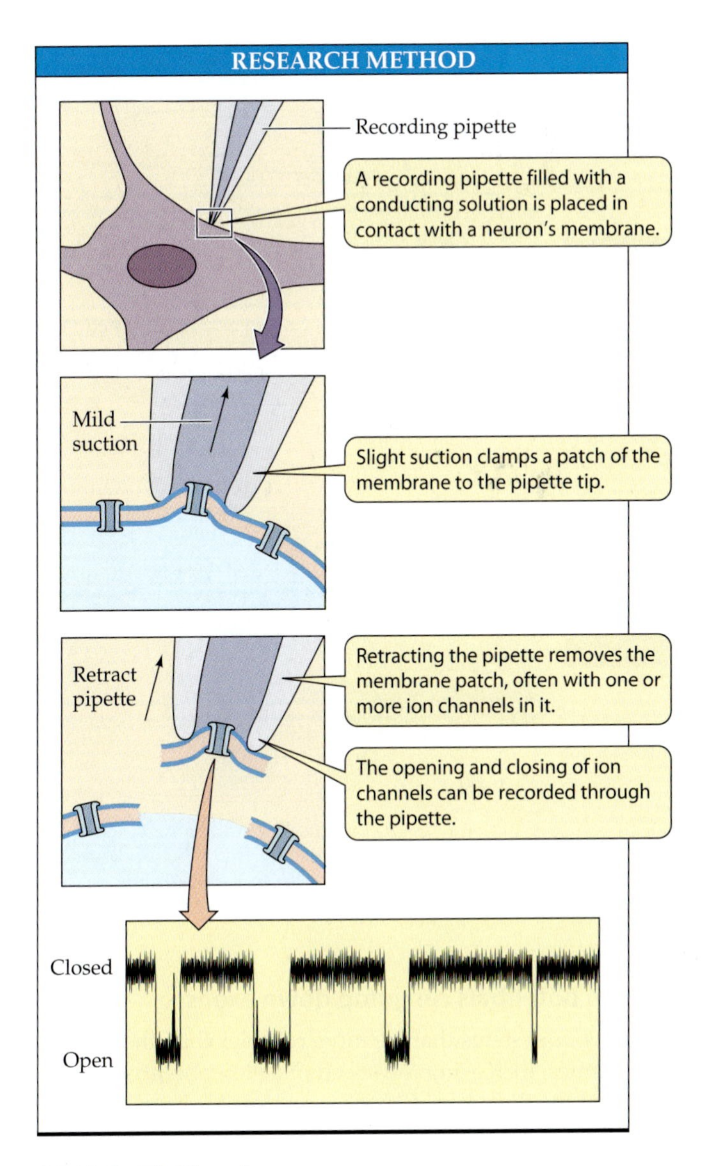

44.12 Patch Clamping
A patch clamp electrode can record the opening and closing of a single ion channel.

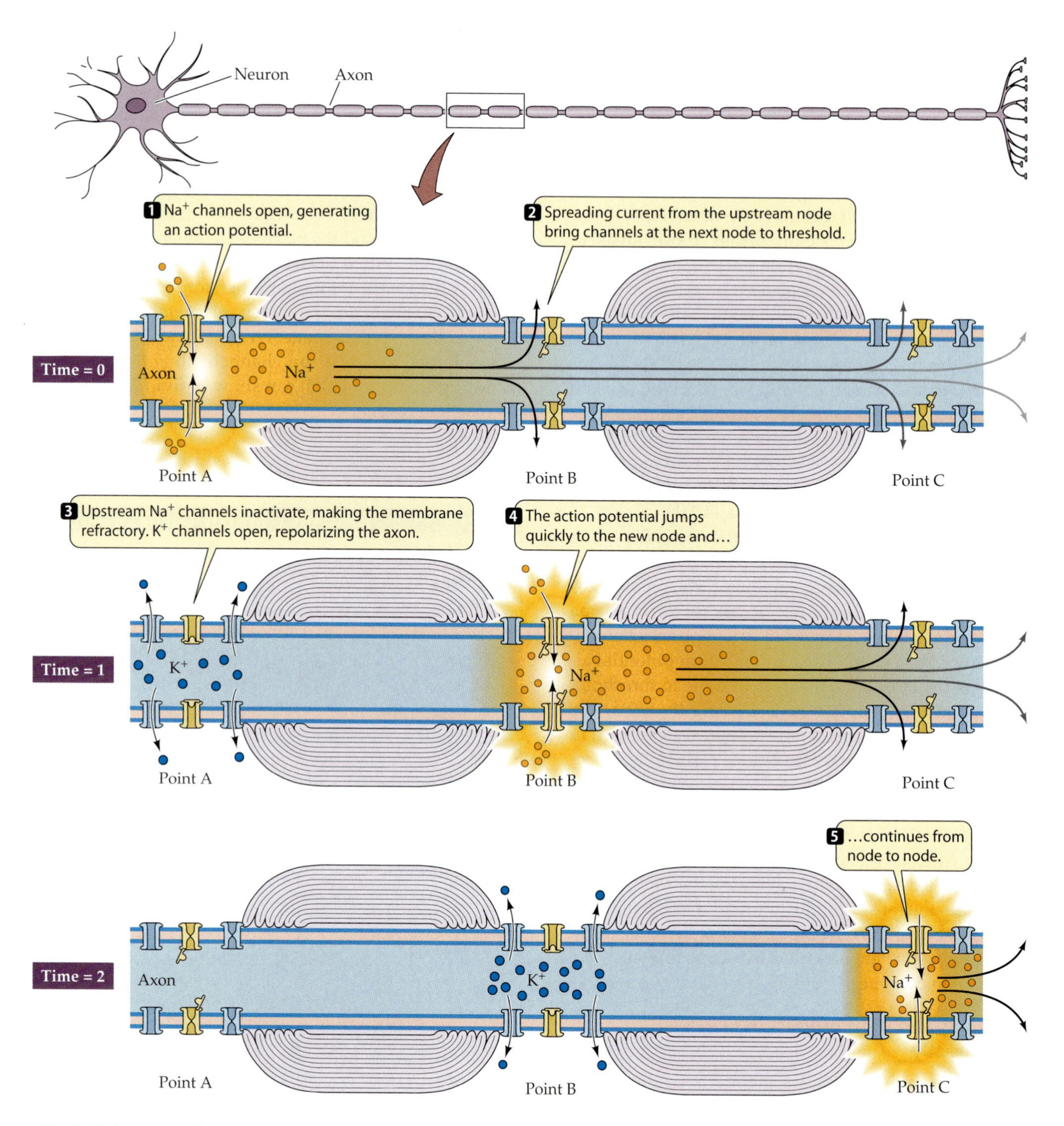

44.13 Saltatory Action Potentials
Action potentials appear to jump from node to node in myelinated axons.

Action potentials can jump down axons

In nervous systems that are more complex than those of invertebrates, increasing the speed of action potentials by increasing the diameter of axons is not feasible because of the huge number of axons that are required. Each of our eyes, for example, has about a million axons extending from it. Evolution has increased propagation velocity in vertebrate axons in a way that does not require large size.

When we discussed glial cells earlier in the chapter, we saw that glial cells of one type, called Schwann cells, send out projections that wrap around axons, covering them with concentric layers of plasma membrane (see Figure 44.4). These myelin wrappings are not continuous, but have regularly spaced gaps called **nodes of Ranvier** where the axon is not covered (Figure 44.13).

Myelin electrically insulates the axon; that is, charged ions cannot cross the regions of the plasma membrane that are wrapped in myelin. Additionally, ion channels are clustered at the nodes. Thus an axon can fire an action potential only at a node of Ranvier, but that action potential cannot be conducted through the adjacent patch of axon covered with myelin. The positive charges that flow into the axon at the node, however, spread down the axon. When the spread of current causes the plasma membrane at the next node to depolarize to threshold, an action potential is fired at that node. Action potentials therefore appear to jump from node to node down the axon.

The speed of conduction is increased in these myelin-wrapped axons because electric current flows very fast through the cytoplasm in comparison to the time required for channels to open and close. This form of impulse propagation is called **saltatory** (jumping) **conduction** and is much quicker than continuous propagation of action potentials down an unmyelinated axon.

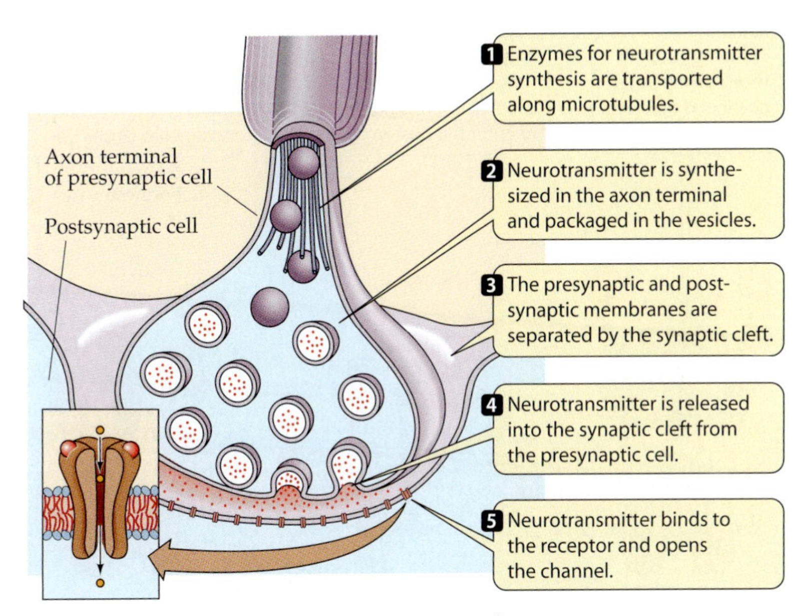

44.14 The Neuromuscular Junction Is a Chemical Synapse
A motor neuron communicates chemically with a muscle cell at a neuromuscular junction when a neurotransmitter called acetylcholine (red dots) crosses the synaptic cleft.

Neurons, Synapses, and Communication

The most remarkable abilities of nervous systems stem from interactions among connected neurons. These interactions process and integrate information. Our nervous systems can orchestrate complex behaviors, deal with complex concepts, and learn and remember because large numbers of neurons interact with one another. The mechanisms of these interactions depend on synapses between cells. **Synapses** are structurally specialized junctions where one cell influences another cell directly through the transfer of an electrical or chemical message. The cell that sends the message is called the *presynaptic cell*, and the cell that receives it is called the *postsynaptic cell*. The most common type of synapse in the nervous system is the chemical synapse—one in which chemical messages released by a presynaptic cell induce changes in a postsynaptic cell.

In this section, we examine the specializations and functions of presynaptic and postsynaptic cells. We discover how synapses can integrate information, and at the end of the section we examine the diversity of chemical messages released by neurons.

The neuromuscular junction is a classic chemical synapse

Synapses between motor neurons and muscle cells are called **neuromuscular junctions**, and they are excellent models for how chemical synaptic transmission works. Like other neurons, the motor neuron that innervates a muscle has only one axon, but that axon can have many branches, each with an axon terminal that forms a neuromuscular junction with a muscle cell. At each axon terminal is an enlarged knob or buttonlike structure that contains many spherical vesicles filled with chemical messenger molecules, or **neurotransmitters** (Figure 44.14). The neurotransmitter used by all neurons that innervate vertebrate skeletal muscle is **actetylcholine**. The portion of the axon terminal plasma membrane that forms a synapse with a muscle cell is called the **presynaptic membrane**. Acetylcholine is released by exocytosis when the membrane of a vesicle fuses with the presynaptic membrane.

Where does the neurotransmitter come from? Some neurotransmitters, like acetylcholine, are synthesized in the axon terminal and packaged in vesicles. The enzymes required for acetylcholine biosynthesis, however, are produced in the cell body of the motor neuron and are transported down the axon to the terminals along microtubules. Other kinds of neurotransmitters, such as peptide neurotransmitters, are produced in the cell body and transported down the axon to the terminals.

The **postsynaptic membrane** of the neuromuscular junction is a modified part of the muscle cell plasma membrane called a **motor end plate**. The space between the presynaptic membrane and the postsynaptic membrane is the **synaptic cleft**. In chemical synapses, the synaptic cleft is, on average, about 20–40 nm wide, and the neurotransmitter released into the cleft diffuses across to the postsynaptic membrane (see Figure 44.14).

The patches of muscle cell plasma membrane that form a motor end plate contain acetylcholine receptor molecules. These receptors are chemically gated channels that allow both Na^+ and K^+ to pass through. Since the resting membrane is already permeable to K^+, the major change that occurs when these channels open is the movement of Na^+ into the cell. When a receptor binds acetylcholine, its channel opens, and Na^+ moves across the membrane, depolarizing the motor end plate (Figure 44.15).

Acetylcholine action is limited by the enzyme **acetylcholinesterase**, found in the synaptic cleft. This powerful enzyme cleaves any acetylcholine molecules it encounters.

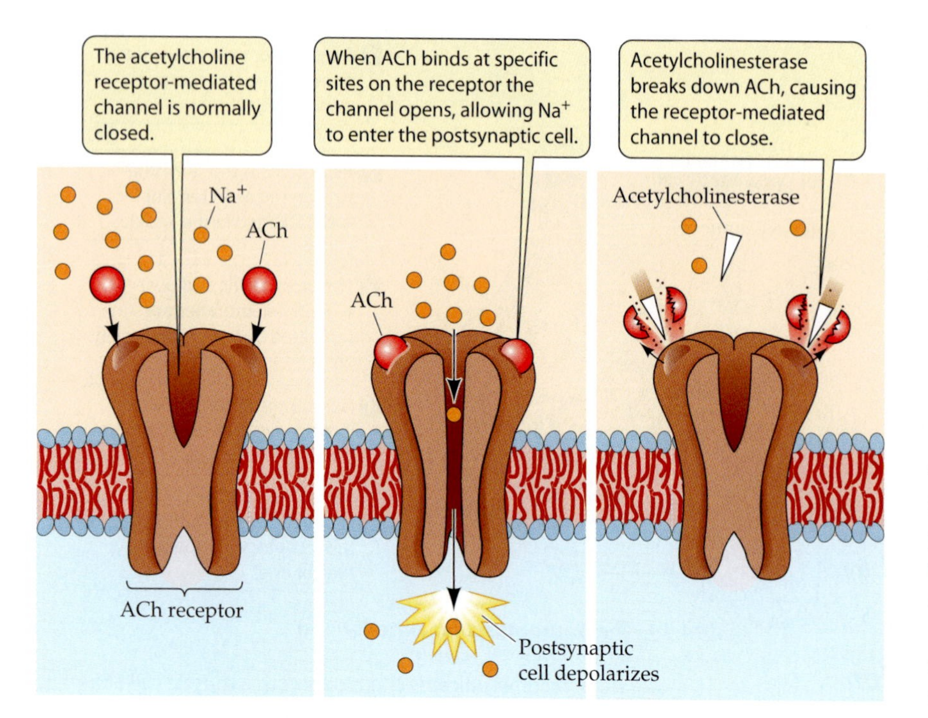

44.15 The Actions of Acetylcholine and Acetylcholinesterase Balance Out When a receptor on the motor end plate binds acetylcholine (ACh), a chemically gated channel opens, and Na^+ ions move into the postsynaptic cell, making the cell more positive inside (depolarization). Acetylcholinesterase in the synapse destroys ACh, closing the channel; the breakdown products are then resynthesized into more ACh.

Thus, the activity of the neuromuscular junction is a balance between the release of acetylcholine by the presynaptic membrane and its destruction by acetylcholinesterase in the synaptic cleft. The breakdown products of neurotransmitter degradation are taken up by the presynaptic terminal and resynthesized into acetylcholine.

The arrival of an action potential causes the release of neurotransmitter

Neurotransmitter is released from the presynaptic membrane when an action potential arrives at the axon terminal. The presynaptic membrane has a type of voltage-gated ion channel found nowhere else on the axon: a voltage-gated calcium channel. When the action potential reaches the axon terminal, it causes these calcium channels to open (Figure 44.16). Because Ca^{2+} concentration is greater outside the cell than inside the cell, Ca^{2+} rushes in.

The increase in Ca^{2+} inside the presynaptic cell causes the vesicles containing acetylcholine to fuse with the presynaptic membrane and empty their contents into the synaptic cleft. The acetylcholine molecules diffuse across the cleft and bind to the receptors on the motor end plate, causing the sodium channels to open briefly and depolarize it.

The postsynaptic membrane integrates synaptic input

The postsynaptic membranes of neuromuscular junctions differ from the presynaptic membranes in an important way. Motor end plates have very few voltage-gated sodium channels; therefore, they do not fire action potentials. This is true not only of motor end plates, but also of dendrites and of most regions of neuronal cell bodies. The binding of neurotransmitter to receptors at the motor end plate and the resultant opening of chemically gated ion channels perturb the resting potential of the postsynaptic membrane. This local change in membrane potential spreads to neighboring regions of the plasma membrane of the postsynaptic cell.

Eventually, the spreading depolarization may reach an area of membrane that does contain voltage-gated channels. The entire plasma membrane of a skeletal muscle fiber, except for the motor end plates, has voltage-gated sodium channels. If the axon terminal of a motor neuron releases sufficient amounts of neurotransmitter to depolarize a motor end plate enough to bring the surrounding membrane to threshold, an action potential is fired. This action potential is then conducted throughout the muscle cell's system of membranes, causing the cell to contract. (We'll learn more about muscle membrane action potentials and the contraction of muscle cells in Chapter 47.)

How much neurotransmitter is enough? Neither a single acetylcholine molecule nor the contents of an entire vesicle (about 10,000 acetylcholine molecules) are enough to bring the plasma membrane of a muscle cell to threshold. However, a single action potential in an axon terminal releases about 100 vesicles, which is enough to fire an action potential in the muscle cell and cause it to contract.

Synapses between neurons can be excitatory or inhibitory

In vertebrates, the synapses between motor neurons and skeletal muscle are always **excitatory**; that is, motor end plates always respond to acetylcholine by depolarizing the postsynaptic membrane. However, synapses between neurons can be either excitatory or inhibitory.

Recall that a neuron may have many dendrites. Axon terminals from many neurons may make synapses with those dendrites and with the cell body. The axon terminals of different presynaptic neurons may store and release different neurotransmitters, and membranes of the dendrites and cell body of a postsynaptic neuron may have receptors for a variety of neurotransmitters. Thus a postsynaptic neuron can receive various chemical messages. If the postsynaptic neuron's response to a neurotransmitter is depolarization, as at the neuromuscular junction, the synapse is

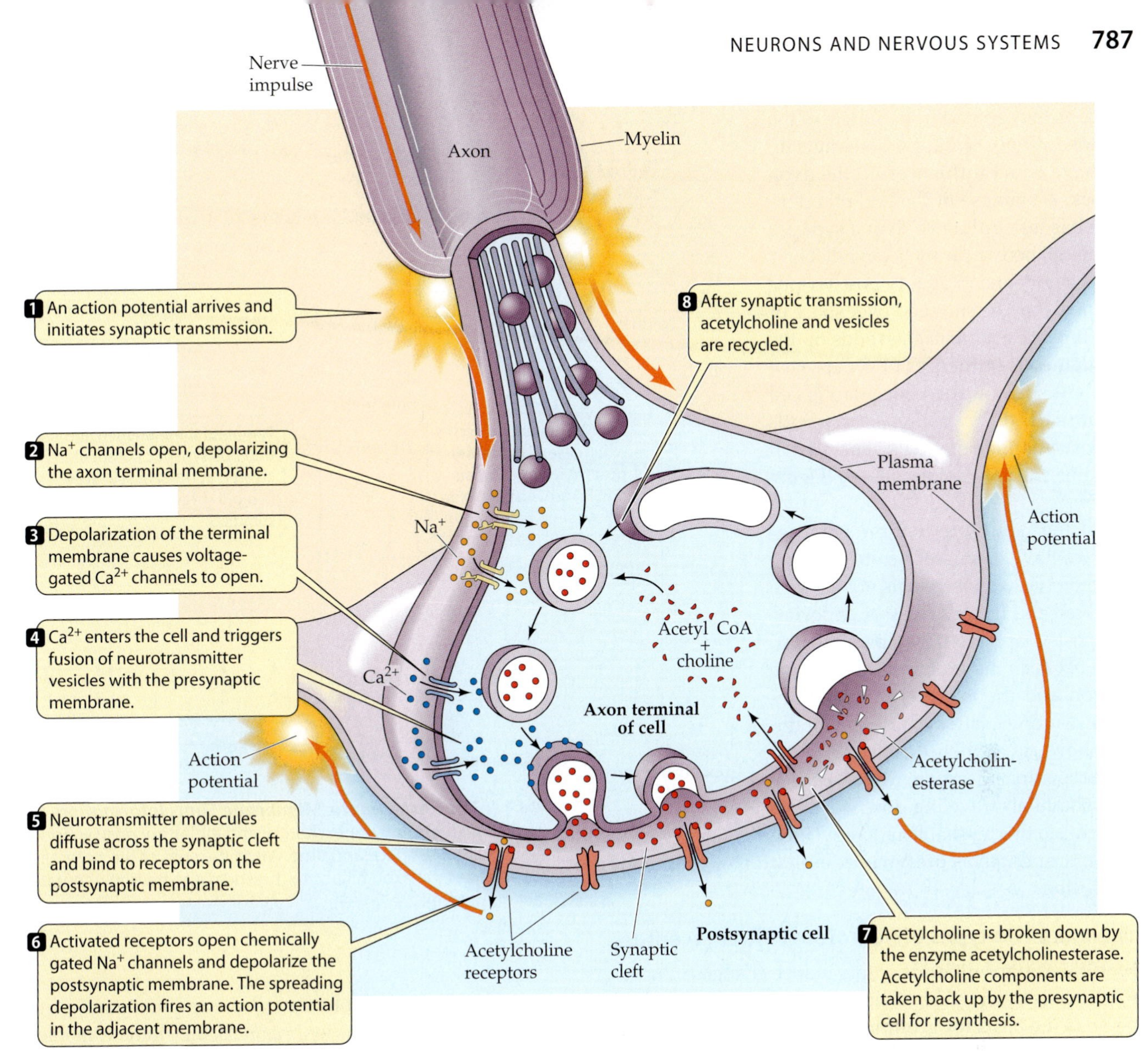

1. An action potential arrives and initiates synaptic transmission.
2. Na^+ channels open, depolarizing the axon terminal membrane.
3. Depolarization of the terminal membrane causes voltage-gated Ca^{2+} channels to open.
4. Ca^{2+} enters the cell and triggers fusion of neurotransmitter vesicles with the presynaptic membrane.
5. Neurotransmitter molecules diffuse across the synaptic cleft and bind to receptors on the postsynaptic membrane.
6. Activated receptors open chemically gated Na^+ channels and depolarize the postsynaptic membrane. The spreading depolarization fires an action potential in the adjacent membrane.
7. Acetylcholine is broken down by the enzyme acetylcholinesterase. Acetylcholine components are taken back up by the presynaptic cell for resynthesis.
8. After synaptic transmission, acetylcholine and vesicles are recycled.

Nerve impulse
Axon
Myelin
Plasma membrane
Action potential
Na^+
Ca^{2+}
Acetyl CoA + choline
Axon terminal of cell
Acetylcholinesterase
Action potential
Acetylcholine receptors
Synaptic cleft
Postsynaptic cell

44.16 Synaptic Transmission Begins with the Arrival of an Action Potential The figure diagrams the sequence of events involved in transmission at the motor end plate, a typical chemical synapse.

excitatory; if the response is hyperpolarization, the synapse is **inhibitory** (see Figure 44.9).

How do inhibitory synapses work? In vertebrates, the two most common inhibitory neurotransmitters are γ-amino butyric acid (GABA) and glycine. The postsynaptic cells at inhibitory synapses have chemically gated chloride channels. When these channels are activated by a neurotransmitter, they can hyperpolarize the postsynaptic membrane. Thus the release of neurotransmitter at an inhibitory synapse makes the postsynaptic cell *less* likely to fire an action potential.

Neurotransmitters that depolarize the postsynaptic membrane are excitatory; they bring about an **excitatory postsynaptic potential** (**EPSP**). Neurotransmitters that hyperpolarize the postsynaptic membrane are inhibitory; they bring about an **inhibitory postsynaptic potential** (**IPSP**).

The postsynaptic membrane sums excitatory and inhibitory input

Individual neurons "decide" whether or not to fire an action potential by summing excitatory and inhibitory postsynaptic potentials. This summation ability of neurons is the major mechanism by which the nervous system integrates information. Each neuron may receive 1,000 or more synaptic inputs, but it has only one output: an action potential in a single axon. All the information contained in the thousands of inputs a neuron receives is reduced to the rate at which that neuron generates action potentials in its axon.

For most neurons, the critical area for "decision making" is the **axon hillock**, the region of the cell body at the base of the axon. The plasma membrane of the axon hillock is not insulated by glial cells and has many voltage-gated channels. Excitatory and inhibitory postsynaptic potentials from synapses anywhere on the dendrites or the cell body spread to the axon hillock by current flow. If the resulting combined potential depolarizes the axon hillock to threshold, the axon fires an action potential. Because postsynaptic potentials decrease as they spread from the site of the synapse, all postsynaptic potentials do not have equal influences on the axon hillock. A synapse at the end of a dendrite has less influence than a synapse on the cell body near the axon hillock.

Excitatory and inhibitory postsynaptic potentials can be summed over space or over time. *Spatial summation* adds up the simultaneous influences of synapses at different sites on the postsynaptic cell (Figure 44.17*a*). *Temporal summation* adds up postsynaptic potentials generated at the same site in a rapid sequence (Figure 44.17*b*).

All the neuron-to-neuron synapses that we have discussed up to now are between the axon terminals of a presynaptic cell and the cell body or dendrites of a postsynaptic cell. Synapses can also form between the axon terminals of one neuron and the axon terminals of another neuron. Such a synapse can modulate how much neurotransmitter the second neuron releases in response to action potentials traveling down its axon. We refer to this mechanism of regulating synaptic strength as **presynaptic excitation** or **presynaptic inhibition**.

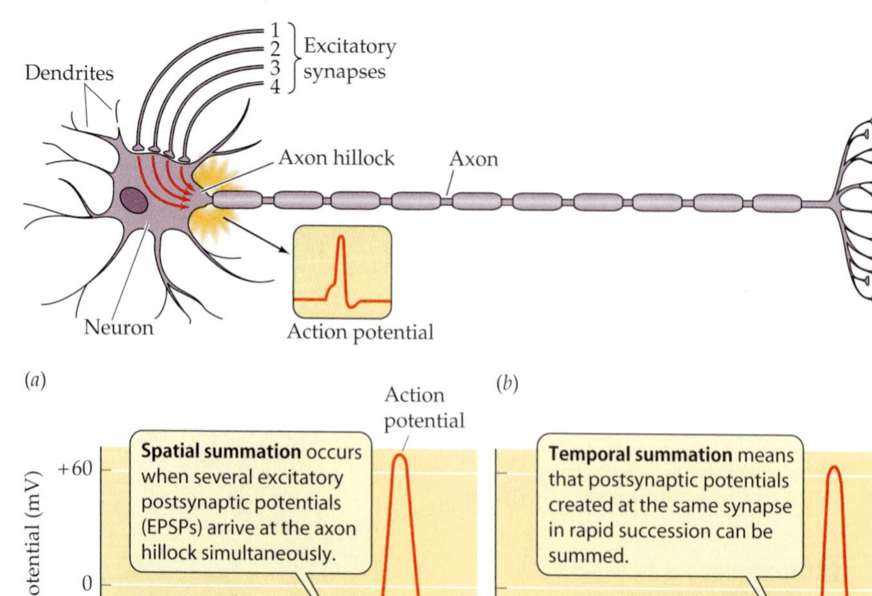

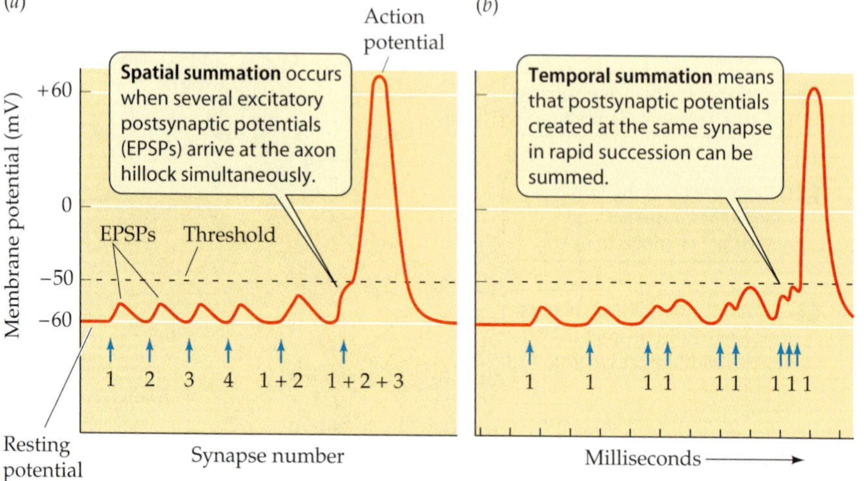

44.17 The Postsynaptic Membrane Sums Information Individual neurons sum excitatory and inhibitory postsynaptic potentials over space and time. When the sum of the potentials depolarizes the axon hillock to threshold, the neuron generates an action potential.

There are two types of neurotransmitter receptors

Most neurotransmitter receptors induce changes in postsynaptic cells by opening or closing ion channels. How they do so is the basis for grouping receptors into two general families.

- **Ionotropic receptors** are themselves ion channels. Neurotransmitter binding by an ionotropic receptor causes a direct change in ion movements across the postsynaptic cell membrane.
- **Metabotropic receptors** are not ion channels, but they induce changes in the postsynaptic cell that can secondarily lead to changes in ion channels.

Postsynaptic cell responses mediated by metabotropic receptors are generally slower and longer-lasting than those induced by ionotropic receptors.

The acetylcholine (ACh) receptor of the motor end plate of muscle cells is an example of an ionotropic receptor. This receptor is called the *nicotinic ACh receptor* because it also binds the drug nicotine. The receptor consists of five subunits, each of which extends through the plasma membrane (see Figure 44.15). When assembled, the subunits create a central pore that is the ion channel. There are several different kinds of subunits, and only one kind has the ability to bind ACh or nicotine. Each functional receptor–ion channel has two of the nicotine-binding subunits and three other subunits.

Metabotropic receptors are also transmembrane proteins, but instead of acting as ion channels, they initiate an intracellular signaling process that can result in the opening or closing of an ion channel. These receptors have seven transmembrane domains, and they are coupled to G proteins (Figure 44.18; see Chapter 15). When a neurotransmitter binds to the extracellular domain of a metabotropic receptor, the intracellular domain activates a G protein. In its inactive state, the G protein has three subunits, one of which (the α subunit) is bound to a molecule of GDP. When the receptor binds its neurotransmitter, the GDP is replaced with a GTP molecule, and the α subunit separates from the other two subunits (called β and γ). The separated subunits move across the membrane to where they can either directly activate an ion channel or activate an intermediate effector protein that eventually activates the ion channel.

An example of a metabotropic receptor is the muscarinic ACh receptor, so called because it binds both ACh and a molecule called muscarine that is a toxin in some poisonous mushrooms. Muscarinic ACh receptors are found in heart muscle. When activated, they cause the opening of potassium channels, which slows the heartbeat.

Metabotropic receptors can induce intracellular changes other than the opening or closing of ion channels. G proteins can activate a number of intracellular second messenger sys-

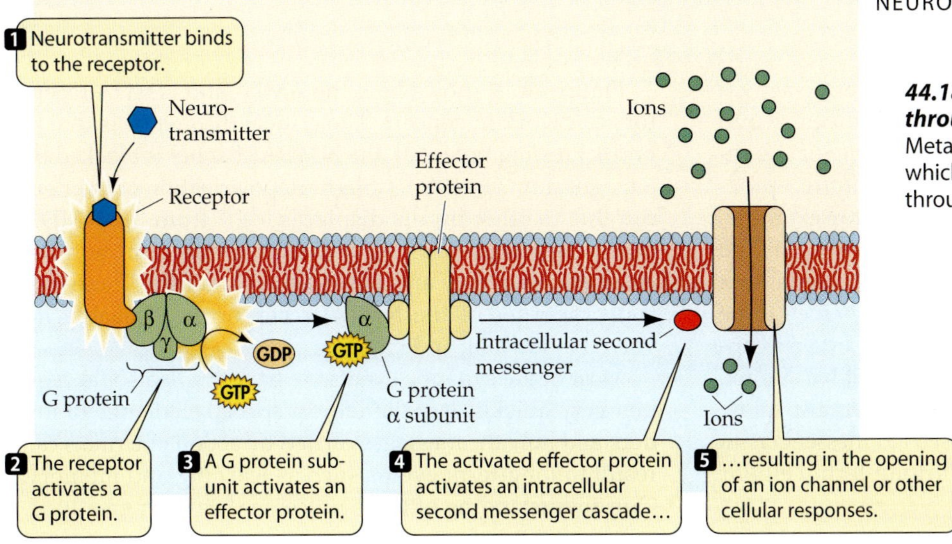

44.18 Metabotropic Receptors Act through G Proteins
Metabotropic receptors activate G proteins, which can influence ion channels directly or through second messengers.

tems (see Chapter 15). Through these signaling pathways, the metabotropic receptor can activate protein kinases, activate or inactivate enzymes, and alter gene expression.

Electrical synapses are fast but do not integrate information well

Electrical synapses, or *gap junctions*, are different from chemical synapses because they couple neurons electrically. At gap junctions, the presynaptic and postsynaptic cell membranes are separated by a space of only 2 to 3 nm, and specific membrane proteins called *connexons* link the two neurons by forming molecular tunnels between the two cells. Ions and small molecules can pass directly from cell to cell through the connexons (see Figure 15.19). Electrical transmission across gap junctions is very fast and can proceed in either direction; that is, stimulation of either neuron can result in an action potential in the other. In contrast, chemical synapses are slower, and they are unidirectional.

Gap junctions are less common in the complex nervous systems of vertebrates than are chemical synapses for several reasons. First, electrical continuity between neurons does not allow temporal summation of synaptic inputs, which is one way that complex nervous systems integrate information. Second, an effective electrical synapse requires a large area of contact between the presynaptic and postsynaptic cells. This condition rules out the possibility of thousands of synaptic inputs to a single neuron—which is the norm in complex nervous systems. Third electrical synapses cannot be inhibitory. Finally, there appears to be little plasticity (modifiability) in electrical synapses. Thus, electrical synapses are good for rapid communication, but not for processes of integration and learning.

The action of a neurotransmitter depends on the receptor to which it binds

More than 25 neurotransmitters are now recognized, and more will surely be discovered. Table 44.1 gives some examples. Acetylcholine, as we have seen, is an important neurotransmitter because it is the means whereby nerves command skeletal muscles to contract. Acetylcholine also plays roles in certain synapses between neurons in the central nervous system, but it accounts for only a small percentage of the total neurotransmitter content of the CNS. The workhorse neurotransmitters of the CNS are simple amino acids: glutamate (excitatory) and glycine and GABA (inhibitory). Another important group of neurotransmitters in the CNS is the *monoamines*, which are derivatives of amino acids. They include dopamine and norepinephrine (derivatives of tyrosine) and serotonin (a derivative of tryptophan). Peptides also function as neurotransmitters. An exciting recent discovery revealed that two gases, carbon monoxide and nitric oxide, are used by neurons as intercellular messengers even though they do not have the characteristics of classic neurotransmitters (that is, they do not have receptors).

The complexity of neurotransmission is increased by the fact that each neurotransmitter has multiple receptor types. We have already seen that acetylcholine has two well-known receptor types, nicotinic and muscarinic, one being ionotropic and the other metabotropic. Both types of ACh receptors are found in the CNS, where nicotinic receptors tend to be excitatory and muscarinic receptors tend to be inhibitory. ACh actions can differ outside of the CNS as well. ACh acting through nicotinic receptors causes the smooth muscle of the gut to depolarize and therefore increases its motility, but ACh acting through muscarinic receptors causes cardiac muscle to hyperpolarize and therefore decreases the contractility of the heart.

We could give many more examples of neurotransmitters that have different effects in different tissues, but the important thing to remember is that *the action of a neurotransmitter depends on the receptor to which it binds.*

Glutamate receptors may be involved in learning and memory

Glutamate can bind to a variety of receptors, including metabotropic receptors that activate G proteins and ionotropic receptors that are directly linked to ion channels that open when the receptor binds glutamate. The glutamate receptors are divided into several classes because they are differentially activated by other chemicals that mimic the action of glutamate. One class of ionotropic glutamate receptor is the *NMDA receptor;* they can be activated by the chemical *N*-methyl-D-aspartate. Another class of ionotropic glutamate receptors is activated by a different chemical, abbreviated as AMPA.

Glutamate is an excitatory neurotransmitter, so activation of all glutamate receptors results in sodium entry into the neuron and depolarization. But the *timing* of the response to activation of these different types of receptors differs significantly. The AMPA receptors, for example, allow a rapid influx of Na^+ into the postsynaptic cell. The NMDA receptors allow a slower and longer-lasting influx of Na^+. The NMDA receptors also require the cell to be somewhat depolarized through the action of other receptors before they will open channels and permit Na^+ influx. When they do respond, the channels they open also allow Ca^{2+} to enter the cell. Ca^{2+} ions act as second messengers in the cell and can trigger a variety of long-term cellular changes.

Figure 44.19 shows how the AMPA and NMDA receptors can work in concert. The critical difference is that at normal resting potential, the NMDA channel is blocked by a magnesium ion (Mg^{2+}). Slight depolarization of the neuron due to other inputs displaces Mg^{2+} from the NMDA channels and allows Na^+ and Ca^{2+} to pass through the channels when they are activated by glutamate. These special properties of the NMDA receptor are probably involved in learning and memory.

Most of the synaptic events we have studied so far happen very quickly. It is therefore a special challenge to understand how the messages carried by action potentials can result in long-term events such as learning and memory. Our understanding of these processes has been greatly af-

44.1 *Some Well-Known Neurotransmitters*

NEUROTRANSMITTER	ACTIONS	COMMENTS
Acetylcholine	The neurotransmitter of vertebrate motor neurons and of some neural pathways in the brain	Broken down in the synapse by acetylcholinesterase; blockers of this enzyme are powerful poisons
Monoamines		
Norepinephrine	Used in certain neural pathways in the brain. Also found in the peripheral nervous system, where it causes gut muscles to relax and the heart to beat faster	Related to epinephrine and acts at some of the same receptors
Dopamine	A neurotransmitter of the central nervous system	Involved in schizophrenia. Loss of dopamine neurons is the cause of Parkinson's disease
Histamine	A minor neurotransmitter in the brain	Thought to be involved in maintaining wakefulness
Serotonin	A neurotransmitter of the central nervous system that is involved in many systems, including pain control, sleep/wake control, and mood	Certain medications that elevate mood and counter anxiety act by increasing serotonin levels
Purines		
ATP	Co-released with many neurotransmitters	Large family of receptors may shape postsynaptic responses to classical neurotransmitters
Adenosine	Transported across cell membranes; not synaptically released	Largely inhibitory effects on postsynaptic cells
Amino acids		
Glutamate	The most common excitatory neurotransmitter in the central nervous system	Some people have reactions to the food additive monosodium glutamate because it can affect the nervous system
Glycine Gamma-aminobutyric acid (GABA)	Common inhibitory neurotransmitters	Drugs called benzodiazepines, used to reduce anxiety and produce sedation, mimic the actions of GABA
Peptides		
Endorphins Enkephalins Substance P	Used by certain sensory nerves, especially in pain pathways	Receptors are activated by narcotic drugs: opium, morphine, heroin, codeine
Gas		
Nitric oxide	Widely distributed in the nervous system	Not a classic neurotransmitter, it diffuses across membranes rather than being released synaptically. A means whereby a postsynaptic cell can influence a presynaptic cell

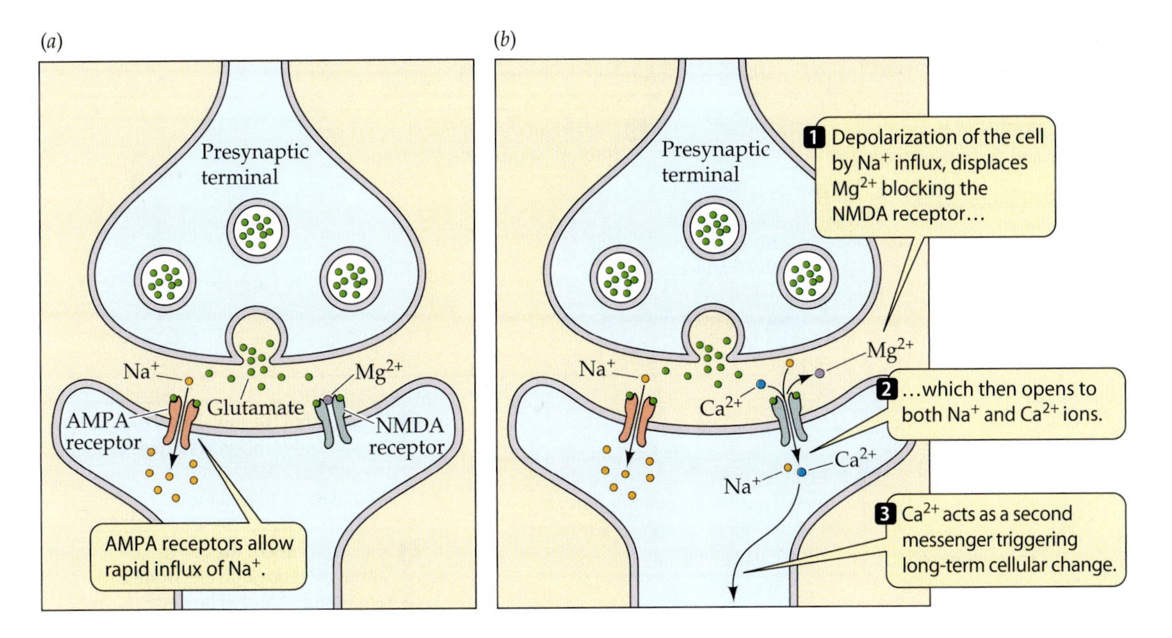

44.19 Two Ionotropic Glutamate Receptors
(*a*) AMPA receptors allow rapid influx of Na^+ into the postsynaptic cell. (*b*) NMDA receptors allow both Na^+ and Ca^{2+} to enter the cell.

fected by a phenomenon called **long-term potentiation**, or **LTP**, that was discovered by neurobiologists working with slices of brain that they kept alive in dishes of culture medium. Using these brain slice preparations, it is possible to stimulate and record from specific brain regions, or even specific cells.

In the studies leading to the discovery of LTP, experimenters repeatedly stimulated the synaptic inputs to a particular neuron and observed the usual action potential response. When the neuron was stimulated repeatedly in rapid succession, however, they found that the properties of the responding neuron changed. The size of the postsynaptic response was enhanced, or *potentiated*, and this change lasted for days or weeks (Figure 44.20).

How does this potentiation occur? The answer in some areas of the brain now seems quite clear. With low levels of stimulation, the glutamate released by presynaptic cells activates only the AMPA receptors, and the postsynaptic cell simply responds with action potentials. With higher levels of stimulation, however, the NMDA receptor is activated, allowing both Na^+ and Ca^{2+} ions to enter the postsynaptic neuron. The Ca^{2+} ions induce long-term changes in the postsynaptic neuron that make it more sensitive to synaptic input.

Exploiting the LTP system, Dr. Joe Tsien and his students and collaborators at Princeton University have made mice smarter by increasing the ability of their NMDA receptors to induce long-term changes in synaptic transmission. The experimenters genetically engineered mice so that their NMDA receptors had a slightly altered structure and were activated for a longer time whenever they bound a molecule of glutamate. The mice with these altered NMDA receptors learned tasks better, ran mazes faster, and remembered the mazes longer than normal mice. Maybe the world does not need smarter mice, but these exciting experiments confirm that we are on the right track to understanding how the brain achieves learning and memory.

To turn off responses, synapses must be cleared of neurotransmitter

Turning off the action of neurotransmitters is as important as turning it on. If released neurotransmitter molecules simply remained in the synaptic cleft, the postsynaptic membrane would become saturated with neurotransmitter, and receptors would be constantly bound. As a result, the postsynaptic cell would remain hyperpolarized or depolarized and would be unresponsive to short-term changes in the presynaptic cell. The more discrete each separate neural signal is, the more information can be processed in a given time. Thus neurotransmitter must be cleared from the synaptic cleft shortly after it is released by the axon terminal.

Neurotransmitter action may be terminated in several ways. First, enzymes may destroy the neurotransmitter. As we have seen, acetylcholine is rapidly destroyed by the enzyme acetylcholinesterase, which is present in the synaptic cleft in close association with the acetylcholine receptors on the postsynaptic membrane. Some of the most deadly nerve gases that were developed for chemical warfare work by inhibiting acetylcholinesterase. As a result, acetylcholine lingers in the synaptic clefts, causing the victim to die of spastic (contracted) muscle paralysis. Some agricultural insecticides, such as malathion, also inhibit acetylcholinesterase and can poison farm workers if used without safety precautions.

Second, neurotransmitter may simply diffuse away from the cleft. Third, neurotransmitter may be taken up via ac-

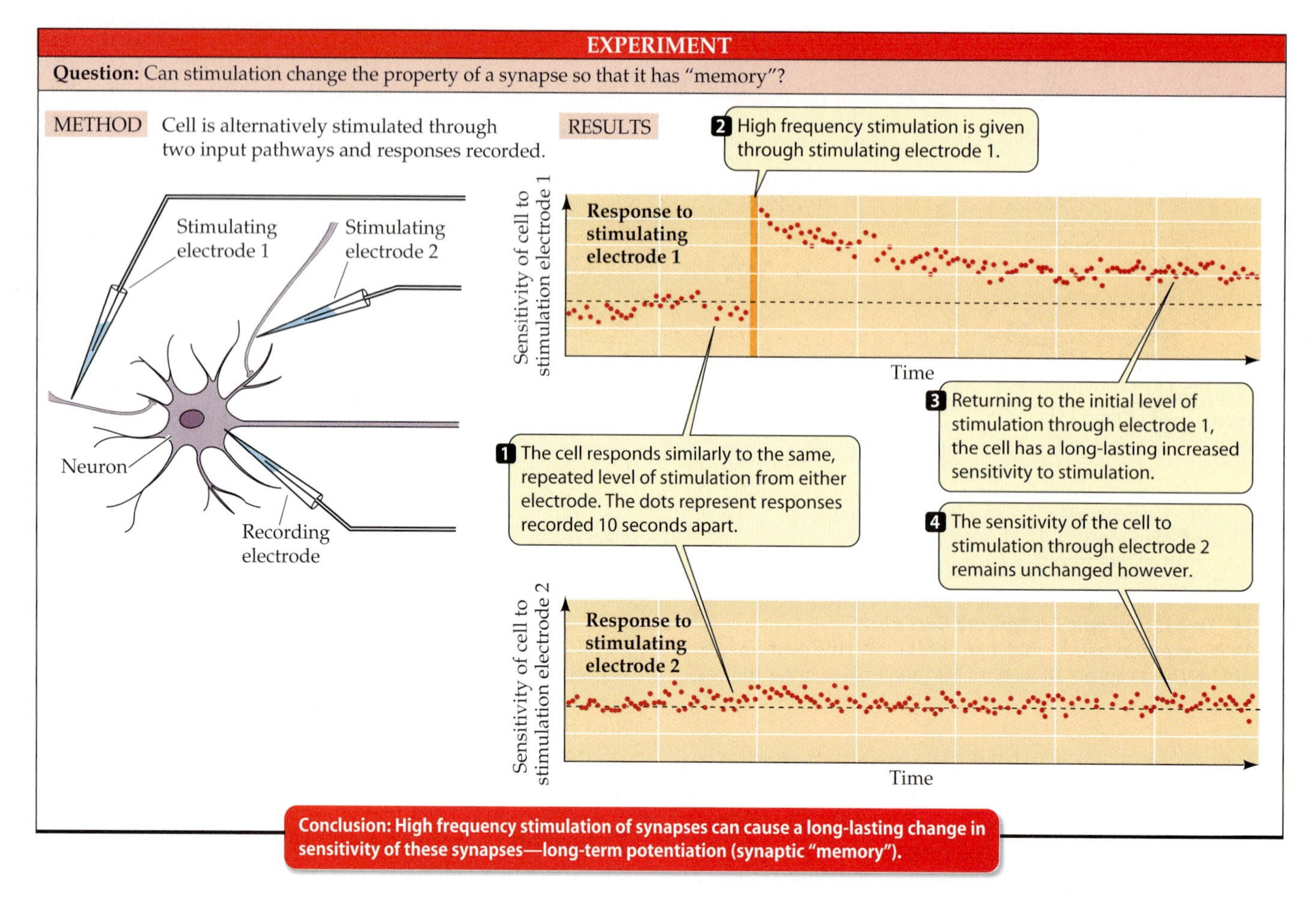

44.20 Repeated Stimulation Can Cause Long-Term Potentiation
When a cell receives regular synaptic input, the resulting postsynaptic potential remains constant. If, however, that same synaptic pathway is stimulated briefly at a high frequency, the subsequent sensitivity of the postsynaptic cell to the original level of synaptic input is potentiated for a long time.

tive transport by nearby cell membranes. The mode of action of Prozac, a commonly prescribed drug for depression, is to *slow* the reuptake of the neurotransmitter serotonin, thus enhancing its activity at the synapse.

Neurons in Networks

Because neurons can interact in the complex ways we have just discussed, networks of neurons can process and integrate information. Multiple neuronal networks constitute the nervous systems of animals. In subsequent chapters, we will see many examples of how neurons work together in networks to accomplish specific tasks. These networks use all of the mechanisms we have discussed: excitatory and inhibitory synapses, presynaptic excitation and inhibition, and mechanisms of long-term potentiation (and depression). Through these operations, our brains solve puzzles, create inventions, remember experiences, fall in love, and learn about biology. The challenge for the future is to understand how these networks work.

Chapter Summary

Nervous Systems: Cells and Functions

▶ Nervous systems consist of cells that process and transmit information.

▶ Sensory cells transduce information from the environment and the body and communicate commands to effectors such as muscles or glands.

▶ The nervous systems of different species vary, but all are composed of cells called neurons. **Review Figures 44.1, 44.2**

▶ In vertebrates, the brain and spinal cord form the central nervous system, which communicates with other tissues of the body via the peripheral nervous system.

▶ Neurons receive information mostly via their dendrites and transmit information over their axons. Neurons function in networks. **Review Figure 44.3**

▶ The information that neurons process is in the form of electrical events in their plasma membranes. Where neurons and other cells meet, information is transmitted between them, mostly by the release of chemical signals called neurotransmitters.

▶ Glial cells physically support neurons and perform many housekeeping functions. Schwann cells and oligodendrocytes produce myelin, which insulates neurons. Astrocytes create the blood–brain barrier. **Review Figure 44.4**

Neurons: Generating and Conducting Nerve Impulses

▶ Neurons have an electric charge difference across their plasma membranes. This resting potential is created by ion pumps and ion channels. **Review Figure 44.5**

▶ The sodium–potassium pump concentrates K^+ on the inside of neurons and Na^+ on the outsides. Ion channels allow K^+ to leak out, leaving behind unbalanced negative charges and leading to the resting potential. **Review Figures 44.6, 44.7**

▶ A potassium equilibrium potential exists when the electric charge that develops across the membrane is sufficient to prevent a net diffusion of K^+ down its concentration gradient. This potential can be calculated with the Nernst equation. **Review Figure 44.8**

▶ The resting potential is perturbed when ion channels open or close, thus changing the permeability of the plasma membrane to charged ions. Through this mechanism, neurons become depolarized or hyperpolarized in response to stimuli. **Review Figure 44.9**

▶ Rapid reversals in charge across portions of the plasma membrane resulting from the opening and closing of voltage-gated sodium and potassium channels produce action potentials. These changes in voltage-gated channels occur when the plasma membrane depolarizes to a threshold level. **Review Figure 44.10**

▶ Action potentials are conducted down axons because of local current flow, which depolarizes adjacent regions of membrane and brings them to threshold for the opening of voltage-gated sodium channels. **Review Figure 44.11**

▶ Patch clamping allows us to study single ion channels. **Review Figure 44.12**

▶ In myelinated axons, the action potentials appear to jump between nodes of Ranvier, patches of plasma membrane that are not covered by myelin. **Review Figure 44.13**

Neurons, Synapses, and Communication

▶ Neurons communicate with each other and with other cells at specialized junctions called synapses, where the plasma membranes of two cells come close together.

▶ The classic chemical synapse is the neuromuscular junction, a synapse between a motor neuron and a muscle cell. Its neurotransmitter is acetylcholine, which causes a depolarization of the postsynaptic membrane when it binds to its receptor. **Review Figure 44.14**

▶ When an action potential reaches an axon terminal of the presynaptic cell, it causes the release of neurotransmitters, chemical signals that diffuse across the synaptic cleft and bind to receptors on the postsynaptic membrane. **Review Figures 44.15, 44.16**

▶ Synapses between neurons can be either excitatory or inhibitory. Synapses can also form on presynaptic membranes and thereby influence the release of neurotransmitter by the presynaptic cell.

▶ A postsynaptic neuron integrates information by summing its synaptic inputs in both space and time. **Review Figure 44.17**

▶ Ionotropic neurotransmitter receptors are ion channels. Metabotropic receptors influence the postsynaptic cell through various signal transduction pathways that involve G proteins. These pathways can result in changes in ion channels, alterations of enzyme activity, and even gene expression. The actions of ionotropic synapses are generally faster than those of metabotropic synapses. **Review Figures 44.18, 44.19**

▶ Electrical synapses pass electric signals between cells without the use of neurotransmitters.

▶ There are many different neurotransmitters and even more receptors. The action of a neurotransmitter depends on the receptor to which it binds. **Review Table 44.1**

▶ Glutamate binds to both ionotropic and metabotropic receptors, and may be involved in learning and memory. **Review Figure 44.19**

▶ With repeated stimulation, a neuron can become more sensitive to its inputs. Since this increased sensitivity can last a long time, it is called long-term potentiation, or LTP. The properties of the NMDA glutamate receptor appear to explain LTP. **Review Figure 44.20**

▶ In chemical synapses, the transmitter must be cleared rapidly from the synapse. Some poisons and some drugs act by blocking or slowing the clearance of transmitter from the synapse.

Neurons in Networks

▶ Neurons work together in networks to accomplish specific tasks. These networks use all of the mechanisms we have discussed in this chapter.

For Discussion

1. The language of the nervous system consists of one "word," the action potential. How can this single message convey a diversity of information, how can that information be quantitative, and how can it be integrated?
2. If you stimulate an axon in the middle, action potentials are conducted in both directions. Yet when an action potential is generated at the axon hillock, it goes only *toward* the axon terminals and does not backtrack. Explain why action potentials are bidirectional in the first example and unidirectional in the second.
3. The nature of synapses presents various opportunities for plasticity in the nervous system. Discuss at least four synaptic mechanisms that could be altered to change the response of a neuron to a specific input.
4. If Dr. Tsien had genetically engineered the AMPA receptor to remain open longer when activated, would it have made his mice smarter? Why or why not?

45 Sensory Systems

Animals perceive the world through their senses. Different species look through different sensory windows, so their views of the world are not the same. Dogs, for example, do not see color well, but they have keener senses of hearing and smell than humans do. While you enjoy a beautiful sunset, your dog might be enjoying sniffing around in the bushes and listening for the sounds of small animals in the underbrush.

Human hunters have exploited the remarkable sensory abilities of dogs for thousands of years. Most recently that hunt has extended to illicit drugs, smuggled contraband, bombs, and firearms. Dogs can be trained to detect the signature odors of such items, so they are used by police, customs agents, and other investigators to identify those odors wherever suspicious activities are occurring.

A black Labrador named Charlie (badge K9-001) was the first dog trained by the U.S. Treasury Department's Bureau of Alcohol, Tobacco, and Firearms to sniff out firearms and explosives. Charlie has sniffed out more than 200 illegal guns and 500 pounds of hidden explosives. With a nose that outperforms electronic sensors, Charlie helped solve a terrorist bombing case by discovering a tiny fragment of the bomb hundreds of yards from the site of the explosion. Charlie's nose is never off duty; on a recreational visit to a U.S. Civil War battlefield, it smelled out cannonball fragments that had been buried for 130 years. ATF dogs receive a lot of expert training, but their careers are based on their remarkable sense of smell.

In this chapter, we look at the general properties of sensory cells and see how they convert environmental stimuli to neural information. Sensory cells are generally called *receptors*, which creates some confusion with the *receptor proteins* that bind signaling molecules. To avoid this confusion, we use the terms *sensory cells* or *receptor cells* in this chapter. We examine in detail the cells responsible for chemoreception, mechanoreception, and photoreception and see how they are incorporated into sensory systems that provide the CNS with information about the world around and within us. In the course of our study of sensory systems, we will learn about the unusual sensory abilities of many animals.

Special Agent K9-001
Charlie's remarkable sense of smell enables him and his partner to discover illicit firearms and explosives.

Sensory Cells, Sensory Organs, and Transduction

Sensory cells *transduce* (convert) physical or chemical stimuli into signals that are transmitted to other parts of the nervous system for processing and interpretation. Most sensory cells are modified neurons, but some are other types of cells closely associated with neurons. Sensory cells are specialized for detecting specific kinds of stimuli, such as pressure, heat, or light.

Most sensory cells possess a membrane receptor protein that detects a stimulus and responds by altering the flow of ions across the plasma membrane (Figure 45.1). The resulting change in membrane potential causes the sensory cell either to fire action potentials itself or to change its secretion of neurotransmitter onto an associated cell that fires action potentials. The intensity of the stimulus is encoded in the frequency of action potentials.

Sensation depends on which neurons in the CNS receive action potentials from sensory cells

If the messages derived from all sensory cells are the same—action potentials—how can we perceive different sensations? Sensations such as heat, itch, pressure, pain, light, smell, and sound differ because the messages from sensory cells arrive at different places in the central nervous

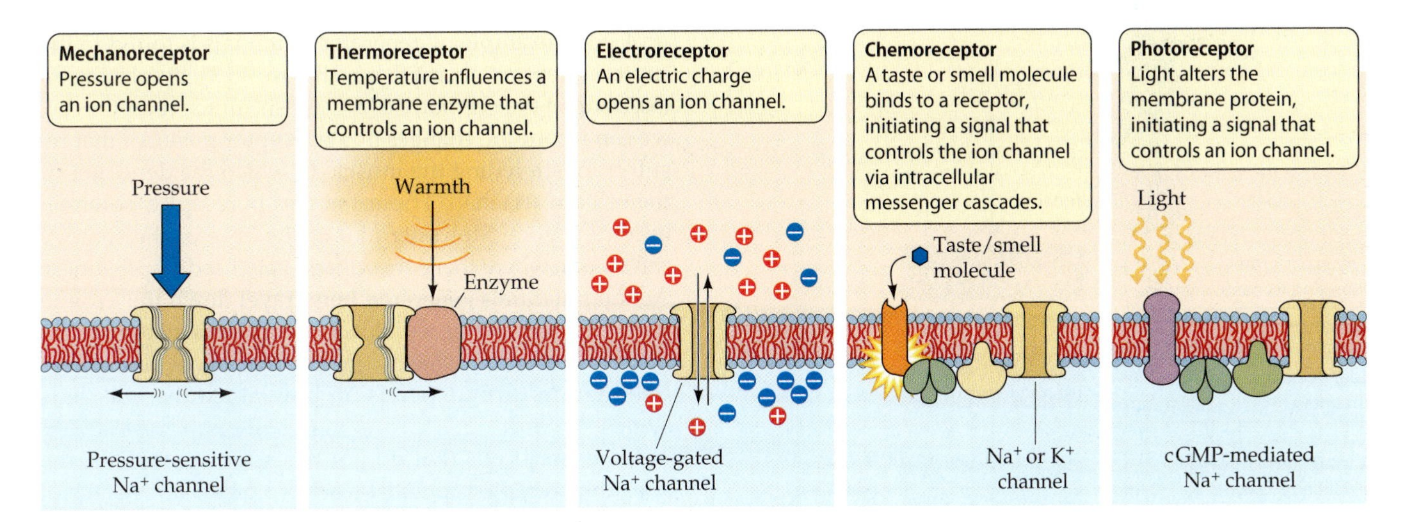

45.1 Sensory Cell Membrane Receptor Proteins Respond to Stimuli
The receptors in mechanoreceptors, thermoreceptors, and electroreceptors are themselves ion channels. The activated receptor proteins of chemoreceptors and photoreceptors initiate biochemical cascades that eventually open or close ion channels.

system (CNS). Action potentials arriving in the visual cortex are interpreted as light, in the auditory cortex as sound, in the olfactory bulb as smell, and so forth.

A small patch of skin on your arm contains sensory cells that increase their firing rates when the skin is warmed. Others increase their activity when the skin is cooled. Other types of sensory cells in the same patch of skin respond to touch, movement of hairs, irritants such as mosquito bites, and pain from cuts or burns. These sensory cells in your arm transmit their messages through axons that enter the CNS through the spinal cord. The synapses made by those axons in the spinal cord and the subsequent pathways of transmission determine whether the stimulation of the patch of skin on your arm is perceived as warmth, cold, pain, touch, itch, or tickle.

The specificity of these sensory circuits is dramatically illustrated in persons who have had a limb or part of a limb amputated. Although the sensory cells from that region are gone, the axons that communicated information from those sensory cells to the CNS may remain. If those axons are stimulated, the person feels specific sensations as if they were coming from the limb that is no longer there—a phantom limb.

The messages from some sensory cells communicate information about internal conditions in the body, but we may not be consciously aware of that information. The brain receives continuous information about body temperature, blood sugar, blood carbon dioxide and oxygen concentration, arterial pressure, muscle tension, and the position of the limbs. All this information is important for the maintenance of homeostasis, but we don't have to think about it. If we did, there would be no time to think about anything else. All sensory cells produce information that the nervous system can use, but that information does not always result in conscious sensation.

Sensory organs are specialized for detecting specific stimuli

Some sensory cells are assembled with other types of cells into *sensory organs*, such as eyes, ears, and noses, that enhance the ability of the sensory cells to collect, filter, and amplify stimuli. A photoreceptor cell, for example, detects electromagnetic radiation of only a particular range of wavelengths. This selectivity is the basis for color vision and explains why some insects can see ultraviolet light. In some simple organisms photoreceptors sense only the presence of light, but in more complex animals, photoreceptors are combined with other cell types into eyes. We'll learn how eyes collect light and focus it onto sheets of photoreceptors so that patterns of light can be detected.

Sensory transduction involves changes in membrane potentials

In this chapter we examine several sensory cell types and the sensory organs with which they are associated. In each case we can ask the same general question: How does the sensory cell transduce energy from a stimulus into action potentials? The details differ for different sensory cells, but those details all fit into a general pattern.

We have already seen the first step of sensory transduction in Figure 45.1: the activation of a receptor protein. A receptor protein in the plasma membrane of the sensory cell is activated by a specific stimulus. The activated receptor protein opens or closes ion channels in the membrane by one of several mechanisms. The receptor protein may be part of an ion channel, much like an ionotropic synaptic receptor, and by changing its conformation it may open or close the channel directly. Alternatively, the activated receptor protein may function like a metabotropic synaptic receptor coupled to a G protein, setting in motion intracellular events that eventually affect ion channels (see Figures 15.8 and 15.10). Figure 45.2 reviews these first steps of sensory transduction and outlines the subsequent steps.

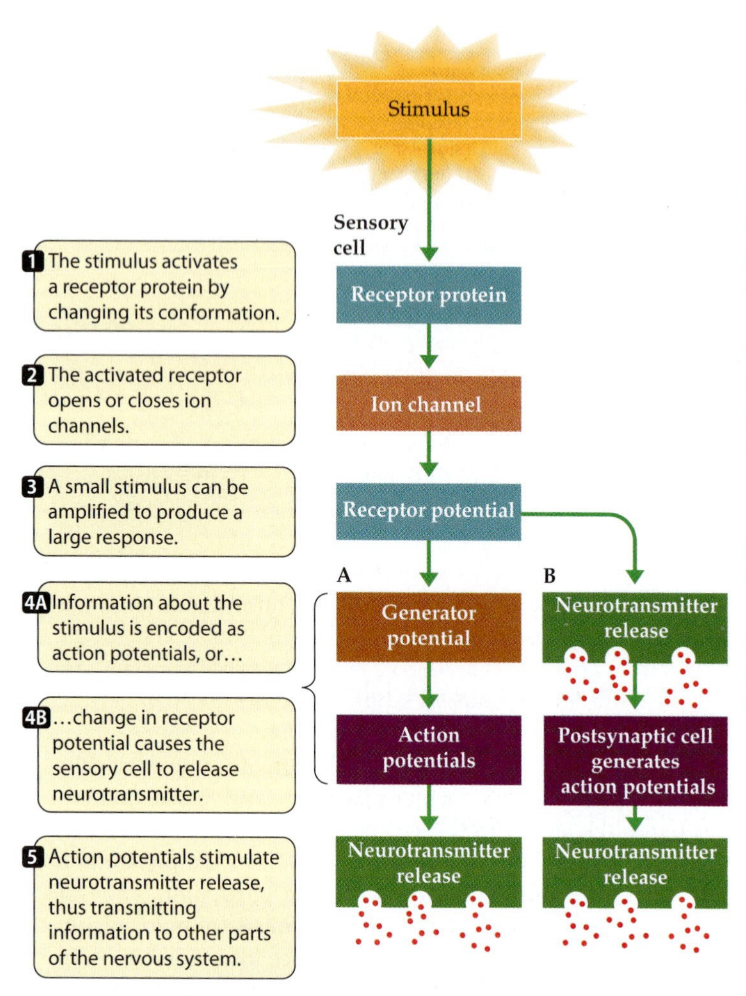

45.2 Sensory Transduction Is a Series of Steps
The sensory cell itself may generate action potentials as a result of stimulation, or it may vary its release of neurotransmitter in response to changes in its membrane potential, and the neurotransmitter may induce another cell to generate action potentials.

The opening or closing of ion channels in response to a stimulus changes the membrane potential of the sensory cell, which is called the **receptor potential**. Such changes in membrane potential can spread by current flow over short distances, but to travel long distances in the nervous system, receptor potentials must be converted into action potentials. The intracellular events involved in the conversion of the original stimulus-induced alteration of ion channels to action potentials can *amplify* the signal. In other words, the energy in the output of the sensory cell can be greater than the energy in the stimulus.

Receptor potentials produce action potentials in two ways: by generating action potentials within the sensory cell itself, or by releasing a neurotransmitter that induces an associated neuron to generate action potentials (see Figure 45.2). In the first case, the sensory cell has a region of plasma membrane with voltage-gated sodium channels. A receptor potential that spreads to this region is called a **generator potential** because it generates action potentials by causing the voltage-gated sodium channels to open.

A good example of generator potentials is found in the stretch receptors of crayfishes (Figure 45.3). By placing an electrode in the cell body of a crayfish stretch receptor cell, we can record the changes in the receptor potential that result from stretching the muscle to which the dendrites of the cell are attached. These changes in receptor potential become a generator potential at the base of the sensory cell's axon, where there are voltage-gated sodium channels. Action potentials generated here travel down the axon to the CNS. The rate at which action potentials are fired by the axon depends on the magnitude of the generator potential; that, in turn, depends on how much the muscle is stretched.

In sensory cells that do not fire action potentials, the spreading receptor potential reaches a presynaptic patch of plasma membrane and induces the release of a neurotransmitter.

Whether or not the sensory cell itself fires action potentials, ultimately the stimulus is transduced into action potentials and the intensity of the stimulus is encoded by the frequency of action potentials.

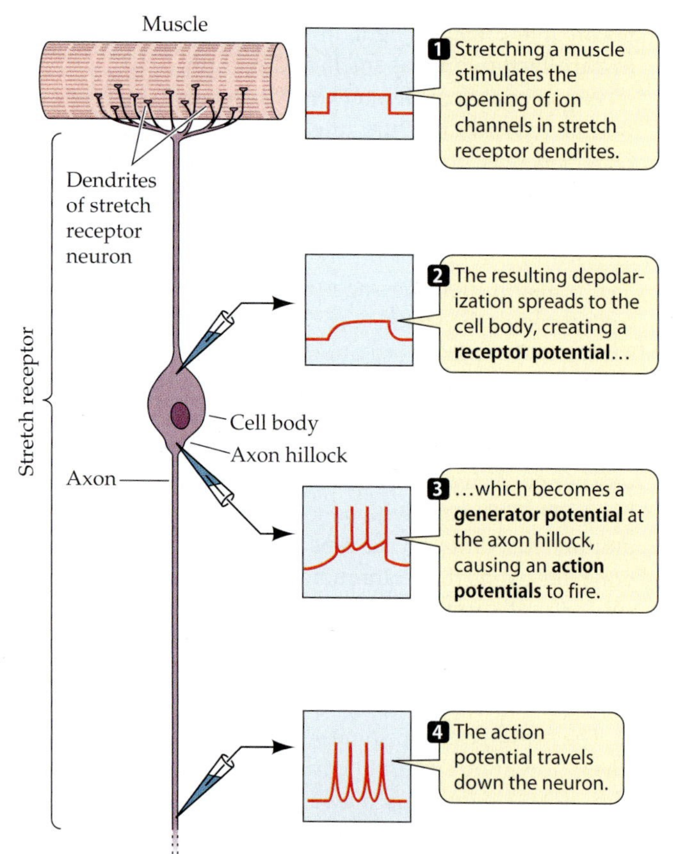

45.3 Stimulating a Sensory Cell Produces a Generator Potential
The stretch receptor of a crayfish produces a generator potential when the muscle is stretched. At the axon hillock, the receptor potential becomes a generator potential, firing action potentials that travel down the axon.

Many receptors adapt to repeated stimulation

Some sensory cells give gradually diminishing responses to maintained or repeated stimulation. This phenomenon is known as **adaptation**, and it enables an animal to ignore background or unchanging conditions while remaining sensitive to changes or to new information. (Note that this use of the term "adaptation" is different from its application in an evolutionary context.) When you dress, you feel each item of clothing touch your skin, but the sensation of clothes touching your skin is not constantly on your mind throughout the day. You are immediately aware, however, when a seam rips, your shoe comes untied, or someone lightly touches your back.

The ability of animals to discriminate between continuous and changing stimuli is due partly to the fact that some sensory cells adapt; it is also due to information processing by the CNS. Some sensory cells adapt very little or very slowly; examples are pain receptors and mechanoreceptors for balance.

In the rest of this chapter we will learn how sensory systems gather and filter stimuli, transduce specific stimuli into action potentials, and transmit action potentials to the CNS.

Chemoreceptors: Responding to Specific Molecules

Animals receive information about chemical stimuli through **chemoreceptors**. Chemoreceptors are responsible for smell, taste, and the monitoring of aspects of the internal environment such as the level of carbon dioxide in the blood. Chemosensitivity is universal among animals.

A colony of corals responds to a small amount of meat extract in the seawater around it by extending bodies and tentacles and searching for food. A solution of a single amino acid can stimulate this response. Humans have similar reactions to chemical stimuli. When we smell freshly baked bread, we salivate and feel hungry, but we gag and retch when we smell diamines from rotting meat. Information from chemoreceptors can cause powerful behavioral and physiological responses.

Arthropods provide good examples for studying chemosensation

Arthropods use chemical signals to attract mates. These signals, called *pheromones*, demonstrate the sensitivity of chemosensory systems. One of the best-studied examples of this phenomenon is the silkworm moth.

To attract a mate, the female silkworm moth releases a pheromone called bombykol from a gland at the tip of her abdomen. The male silkworm moth has receptors for this molecule on his antennae (Figure 45.4). Each feathery antenna carries about 10,000 bombykol-sensitive hairs. A single molecule of bombykol is sufficient to generate action potentials in the antennal nerve that transmits the signal to the CNS. Because of the male's high degree of sensitivity, the sexual message of a female moth is likely to reach any male that happens to be within a downwind area stretching over several kilometers. When approximately 200 hairs per second are activated, the male flies upwind in search of the female. Because the rate of firing in the male's sensory nerves is proportional to bombykol concentrations in the air, he can follow a concentration gradient and home in on the signaling female.

Many arthropods have chemoreceptor hairs, each containing one or more specific types of receptors. Crabs and flies, for example, have chemoreceptor hairs on their feet; these hairs have receptors for sugars, amino acids, salts, and other molecules. A fly tastes a potential food by stepping in it.

Olfaction is the sense of smell

The sense of smell, known as **olfaction**, also depends on chemoreceptors. In vertebrates, the olfactory sensors are neurons embedded in a layer of epithelial cells at the top of the nasal cavity. The axons of these neurons project to the olfactory bulb of the brain, and their dendrites end in olfac-

45.4 Some Scents Travel Great Distances
Mating in silkworm moths of the genus *Bombyx* is coordinated by a pheromone called bombykol.

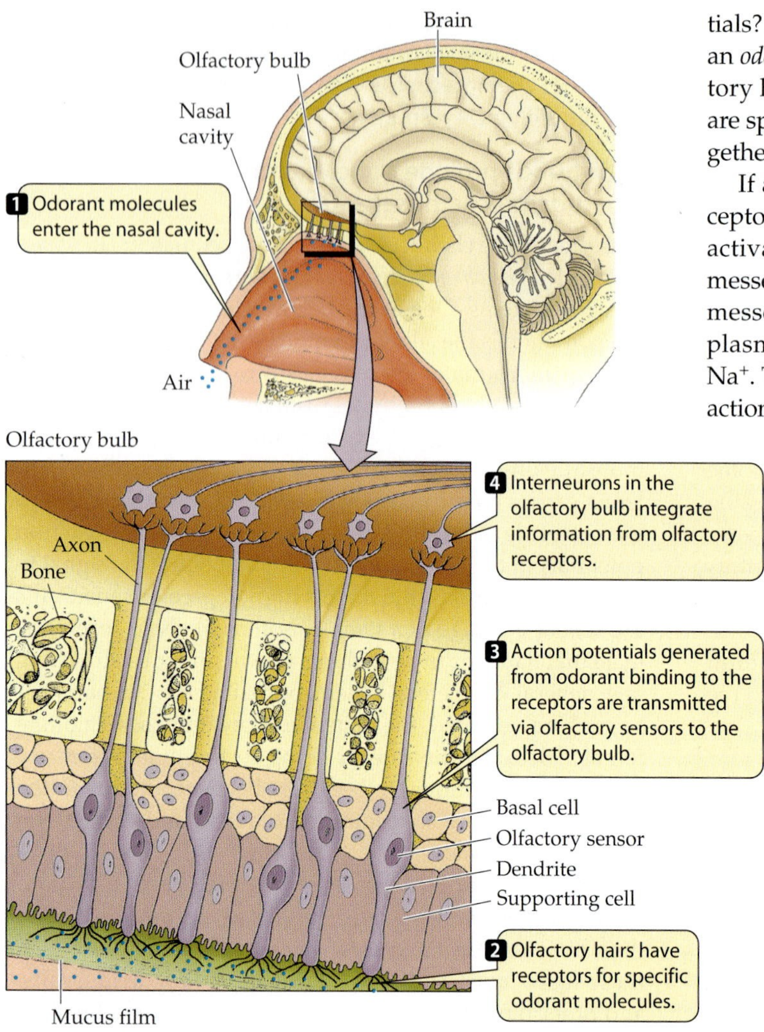

45.5 Olfactory Receptors Communicate Directly with the Brain
The receptors of the human olfactory system are embedded in tissues lining the nasal cavity and send their axons to the olfactory bulb of the brain.

tory hairs at the surface of the nasal epithelium. A protective layer of mucus covers the epithelium. Molecules from the environment must diffuse through this mucus to reach the receptor proteins on the olfactory hairs. When you have a cold, the amount of mucus increases and the epithelium swells. With this in mind, study Figure 45.5 and you will easily understand why respiratory infections can cause you to lose your sense of smell.

A dog has up to 40 million nerve endings per square centimeter of nasal epithelium, many more than we do. Humans have a fairly sensitive olfactory system, but we are unusual among mammals in that we depend more on vision than on olfaction (we tend to join bird-watching societies more often than mammal-smelling societies). Whales and porpoises have no olfactory receptors and hence no sense of smell.

How does an olfactory sensory cell transduce the structure of a molecule from the environment into action potentials? A molecule that triggers an olfactory receptor is called an *odorant*. Odorants bind to receptor proteins on the olfactory hairs of the sensory cells. Olfactory receptor proteins are specific for particular odorant molecules—the two fit together like a lock and key.

If a "key" (an odorant molecule) fits the "lock" (the receptor protein), then a G protein is activated, which in turn activates an enzyme that causes an increase of a second messenger in the cytoplasm of the sensory cell. The second messenger binds to sodium channels in the sensory cell's plasma membrane and opens them, causing an influx of Na^+. The sensory cell thus depolarizes to threshold and fires action potentials (see Figure 15.16).

The olfactory world has an enormous number of "keys"—molecules that produce distinct smells. The number of "locks"—receptor proteins—is large, but not nearly as large as the number of possible odorants. A family of about a thousand genes codes for olfactory receptor proteins. Each receptor protein is expressed in a limited number of sensory cells in the olfactory epithelium, and those cells all project to the same regions in the olfactory bulb. A given odorant molecule may bind to one or to more than one receptor protein. Therefore, each odorant molecule can excite a unique selection of cells in the olfactory bulb, so an olfactory system with a thousand different receptor proteins can discriminate a large number of smells.

How does the sensory cell signal the intensity of a smell? It responds in a graded fashion to the concentration of odorant molecules: The more odorant molecules that bind to receptors, the more action potentials are generated and the greater the intensity of the perceived smell.

Gustation is the sense of taste

The sense of taste, or **gustation**, in humans and other vertebrates depends on clusters of sensory cells called **taste buds**. The taste buds of terrestrial vertebrates are confined to the mouth cavity, but some fishes have taste buds in the skin that enhance their ability to sense their environment. Some fishes living in murky water are very sensitive to small amounts of amino acids in the water around them and can find food without the use of vision. The duck-billed platypus, a monotreme mammal (see Figure 33.22*b*), has similar talents as a result of taste buds on the sensitive skin of its bill. What is a taste bud and how does it work?

A taste bud is a cluster of sensory cells (Figure 45.6). A human tongue has approximately 10,000 taste buds. The taste buds are embedded in the epithelium of the tongue, and many are found on the raised papillae of the tongue. (Look at your tongue in a mirror—the papillae make it look fuzzy.) Each papilla has many taste buds. The outer surface of a taste bud has a pore that exposes the tips of the sensory cells. Microvilli (tiny hairlike projections) increase the surface area of the sensory cells where their tips converge at the pore. These sensory cells, unlike olfactory receptors, are not

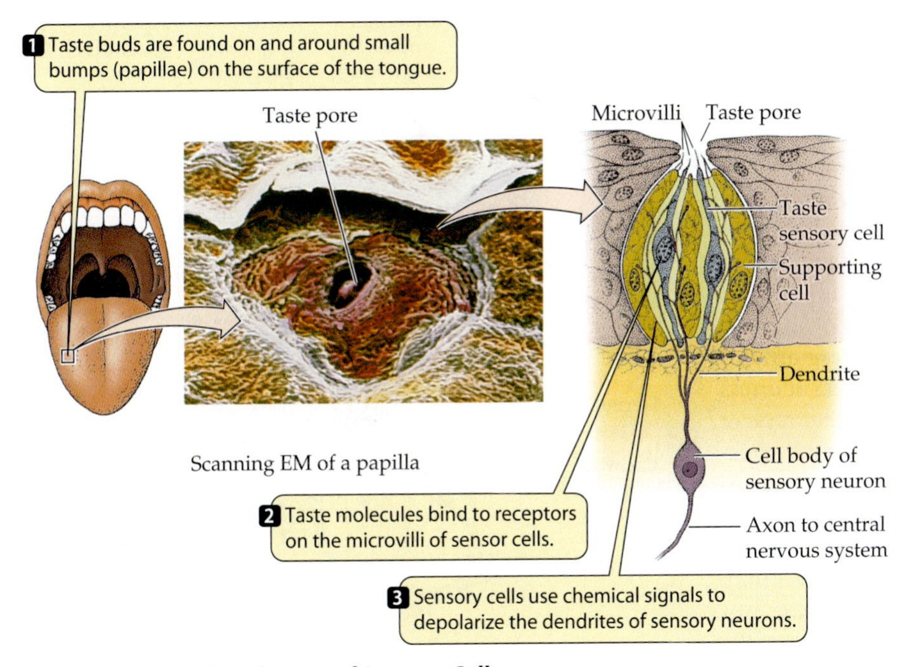

45.6 Taste Buds Are Clusters of Sensory Cells
Each taste bud contains a number of sensory cells that are not neurons.

neurons. At their bases, they form synapses with dendrites of sensory neurons.

Gustation begins with receptor proteins in the membranes of the microvilli. As with olfactory transduction, receptor proteins on the sensory cells bind specific molecules (such as sugar). This binding causes changes in the membrane potential of the sensory cells, which release neurotransmitters onto the dendrites of the sensory neurons. The sensory neurons fire action potentials that are conducted to the CNS.

The tongue does a lot of hard work, so its epithelium is shed and replaced at a rapid rate. Individual taste buds last only a few days before they are replaced, but the sensory neurons associated with them live on, always forming new synapses as new taste buds form.

You have probably heard that humans can perceive only four tastes: sweet, salty, sour, and bitter. In actuality, taste buds can distinguish among a variety of sweet-tasting molecules and a variety of bitter-tasting molecules. The full complexity of the chemosensitivity that enables us to enjoy the subtle flavors of food comes from the combined activation of gustatory and olfactory receptors; that is the reason you lose some of your sense of taste when you have a cold.

Why does a snake continually sample the air by darting its forked tongue in and out? The forks of the snake's tongue fit into cavities in the roof of its mouth that are richly endowed with olfactory receptors. The tongue samples the air and presents the sample directly to the olfactory receptors. Thus the snake is really using its tongue to smell its environment, not to taste it. Why doesn't the snake simply use the flow of air to and from its lungs, as we do, to smell the environment? In reptiles, air flows to and from the lungs slowly (and can even stop entirely for long periods of time), but the tongue can dart in and out many times in a second. It is a quick source of olfactory information.

Mechanoreceptors: Detecting Stimuli that Distort Membranes

Mechanoreceptors are cells that are sensitive to mechanical forces. In the skin, different kinds of mechanoreceptors are responsible for the perception of touch, pressure, and tickle. Stretch receptors in muscles, tendons, and joints provide information about the position of the parts of the body in space and the forces acting on them. Stretch receptors in the walls of blood vessels signal changes in blood pressure. "Hair" cells with extensions that are sensitive to bending are incorporated into mechanisms for hearing and for signaling the body's position in space.

Physical distortion of a mechanoreceptor's plasma membrane causes ion channels to open, altering the membrane potential of the cell, which in turn leads to the generation of action potentials. The rate of action potentials tells the CNS the strength of the stimulus exciting the mechanoreceptor.

Many different sensory cells respond to touch and pressure

Objects touching the skin generate varied sensations because skin is packed with diverse mechanoreceptors (Figure 45.7). The outer layers of skin contain whorls of nerve endings enclosed in connective tissue capsules. These very sensitive mechanoreceptors, called **Meissner's corpuscles**, respond to objects that touch the skin even lightly. Meissner's corpuscles adapt very rapidly. That is why you roll a small object between your fingers, rather than holding it still, to discern its shape and texture. As you roll it, you continue to stimulate these receptors anew.

Also in the outer regions of the skin are **expanded-tip tactile receptors** of various kinds. They differ from Meissner's corpuscles in that they adapt only partly and slowly. They are useful for providing steady-state information about objects that continue to touch the skin.

The density of these tactile mechanoreceptors varies across the surface of the body. A two-point discrimination test demonstrates this fact. If you lightly touch someone's back with two toothpicks, you can determine how far apart the two stimuli have to be before the person can distinguish whether he or she was touched with one or two points. On the back, the stimuli have to be rather far apart. The same test applied to the person's lips or fingertips reveals finer spatial discrimination; that is, the person can identify as separate two stimuli that are close together.

Deep in the skin, the dendrites of neurons wrap around hair follicles. When the hairs are displaced, those neurons are stimulated. Also deep within the skin is another type of mechanoreceptor, the **Pacinian corpuscle**. Pacinian corpuscles look like onions because they are made up of concentric layers of connective tissue, which encapsulate an extension of a sensory neuron. Pacinian corpuscles respond especially well to vibrations applied to the skin, but they adapt rapidly to steady pressure. The connective tissue capsule is important in this adaptation. An initial pressure distorts the corpuscle and the plasma membrane of the neuron at its core, but the layers of the capsule rapidly rearrange to redistribute the force, thus eliminating the distortion of the neuronal plasma membrane.

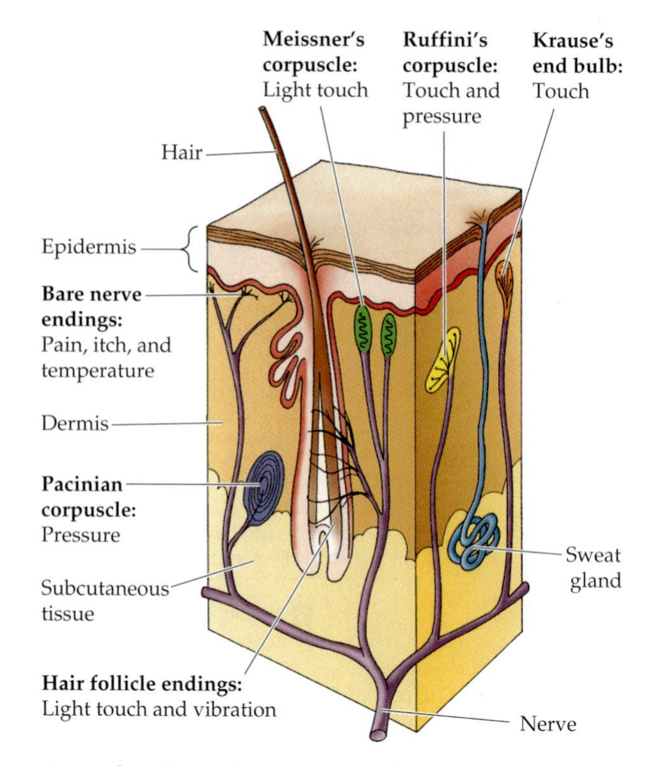

45.7 The Skin Feels Many Sensations
Even a very small patch of skin contains a diversity of sensory cells.

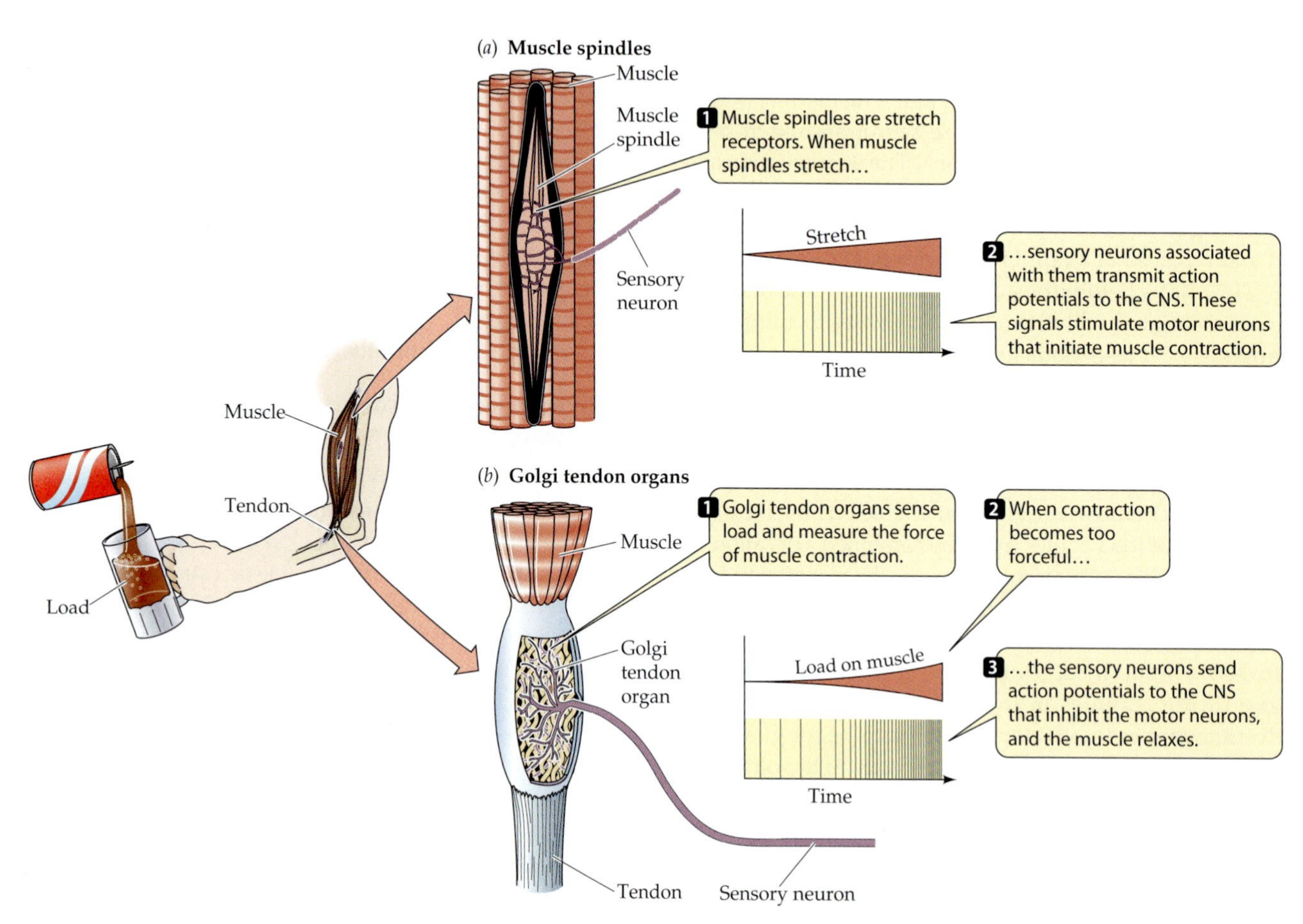

45.8 Stretch Receptors Are Activated When Limbs Are Stretched
Stretch receptors provide information about the stresses on muscles and joints in an animal's limbs. (*a*) Signals from muscle spindles to the CNS initiate muscle contraction. (*b*) Golgi tendon organs in tendons and ligaments inhibit a contraction that becomes too forceful, triggering relaxation and protecting the muscle from tearing.

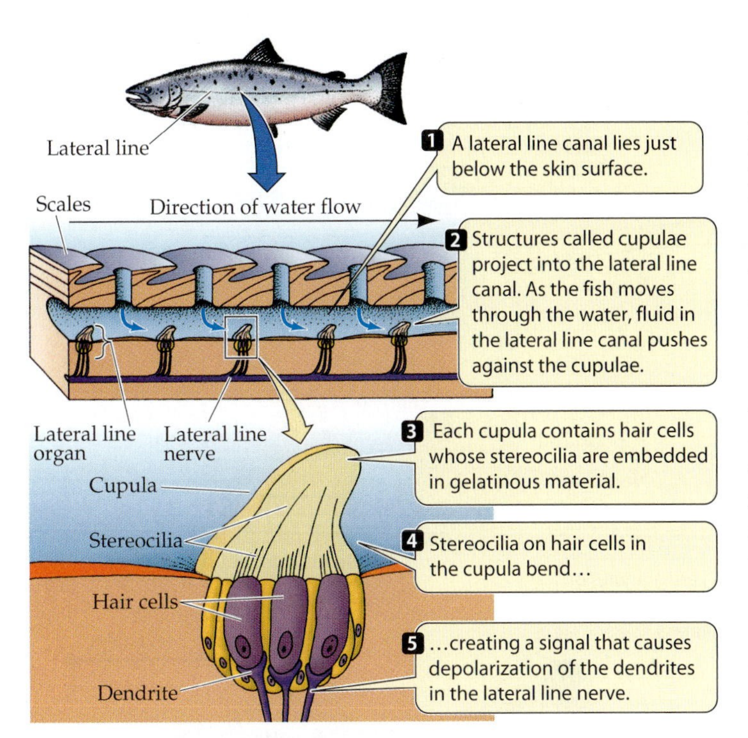

45.9 The Lateral Line System Contains Mechanoreceptors
Hair cells in the lateral line of a fish detect movement of the water around the animal, giving the fish information about its own movements and the movements of objects nearby.

Stretch receptors are found in muscles, tendons, and ligaments

An animal receives information from **stretch receptors** about the position of its limbs and the stresses on its muscles and joints. These mechanoreceptors are activated by being stretched. They continuously feed information to the CNS, and that information is essential for the coordination of movements.

The stretch receptors in skeletal muscle are called **muscle spindles**. They are embedded in connective tissue within skeletal muscle. They consist of modified muscle fibers that are innervated in the center with extensions of sensory neurons. Whenever the muscle stretches, muscle spindles are also stretched and the neurons transmit action potentials to the CNS (Figure 45.8*a*). Earlier in this chapter, we saw how crayfish stretch receptors transduce physical force into action potentials (see Figure 45.3). The actions of muscle spindles are similar.

Another stretch receptor, the **Golgi tendon organ**, is found in tendons and ligaments. It provides information about the force generated by a contracting muscle. When a contraction becomes too forceful, the information from the Golgi tendon organ feeds into the spinal cord, inhibits the motor neuron, and causes the contracting muscle to relax, thus protecting the muscle from tearing (Figure 45.8*b*).

Hair cells provide information about balance, orientation in space, and motion

Hair cells are also mechanoreceptors. Projecting from the surface of each hair cell is a set of **stereocilia**, which looks like a set of organ pipes. When these stereocilia (which are really microvilli) are bent, they alter receptor proteins in the hair cell's plasma membrane. When the stereocilia of some hair cells are bent in one direction, the receptor potential becomes more negative; when they are bent in the opposite direction, it becomes more positive. When the receptor potential becomes more positive, the hair cell releases neurotransmitter to the sensory neuron associated with it and the sensory neuron sends action potentials to the CNS.

Hair cells are found in the **lateral line** sensory system of fishes. The lateral line consists of a canal just under the surface of the skin that runs down each side of the fish (Figure 45.9). The lateral line provides information about movements of the fish through the water, as well as about moving objects, such as predators or prey, that cause pressure waves in the surrounding water.

Many invertebrates have equilibrium organs called **statocysts** that use hair cells to signal the position of the animal with respect to gravity (Figure 45.10). A statocyst is a chamber lined with hair cells that contains a dense object called a *statolith*. As the animal changes its position, the statolith moves in response to gravity, stimulating the hair cells below it. Replacing the statoliths of a lobster with iron filings and holding a magnet over the animal causes it to swim upside down. When a magnet is held to the lobster's side, it swims on its side.

Vertebrates also have equilibrium organs. The mammalian inner ear has two equilibrium organs that use hair cells to detect the position of the body with respect to gravity: semicircular canals and the vestibular apparatus. The

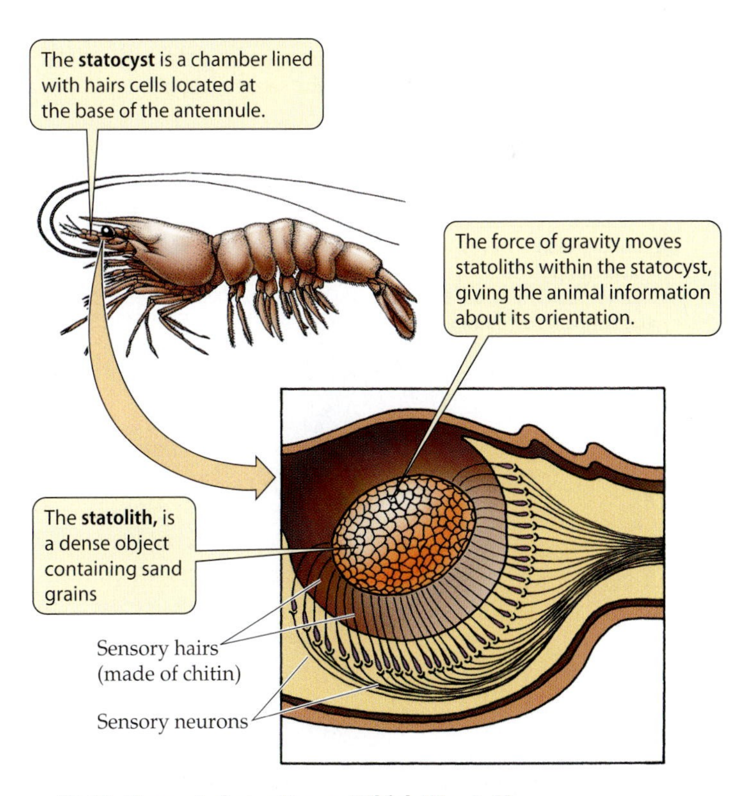

45.10 How a Lobster Knows Which Way Is Up
The statocyst is an equilibrium-sensing organ found in many invertebrates.

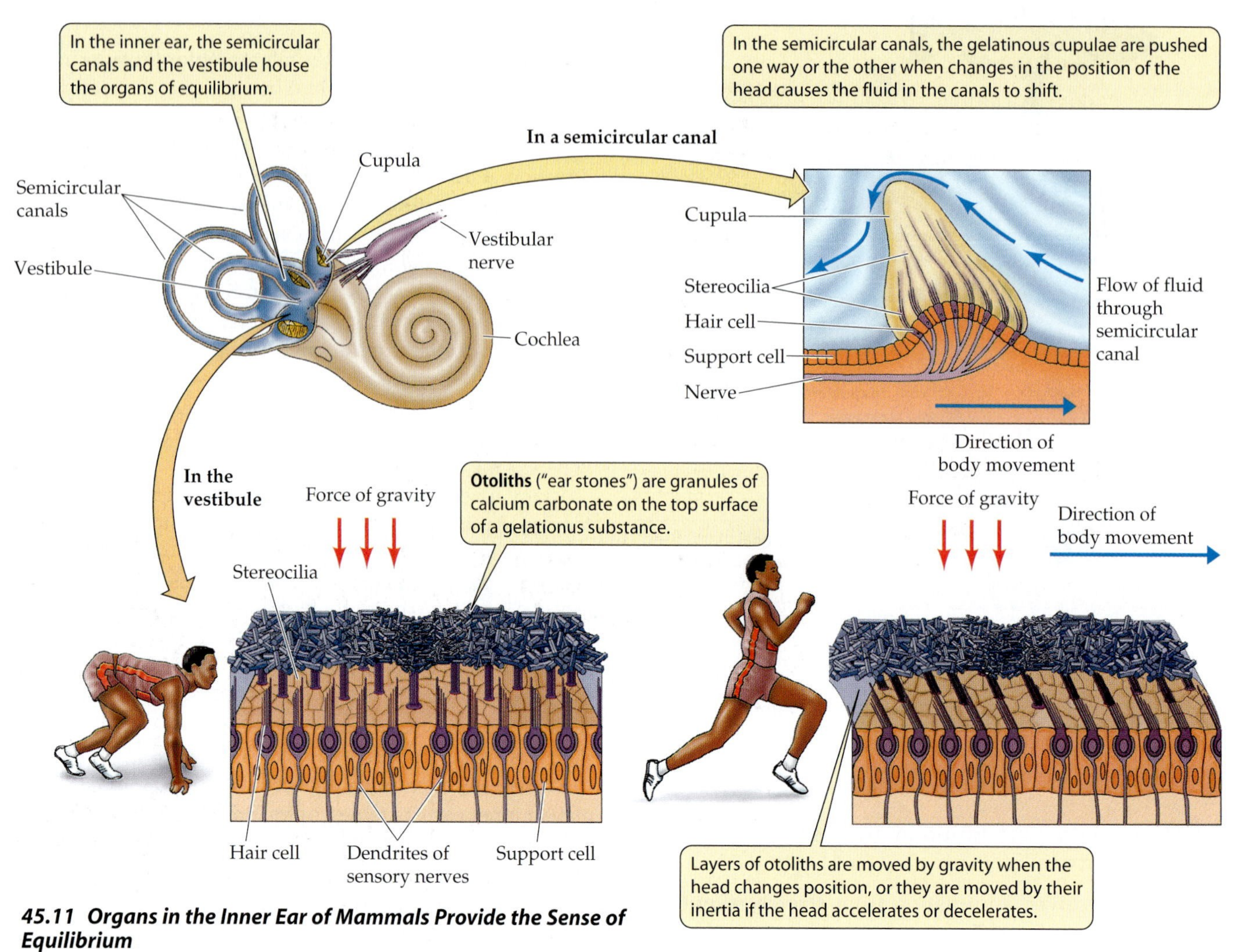

45.11 Organs in the Inner Ear of Mammals Provide the Sense of Equilibrium
The bony inner ear has three parts: the snail-shaped cochlea, the semicircular canals, and the vestibule. The cochlea is for hearing; the semicircular canals and the vestibular apparatus provide the sense of equilibrium.

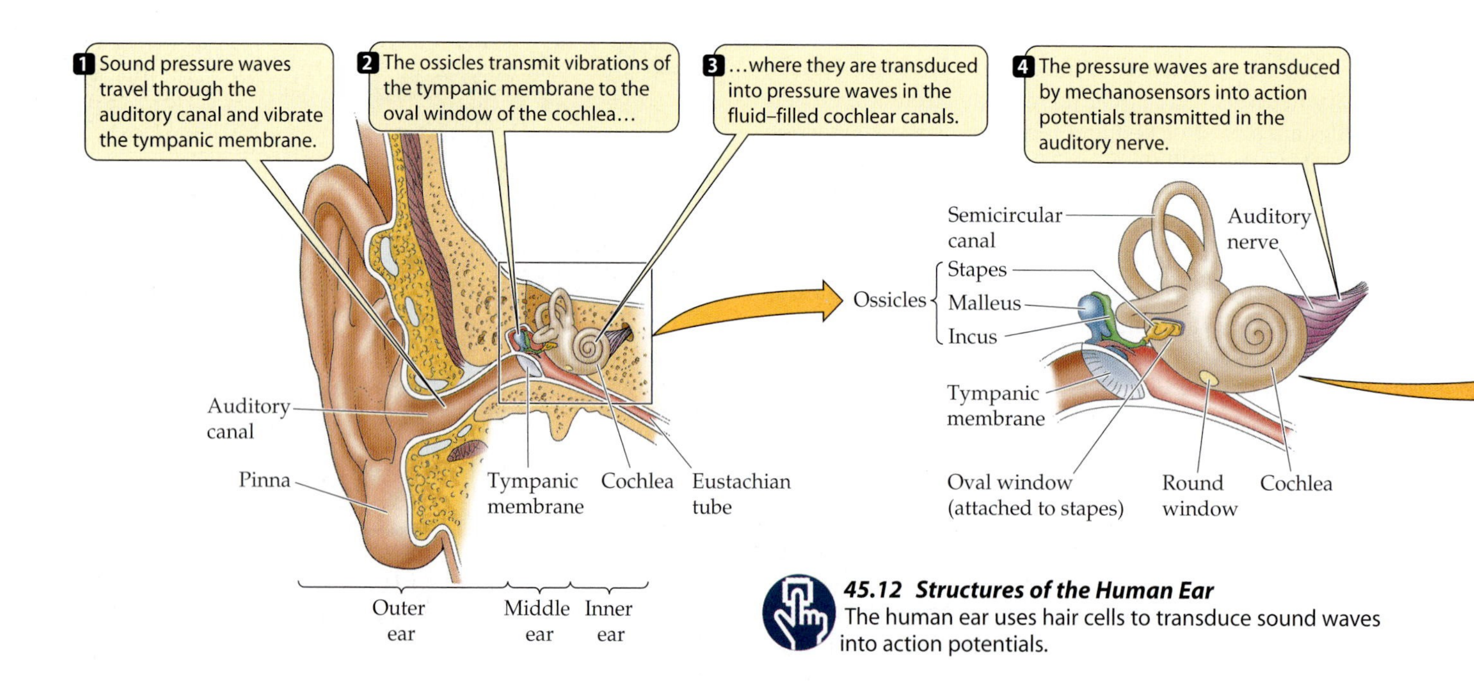

45.12 Structures of the Human Ear
The human ear uses hair cells to transduce sound waves into action potentials.

inner ear contains three **semicircular canals** at right angles to one another. The structure and function of these organs are described in Figure 45.11. The **vestibular apparatus** has two chambers that perform a function similar to that of the statocysts of invertebrates.

Auditory systems use hair cells to sense sound waves

The stimuli that animals perceive as sounds are pressure waves. **Auditory systems** use mechanoreceptors to transduce pressure waves into action potentials. Auditory systems include special structures to gather sound waves, direct them to the sensory organ, and amplify their effect on the mechanoreceptors.

Human hearing provides a good example of an auditory system. The organs of hearing are the ears. The two prominent structures on the sides of our heads usually thought of as ears are the **ear pinnae**. The pinna of an ear collects sound waves and directs them into the auditory canal, which leads to the actual hearing apparatus in the middle ear and the inner ear (Figure 45.12). If you have ever watched a rabbit, a horse, or a dog change the orientation of its ear pinnae to focus on a particular sound, then you have witnessed the role of ear pinnae in hearing.

The eardrum, or **tympanic membrane**, covers the end of the auditory canal. The tympanic membrane vibrates in response to pressure waves traveling down the auditory canal. The middle ear, an air-filled cavity, lies on the other side of the tympanic membrane.

The middle ear is open to the throat at the back of the mouth through the *eustachian tube*. Because air flows through the eustachian tube, pressure equilibrates between the middle ear and the outside world. When you have a cold or allergy, the tube can become blocked by mucus or by tissue swelling, so you have difficulty "clearing your ears," or equilibrating the pressure in the middle ear with the outside air pressure. As a result, the flexible tympanic membrane bulges in or out, dampening your hearing and sometimes causing earaches.

The middle ear contains three delicate bones called the **ear ossicles**, individually named the *malleus* (hammer), *incus* (anvil), and *stapes* (stirrup). The ossicles transmit the vibrations of the tympanic membrane to another flexible membrane called the **oval window**. The leverlike action of the ossicles amplifies the vibrations of the tympanic membrane about 20-fold in transmitting them to the oval window membrane. Behind the oval window lies the fluid-filled inner ear. Movements of the oval window result in pressure changes in the inner ear. These pressure waves are transduced into action potentials.

The inner ear is a long, tapered, coiled chamber called the **cochlea** (from Latin and Greek words for "snail" or "shell"). A cross section of this chamber reveals that it is composed of three parallel canals separated by two membranes: **Reissner's membrane** and the **basilar membrane** (see Figure 45.12). Sitting on the basilar membrane is the **organ of Corti**, the apparatus that transduces pressure waves into action potentials in the auditory nerve, which in turn conveys information from the ear to the brain. The organ of Corti contains hair cells whose stereocilia are in contact with an overhanging, rigid shelf called the **tectorial membrane**. Whenever the basilar membrane flexes, the tectorial membrane bends the hair cell stereocilia. As a consequence, the hair cells depolarize or hyperpolarize, altering the rate of action potentials transmitted to the brain by their associated sensory neurons.

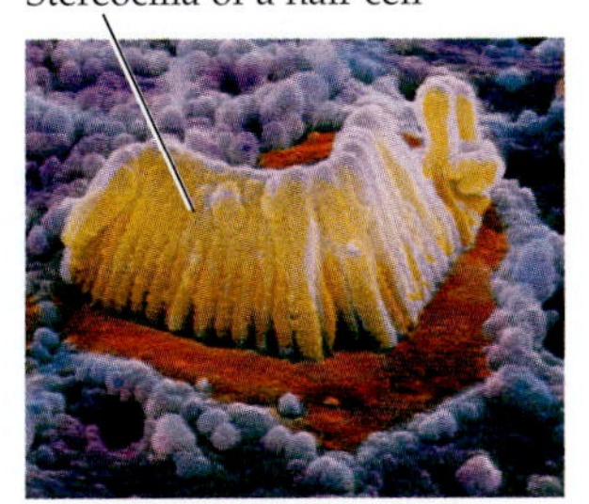

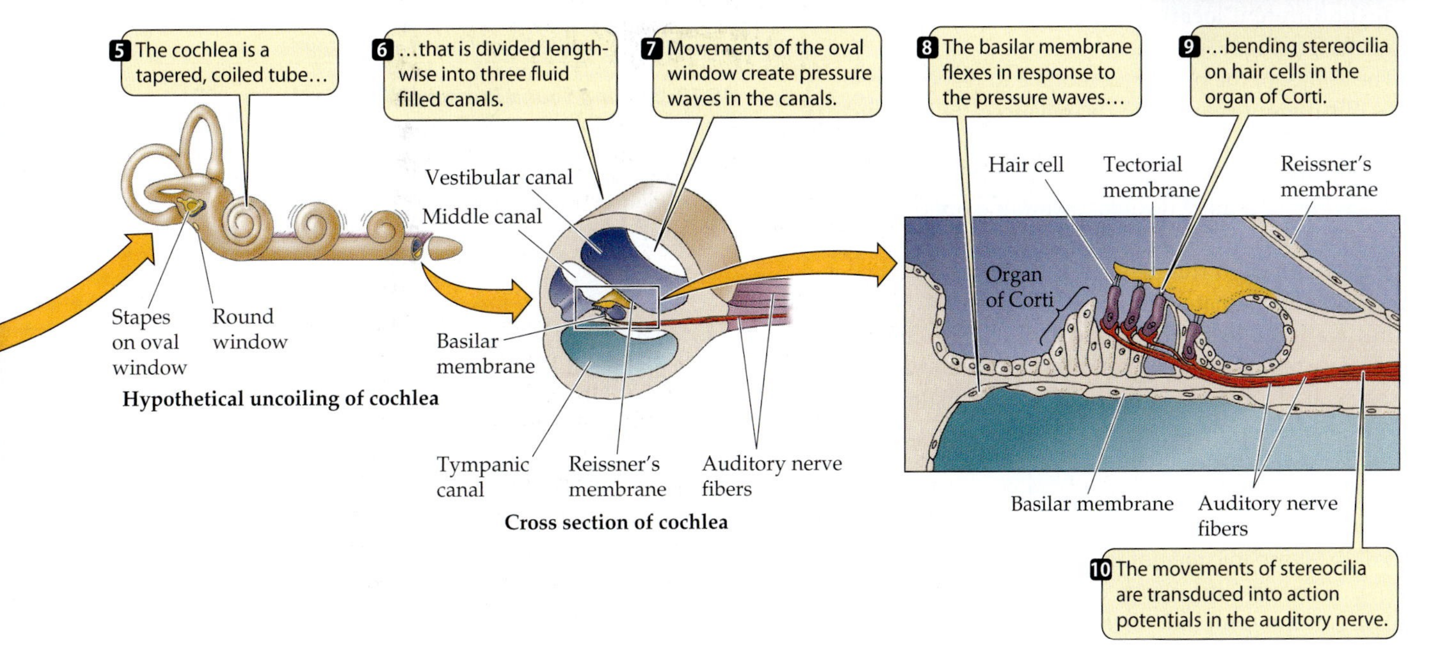

What causes the basilar membrane to flex, and how does this mechanism distinguish sounds of different frequencies? In Figure 45.13, the cochlea is shown uncoiled to make it easier to understand its structure and function. To simplify matters, we have left out Reissner's membrane, thus combining the upper and the middle canals into one upper canal. The purpose of Reissner's membrane is to contain a specific aqueous environment for the organ of Corti separate from the aqueous environment in the rest of the cochlea.

The simplified diagram of the cochlea shown in Figure 45.13 reveals two additional features that are important to its function. First, the upper and lower chambers separated by the basilar membrane are joined at the distal end of the cochlea (the end farthest from the oval window), making one continuous canal that folds back on itself. Second, just as the oval window is a flexible membrane at the beginning of the cochlea, the **round window** is a flexible membrane at the end of the long cochlear canal.

Air is highly compressible, but fluids are not. Therefore, a sound pressure wave can travel through air without much displacement of the air, but a sound pressure wave in fluid causes displacement of the fluid. When the stapes pushes the oval window in, the fluid in the upper canal of the cochlea is displaced. The cochlear fluid displacement travels down the upper canal, around the bend, and back through the lower canal. At the end of the lower canal, the displacement is absorbed by the outward bulging of the round window.

If the oval window vibrates in and out rapidly, the waves of fluid displacement do not have enough time to travel all the way to the end of the upper canal and back through the lower canal. Instead, they take a shortcut by crossing the basilar membrane, causing it to flex. The more rapid the vibration, the closer to the oval and round windows the wave of displacement will flex the basilar membrane. Thus different pitches of sound flex the basilar membrane at different locations and activate different sets of hair cells (see Figure 45.13).

The ability of the basilar membrane to respond to vibrations of different frequencies is enhanced by its structure. Near the oval and round windows, at the proximal end, the basilar membrane is narrow and stiff, but it gradually becomes wider and more flexible toward the opposite (distal) end. So it is easier for the proximal basilar membrane to resonate with high frequencies and for the distal basilar membrane to resonate with lower frequencies. A complex sound made up of many frequencies distorts the basilar membrane at many places simultaneously and activates a unique subset of hair cells. Action potentials stimulated by the mechanoreceptors at different positions along the organ of Corti travel to the brain stem along the auditory nerve.

Deafness, the loss of the sense of hearing, has two general causes. *Conduction deafness* is caused by the loss of function of the tympanic membrane and the ossicles of the middle ear. Repeated infections of the middle ear can cause scarring of the tympanic membrane and stiffening of the connections between the ossicles. The consequence is less efficient conduction of sound waves from the tympanic membrane to the oval window. With increasing age, the ossicles progressively stiffen, resulting in a gradual loss of the ability to hear high-frequency sounds. *Nerve deafness* is caused by damage to the inner ear or the auditory pathways. A common cause of nerve deafness is damage to the hair cells of the delicate organ of Corti by exposure to loud sounds such as jet engines, pneumatic drills, or highly amplified rock music. This damage is cumulative and permanent.

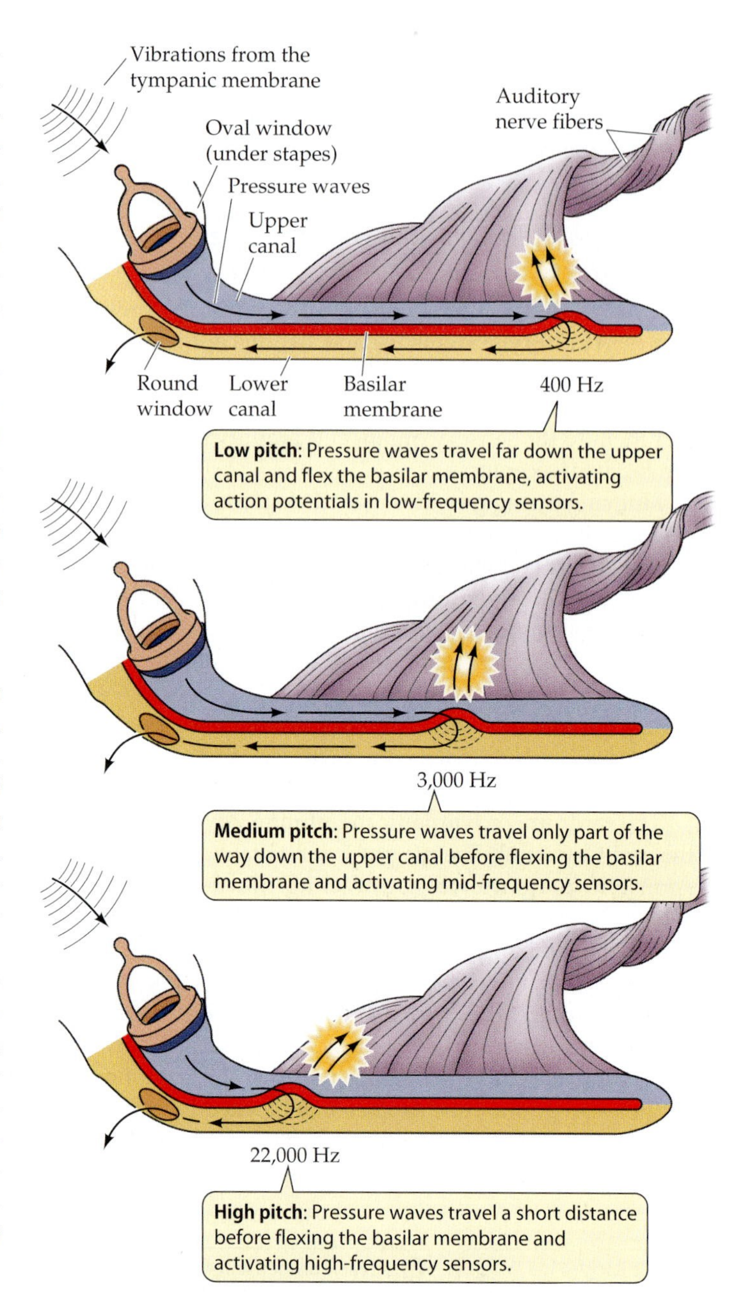

45.13 Sensing Sound Pressure Waves in the Inner Ear
For simplicity, this diagram illustrates the cochlea as uncoiled, and leaves out Reissner's membrane. Pressure waves of different frequencies flex the basilar membrane at different locations. Information about sound frequency is specified by which hair cells are activated.

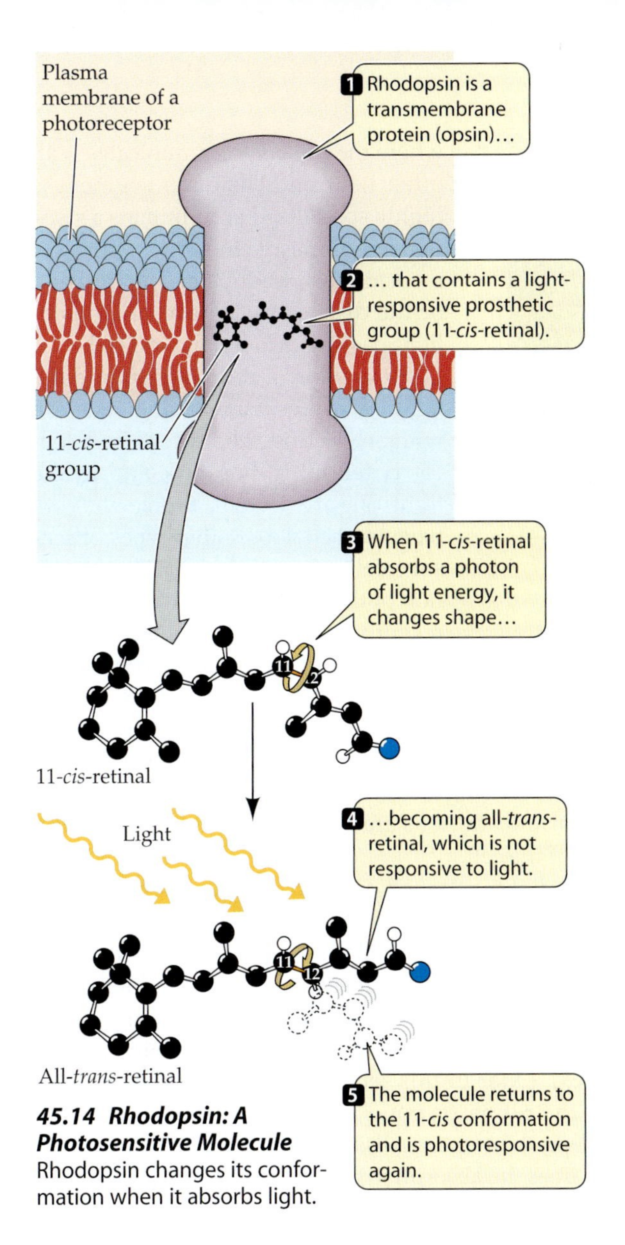

45.14 Rhodopsin: A Photosensitive Molecule
Rhodopsin changes its conformation when it absorbs light.

Photoreceptors and Visual Systems: Responding to Light

Sensitivity to light—**photosensitivity**—confers on the simplest animals the ability to orient to the sun and sky and gives more complex animals rapid and extremely detailed information about objects in their environment. It is not surprising that both simple and complex animals can sense and respond to light. What is remarkable is that across the entire range of animal species, evolution has conserved the same basis for photosensitivity: the family of pigments called **rhodopsins**.

In this section we will learn how rhodopsin molecules respond when stimulated by light energy and how that response is transduced into neural signals. We will also examine the structures of eyes, the organs that gather and focus light energy onto photoreceptor cells.

Rhodopsin is responsible for photosensitivity

Photosensitivity depends on the ability of rhodopsins to absorb photons of light and to undergo a change in conformation. A rhodopsin molecule consists of a protein, **opsin** (which by itself does not absorb light), and a light-absorbing prosthetic group, **11-*cis*-retinal**. The light-absorbing group is cradled in the center of the opsin and the entire rhodopsin molecule sits within the plasma membrane of a photoreceptor cell (Figure 45.14).

When the 11-*cis*-retinal absorbs a photon of light energy, its shape changes into a different isomer of retinal—all-*trans*-retinal. This change puts a strain on the bonds between retinal and opsin, changing the conformation of opsin. This change in conformation signals the detection of light. In vertebrate eyes, the retinal and the opsin eventually separate from each other—a process called *bleaching*, which causes the molecule to lose its photosensitivity. When the retinal spontaneously returns to its 11-*cis* isomer and recombines with opsin, it once again becomes photosensitive rhodopsin.

How does the light-induced conformational change of rhodopsin transduce light into a cellular response? After retinal converts from the 11-*cis* into the all-*trans* form, its interactions with opsin pass through several unstable intermediate stages. One of these stages is known as

45.15 The Rod Cell: A Vertebrate Photoreceptor
(*a*) The rod cell of the vertebrate retina is a neuron modified for photosensitivity. The membranes of a rod cell's discs are densely packed with rhodopsin. (*b*) A transmission electron micrograph of a section through a photoreceptor.

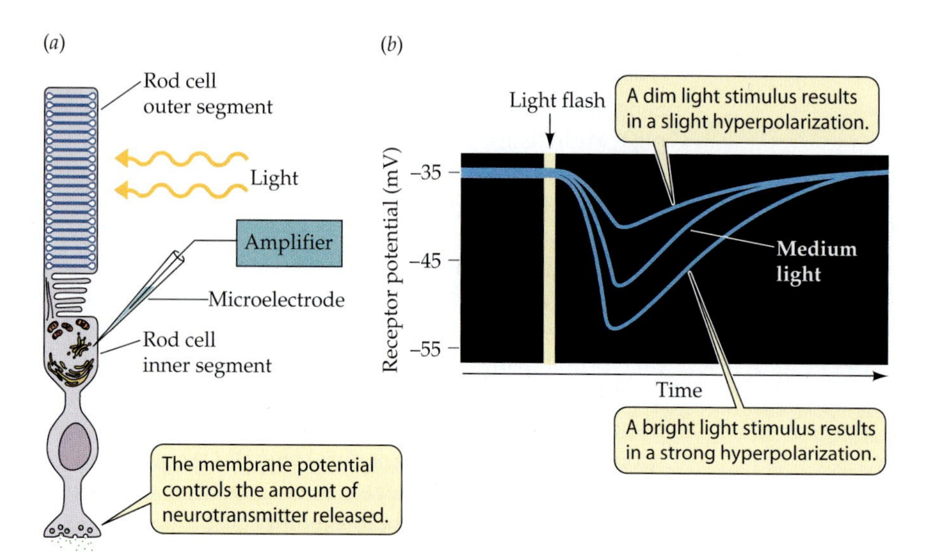

45.16 A Rod Cell Responds to Light
The plasma membrane of a rod cell hyperpolarizes—becomes more negative—in response to a flash of light.

photoexcited rhodopsin because it triggers a cascade of reactions that results in the alteration of membrane potential that is the photoreceptor cell's response to light.

To get a better idea of how rhodopsin alters the membrane potential of a photoreceptor cell and how that photoreceptor cell signals that it has been stimulated by light, let's look at a vertebrate photoreceptor cell, the **rod cell**, Like other vertebrate photoreceptor cells, the rod cell is a modified neuron (Figure 45.15). At the back of the vertebrate eye is the **retina**, which consists of several layers of neurons. One of these layers contains the photoreceptor cells. The other layers of the retina transduce the visual world into action potentials.

Each rod cell in the retina has an outer segment, an inner segment, and a synaptic terminal. The inner segment contains the usual organelles of a cell. The synaptic terminal is where the rod cell communicates with other neurons. The outer segment is highly specialized and contains a stack of discs of plasma membrane densely packed with rhodopsin. The function of the discs is to capture photons of light passing through the rod cell.

To see how a rod cell responds to light, we can penetrate a single rod cell with an electrode and record its receptor potential in the dark and in the light (Figure 45.16*a*). From what we have learned about other types of sensory cells, we might expect stimulation of the rod cell by light to make its receptor potential less negative. But photoreceptor cells are atypical, and the opposite is true. When a rod cell is kept in the dark, it already has a relatively depolarized resting potential in comparison with other neurons. In fact, the plasma membrane of the rod cell is almost as permeable to Na^+ ions as to K^+ ions, and Na^+ ions are continually entering the outer segment of the cell.

When a light is flashed on the dark-adapted rod cell, its receptor potential becomes more negative—it hyperpolarizes (Fig 45.16*b*). The rod cell itself does not generate action potentials. However, the rod cell changes its rate of neurotransmitter release as its membrane potential changes (since the rod cell hyperpolarizes, neurotransmitter release decreases). Later in this section we will learn how other cells in the retina respond to neurotransmitter released from the photoreceptor cells.

How does the absorption of light by rhodopsin hyperpolarize the rod cell? When rhodopsin is excited by light, it initiates a cascade of events. The photoexcited rhodopsin combines with and activates another protein, a G protein

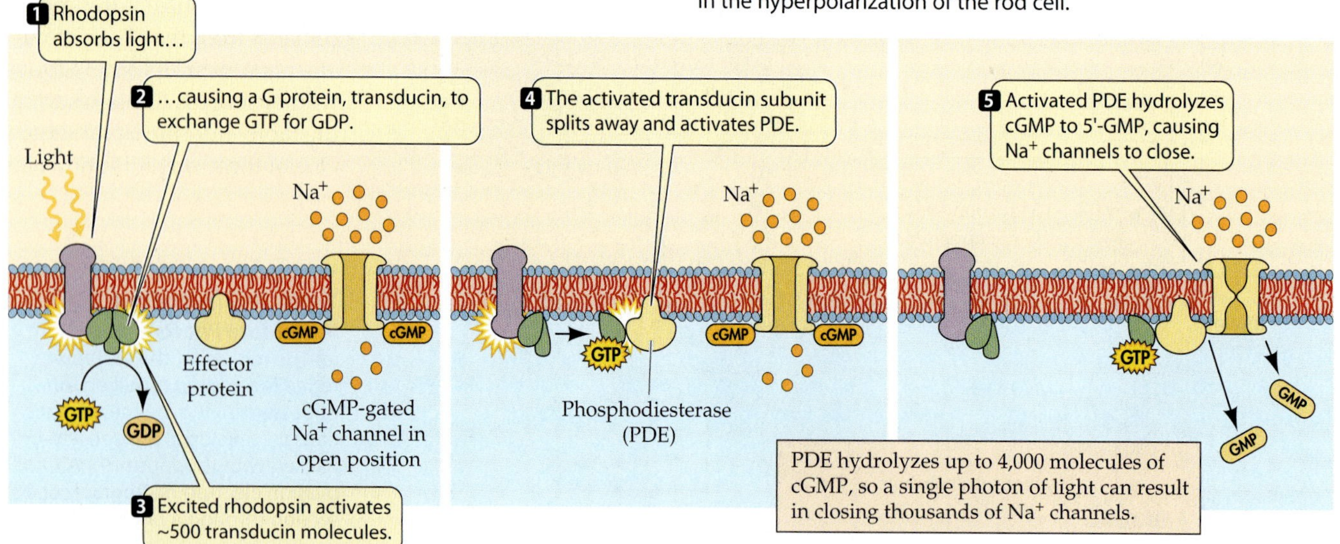

45.17 Light Absorption Closes Sodium Channels
The absorption of light by rhodopsin initiates a cascade resulting in the hyperpolarization of the rod cell.

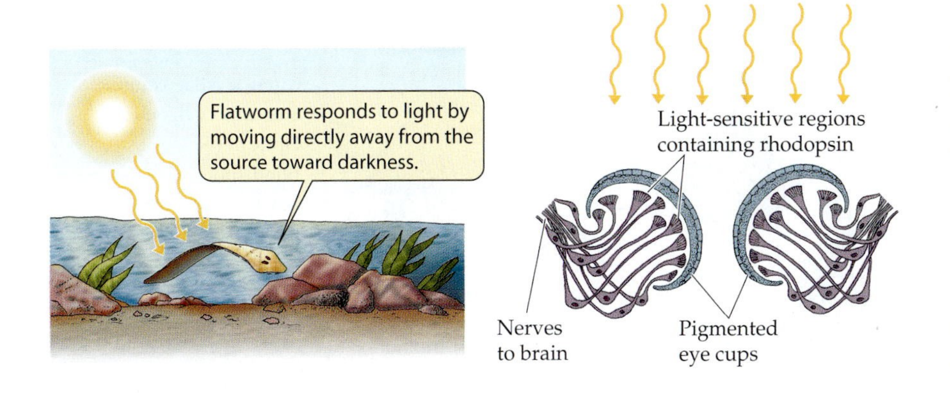

45.18 A Simple Photosensory System
Although flatworms do not "see" as we do, their eye cups enable them to move away from a light source to an area where they may be less visible to predators.

called *transducin*. Activated transducin in turn activates a phosphodiesterase, which converts cyclic GMP (cGMP) to 5′-GMP. This reaction plays a central role in phototransduction. In the dark, the cGMP in the outer segment binds to sodium channels, keeping them open and allowing Na^+ to enter the outer segment. As cGMP is converted to 5′-GMP, the sodium channels close, and the cell hyperpolarizes.

This may seem like a roundabout way of doing business, but its advantage is its enormous amplification ability. Each molecule of photoexcited rhodopsin can activate as many as 500 transducin molecules, thus activating a large number of phosphodiesterase molecules. The catalytic capacity of a molecule of phosphodiesterase is great: it can hydrolyze more than 4,000 molecules of cGMP per second. The bottom line is that a single photon of light can cause more than a million sodium channels to close, thereby changing the rod cell's receptor potential (Figure 45.17).

Now let's see how photoreceptors of different types are incorporated into different kinds of visual systems.

Invertebrates have a variety of visual systems

Flatworms obtain directional information about light from photoreceptor cells that are organized into *eye cups* (Figure 45.18). The eye cups are bilateral structures, each partly shielded from light by a layer of pigmented cells lining the cup. The photoreceptors on the two sides of the animal are unequally stimulated unless the animal is facing directly toward or away from a light source. The flatworm generally uses directional information about light sources to move away from light.

Arthropods (crustaceans, spiders, and insects) have evolved **compound eyes** that provide them with information about patterns or images in the environment. Each compound eye consists of many optical units called **ommatidia** (singular ommatidium) (Figure 45.19). The number of ommatidia in a compound eye varies from only a few in some ants, to 800 in fruit flies, to 10,000 in some dragonflies.

Each ommatidium has a lens structure that directs light onto photoreceptors called *retinula cells*. Flies, for example, have seven elongated retinula cells in each ommatidium. The inner borders of the retinula cells are covered with microvilli that contain rhodopsin and thus trap light. Since the microvilli of the different retinula cells overlap, they appear to form a central rod, called a *rhabdom*, down the center of the ommatidium.

Axons from the retinula cells communicate with the nervous system. Since each ommatidium of a compound eye is directed at a slightly different part of the visual world, only a crude, or perhaps a broken-up, image can be communicated from the compound eye to the CNS.

Image-forming eyes evolved independently in vertebrates and cephalopods

Both vertebrates and cephalopod mollusks have evolved eyes with exceptional abilities to form images of the visual world. Like cameras, these eyes focus images on a surface sensitive to light. Considering that they evolved independently of each other, their high degree of similarity is remarkable (Figure 45.20).

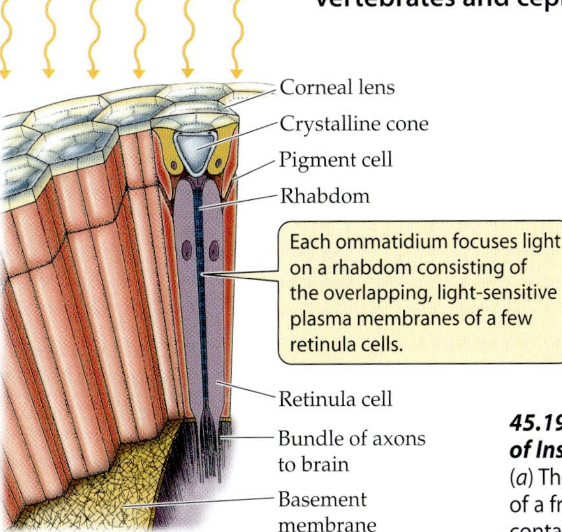

45.19 Ommatidia: The Functional Units of Insect Eyes
(*a*) The micrograph shows the compound eye of a fruit fly (*Drosophila*). (*b*) The rhodopsin-containing retinula cells are the photoreceptors in ommatidia.

45.20 Eyes Like Cameras The lenses of cephalopod and vertebrate eyes focus images on layers of photoreceptor cells, just as a camera's lens focuses images on film.

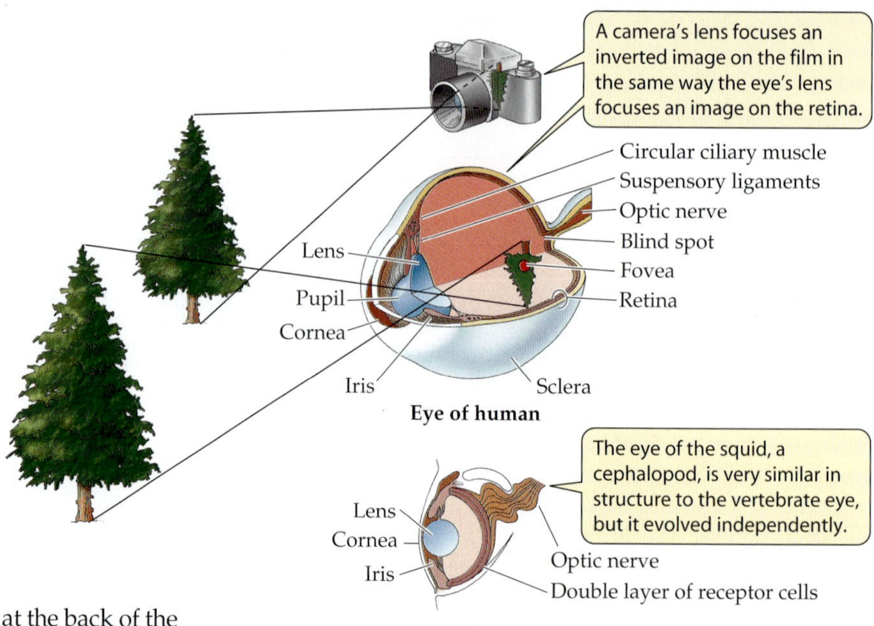

The vertebrate eye is a spherical, fluid-filled structure bounded by a tough connective tissue layer called the *sclera*. At the front of the eye, the sclera forms the transparent **cornea** through which light passes to enter the eye. Just inside the cornea is the pigmented **iris**, which gives the eye its color. The function of the iris is to control the amount of light that reaches the photoreceptor cells at the back of the eye, just as the diaphragm of a camera controls the amount of light reaching the film. The central opening of the iris is the **pupil**. The iris is under neural control. In bright light the iris constricts and the pupil is very small. As light levels fall, the iris relaxes and the pupil enlarges.

Behind the iris is the crystalline protein **lens**, which helps focus images on the photosensitive layer, the *retina*, at the back of the eye. The cornea and the fluids of the eye chambers are mostly responsible for focusing light on the retina, but the lens allows the eye to *accommodate*—that is, to focus on objects at various locations in the near visual field. To focus a camera on objects close at hand, you adjust the distance between the lens and the film. Fishes, amphibians, and reptiles accommodate in a similar manner, moving the lenses of their eyes closer to or farther from their retinas. Mammals and birds use a different method: they alter the shape of the lens.

The lens is contained in a connective tissue sheath that tends to keep it in a spherical shape, but it is attached to suspensory ligaments that pull it into a flatter shape. Circular muscles called the *ciliary muscles* counteract the pull of the suspensory ligaments and permit the lens to round up. With the ciliary muscles at rest, the flatter lens has the correct optical properties to focus distant images on the retina, but not close images. Contracting the ciliary muscles rounds up the lens, changing its light-bending properties to bring close images into focus (Figure 45.21). As we age, our lenses become less elastic and we lose the ability to focus on objects close at hand without the help of corrective lenses. As a consequence, most adults over the age of 45 need the assistance of bifocal lenses or reading glasses to compensate for their lost ability to accommodate.

The vertebrate retina receives and processes visual information

During development, neural tissue grows out from the brain to form the retina. In addition to a layer of photoreceptor cells, the retina includes layers of cells that process the visual information from the photoreceptors and produce an output signal that is transmitted to the brain via the optic nerve. A curious feature of the anatomy of the retina is that the light-absorbing outer segments of the photoreceptor cells are all the way at the back of the retina. Light must pass through all the layers of retinal cells before reaching the place where photons are captured by rhodopsin. We will examine in detail how the cells of the retina process information, but first let's describe some general features of retinal organization.

THE CELLULAR STRUCTURE OF THE RETINA. The density of photoreceptor cells is not the same across the entire retina. Light coming from the center of the visual field falls on an area of the retina called the **fovea**, where the density of photoreceptor cells is the highest. The human fovea has about 160,000 photoreceptors per square millimeter. A hawk has about 1 million photoreceptors per square millimeter of fovea, making its vision sharper than ours. In addition, the hawk has two foveas in each eye: one receives light from

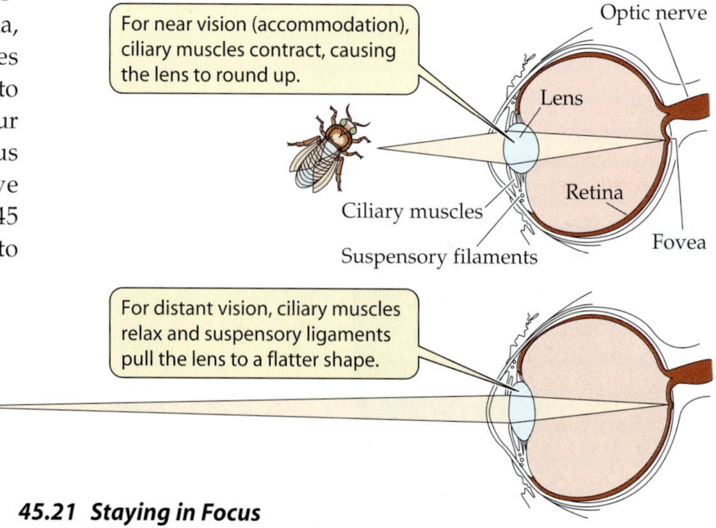

45.21 Staying in Focus Mammals and birds focus their eyes by changing the shape of the lens.

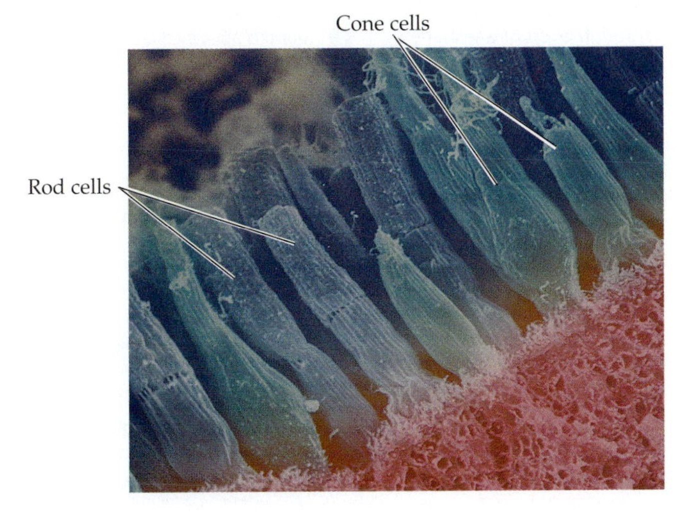

45.22 Rods and Cones
This scanning electron micrograph of photoreceptors in the retina of a mud puppy (an amphibian) shows cylindrical rods and tapered cones.

straight ahead, while the other receives light from below. Thus while the hawk is flying, it sees both its projected flight path and the ground below, where it might detect a mouse scurrying in the grass.

The fovea of a horse is a long, vertical patch of retina. The horse's lens is not good at accommodation, but it focuses distant objects that are straight ahead on one part of this long fovea and close objects that are below the head on another part. When horses are startled by an object close at hand, they pull their heads back and rear up to bring the object into focus on the close-vision part of the fovea.

Where blood vessels and the bundle of axons going to the brain (the optic nerve) pass through the back of the eye, there are no photoreceptors, so there is a blind spot on the retina. You are normally not aware of your blind spot, but you can find it. Stare straight ahead, holding a pencil in your outstretched hand so that the eraser is in the center of your field of vision. While continuing to stare straight ahead, slowly move the pencil to the side until the eraser disappears. When this happens, the light from the eraser is focused directly on your blind spot.

Until now we have referred to only one kind of photoreceptor cell, the rod cell. But there are two major kinds of vertebrate photoreceptors, both named for their shapes—rod cells and cone cells (Figure 45.22). A human retina has about 3 million cones and about 100 million rods. Rod cells are more sensitive to light, but do not contribute to color vision. **Cone cells** are responsible for color vision, but are less sensitive to light. Cones are also responsible for our sharpest vision. Even though there are many more rods than cones in human retinas, our foveas contain only cones.

Because cones have low sensitivity to light, they are of no use in dim light. At night our vision is not very sharp and we see mostly in shades of gray. You may have trouble seeing a small object such as a keyhole at night when you are looking straight at it—that is, when its image is falling on your fovea. If you look a little to the side, so that the image falls on a rod-rich area of your retina, you can see the object better. Astronomers looking for faint objects in the sky learned this trick a long time ago. Animals that are nocturnal (such as mice or flying squirrels) may have retinas made up almost entirely of rods and have little or no color vision. By contrast, some animals that are active only during the day (such as chipmunks and ground squirrels) have only cones in their retinas.

How do cone cells enable us to see color? The human retina has three kinds of cone cells, each containing slightly different types of opsin molecules. These opsin molecules differ in the wavelengths of light they absorb best. Although the same 11-*cis*-retinal group is the light absorber (see Figure 45.14), its molecular interactions with opsin tune the spectral sensitivity of the rhodopsin molecule as a whole. Some opsins cause retinal to absorb most efficiently in the blue region, some in the green, and some in the yellow and red (Figure 45.23). Intermediate wavelengths of light excite these different classes of cones in different proportions. The genes that encode the different opsins of humans have been cloned: One codes for blue-sensitive opsin, one for red-sensitive opsin, and several for green-sensitive opsin.

The human retina is organized into five layers of neurons that receive visual information and process it before sending it to the brain (Figure 45.24). As mentioned earlier, the layer of photoreceptors is all the way at the back of the retina. The outer segments of the rods and cones are partly buried in a layer of pigmented epithelium that absorbs photons not captured by rhodopsin and prevents any backscattering of light that might decrease visual sharpness.

INFORMATION FLOW IN THE RETINA. A first step in investigating how the human retina tells the brain what it is seeing is to study how its five layers of neurons are interconnected and how they influence one another. As we know, the photoreceptor cells at the back of the retina hyperpolarize in response to light and do not generate action poten-

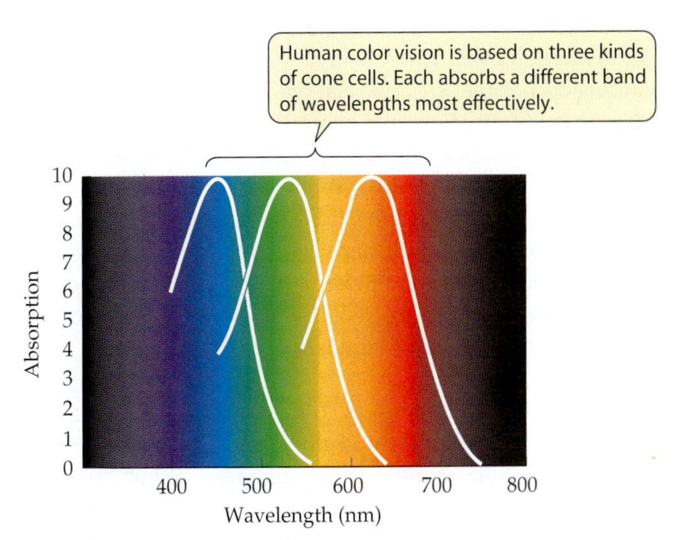

45.23 Absorption Spectra of Cone Cells
The three kinds of cone cells contain slightly different opsin molecules, which absorb different wavelengths of light.

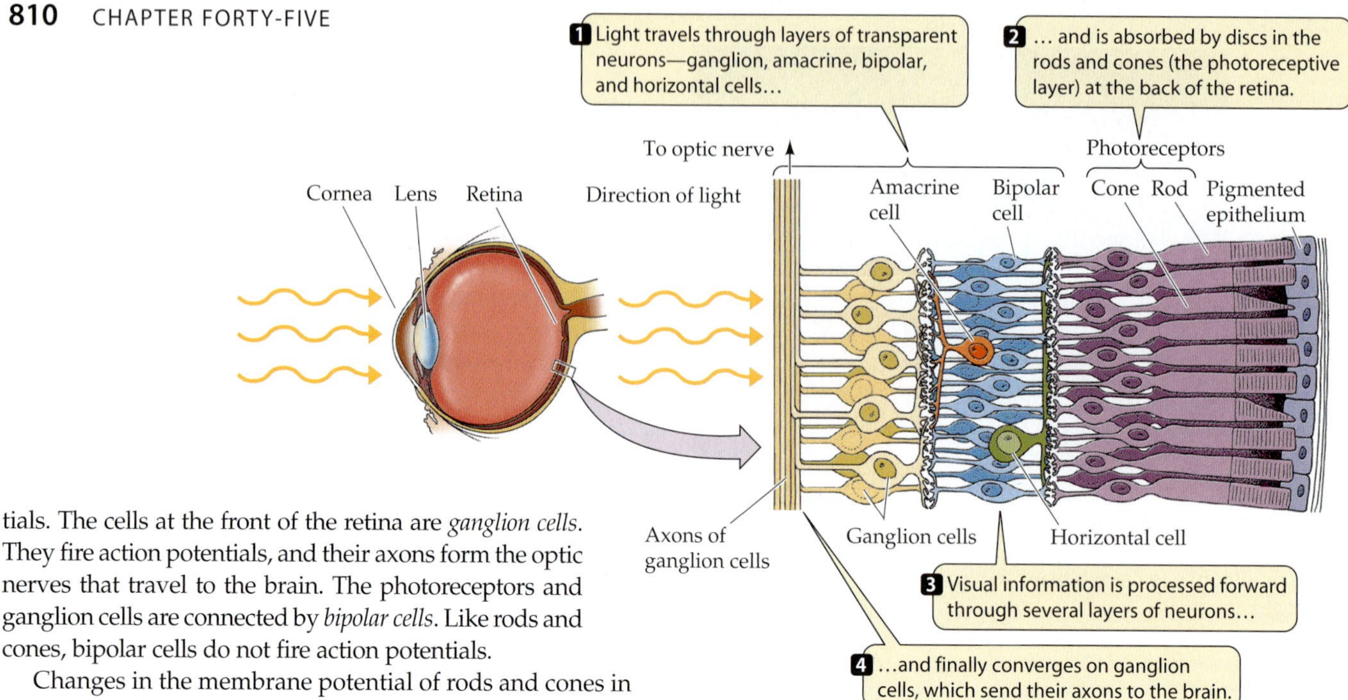

45.24 The Retina
The human retina has five layers of neurons that receive and process visual information.

tials. The cells at the front of the retina are *ganglion cells*. They fire action potentials, and their axons form the optic nerves that travel to the brain. The photoreceptors and ganglion cells are connected by *bipolar cells*. Like rods and cones, bipolar cells do not fire action potentials.

Changes in the membrane potential of rods and cones in response to light alter the rate at which the rods and cones release neurotransmitter at their synapses with the bipolar cells. In response to neurotransmitter from the photoreceptors, the membrane potentials of the bipolar cells change, altering the rate at which they release neurotransmitter onto ganglion cells. The ganglion cells generate action potentials, and the rate of neurotransmitter release from the bipolar cells determines the rate at which they do so. Thus the direct flow of information in the retina is from photoreceptor to bipolar cell to ganglion cell. Ganglion cells send the information to the brain.

The other two cell layers, the horizontal cells and the amacrine cells, communicate laterally across the retina. *Horizontal cells* connect neighboring pairs of photoreceptors and bipolar cells. Thus the communication between a photoreceptor and its bipolar cell can be influenced by the amount of light absorbed by neighboring photoreceptors. This lateral flow of information enables the retina to sharpen the perception of contrast between light and dark patterns.

Amacrine cells connect neighboring pairs of bipolar cells and ganglion cells. The role of amacrine cells is still poorly understood. Some amacrine cell types are highly sensitive to changing illumination or to motion. Others assist in adjusting the sensitivity of the eyes according to the overall level of light falling on the retina. When background light levels change, amacrine cell connections to the ganglion cells help the ganglion cells remain sensitive to temporal changes in stimulation. Thus even with large changes in background illumination, the eyes are sensitive to smaller, more rapid changes in the pattern of light falling on the retina.

INFORMATION PROCESSING IN THE RETINA. Knowing the paths of information flow through the retina still doesn't tell us how that information is processed. What does the eye tell the brain in response to a pattern of light falling on the retina? One aspect of information processing in the retina is *convergence of information*. There are more than 100 million photoreceptors in each retina, but only about 1 million ganglion cells sending messages to the brain. How is the information from all those photoreceptors integrated by the ganglion cells?

This question was addressed in some elegant, classic experiments in which electrodes were used to record the activity of single ganglion cells in living animals while their retinas were stimulated with spots of light. These studies revealed that each ganglion cell has a well-defined **receptive field** that consists of a specific group of photoreceptor cells. Stimulating these photoreceptors with light activates the ganglion cell (Figure 45.25). Information from many photoreceptor cells is integrated in this way to produce a single message.

The receptive fields of many ganglion cells are circular, but the way a spot of light influences the activity of the ganglion cell depends on where in the receptive field it falls. The receptive field of a ganglion cell can be divided into two concentric areas, called the *center* and the *surround*. There are two kinds of receptive fields, *on-center* and *off-center*. Stimulating the center of an on-center receptive field excites the ganglion cell, and stimulating the surround inhibits it. Stimulating the center of an off-center receptive field inhibits the ganglion cell, and stimulating the surround excites it. Center effects are always stronger than surround effects.

The response of a ganglion cell to stimulation of the center of its receptive field depends on how much of the surround area is also stimulated. A small dot of light directly on the center has the maximal effect. A bar of light hitting the center and pieces of the surround has less of an effect, and a large, uniform patch of light falling equally on center and surround has very little effect. Ganglion cells communicate information about contrasts between light and dark that fall on different regions of their receptive fields.

How are receptive fields related to the connections among the neurons of the retina? The photoreceptors in the center of the receptive field of a ganglion cell are connected to that ganglion cell by bipolar cells. The photoreceptors in the surround send information to the center photoreceptors, and thus to the ganglion cell, through the lateral connections of horizontal cells. Thus the receptive field of a ganglion cell consists of a pattern of synapses among photoreceptors, horizontal cells, bipolar cells, and ganglion cells.

The receptive fields of neighboring ganglion cells can overlap greatly; a given photoreceptor can be effectively connected to several ganglion cells. Thus the ganglion cells send simple messages to the brain about the pattern of light

45.25 What Does the Eye Tell the Brain?
When the retina is stimulated with dots and rings of light, individual ganglion cells show different responses.

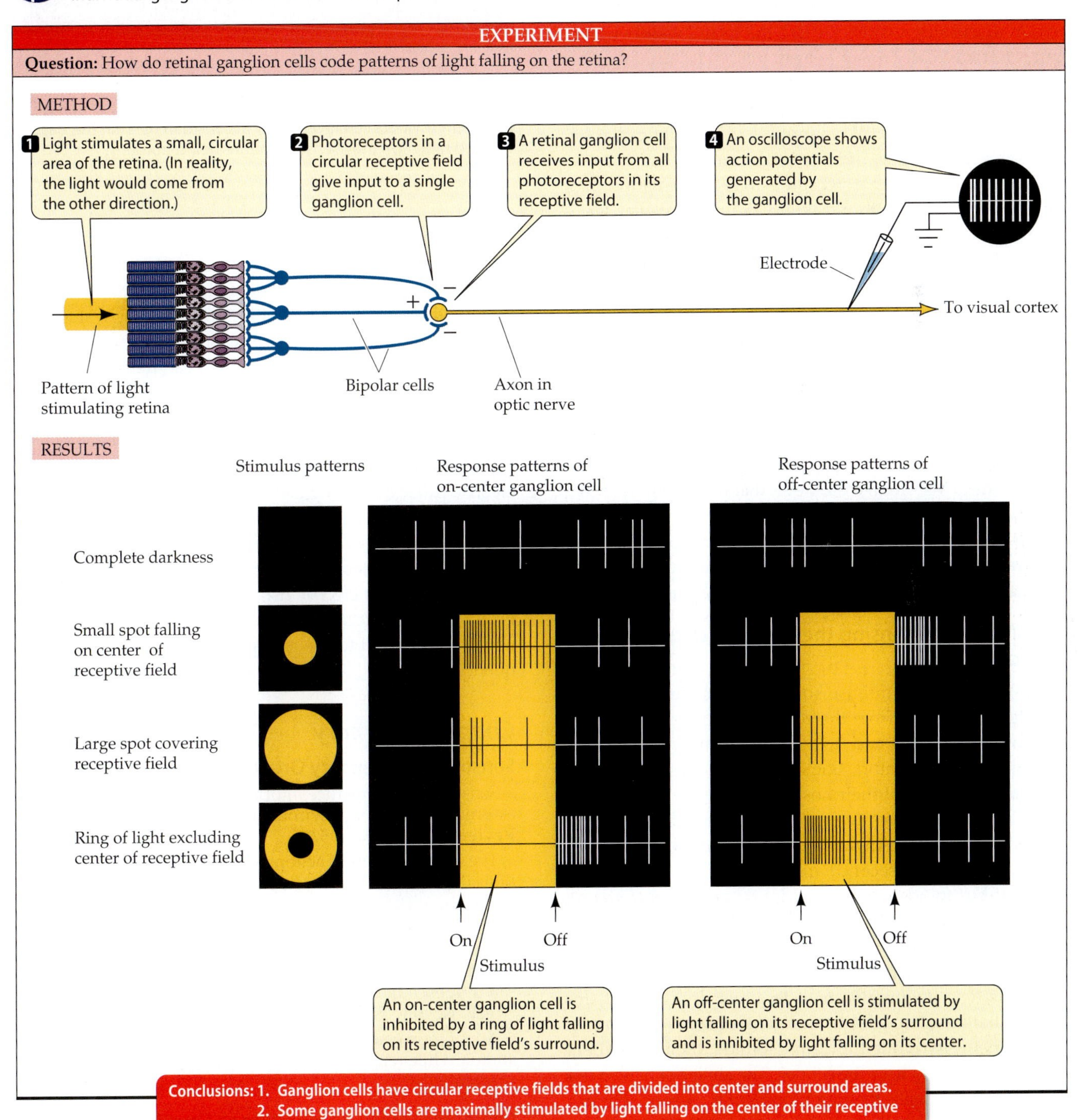

intensities falling on small, circular patches of retina. In Chapter 47 we will see how the brain reassembles that information into our view of the world.

Sensory Worlds Beyond Our Experience

Humans make use of only a subset of the information available to us in the environment. Other animals have sensory systems that enable them to use different subsets and different types of information.

Some species can see infrared and ultraviolet light

When discussing vision, we use the term "visible light," but what we really mean is light visible to humans. Our visible spectrum is a very narrow region of the entire, continuous range of electromagnetic radiation in the environment (see Figure 8.5). We cannot see ultraviolet radiation, for example, but many other animals can.

One of the seven photoreceptors in each ommatidium of a fruit fly is sensitive to ultraviolet light. The visual sensitivity of many pollinating insects includes the ultraviolet part of the spectrum. Some flowers have patterns that are invisible to us but show up if we photograph them with film that is sensitive to ultraviolet light. Those patterns provide information to prospective pollinators, but humans are not equipped to receive that information.

At the other end of the spectrum is infrared radiation, which we sense as heat. Other animals extract much more information from infrared radiation—especially that emitted by potential prey. Pit vipers such as rattlesnakes have *pit organs*, one just in front of each eye, that use highly sensitive heat detectors and a simple pinhole camera arrangement to sense and locate infrared radiation (Figure 45.26). In total darkness, these snakes can locate a prey animal such as a mouse, orient to it, and strike it with great accuracy.

Echolocation is sensing the world through reflected sound

Some species emit intense sounds and create images of their environments from the echoes of those sounds. Bats, porpoises, dolphins, and (to a lesser extent) whales are accomplished echolocators. Some species of bats have elaborate modifications of their noses to direct the sounds they emit, as well as impressive ear pinnae to collect the returning echoes. The high-frequency sounds they emit as pulses (about 20 to 80 per second) are above the range of human hearing, but they are extremely loud in contrast to the resulting faint echoes bouncing off small insects. An echolocating bat is similar to a construction worker who is trying to overhear a whispered conversation on a street corner while using a pneumatic drill. To avoid deafening themselves, bats use muscles in their middle ears to dampen their sensitivity while they emit sounds, then relax them quickly enough to hear the echoes. The ability of bats to use echolocation to sense their environment is so good that in a totally dark room strung with fine wires, they can capture tiny flying insects while navigating around the wires.

45.26 Stalking in the Dark
The black-tailed rattlesnake of the southwestern United States is a pit viper. Pit vipers can locate prey in total darkness on the basis of infrared radiation they sense through their pit organs.

Some fish can sense electric fields

We discussed the mechanoreceptors in the lateral lines of fishes (see Figure 45.9). The lateral lines of some species, especially those such as catfish that live in murky waters, also contain electroreceptors. These sensory cells enable the fish to detect weak electric fields, which can help them locate prey.

The use of electroreceptors is even more sophisticated in species called electric fishes. These fishes have evolved electric organs in their tails that generate a continuous series of electric pulses, creating a weak electric field around their bodies. Any objects in the environment, such as rocks, plants, or other fish, disrupt the electric fish's electric field, and the electroreceptors of the lateral line detect those disruptions. In some electric fish species, each individual in a group emits its electric pulses at a different frequency. If a new fish is added to the group, they all readjust their frequencies.

Chapter Summary

Sensory Cells, Sensory Organs, and Transduction

- Sensory cells transduce information about an animal's external and internal environment into action potentials. **Review Figures 45.1, 45.2**
- The interpretation of action potentials as particular sensations depends on which neurons in the CNS receive them.
- Sensory cells have membrane receptor proteins that cause ion channels to open or close, generating receptor potentials. Receptor potentials can spread to regions of the sensory cell plasma membrane that generate action potentials, or they can influence the release of neurotransmitter from the sensory cell. **Review Figure 45.3**
- Adaptation enables the nervous system to ignore irrelevant stimuli while remaining responsive to relevant or to new stimuli.

Chemoreceptors: Responding to Specific Molecules

- Smell, taste, and the sensing of pheromones are examples of chemosensation. Chemoreceptor cells have receptor proteins that can bind to specific molecules that come into contact with the sensory cell membrane. **Review Figure 45.5**

▶ The binding of an odorant molecule to a receptor protein causes the production of a second messenger in the chemoreceptor cell. The second messenger alters ion channels and creates a receptor potential. **Review Figure 15.17**

▶ Chemoreceptors in the mouth cavities of vertebrates are responsible for the sense of taste. **Review Figure 45.6**

Mechanoreceptors: Detecting Stimuli that Distort Membranes

▶ In the skin there are a diversity of mechanoreceptors that respond to touch and pressure. The density of mechanoreceptors in any skin area determines the sensitivity of that area to touch and pressure. **Review Figure 45.7**

▶ Stretch receptors in muscles, tendons, and ligaments inform the CNS of the positions of and the loads on parts of the body. **Review Figure 45.8**

▶ Hair cells are mechanoreceptors that are not neurons. The bending of their stereocilia alters their membrane proteins and therefore their receptor potentials. Hair cells are found in organs of equilibrium and orientation such as the lateral line system of fishes, the statocysts of invertebrates, and the semicircular canals and vestibular apparatus of mammals. **Review Figures 45.9, 45.10, 45.11**

▶ Hair cells are responsible for mammalian auditory sensitivity. Ear pinnae collect and direct sound waves to the tympanic membrane, which vibrates in response to sound waves. The movements of the tympanic membrane are amplified through a chain of ossicles that conduct the vibrations to the oval window. Movements of the oval window create pressure waves in the fluid-filled cochlea. **Review Figure 45.12**

▶ The basilar membrane running down the center of the cochlea is distorted at specific locations that depend on the frequency of the pressure wave. These distortions cause the bending of hair cells in the organ of Corti, which rests on the basilar membrane. Changes in hair cell receptor potentials create action potentials in the auditory nerve, which conducts the information to the CNS. **Review Figure 45.13**

Photoreceptors and Visual Systems: Responding to Light

▶ Photosensitivity depends on the capture of photons of light by rhodopsin, a photoreceptor molecule that consists of a protein called opsin and a light-absorbing prosthetic group called retinal. Absorption of light by retinal is the first step in a cascade of intracellular events leading to a change in the receptor potential of the photoreceptor cell. **Review Figure 45.14**

▶ When excited by light, vertebrate photoreceptor cells hyperpolarize and release less neurotransmitter onto the neurons with which they form synapses. They do not fire action potentials. **Review Figures 45.15, 45.16, 45.17**

▶ Vision results when eyes focus patterns of light onto layers of photoreceptors. Eyes vary from the simple eye cups of flatworms, which enable the animal to sense the direction of a light source, to the compound eyes of arthropods, which enable the animal to detect shapes and patterns, to the lensed eyes of cephalopods and vertebrates. **Review Figures 45.18, 45.19**

▶ The eyes of vertebrates and cephalopods focus detailed images of the visual field onto dense arrays of photoreceptors that transduce the visual image into neural signals. **Review Figures 45.20, 45.21**

▶ The vertebrate photoreceptors are rod cells, responsible for dim light and black-and-white vision, and cone cells, responsible for color vision by virtue of their spectral sensitivities. **Review Figure 45.23**

▶ The vertebrate retina is a dense array of neurons lining the back of the eyeball. It consists of five layers of cells. The outermost layer consists of the rods and cones. The innermost layer consists of the ganglion cells, which send their axons in the optic nerve to the brain. Between the photoreceptors and the ganglion cells are neurons that process the information from the photoreceptors. **Review Figure 45.24**

▶ The area of the retina that receives light from the center of the visual field, the fovea, has the greatest density of photoreceptors. In humans the fovea contains almost exclusively cone cells, which are responsible for color vision but are not very sensitive in dim light.

▶ Each ganglion cell is stimulated by light falling on a small circular patch of photoreceptors called a receptive field. Receptive fields have a center and a surround, which have opposing effects on the ganglion cell. If the center is excitatory, the surround is inhibitory, and vice versa. **Review Figure 45.25**

Sensory Worlds Beyond Our Experience

▶ Many animals have sensory abilities that we do not share. Bats echolocate, insects see ultraviolet radiation, pit vipers "see" infrared radiation, and fish sense electric fields.

For Discussion

1. Compare and contrast the functioning of olfactory receptors and photoreceptors. How do these sensory cells enable the CNS to discriminate between an apple and an orange?
2. Amplification of signal is an important feature of sensory systems. Compare mechanisms of amplification in olfactory, visual, and auditory systems.
3. If you were blindfolded and placed in a wheelchair, how would you know if you were being pushed forward or backward?
4. Describe and contrast two sensory systems that enable animals to "see" in the dark. What problems or limitations are inherent in these systems in comparison with vision?
5. Communication is the transfer of information from one animal to another. Animals can use visual, olfactory, tactile, and auditory signals to communicate. From what you know about these sensory systems, discuss the relative advantages and disadvantages of these systems for communication.

46 The Mammalian Nervous System: Structure and Higher Functions

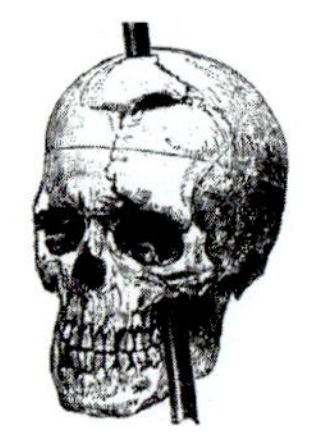

PHINEAS GAGE WAS AN INDUSTRIOUS, responsible, considerate young man. He was 25 years old in 1848, working as a railroad construction foreman. He had the respect of his men, and he looked out for them to the extent that he took on himself the most dangerous tasks associated with blasting the rocks in the path of the railroad.

Late one afternoon the last hole had been drilled for the day. Gage poured blasting powder into the hole and tamped it with a meter-long, 3-cm wide iron rod. The tamping iron hit the side of the hole, struck a spark, and ignited the powder. The explosion shot the rod out of the hole like a bullet. It struck Phineas below his left eye, penetrated his skull, passed through the part of his brain behind his forehead, and exited out the top of his head. Was this the end of Phineas Gage?

Gage regained consciousness within minutes and could speak. He was taken to a hotel, where a physician dressed his wounds, but the doctor could do little else. Infections were a problem, but Gage's senses and memory were intact. In 3 weeks he left his bed, but he did not return to his work at the railroad. The recovered Phineas Gage was quarrelsome, bad-tempered, lazy, and irresponsible. He was impatient and obstinate, and he used profane language, which he had never done before.

The body of Phineas Gage survived the accident, but he was an entirely different person. He spent the rest of his days as a drifter, earning money by telling his story, exhibiting his scars and his tamping iron. If you are in Cambridge, Massachusetts, you can pay him a visit. His skull, death mask, and the tamping iron are on display in the Museum of the Medical College of Harvard University.

The sad story of Phineas Gage reveals that the essence of individuality and personality resides in the brain. What is this miraculous organ, and what does it do? The human brain weighs about 1.5 kg, is mostly water, and has the consistency of custard. Yet the complexity of this small mass of tissue exceeds that of any other known matter. The work of the brain is to process and store information and to control the physiology and behavior of the body. The brain is constantly receiving, integrating, and interpreting information from all the senses. To respond to that information, it generates commands to the muscles and organs of the body.

The unit of function of the brain is the neuron. The human brain consists of about 10 billion neurons, which account for its ability to handle vast amounts of information. In the previous two chapters we learned about the cellular properties of neurons: how they generate and conduct action potentials, how they transduce sensory information, how they communicate with each other at synapses, and how information is integrated at synapses. In this chapter we take on the challenge of understanding the functions of the human nervous system in terms of these cellular mechanisms.

The Nervous System: Structure, Function, and Information Flow

The human nervous system is more than the brain. The brain and spinal cord together constitute the **central nervous system** (**CNS**). Information is brought to and from the

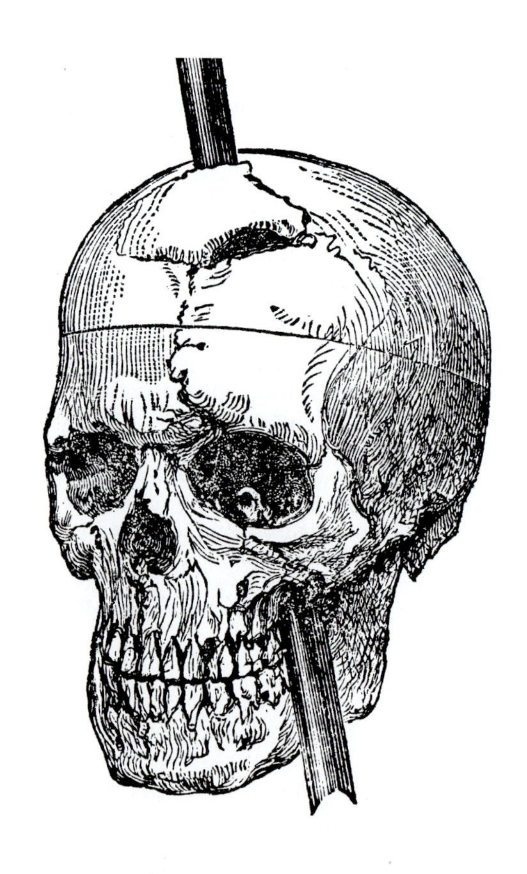

An Unintentional Experiment
In a nineteenth-century railroad construction accident, an explosion blew a tamping iron through the brain of Phineas Gage. Unbelievably, Gage survived, but his personality was radically changed. This drawing of Gage's skull was made at the time of his death.

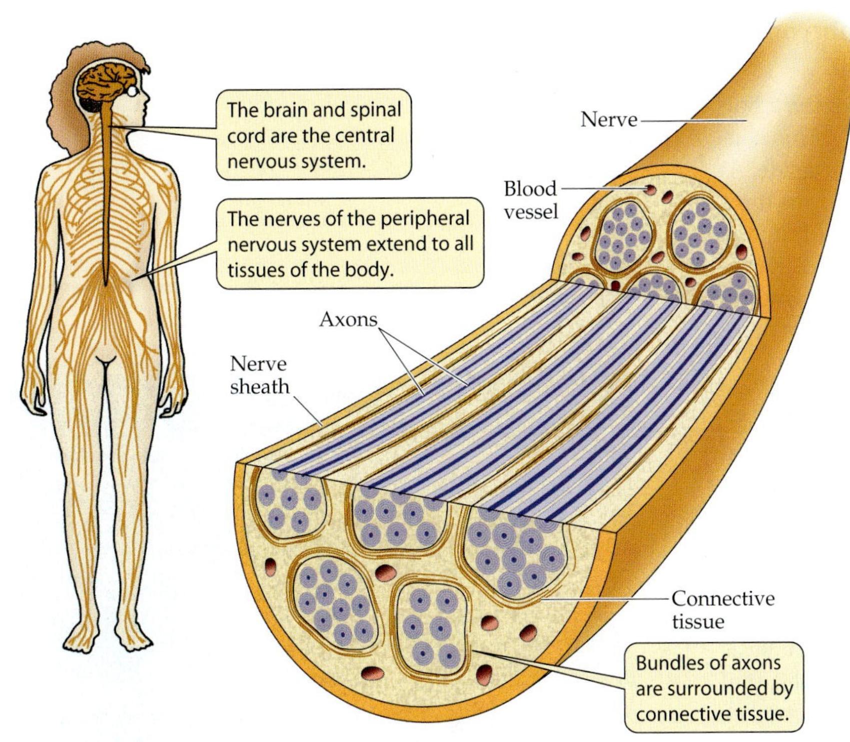

46.1 *Anatomy of the Human Nervous System*
(*a*) Information is communicated between the central nervous system and the other tissues of the body through the peripheral nervous system. (*b*) A nerve contains the axons of many neurons. Some of these neurons conduct information to and others from the central nervous system.

CNS by means of an enormous network of nerves that make up the **peripheral nervous system** (Figure 46.1*a*). The peripheral nervous system reaches every tissue of the body. It connects to the CNS via spinal nerves and cranial nerves.

A **nerve** is a bundle of axons (Figure 46.1*b*) that carries information about many things simultaneously. It is important to distinguish between the axon of a single neuron and a nerve. Some axons in a nerve may be carrying information to the CNS while other axons in the same nerve are carrying information from the CNS to the organs of the body.

A conceptual diagram of the nervous system traces information flow

The nervous system is an information processing system—a very complex one that handles many tasks simultaneously. It will help to organize our thinking about the nervous system by beginning with a conceptual diagram of information flow (Figure 46.2). We can then plug anatomical and functional details into this general model.

The **afferent** portion of the peripheral nervous system carries information *to the CNS*. We are consciously aware of much of the information that moves through these afferent pathways (for example, vision, hearing, temperature, pain, the position of limbs), but we are not consciously aware of other afferent information that is important for physiological regulation (for example, blood pressure, deep body temperature, blood oxygen supply).

The **efferent** portion of the peripheral nervous system carries information *from the CNS* to the muscles and glands of the body. Efferent pathways can be divided into a **voluntary** division, which executes our conscious movements, and an **involuntary**, or **autonomic**, division, which controls physiological functions.

In addition to neural information, the CNS receives chemical information in the form of hormones circulating in the blood. Neurohormones released by neurons into the extracellular fluids of the brain can send chemical information to other neurons in the brain or can leave the brain and enter the circulation. In Chapter 41 we learned of the important role of neurohormones in the control of the anterior pituitary and saw that other neurohormones are released from the posterior pituitary into the circulation.

Now we can begin to translate our conceptual scheme of information flow into an anatomical view of the nervous system. It can be difficult to learn the relationships between the different structures of the adult nervous system, but the task is much easier if we begin with the development of the nervous system from a simple tubular structure that forms in the embryo.

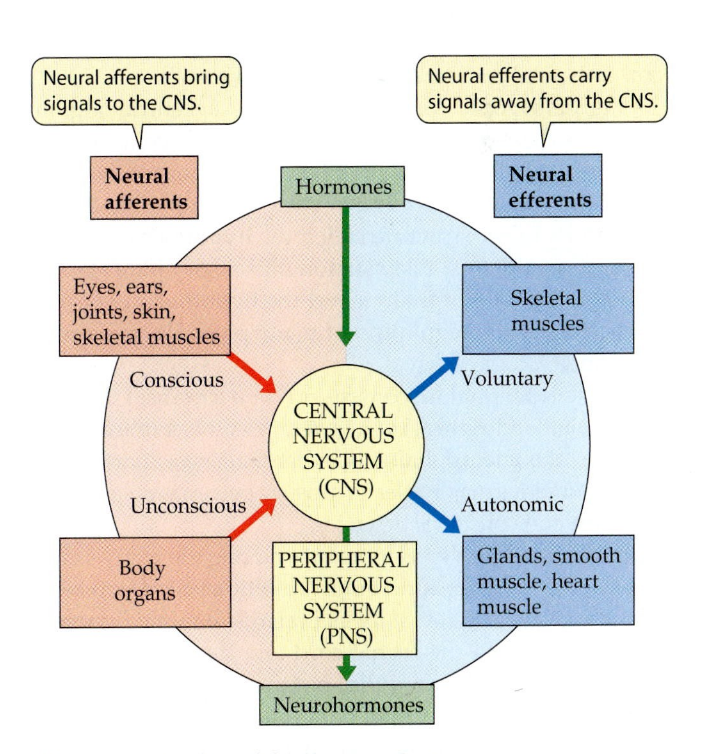

46.2 *Organization of the Nervous System*
The peripheral nervous system carries information both to and from the central nervous system. The CNS also receives hormonal inputs and produces hormonal outputs.

The vertebrate CNS develops from the embryonic neural tube

Early in the development of all vertebrate embryos, a hollow tube of neural tissue forms (see Chapter 43). This neural tube runs the length of the embryo on its dorsal side. At the anterior end of the embryo, the neural tube forms three swellings that become the basic divisions of the brain: the **hindbrain**, the **midbrain**, and the **forebrain**. The rest of the neural tube becomes the spinal cord (Figure 46.3). The cranial and spinal nerves, which make up the peripheral nervous system, sprout from the neural tube and grow throughout the embryo.

Each of the three regions of the embryonic brain develops into several structures in the adult brain. From the hindbrain come the **medulla**, the **pons**, and the **cerebellum**. The medulla is continuous with the spinal cord. The pons is in front of the medulla, and the cerebellum is a dorsal outgrowth of the pons. The medulla and pons contain distinct groups of neurons that are involved in the control of physiological functions such as breathing and circulation or basic motor patterns such as swallowing and vomiting. All neural information traveling between the spinal cord and higher brain areas must pass through the pons and the medulla.

The cerebellum is like the conductor of an orchestra; it receives "copies" of the commands going to the muscles from higher brain areas, and it receives information coming up the spinal cord from the joints and muscles. Thus it can compare the motor "score" with the actual behavior of the muscles and refine the motor commands.

From the embryonic midbrain come structures that process aspects of visual and auditory information. In addition, all information traveling between higher brain areas and the spinal cord must pass through the midbrain.

The embryonic forebrain develops a central region called the **diencephalon** and a surrounding structure called the **telencephalon**. The diencephalon is the core of the forebrain and consists of an upper structure called the **thalamus** and a lower structure called the **hypothalamus**. The thalamus is the final relay station for sensory information going to the telencephalon, and the hypothalamus is responsible for the regulation of many physiological functions and biological drives.

The telencephalon consists of two **cerebral hemispheres**, left and right (also referred to as the **cerebrum**). In humans, the telencephalon is by far the largest part of the brain and plays major roles in sensory perception, learning, memory, and conscious behavior.

Understanding the relationships among the many structures of the complex adult brain is a little easier if you keep this linear organization of the neural axis in mind: Communication between the spinal cord and the telencephalon travels through the medulla, pons, midbrain, and diencephalon. The medulla, pons, and midbrain are referred to collectively as the **brain stem**. In general, more primitive and autonomic functions are localized farther down this neural axis, while more complex and evolutionarily advanced functions are found higher on the axis.

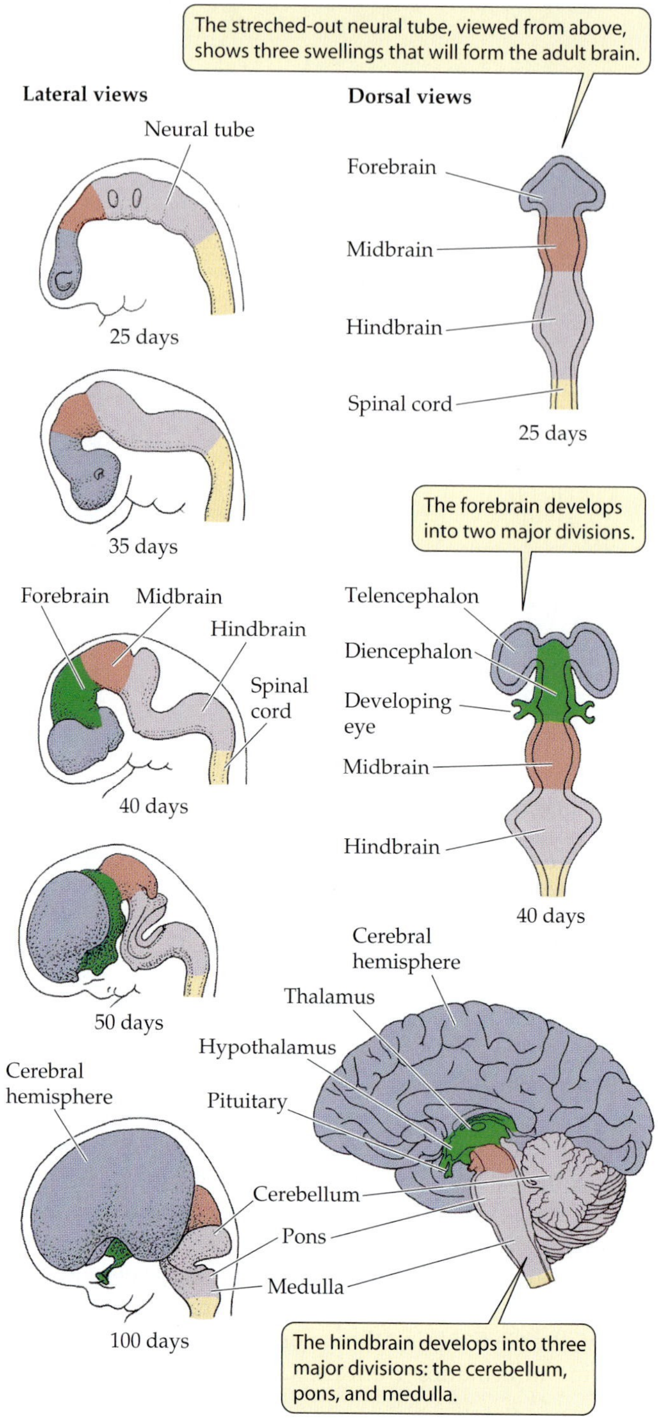

46.3 Development of the Human Nervous System
Three swellings at the anterior end of the hollow neural tube in early vertebrate embryos develop into the parts of the adult brain. The final view is an adult human brain section cut in half through the midline.

As we go up the vertebrate phylogenetic scale from fish to mammals, the telencephalon increases in size, complexity, and importance. The forebrain dominates the nervous systems of mammals, and damage to this region results in severe impairment of sensory, motor, or cognitive functions, and even coma. In contrast, a shark with its telencephalon removed can swim almost normally.

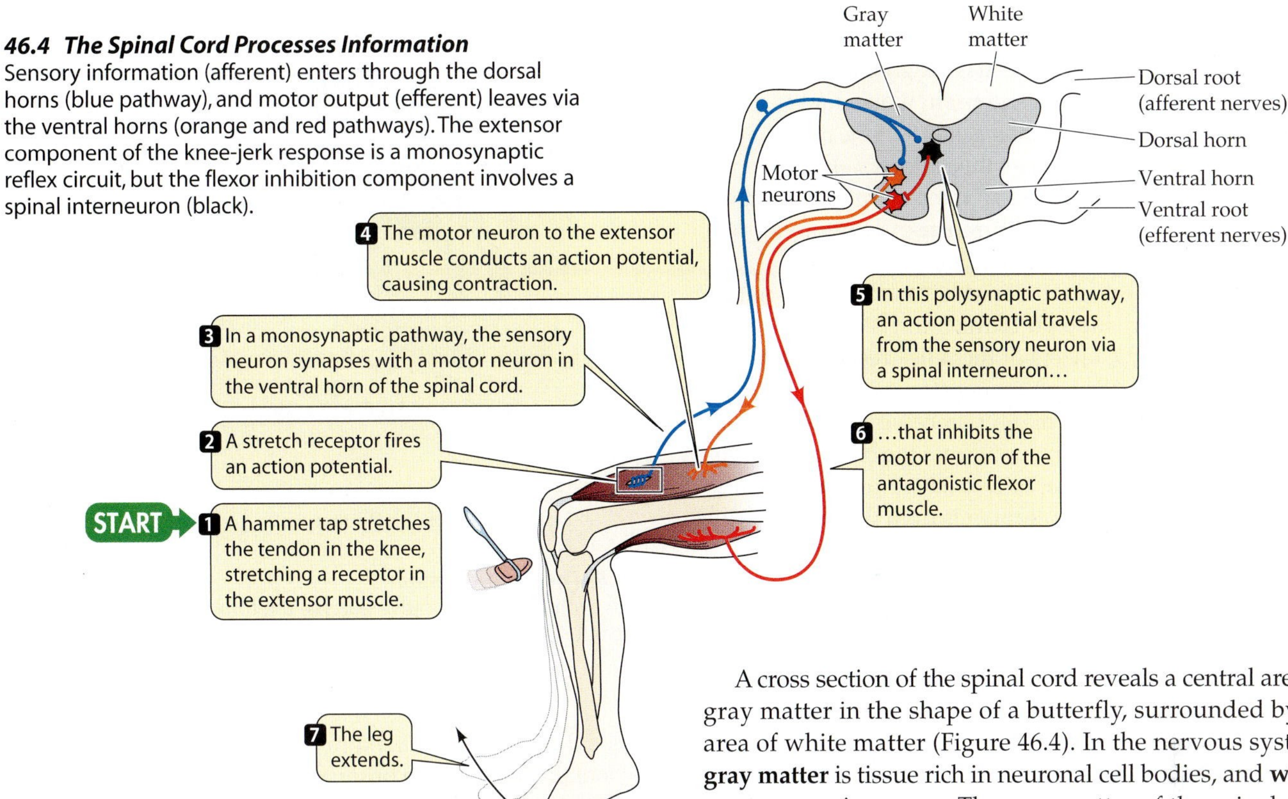

46.4 The Spinal Cord Processes Information
Sensory information (afferent) enters through the dorsal horns (blue pathway), and motor output (efferent) leaves via the ventral horns (orange and red pathways). The extensor component of the knee-jerk response is a monosynaptic reflex circuit, but the flexor inhibition component involves a spinal interneuron (black).

Functional Subsystems of the Nervous System

When we talk about the development of the nervous system, we describe it in terms of anatomically distinct structures. However, the nervous system is always engaged in many tasks at the same time—a phenomenon called *parallel processing of information*. Any one task usually involves many different anatomical structures. Understanding the nervous system is made simpler if we recognize its functional subsystems, such as the spinal cord, reticular system, limbic system, and cerebrum. Any one anatomical structure may play roles in several functional subsystems.

The spinal cord receives and processes information from the body

The spinal cord conducts information in both directions between the brain and the organs of the body. It also integrates a great deal of the information coming from the peripheral nervous system and responds to that information by issuing motor commands.

The conversion of afferent to efferent information in the spinal cord without participation of the brain is called a **spinal reflex**. The simplest type of spinal reflex involves only two neurons and one synapse and is therefore called a **monosynaptic reflex**. An example is the knee-jerk reflex, which your physician checks by tapping just below your knee with a small mallet. We can diagram the wiring of a monosynaptic reflex by following the flow of information through the spinal cord.

A cross section of the spinal cord reveals a central area of gray matter in the shape of a butterfly, surrounded by an area of white matter (Figure 46.4). In the nervous system, **gray matter** is tissue rich in neuronal cell bodies, and **white matter** contains axons. The gray matter of the spinal cord contains the cell bodies of the spinal neurons; the white matter contains the axons that conduct information up and down the spinal cord. Spinal nerves leave the spinal cord at regular intervals on each side. Each spinal nerve has two roots connecting to the gray matter—one connecting with the *dorsal horn*, the other with the *ventral horn*. Each spinal nerve carries both afferent and efferent information. The afferent axons enter the spinal cord through the dorsal roots and the efferent axons leave the spinal cord through the ventral roots.

In the case of the knee-jerk reflex, sensory information comes from stretch receptors in the leg muscle that is suddenly stretched when the mallet strikes the tendon that runs over the knee. Each stretch receptor initiates action potentials that are conducted by the axon of a sensory neuron in through the dorsal horn of the spinal cord and all the way to the ventral horn. In the ventral horn, the sensory neuron synapses with motor neurons, causing them to fire action potentials that are then conducted back to the leg extensor muscle, causing it to contract. The function of this simple circuit is to sense an increased load on the limb, and to cause the muscle to increase its strength of contraction to compensate for the added load.

Most spinal circuits are more complex than this monosynaptic reflex. We can demonstrate that by building on the circuit we have just traced. Limb movement is controlled by *antagonistic* sets of muscles—muscles that work against each other. When one member of an antagonistic set of muscles contracts, it bends or flexes the limb; it is therefore called a *flexor*. The antagonist to this muscle straightens or extends the limb and is called an *extensor*. For a limb to move, one muscle of the pair must relax while the other contracts. Thus sensory input that activates the motor neuron of one muscle also inhibits its antagonist. This coordination is achieved by an **interneuron**, which makes an inhibitory synapse onto the motor neuron of the antagonistic muscle (see Figure 46.4). Thus the reciprocal inhibition of antagonistic muscles involves an interneuron between the sensory cell and the motor neuron and therefore at least two synapses.

Information entering the dorsal horn is also transmitted by axons up the spinal cord to the brain. We are aware of the mallet hitting the knee, but the reflex response actually begins before that information registers in our consciousness. A great deal of information processing takes place in the spinal cord without any input from the brain. Spinal circuits can even generate repetitive motor patterns such as those of walking without commands from the brain.

The reticular system alerts the forebrain

The **reticular system** of the brain stem is a highly complex network of neuronal axons and dendrites. Within the reticular system are many discrete groups of neurons. Such an anatomically distinct group of neurons in the CNS is called a **nucleus**.

The reticular system is distributed through the core of the medulla, pons, and midbrain (Figure 46.5). Afferent information coming up the neural axis passes through the reticular system, where many connections are made to neurons involved in controlling many functions of the body. Information from joints and muscles, for example, is directed to nuclei in the pons and cerebellum that are involved in balance and coordination, whereas information from pain receptors is directed to nuclei in the brain stem that control sensitivity to pain. This information continues up the neural axis to the forebrain, where it results in conscious sensations that can be localized to the specific sites in the body where the information originated.

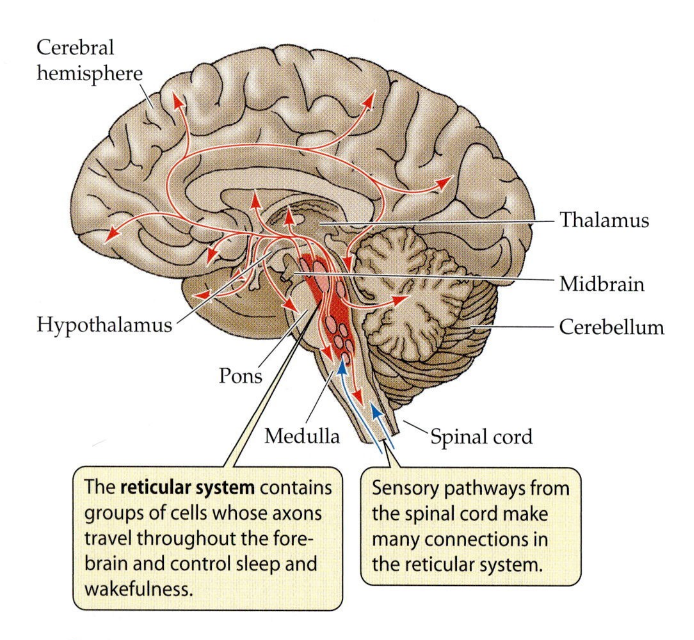

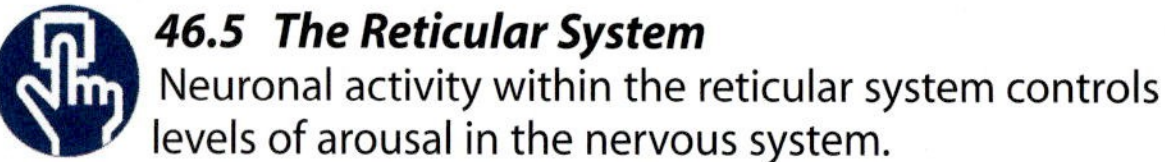

46.5 The Reticular System
Neuronal activity within the reticular system controls levels of arousal in the nervous system.

46.6 The Limbic System
The evolutionarily primitive parts of the forebrain (shown in blue) are referred to as the limbic system.

The information routed through the reticular system also influences the level of arousal of the nervous system. Nuclei in the reticular system are involved in the control of sleep and waking. High levels of activity in the reticular system influence these nuclei to maintain the brain in a waking condition; low levels of activity enable sleep. Because of the alerting function of the reticular core of the brain stem, it has been called the *reticular activating system*.

If the brain stem of a person is damaged at midbrain or higher levels, and the alerting action of the reticular system cannot reach the forebrain, the person loses the ability to be in a conscious, waking state and becomes comatose. Damage to the brain stem or the spinal cord below the reticular system does not interfere with the ascending alerting actions of the reticular system and leaves the person with normal patterns of sleep and waking, although it can cause lack of sensation (*paresthesia*) and loss of motor function (*paralysis*).

The limbic system supports basic functions of the forebrain

The telencephalon of fishes, amphibians, and reptiles consists of only a few structures surrounding the diencephalon. In birds and mammals, these primitive forebrain structures are completely covered by the evolutionarily more recent elaborations of the telencephalon called the **neocortex**, but they still have important functions. These primitive parts of the forebrain are collectively referred to as the **limbic system** (Figure 46.6).

The limbic system is responsible for basic physiological drives, instincts, and emotions. Within the limbic system are areas that when stimulated with small electric currents can cause intense sensations of pleasure, pain, or rage. If a

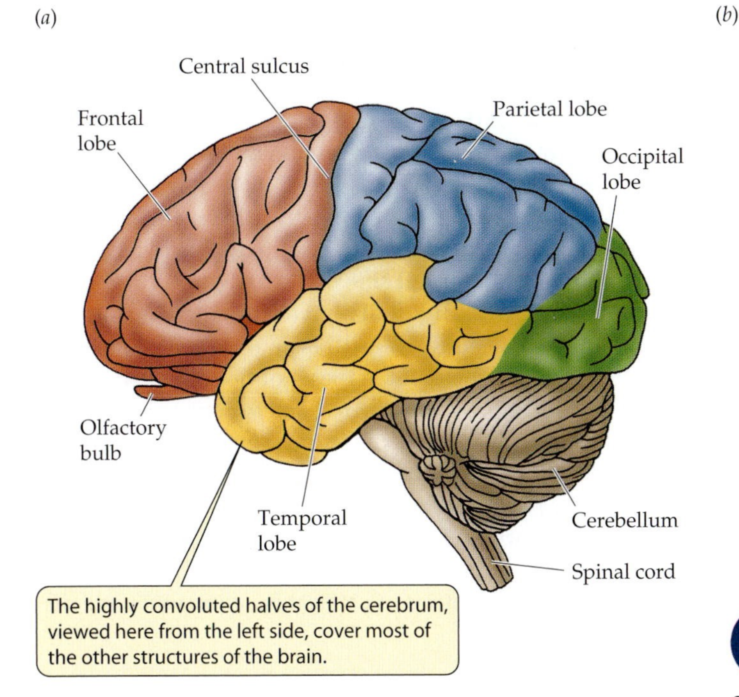

46.7 The Human Cerebrum
(*a*) Each cerebral hemisphere is divided into four lobes. (*b*) Different functions are localized in particular areas of the cerebral lobes.

rat is given the opportunity to stimulate its own pleasure centers by pressing a switch, it will ignore food, water, and even sex, pushing the switch until it is exhausted. Pleasure and pain centers in the limbic system are believed to play roles in learning and in physiological drives.

One part of the limbic system, the **hippocampus**, is necessary in humans for the transfer of short-term memory to long-term memory. If you are told a new telephone number, you may be able to hold it in short-term memory for a few minutes, but within half an hour it is forgotten unless you make a real effort to remember it. The phenomenon of remembering something for more than a few minutes requires its transfer to long-term memory.

Regions of the cerebrum interact to produce consciousness and control behavior

The cerebral hemispheres are the dominant structures in the mammalian brain. In humans they are so large that they cover all other parts of the brain except the cerebellum (Figure 46.7). A sheet of gray matter called the **cerebral cortex** covers each cerebral hemisphere. The cortex is about 4 mm thick and covers a total surface area over both hemispheres of 1 square meter. The cerebral cortex is convoluted, or folded, into ridges called *gyri* and valleys called *sulci*. These convolutions allow it to fit into the skull, which is of limited size and volume. Under the cerebral cortex is white matter, made up of the axons that connect the cell bodies in the cortex with one another. They also connect with other areas of the brain.

Different regions of the cerebral cortex have specific functions. Some of those functions are easily defined, such as receiving and processing sensory information, but most of the cortex is involved in higher-order information processing that is less easy to define. These latter areas are given the general name of *association cortex*.

To understand the cerebral cortex, it helps to have an anatomical road map. As viewed from the left side, a left cerebral hemisphere looks like a boxing glove for the right hand with the fingers pointing forward, the thumb pointing out, and the wrist at the rear (see Figure 46.7*a*). The "thumb" area is the **temporal lobe**, the fingers the **frontal lobe**, the back of the hand the **parietal lobe**, and the wrist the **occipital lobe**. A mirror image of this arrangement characterizes the right cerebral hemisphere. Let's look at each lobe of the cerebrum separately.

THE TEMPORAL LOBE. The upper region of the temporal lobe receives and processes auditory information. The association areas of the temporal lobe are involved in the recognition, identification, and naming of objects. Damage to the temporal lobe results in disorders called *agnosias* in which the individual is aware of a stimulus but cannot identify it. Some of these deficits can be quite specific.

Damage to one area of the temporal lobe results in the inability to recognize faces. Even old acquaintances cannot be identified by facial features, although they may be identified by other attributes such as voice, body features, and characteristic style of walking. Using monkeys, it has been possible to record from neurons in this region that respond selectively to faces in general. These neurons do not respond to other stimuli in the visual field, and their responsiveness decreases if some of the features of the face are missing or appear in inappropriate locations (Figure 46.8). Damage to other association areas of the temporal lobe causes deficits in understanding spoken language, even though speaking, reading, and writing abilities may be intact.

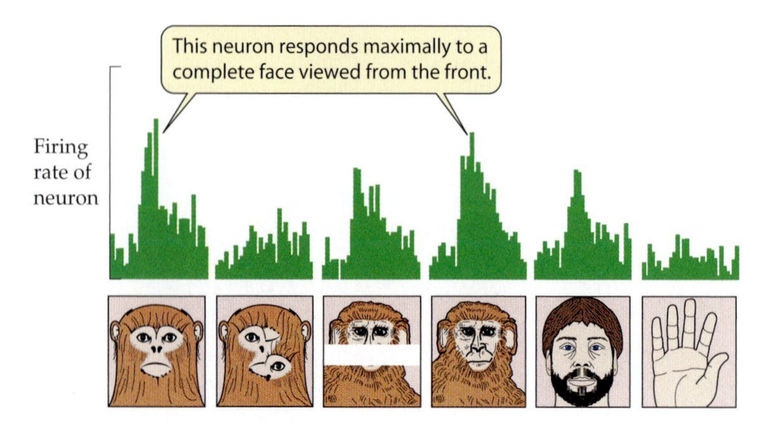

46.8 *Neurons in One Region of the Temporal Lobe Respond to Faces*
The traces represent the firing rate of a neuron in the temporal lobe of a monkey in response to the pictures shown below them.

THE FRONTAL LOBE. A strip of the frontal lobe cortex just in front of the central sulcus is called the *primary motor cortex* (see Figure 46.7*b*). The neurons in this region have axons that project to muscles in specific parts of the body. The parts of the body can be mapped onto the primary motor cortex, from the head region on the lower side to the lower part of the body at the top. Areas with fine motor control, such as the face and hands, have the greatest representation (Figure 46.9). If a neuron in the primary motor cortex is electrically stimulated, the response is the twitch of a muscle, but not a coordinated, complex behavior.

The association functions of the frontal lobe are diverse and are best described as having to do with planning. The story of Phineas Gage at the beginning of this chapter demonstrates the effects of damage to these association areas. People with such deficits have drastic alterations of personality because they cannot create an accurate view of themselves in the context of the world around them and cannot plan for future events.

THE PARIETAL LOBE. The frontal and parietal lobes are separated by a deep valley called the *central sulcus*. The strip of parietal lobe cortex just behind the central sulcus is the *primary somatosensory cortex* (see Figure 46.7*b*). This area receives information through the thalamus about touch and pressure sensations.

The whole body surface is represented in the primary somatosensory cortex—the head at the bottom and the legs at the top (see Figure 46.9). Areas of the body that have lots of sensory neurons and are capable of making fine distinctions in touch (such as the lips and the fingers) have disproportionately large representation. If a very small area of the primary somatosensory cortex is stimulated electrically, the subject reports feeling specific sensations, such as touch, from a very localized part of the body.

A major association function of the parietal lobe is attending to complex stimuli. Damage to the right parietal lobe causes a condition called *contralateral neglect syndrome*, in which the individual tends to ignore stimuli from the left side of the body or the left visual field. Such individuals have difficulty performing complex tasks such as dressing the left side of the body; an afflicted man may not be able to shave the left side of his face. When asked to copy simple drawings, a person who exhibits this syndrome can do well with the right side of the drawing, but not the left (Figure 46.10). The parietal cortex is not symmetrical with respect to its role in attention. Damage to the left parietal cortex does not cause neglect of the right side of the body. We will see similar asymmetries in cortical function when we discuss language.

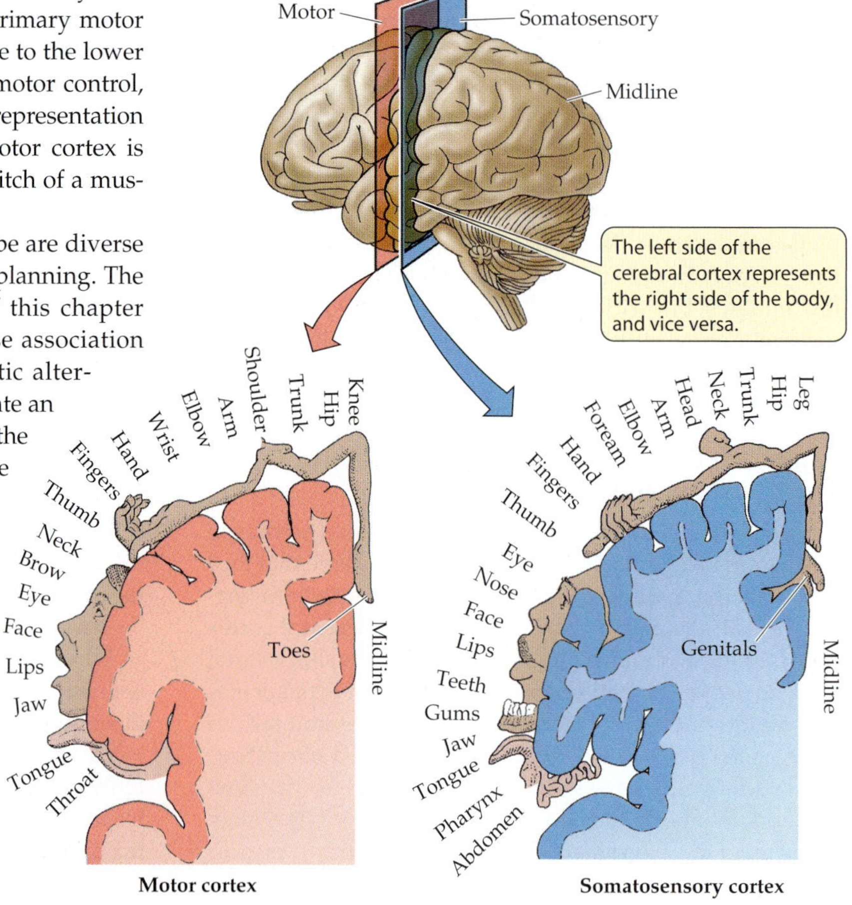

46.9 *The Body Is Represented in the Primary Motor Cortex and the Primary Somatosensory Cortex*
Cross sections through the primary motor and primary somatosensory cortexes can be represented as maps of the human body. Body parts are shown in relation to the brain area devoted to them.

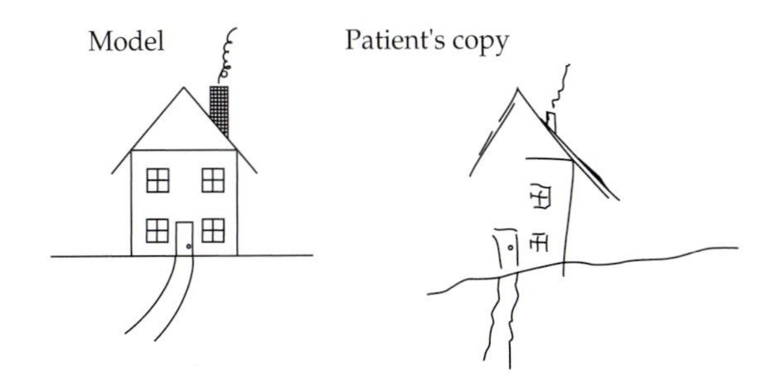

46.10 Contralateral Neglect Syndrome
A person with damage to the right parietal association cortex will neglect the left side of a drawing when asked to copy a model.

THE OCCIPITAL LOBE. The occipital lobes receive and process visual information. We'll learn more about the details of that process later in this chapter. The association areas of the occipital cortex are essential for making sense out of the visual world and translating visual experience into language. Some deficits resulting from damage to association areas of the occipital cortex are quite specific. In one case, a woman with limited damage was unable to see motion. Her vision was intact, but she could see a waterfall only as a still image, and a car approaching only as a series of scenes of a stationary object at different distances.

The cerebrum has increased in size and complexity

As mentioned earlier, the size of the telencephalon relative to the rest of the brain increases substantially as we go from fishes to amphibians, to reptiles, to birds and mammals. Even when we consider only mammals, the cerebral cortex increases in size and complexity when we compare animals such as rodents, whose behavioral repertoires are relatively simple, with animals such as primates that have much more complex behavior.

The most dramatic increase in the size of the cerebral cortex took place during the last several million years of human evolution. The incredible intellectual capacities of *Homo sapiens* are associated with enlargement of the cerebral cortex. Humans do not have the largest brains in the animal kingdom; elephants, whales, and porpoises have larger brains in terms of mass. If we compare brain size to body size, however, humans and dolphins top the list. Humans have the largest ratio of brain size to body size, and they have the most highly developed cerebral cortex. Another feature of the cerebral cortex that reflects increasing behavioral and intellectual capabilities is the ratio of association cortex to primary somatosensory and motor cortexes. Humans have the largest relative amount of association cortex.

Information Processing by Neuronal Networks

In Chapter 44 we learned how neurons interact to process information. A goal of neurobiology is to understand the complex functions of the nervous system in terms of the properties of neurons and synapses between them. We will use two subsystems as examples to show how the functions of the nervous system can be understood in terms of neuronal networks. The first example, the autonomic nervous system, consists of efferent pathways; the second, the visual system, consists of afferent and integrative pathways. Techniques that have allowed neurobiologists to trace neuronal connections, chemically characterize synapses, and record the activities of single cells and groups of cells have advanced our understanding of how certain subsystems of the nervous system work.

The autonomic nervous system controls organs and organ systems

The autonomic nervous system is divided into two parts: the **sympathetic** and **parasympathetic** divisions. These two divisions work in opposition to each other in their effects on most organs, one causing an increase in activity and the other causing a decrease. The best-known functions of the autonomic nervous system are those of the sympathetic division called the "fight-or-flight" mechanisms, which increase heart rate, blood pressure, and cardiac output and prepare the body for emergencies (see Chapter 41). In contrast, the parasympathetic division slows the heart and lowers blood pressure.

It is tempting to think of the sympathetic division as the one that speeds things up and the parasympathetic division as the one that slows things down, but that is not always a correct distinction. The sympathetic division slows the digestive system and the parasympathetic division accelerates it. The two divisions of the autonomic nervous system are easily distinguished from each other by their anatomy, their neurotransmitters, and their actions (Figure 46.11).

Both divisions of the autonomic nervous system are efferent pathways. Each autonomic efferent pathway begins with a neuron that has its cell body in the brain stem or spinal cord and uses acetylcholine as its neurotransmitter. These cells are called **preganglionic neurons** because the second neuron in the pathway with which they synapse resides in a **ganglion** (a collection of neuronal cell bodies that is outside of the CNS). The second neuron is called a **postganglionic neuron** because its axon extends out from the ganglion. The axons of the postganglionic neurons end on the cells of the target organs.

The postganglionic neurons of the sympathetic division use norepinephrine as their neurotransmitter; those of the parasympathetic division use acetylcholine. In organs that receive both sympathetic and parasympathetic input, the target cells respond in opposite ways to norepinephrine and to acetylcholine. A region of the heart called the *pacemaker*, which generates the heartbeat, is an example. Stimulating the sympathetic nerve to the heart or dripping norepinephrine onto the pacemaker region depolarizes the pacemaker cells, increases their firing rate, and causes the heart to beat faster. Stimulating the parasympathetic nerve to the heart or dripping acetylcholine onto the pacemaker region hyperpolarizes the pacemaker cells, decreases their firing rate, and causes the heart to beat slower. In contrast,

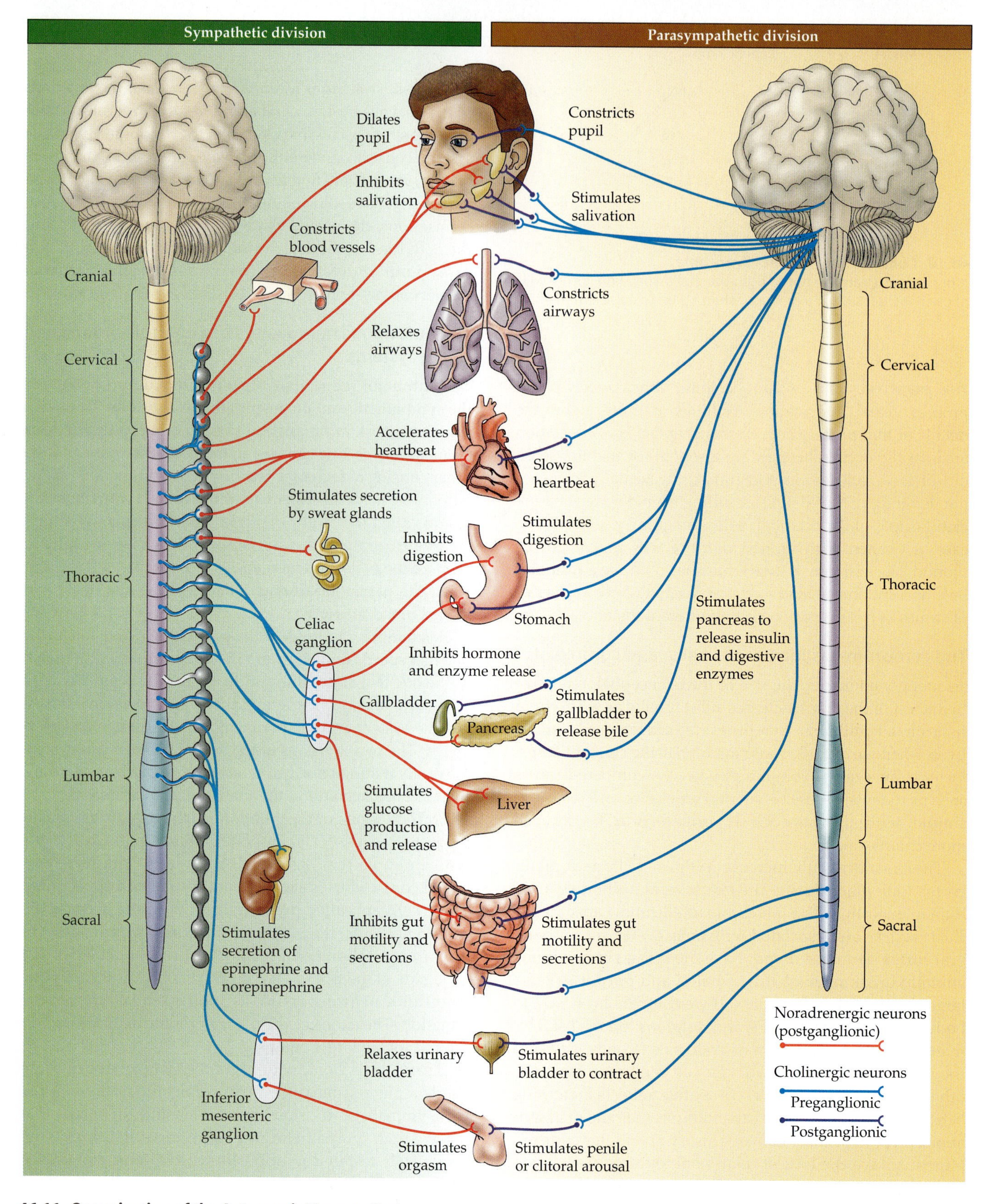

46.11 Organization of the Autonomic Nervous System
The autonomic nervous system is divided into the sympathetic and parasympathetic divisions, which work in opposition to each other in their effects on most organs (one causing an increase and the other a decrease in activity).

in the digestive tract, norepinephrine hyperpolarizes muscle cells, which slows digestion, and acetylcholine depolarizes muscle cells, which accelerates digestion.

The sympathetic and parasympathetic divisions of the autonomic nervous system can also be distinguished by anatomy (see Figure 46.11). The preganglionic neurons of the parasympathetic division come from the brain stem and the last segment of the spinal cord. The preganglionic neurons of the sympathetic division come from the upper regions of the spinal cord below the neck—the thoracic and lumbar regions. The ganglia of the sympathetic division are mostly lined up in two chains, one on either side of the spinal cord. The parasympathetic ganglia are close to—sometimes sitting on—the target organs.

The autonomic nervous system is an important link between the CNS and many physiological functions of the body. Its control of diverse organs and tissues is crucial to homeostasis. In spite of its complexity, work by neurobiologists and physiologists over many decades has made it possible to understand its functions in terms of neuronal properties and circuits. In Chapter 49, for example, we will see how information from pressure receptors in the blood vessels is transmitted to the CNS, where it produces autonomic signals that control the rate of the heartbeat.

Neurons and circuits in the occipital cortex integrate visual information

In Chapter 45 we learned that the information conveyed to the brain in the optic nerve consists of action potentials that are stimulated by light falling on small circular areas of the retina called receptive fields. A receptive field contains many photoreceptor cells connected together in a circuit in such a way that the signals they produce are integrated and transmitted to the brain by a single retinal ganglion cell. The axon of each ganglion cell travels to the brain in the optic nerve. How does the brain construct visual images from information about circular patches of light falling on the retina?

Information from the retina is transmitted through the optic nerve to a relay station in the thalamus, and then to the brain's visual processing area, in the occipital cortex at the back of the cerebral hemispheres (see Figure 46.7*b*). David Hubel and Torsten Wiesel of Harvard University studied the activity of neurons in this *visual cortex*. They recorded the activities of single cells in the brains of living animals while they stimulated the animals' retinas with spots and bars of light. They found that cells in the visual cortex, like retinal ganglion cells, have receptive fields—specific areas of the retina that, when stimulated by light, influence the rate at which the cells fire action potentials.

Cells in the visual cortex, however, have receptive fields that differ from the simple circular receptive fields of retinal ganglion cells. Cortical cells called **simple cells** are maximally stimulated by bars of light that have specific orientations. Simple cells probably receive input from several ganglion cells whose circular receptive fields are lined up in a row.

Complex cells in the visual cortex are also maximally stimulated by a bar of light with a particular orientation, but the bar may fall anywhere on a large area of retina described as that cell's receptive field. Complex cells receive input from several simple cells that share a certain stimulus orientation, but have receptive fields in different places on the retina (Figure 46.12). Some complex cells respond most strongly when the bar of light moves in a particular direction.

The concept that emerges from these experiments is that the brain assembles a mental image of the visual world by analyzing edges of patterns of light falling on the retina.

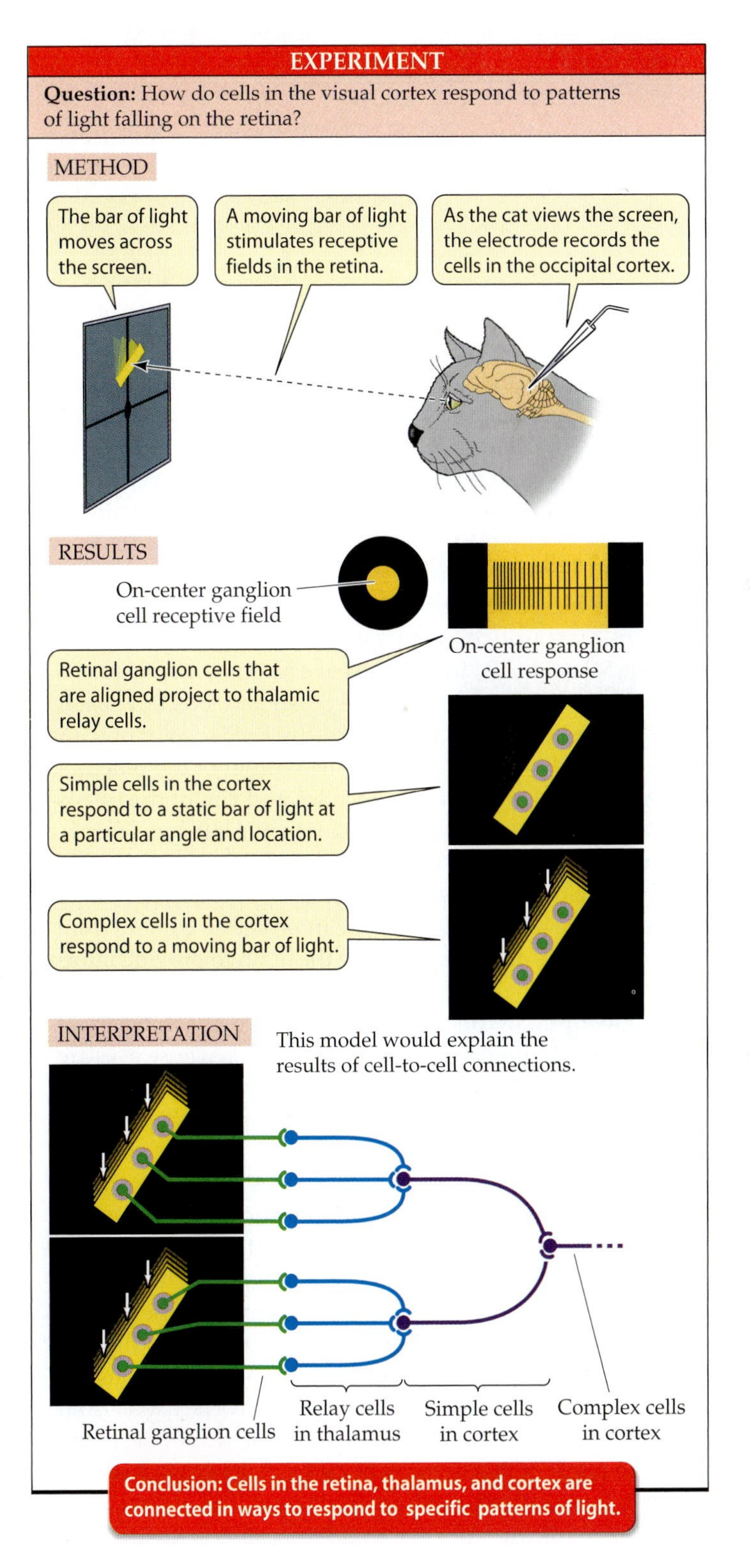

46.12 Receptive Fields of Cells in the Visual Cortex
Cells in the visual cortex respond to specific patterns of light falling on the retina. Ganglion cells that project information about circular receptive fields converge on simple cells in the cortex in such a way that the simple cells have linear receptive fields. Simple cells project to complex cells in such a way that the complex cells can respond to linear stimuli falling on different areas of the retina.

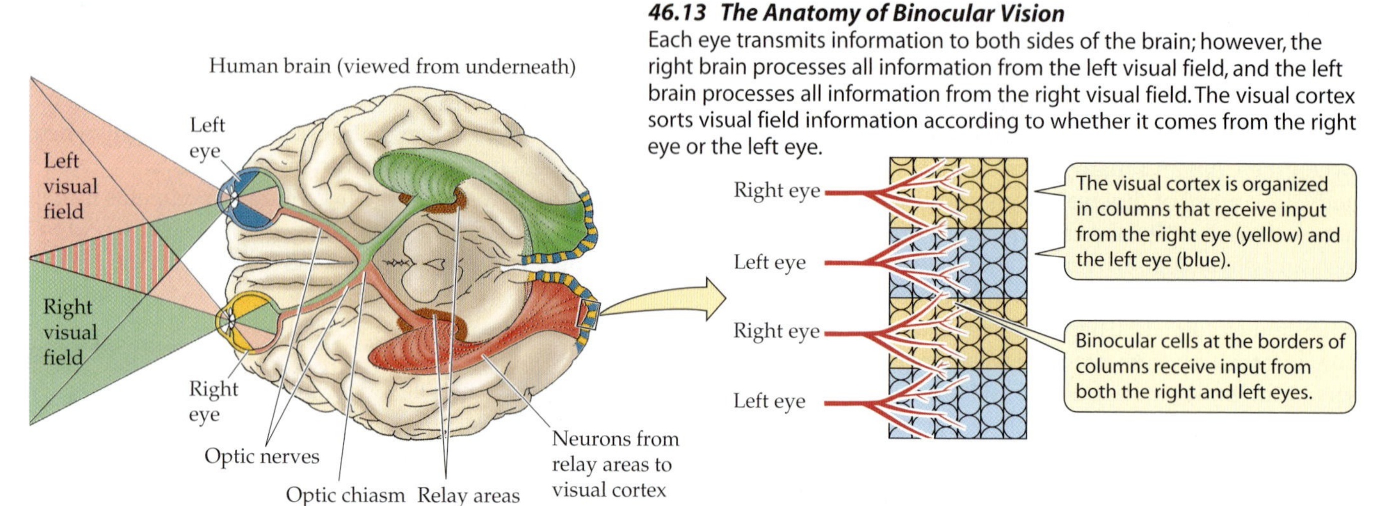

46.13 The Anatomy of Binocular Vision
Each eye transmits information to both sides of the brain; however, the right brain processes all information from the left visual field, and the left brain processes all information from the right visual field. The visual cortex sorts visual field information according to whether it comes from the right eye or the left eye.

This analysis is conducted in a massively parallel fashion. Each retina sends a million axons to the brain, but there are hundreds of millions of neurons in the visual cortex. Each bit of information from a retinal ganglion cell is received by hundreds of cortical cells, each responsive to a different combination of orientation, position, and even movement of contrasting lines in the pattern of light falling on the retina.

Cortical cells receive input from both eyes

How do we see objects in three dimensions? The quick answer is that our two eyes see overlapping, yet slightly different, visual fields; that is, we have *binocular vision*. Turn a typical conical flowerpot upside down and look down at it so that the bottom of the pot is exactly in the center of your overall field of vision. You see the bottom of the pot, and you see equal amounts of the sides and rim of the pot as concentric circles. Now, if you close your left eye, you see more of the right side and right rim of the pot. With your right eye closed, you see more of the left side and left rim of the pot. The discrepancies in the information coming from your two eyes are interpreted by the brain to provide information about the depth and the three-dimensional shape of the flowerpot. If you are blind in one eye, you have great difficulty discriminating distances. Animals whose eyes are on the sides of their heads have nonoverlapping fields of vision and, as a result, poor depth vision, but they can see predators creeping up from behind.

The story of how the brain integrates information from two eyes begins with the paths of the optic nerves. If you look at the underside of the brain, the optic nerves from the two eyes appear to join together just under the hypothalamus and then separate again. The place where they join is called the **optic chiasm** (Figure 46.13). Axons from the half of each retina closest to your nose cross in the optic chiasm and go to the opposite side of your brain. The axons from the other half of each retina go to the same side of the brain.

The result of this division of axons in the optic chiasm is that all visual information from your left visual field (everything left of straight ahead) goes to the right side of your brain, as shown in red in Figure 46.13. All visual information from your right visual field goes to the left side of your brain, as indicated in green in the figure. Both eyes transmit information about a specific spot in your right visual field, for example, to the same place in the left visual cortex. How are the two sources of information integrated?

Cells in the visual cortex are organized in columns. These columns alternate: left eye, right eye, left eye, right eye, and so on. Cells closest to the border between two columns receive input from both eyes and are therefore called **binocular cells**. Binocular cells interpret distance by measuring the disparity between where the same stimulus falls on the two retinas.

What is disparity? Hold your finger out in front of you and look at it, closing one eye and then the other. Your finger appears to jump back and forth because its image falls on a different position on each retina. Repeat the exercise with an object at a distance. It doesn't appear to jump back and forth as much because there is less disparity in the positions of the image on the two retinas. Certain binocular cells respond optimally to a stimulus falling on both retinas with a particular disparity. Which set of binocular cells is stimulated depends on how far away the stimulus is.

When we look at something, we can detect its shape, color, depth, and movement. Where does all this information come together? Is there a single cell that fires only when a red sports car drives by? Probably not. A specific visual experience comes from simultaneous activity in a large collection of cells. In addition, most visual experiences are enhanced by information from the other senses and from memory as well. This realization helps explain why about 75 percent of the cerebral cortex is association cortex.

Understanding Higher Brain Functions in Cellular Terms

Very few functions of the nervous system have been worked out to the point of identifying the underlying neuronal networks. The processes responsible for the higher brain functions discussed in the remaining pages are undeniably complex. Nevertheless, neurobiologists, using a wide range of techniques, are making considerable progress in understanding some of the mechanisms involved. The following discussion presents several complex aspects of brain and behavior that present challenges to neurobiologists: sleep and dreaming, learning and memory, language use, and consciousness.

Sleeping and dreaming involve electrical patterns in the cerebrum

A dominant feature of our behavior is the daily cycle of sleep and waking. All birds and mammals, and probably all other vertebrates, sleep. We spend one-third of our lives sleeping, yet we do not know why or how.

We do know, however, that we need to sleep. Loss of sleep impairs alertness and performance. Most people in our society—certainly most college students—are chronically sleep-deprived. Large numbers of accidents and serious mistakes that endanger lives can be attributed to impaired alertness due to sleep loss. Yet insomnia (difficulty in falling or staying asleep) is one of the most common medical complaints. To discover ways of dealing with these problems, it is important to learn more about the neural control of sleep.

THE ELECTROENCEPHALOGRAM. A common tool of sleep researchers is the *electroencephalogram* (*EEG*). To record an EEG, electrodes are placed at different locations on the scalp, and changes in the electric potential differences between electrodes are recorded through time. These electric potential differences reflect the electrical activity of the neurons in the brain regions under the electrodes, primarily regions of the cerebral cortex. Pens writing on a moving chart are used to record the patterns of electric potential differences between electrodes (Figure 46.14*a*,*b*). Usually, the electrical activity of one or more skeletal muscles is also recorded on the chart; this record is called an *electromyogram* (*EMG*).

EEG and EMG patterns reveal the transition from being awake to being asleep. They also reveal that there are different states of sleep. In mammals other than humans, two major sleep states are easily distinguished. They are called **slow-wave sleep** and **rapid-eye-movement (REM) sleep**. In humans, we characterize sleep states as **non-REM sleep** and **REM sleep**. Human non-REM sleep is divided into four stages. Only the two deepest stages are considered true slow-wave sleep.

When a person falls asleep at night, the first sleep state entered is non-REM sleep, which progresses from stage 1 to stage 4. Stages 3 and 4 are deep, restorative, slow-wave sleep. After this first episode of non-REM sleep follows an episode of REM sleep. Throughout the night, we have four or five cycles of non-REM and REM sleep (Figure 46.14*c*). About 80 percent of our sleep is non-REM sleep and 20 percent is REM sleep.

We have vivid dreams and nightmares during REM sleep, which gets its name from the jerky movements of the eyeballs that occur during this state. The most remarkable feature of REM sleep is that inhibitory commands from the brain almost completely paralyze the skeletal muscles. Occasional muscle twitches break through the paralysis, as can be seen in a dog that appears to be trying to run in its sleep. If you look closely at a sleeping dog when its legs and paws are twitching, you will be able to see the rapid eye movements as well. Probably the function of muscle paralysis during REM sleep is to prevent the acting out of dreams. Sleepwalking, however, occurs during non-REM sleep.

The EEG characterization of sleep raises many questions. Why do we have such very different states of sleep?

(*a*)

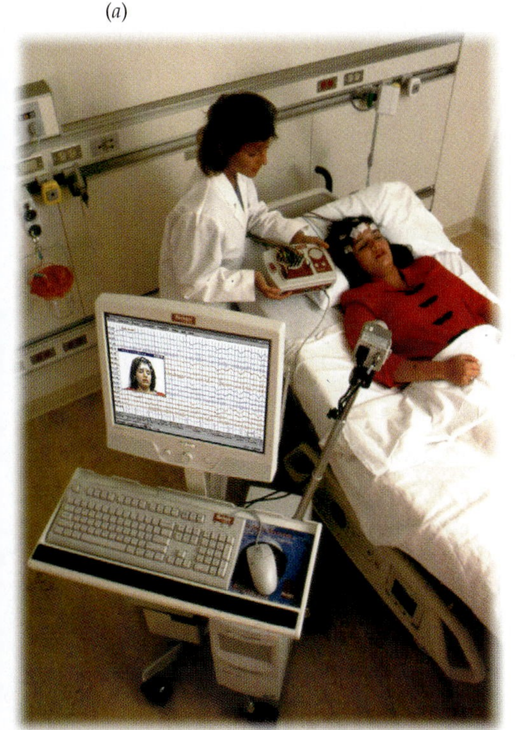

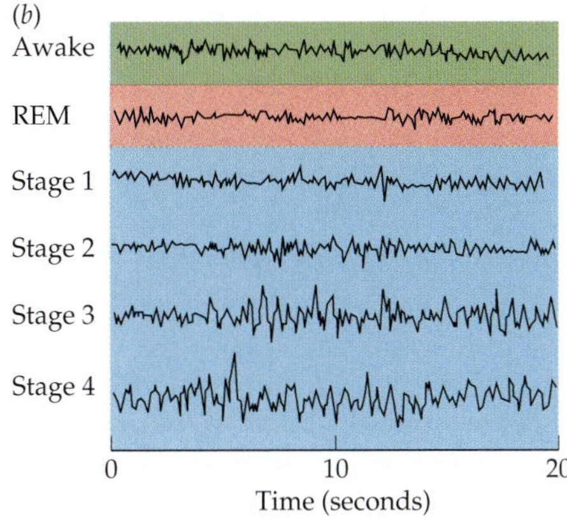

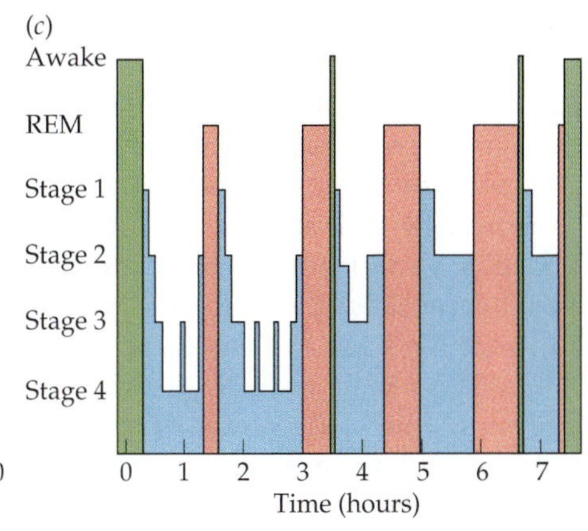

46.14 Patterns of Electrical Activity in the Cerebral Cortex Characterize Stages of Sleep
(*a*) Electrical activity in the cerebral cortex is detected by electrodes placed on the scalp and recorded on moving chart paper by a polygraph. (*b*) The resulting record is an electroencephalogram (EEG). (*c*) During a night, we cycle through the different stages of sleep.

Why does non-REM sleep precede and cycle with REM sleep? Why do we dream in REM sleep? Why do we have deeper non-REM sleep (stages 3 and 4) earlier in the night and more REM sleep later in the night? The answers to these questions are beginning to be revealed as we improve our understanding of the cells and circuits that underlie sleep.

CELLULAR CHANGES DURING SLEEP. There are striking neurophysiological differences between non-REM and REM sleep. Non-REM sleep is characterized by a decrease in the responsiveness of neurons in the thalamus and cerebral cortex. Remember that neurons have a resting membrane potential that is negative (the inside of the cell is negative relative to the outside), and a threshold potential for firing action potentials. Usually the resting potential is below the threshold potential, so the neuron is not firing. When synaptic input causes the plasma membrane to become less negative (depolarized), the cell can reach the threshold potential and fire action potentials.

During waking, several nuclei in the brain stem are continuously active. Many axons from these nuclei extend to the thalamus and the cortex, and the neurotransmitters released by the terminals of these axons are generally depolarizing. Therefore, these broadly distributed neurotransmitters keep the resting potential of the neurons of the thalamus and cortex close to threshold and sensitive to synaptic inputs, and thereby maintain waking.

With the onset of sleep, the activity in these brain stem nuclei decreases, and their terminals in the thalamus and cerebral cortex release less neurotransmitter. With the withdrawal of these depolarizing neurotransmitters, the resting potentials of the cells of the thalamus and cerebral cortex become more negative (hyperpolarized), and the cells are less sensitive to synaptic input. Their processing of information is inhibited, and consciousness is lost.

An interesting neuronal event happens as a result of thalamo-cortical hyperpolarization: The cells begin to fire in bursts. The synchronization of these bursts over broad areas of cerebral cortex results in the EEG slow-wave pattern that characterizes non-REM sleep. Studies of neurons of the thalamus and the cortex using intracellular recording techniques have shown that their hyperpolarization during non-REM sleep is due to increased opening of K^+ channels, and the bursting is due to Ca^{2+} channels that rapidly deactivate and require hyperpolarization to be reactivated. We can therefore explain the EEG pattern of non-REM sleep in terms of the properties of neurons and ion channels.

In addition to the brain stem nuclei that bring on sleep by decreasing their activity and causing general hyperpolarization, a substance that is broadly distributed in the extracellular fluid of the brain has a strong hyperpolarizing influence. This substance is adenosine—part of the molecule ATP, which supplies energy for most cellular processes. When cells cannot maintain an adequate supply of ATP, they release adenosine. It has therefore been suggested that increased concentrations of adenosine in the brain reflect the depletion of brain energy reserves, and that due to its hyperpolarizing influence, adenosine contributes to sleepiness and to the depth of non-REM sleep. A corollary of this hypothesis is that one function of non-REM sleep could be to restore brain energy reserves. There are many other molecules in the brain that influence sleep. Understanding their role in sleep control will provide new information on the possible functions of sleep.

At the transition from non-REM to REM sleep, dramatic changes occur. Some of the brain stem nuclei that were inactive during non-REM sleep become active again, causing a general depolarization of cortical neurons. Thus the bursts of firing cease, the slow waves in the EEG disappear, and the EEG resembles that of the waking brain. Because the resting potentials of the neurons return to near threshold levels, the cortex can process information, and vivid dreams occur.

During REM sleep, however, the brain inhibits both afferent (sensory) and efferent (motor) pathways; therefore the activity in the cortex is unconstrained by its usual sources of information. One example of the effect of this loss of motor output and sensory input is the frequently reported dream in which a person is trying to run but cannot move. We do not know the function of REM sleep, but since a wide variety of mammals have about the same percentage of total sleep time that is REM sleep, it is probably a rather basic, cellular function.

One prominent hypothesis about the functions of sleep is that it is essential for the maintenance and repair of neural connections, and for the neural changes that are involved in learning and memory. However, evidence for such functions is still meager.

Some learning and memory can be localized to specific brain areas

Learning is the modification of behavior by experience. *Memory* is the ability of the nervous system to retain what is learned and what is experienced. Even very simple animals can learn and remember, but these two abilities are most highly developed in humans.

Consider the amount of information associated with learning a language. The capacity of memory and the rate at which items can be retrieved are remarkable features of the human nervous system. A major challenge in neurobiology is to understand these phenomena in terms of the cells and molecules that make up the brain. Such knowledge could help us find ways to prevent the tragic loss of learning abilities and memory that occurs in the common condition of the elderly called *Alzheimer's disease*.

LEARNING. Learning that leads to long-term memory and modification of behavior must involve long-lasting synaptic changes. Synaptic changes that last for weeks have been discovered, as we saw in Chapter 44. High-frequency electrical stimulation of certain identifiable circuits of the mammalian hippocampus makes them more sensitive to subsequent stimulation. This phenomenon, called long-term potentiation (LTP), is an effect of ion channels controlled by the neurotransmitter glutamate. These channels allow both Na^+ and Ca^{2+} ions to enter the neurons, and the increase in intracellular Ca^{2+} activates enzymes that cause modifications in the cell that enhance its responsiveness.

In contrast, continuous, repetitive, low-level stimulation of the hippocampal circuit reduces its responsiveness, a phenomenon that has been called **long-term depression** (**LTD**). LTP and LTD have been demonstrated in circuits other than hippocampal circuits, and they may be fundamental cellular or molecular mechanisms involved in learning and memory.

A form of learning that is widespread among animal species is **associative learning**, in which two unrelated stimuli become linked to the same response. The simplest example of associative learning is the **conditioned reflex**, discovered by the Russian physiologist Ivan Pavlov. Pavlov was studying the control of digestive functions in dogs and observed that a dog salivates at the sight or smell of food—a simple autonomic reflex. He discovered that if he rang a bell just before food was presented to the dog, after a few trials the dog would salivate at the sound of the bell, even if no food followed. The salivation reflex was conditioned to be associated with the sound of a bell, which normally is unrelated to feeding and digestion.

This simple form of learning has been studied extensively in efforts to understand its underlying neural mechanisms. In a series of studies led by Richard Thompson, the eye-blink reflex of a rabbit in response to a puff of air directed at its eyes was conditioned to be associated with a tone stimulus. After conditioning, the rabbit blinked when presented with just the tone (Figure 46.15). A small and specific area of the cerebellum was discovered to be necessary for this conditioned reflex. Thus it was possible to localize learning to an identifiable set of synapses in the mammalian brain.

MEMORY. Attempts to treat human neurological diseases have led to the localization of areas of the brain involved in the formation and recall of memories. *Epilepsy* is a disorder characterized by uncontrollable increases in neural activity in specific parts of the brain. The resulting *seizures*, or "epileptic fits," can endanger the afflicted individual. In the past, serious cases of epilepsy were sometimes treated by destroying the part of the brain from which the surge of activity originated.

To find the right area, the surgery was done under local anesthesia, and different regions of the brain were electrically stimulated with fine electrodes while the patient reported on the resulting sensations. When some regions of association cortex were stimulated, patients reported vivid

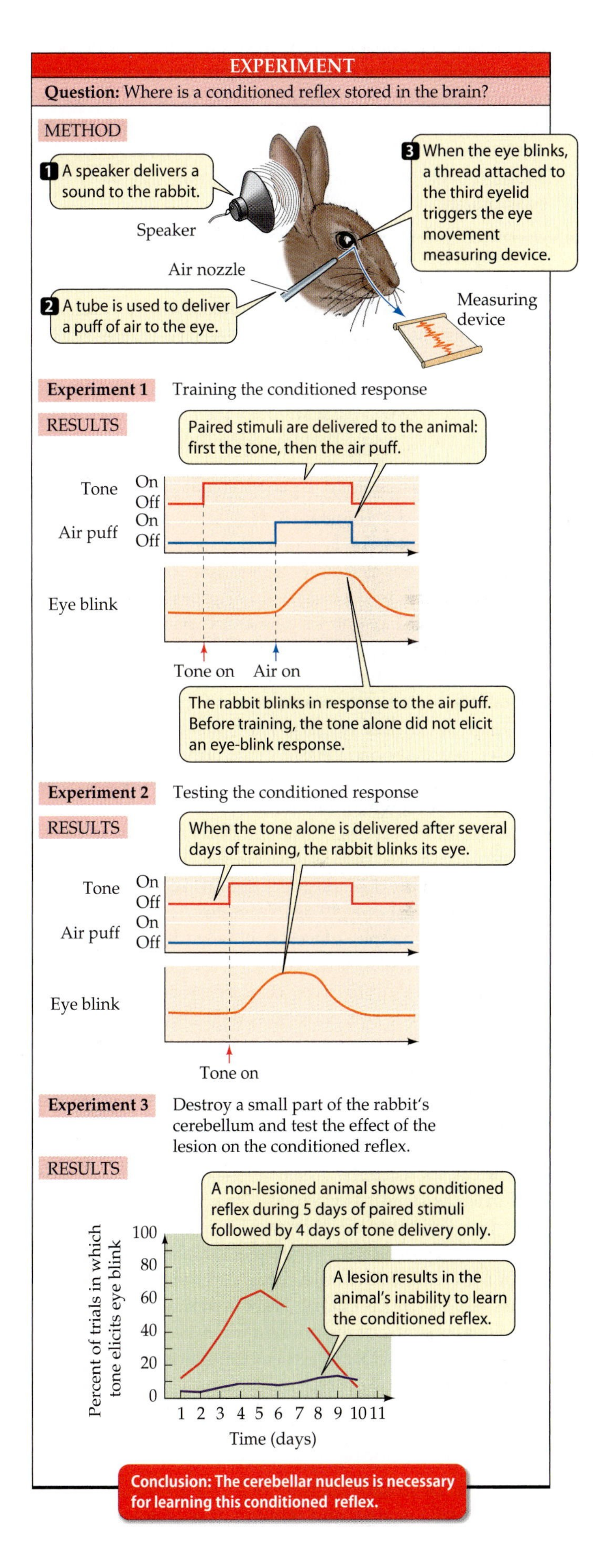

46.15 The Conditioned Eye-Blink Reflex Depends on a Cerebellar Circuit
A small and specific area of the cerebellum is necessary for a rabbit to form a conditioned reflex.

memories. Such observations were the first evidence that memories have anatomical locations in the brain and exist as properties of neurons and networks of neurons. Yet the destruction of a small area does not completely erase a memory, so it is postulated that memory is a function distributed over many brain regions and that a memory may be stimulated via many different routes.

You can recognize several forms of memory from your own experience. There is *immediate memory* for events that are happening now. Immediate memory is almost perfectly photographic, but it lasts only seconds. *Short-term memory* contains less information, but it lasts longer—on the order of 10 to 15 minutes. If you are introduced to a group of new people, you may remember most of their names for 5 or 10 minutes, but you will have forgotten them in an hour or so if you have not repeated them, written them down, or used them in a conversation. Repetition, use, or reinforcement by something that gets your attention (such as the title President) facilitates the transfer of short-term memory to *long-term memory*, which can last for days, months, or years.

Knowledge about neural mechanisms for the transfer of short-term memory to long-term memory has come from observations of persons who have lost parts of the limbic system, notably the hippocampus. A famous case is that of a man identified as H.M., whose hippocampus on both sides of the brain was removed in an effort to control severe epilepsy. Since that surgery, H.M. has not been able to transfer information to long-term memory. If someone is introduced to him, has a conversation with him, and then leaves the room, when that person returns he or she is unknown to H.M.—it is as if the previous conversation had never taken place. H.M. retains memories of events that happened before his surgery, but he remembers post-surgery events for only 10 or 15 minutes.

Memory of people, places, events, and things is called *declarative memory* because you can consciously recall and describe them. Another type of memory, called *procedural memory*, cannot be consciously recalled and described: It is the memory of how to perform a motor task. When you learn to ride a bicycle, ski, or use a computer keyboard, you form procedural memories. Although H.M. is incapable of forming declarative memories, he is capable of forming procedural memories. When taught a motor task day after day, he cannot recall the lessons of the previous day, yet his performance steadily improves. Thus procedural learning and memory must involve mechanisms different from those used in declarative learning and memory.

Our understanding of learning and memory in cellular terms is very rudimentary. New techniques that enable functional imaging of the brain in ways that reveal changes in the metabolic activity of specific regions and structures are greatly enhancing progress in this area.

Language abilities are localized in the left cerebral hemisphere

No aspect of brain function is as integrally related to human consciousness and intellect as is language. Therefore, studies of the brain mechanisms that underlie the acquisition and use of language are extremely interesting to neuroscientists. A curious observation about language abilities is that they are usually located in only one cerebral hemisphere—which in 97 percent of people is the left hemisphere. This phenomenon is referred to as the **lateralization** of language functions.

Some of the most fascinating research on this subject was conducted by Roger Sperry and his colleagues at the California Institute of Technology; Sperry received the Nobel prize in medicine for this work. The two cerebral hemispheres are connected by a tract of white matter called the **corpus callosum**. In one severe form of epilepsy, bursts of action potentials travel from hemisphere to hemisphere across the corpus callosum. Cutting the tract eliminates the problem, and patients function nearly normally following the surgery. But experiments revealed interesting deficits in the language abilities of these "split-brain" persons. Without the connection between the two hemispheres, the knowledge or experience of the right hemisphere could no longer be expressed in language, nor could language be used to communicate with the right hemisphere.

Another curious feature of our nervous systems is that the left side of the body is served (in both sensory and motor aspects) mostly by the right side of the brain, and the right side of the body is served mostly by the left side of the brain. Thus, sensory input from the right hand goes to the left cerebral hemisphere, and sensory input from the left hand goes to the right cerebral hemisphere. Language abilities reside predominantly in the left hemisphere.

The mechanisms of language in the left hemisphere have been the focus of much research. Again, the experimental subjects are persons who have suffered damage to the left hemisphere and are left with one of many forms of *aphasia*, a deficit in the ability to use or understand words. These studies have identified several language areas in the left hemisphere (Figure 46.16).

Broca's area, located in the frontal lobe just in front of the motor cortex, is essential for speech. Damage to Broca's area results in halting, slow, poorly articulated speech or even complete loss of speech, but the patient can still read and understand language. In the temporal lobe, close to its border with the occipital lobe, is *Wernicke's area*, which is more involved with sensory than with motor aspects of language. Damage to Wernicke's area can cause a person to lose the ability to speak sensibly while retaining the abilities to form the sounds of normal speech and to imitate its cadence. Moreover, such a patient cannot understand spoken or written language. Near Wernicke's area is the *angular gyrus*, which is believed to be essential for integrating spoken and written language.

Normal language ability depends on the flow of information among various areas of the left cerebral cortex. Input from spoken language travels from the primary auditory cortex to Wernicke's area (see Figure 46.16*a*). Input from written language travels from the primary visual cortex to the angular gyrus to Wernicke's area (see Figure

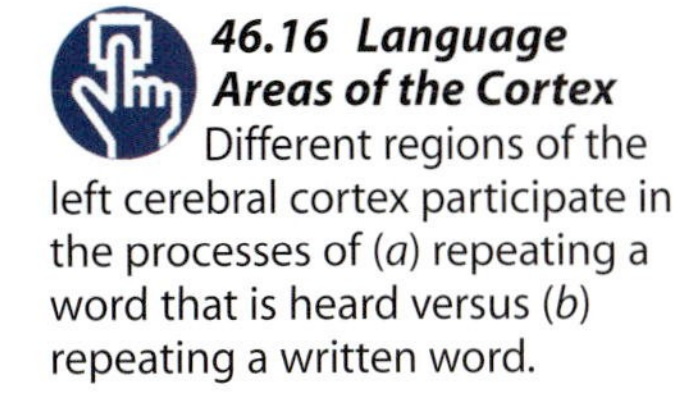

46.16 Language Areas of the Cortex
Different regions of the left cerebral cortex participate in the processes of (*a*) repeating a word that is heard versus (*b*) repeating a written word.

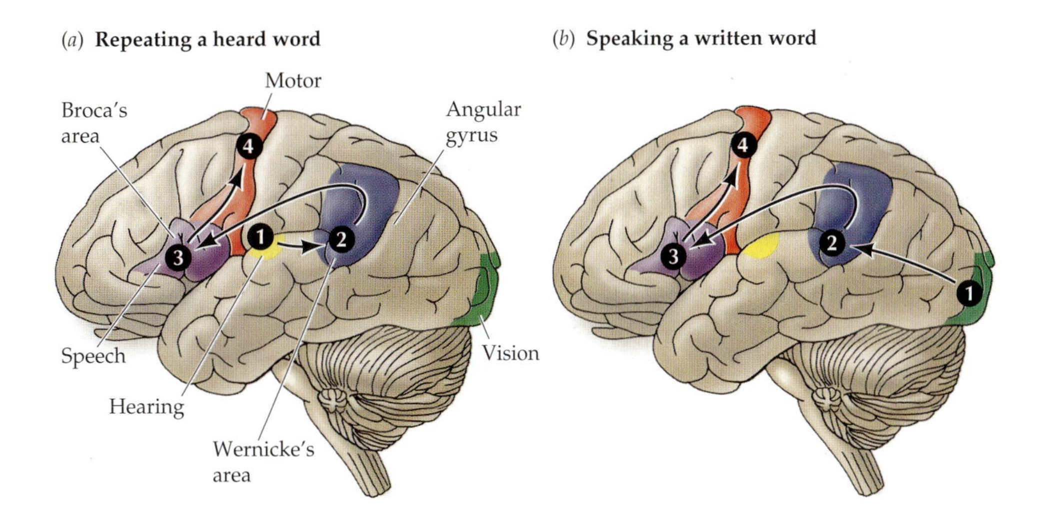

46.16*b*). Commands to speak are formulated in Wernicke's area and travel to Broca's area and from there to the primary motor cortex. Damage to any one of those areas or the pathways between them can result in aphasia. Using modern methods of functional brain imaging, it is possible to see the metabolic activity in different brain areas when the brain is using language (Figure 46.17).

What is consciousness?

This chapter has only scratched the surface of our knowledge about the organization and functions of the human brain, but it may give you some idea of the incredible challenge that neurobiologists face in trying to understand their own brains. Progress is being aided by powerful new technologies, such as patch clamping (see Chapter 44), functional imaging (see Figure 46.17), and neurochemical and molecular methods. However, even these sophisticated new research tools may not allow us to answer the question "What is consciousness?"

If you look at a black dog, you are conscious of the fact that it is a dog, it is black, and it is a Labrador retriever, and you may remember that its name is Sarina, it is 3 years old, it belongs to your friend Meera, and so on. From what you have learned in this chapter, imagine how many neurons would be active during this experience: neurons in the visual system, the language areas, and in different regions of association cortex. But is being conscious of the black dog simply a result of the fact that all of these neurons are firing at the same time? Your brain is simultaneously processing many other sensory inputs, but you are not necessarily conscious of those inputs. What makes you conscious of the black dog and associated memories and not conscious of other information the brain is processing at the same time?

If we could describe all the neurons and all the synapses involved in the conscious experience of seeing and naming a black dog, and then build a computer with devices that modeled all these neurons and connections, would that computer be conscious? It has been said that the question of consciousness resolves into two types of problems: "easy" and "hard." The easy problems deal with all the cells and circuits that process the information that is involved in conscious experience. The implication of "easy" is that we seem to have the tools to solve these kinds of problems, as complex as they may be. The hard problems involve explaining how properties of cells and networks result in consciousness, and we seem to lack the proper tools or concepts even to begin to solve these problems.

Passively viewing words | Listening to words | Speaking words | Generating words

46.17 Imaging Techniques Reveal Active Parts of the Brain
Positron emission tomography (PET) scanning reveals the brain regions that are activated by different aspects of language use. A radioactive form of glucose is given to the subject. Radioactivity accumulates in brain areas in proportion to their metabolic use of glucose. The PET scan visualizes levels of radioactivity in specific brain regions.

Chapter Summary

The Nervous System: Structure, Function, and Information Flow

▶ The brain and spinal cord make up the central nervous system; the cranial and spinal nerves make up the peripheral nervous system. A nerve is a bundle of many axons carrying information to and from the central nervous system. **Review Figure 46.1**

▶ The nervous system can be modeled conceptually in terms of the direction of information flow and whether or not we are conscious of the information. **Review Figure 46.2**

▶ The vertebrate nervous system develops from a hollow dorsal neural tube. The brain forms from three swellings at the anterior end of this neural tube, which become the hindbrain, the midbrain, and the forebrain. **Review Figure 46.3**

▶ The forebrain develops into the cerebral hemispheres (the telencephalon) and the underlying thalamus and hypothalamus (the diencephalon). The midbrain and hindbrain develop into the brain stem. More primitive and autonomic functions are localized in the brain stem, and conscious experience depends on the cerebrum.

Functional Subsystems of the Nervous System

▶ The nervous system is composed of many subsystems that function simultaneously. Some important subsystems are the spinal cord, the reticular system, the limbic system, and the cerebrum.

▶ The spinal cord communicates information between the brain and the body. It also processes and integrates much information, and can issue some commands to the body without input from the brain. **Review Figure 46.4**

▶ The reticular system of the brain stem is a complex network that directs incoming information to appropriate brain stem nuclei that control autonomic functions, as well as transmitting the information to the forebrain that results in conscious sensation. The reticular system controls the level of arousal of the nervous system. **Review Figure 46.5**

▶ The limbic system is an evolutionarily primitive part of the forebrain that is involved in emotions, physiological drives, instincts, and memory. **Review Figure 46.6**

▶ The cerebral hemispheres are the dominant structures of the human brain. Their surfaces consist of a layer of neurons called the cerebral cortex.

▶ Most of the cerebral cortex is involved in higher-order information processing, and these areas are generally called association cortex.

▶ The cerebral hemispheres can be divided into temporal, frontal, parietal, and occipital lobes. Many motor functions are localized in parts of the frontal lobe, information from many receptors around the body projects to a region of the parietal lobe, visual information projects to the occipital lobe, and auditory information projects to a region of the temporal lobe. **Review Figures 46.7, 46.8, 46.9, 46.10**

Information Processing by Neuronal Circuits

▶ The functions of the nervous system are beginning to be understood in terms of the properties of cells organized in neuronal circuits.

▶ The autonomic nervous system consists of efferent pathways that control the organs and organ systems of the body. Its sympathetic and parasympathetic divisions normally work in opposition to each other. These divisions are characterized by their anatomy, neurotransmitters, and effects on target tissues. **Review Figure 46.11**

▶ Neuronal circuits in the occipital cortex integrate visual information. Receptive field responses of retinal ganglion cells are communicated to the brain in the optic nerves. This information is projected to the visual cortex in such a way as to create receptive fields for cortical cells.

▶ A simple cell is stimulated by a bar of light with a specific orientation falling at a specific location on the retina. A complex cell is maximally stimulated by such a stimulus moving across the retina. The visual cortex seems to assemble a mental image of the visual world by analyzing edges of patterns of light. **Review Figure 46.12**

▶ Binocular vision results from circuits that communicate information from both eyes to binocular cells in the visual cortex. These cells interpret distance by measuring the disparity between where the same stimulus falls on the two retinas. **Review Figure 46.13**

Understanding Higher Brain Functions in Cellular Terms

▶ Humans have a daily cycle of sleep and waking. Sleep can be divided into slow-wave (non-REM) sleep and rapid-eye-movement (REM) sleep. Human non-REM sleep is divided into four stages of increasing depth. **Review Figure 46.14**

▶ Some learning and memory processes have been localized to specific brain areas. Repeated activation of identified circuits in brain regions such as the hippocampus have revealed long-lasting changes in synaptic properties referred to as long-term potentiation and long-term depression, which may be involved in learning and memory. **Review Figure 46.15**

Complex memories can be elicited by stimulating small regions of association cortex. Damage to the hippocampus can destroy the ability to form long-term declarative memories, but not procedural memories.

▶ Language abilities are localized mostly in the left cerebral hemisphere, a phenomenon known as lateralization.

▶ Different areas of the left hemisphere—including Broca's area, Wernicke's area, and the angular gyrus—are responsible for different aspects of language. **Review Figure 46.16**

For Discussion

1. The mammalian nervous system begins as a hollow tubular structure, but just before the three regions of the brain begin to form, this hollow tube constricts at the site that will become the junction between the brain and the spinal cord. What could be the significance of this constriction? What does it suggest about the mechanism by which the brain develops?
2. The stretch receptors in muscles are modified muscle fibers, and they have their own motor neurons. What is the function of these motor neurons? To think about this question, remember that the function of the monosynaptic reflex is to adjust muscle tension to a change in load so that the position of the limb does not change.
3. A patient is unable to speak coherently. He can read and write, and he has no obvious loss of muscle function. Where would you expect to find an abnormality if you did brain scans of this patient?
4. We described the organization of the visual cortex as columns of cells that alternately receive input from the left eye and the right eye. If a young kitten is allowed to see light out of only one eye for a day, more synapses begin to form in the cortical columns receiving input from that eye, while synapses decrease in the intervening columns. This redistribution of synapses does not occur, however, if the kitten is not allowed to sleep. What hypotheses could you propose on the basis of these results?

47 Effectors: Making Animals Move

THE CENTRAL NERVOUS SYSTEM IS MORE than a processor and storage medium for information. It also allows animals to respond to that information. **Effectors** are tissues and organs commanded by the CNS to carry out these responses, and most responses of animals involve movement.

A fascinating array of adaptations enable animals to move. Consider the act of jumping. When you jump, neural signals from the visual cortex of your brain are routed through spinal circuits that tell certain leg muscles to contract, extending your legs, and hence you jump. Highly skilled and trained athletes can actually outleap their own body height.

But as "record jumpers" go, many other animals—cats, spiders, kangaroos, and fleas to name just a few—far surpass even the Olympian feats of humans. A flea, for example, can jump more than 200 times its body length. Unlike a human jumper, the flea's jumping mechanism doesn't involve muscles but works like a slingshot. The flea is so small, and its initial acceleration is so great, that no muscle can contract fast enough to cause such a movement. Instead, at the base of its jumping legs is an elastic material that is compressed by muscles while the flea is resting. When a trigger mechanism is released, the elastic material recoils and "fires" the flea up and over to its target (or away from an enemy).

Jumping is just one adaptation an animal can use to respond to information received by its sensory receptors. Effectors include the internal organs and organ systems that the animal uses to control its internal environment; these effectors are the subjects of subsequent chapters. In this chapter, we focus on cilia, flagella, muscles, and skeletons—the mechanisms that create mechanical forces and use those forces to change shape and move, and which are the basis for most animal behavior. At the end we will briefly consider a few effectors other than those that create movement.

A Champion Jumps
Sergey Kliugin of Russia won the gold medal for high jumping in the 2000 Summer Olympics. His winning jump was 2.35 meters, or about 1.3 times his height.

Cilia, Flagella, and Cell Movement

Two subcellular structures, microtubules and microfilaments (see Figure 4.21), generate cell movement. Both of these structures consist of long protein molecules that can change length or shape. Microtubules generate the small-scale movements of cilia and flagella. Microfilaments reach their highest level of organization in muscle cells, which generate large-scale movements.

Cilia are tiny, hairlike appendages of cells

Certain protists are covered by dense patches of **cilia** that propel them through their aqueous environment. Each cilium is tiny, about 0.25 μm in diameter. Multicellular animals use ciliated cells to move liquids and particles over cell surfaces. Many invertebrates use ciliated cells to obtain food and oxygen. Some mollusks, for example, use cilia to circulate a current of water across their gas exchange and feeding surfaces (Figure 47.1).

The airways of many animals are lined with and cleaned by ciliated cells (Figure 47.2). In humans, the cilia continuously sweep a layer of mucus from deep down in the lungs, up through the windpipe, and into the throat. The mucus carries particles of dirt and dead cells. We can then either swallow or spit out the mucus, and with it the trapped detritus. Ciliated cells lining the female reproductive tract create currents that sweep eggs from the ovaries into the oviducts and all the way down to the uterus.

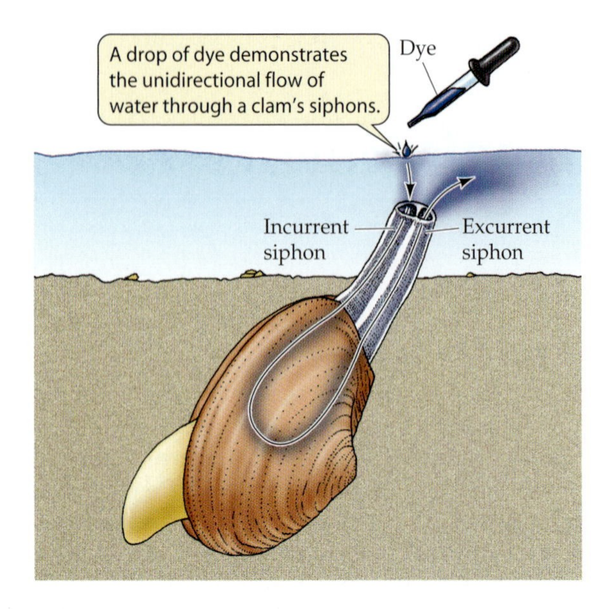

47.1 Cilia Create Water Currents in a Clam's Siphons
In burrowing mollusks such as clams, cilia lining the siphons maintain a unidirectional flow of water: in one siphon, over the gills, and out the other siphon. The gills extract oxygen and food from this flow of water.

47.2 Cilia Line Respiratory Passages
A scanning electron micrograph of a rabbit's airway shows many cilia.

A cilium moves with the same basic motion as a swimmer's arms during the breaststroke (Figure 47.3*a*). During the *power stroke*, the cilium projects stiffly outward and moves backward, propelling the cell forward (or the medium backward). During the *recovery stroke*, the cilium folds as it returns to its original position. The power stroke is fast, the recovery stroke slow. As you know from moving your arms or legs in water, there is less resistance the slower you move. The resistance of the medium to the recovery stroke is thus slight compared with its resistance to the power stroke. Groups of cilia typically beat in coordinated waves. At any particular moment, some cilia of a cell are moving through the power stroke and others are recovering.

Flagella are like long cilia

The **flagella** of eukaryotes are identical to cilia except that they are longer and occur singly, or in groups of only a few, on any one cell. Flagellated cells maintain a flow of water through the bodies of sponges, bringing in food and oxygen and removing carbon dioxide and wastes. Flagella power the movement of the sperm of most species. Because of their greater length, flagella have a whiplike stroke pattern rather than the "swimming" stroke pattern of cilia (Figure 47.3*b*).

Cilia and flagella are moved by microtubules

The core of a cilium or a flagellum is called the **axoneme**. The axoneme contains a ring of nine pairs of microtubules. In the center of the ring may be an additional pair of microtubules, a single microtubule, or no microtubule. As discussed in Chapter 4, microtubules are hollow tubes formed from polymerization of the globular polypeptide tubulin. Other proteins in the axoneme form spokes, side arms, and cross-links. Side arms composed of the protein dynein generate force (see Figures 4.24 and 4.25). Dynein is an enzyme that catalyzes the hydrolysis of ATP and uses the released energy to change its shape, thereby generating mechanical force.

When the dynein arms on one microtubule pair contact a neighboring microtubule pair and bind to it, ATP is broken down, and the resulting conformational changes in the dynein molecules cause the arms to point toward the base of the axoneme (Figure 47.4). This action pushes the microtubule pair ahead in relation to its neighbor. The dynein arms then detach from the neighboring pair and reorient to their starting horizontal position. As the cycle is repeated,

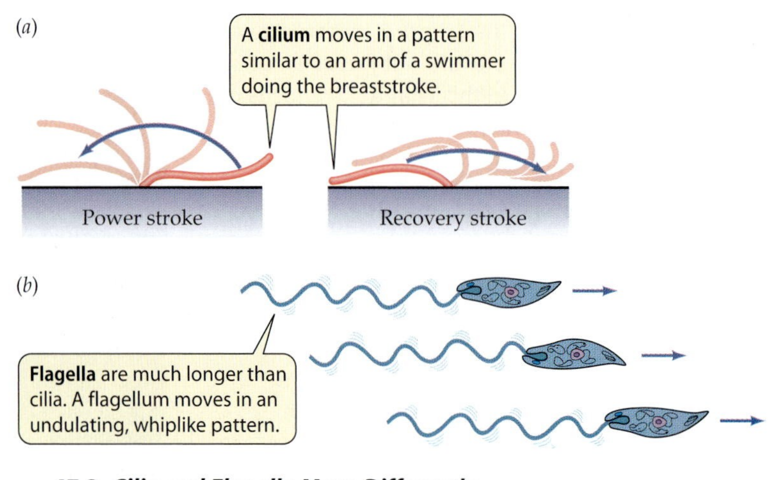

47.3 Cilia and Flagella Move Differently
Cilia (*a*) have a "swimmer's stroke," whereas flagella (*b*) have a whiplike motion.

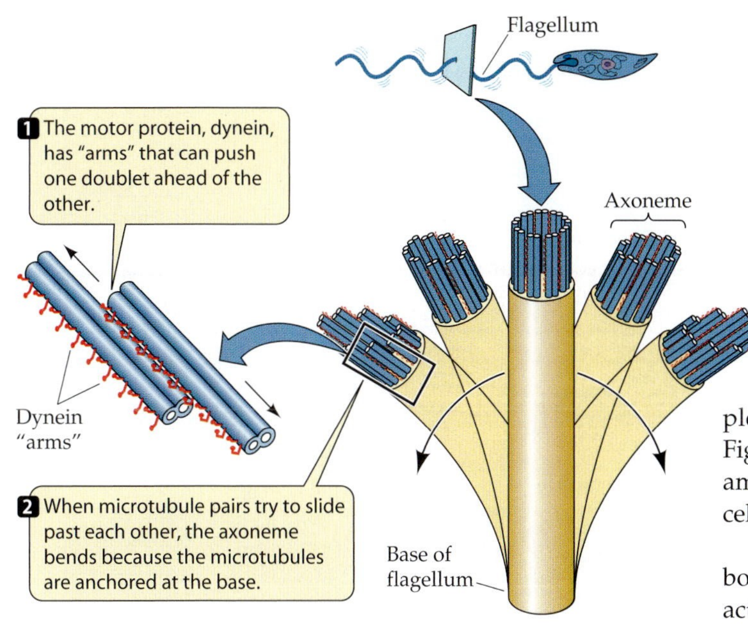

47.4 Microtubules Create Motion by Pushing Against Each Other
Cilia and flagella move because of interactions between microtubules in the axoneme.

adjacent microtubule pairs try to "row" past each other, with the dynein arms acting as "oars." Axonemes severed from cells continue to flex in a normal pattern if exposed to calcium ions (Ca^{2+}) and ATP, demonstrating that the motile mechanism is part of the axoneme itself and not derived from the associated cell.

Microtubules are intracellular effectors

Microtubules, as components of the cytoskeleton, contribute to the shape of eukaryotic cells. Microtubules are important intracellular effectors for changing cell shape, moving organelles, and enabling cells to respond to their environment. Microtubules change the shapes of cells and move cells by polymerizing and depolymerizing the protein tubulin.

As we saw in Chapter 9, the spindle that moves chromosomes to the mitotic poles at anaphase is made up of microtubules. Another example of microtubule involvement in cell movement is the growth of the axons of neurons in the developing nervous system. Neurons find and make their appropriate connections by sending out long extensions that search for the correct contact cells. If polymerization of tubulin is chemically inhibited, the neurons do not extend.

Microfilaments change cell shape and cause cell movements

Microfilaments are proteins that change conformation as a means of generating forces. The dominant microfilament in animal cells is the protein **actin**. Bundles of cross-linked actin strands form important structural components of cells. The microvilli that increase the absorptive surface area of the cells lining the gut are stiffened by actin microfilaments (see Figure 4.23), as are the stereocilia of the sensory hair cells in the mammlian ear (see Figure 45.12). Actin microfilaments can change the shape of a cell simply by polymerizing and depolymerizing.

Together with the protein **myosin**, actin microfilaments generate the contractile forces responsible for many aspects of cell movement and changes in cell shape. The contractile ring that divides an animal cell undergoing mitosis into two daughter cells is composed of actin microfilaments in association with myosin. The mechanisms that many cells employ to engulf materials (endocytosis; see Chapter 5 and Figure 19.4) also rely on interactions between actin microfilaments and myosin. Nets of actin and myosin beneath the cell membrane change a cell's shape during endocytosis.

Certain cells in multicellular animals travel within the body by **amoeboid movement**, which is generated by the activity of actin microfilaments and myosin. During development, many cells migrate by amoeboid movement. Throughout an animal's life, phagocytic cells circulate in the blood, squeeze through the walls of the blood vessels, and wander through the tissues by amoeboid movement. The mechanisms of amoeboid movement have been studied extensively in the protist for which this type of movement was named—the amoeba, which lives in freshwater streams and ponds (see the photograph on page 476).

Amoebas move by extending lobe-shaped projections called *pseudopods* and then seemingly squeezing themselves into those pseudopods. The cytoplasm in the core of the amoeba, called *plasmasol*, is relatively liquid, but just beneath the plasma membrane the cytoplasm is much thicker, and is called *plasmagel*. Reversible changes between plasmasol and plasmagel move the cytoplasm.

To form a pseudopod, the thick plasmagel in one area of the cell thins, allowing a bulge to form. Just under the cell surface, in the plasmagel, is a network of actin microfilaments that interacts with myosin to squeeze plasmasol into the bulge. As the microfilament network continues to contract, cytoplasm streams in the direction of the pseudopod. Eventually the cytoplasm at the leading edge of the pseudopod converts to plasmagel, and the pseudopod stops forming. Thus the basis for amoeboid motion is the ability of the cytoplasm to cycle through sol and gel states and the ability of the microfilament network under the cell membrane to contract and cause the cytoplasmic streaming that pushes out a pseudopod.

Muscle Contraction

Most behavioral and many physiological responses depend on muscle cells. Muscle cells are specialized for contraction and have high densities of actin and myosin. Such cells are found throughout the animal kingdom. Wherever whole tissues contract in animals, muscle cells are responsible.

In muscle cells, actin and myosin molecules are organized into **filaments** consisting of two or more molecules.

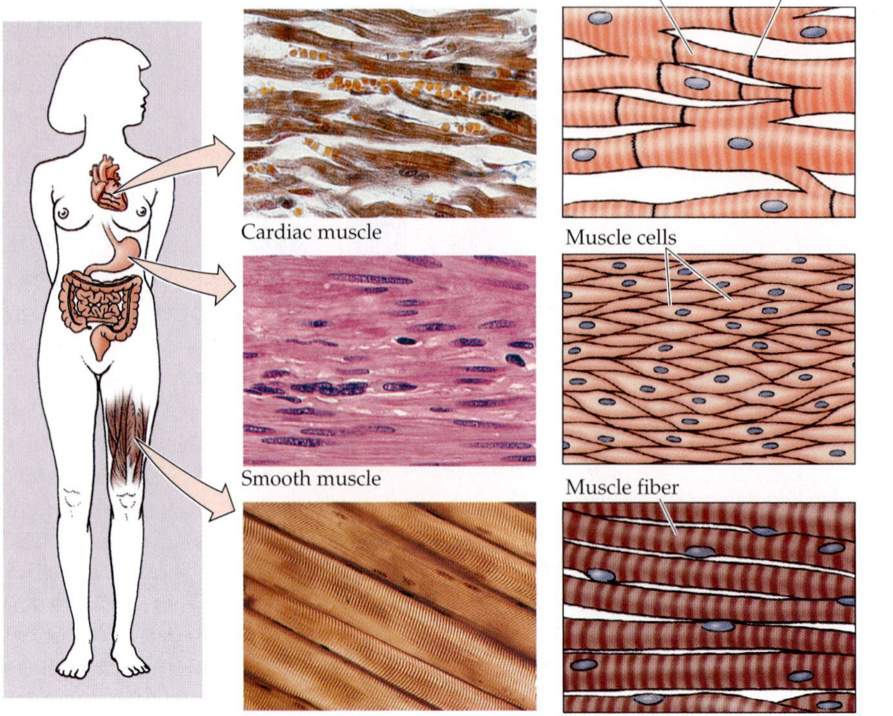

47.5 Vertebrate Muscle Tissue
The fibers of cardiac, or heart muscle (*top*), branch and create a meshwork that resists tearing or breaking. Intercalated discs provide strong mechanical adhesions between the cells. In smooth muscle (*center*), the cells are usually arranged in sheets. Skeletal muscle (*bottom*) appears striped, or striated. The individual cells, called muscle fibers, are very large and are multinucleated.

Actin filaments consist of two actin molecules twisting around each other, and myosin filaments are bundles of many myosin molecules. The actin and myosin filaments lie parallel to each other. When contraction is triggered, the actin and myosin filaments slide past each other in a telescoping fashion.

There are three types of vertebrate muscle: smooth muscle, cardiac (heart) muscle, and skeletal muscle (Figure 47.5). Contraction in all three types is triggered by action potentials moving along their membranes. Although they all use the same contractile mechanism, these three muscle types have important differences that adapt them to their particular functions.

Smooth muscle causes slow contractions of many internal organs

Smooth muscle provides the contractile force for most of our internal organs, which are under the control of the autonomic nervous system. Smooth muscle moves food through the digestive tract, controls the flow of blood through blood vessels, and empties the urinary bladder. Structurally, smooth muscle cells are the simplest muscle cells. They are usually long and spindle-shaped, and each cell has a single nucleus. Because the filaments of actin and myosin in smooth muscle are not as regularly arranged as those in the other muscle types, the contractile machinery is not obvious when the cells are viewed under the light microscope (Figure 47.5, center).

If we study smooth muscle tissue from a particular organ, such as the wall of the digestive tract, we find that it has some interesting properties. The cells are arranged in sheets, and individual cells in the sheets are in electrical contact with one another through gap junctions. As a result, an action potential generated in the membrane of one smooth muscle cell can spread to all the cells in the sheet of tissue.

Another interesting property of a smooth muscle cell is that the resting potential of its membrane is sensitive to being stretched. If the wall of the digestive tract is stretched in one location (as by receiving a mouthful of food), the membranes of the stretched cells depolarize, reach threshold, and fire action potentials that cause the cells to contract. Thus smooth muscle contracts after being stretched, and the harder it is stretched, the stronger the contraction. (Later in this chapter we will see how membrane depolarization triggers contraction.)

Other factors that alter the membrane potential of smooth muscle cells are the neurotransmitters of the autonomic nervous system (see Figure 46.11). In the case of the digestive tract, acetylcholine causes smooth muscle cells to depolarize and thus makes them more likely to fire action potentials and contract. Norepinephrine causes these muscle cells to hyperpolarize and therefore makes them less likely to fire action potentials and contract (Figure 47.6).

Cardiac muscle causes the heart to beat

Cardiac muscle looks different from smooth muscle or skeletal muscle when viewed under the microscope (Figure 47.5, top). The cells appear striped, or *striated*, because of the regular arrangement of bundles of actin and myosin filaments within them. Actin and myosin are arranged in a similar way in skeletal muscle, as we'll see below.

The unique feature of cardiac muscle cells is that they branch. The branches of adjoining cells are interdigitated into a meshwork that gives cardiac muscle an ability to resist tearing. As a result, the heart walls can withstand high pressures while pumping blood without the danger of developing leaks. Also adding to the strength of cardiac muscle are *intercalated discs* that provide strong mechanical adhesions between adjacent cells.

As is true of smooth muscle, the individual cells in a sheet of cardiac muscle are in electrical contact with one another. Gap junctions present in the intercalated discs present low resistance to ions or electric currents. Therefore, a

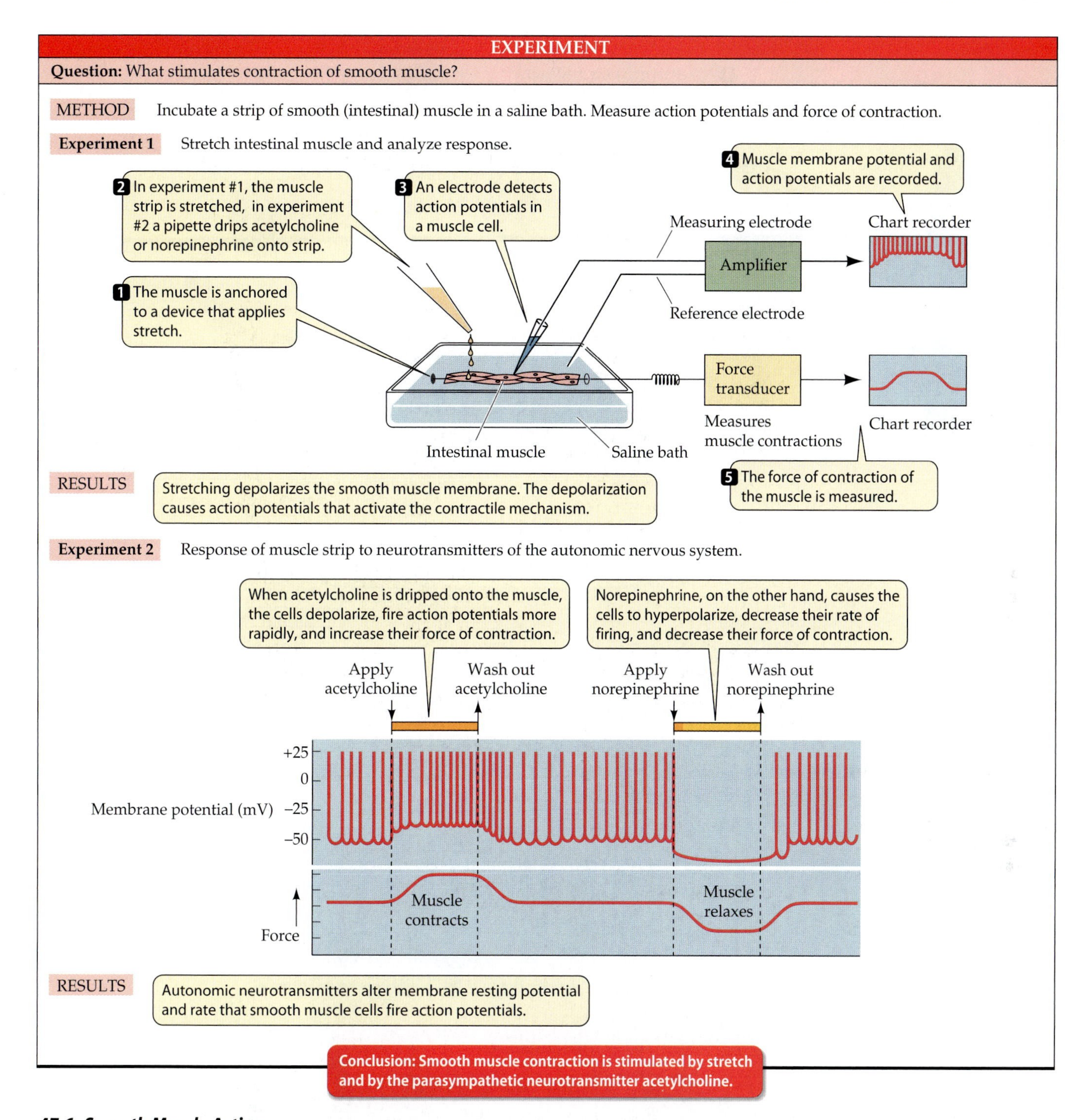

47.6 Smooth Muscle Action
Stretching depolarizes the membrane of smooth muscle cells, and this depolarization causes action potentials that activate the contractile mechanism. The neurotransmitters acetylcholine and norepinephrine also alter the membrane potential of smooth muscle, making it more or less likely to contract.

depolarization initiated at one point in the heart spreads rapidly through the mass of cardiac muscle.

An interesting feature of vertebrate cardiac muscle is that certain specialized muscle cells, called **pacemaker cells**, initiate the rhythmic contractions of the heart. We'll learn about the molecular basis for this pacemaking function in Chapter 49. Because of these specialized pacemaker cells, the heartbeat is *myogenic*—generated by the heart muscle itself. The autonomic nervous system modifies the rate of the pacemaker cells, but is not essential for their continued rhythmic function. A heart removed from an animal continues to beat with no input from the nervous system. The myogenic nature of the heartbeat is a major factor in making heart transplants possible.

Skeletal muscle carries out behavior

Skeletal muscle carries out, or *effects*, all voluntary movements, such as running or playing a piano, and generates the movements of breathing. Skeletal muscle is also called

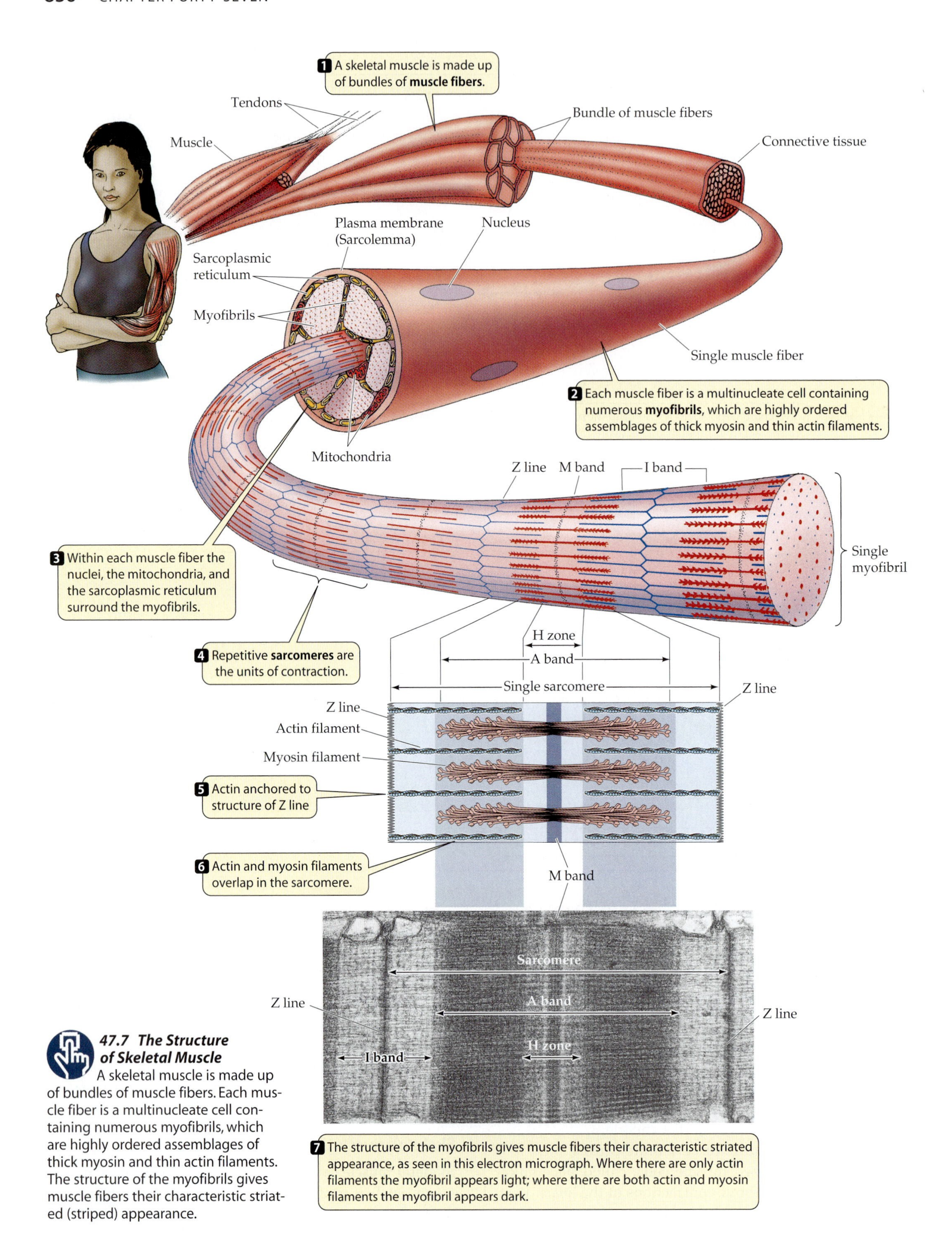

47.7 The Structure of Skeletal Muscle
A skeletal muscle is made up of bundles of muscle fibers. Each muscle fiber is a multinucleate cell containing numerous myofibrils, which are highly ordered assemblages of thick myosin and thin actin filaments. The structure of the myofibrils gives muscle fibers their characteristic striated (striped) appearance.

striated muscle because the highly regular arrangement of its actin and myosin filaments gives it a striped appearance. Skeletal muscle cells, called **muscle fibers**, are large. Unlike smooth muscle and cardiac muscle cells, each of which has a single nucleus, skeletal muscle fibers have many nuclei because they develop through the fusion of many individual cells. A muscle such as your biceps (which bends your arm) is composed of many muscle fibers bundled together by connective tissue.

What is the relation between a muscle fiber and the actin and myosin filaments responsible for its contraction? Each muscle fiber is composed of **myofibrils**—bundles of contractile filaments made up of actin and myosin (Figure 47.7). Within each myofibril are thin actin filaments and thick myosin filaments. If we cut across the myofibril at certain locations, we see only thick filaments; if we cut at other locations, we see only thin filaments. But, in most regions of the myofibril, each thick myosin filament is surrounded by six thin actin filaments.

A longitudinal view of a myofibril reveals the reason for the striated appearance of skeletal muscle. The band pattern of the myofibril is due to repeating units called **sarcomeres**, which are the units of contraction (see Figure 47.7). Each sarcomere is made of overlapping filaments of actin and myosin. As the muscle contracts, the sarcomeres shorten, and the appearance of the band pattern changes.

The observation that the widths of the bands in the sarcomeres change when a muscle contracts led two British biologists, Hugh Huxley and Andrew Huxley, to propose a molecular mechanism of muscle contraction. Let's look at the band pattern of the myofibril in detail (see the micrograph in Figure 47.7). Each sarcomere is bounded by *Z lines*, which are structures that anchor the thin actin filaments. Centered in the sarcomere is the *A band*, which contains all the myosin filaments. The *H zone* and the *I band*, which appear light, are regions where actin and myosin filaments do not overlap in the relaxed muscle. The dark stripe within the H zone is called the *M band*; it contains proteins that help hold the myosin filaments in their regular hexagonal arrangement.

When the muscle contracts, the sarcomere shortens. The H zone and the I band become much narrower, and the Z lines move toward the A band as if the actin filaments were sliding into the region occupied by the myosin filaments. This observation led Huxley and Huxley to propose the **sliding filament theory** of muscle contraction: Actin and myosin filaments slide past each other as the muscle contracts.

To understand what makes the filaments slide, we must examine the structure of actin and of myosin (Figure 47.8). Each myosin molecule consists of two long polypeptide chains coiled together, each ending in a large globular head. A myosin filament is made up of many myosin molecules arranged in parallel, with their heads projecting laterally from one or the other end of the filament. The actin filament consists of a helical arrangement of two chains of monomers twisted together like two strands of pearls. Twisting around the actin chains is another protein, tropomyosin, and attached to it at intervals are molecules of troponin. We'll discuss the roles of these last two proteins in the following section.

The myosin heads have sites that can bind to actin and thereby form bridges between the myosin and the actin filaments. The myosin heads also have ATPase activity; that is, they bind and hydrolyze ATP. The energy released when this happens changes the orientation of the myosin head.

Together, these details explain the cycle of events that cause the actin and myosin filaments to slide past each other and shorten the sarcomere. A myosin head binds to an actin filament (see Figure 47.8). Upon binding, the head changes its orientation with respect to the myosin filament, thus exerting a force that causes the actin and myosin filaments to slide about 5 to 10 nm relative to each other. Next, the myosin head binds a molecule of ATP, which causes it to release the actin. When the ATP is hydrolyzed, the energy released causes the myosin head to return to its original conformation, in which it can bind again to actin. It is as if the energy from ATP hydrolysis is being used to cock the hammer of a pistol, and contact of the myosin head with an actin binding site pulls the trigger.

We have been discussing the cycle of contraction in terms of a single myosin head. Don't forget that each myosin filament has many myosin heads at both ends and is surrounded by six actin filaments; thus the contraction of the sarcomere involves a great many cycles of interaction between actin and myosin molecules. That is why when a single myosin head breaks its contact with actin, the actin filaments do not slip backward.

An interesting aspect of this contractile mechanism is that ATP is needed to break the actin–myosin bonds, but not to form them. Thus muscles require ATP to *stop* con-

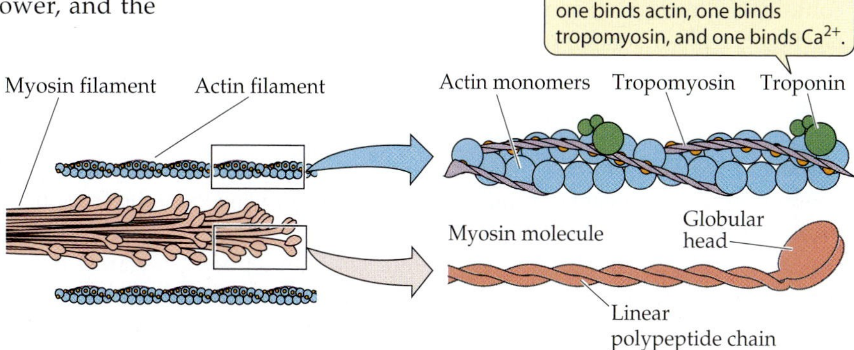

47.8 Actin and Myosin Filaments Overlap to Form Myofibrils
Myosin filaments are bundles of molecules with globular heads and polypeptide tails. Actin filaments consist of two chains of actin monomers twisted together. They are wrapped by chains of the polypeptide tropomyosin and studded at intervals with another protein, troponin.

tracting. This fact explains why muscles stiffen soon after animals die, a condition known as *rigor mortis*. Death stops the replenishment of the ATP stores of muscle cells, so the actin–myosin bonds cannot be broken, and the muscles stiffen. Eventually the proteins begin to lose their integrity, and the muscles soften. These events have regular time courses that differ somewhat for different regions of the body; therefore, an examination of the stiffness of the muscles of a corpse sometimes can help a coroner estimate the time of death.

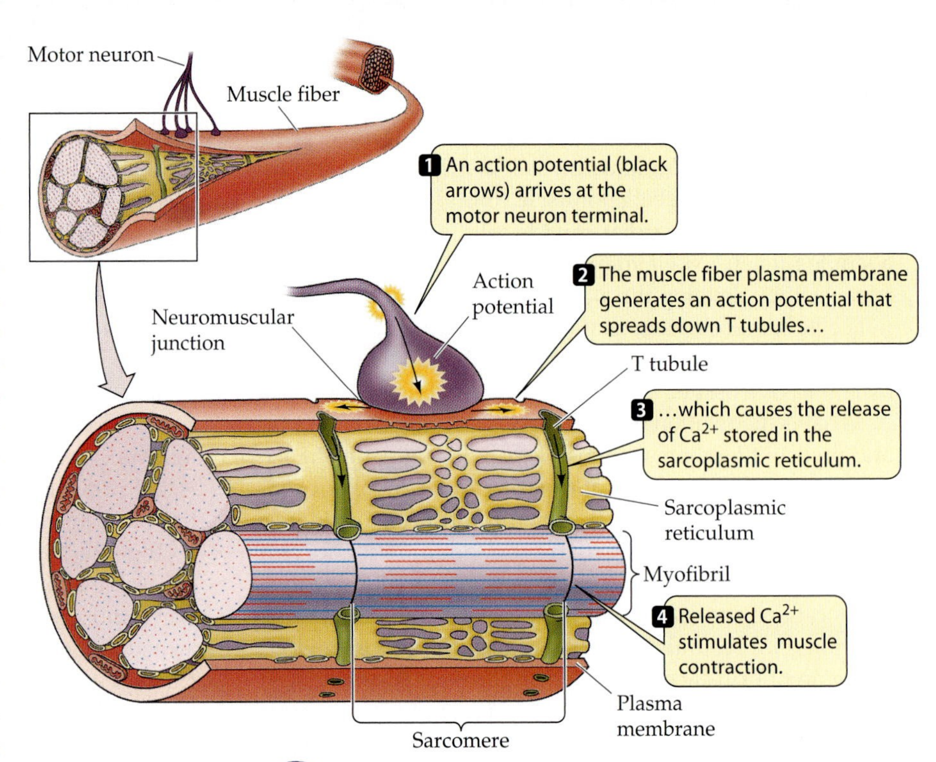

47.9 T Tubules in Action
An action potential at the neuromuscular junction spreads throughout the muscle fiber via a network of T tubules, triggering the release of Ca^{2+} from the sarcoplasmic reticulum.

Actin–myosin interactions are controlled by Ca^{2+}

Muscle contractions are initiated by action potentials from motor neurons arriving at the neuromuscular junction (see Figure 44.16). Motor neurons are generally highly branched and can synapse with up to a hundred muscle fibers each. All the fibers activated by a single motor neuron constitute a **motor unit** and contract simultaneously in response to the action potentials fired by that motor neuron. To understand the fine control the nervous system has over muscle contraction, we must examine the membrane system of the muscle fiber and some additional protein components of the actin filaments.

Like neurons, vertebrate skeletal muscle fibers are excitable cells: When they are depolarized to a threshold that opens their voltage-gated sodium channels, their plasma membranes generate action potentials, just as the membranes of axons do. When an action potential arrives at the neuromuscular junction, neurotransmitter from the motor neuron binds to receptors in the postsynaptic membrane, causing ion channels in the motor end plate to open. Most of the ions that flow through these channels are K^+, and therefore the motor end plate is depolarized.

The depolarization of the motor end plate spreads to the surrounding plasma membrane of the muscle fiber, which contains voltage-gated ion channels. When threshold is reached, the plasma membrane fires an action potential that is conducted rapidly to all points on the surface of the muscle fiber.

The action potential in a muscle fiber also travels deep within the cell. The plasma membrane of the muscle fiber is continuous with a system of tubules that descends into and branches throughout the cytoplasm (also called the **sarcoplasm**) of the muscle fiber (Figure 47.9). The action potential that spreads over the plasma membrane of the muscle fiber also spreads through this system of transverse tubules, or **T tubules**.

The T tubules come into very close contact with a network of intracellular membranes called the **sarcoplasmic reticulum** that extends throughout the sarcoplasm, surrounding every myofibril. Calcium pumps in the sarcoplasmic reticulum cause this membrane-enclosed compartment of the cell to take up Ca^{2+} ions from the sarcoplasm. Therefore, when the muscle fiber is at rest, there is a high concentration of Ca^{2+} in the sarcoplasmic reticulum and a low concentration of Ca^{2+} in the sarcoplasm surrounding the myofibrils.

When an action potential spreads through the T tubule system, it causes calcium channels in the sarcoplasmic reticulum to open, resulting in the diffusion of Ca^{2+} ions out of the sarcoplasmic reticulum and into the sarcoplasm surrounding the myofibrils. The Ca^{2+} stimulates the interaction of actin and myosin and the sliding of the filaments. How does this work?

An actin filament, as we have seen, is a helical arrangement of two strands of actin monomers. Lying in the grooves between the two actin strands is the two-stranded protein **tropomyosin** (see Figure 47.8). At regular intervals, the filament also includes another globular protein, **troponin**. The troponin molecule has three subunits: one binds actin, one binds tropomyosin, and one binds Ca^{2+}.

When Ca^{2+} is sequestered in the sarcoplasmic reticulum, the tropomyosin strands block the sites on the actin filament where myosin heads can bind. When the T tubule system depolarizes, Ca^{2+} is released into the sarcoplasm, where it binds to the troponin, changing the shape of the troponin molecule. Because the troponin is bound to the tropomyosin, this conformational change of the troponin twists the tropomyosin enough to expose the actin–myosin binding sites. Thus the cycle of making and breaking actin–myosin bonds is initiated, the filaments are pulled past one another, and the muscle fiber contracts. When the

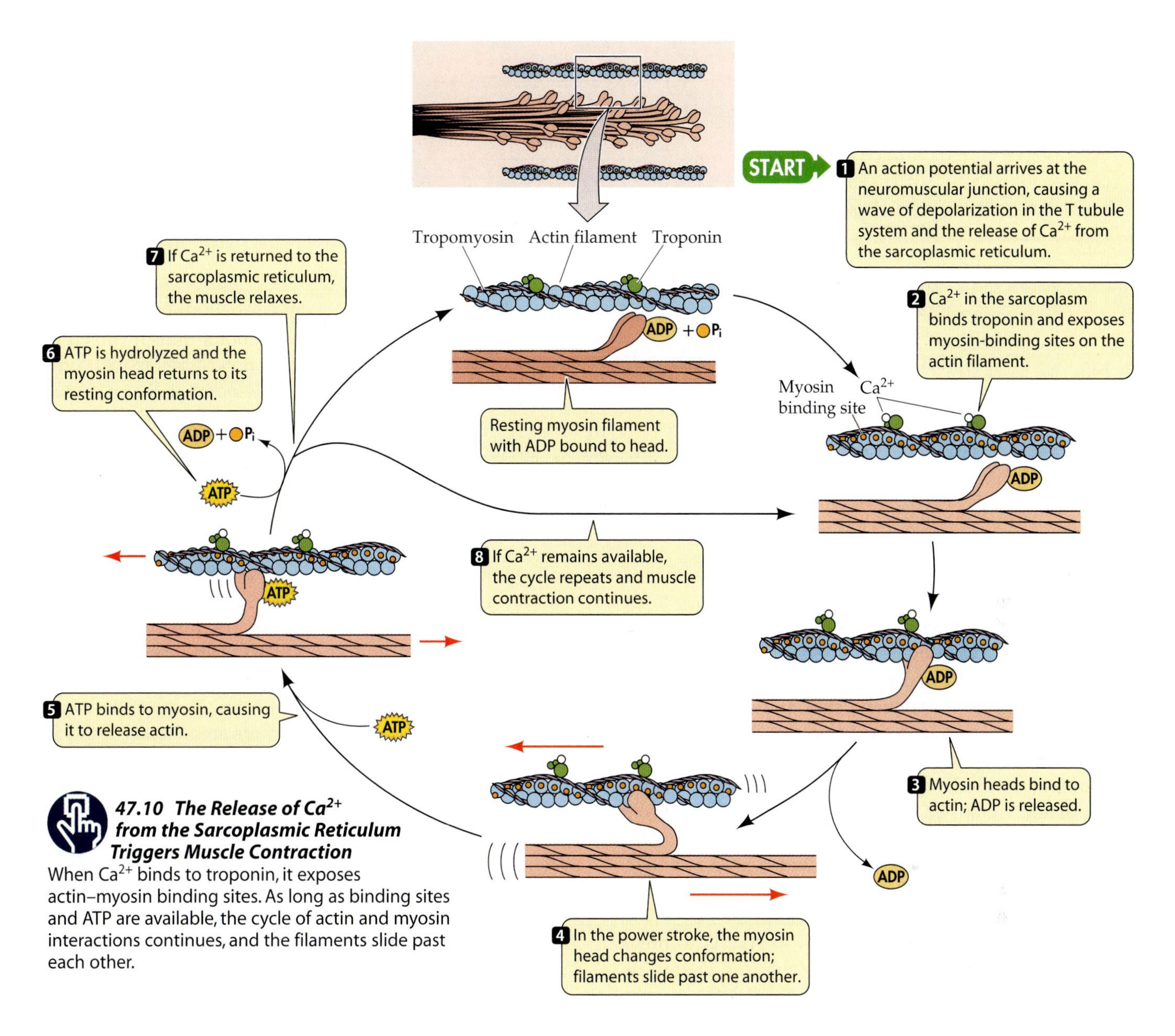

47.10 The Release of Ca^{2+} from the Sarcoplasmic Reticulum Triggers Muscle Contraction

When Ca^{2+} binds to troponin, it exposes actin–myosin binding sites. As long as binding sites and ATP are available, the cycle of actin and myosin interactions continues, and the filaments slide past each other.

T tubule system repolarizes, the calcium pumps remove the Ca^{2+} ions from the sarcoplasm, causing the tropomyosin to return to the position in which it blocks the binding of myosin heads to actin, and the muscle fiber returns to its resting condition. Figure 47.10 summarizes this cycle.

Calmodulin mediates Ca^{2+} control of contraction in smooth muscle

Smooth muscle cells do not have the troponin–tropomyosin mechanism for controlling contraction, but Ca^{2+} still plays a critical role. The Ca^{2+} influx into the sarcoplasm of a smooth muscle cell can be stimulated by action potentials, by hormones, or by stretching. The Ca^{2+} that enters the sarcoplasm combines with a protein called **calmodulin**. The calmodulin–Ca^{2+} complex activates an enzyme called *myosin kinase*, which can phosphorylate myosin heads. When the myosin heads in smooth muscle are phosphorylated, they can undergo cycles of binding and releasing actin, causing muscle contraction. As Ca^{2+} is removed from the sarcoplasm, it dissociates from calmodulin, and the activity of myosin kinase falls. In addition, another enzyme, *myosin phosphatase*, dephosphorylates the myosin and helps stop the actin–myosin interactions.

Single muscle twitches are summed into graded contractions

In skeletal muscle, the arrival of an action potential at a neuromuscular junction causes an action potential in a muscle fiber. The spread of the action potential through the T tubule system of the muscle fiber causes a minimum unit of contraction, called a **twitch**. A twitch can be measured in terms of the *tension*, or force, it generates (Figure 47.11*a*).

If action potentials in the muscle fiber are adequately separated in time, each twitch is a discrete, all-or-none phenomenon. If action potentials are fired more rapidly, however, new twitches are triggered before the myofibrils have had a chance to return to their resting condition. As a result, the twitches sum, and the tension generated by the

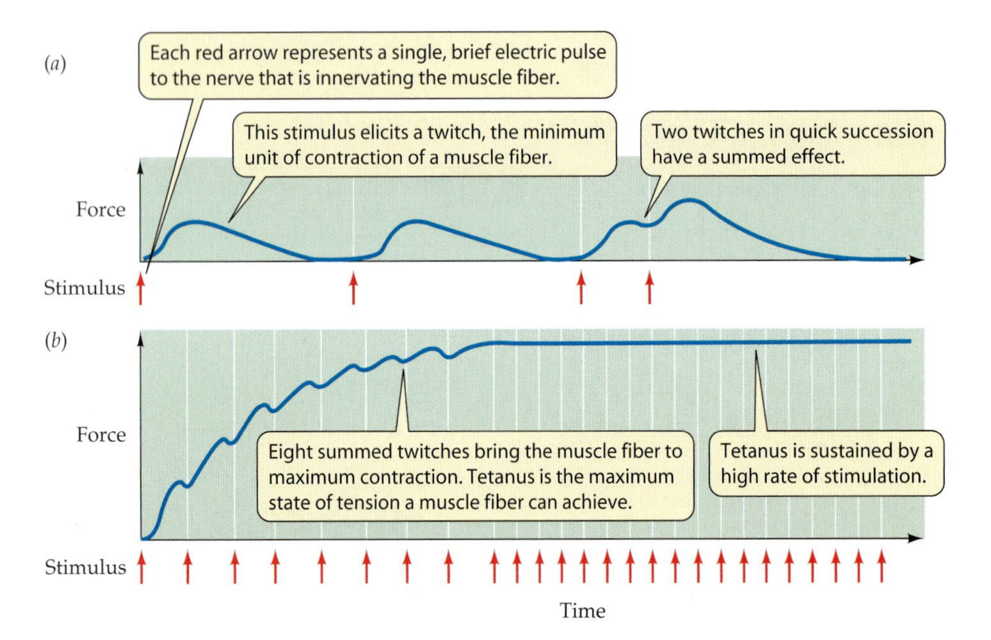

47.11 Twitches and Tetanus
(*a*) Action potentials from a motor neuron cause a muscle fiber to twitch. Twitches in quick succession can be summed. (*b*) Summation of many twitches can bring the muscle fiber to the maximum level of contraction, known as tetanus.

fiber increases and becomes more sustained. Thus an individual muscle fiber can show a graded response to increased levels of stimulation by its motor neuron.

At high levels of stimulation, the calcium pumps in the sarcoplasmic reticulum can no longer remove Ca^{2+} ions from the sarcoplasm between action potentials, and the contractile machinery generates maximum tension—a condition known as **tetanus** (Figure 47.11*b*). (Do not confuse this condition with the disease *tetanus*, which is caused by a bacterial toxin and is characterized by spastic contractions of skeletal muscles.)

How long a muscle fiber can maintain a tetanic contraction depends on its supply of ATP. Eventually the fiber will become fatigued. It may seem paradoxical that the *lack* of ATP causes fatigue, since the action of ATP is to break actin–myosin bonds. But remember that the energy released from the hydrolysis of ATP "recocks" the myosin heads, allowing them to cycle through another power stroke. When a muscle is contracting against a load, the cycle of making and breaking actin–myosin bonds must continue to prevent the load from stretching the muscle. The situation is like rowing a boat upstream: You cannot maintain your position relative to the stream bank by just holding the oars out against the current; you have to keep rowing. Likewise, actin–myosin bonds have to keep cycling to maintain tension in the muscle.

The ability of a whole muscle to generate different levels of tension depends on how many fibers in that muscle are activated. Whether a muscle contraction is strong or weak depends both on how many of the motor neurons that synapse with that muscle are firing and on the rate at which those neurons are firing. These two factors can be thought of as *spatial summation* and *temporal summation*, respectively.

Both types of summation increase the strength of contraction of the muscle as a whole. Faster twitching of individual fibers causes temporal summation (see Figure 47.11*b*), and an increase in the number of motor units involved in the contraction causes spatial summation. (Remember that a motor unit consists of all the muscle fibers innervated by a single neuron, and that a single muscle consists of many motor units.)

Many muscles of the body maintain a low level of tension called **tonus** even when the body is at rest. For example, the muscles of the neck, trunk, and limbs that maintain our posture against the pull of gravity are always working, even when we are standing or sitting still. Muscle tonus comes from the activity of a small but changing number of motor units in a muscle; at any one time, some of the muscle's fibers are contracting and others are relaxed. Tonus is constantly being readjusted by the nervous system.

Muscle fiber types determine endurance and strength

Not all skeletal muscle fibers are alike, and a single muscle may contain more than one type of fiber. The two major types of skeletal muscle fibers are slow-twitch fibers and fast-twitch fibers (Figure 47.12*a*). **Slow-twitch fibers** are also called *red muscle* because they have lots of the oxygen-binding molecule myoglobin, they have lots of mitochondria, and they are well supplied with blood vessels. A single twitch of a slow-twitch fiber produces low tension.

The maximum tension a slow-twitch fiber can produce is low and develops slowly, but these fibers are highly resistant to fatigue. Because slow-twitch fibers have substantial reserves of fuel (glycogen and fat), their abundant mitochondria can maintain a steady, prolonged production of ATP if oxygen is available. Muscles with high proportions of slow-twitch fibers are good for long-term aerobic work (that is, work that requires lots of oxygen). Champion long-distance runners, cross-country skiers, swimmers, and bicyclists have leg and arm muscles consisting mostly of slow-twitch fibers (Figure 47.12*b*).

Fast-twitch fibers are also called *white muscle* because, in comparison to slow-twitch fibers, they have fewer mitochondria, little or no myoglobin, and fewer blood vessels. The white meat of domestic chickens is composed of fast-twitch fibers. Fast-twitch fibers can develop maximum tension more rapidly than slow-twitch fibers can, and that maximum tension is greater, but fast-twitch fibers become

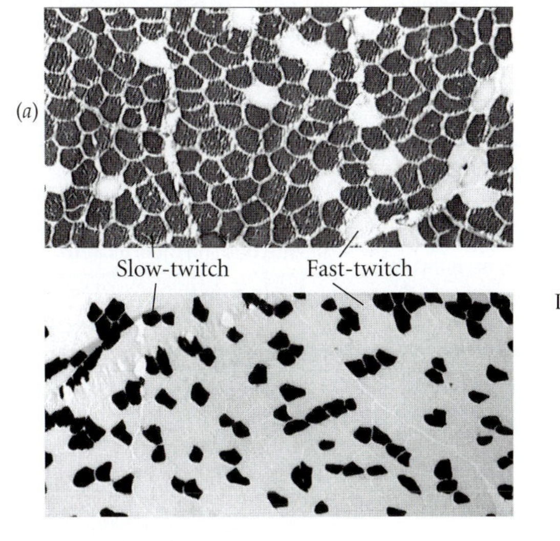

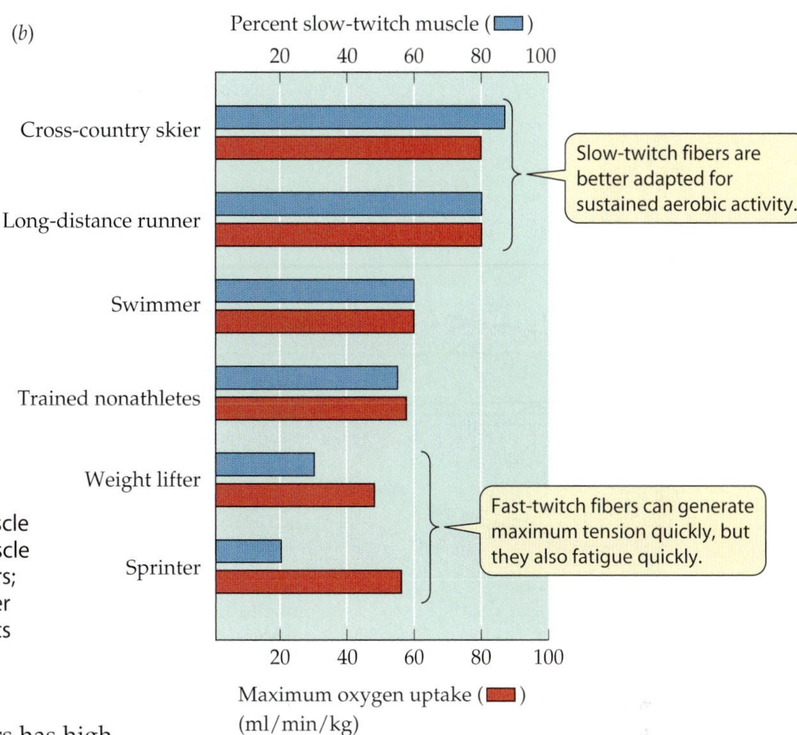

47.12 Two Types of Muscle Fibers
(*a*) Skeletal muscles stained with a reagent that shows slow-twitch fibers as dark. The upper photo shows muscle from a professional cyclist. The lower photo shows muscle from a nonathlete who has about 75% fast-twitch fibers; this person would probably perform better as a sprinter than as a distance runner. (*b*) Athletes in different sports have different distributions of muscle fiber types.

fatigued rapidly. The myosin of fast-twitch fibers has high ATPase activity, so they can put the energy of ATP to work very rapidly, but they cannot replenish it quickly enough to sustain contraction for a long time. Fast-twitch fibers are especially good for short-term work that requires maximum strength. Champion weight lifters and sprinters have leg and arm muscles with high proportions of fast-twitch fibers.

What determines the proportion of fast- and slow-twitch fibers in your skeletal muscles? The most important factor is your genetic heritage, so there is some truth to the statement that champions are born, not made. To a certain extent, however, you can alter the properties of your muscle fibers through training. With aerobic training, the oxidative capacity of fast-twitch fibers can improve substantially. But a person born with a high proportion of fast-twitch fibers will never become a champion marathon runner, and a person born with a high proportion of slow-twitch fibers will never become a champion sprinter.

Skeletal Systems Provide Support for Muscles

Muscles can only contract and relax. Without something rigid to pull against, a muscle would just be a formless mass that twitches and changes shape. Skeletal systems provide rigid supports against which muscles can pull, creating directed movements. In this section, we'll examine the three types of skeletal systems found in animals: hydrostatic skeletons, exoskeletons, and endoskeletons.

A hydrostatic skeleton consists of fluid in a muscular cavity

The simplest type of skeleton is the **hydrostatic skeleton** of cnidarians, annelids, and many other soft-bodied invertebrates. It consists of a volume of incompressible fluid (water) enclosed in a body cavity surrounded by muscle. When muscles oriented in a certain direction contract, the fluid-filled body cavity bulges out in the opposite direction.

The sea anemone, a cnidarian (see Figure 31.6*c*), has a hydrostatic skeleton. Its body cavity is filled with seawater. To extend its body and its tentacles, the anemone closes its mouth and constricts muscle fibers that are arranged in circles around its body. Contraction of these circular muscles puts pressure on the water in the body cavity, and that pressure forces the body and tentacles to extend. The anemone retracts its tentacles and body by contracting muscle fibers that are arranged longitudinally in the body wall and along the tentacles.

An earthworm uses its hydrostatic skeleton to crawl. The earthworm's body cavity is divided into many separate, fluid-filled segments. The body wall surrounding each segment has two muscle layers: one in which the muscle fibers are arranged in circles around the body cavity, and another in which the muscle fibers run lengthwise (Figure 47.13*a*). If the circular muscles in a segment contract, the compartment in that segment narrows and elongates. If the lengthwise (longitudinal) muscles of a segment contract, the compartment shortens and bulges outward.

Alternating contractions of the earthworm's circular and longitudinal muscles create waves of narrowing and widening, lengthening and shortening, that travel down the body. The bulging, short segments serve as anchors as the narrowing, expanding segments project forward, and longitudinal contractions pull other segments forward. Bristles help the widest parts of the body to hold firm against the substrate (Figure 47.13*b*).

Another type of locomotion made possible by hydrostatic skeletons is the jet propulsion used by squid and oc-

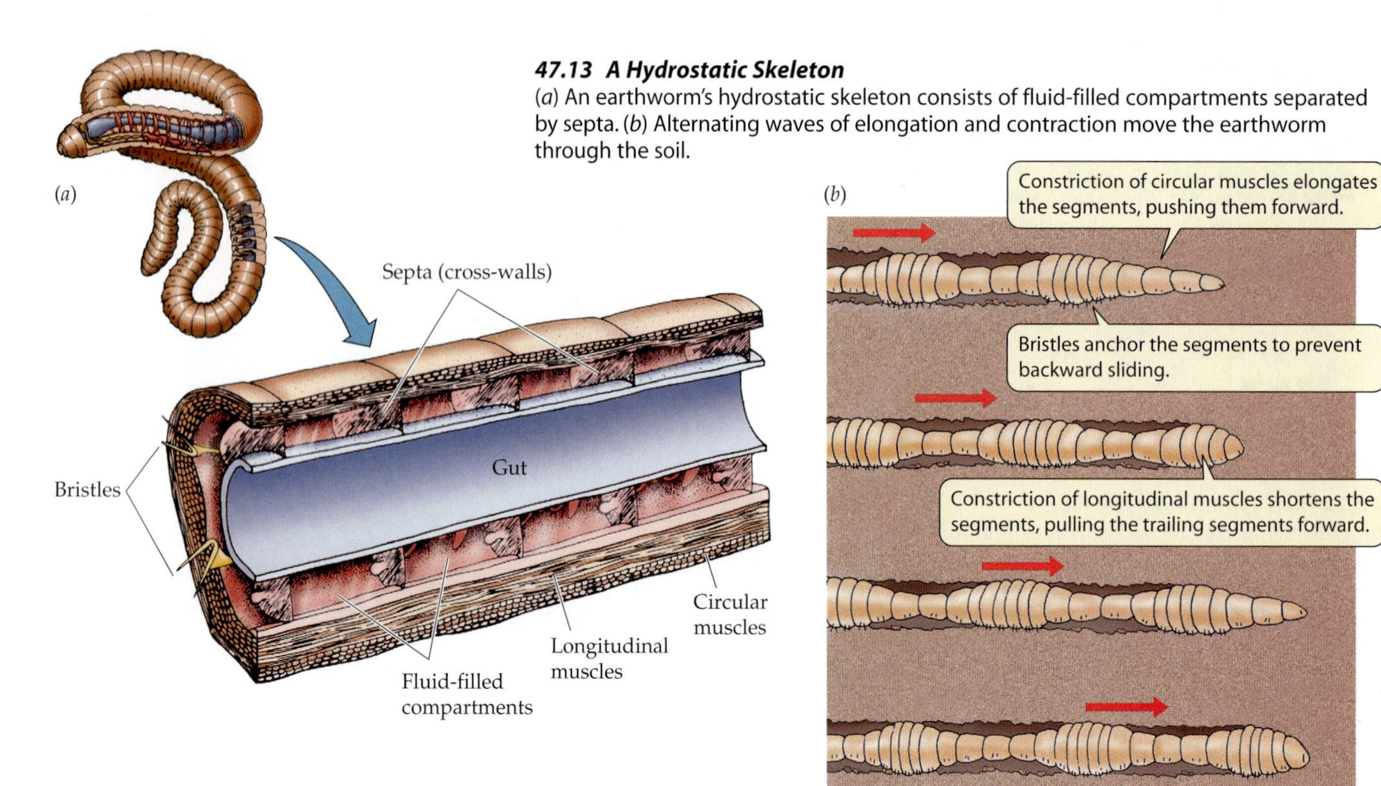

47.13 A Hydrostatic Skeleton
(*a*) An earthworm's hydrostatic skeleton consists of fluid-filled compartments separated by septa. (*b*) Alternating waves of elongation and contraction move the earthworm through the soil.

topuses. Muscles surrounding a water-filled cavity in these cephalopods contract, forcefully expelling water from the animal's body. As the water shoots out under pressure, the animal is propelled in the opposite direction.

Exoskeletons are rigid outer structures

An **exoskeleton** is a hardened outer surface to which muscles can be attached. Contractions of the muscles cause jointed segments of the exoskeleton to move relative to each other. The simplest example of an exoskeleton is the shell of a mollusk. Some marine mollusks, such as clams and snails, have shells composed of protein strengthened by crystals of calcium carbonate (a rock-hard material). These shells can be massive, affording significant protection against predators. The shells of land snails generally lack the hard mineral component and are much lighter. Molluscan shells can grow as the animal grows, and growth rings are usually apparent on the shells.

The most complex exoskeletons are found among the arthropods. An exoskeleton, or **cuticle**, covers all the outer surfaces of the arthropod's body and all its appendages. It is made up of plates secreted by a layer of cells just below the exoskeleton. A continuous, layered, waxy coating covers the entire body. The cuticle contains stiffening materials everywhere except at the joints, where flexibility must be retained. Muscles attached to the inner surfaces of the arthropod exoskeleton move its parts around the joints (Figure 47.14).

The layers of the cuticle include an outer, thin, waxy *epicuticle* that protects the bodies of terrestrial arthropods from drying out, and a thicker, inner *endocuticle* that forms most of the structure. The endocuticle is a tough, pliable material found only in arthropods. It consists of a complex of protein and **chitin**, a nitrogen-containing polysaccharide. In marine crustaceans the endocuticle is further toughened by insoluble calcium salts. The thickness of the cuticle varies, but it can be thick enough to form a protective armor.

An exoskeleton protects all the soft tissues of the animal, but is itself subject to damage by abrasion and crushing. The greatest drawback of the arthropod exoskeleton is that

47.14 An Insect's Exoskeleton
Muscles attached to the exoskeleton of this lubber grasshopper move parts of the body around flexible joints.

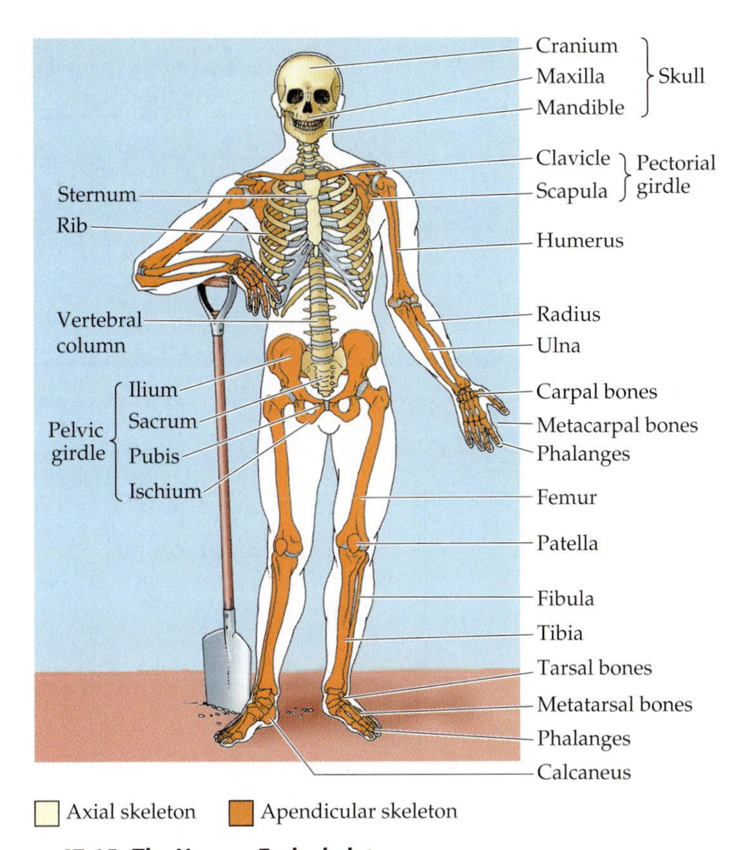

47.15 The Human Endoskeleton
Cartilage and bone make up the internal skeleton of a human being.

it cannot grow. Therefore, if the animal is to become larger, it must *molt*, shedding its exoskeleton and forming a new, larger one. A molting animal is vulnerable because the new exoskeleton takes time to harden. The animal's body is temporarily unprotected, and without a firm exoskeleton against which its muscles can exert maximum tension, it is unable to move rapidly. Soft-shelled crabs, a gourmet delicacy, are crabs caught when they are molting.

Vertebrate endoskeletons provide supports for muscles

The **endoskeleton** of vertebrates is an internal scaffolding to which muscles attach and against which they can pull. It is composed of rodlike, platelike, and tubelike bones, which are connected to each other at a variety of joints that allow a wide range of movements. Endoskeletons do not provide the protection that exoskeletons do, but their advantage is that they can grow. Because bones are inside the body, the body can enlarge without shedding its skeleton.

The human skeleton consists of 206 bones, some of which are shown in Figure 47.15. It can be divided into an *axial skeleton*, which includes the skull, vertebral column, and ribs, and an *appendicular skeleton*, which includes the pectoral girdle, the pelvic girdle, and the bones of the arms, legs, hands, and feet.

Two kinds of connective tissue cells produce large amounts of extracellular matrix material to create the vertebrate endoskeleton. The matrix material produced by **cartilage** cells is a rubbery mixture of proteins and polysaccharides. The principal protein in cartilage is collagen. Collagen fibers run in all directions through the gel-like matrix and give it the well-known strength and resiliency of "gristle."

Cartilage is found in parts of the endoskeleton where both stiffness and resiliency are required, such as on the surfaces of joints, where bones move against each other. Cartilage is also the supportive tissue in stiff but flexible structures such as the larynx (voice box), the nose, and the ear pinnae. Sharks and rays are called cartilaginous fishes (see Figure 33.11) because their skeletons are composed entirely of cartilage. In all other vertebrates, cartilage is the principal component of the embryonic skeleton, but during development most of it is gradually replaced by bone.

Bone consists mostly of extracellular matrix material that contains collagen fibers as well as crystals of insoluble calcium phosphate, which give bone its rigidity and hardness. The skeleton serves as a reservoir of calcium for the rest of the body and is in dynamic equilibrium with soluble calcium in the extracellular fluids of the body. This equilibrium is under the control of calcitonin and parathyroid hormone (see Figure 41.10). If too much calcium is taken from the skeleton, the bones are seriously weakened.

The living cells of bone—called osteoblasts, osteocytes, and osteoclasts—are responsible for the dynamic remodeling of bone that is constantly under way (Figure 47.16). **Osteoblasts** lay down new matrix material on bone surfaces. These cells gradually become surrounded by matrix and eventually become enclosed within the bone, at which point they cease laying down matrix but continue to exist within small lacunae (cavities) in the bone. In this state they are called **osteocytes**. In spite of the vast amounts of matrix between them, osteocytes remain in contact with one another through long cellular extensions that run through tiny channels in the bone. Communication between osteocytes is important in controlling the activities of the cells that are laying down new bone or eroding it away.

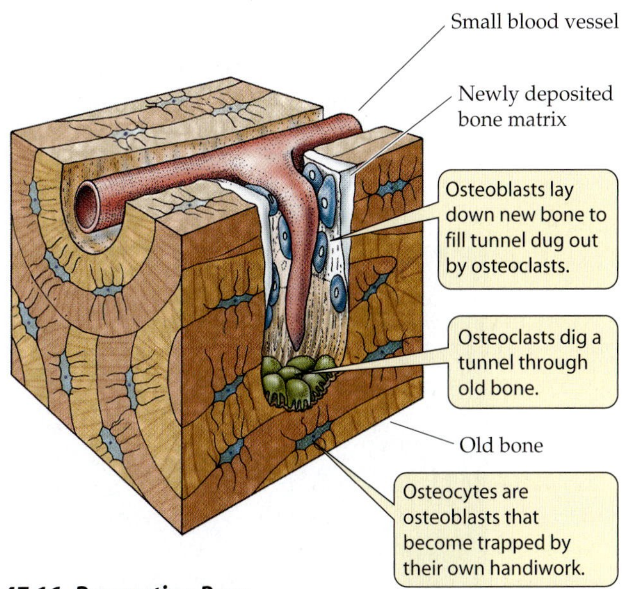

47.16 Renovating Bone
Bones are constantly being remodeled by osteoblasts, which lay down bone, and osteoclasts, which dissolve bone.

The cells that erode or reabsorb bone are the **osteoclasts**. They are derived from the same cell lineage that produces the white blood cells. Osteoclasts burrow into bone, forming cavities and tunnels. Osteoblasts follow osteoclasts, depositing new bone. Thus the interplay of osteoblasts and osteoclasts constantly replaces and remodels the bones.

How the activities of the bone cells are coordinated is not understood, but stress placed on bones somehow provides them with information. A remarkable finding in studies of astronauts who had spent long periods in zero gravity was that their bones had decalcified. Conversely, certain bones of athletes can thicken during training, becoming considerably thicker than the same bones in nonathletes. Both thickening and thinning of bones are experienced by someone who has a leg in a cast for a long time. The bones of the uninjured leg carry the person's weight and thicken, while the bones of the inactive leg in the cast thin. The jawbones of people who lose their teeth experience less compressional force during chewing and become considerably reduced.

Bones develop from connective tissues

Bones are divided into two types on the basis of how they develop. **Membranous bone** forms on a scaffolding of connective tissue membrane. **Cartilage bone** forms first as a cartilaginous structure and is gradually hardened (ossified) to become bone. The outer bones of the skull are membranous bones; the bones of the limbs are cartilage bones.

Cartilage bones can grow throughout the ossification process. The long bones of the legs and arms, for example, ossify first at the centers and later at each end (Figure 47.17). Growth can continue until these areas of ossification join. The membranous bones forming the skull cap grow until their edges meet. The soft spot on the top of a baby's head is the point at which the skull bones have not yet joined.

The structure of bone may be **compact** (solid and hard) or **cancellous** (having numerous internal cavities that make it appear spongy, even though it is rigid). The architecture of a specific bone depends on its position and function, but most bones have both compact and cancellous regions. The shafts of the long bones of the limbs, for example, are cylinders of compact bone surrounding central cavities that contain the bone marrow, where the cellular elements of the blood are made. The ends of the long bones are cancellous (Figure 47.18*a*). Cancellous bone is lightweight because of its numerous cavities, but it is also strong because its internal meshwork constitutes a support system. It can with-

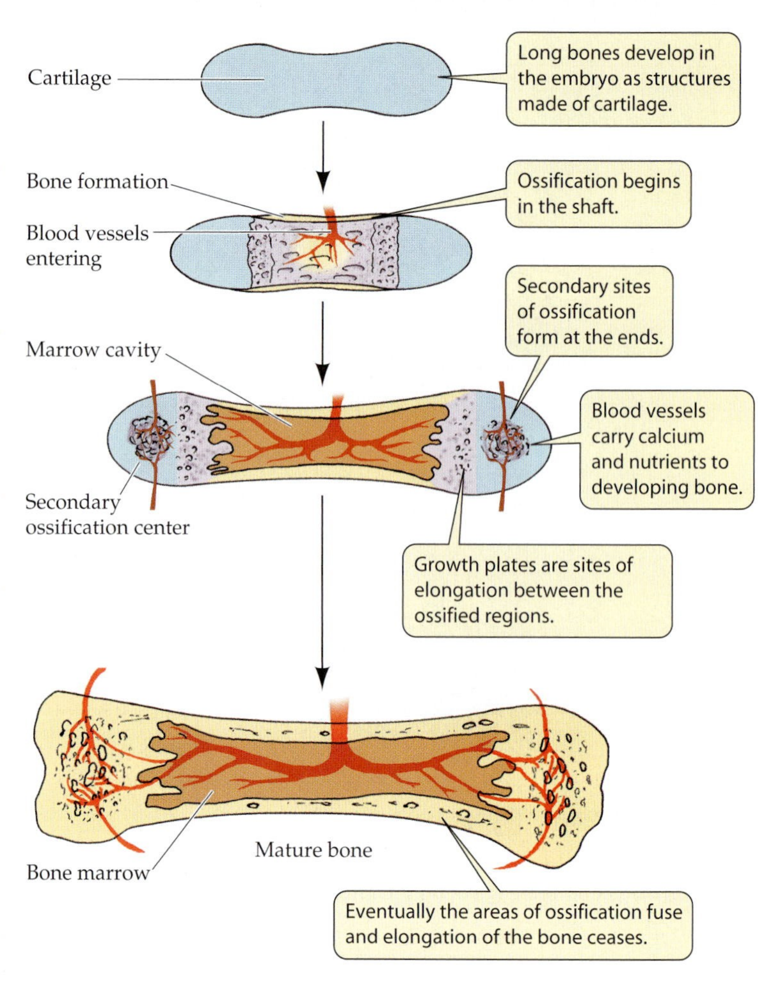

47.17 The Growth of Long Bones
In the long bones of human limbs, ossification occurs first at the centers and later at each end.

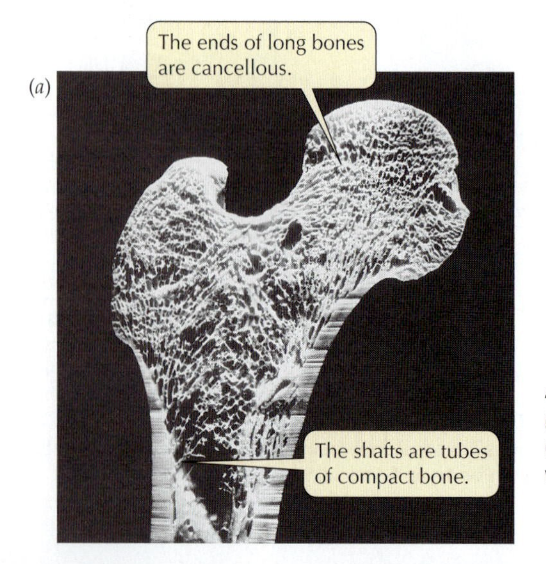

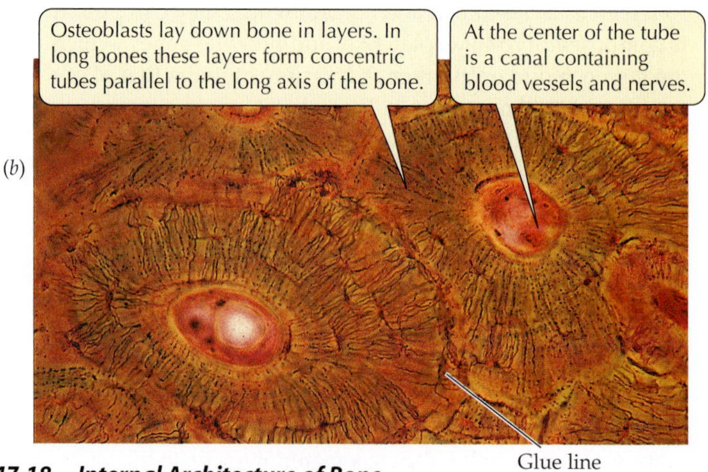

47.18 Internal Architecture of Bone
(*a*) Bone may have cancellous ("with holes") and compact (solid) regions. (*b*) A micrograph of a section of a long bone shows Haversian systems with their central channels. Glue lines separate Haversian systems.

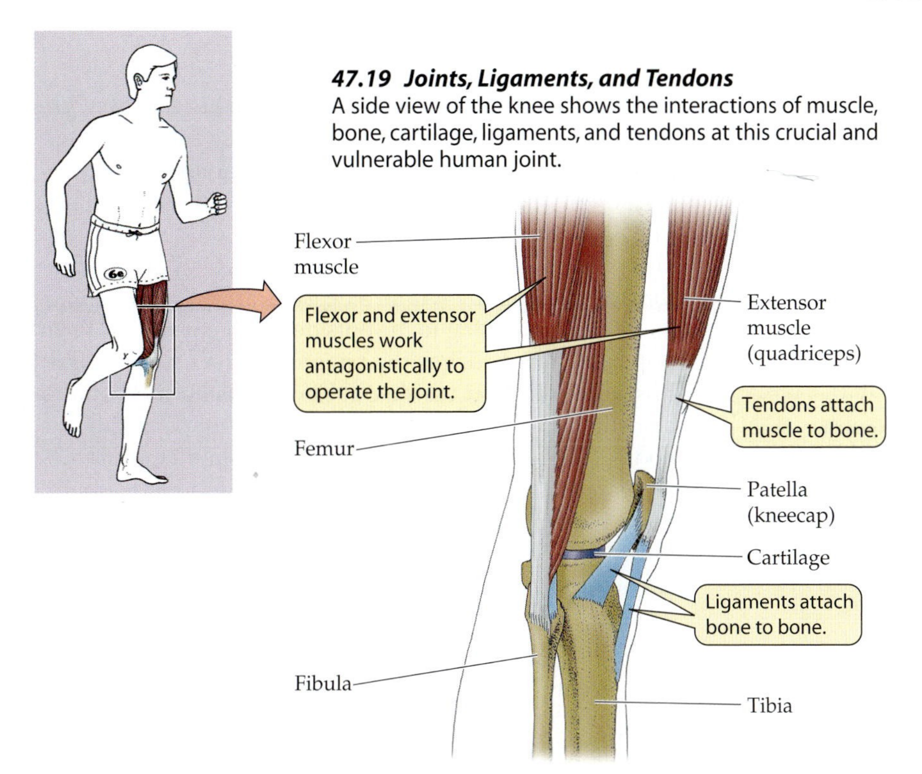

47.19 Joints, Ligaments, and Tendons
A side view of the knee shows the interactions of muscle, bone, cartilage, ligaments, and tendons at this crucial and vulnerable human joint.

stand considerable forces of compression. The rigid, tubelike shaft of compact bone can withstand compression and bending forces. Architects and nature alike use hollow tubes as lightweight structural elements.

Most of the compact bone in mammals is called *Haversian bone* because it is composed of structural units called **Haversian systems** (Figure 47.18*b*). Each Haversian system is a set of thin, concentric bony cylinders, between which are the osteocytes in their lacunae. Through the center of each Haversian system runs a narrow canal containing blood vessels (see Figure 47.16). Adjacent Haversian systems are separated by boundaries called *glue lines*. Haversian bone is resistant to fracturing because cracks tend to stop at glue lines.

Bones that have a common joint can work as a lever

Muscles and bones work together around **joints**, where two or more bones come together. Since muscles can only contract and relax, they create movement around joints by working in antagonistic pairs: When one contracts, the other relaxes. With respect to a particular joint, such as the knee, we can refer to the muscle that bends or flexes the joint as the **flexor** and the muscle that straightens or extends the joint as the **extensor**. The bones that meet at the joint are held together by **ligaments**, which are flexible bands of connective tissue. Other straps of connective tissue, called **tendons**, attach the muscles to the bones (Figure 47.19). Ligaments help direct the forces generated by muscles by holding tendons in place. In many kinds of joints, only the tendon spans the joint, sometimes moving over the surfaces of the bone like a rope over a pulley. The tendon of the quadriceps muscle traveling over the knee joint is what is tapped to elicit the knee-jerk reflex (see Figure 46.4). The human skeleton has a wide variety of joints with different ranges of movement (Figure 47.20).

Bones move around joints and the muscles that work with those bones can be thought of as systems of levers. A lever has a *power arm* and a *load arm* that work around a *fulcrum* (pivot). The length ratio of the two arms determines whether a particular lever can exert a lot of force over a short distance or is better

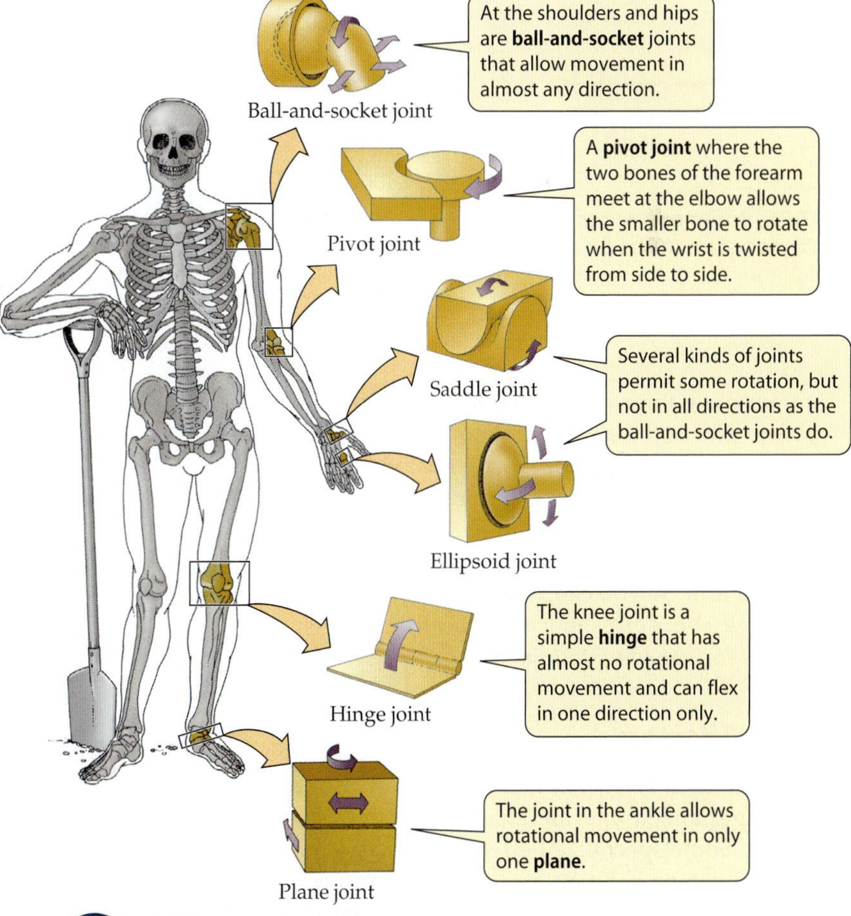

47.20 Types of Joints
The designs of joints are similar to mechanical counterparts and enable a variety of movements.

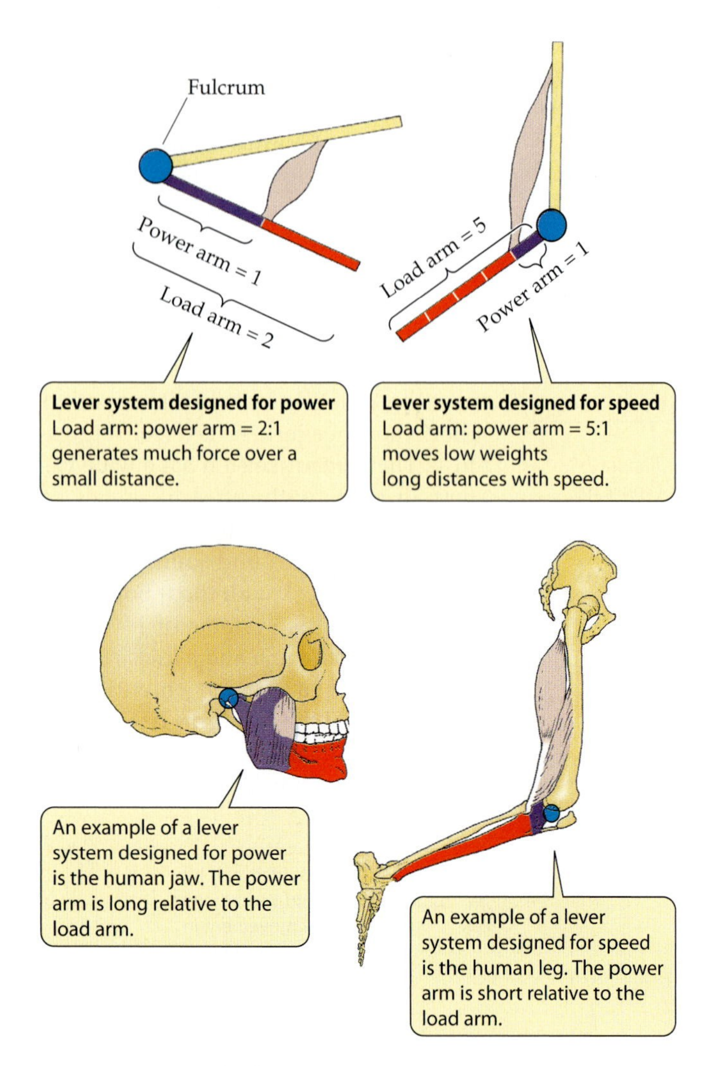

47.21 Bones and Joints Work Like Systems of Levers
A lever system can be designed for power or speed.

at translating force into large or fast movements. Compare the jaw joint and the knee joint, for example (Figure 47.21). The power arm of the jaw is long relative to the load arm, allowing the jaw to apply great pressures over a small distance, as when you crack a nut with your teeth. The power arm of the lower leg, on the other hand, is short relative to the load arm, so you can run fast, jump high, and deliver swift kicks, but you cannot apply nearly the pressure with a leg that you can with your jaws.

Other Effectors

Muscles are universal in animals, but other effectors are more specialized and are shared by only a few animal species. Some specialized effectors are used for defense, some for communication, and some for capture of prey or avoidance of predators. In this section we mention only a few specialized effectors to give a sampling of their evolutionary diversity.

Nematocysts capture prey and repel predators

Some animals possess highly specialized organs that are fired like miniature missiles to capture prey and repel predators. **Nematocysts** are elaborate cellular structures produced only by hydras, jellyfishes, and other cnidarians. They are concentrated in huge numbers on the outer surface of the tentacles. Each nematocyst consists of a slender thread coiled tightly within a capsule, armed with a spine-like trigger projecting to the outside (see Figure 31.7). When potential prey brushes the trigger, the nematocyst fires, turning the thread inside out and exposing little spines along its base. The thread either entangles or penetrates the body of the victim, and a poison may be simultaneously released around the point of contact. Once the prey is subdued, it is pulled into the mouth of the cnidarian and swallowed. A jellyfish called the Portuguese man-of-war has tentacles that can be several meters long. These animals can capture, subdue, and devour full-grown mackerel, and the poison of their nematocysts is so potent that it can kill a human who becomes tangled in the tentacles.

Chromatophores enable animals to change color

A change in body color is a response that some animals use to camouflage themselves in a particular environment or to communicate with other animals. **Chromatophores** are pigment-containing cells in the skin that can change the color and pattern of the animal. Chromatophores are under nervous or hormonal control, or both; in most cases they can effect a change within minutes or even seconds.

In squid, sole, and flounder, all of which spend much time on the seafloor, the famous chameleons (a group of African lizards; see Figure 33.19*b*), and a few other animals, chromatophores enable the animal to blend in with the background on which it is resting and thus escape discovery by predators. Chromatophores with different pigments enable animals to assume different hues or to become mottled to match the background more precisely. In other mollusks, fishes, and lizards, a color change sends a signal to potential mates and territorial rivals of the same species.

There are three principal types of chromatophore cells. The most common type has fixed cell boundaries, within which pigmented granules may be moved about by microfilaments. When the pigment is concentrated in the center of each chromatophore, the animal is pale; the animal turns darker when the pigment is dispersed throughout the cell. Another type of chromatophore is capable of amoeboid movement. These cells can mold themselves into shapes with a minimal surface area, leaving the tissue relatively pale, or they can flatten out to make the tissue appear darker.

The third type of chromatophore changes shape as a result of the action of muscle fibers radiating outward from the cell (Figure 47.22*a*). When the muscle fibers are relaxed, the chromatophores are small and compact, and the animal is pale. To darken the animal, the muscle fibers contract and spread the chromatophores over more of the body surface. These chromatophores can change so rapidly that they are

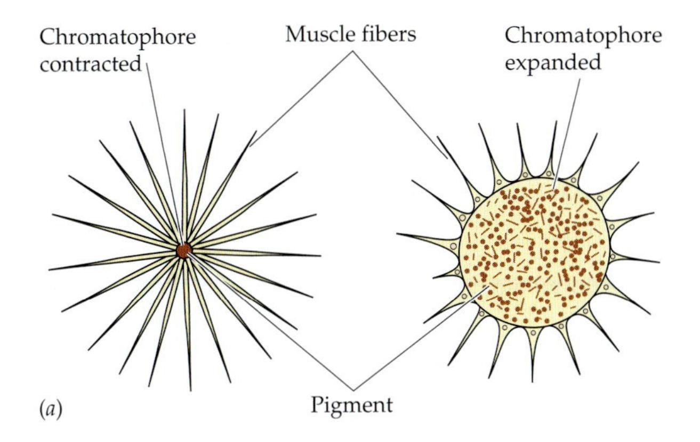

47.22 Chromatophores Help Animals Camouflage Themselves or Communicate
(*a*) Muscle fibers around the chromatophores cause chromatophores to contract. (*b*) Cuttlefish are cephalopod mollusks that can change color patterns so fast that these changes can be used for rapid communication.

used in some species for communication during courtship and aggressive interactions. For example, the cuttlefish, a cephalophod, can signal courtship intentions to a potential mate on one side of its body while signaling aggressive threats to a rival on the other side of its body (Figure 47.22*b*).

Glands can be effectors

Glands are effector organs that produce and release chemicals. Some glands produce hormones for internal signaling. Other glands secrete substances into the gut or onto the body surface. Some of these secretions are used defensively or to capture prey. Others are *pheromones*, chemical signals released into the environment for communication with other individuals.

Certain snakes, frogs, salamanders, spiders, mollusks, and fishes have poison glands. Many of the poisons produced by these glands are extremely specific in their modes of action. For example, the poison dendrotoxin, which certain tribes of the Amazonian rainforest use on the tips of their arrows for hunting, comes from the skin of a frog and blocks certain potassium channels. The snake venom bungarotoxin inactivates the neuromuscular acetylcholine receptors. The puffer fish poison tetrodotoxin blocks voltage-gated sodium channels. A poison from a mollusk, conotoxin, blocks calcium channels. Not all defensive secretions are poisonous, however. A well-known example is the odoriferous chemical mercaptan sprayed by skunks.

Electric organs can be shocking

Various fishes can generate electricity, as we saw in Chapter 45. These species include the electric eel, the knife fish, the torpedo (a type of ray), and the electric catfish. The electric fields they generate are used for sensing the environment, for communication, and also for stunning potential predators or prey. The electric organs of these animals evolved from muscles, and they produce electric potentials in the same general way as nerves and muscles do.

Electric organs consist of very large, disc-shaped cells arranged in long rows like stacks of batteries. When the cells discharge simultaneously, the electric organ can generate far more voltage and current than can nerve or muscle tissue. Electric eels, for example, can produce up to 600 volts with an output of approximately 100 watts—which is enough to light a row of light bulbs or to temporarily stun a person.

Chapter Summary

▶ Effectors enable animals to respond to information from their internal and external environments. Most effector mechanisms generate mechanical forces and cause movement.

Cilia, Flagella, and Cell Movement

▶ Cell movement is generated by two structures, microtubules and microfilaments, both of which consist of long protein molecules that can change their length or shape.

▶ The movements of cilia and flagella depend on microtubules. **Review Figures 47.1, 47.3, 47.4**

▶ Microfilaments allow animal cells to change their shape and move.

Muscle Contraction

▶ The three types of vertebrate muscle are smooth, cardiac, and skeletal (striated). **Review Figure 47.5**

▶ Smooth muscle provides contractile force for internal organs. Smooth muscle cells are electrically coupled through gap junctions, so action potentials that cause contraction spread rapidly throughout smooth muscle tissue. Autonomic neurotransmitters alter the membrane potential of smooth muscle cells. **Review Figure 47.6**

▶ The walls of the heart consist of sheets of branching cardiac muscle cells. The cells are electrically coupled through gap junctions, so that action potentials spread rapidly throughout sheets of cardiac muscle and cause coordinated contractions. Some cardiac muscle cells are pacemaker cells that generate the heartbeat.

▶ Skeletal, or striated, muscle consists of bundles of muscle fibers. Each muscle fiber is a huge cell containing multiple nuclei and numerous myofibrils, which are bundles of actin and myosin filaments. The regular, overlapping arrangement of the actin and myosin filaments into sarcomeres gives skeletal muscle its striated appearance. During contraction, the actin and myosin filaments slide past each other in a telescoping fashion. **Review Figure 47.7**

▶ The molecular mechanism of muscle contraction involves the binding of the globular heads of myosin molecules to actin. Upon binding, the myosin head changes conformation, causing the two filaments to move relative to each other. Release of the myosin heads from actin and their return to their original conformation requires ATP. **Review Figure 47.8**

▶ The plasma membrane of the muscle fiber is continuous with a system of T tubules that extends deep into the sarcoplasm (muscle cell cytoplasm). **Review Figure 47.9**

▶ When an action potential spreads across the plasma membrane and through the T tubules, it causes Ca^{2+} ions to be released from the sarcoplasmic reticulum. The Ca^{2+} ions bind to troponin and change its conformation, pulling the tropomyosin strands away from the myosin binding sites on the actin filament. Cycles of actin–myosin binding and release occur, and the muscle fiber contracts until the Ca^{2+} is returned to the sarcoplasmic reticulum. **Review Figure 47.10**

▶ In striated muscle, a single action potential causes a minimum unit of contraction called a twitch. Twitches occurring in rapid succession can be summed, thus increasing the strength of contraction. **Review Figure 47.11**

▶ Slow-twitch muscle fibers are adapted for extended, aerobic work; fast-twitch fibers are adapted for generating maximum forces for short periods of time. The ratio of slow-twitch to fast-twitch fibers in the muscles of an individual is genetically determined. **Review Figure 47.12**

Skeletal Systems Provide Support for Muscles

▶ Skeletal systems provide rigid supports against which muscles can pull.

▶ Hydrostatic skeletons are fluid-filled cavities that can be squeezed by muscles. **Review Figure 47.13**

▶ Exoskeletons are hardened outer surfaces to which internal muscles are attached. **Review Figure 47.14**

▶ Endoskeletons are internal, articulated systems of rigid rodlike, platelike, and tubelike supports consisting of bone and cartilage to which muscles are attached. **Review Figure 47.15**

▶ Bone is continually being remodeled by osteoblasts, which lay down new bone, and osteoclasts, which erode bone. **Review Figure 47.16**

▶ Bones develop from connective tissue membranes or from cartilage through ossification. Cartilage bone can grow until centers of ossification meet. **Review Figure 47.17**

▶ Bone can be solid and hard (compact bone), or it can contain numerous internal spaces (cancellous bone).

▶ Tendons connect muscles to bones; ligaments connect bones to each other and also help direct the forces generated by muscles by holding tendons in place. **Review Figure 47.19**

▶ Muscles and bones work together around joints as systems of levers. **Review Figures 47.20, 47.21**

Other Effectors

▶ Effector organs other than muscles include nematocysts, chromatophores, glands, and structures that produce electric pulses.

For Discussion

1. The amount of force a skeletal muscle can generate depends on its initial length. If the muscle is stretched or compressed prior to stimulation, it cannot generate maximum force of contraction. Why? Explain in terms of the molecular structure of the contractile mechanism.
2. If an intact axoneme is stimulated with Ca^{2+} and ATP, it flexes back and forth. However, if all of the proteins of the axoneme are enzymatically removed except for the microtubules and dynein, and the axoneme is then stimulated with Ca^{2+} and ATP, the microtubules telescope apart and the structure gets longer. Why?
3. Wombats are powerful digging animals, and kangaroos are powerful jumping animals. How do you think the structures of their legs would compare in terms of their designs as lever systems?
4. Why are ducks better long-distance fliers than chickens?
5. If an adolescent breaks a leg bone close to the ankle joint, after the break heals, that leg may not grow as long as the other one. Why?

48 Gas Exchange in Animals

In his book about the first ascent of Mt. Everest, Sir John Hunt relates the following observation made at 26,000 feet, on the South Col, the last camp before the summit attempt. At this altitude climbers are almost totally incapacitated if they do not breathe supplemental oxygen from pressurized bottles:

> And so back up the gradual slopes, the wind behind me. A much greater effort this, stopping every few yards with a slight anxiety lest I should not make the distance. As I approached the tents, I was astonished to see a bird, a chough, strutting about on the stones near me. … During this day, too, Charles Evans saw what must have been a migration of small grey birds across the Col. Neither of us had thought to find any signs of life as high as this.

Birds in flight can consume oxygen at a rate that a well-trained human athlete cannot sustain for more than a few minutes. Yet birds fly over the summit of Everest, where human climbers must breathe supplemental oxygen just to plod along at a slow pace. How do they do it? Fish breathing water, with an oxygen content less than 5 percent that of air, can swim much faster, farther, and longer than the best human swimmer. How do they do it? The abilities of some animals to maintain high rates of metabolism depend in part on the capacities of their respiratory gas exchange systems.

Some animals carry out their respiratory gas exchange with water and others with air. Both water breathers and air breathers have respiratory systems with adaptations that maximize exchanges of oxygen and carbon dioxide with the environment. These adaptations include specialized surface areas where gas exchange takes place, breathing mechanisms that bring fresh air or water to those surfaces, and circulatory mechanisms for transporting respiratory gases to and from the internal sides of the gas exchange surfaces. Since gases cross the gas exchange surfaces by diffusion only, physical factors that limit diffusion determine the maximum capacities of gas exchange systems.

In this chapter we first describe the physical factors that influence respiratory gas exchange. Then we examine the respiratory gas exchange systems of a variety of species, including some highly efficient ones such as fish gills and bird lungs, and less efficient ones such as our own. We also look at the adaptations of the blood for transporting respiratory gases. Finally, we see how respiratory gas exchange systems are controlled and regulated.

Respiratory Gas Exchange

The **respiratory gases** are oxygen (O_2) and carbon dioxide (CO_2). Cells need to obtain O_2 from the environment to produce an adequate supply of ATP through the oxidation of nutrient molecules (see Chapter 7). An end product of the oxidative metabolism of nutrients is CO_2, which must be lost to the environment to prevent toxic effects.

Diffusion is the only means by which respiratory gases are exchanged between the internal body fluids of an animal and the outside medium—air or water. There are no active transport mechanisms to move respiratory gases across biological membranes. Because diffusion is a physical process, knowing the physical factors that influence rates of diffusion helps us understand the diverse adaptations of gas exchange systems. You might want to review what you learned about diffusion as a physical phenomenon in Chapter 5. Here, we discuss environmental factors that in-

Flying High
Many birds can sustain the high metabolic costs of flight even at very high altitudues, where oxygen is scarce.

(*a*)

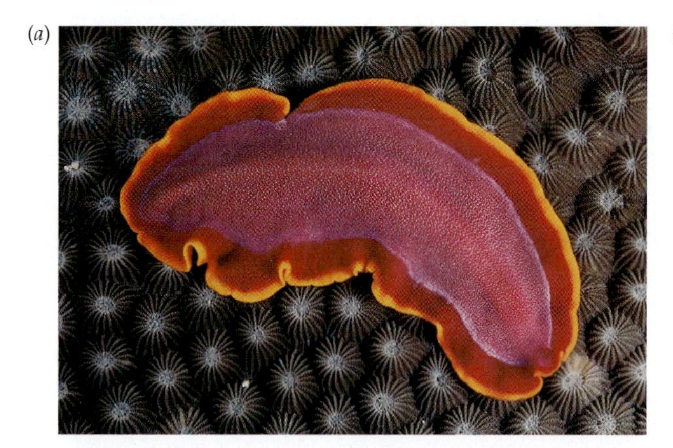

(*b*)

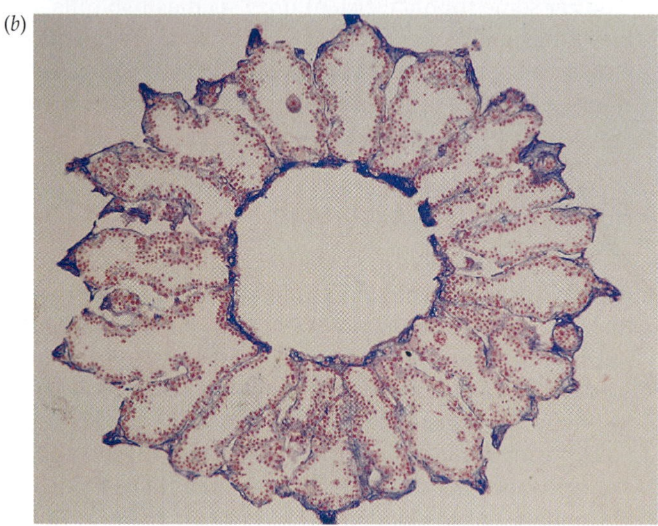

(*c*)

48.1 Keeping in Touch with the Medium
(*a*) No cell in the leaflike body of this marine flatworm is more than a millimeter away from seawater. (*b*) The same is true of sponges, which have body walls perforated by many channels lined with flagellated cells. These channels communicate with the outside world and with a central cavity. The flagella maintain currents of water through the channels, through the central cavity, and out of the animal. Every cell in the sponge is very close to the respiratory medium. (*c*) The gills of this newt project like a feathery fringe and provide a large surface area for gas exchange. Blood circulating through the gills comes into close contact with the respiratory medium.

fluence diffusion rates, and then describe the adaptations of respiratory systems for facilitating the diffusion of respiratory gases.

Air is a better respiratory medium than water

O_2 can be obtained more easily from air than from water for several reasons.

- The oxygen content of air is much higher than the oxygen content of an equal volume of water. The maximum O_2 content of a rapidly flowing stream splashing over rocks and tumbling over waterfalls is less than 10 ml of O_2 per liter of water. The O_2 content of fresh air is about 200 ml of O_2 per liter of air.
- O_2 diffuses about 8,000 times more rapidly in air than in water. In a still pond, the O_2 content of the water can be zero only a few millimeters below the surface.
- When an animal breathes, it does work to move water or air over its specialized gas exchange surfaces. More energy is required to move water than to move air because water is 800 times more dense than air and about 50 times more viscous.

The slow diffusion of O_2 molecules in water is a problem for air-breathing animals as well as for water-breathing animals. Eukaryotic cells carry out cellular respiration in their mitochondria, which are located in the cytoplasm—an aqueous medium. Cells are bathed in extracellular fluid—also an aqueous medium. The slow rate of O_2 diffusion in water limits the efficiency of O_2 distribution from gas exchange surfaces to the sites of cellular respiration in both air-breathing and water-breathing animals.

Diffusion of O_2 in water is so slow that even animal cells with low rates of metabolism can be no more than a couple of millimeters away from a good source of environmental O_2. Therefore, in animals that lack an internal system for transporting O_2, no cell may be more than about 1 or 2 mm from the outside world—a severe size limit. One way some simple invertebrate animals have grown bigger in spite of this limit is to have a flat, leaflike body (Figure 48.1*a*). Another way is to have a very thin body built around a central cavity through which water circulates (Figure 48.1*b*). Otherwise, an animal must have specialized structures to provide an increased surface area for diffusion, an internal circulatory system to carry respiratory gases to and from these structures, and a way in which the surfaces of these gas exchange structures can be continuously bathed with fresh air or water (Figure 48.1*c*).

High temperatures create respiratory problems for aquatic animals

Animals that breathe water are in a double bind when environmental temperatures rise. Most water breathers are ectothermic—their body temperatures are closely tied to the temperature of the water around them. As the temperature of the water rises, so does their body temperature and metabolic rate (see Chapter 40 for a discussion of Q_{10} relationships). Thus, with rising temperatures, water breathers

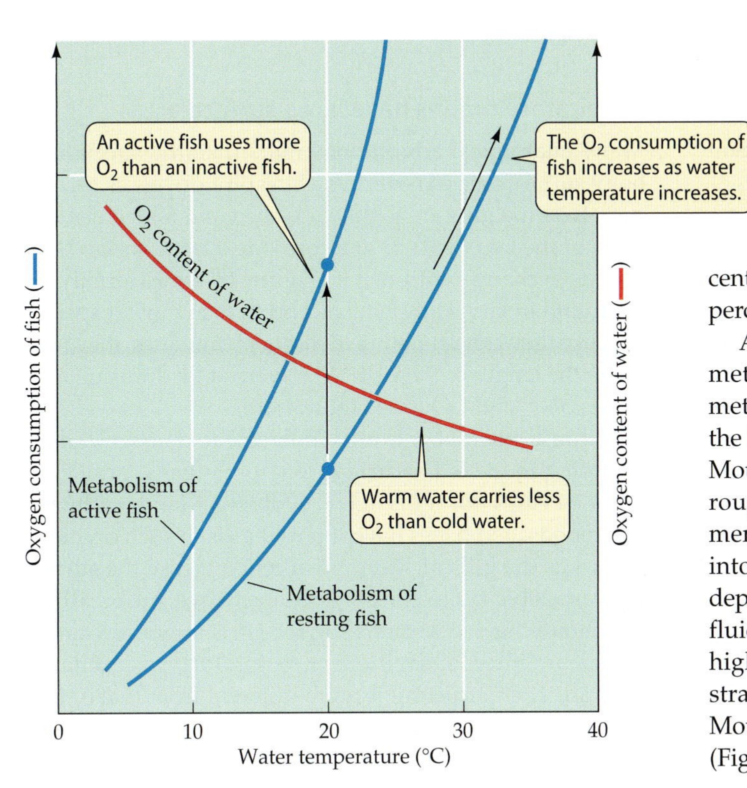

48.2 The Double Bind of Water Breathers
Fish need *more* oxygen in warm water, but warm water carries *less* oxygen than cold water.

need more O_2. But warm water holds less dissolved oxygen gas than cold water does (just think of what happens when you open a warm bottle of soda). So, under conditions that increase the need these animals have for O_2, there is less O_2 in their respiratory medium (Figure 48.2). In addition, if the animal performs work to move water across its gas exchange surfaces (as fish do, for example), the energy the animal must expend to breathe increases as water temperature rises. Therefore, as water temperature goes up, the water breather must extract more and more O_2 from its environment, and a lower percentage of that O_2 is available to support activities other than breathing.

O_2 availability decreases with altitude

Just as a rise in temperature reduces the supply of O_2 available to aquatic animals, an increase in altitude reduces the O_2 supply for air breathers, because the amount of O_2 in the atmosphere decreases with altitude.

One of the ways to express the amounts of gases in air and in water is by their **partial pressures**. At sea level, the pressure exerted by the atmosphere is equivalent to the pressure produced by a column of mercury 760 mm high. Therefore, *barometric pressure* (atmospheric pressure) at sea level is 760 mm of mercury (Hg). Because dry air is 20.9 percent O_2, the *partial pressure of oxygen* (P_{O_2}) at sea level is 20.9 percent of 760 mm Hg, or about 159 mm Hg.

At higher elevations, where there is less air above, barometric pressure declines. At an altitude of 5,300 m, barometric pressure is only half as much as it is at sea level, so the P_{O_2} at that altitude is only 80 mm Hg. At the summit of Mount Everest (8,848 m), the P_{O_2} is only about 50 mm Hg, roughly one-third what it is at sea level. Since the movement of O_2 across respiratory gas exchange surfaces and into the body depends on diffusion, its rate of movement depends on the P_{O_2} difference between the air and the body fluids. Therefore, the drastically reduced P_{O_2} in the air at a high altitude constrains O_2 uptake. Because of these constraints, mountain climbers who venture to the heights of Mount Everest usually breathe O_2 from pressurized bottles (Figure 48.3).

Carbon dioxide is lost by diffusion

Respiratory gas exchange is a two-way process. CO_2 diffuses out of the body as O_2 diffuses in. Given the same partial pressure gradient, CO_2 and O_2 molecules diffuse at about the same rate, whether in air or in water. However, the partial pressure gradients for the diffusion of O_2 and CO_2 across gas exchange surfaces are not the same. The

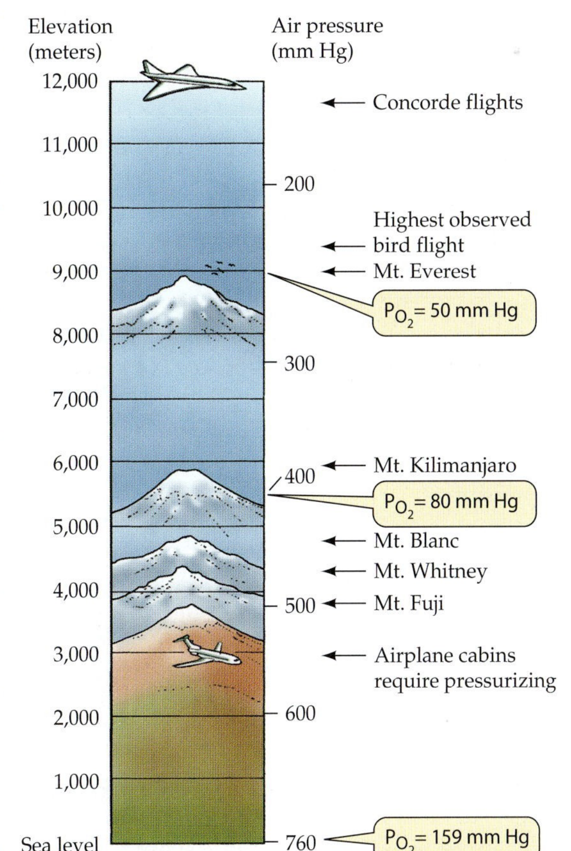

48.3 Scaling Heights
The partial pressure of oxygen in the atmosphere decreases with altitude. Therefore, airplane cabins must be pressurized, and mountain climbers must carry pressurized containers of oxygen, at high altitudes. Birds, however, have been observed flying over even the highest peaks.

amount of CO_2 in the atmosphere is extremely low (0.03 percent), so there is always a good partial pressure gradient for loss of CO_2 from air-breathing animals.

Water-breathing animals are much more likely than air breathers to experience high partial pressures of CO_2 in their environments. If water is well aerated, well mixed, and does not contain a lot of dead organic material, the diffusion of CO_2 from an aquatic animal is not a problem. Stagnant waters that are home to much biological activity or rotting vegetation, however, can have high levels of CO_2 and not be able to support animal life. In both air breathers and water breathers, the need to transport CO_2 from where it is produced in the cells of the body to where it diffuses into the environment can be a limiting factor in gas exchange (and hence in metabolism).

Fick's law applies to all systems of gas exchange

All adaptations that maximize respiratory gas exchange influence one or more components of a simple equation called **Fick's law of diffusion**,

$$Q = DA\frac{C_1 - C_2}{L}$$

where

- Q is the rate at which a substance such as O_2 diffuses between two locations
- D is the *diffusion coefficient*, which is a characteristic of the diffusing substance, the medium, and the temperature (for example, perfume has a higher D than motor oil, and substances diffuse faster in air than in water)
- A is the cross-sectional area over which the substance is diffusing
- C_1 and C_2 are the concentrations of the substance at two locations
- L is the distance between those locations

Therefore, $(C_1 - C_2)/L$ is a concentration gradient. In discussing respiratory gas exchange, we will use partial pressures rather than concentrations; this term will therefore become a partial pressure gradient.

Animals can maximize D for respiratory gases by using air rather than water as their gas exchange medium whenever possible. All other adaptations for maximizing respiratory gas exchange must influence the surface area for exchange or the partial pressure gradient across that surface area.

Respiratory Adaptations for Gas Exchange

Now that we know the factors that determine the rates of diffusion of respiratory gases, let's take a look at the fascinating array of adaptations that animals have evolved for respiratory gas exchange.

Respiratory organs have large surface areas

Many anatomical adaptations maximize the specialized body surface area (A) over which respiratory gases can diffuse. **External gills** are highly branched and folded elaborations of the body surface that provide a large surface area for gas exchange with water (Figure 48.4*a*). External gills are found in larval amphibians and in many insect species. Because they consist of thin, delicate membranes, they minimize the length of the path (L) traversed by diffusing molecules of O_2 and CO_2 (see Figure 48.1*c*).

Because external gills are vulnerable to damage and are tempting morsels for carnivorous organisms, protective body cavities for gills have evolved. Many mollusks, arthropods, and fishes have **internal gills** in such cavities.

Just as the gills of water breathers increase the surface area available for respiratory gas exchange, air-breathing vertebrates have enormous surface areas for gas exchange. **Lungs** are internal cavities for respiratory gas exchange with air. Their structure is quite different from that of gills (Figure 48.4*b*). Lungs have a large surface area because they are highly divided, and they are elastic so they can be inflated and deflated with air.

Most air-breathing invertebrates are insects, which have a unique respiratory gas exchange system consisting of a highly branched network of air-filled tubes called *tracheae* that branch through all the tissues of the insect's body. The terminal branches of these tubes are so numerous that they have an enormous surface area.

Ventilation and perfusion maximize partial pressure gradients

Fick's law of diffusion points to other possible adaptations besides increasing surface area that can increase respira-

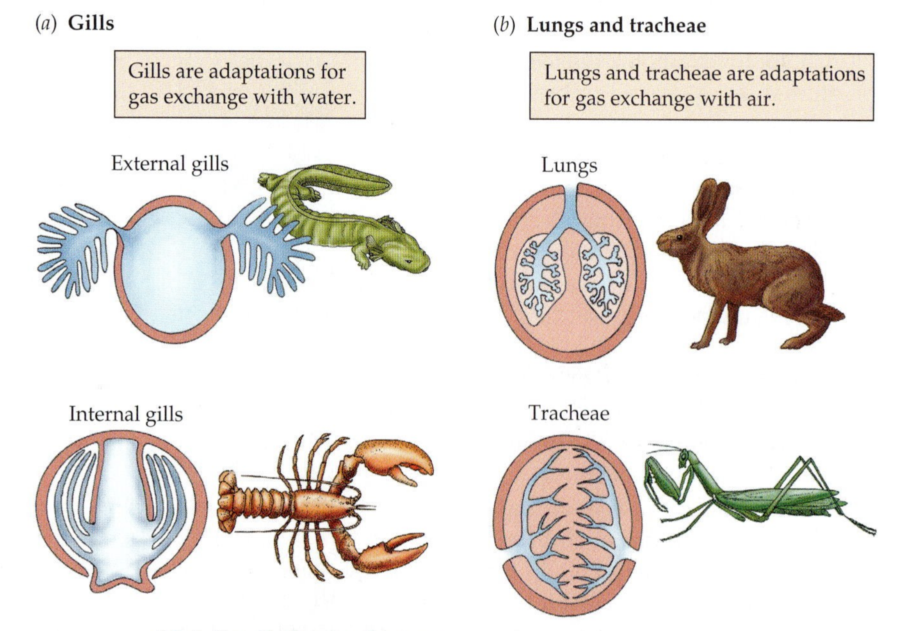

48.4 Gas Exchange Systems
A large surface area for the diffusion of respiratory gases is a common feature of animals.

tory gas exchange. Animals maximize the partial pressure gradients ($C_1 - C_2/L$) that drive the diffusion of respiratory gases across their gas exchange membranes in several ways:

- Gill and lung membranes are very thin so that the path length for diffusion (L) is small.
- Breathing mechanisms **ventilate** the environmental side of the exchange surfaces so that they are exposed to fresh respiratory medium (air or water) with the highest possible partial pressure of O_2 and the lowest possible partial pressure of CO_2.
- Circulatory systems **perfuse** the internal side of the exchange surfaces with a respiratory gas transport medium that helps maintain the lowest possible partial pressure of O_2 and highest possible partial pressure of CO_2 on the inside of the exchange membranes.

An animal's **gas exchange system** is made up of its gas exchange surfaces and the mechanisms it uses to ventilate and perfuse those surfaces. The following sections describe four gas exchange systems. First we look at the unique gas exchange system of insects. Then we describe two remarkably efficient systems: fish gills and bird lungs. Finally, we discuss mammalian lungs.

INSECT TRACHEAE. Respiratory gases diffuse through air most of the way to and from every cell of an insect's body. This diffusion is achieved through a system of air tubes, or **tracheae**, that open to the outside environment through holes called *spiracles* in the sides of the abdomen (Figure 48.5*a,b*). The tracheae branch into even finer tubes, or *tracheoles*, until they end in tiny *air capillaries* (Figure 48.5*c*). In the insect's flight muscles and other highly active tissues, no mitochondrion is more than a few micrometers away from an air capillary.

Because the diffusion rate of oxygen is so much higher in air than in water, air capillaries enable insects to supply oxygen to their cells at high rates. Many insects metabolize at high rates, but their relatively simple gas exchange systems are able to provide them with the oxygen they need. However, the rate of diffusion in insect tracheae and air capillaries is limited by their small diameter (A) and by the length (L) of these dead-end airways, so insects must be relatively small animals.

Some species of insects that dive and stay underwater for long periods make use of an interesting variation on diffusion. These insects carry with them a bubble of air. A small bubble may not seem like a very large reservoir of oxygen, yet these insects can stay underwater almost indefinitely with their small air supplies. The secret has to do with the P_{O_2} in the bubble. When the insect dives, the air bubble contains about 80 percent nitrogen and 20 percent O_2. As the insect consumes the O_2 in its bubble, the bubble shrinks a little. The bubble doesn't disappear, however, because it consists mostly of nitrogen, which the insect does not consume. When the P_{O_2} in the bubble falls below the P_{O_2} in the surrounding water, O_2 diffuses from the water into the bubble. For these small animals, the rate of O_2 diffusion into the bubble is enough to meet their O_2 demand while they are underwater.

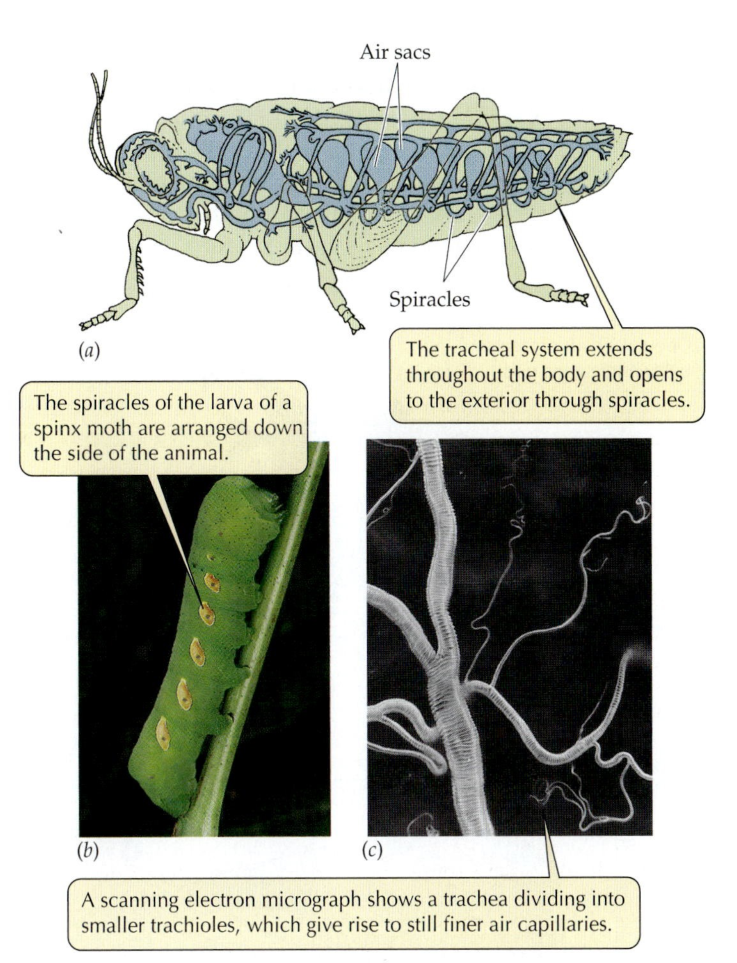

48.5 The Tracheal Gas Exchange System of Insects
In insects, respiratory gases diffuse through a system of air tubes (tracheae) that open to the external environment through holes called spiracles.

FISH GILLS. The internal gills of fishes are supported by usually four *gill arches* on either side of the fish and lie between the mouth cavity and the protective *opercular flaps* (Figure 48.6*a*). Water flows unidirectionally into the fish's mouth, over the gills, and out from under the opercular flaps, so that the gills are continuously bathed with fresh water. This constant flow of water moving over the gills maximizes the P_{O_2} on the external surfaces. On the internal side, the circulation of blood minimizes the P_{O_2} by sweeping the O_2 away as rapidly as it diffuses across.

The gills have an enormous surface area for gas exchange because they are so highly divided. Each gill consists of hundreds of leaf-shaped *gill filaments* (Figure 48.6*b*). The upper and lower flat surfaces of each gill filament have rows of evenly spaced folds, or *lamellae*. The lamellae are the gas exchange surfaces. Their delicate structure minimizes the path length for diffusion of gases between blood and water. The surfaces of the lamellae consist of highly flattened epithelial cells, so the water and the red blood cells are separated by little more than 1 or 2 μm.

The flow of blood perfusing the inner surfaces of the lamellae, like the flow of water over the gills, is unidirectional. *Afferent* blood vessels bring blood to the gills, while *efferent* blood vessels take blood away from the gills (Figure 48.6*c*). Blood flows through the lamellae in the direction opposite to the flow of water over the lamellae. This **countercurrent flow** maximizes the P_{O_2} gradient between water and blood, making gas exchange more efficient than it would be in a system using concurrent (parallel) flow (Figure 48.7).

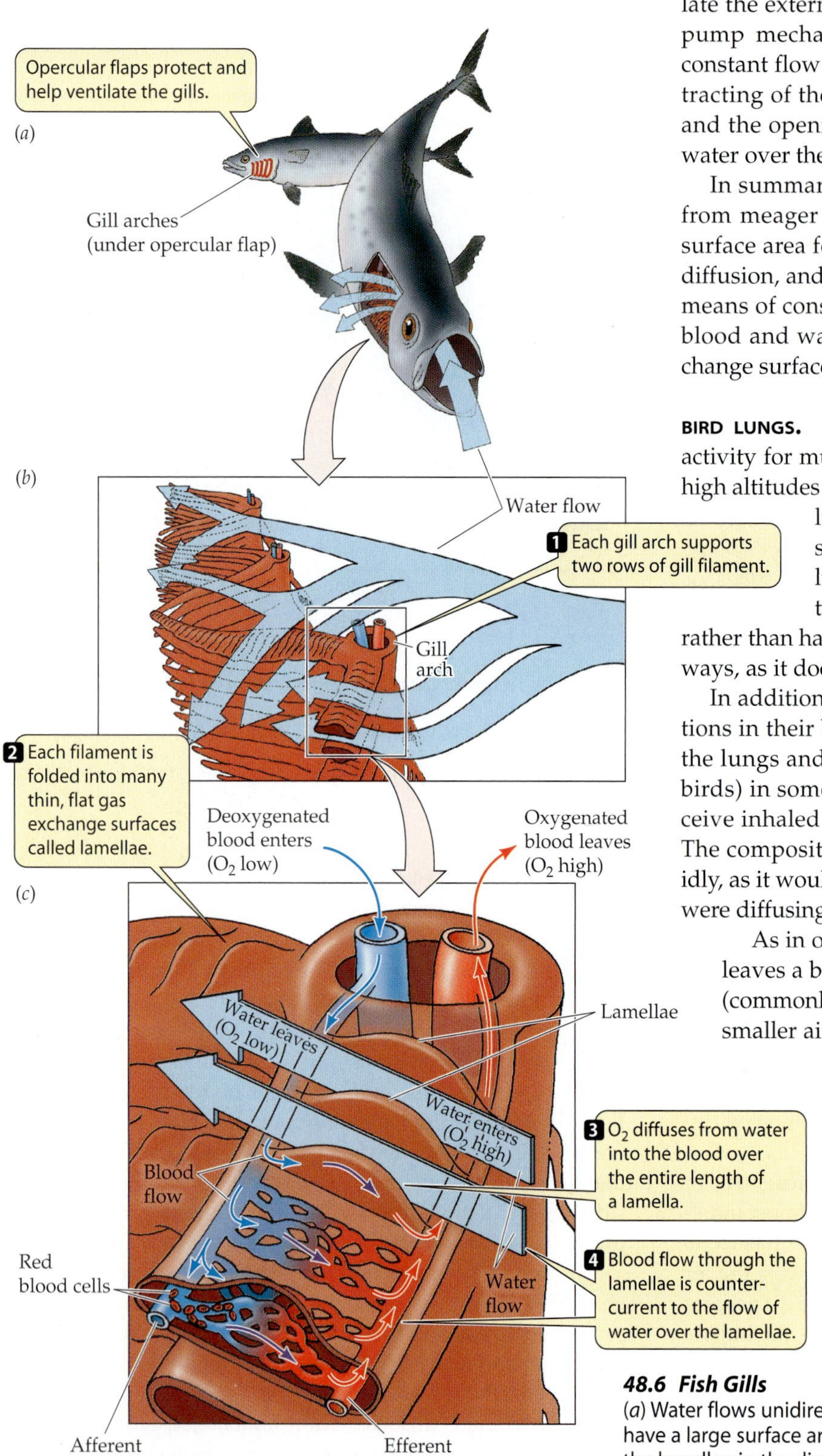

48.6 Fish Gills
(*a*) Water flows unidirectionally over the gills of a fish. (*b*) Gill filaments have a large surface area and thin membranes. (*c*) Blood flows through the lamellae in the direction opposite (left to right, in this depiction) to the flow of water (right to left) over the lamellae.

Some fish, including anchovies, tuna, and certain species of sharks, ventilate their gills by swimming almost constantly with their mouths open. Most fish, however, ventilate the external surfaces of their gills by means of a two-pump mechanism that maintains a unidirectional and constant flow of water over the gills. The closing and contracting of the mouth cavity pushes water over the gills, and the opening and closing of the opercular flaps pulls water over the gills.

In summary, fish can extract an adequate supply of O_2 from meager environmental sources by maximizing the surface area for diffusion, minimizing the path length for diffusion, and maximizing oxygen extraction efficiency by means of constant, unidirectional, countercurrent flow of blood and water over the opposite sides of their gas exchange surfaces.

BIRD LUNGS. Birds can sustain extremely high levels of activity for much longer than mammals can—even at very high altitudes where mammals cannot even survive. Yet the lungs of a bird are smaller than the lungs of a similar-sized mammal. How can this be? Bird lungs have a unique structure that allows air to flow unidirectionally through the lungs, rather than having to flow in and out through the same airways, as it does in mammalian lungs.

In addition to lungs, birds have **air sacs** at several locations in their bodies. The air sacs are interconnected with the lungs and with air spaces (another unique feature of birds) in some of the bones (Figure 48.8). The air sacs receive inhaled air, but they are not gas exchange surfaces. The composition of air in an air sac does not change rapidly, as it would if O_2 were diffusing into the blood and CO_2 were diffusing into the air sac.

As in other air-breathing vertebrates, air enters and leaves a bird's gas exchange system through a **trachea** (commonly known as the *windpipe*), which divides into smaller airways called **bronchi** (singular bronchus). In air-breathing vertebrates other than birds, the bronchi generate trees of branching airways that become finer and finer until they dead-end in clusters of microscopic, membrane-enclosed air sacs, where gases are exchanged. In bird lungs, however, there are no dead ends; air flows unidirectionally through the lungs.

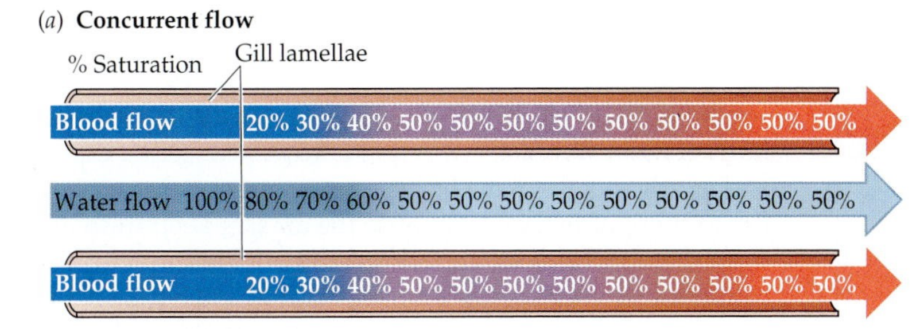

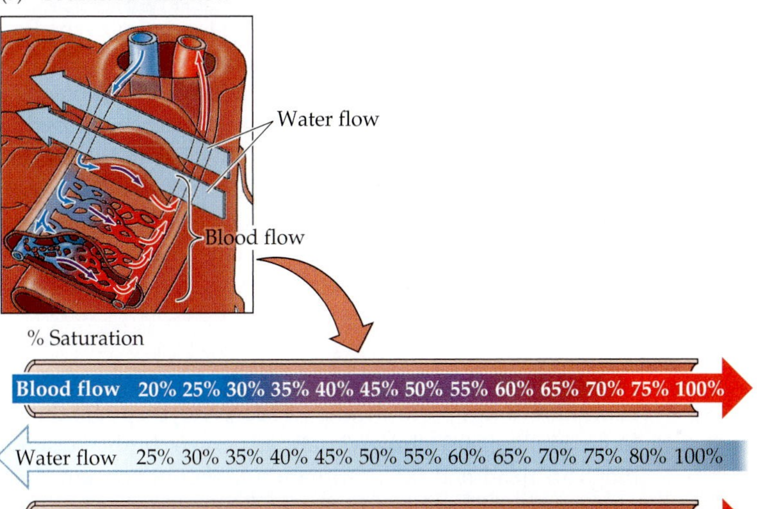

48.7 Countercurrent Exchange Is More Efficient than Concurrent Exchange
In these models of concurrent and countercurrent exchange, the numbers represent the oxygen saturation of blood and water. (*a*) In the concurrent exchanger, the percentages of saturation of blood and water reach equilibrium even before the water has flowed halfway across the exchange surface. (*b*) There is more complete exchange in the case of countercurrent flow because the water is *always* more saturated than the blood as it passes over the exchange surface, so that a gradient of O_2 saturation is always maintained.

In bird lungs, the bronchi divide into tubelike **parabronchi** (Figure 48.9). Running between the parabronchi are tiny airways called *air capillaries*. Air flows through the lungs in the parabronchi, but crosses between parabronchi through the air capillaries. The air capillaries are the gas exchange surfaces. They are tiny but numerous, so they provide an enormous surface area for gas exchange.

Another unusual feature of bird lungs is that they expand and contract less during a breathing cycle than mammalian lungs do. To make things even more puzzling, bird lungs contract during inhalation and expand during exhalation!

The puzzle of how birds breathe was solved by an experiment that placed small oxygen sensors at different locations in the air sacs and airways of birds. The bird could then be exposed to pure oxygen for just a single breath, and the progress of that single breath through the bird's gas exchange

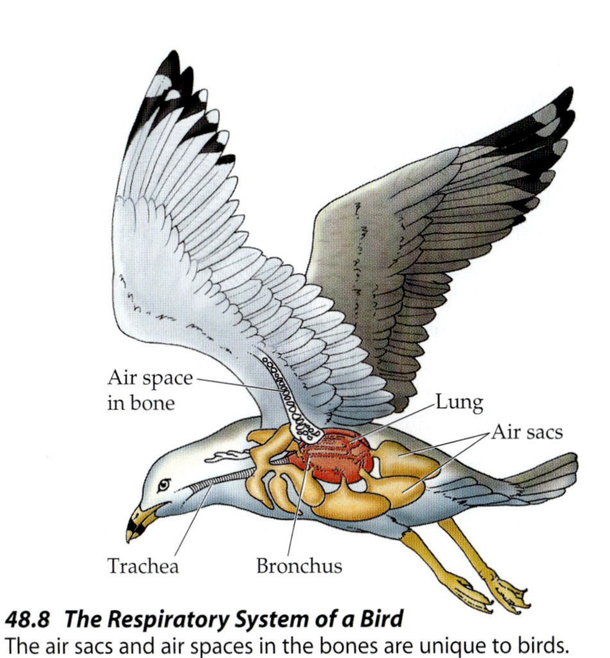

48.8 The Respiratory System of a Bird
The air sacs and air spaces in the bones are unique to birds.

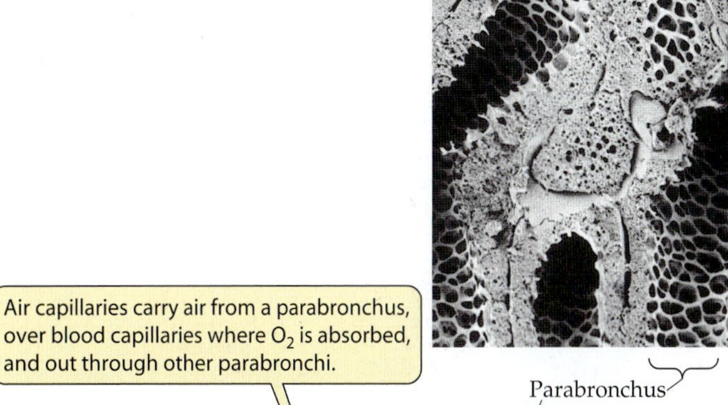

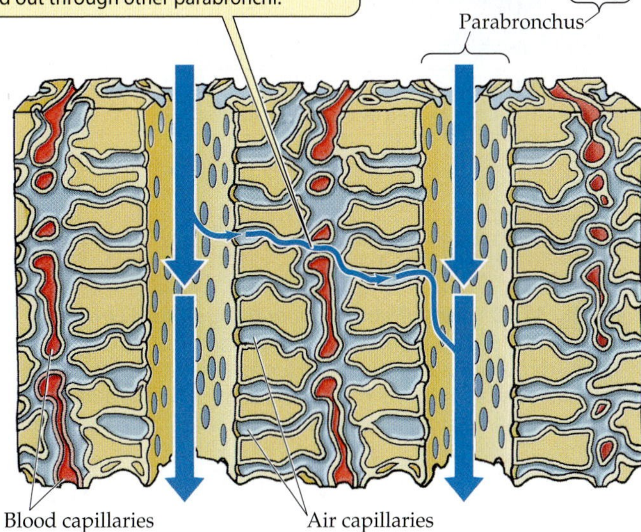

48.9 Air Flows through Bird Lungs Constantly and Unidirectionally
The gas exchange surfaces of birds are air capillaries branching off the parabronchi, which run through the lungs.

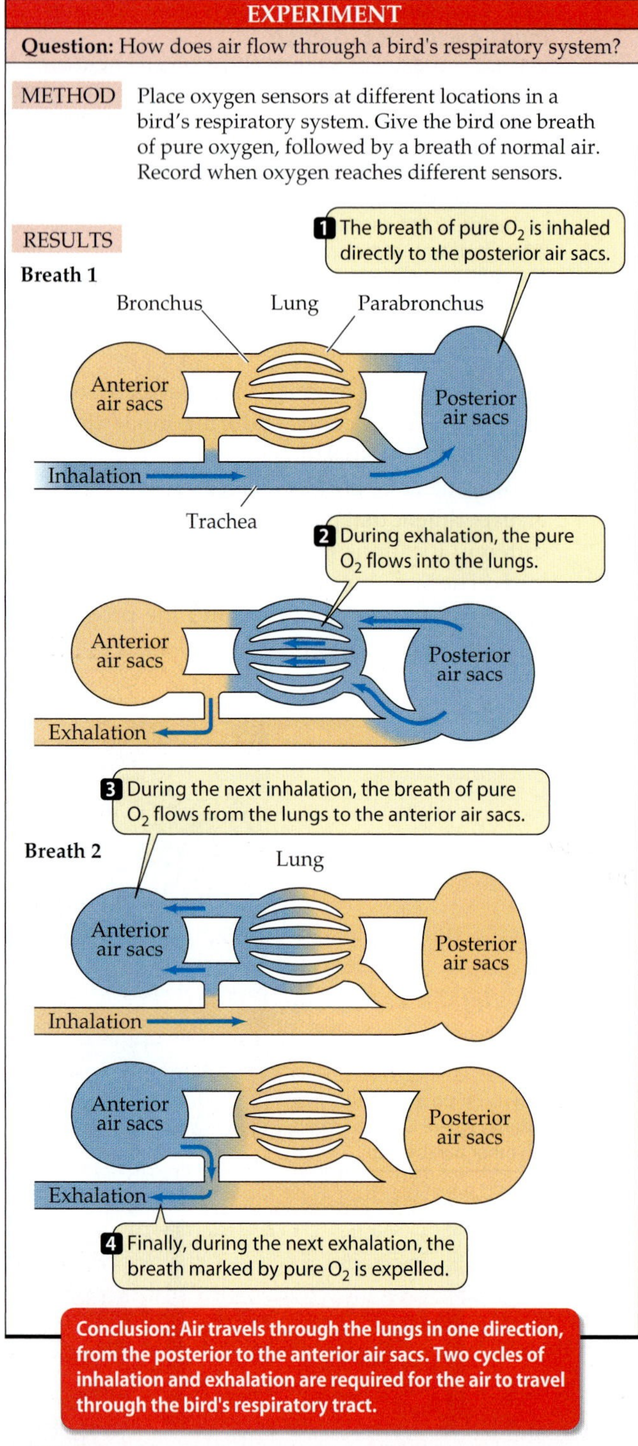

48.10 The Path of Air Flow through Bird Lungs The fresh air a bird takes in with one breath (blue) travels through the lungs in one direction, from the posterior to the anterior air sacs. Two cycles of inhalation and exhalation are required for the air to travel the full length of the bird's respiratory system.

system could be followed by the oxygen sensors. This experiment showed that a single breath remains in the bird's gas exchange system for two cycles of inhalation and exhalation, and that the air sacs work as bellows maintaining a continuous and unidirectional flow of fresh air through the lungs (Figure 48.10).

The advantages of the bird gas exchange system are similar to those of fish gills. Because the air sacs keep fresh air from the outside flowing unidirectionally and practically continuously over the gas exchange surfaces, the P_{O_2} on the environmental side of those surfaces is maximized. Furthermore, the unidirectional flow of air through the system makes possible a pattern of blood flow to minimize the P_{O_2} on the internal side of the exchange surfaces.

It is now clear how birds can fly over Mount Everest. A bird is able to supply its gas exchange surfaces with a continuous flow of fresh air that has a P_{O_2} close to that of the ambient air. Even when the P_{O_2} of the ambient air is only slightly above the P_{O_2} of the blood, O_2 can diffuse from air to blood. Next we will see why humans find it difficult to sustain even low levels of metabolic activity at such high altitudes.

TIDAL BREATHING IN MAMMALS. At the beginning of their evolution, lungs were dead-end sacs, and they remain so today in all air-breathing vertebrates except birds. Because lungs are dead-end sacs, ventilation cannot be constant and unidirectional, but must be **tidal**: Air flows in and out by the same route.

A *spirometer* shows how we use our lung capacity in breathing (Figure 48.11). When we are at rest, the amount of air that our normal breathing cycle moves per breath is called the *tidal volume* (about 500 ml for an average human adult). We can breathe much more deeply and inhale more air than our resting tidal volume; the additional volume of air we can take in above normal tidal volume is our *inspiratory reserve volume*. Conversely, we can forcefully exhale more air than we normally do during a resting exhalation. This additional amount of air that can be forced out of the lungs is the *expiratory reserve volume*. But even after the most extreme exhalation possible, some air remains in the lungs. The lungs and airways cannot be collapsed completely; they always contain a *residual volume*. The *total lung capacity* is the sum of the residual volume, expiratory reserve volume, tidal volume, and inspiratory reserve volume.

Tidal breathing severely limits the partial pressure gradient available to drive the diffusion of O_2 from air into the blood. Fresh air is not moving into the lungs during half of the respiratory cycle; therefore, the average P_{O_2} of air in the lungs is considerably less than it is in the air outside the lungs. Furthermore, the incoming air mixes with the stale air that was not expelled by the previous exhalation. The lung volume that is not ventilated with fresh air is called *dead space*. This dead space consists of the residual volume and, depending on the depth of breathing, some or all of the expiratory reserve volume.

The scale in Figure 48.11 tells us that a tidal volume of 500 ml of fresh air mixes with up to 2,000 ml of stale air before reaching the gas exchange surfaces in our lungs. When the P_{O_2} in the ambient air is 150 mm Hg, the P_{O_2} of the air that reaches our gas exchange surfaces is only about 100

RESEARCH METHOD

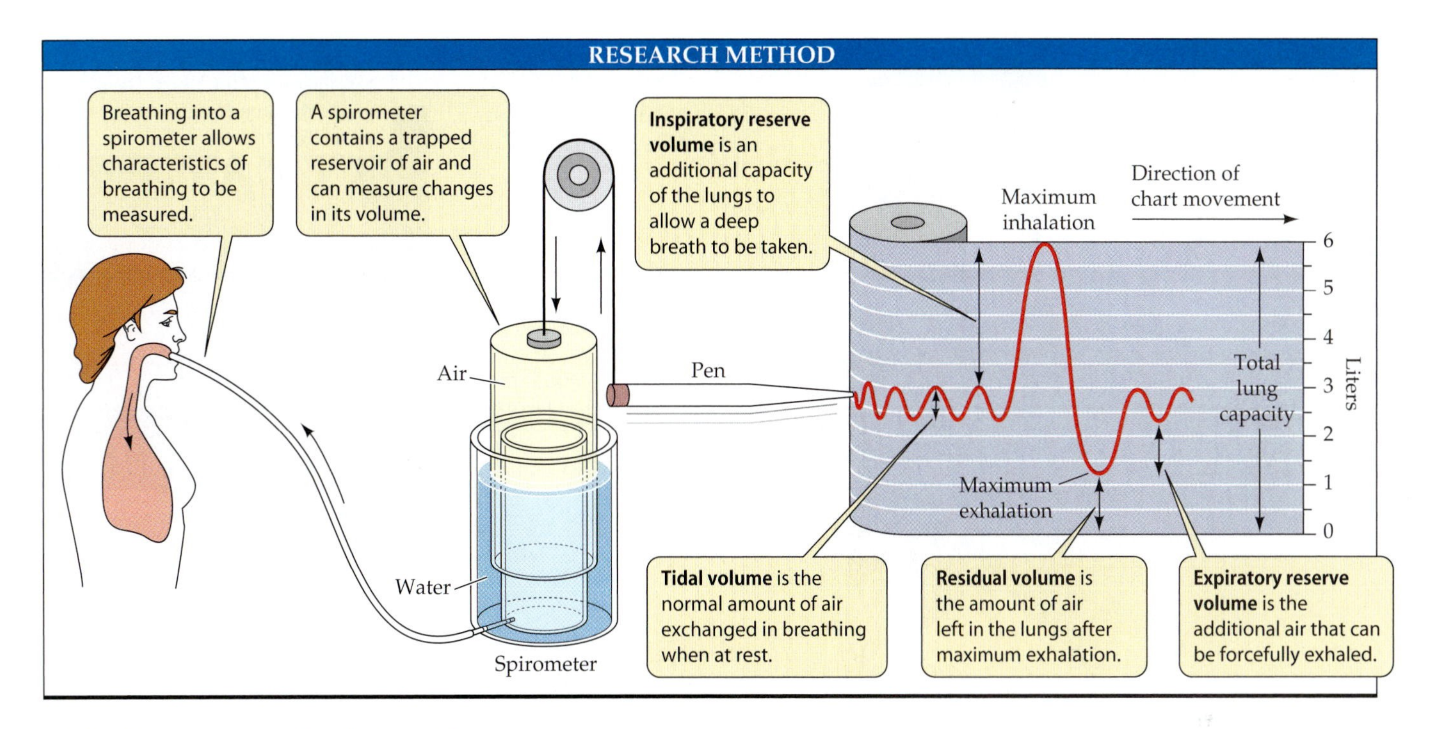

48.11 Measuring Lung Ventilation with a Spirometer
Breathing from a closed reservoir of air and measuring the changes in the volume of that reservoir demonstrates the characteristics of mammalian tidal breathing.

mm Hg. By contrast, the P_{O_2} in the water that bathes the lamellae of fish gills or in the air that flows through the air capillaries of bird lungs is the same as the P_{O_2} in the outside water or air.

In addition to reducing the partial pressure gradient, tidal breathing reduces the efficiency of gas exchange in another way: It does not allow countercurrent gas exchange between air and blood. Because air enters and leaves the gas exchange structures by the same route, there is no anatomical way that blood can flow countercurrent, or even crosscurrent, to the air flow.

Mammalian Lungs and Gas Exchange

To offset the inefficiencies of tidal breathing, mammalian lungs have some design features that maximize the rate of gas exchange: an enormous surface area, and a very short path length for diffusion. Mammalian lungs serve the respiratory needs of mammals well, considering the ecologies and lifestyles of these animals.

Air enters the lungs through the oral cavity or nasal passage, which join together in the *pharynx* (Figure 48.12). From the pharynx, the esophagus conducts food to the stomach and a single airway leads to the lungs. At the beginning of this airway is the *larynx*, or voice box, which houses the vocal cords. The larynx is the "Adam's apple" that you can see or feel on the front of your neck. The major airway, the *trachea*, is about 2 cm in diameter. The thin walls of the trachea are prevented from collapsing by rings of cartilage that support them as air pressure changes during the breathing cycle. If you run your fingers down the front of your neck just below your larynx, you can feel a couple of these rings of cartilage.

The trachea branches into two smaller *bronchi*, one leading to each lung. The bronchi branch repeatedly to generate a treelike structure of progressively smaller airways extending to all regions of the lungs (see Figure 48.12). As the branching of the bronchial tree continues to produce still smaller airways, the cartilage supports eventually disappear, marking the transition to **bronchioles**. The branching continues until the bronchioles are smaller than the diameter of a pencil lead, at which point tiny, thin-walled air sacs called **alveoli** begin to appear.

The alveoli are the sites of gas exchange. Because the airways only conduct air to and from the alveoli and do not themselves conduct gas exchange, their volume is physiological dead space. The number of alveoli in human lungs is about 300 million. Even though each alveolus is very small, the combined surface area for diffusion of respiratory gases is about 70 m^2—the size of a badminton court.

Each alveolus is made of very thin cells. Between and surrounding the alveoli are networks of the smallest of blood vessels, the *capillaries*, whose walls are also made up of exceedingly thin endothelial cells. Where capillary meets alveolus, very little space separates them (see Figure 48.12), so the length of the diffusion path between air and blood is less than 2 μm. Even the diameter of a red blood cell is greater—about 7 μm.

Respiratory tract secretions aid breathing

Mammalian lungs have two other important adaptations that do not directly influence their gas exchange properties:

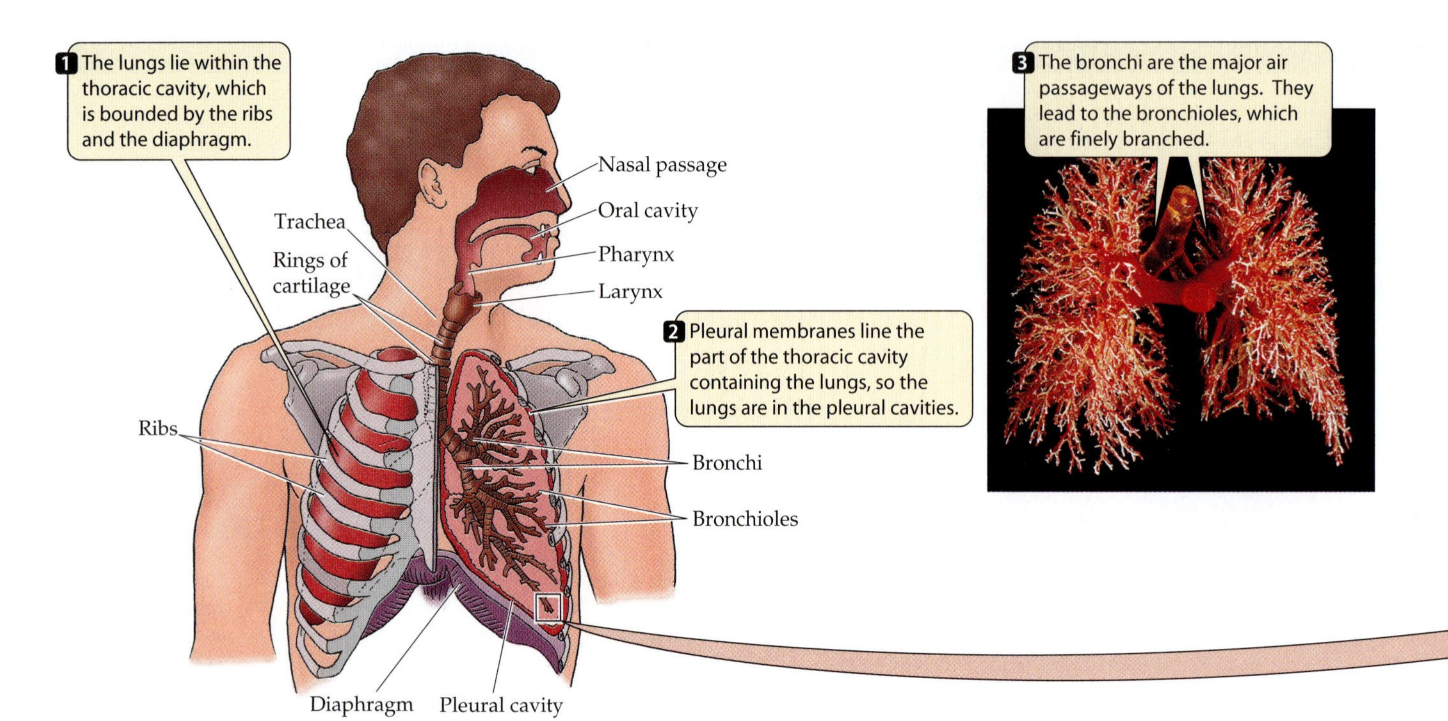

48.12 The Human Respiratory System
The diagram traces the hierarchy of human respiratory structures from the lungs down to the minuscule alveoli.

the production of mucus and the production of surfactant.

A **surfactant** is a substance that reduces the surface tension of a liquid. Lung surfactant reduces the surface tension of the film of fluid lining the insides of the alveoli. What is surface tension and how does it affect lung function? Surface tension is a result of cohesion between the molecules of a liquid. Cohesion gives the surface of the liquid the properties of an elastic membrane. Surface tension is what allows some insects to walk on the surface of water (see Figure 2.16). A surfactant interferes with the cohesive forces that create surface tension. Detergent is a surfactant, and when added to water, it makes walking on water difficult for the water strider.

The thin, aqueous layer that lines the alveoli has surface tension, which must be overcome to inflate the lungs. Surface tension normally is reduced by surfactant molecules produced by certain cells in the alveoli. If a baby is born more than a month prematurely, however, these cells may not yet be producing surfactant. Such a premature baby has great difficulty breathing because an enormous effort is required to stretch the alveoli against the surface tension. This condition, known as *respiratory distress syndrome*, may cause the baby to die from exhaustion and suffocation. Common treatments have been to put the baby on a respirator to assist its breathing and to give the baby hormones to speed its lung development. A new approach, however, is to apply surfactant to the lungs via an aerosol.

Many cells lining the airways produce a sticky *mucus* that captures bits of dirt and microorganisms that are inhaled. This mucus must be continually cleared from the airways. Other cells lining the airways have cilia (see Figure 47.2) whose beating moves the mucus with its trapped debris up toward the pharynx, where it is swallowed. This phenomenon, called the **mucus escalator**, can be adversely affected by inhaled pollutants. Smoking one cigarette can immobilize the cilia of the airways for hours. A smoker's cough results from the need to clear the obstructing mucus from the airways when the mucus escalator is out of order.

Lungs are ventilated by pressure changes in the thoracic cavity

As Figure 48.12 shows, human lungs are suspended in the **thoracic cavity**, which is bounded on the top by the shoulder girdle, on the sides by the rib cage, and on the bottom by a domed sheet of muscle, the **diaphragm**. The thoracic cavity is lined on the inside by the **pleural membranes**, which divide it into right and left **pleural cavities** enclosing each lung. Because the pleural cavities are closed spaces, any effort to increase their volume creates negative pressure—suction—inside them.

Negative pressure within the pleural cavities causes the lungs to expand as air flows into them from the outside. This is the mechanism of inhalation. The diaphragm contracts to begin an inhalation. This contraction pulls the diaphragm down, increasing the volume of the thoracic and pleural cavities (Figure 48.13). As pressure in the pleural cavities becomes more negative, air moves into the lungs. Exhalation begins when the contraction of the diaphragm ceases. The diaphragm relaxes and moves up, and the elastic recoil of the lungs pushes air out through the airways. During tidal breathing, inhalation is an active process and exhalation is a passive process.

The diaphragm is not the only muscle that changes the volume of the pleural cavities. Between the ribs are two sets

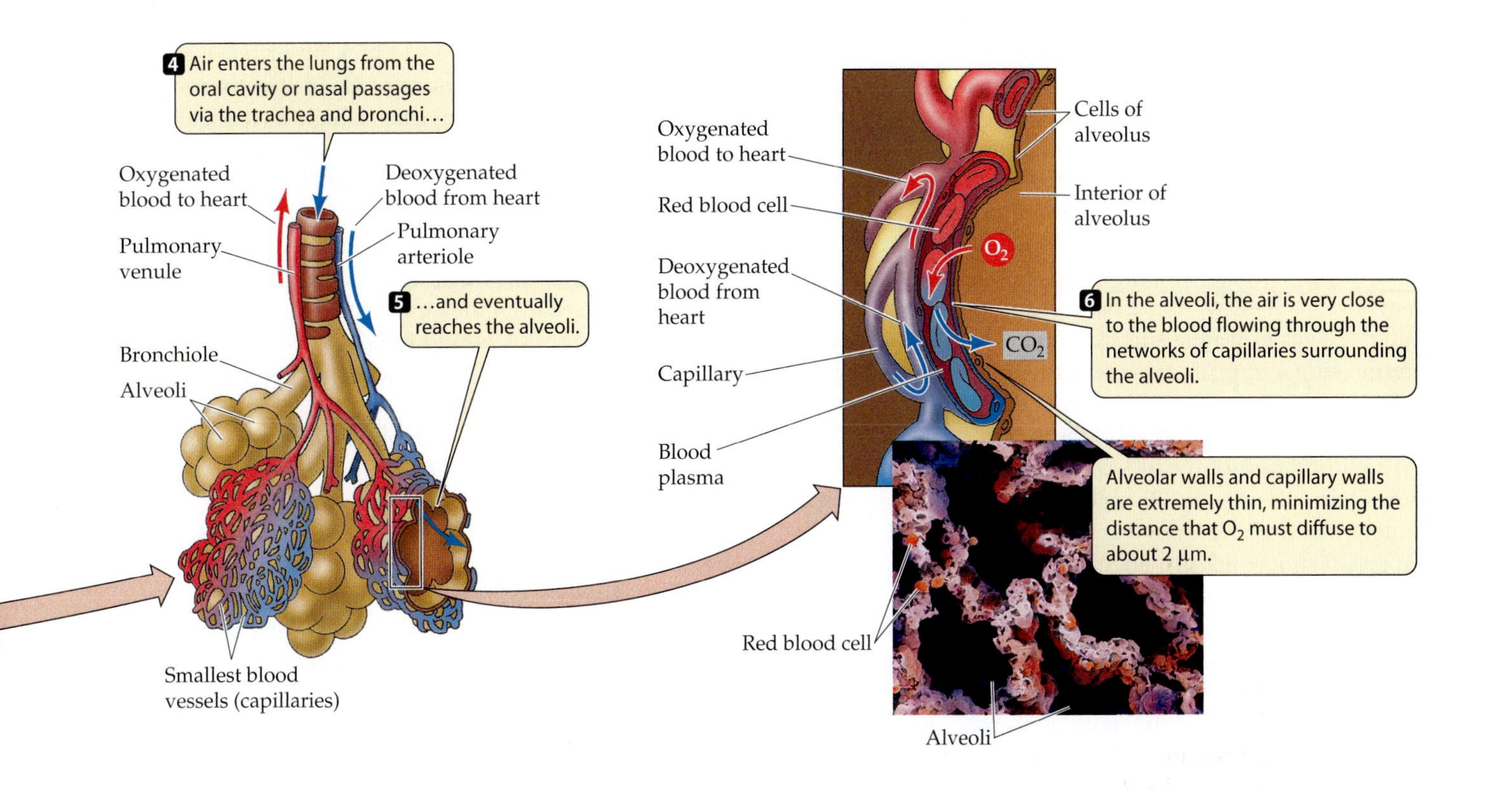

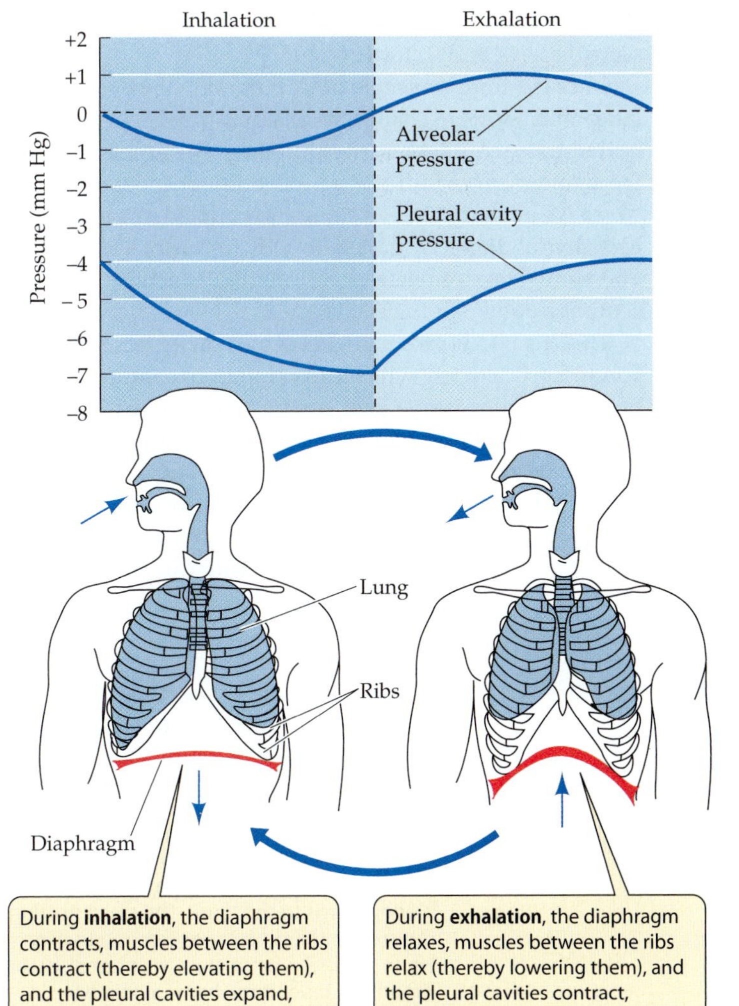

of **intercostal muscles**. The *external intercostal muscles* expand the pleural cavities by lifting the ribs up and outward. The *internal intercostal muscles* decrease the volume of the thoracic cavity by pulling the ribs down and inward. When heavy demands are placed on the gas exchange system, such as during strenuous exercise, the external intercostal muscles increase the volume of air inhaled, making use of the inspiratory reserve volume, and the internal intercostal muscles increase the amount of air exhaled, making use of the expiratory reserve volume.

When the diaphragm is at rest between tidal breaths, the pressure in the pleural cavities is still slightly negative. This slight suction keeps the alveoli partly inflated. If the thoracic wall is punctured—by a knife wound, for example—air leaks into the pleural cavity, and the pressure from this air causes the lung to collapse. If the hole in the thoracic wall is not sealed, the breathing movements of the diaphragm and intercostal muscles pull air into the pleural cavity rather than into the lung, and ventilation of the alveoli in that lung ceases.

48.13 Into the Lungs and Out Again
Inhalation is an active process spurred by the contraction of the diaphragm. Exhalation generally is a passive process as the diaphragm relaxes. During inhalation, the negative pressure in the pleural cavity increases, expanding the elastic lung tissue and sucking air into the lungs. During exhalation, the negative pressure in the pleural cavity decreases, allowing the elastic lung tissue to recoil to create a positive pressure in the lungs and expel air.

Blood Transport of Respiratory Gases

The circulatory system is the subject of the next chapter, but since two of the substances the blood transports are the respiratory gases (O_2 and CO_2), we must discuss blood here. The circulatory system uses a pump (the heart) and a network of blood vessels to transport blood and the substances it carries around the body. As O_2 diffuses across the gas exchange surfaces into the blood vessels, the circulating blood sweeps it away. As we have seen, this internal perfusion of the gas exchange surfaces minimizes the P_{O_2} on the internal side and promotes the diffusion of O_2 across the surface at the highest possible rate. The blood then delivers this O_2 to the cells and tissues of the body.

The liquid part of the blood, the **blood plasma**, carries some O_2 in solution, but its ability to transport O_2 is quite limited. Blood plasma can carry only about 0.3 ml of oxygen per 100 ml, which is inadequate to support the metabolism of a person at rest. Fortunately, the blood also contains **red blood cells**, which are red because they are loaded with the oxygen-binding pigment **hemoglobin**. Hemoglobin increases the capacity of blood to transport oxygen by about 60-fold. There is quite a diversity of O_2-binding pigments among the animals; there is even considerable diversity of hemoglobin molecules. The discussion that follows focuses on human hemoglobin.

Hemoglobin combines reversibly with oxygen

Red blood cells contain enormous numbers of hemoglobin molecules. Hemoglobin is a protein consisting of four polypeptide subunits (see Figure 3.7). Each of these polypeptides surrounds a heme group—an iron-containing ring structure that can reversibly bind a molecule of O_2.

As O_2 diffuses into the red blood cells, it binds to hemoglobin. Once O_2 is bound, it cannot diffuse back across the red cell plasma membrane. By mopping up O_2 molecules as they enter the red blood cells, hemoglobin maximizes the partial pressure gradient driving the diffusion of O_2 into the cells. In addition, it enables the red blood cells to carry a large amount of O_2 to the tissues of the body.

The ability of hemoglobin to pick up or release O_2 depends on the partial pressure of O_2 in its environment. When the P_{O_2} of the blood plasma is high, as it usually is in the lung capillaries, each molecule of hemoglobin can carry its maximum load of four molecules of O_2. As the blood circulates through the rest of the body, it encounters lower P_{O_2} values. At these lower P_{O_2} values, the hemoglobin releases some of the O_2 it is carrying (Figure 48.14).

As you can see from the figure, the relation between P_{O_2} and the amount of O_2 bound to hemoglobin is not linear, but S-shaped (sigmoid). The sigmoid hemoglobin–O_2 binding curve reflects interactions between the four subunits of the hemoglobin molecule, each of which can bind one molecule of O_2. At low P_{O_2} values, only one subunit will bind an O_2 molecule.When it does so, the shape of this subunit changes, causing an alteration in the quaternary structure of the whole hemoglobin molecule (see Chapter 3). This structural change makes it easier for the other subunits to bind a molecule of O_2; that is, their O_2 affinity is increased. Therefore a smaller increase in P_{O_2} is necessary to get most of the hemoglobin molecules to bind two O_2 molecules (that is, to become 50 percent saturated) than it was to get them to bind one molecule of O_2 (25 percent saturated). The influence of the binding of O_2 by one subunit on the O_2 affinity of the other subunits is called **positive cooperativity**, because binding of the first molecule makes binding of the second easier, and so forth.

Once the third molecule of O_2 is bound, however, the relationship seems to change, as a larger increase in P_{O_2} is required to reach 100 percent saturation. This upper bend of the sigmoid curve is due to a probability phenomenon: The closer we get to having all subunits occupied, the less likely it is that a single O_2 molecule will find a place to bind. Therefore it takes a relatively greater P_{O_2} to achieve 100 percent saturation.

This is a good place to mention the danger posed by carbon monoxide (CO), which can come from a faulty furnace or from combusting a fuel such as charcoal or kerosene without adequate ventilation. CO binds to hemoglobin with a higher affinity than does O_2. Thus, CO destroys the ability of hemoglobin to transport and release O_2 to the tissues of the body. The victim loses consciousness and can die because the brain lacks O_2.

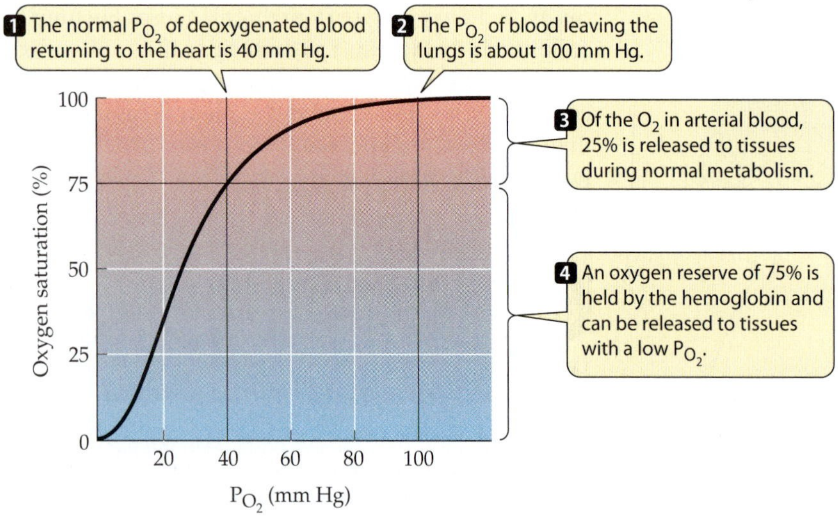

1 The normal P_{O_2} of deoxygenated blood returning to the heart is 40 mm Hg.

2 The P_{O_2} of blood leaving the lungs is about 100 mm Hg.

3 Of the O_2 in arterial blood, 25% is released to tissues during normal metabolism.

4 An oxygen reserve of 75% is held by the hemoglobin and can be released to tissues with a low P_{O_2}.

Oxygen saturation (%)

100 75 50 25 0

20 40 60 80 100

P_{O_2} (mm Hg)

48.14 The Binding of Oxygen to Hemoglobin Depends on the P_{O_2}
Hemoglobin in blood leaving the lungs is 100 percent saturated (four molecules of O_2 are bound to each hemoglobin). Most hemoglobin molecules will drop only one of their four O_2 molecules as they circulate through the body, and are still 75 percent saturated when the blood returns to the lungs. The steep portion of this oxygen-binding curve comes into play when tissue P_{O_2} falls below the normal 40 mm Hg, at which point the hemoglobin will "unload" its oxygen reserves.

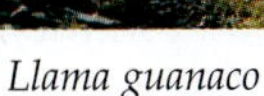
Llama guanaco

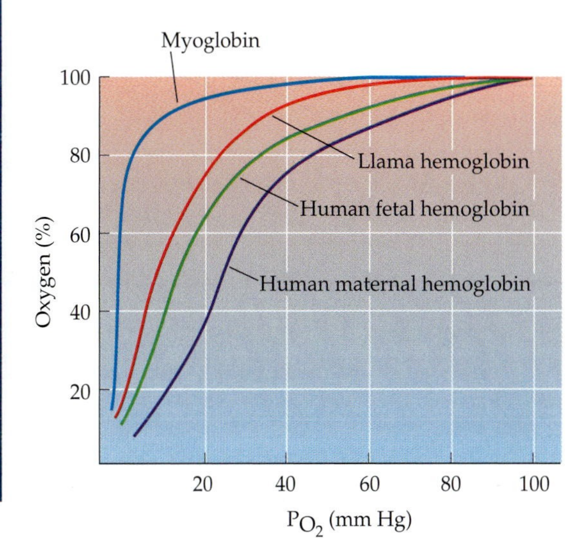

48.15 Oxygen-Binding Adaptations
Evolution has adapted the oxygen-binding properties of different hemoglobins and of myoglobin. The hemoglobin of llamas, for example, is adapted for binding oxygen at high altitudes, where P_{O_2} is low.

The O_2-binding properties of hemoglobin help get O_2 to the tissues that need it most. In the lungs, where the P_{O_2} is about 100 mm Hg, the hemoglobin is 100 percent saturated. The P_{O_2} in blood returning to the heart from the body is usually about 40 mm Hg. You can see from Figure 48.14 that at this P_{O_2} the hemoglobin is still about 75 percent saturated. This means that as the blood circulates around the body, only about 1 of 4 O_2 molecules it carries is released to the tissues. That seems inefficient, but it is really quite adaptive, because the hemoglobin keeps 75 percent of its oxygen in reserve to meet peak demands.

When a tissue becomes starved of oxygen and its local P_{O_2} falls below 40 mm Hg, the hemoglobin flowing through that tissue is on the steep portion of its sigmoid binding curve. That means that relatively small decreases in P_{O_2} below 40 mm Hg will result in the release of lots of O_2 to the tissue. Thus the cooperativity of O_2 binding by hemoglobin is very effective in making O_2 available to the tissues precisely when and where it is needed most.

Myoglobin holds an oxygen reserve

Muscle cells have their own oxygen-binding molecule, **myoglobin**. Myoglobin consists of just one polypeptide chain associated with an iron-containing ring structure that can bind one molecule of oxygen. Myoglobin has a higher affinity for O_2 than hemoglobin does (see Figure 48.15), so it picks up and holds oxygen at P_{O_2} values at which hemoglobin is releasing its bound O_2.

Myoglobin provides a reserve of oxygen for the muscle cells for times when metabolic demands are high and blood flow is interrupted. Interruption of blood flow in muscles is common because contracting muscles constrict blood vessels. When tissue P_{O_2} values are low and hemoglobin can no longer supply more O_2, myoglobin releases its bound O_2. Diving mammals such as seals have high concentrations of myoglobin in their muscles, which is one reason they can stay underwater for so long. (We will learn more about adaptations for diving in the next chapter.) Even in nondiving animals, muscles called on for extended periods of work frequently have more myoglobin than muscles that are used for short, intermittent periods. This is one of the reasons for the difference in appearance between fast-twitch and slow-twitch muscle (see Figure 47.12).

The affinity of hemoglobin for oxygen is variable

Various factors influence the oxygen-binding properties of hemoglobin, thereby influencing oxygen delivery to tissues. In this section we examine three of these factors: the chemical composition of the hemoglobin, pH, and the presence of 2,3 diphosphoglyceric acid.

HEMOGLOBIN COMPOSITION. As we noted above, there is more than one type of hemoglobin. The chemical composition of the polypeptide chains that form the hemoglobin molecule varies. The normal hemoglobin of adult humans has two each of two kinds of polypeptide chains—two *alpha* chains and two *beta* chains—and the oxygen-binding characteristics shown in Figure 48.14.

Before birth, the human fetus has a different form of hemoglobin, consisting of two *alpha* chains and two *gamma* chains. The functional difference between these two types of hemoglobin is that the fetal hemoglobin has a higher affinity for O_2. Therefore, the hemoglobin–O_2 binding curve of fetal hemoglobin is shifted to the left in comparison to the curve for adult hemoglobin (Figure 48.15). You can see from these curves that if both types of hemoglobin are at the same P_{O_2}, the fetal hemoglobin will pick up oxygen released by the adult hemoglobin. This difference in O_2 affinities facilitates the transfer of O_2 from the mother's blood to the blood of the fetus in the placenta.

Llamas and vicuñas are mammals native to high altitudes in the Andes Mountains of South America. In the natural habitat of these animals, more than 5,000 m above sea

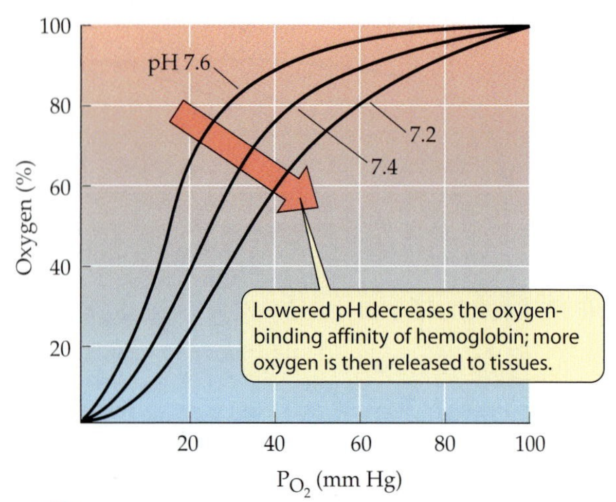

48.16 The Oxygen-Binding Properties of Hemoglobin Can Change
Changes in pH affect the oxygen-binding capacity of hemoglobin.

level, the P_{O_2} is below 85 mm Hg, and the P_{O_2} in their lungs is about 50 mm Hg. Thus, the hemoglobins of these animals must be able to pick up O_2 in an environment that has a low P_{O_2}. The hemoglobins of llamas and vicuñas have oxygen-binding curves to the left of the curves of hemoglobins of most other mammals—in other words, their hemoglobin can become saturated with O_2 at lower P_{O_2} values than those of other animals can.

PH. The oxygen-binding properties of hemoglobin are influenced by physiological conditions. The influence of pH on the function of hemoglobin is known as the **Bohr effect**. As the blood plasma picks up acidic metabolites such as lactic acid, fatty acids, and CO_2 (which combines with water to form carbonic acid) from the tissues, its pH falls. When this happens, the oxygen-binding curve of hemoglobin shifts to the right (Figure 48.16). This shift means that the hemoglobin will release more O_2 to the tissues—another way that O_2 is supplied where and when it is most needed.

2,3 DIPHOSPHOGLYCERIC ACID. 2,3 diphosphoglyceric acid (DPG) is a metabolite in the glycolytic pathway. Mammalian red blood cells have a high concentration of DPG, an important regulator of hemoglobin function. DPG reversibly combines with deoxygenated hemoglobin and changes the shape of the hemoglobin such that it has a lower affinity for O_2. The result is that at any P_{O_2}, hemoglobin releases more of its bound O_2 than it otherwise would. In other words, DPG shifts the oxygen-binding curve of mammalian hemoglobin to the right. When humans go to high altitudes, or when they cease being sedentary and begin to exercise, the level of DPG in their red blood cells goes up and makes it easier for the hemoglobin to deliver more O_2 to the tissues. The reason that fetal hemoglobin has a left-shifted hemoglobin–O_2 binding curve is that it has a lower affinity for DPG than does adult hemoglobin.

Llamas and humans employ opposite adjustments of hemoglobin function as adaptations for life at high altitudes. The llama's hemoglobin has a left-shifted oxygen-binding curve, which means that it can become 100 percent saturated with O_2 at the low P_{O_2} values at high altitudes. As a consequence, the llama's tissues must operate at a lower P_{O_2}. By contrast, human hemoglobin acquires, through acclimation, a right-shifted oxygen-binding curve. The result is that human hemoglobin never becomes fully saturated with O_2 at high altitudes, but more of the O_2 carried by that hemoglobin is released to the tissues.

Carbon dioxide is transported as bicarbonate ions in the blood

Delivering O_2 to the tissues is only half of the respiratory function of the blood. The blood also must take carbon dioxide, a metabolic waste product, away from the tissues. CO_2 is highly soluble and readily diffuses through cell membranes, moving from its site of production in a cell into the blood, where the partial pressure of CO_2 is lower. However, very little dissolved CO_2 is transported by the blood. Most CO_2 produced by the tissues is transported to the lungs in the form of the **bicarbonate ion**, HCO_3^-. How and where CO_2 is converted to HCO_3^-, is transported, and then is converted back to CO_2 is an interesting story.

When CO_2 dissolves in water, some of it slowly reacts with the water molecules to form carbonic acid (H_2CO_3), some of which then dissociates into a proton (H^+) and a bicarbonate ion (HCO_3^-). This reversible reaction is expressed as follows:

$$CO_2 + H_2O \rightleftharpoons H_2CO_3 \rightleftharpoons H^+ + HCO_3^-$$

In the blood plasma, the reaction between CO_2 and H_2O proceeds slowly. But it is a different story in the endothelial cells of the capillaries and in the red blood cells, where the enzyme **carbonic anhydrase** speeds up the conversion of CO_2 to H_2CO_3. The newly formed carbonic acid dissociates and the resulting bicarbonate ions enter the plasma in exchange for Cl^- (Figure 48.17). By converting CO_2 to H_2CO_3, carbonic anhydrase reduces the partial pressure of CO_2 in these cells and in the plasma, facilitating the diffusion of CO_2 from tissue cells to endothelial cells, plasma, and red blood cells. Most CO_2 is transported by the blood plasma as bicarbonate ion produced in endothelial cells and in red blood cells. Some CO_2 is also carried in chemical combination with deoxygenated hemoglobin as carboxyhemoglobin.

In the lungs, the reactions involving CO_2 and bicarbonate ions are reversed. CO_2 diffuses from the pulmonary capillaries to the alveolar air and is exhaled. Since the P_{CO_2} in the alveoli is less than the P_{CO_2} in the plasma and in the endothelial cells, CO_2 leaves the plasma and the endothelial cells and enters the alveoli. As the P_{CO_2} in the plasma falls, CO_2 diffuses from the red blood cells into the plasma and from there into the endothelial cells and into the alveoli. As the P_{CO_2} in the red blood cells falls, more HCO_3^- is converted into CO_2, and more HCO_3^- moves into the red blood

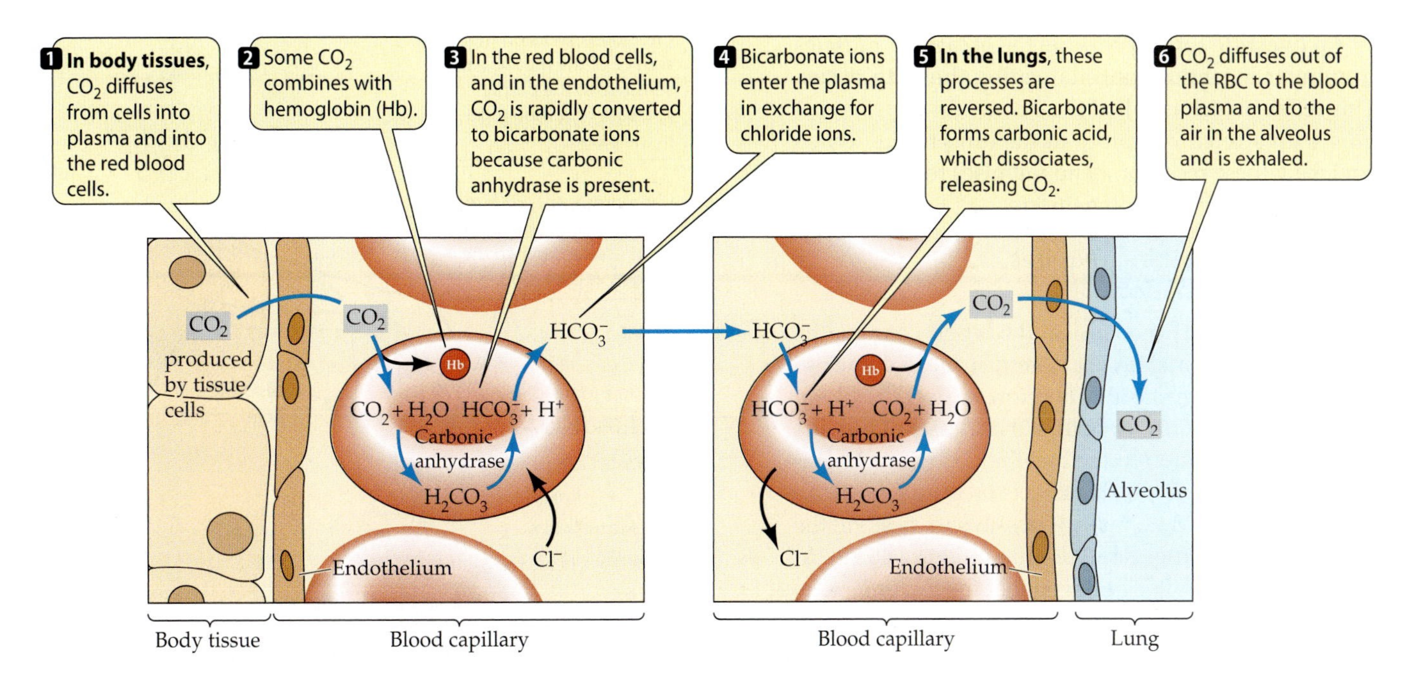

48.17 Carbon Dioxide Is Transported as Bicarbonate Ions
Carbonic anhydrase in capillary endothelial cells and in red blood cells facilitates conversion of CO_2 produced by tissues into bicarbonate ions carried by the plasma. In lungs, the process is reversed as CO_2 is exhaled.

cells from the plasma. Remember that an enzyme like carbonic anhydrase only speeds up a reversible reaction; it does not determine its direction. Direction is determined by concentrations of reactants and products (see Chapter 6).

Regulating Breathing to Supply O_2

We must breathe every minute of our lives, but we don't worry about our need to breathe, or even think about it very often. Breathing is an autonomic function of the nervous system. The breathing pattern easily adjusts itself around other activities (such as speech and eating), and breathing rates change to match the metabolic demands of our bodies. In this section we examine how the regular breathing cycle is generated and controlled.

Breathing is controlled in the brain stem

The autonomic nervous system maintains breathing and modifies its depth and frequency to meet the demands of the body for O_2 supply and CO_2 elimination. Breathing ceases if the spinal cord is severed in the neck region, showing that the breathing pattern is generated in the brain. If the brain stem is cut just above the medulla, the segment of the brain stem just above the spinal cord, an irregular breathing pattern remains (Figure 48.18).

Groups of neurons within the medulla increase their firing rates just before an inhalation begins. As more and more of these neurons fire—and fire faster and faster—the diaphragm contracts. Suddenly the neurons stop firing, the diaphragm relaxes, and exhalation begins. Exhalation is usually a passive process that depends on the elastic recoil of the lung tissues. When breathing demand is high, however, as during strenuous exercise, motor neurons for the intercostal muscles are recruited, which increases both the inhalation and the exhalation volumes. Brain areas above the medulla modify breathing to accommodate speech, ingestion of food, coughing, and emotional states.

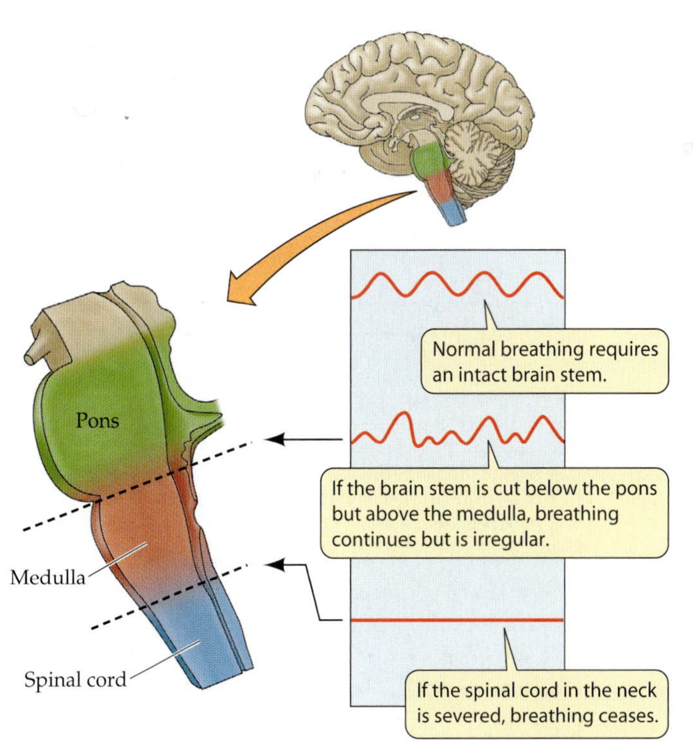

48.18 The Brain Stem Generates and Controls Breathing Rhythm
Severing the brain stem at different levels reveals that the basic breathing rhythm is generated in the medulla and modified by neurons in or above the pons.

An override reflex prevents the breathing muscles from overdistending and damaging the lung tissue. This reflex, which is called the *Hering–Breuer reflex*, begins with stretch receptors in the lung tissue. When stretched, these receptors send impulses to the medulla that inhibit the inhalation neurons.

Regulating breathing requires feedback information

When the P_{O_2} and the P_{CO_2} in the blood change, the breathing rhythm changes to return these values to normal levels. We should therefore expect the blood partial pressure of one or both of these gases to provide feedback information to the breathing rhythm generator. Experiments in which subjects breathe gases with different P_{O_2} and P_{CO_2} make it possible to measure the effect of these changes on breathing (Figure 48.19). In these experiments, it is assumed that the P_{O_2} or P_{CO_2} in inhaled air will be reflected in the blood. The conclusion from such experiments is that humans and other mammals are remarkably insensitive to falling blood levels of O_2, but very sensitive to increases in the P_{CO_2} of the blood.

Where are gas partial pressures in the blood sensed? The major site of CO_2 sensitivity is an area on the ventral surface of the medulla, not far from the groups of neurons that generate the breathing rhythm. Sensitivity to P_{O_2} in the blood resides in nodes of tissue on the large blood vessels leaving the heart: the aorta and the carotid arteries (Figure 48.20). These carotid and aortic bodies receive enormous supplies of blood, and they contain chemoreceptor nerve endings. If the blood supply to these structures decreases, or if the P_{O_2} of the blood falls dramatically, the chemoreceptors are activated and send impulses to the breathing control center. Although we are not very sensitive to changes in blood P_{O_2}, the carotid and aortic bodies can stimulate increases in breathing during exposure to very high altitudes or when blood volume or blood pressure is very low.

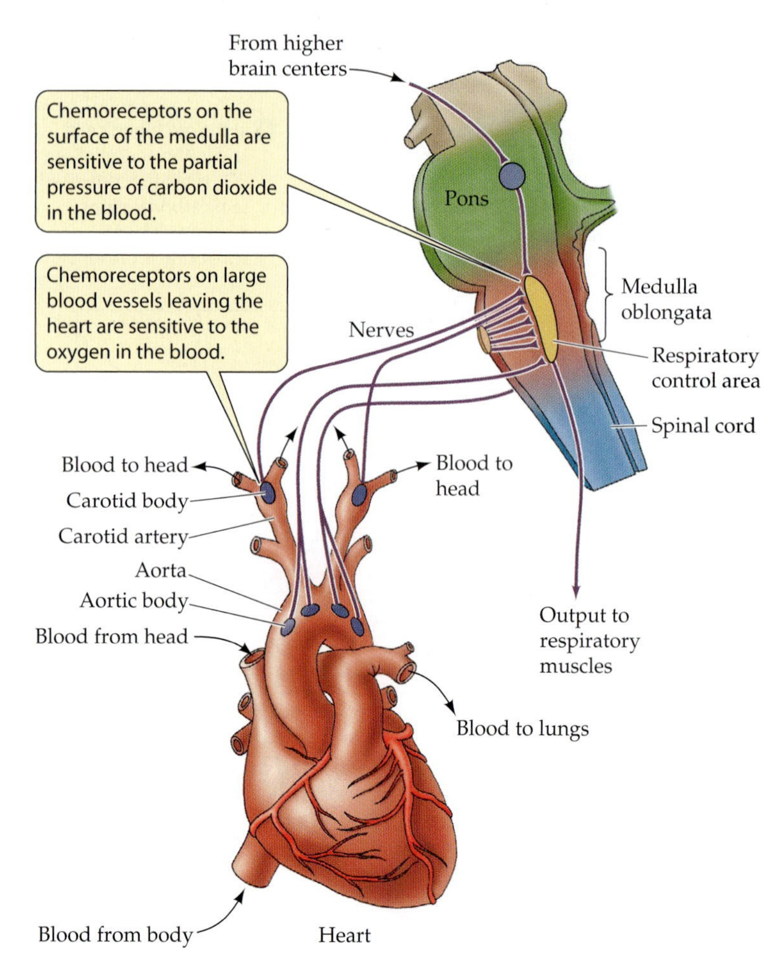

48.20 Feedback Information Controls Breathing
The body uses feedback information from chemosensors in the heart and the brain to match breathing rate to metabolic demand.

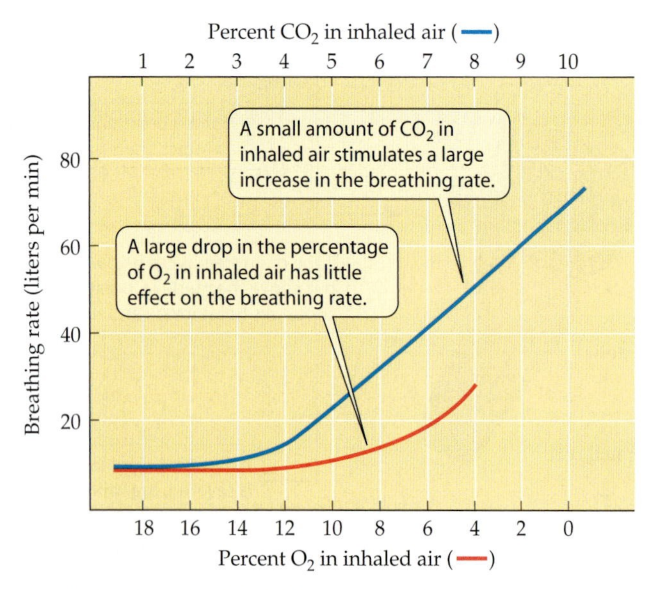

48.19 Carbon Dioxide Affects Breathing Rate
Breathing is more sensitive to increased carbon dioxide content in inhaled air than to decreased oxygen content.

Chapter Summary

Respiratory Gas Exchange

▶ Most cells require a constant supply of O_2 and continuous removal of CO_2. These respiratory gases are exchanged between the body fluids of an animal and its environment by diffusion.

▶ In aquatic animals, gas exchange is limited by the low diffusion rate and low amount of oxygen in water. Aquatic animals face a double bind in that the amount of oxygen in water decreases, but their metabolism and the amount of work required to move water over gas exchange surfaces increase, as water temperature rises. **Review Figure 48.2**

Respiratory Adaptations for Gas Exchange

▶ The evolution of large animals with high metabolic rates required the evolution of adaptations to maximize the rates of diffusion of respiratory gases between animals and their environments. These adaptations involve increasing surface areas for gas exchange and maximizing partial pressure gradients across those exchange surfaces by decreasing their thickness, ventilating the outer surface with respiratory medium, and perfusing the inner surface with blood. **Review Figure 48.4**

• Insects distribute air throughout their bodies in a system of tracheae, tracheoles, and air capillaries. **Review Figure 48.5**

▶ Fish have maximized their rates of gas exchange by having large gas exchange surface areas that are ventilated continuously and unidirectionally with fresh water. Countercurrent blood flow helps increase the efficiency of gas exchange. **Review Figures 48.6, 48.7**

▶ The gas exchange system of birds includes air sacs that communicate with the lungs but are not used for gas exchange. Air flows unidirectionally through bird lungs in parabronchi. Gases are exchanged in air capillaries that run between parabronchi. **Review Figures 48.8, 48.9**

▶ Each breath of air remains in the bird respiratory system for two breathing cycles. The air sacs work as bellows to supply the air capillaries with a continuous, unidirectional flow of fresh air. **Review Figure 48.10**

▶ Breathing in vertebrates other than birds is tidal and is therefore less efficient than gas exchange in fishes or birds. Even though the volume of air exchanged with each breath can vary considerably, the inhaled air is always mixed with stale air. **Review Figure 48.11**

Mammalian Lungs and Gas Exchange

▶ In mammalian lungs, the gas exchange surface area provided by the millions of alveoli is enormous, and the diffusion path length between the air and perfusing blood is very short. **Review Figure 48.12**

▶ Surface tension in the alveoli would make their inflation difficult if the lungs did not produce surfactant.

▶ Inhalation occurs when contractions of the diaphragm and the intercostal muscles create negative pressure in the thoracic cavity. Relaxation of the diaphragm and some intercostal muscles and contraction of other intercostal muscles increases pressure in the thoracic cavity and causes exhalation. **Review Figure 48.13**

Blood Transport of Respiratory Gases

▶ Oxygen is reversibly bound to hemoglobin in red blood cells. Each molecule of hemoglobin can carry a maximum of four molecules of oxygen. Because of positive cooperativity, the affinity of hemoglobin for oxygen depends on the P_{O_2} to which the hemoglobin is exposed. Therefore, hemoglobin gives up oxygen in metabolically active tissues and picks up oxygen as it flows through respiratory exchange structures. **Review Figure 48.14**

▶ Myoglobin has a very high affinity for oxygen and serves as an oxygen reserve in muscle.

▶ There is more than one type of hemoglobin. Fetal hemoglobin has a higher affinity for oxygen than does maternal hemoglobin, allowing fetal blood to pick up oxygen from the maternal blood in the placenta. **Review Figure 48.15**

▶ The affinity of hemoglobin for oxygen is decreased by the presence of hydrogen ions or 2,3 diphosphoglyceric acid. **Review Figure 48.16**

▶ Carbon dioxide is carried in the blood principally as bicarbonate ions. **Review Figure 48.17**

Regulating Breathing to Supply O_2

▶ The breathing rhythm is an autonomic function generated by neurons in the medulla of the brain stem and modulated by higher brain centers. **Review Figure 48.18**

▶ The most important feedback stimulus for breathing is the level of CO_2 in the blood. **Review Figure 48.19**

• The breathing rhythm is sensitive to feedback from chemoreceptors on the ventral surface of the brain stem and in the carotid and aortic bodies on the large vessels leaving the heart. **Review Figure 48.20**

For Discussion

1. A species of fish that lives in Antarctica has no hemoglobin. What anatomical and behavioral characteristics would you expect to find in this fish, and why is its distribution limited to the waters of Antarctica?
2. Blood banks store whole blood for a much shorter period than they store blood plasma. The reason is that when blood that has been stored for too long is infused into a patient, it can actually decrease the oxygen availability to the patient's tissues. Why is this so? Explain in terms of the different physiological functions of 2,3 diphosphoglyceric acid.
3. Explain how llamas and humans can have opposite adaptations for maximizing gas transport at high altitudes.
4. In the disease emphysema, the fine structures of alveoli break down, resulting in the formation of larger air cavities in the lungs. Also, the tissue of the lungs becomes fibrotic and less elastic. Explain at least two reasons why patients with emphysema have a low tolerance for exercise.
5. The disease called "the bends" occurs in scuba divers (persons who spend time underwater by breathing pressurized air) who have come too quickly to the surface after spending an extended period in deep water. The cause of the bends is tiny bubbles of nitrogen coming out of solution in the blood plasma. Seals spend much more time underwater and at deeper depths than scuba divers, yet they do not suffer the bends. Why?

49 Circulatory Systems

YOUR HEART IS A LITTLE BIGGER THAN your fist. This mass of muscle pumps about 70 ml of blood to your lungs and an equal amount to the rest of the organs of your body with each beat, and when you are at rest, it beats about once each second. Even when you are at rest, your heart pumps your total blood volume through your lungs and around your body about once each minute.

Circulating blood has many functions, such as delivery of oxygen and nutrients to cells, removal of waste products of metabolism, and distribution of heat and hormones. It is not surprising, therefore, that when you work or exercise, your heart rate and the amount of blood your heart pumps each minute go up as much as three or four times to match the increasing metabolic demands. Because blood flow can be redistributed to different tissues and organs depending on their needs, blood flow to your active muscles might increase more than 25-fold during exercise.

It makes sense that working muscles should get a greater blood supply. Consider, however, the cardiovascular responses of a seal searching for and pursuing prey underwater for over half an hour. The seal's response to this underwater exercise is very different from yours. Its heart slows dramatically, from about 150 beats per minute to 20, its cardiac output falls proportionally, and blood flow to the muscles propelling its swimming falls practically to zero. Blood flow to the heart is less than a third what it was before the dive began. Only blood flow to its nervous system is maintained at predive levels. Because the circulatory system of the seal responds differently during exercise than yours does, the seal is able to conserve its oxygen supplies and remain underwater for long periods. By the end of this chapter, you will understand the adaptations of circulatory systems that enable you to match blood supply with demand in your exercising muscles and enable the seal to decrease its utilization of oxygen during long dives.

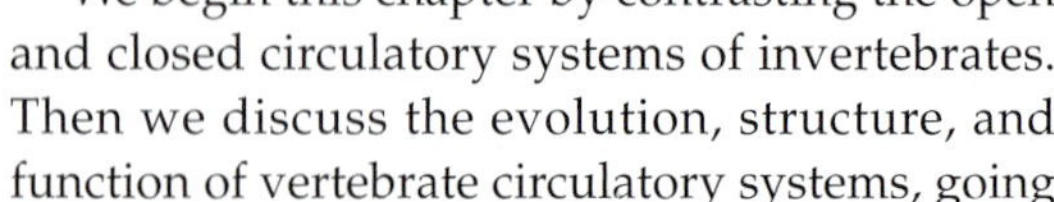

We begin this chapter by contrasting the open and closed circulatory systems of invertebrates. Then we discuss the evolution, structure, and function of vertebrate circulatory systems, going from the two-chambered hearts and single blood circuits of fishes to the four-chambered hearts and double blood circuits of birds and mammals. Taking the human heart as a model, we explore the mechanics of the beating heart and the events of the cardiac cycle that pump blood around the body.

After the heart, we turn to the characteristics of the vascular system: the arteries, capillaries, and veins. We explain how materials are exchanged between the blood and the tissue fluids. The third component of a circulatory system is the blood, and we describe the features of this fluid tissue. The chapter ends with a discussion of the hormonal and neural control and regulation of the human circulatory system and an explanation of the diving adaptations of marine mammals.

Circulatory Systems: Pumps, Vessels, and Blood

A **circulatory system** consists of a pump (heart), a fluid (blood) that can transport materials, and a series of conduits (blood vessels) through which the fluid can be pumped around the body. Heart, blood, and vessels are also known as a *cardiovascular system* (from the Greek *kardia*, "heart," and the Latin *vasculum*, "small vessel"). In this sec-

Champion Divers
A harbor seal (*Phoca vitulina*) can be active underwater for 30 minutes or more without coming to the surface to breathe.

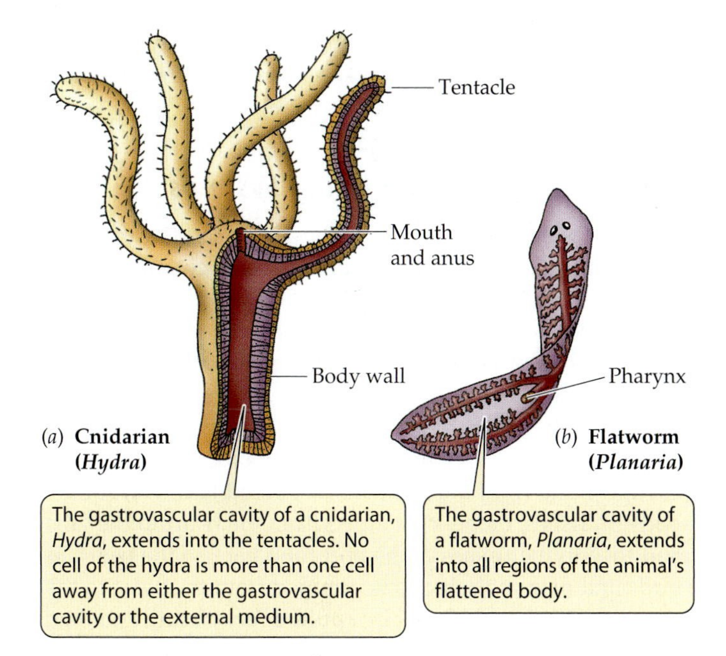

49.1 Gastrovascular Cavities
In small aquatic animals without circulatory systems, a gastrovascular cavity serves the metabolic needs of the innermost cells of the body.

tion, we compare the circulatory systems of different groups of animals.

Some simple aquatic animals do not have circulatory systems

A circulatory system is unnecessary if the cells of an organism are close enough to the external environment that nutrients, respiratory gases, and wastes can diffuse between the cells and the environment. Small aquatic invertebrates have structures and body shapes that permit direct exchanges between cells and environment. The hydra, a cnidarian, is a good example (see Figure 42.1*a*). All cells of the hydra are in contact with, or very close to, the water that either surrounds the animal or circulates through its **gastrovascular cavity**, a dead-end sac that serves both for digestion ("gastro-") and for transport ("vascular") (Figure 49.1*a*). The cells of some other invertebrates are served by highly branched gastrovascular systems, and many have flattened body shapes that maximize the surface area of the animal that is in contact with the external environment (Figure 49.1*b*).

Large surface-to-volume ratios and branched gastrovascular systems cannot satisfy the needs of larger animals with many layers of cells. The cells of these animals are surrounded by internal, but extracellular, fluids—commonly called *tissue fluids*. Circulatory systems carry materials to and from all regions of the body to maintain the optimum composition of the tissue fluids, which in turn serve the needs of the cells. Terrestrial animals require circulatory systems because none of their cells are bathed by an external aqueous medium, and all their cells must be served by tissue fluids.

Open circulatory systems move tissue fluid

The simplest circulatory systems squeeze tissue fluid through intercellular spaces as the animal moves. In these **open circulatory systems** there is no distinction between tissue fluid and blood. Usually a muscular pump, or heart, assists the distribution of the fluid. The contractions of the heart propel the tissue fluid through vessels leading to different regions of the body, but the fluid leaves those vessels to trickle through the tissues and eventually return to the heart. In the arthropod shown in Figure 49.2*a*, the fluid returns to the heart through valved holes called *ostia*. In the mollusk in Figure 49.2*b*, open vessels aid in the return of tissue fluid to the heart.

Closed circulatory systems circulate blood through tissues

In a **closed circulatory system**, a system of vessels keeps circulating blood separate from the tissue fluid. Blood is pumped through this vascular system by one or more muscular hearts, and some components of the blood never leave the vessels.

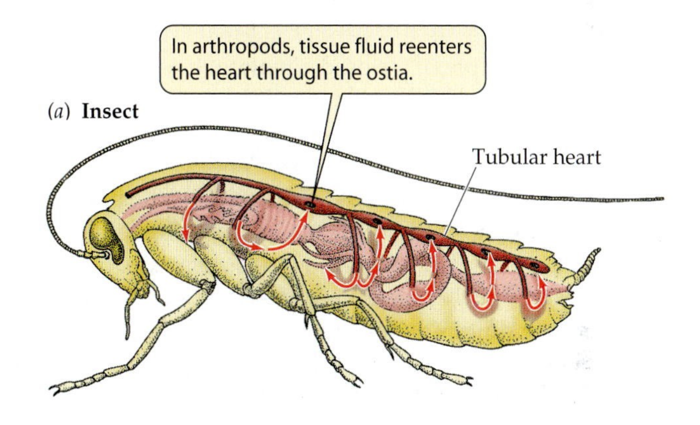

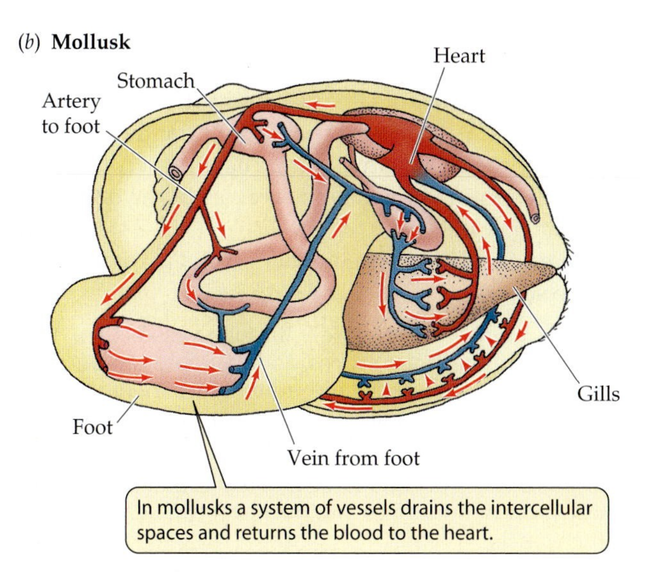

49.2 Open Circulatory Systems
In both arthropods (*a*) and mollusks (*b*), blood is pumped by a tubular heart and directed to regions of the body through vessels that open into intercellular spaces.

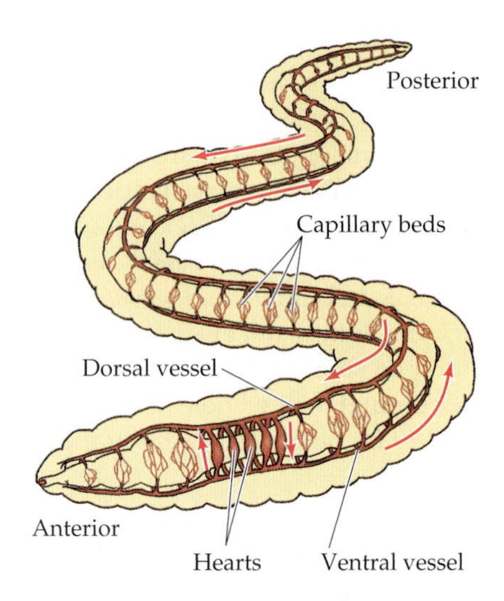

49.3 A Closed Circulatory System
In a closed circulatory system, blood is confined to the blood vessels, kept separate from the tissue fluid, and pumped by one or more muscular hearts. The earthworm, with large dorsal and ventral blood vessels and a branching network of smaller vessels, exemplifies this type of system.

A simple example of a closed circulatory system is that of the common earthworm, an annelid (see Figure 31.23). One large blood vessel on the ventral side of the earthworm carries blood from its anterior end to its posterior end. Smaller vessels branch off and transport the blood to even smaller vessels serving the tissues in each segment of the worm's body. In the smallest vessels, respiratory gases, nutrients, and metabolic wastes diffuse between the blood and the tissue fluid. The blood then flows from these vessels into larger vessels that lead into one large vessel on the dorsal side of the worm. The dorsal vessel carries the blood from the posterior to the anterior end of the body. Five pairs of vessels connect the large dorsal and ventral vessels in the anterior end, thus completing the circuit (Figure 49.3). The dorsal vessel and the five connecting vessels serve as hearts for the earthworm; their contractions keep the blood circulating. The direction of circulation is determined by one-way valves in the dorsal and connecting vessels.

Closed circulatory systems have several advantages over open systems.

- Blood can flow more rapidly through vessels than through intercellular spaces, and therefore can transport nutrients and wastes to and from tissues more rapidly.
- By changing resistances in the vessels, closed systems can selectively direct blood to specific tissues.
- Cellular elements and large molecules that aid in the transport of hormones and nutrients can be kept within the vessels.

Overall, closed circulatory systems can support higher levels of metabolic activity than open systems, especially in larger animals. How, then, do highly active insect species achieve high levels of metabolic output with their open circulatory systems? One answer to this question is can be found in Chapter 48: Insects do not depend on their circulatory systems for respiratory gas exchange (see Figure 48.5).

Vertebrate Circulatory Systems

Vertebrates have closed circulatory systems and hearts with two or more *chambers*. Valves between the chambers, and between the chambers and the vessels, prevent the backflow of blood when the heart contracts.

As we explore the features of the circulatory systems of the different classes of vertebrates, a general evolutionary theme will become apparent: there is a progressively more complete separation of the circulation of blood to the gas exchange organ from the circulation of blood to the rest of the body. In fishes, blood is pumped from the heart to the gills and then to the tissues of the body and back to the heart. In birds and mammals, blood is pumped from the heart to the lungs and back to the heart in a **pulmonary circuit**, and then from the heart to the rest of the body and back to the heart in a **systemic circuit**. We will trace the evolution of the separation of the circulation into two circuits.

The closed vascular system of vertebrates includes **arteries** that carry blood away from the heart and **veins** that carry blood back to the heart. **Arterioles** are small arteries, and **venules** are small veins. **Capillaries** are tiny, thin-walled vessels that connect arterioles and venules. Materials are exchanged between the blood and the tissue fluid only across capillary walls.

Fishes have two-chambered hearts

The fish heart has two chambers. A less muscular chamber, called the **atrium**, receives blood from the body and pumps it into a more muscular chamber, the **ventricle**. The ventricle pumps the blood to the gills, where gases are exchanged. Blood leaving the gills collects in a large dorsal artery, the **aorta**, which distributes blood to smaller arteries and arterioles leading to all the organs and tissues of the body. In the tissues, blood flows through beds of tiny capillaries, collects in venules and veins, and eventually returns to the heart.

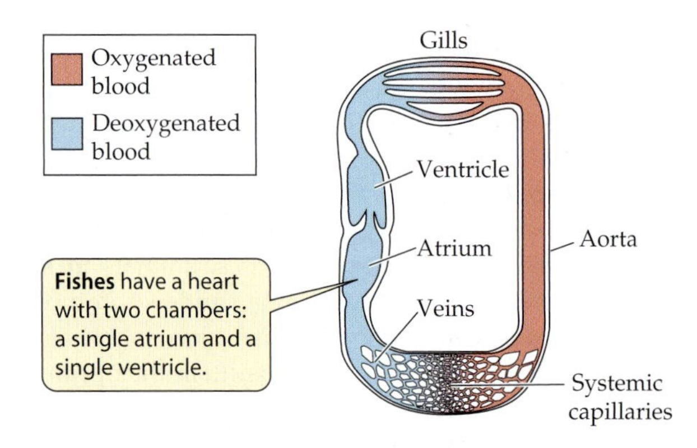

Most of the pressure imparted to the blood by the contraction of the ventricle is dissipated by the high resistance of the narrow spaces through which blood flows in the gills. As a result, blood entering the aorta of the fish is under low pressure, limiting the capacity of the fish circulatory system to supply the tissues with oxygen and nutrients. This limitation on arterial blood pressure does not seem to hamper the performance of many rapidly swimming species, such as tuna and marlin.

The evolutionary transition from breathing water to breathing air had important consequences for the vertebrate circulatory system. An example of how the system changed to serve a primitive lung is seen the African lungfish. These fish are periodically exposed to water with low oxygen content or to situations in which their aquatic environment dries up. Their adaptation for dealing with these conditions is an outpocketing of the gut that serves as a lung. The lung contains many thin-walled blood vessels, so blood flowing through those vessels can pick up oxygen from air gulped into the lung.

How does the circulatory system take advantage of this new organ? The last pair of gill arteries is modified to carry blood to the lung, and a new vessel carries oxygenated blood from the lung back to the heart. In addition, two other gill arches have lost their gills, and their blood vessels deliver blood from the heart directly to the dorsal aorta. Because a few of the gill arches retain gills, the African lungfish can breathe either air or water.

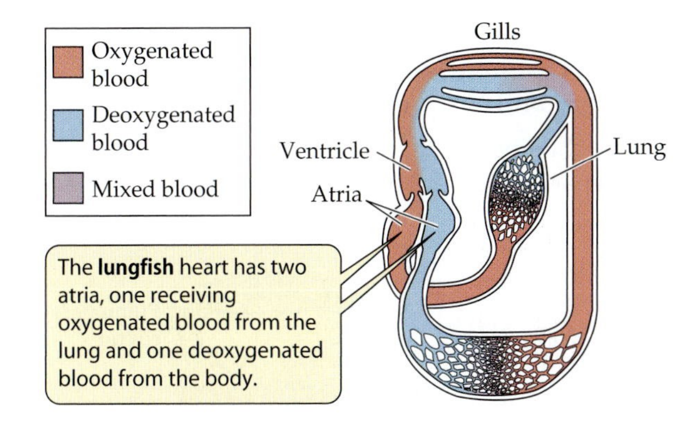

The lungfish heart has adaptations that partially separate the flow of its blood into pulmonary and systemic circuits. Unlike other fishes, the lungfish has a partly divided atrium; the left side receives oxygenated blood from the lungs, and the right side receives deoxygenated blood from the other tissues. These two bloodstreams stay mostly separate as they flow through the ventricle and the large vessel leading to the gill arches, so that the oxygenated blood goes to the gill arteries leading to the dorsal aorta, and the deoxygenated blood goes to the arches with functional gills and to the lung.

Amphibians have three-chambered hearts

Pulmonary and systemic circulation are partly separated in adult amphibians. A single ventricle pumps blood to the lungs and to the rest of the body. Two atria receive blood returning to the heart. One receives oxygenated blood from the lungs, and the other receives deoxygenated blood from the body.

Because both atria deliver blood to the same ventricle, the oxygenated and deoxygenated blood could mix so that blood going to the tissues would not carry a full load of oxygen. Mixing is limited, however, because anatomical features of the ventricle direct the flow of deoxygenated blood from the right atrium to the pulmonary circuit and the flow of oxygenated blood from the left atrium to the aorta.

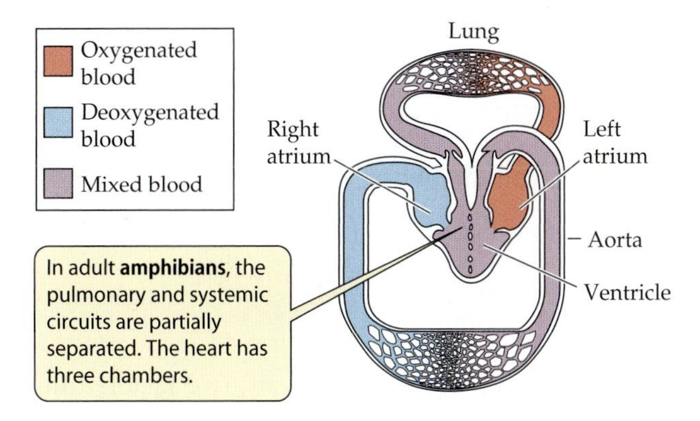

The advantage of this partial separation of pulmonary and systemic circulation is that the high resistance of the gas exchange organ no longer lies between the heart and the tissues. Therefore, the amphibian heart delivers blood to the aorta, and hence to the body, at a higher pressure than the fish heart does.

Reptiles have exquisite control of pulmonary and systemic circulation

Turtles, snakes, and lizards are commonly said to have three-chambered hearts, while crocodilians (crocodiles and alligators) are said to have four-chambered hearts. But this statement is an oversimplification. The hearts of all these animals have two separate atria and a ventricle that is divided in a complex way so that mixing of oxygenated and deoxygenated blood is minimized.

The most important and unusual feature of reptilian and crocodilian hearts is their ability to alter the distribution of blood going to the lungs and to the rest of the body. Consider the behavior, ecology, and physiology of these animals. Despite the common image of turtles as being slow and plodding, reptiles and crocodilians can be fast, active, powerful animals. They can also be inactive for long periods of time, during which they have metabolic rates much lower than the resting metabolic rates of birds and mammals. The enormous range of metabolic demands in these animals means that they do not have to breathe continuously. Some species are also accomplished divers and spend long periods underwater where they cannot breathe.

To understand the wonderful adaptations of the reptilian and crocodilian hearts, you have to realize that there is no benefit in sending blood to the lungs when an animal is not breathing. The hearts of these animals circulate blood through their lungs and then to the rest of their bodies when they are breathing, but when they are not breathing, they can bypass the lung circuit and pump all the blood around the body. How do they accomplish this switching?

Reptiles have two aortas instead of one. This simplified representation of reptilian cardiovascular anatomy shows that the right aorta can receive blood from either the right side or the left side of the ventricle:

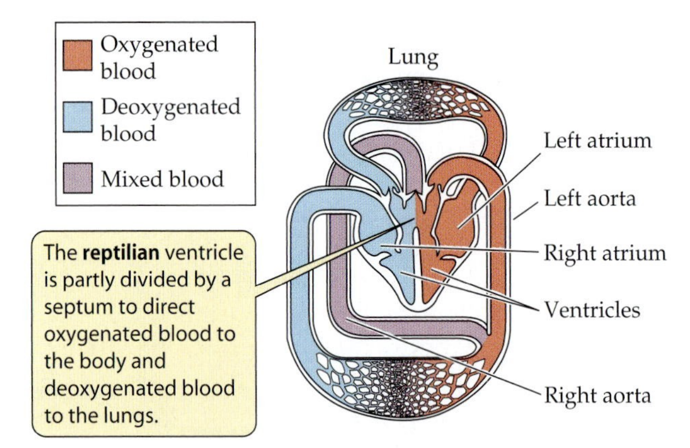

When the animal is breathing air, two factors cause blood from the right side of the ventricle to go preferentially into the pulmonary circuit rather than into the systemic circuit. First, the resistance in the pulmonary circuit is lower than that in the systemic circuit. Second, there is a slight asynchrony in the timing of ventricular contraction, so the blood in the right side of the ventricle tends to be ejected slightly before the blood in the left side. As the ventricle contracts, the deoxygenated blood in the right side of the ventricle moves first into the lung circuit. When the oxygenated blood in the left side of the ventricle starts to move, it encounters resistance in the lung circuit, which is already filled with the deoxygenated blood from the right side. Therefore the blood from the left side tends to flow into the two aortas.

When the reptile stops breathing, blood flow is rerouted by constriction of vessels in the lung circuit. As resistance in the lung circuit increases, the blood from the right side of the ventricle tends to be directed into one of the aortas. As a result, blood from both sides of the ventricle flows through the aortas to the systemic circuit.

The ability of snakes, lizards, and turtles to redirect blood flow from the lung circuit to the systemic circuit depends on the incomplete division of their ventricles. Crocodilians have true four-chambered hearts with completely divided ventricles. Yet the crocodilians have not lost the ability to shunt blood from the lung circuit when they are not breathing. The crocodilians have one aorta originating in the left ventricle and one aorta originating in the right ventricle. However, a short channel connects these two aortas just after they leave the heart.

Because the crocodilians' ventricles are separate, they can generate different pressures when they contract. When the animal is breathing, the pressure in the left ventricle and the left aorta is higher than the pressure in the right ventricle. This higher pressure is communicated through the connecting channel to the right aorta, and this high back pressure prevents right-ventricle blood from entering that aorta. As a result, both aortas carry blood from the left ventricle, and the blood from the right ventricle flows to the lung circuit.

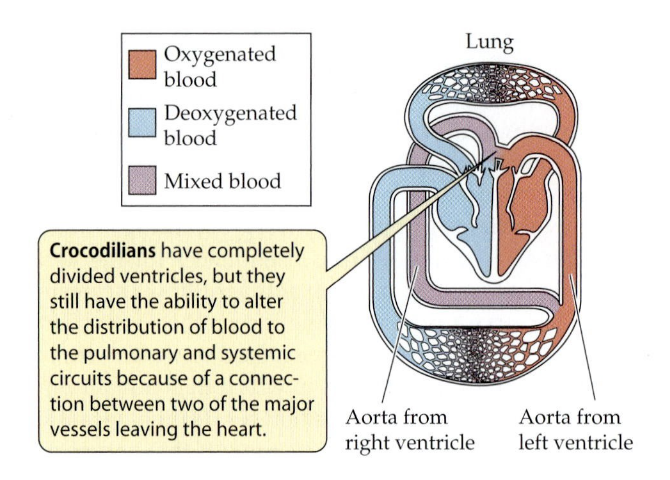

When a crocodilian is not breathing, constriction of vessels in the lung circuit increases the resistance in that circuit. As a result, pressure builds up in the right ventricle to a level that exceeds the back pressure in the right aorta. Under these conditions, blood from both ventricles flows through the two aortas and the systemic circuit, and little blood flows into the lung circuit.

You can now appreciate the fact that reptilian and crocodilian hearts are not primitive. Rather, these hearts and their major vessels are highly adapted to operate efficiently over a wide range of metabolic demands.

Birds and mammals have fully separated pulmonary and systemic circuits

The four-chambered hearts of birds and mammals completely separate their pulmonary and systemic circuits.

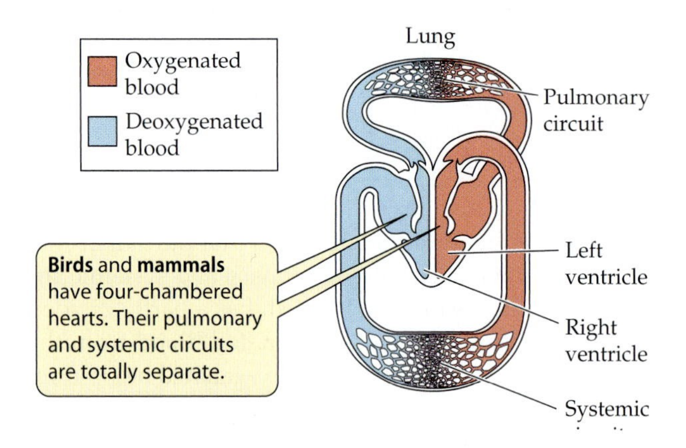

Separate circuits have several advantages:

- Oxygenated and deoxygenated blood cannot mix, and therefore, the systemic circuit is always receiving blood with the highest oxygen content.
- Respiratory gas exchange is maximized because the blood with the lowest oxygen content and highest CO_2 content is sent to the lungs.
- Separate systemic and pulmonary circuits can operate at different pressures.

The tissues of birds and mammals have high nutrient demands and thus a very high density of the smallest vessels, the capillaries. Many small vessels present lots of resistance to the flow of blood. Therefore, high pressure is required in the systemic circuits of birds and mammals. Their pulmonary circuits do not have as many capillaries and as high resistances as their systemic circuits, so the pulmonary circuits of birds and mammals can run at lower pressures.

The Human Heart: Two Pumps in One

Like all other mammalian hearts, the human heart has four chambers: two atria and two ventricles (Figure 49.4). The atrium and ventricle on the right side of your body are called the right atrium and right ventricle. They can be thought of as the right heart. The atrium and ventricle on the left side of your body are called the left atrium and left ventricle. They can be thought of as the left heart. The right heart pumps blood through the pulmonary circuit, and the left heart pumps blood through the systemic circuit.

Valves between the atria and ventricles, the **atrioventricular valves**, prevent backflow of blood into the atria when the ventricles contract. The **pulmonary valve** and the **aortic valve**, positioned between the ventricles and the arteries, prevent the backflow of blood into the ventricles.

In what follows, we'll first focus on the flow of blood through the heart and through the body. Then we'll examine the unique electrical properties of cardiac muscle and see how the heart's electrical activity can be recorded in an EKG (electrocardiogram).

Blood flows from right heart to lungs to left heart to body

Let's follow the circulation of the blood through the heart, starting in the right heart. The right atrium receives deoxygenated blood from the **superior vena cava** and the **inferior vena cava** (see Figure 49.4), large veins that collect blood from the upper and lower body, respectively. The veins of the heart itself also drain into the right atrium. From the right atrium, the blood flows into the right ventricle. Most of the filling of the ventricle is due to passive flow while the heart is relaxed between beats. Just at the end of this period of ventricular filling, the atrium contracts and adds a little more blood to the ventricular volume. The right ventricle then contracts, pumping blood into the **pulmonary artery**, which transports the blood to the lungs.

The **pulmonary veins** return the oxygenated blood from the lungs to the left atrium, from which the blood enters the left ventricle. As with the right side of the heart, most left ventricular filling is passive, and ventricular volume is topped off by contraction of the atrium just at the end of the period of filling.

The walls of the left ventricle are powerful muscles that contract around the blood with a wringing motion starting from the bottom. When pressure in the left ventricle is high enough to push open the aortic valve, the blood rushes into the aorta to begin its circulation throughout the body and eventually back to the right atrium. In Figure 49.4, observe that the left ventricle is more massive than the right ventricle. Because there are many more arterioles and capillaries in the systemic circuit than in the pulmonary circuit, resistance is higher in the systemic circuit, and the left ventricle must squeeze with greater force than the right, even though both are pumping the same volume of blood.

49.4 The Human Heart and Circulation In the human heart, blood flows from right heart to lungs to left heart to body. The atrioventricular valves prevent blood from flowing back into the atria when the ventricles contract. The pulmonary and aortic valves prevent blood from flowing back into ventricles from the arteries when the ventricles relax.

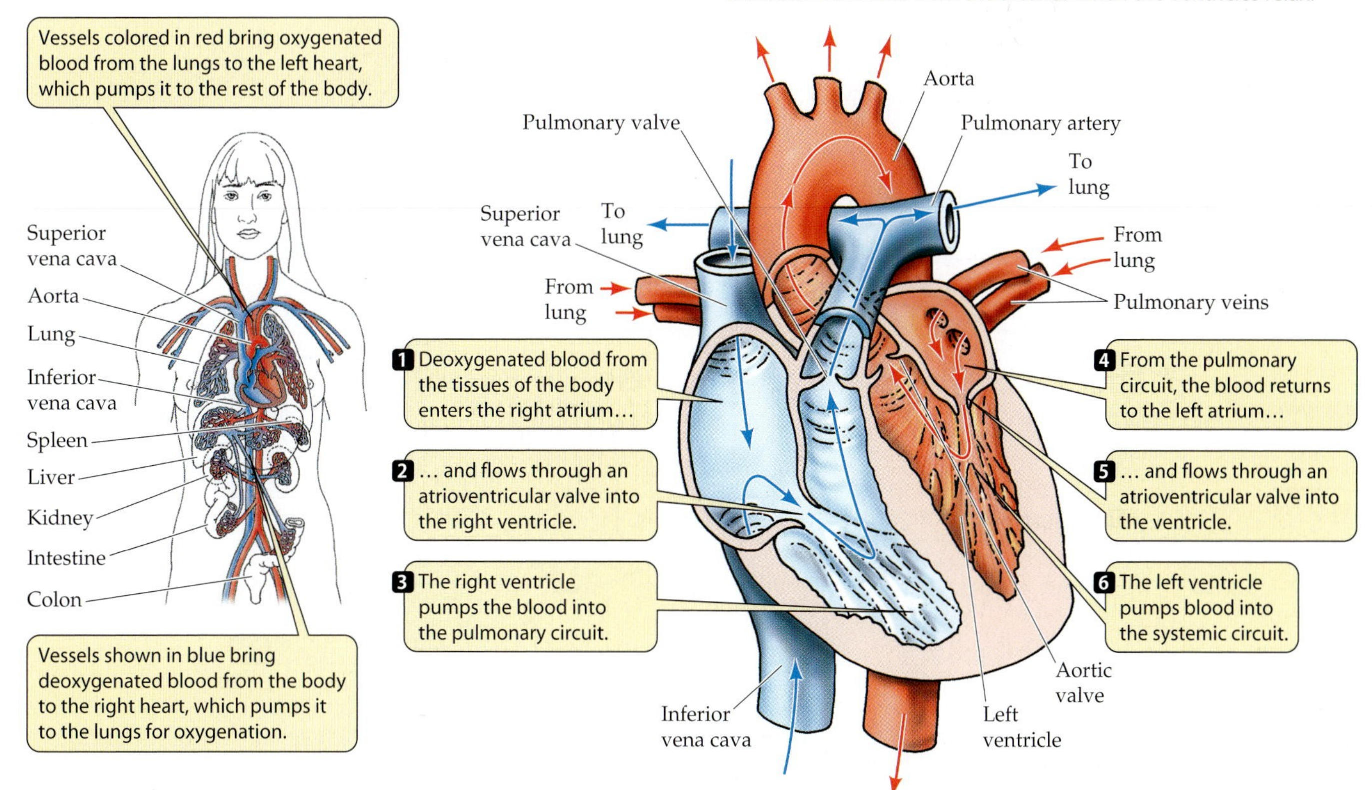

The pumping of the heart—the contraction of the two atria followed by the contraction of the two ventricles, and then relaxation—is the **cardiac cycle**. Contraction of the ventricles is called **systole**, and relaxation of the ventricles **diastole** (Figure 49.5). Just at the end of diastole, the atria contract and top off the volume of blood in the ventricles. The sounds of the cardiac cycle, the "lub-dub" heard through a stethoscope placed on the chest, are created by the slamming shut of the heart valves. The shutting and opening of these valves is simply a mechanical event resulting from pressure differences on the two sides of the valves. As the ventricles begin to contract, the pressure in the ventricles rises above the pressure in the atria, and the atrioventricular valves close ("lub"). When the ventricles begin to relax, the high back pressure in the aorta and pulmonary arteries causes the aortic and pulmonary valves to bang shut ("dub"). Defective valves produce the sounds of *heart murmurs*. For example, if an atrioventricular valve is defective, blood will flow back into the atrium with a "whoosh" sound following the "lub."

The cardiac cycle can be felt in the pulsation of arteries such as the one that supplies blood to your hand. You can feel your pulse by placing two fingers from one hand lightly over the wrist of the other hand just below the thumb. During systole, blood surges through the arteries of your arm and hand, and you can feel the surge as a pulsing of the artery in your wrist.

Blood pressure changes associated with the cardiac cycle can be measured in the large artery in your arm by using an inflatable pressure cuff called a sphygmomanometer and a stethoscope (Figure 49.6). This method measures the minimum pressure necessary to compress an artery so that blood does not flow through it at all (the systolic value) and the minimum pressure that permits intermittent flow through the artery (the diastolic value). In a conventional blood pressure reading, the systolic value is placed over the diastolic value. Normal values for a young adult might be 120 mm of mercury (Hg) during systole and 80 mm Hg during diastole, or 120/80.

The heartbeat originates in the cardiac muscle

Cardiac muscle, as we saw in Chapter 47, has some unique properties that allow it to function as an effective pump. First, the cardiac muscle cells are in electrical continuity with each other. Gap junctions enable action potentials to spread rapidly from cell to cell. Because a spreading action potential stimulates contraction, large groups of cardiac muscle cells contract in unison. This coordinated contraction is important for pumping blood.

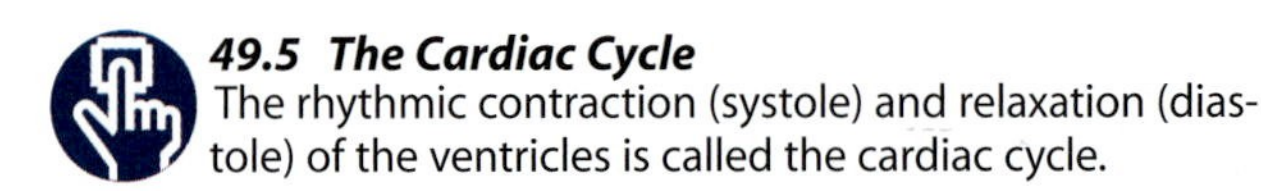

49.5 The Cardiac Cycle
The rhythmic contraction (systole) and relaxation (diastole) of the ventricles is called the cardiac cycle.

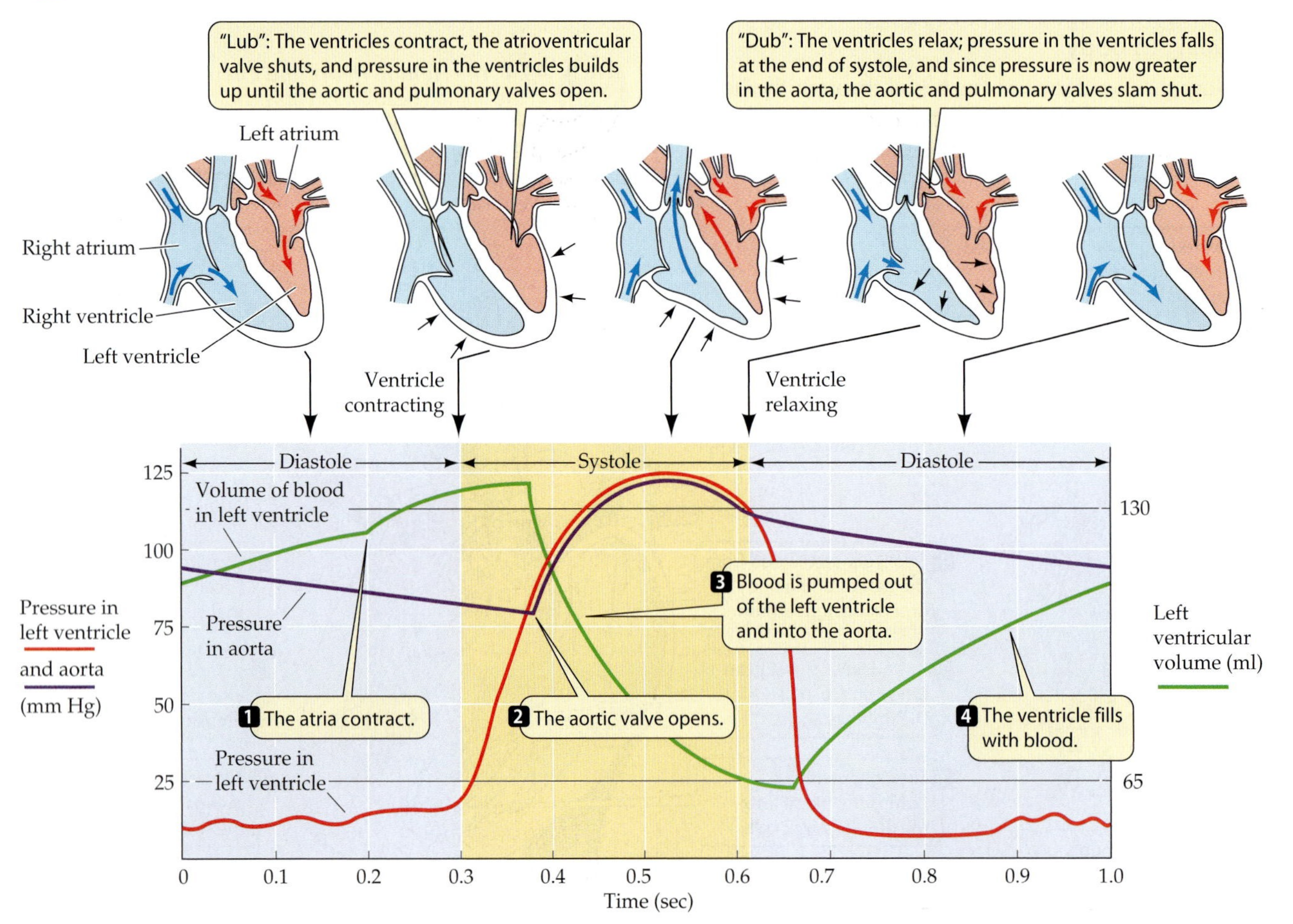

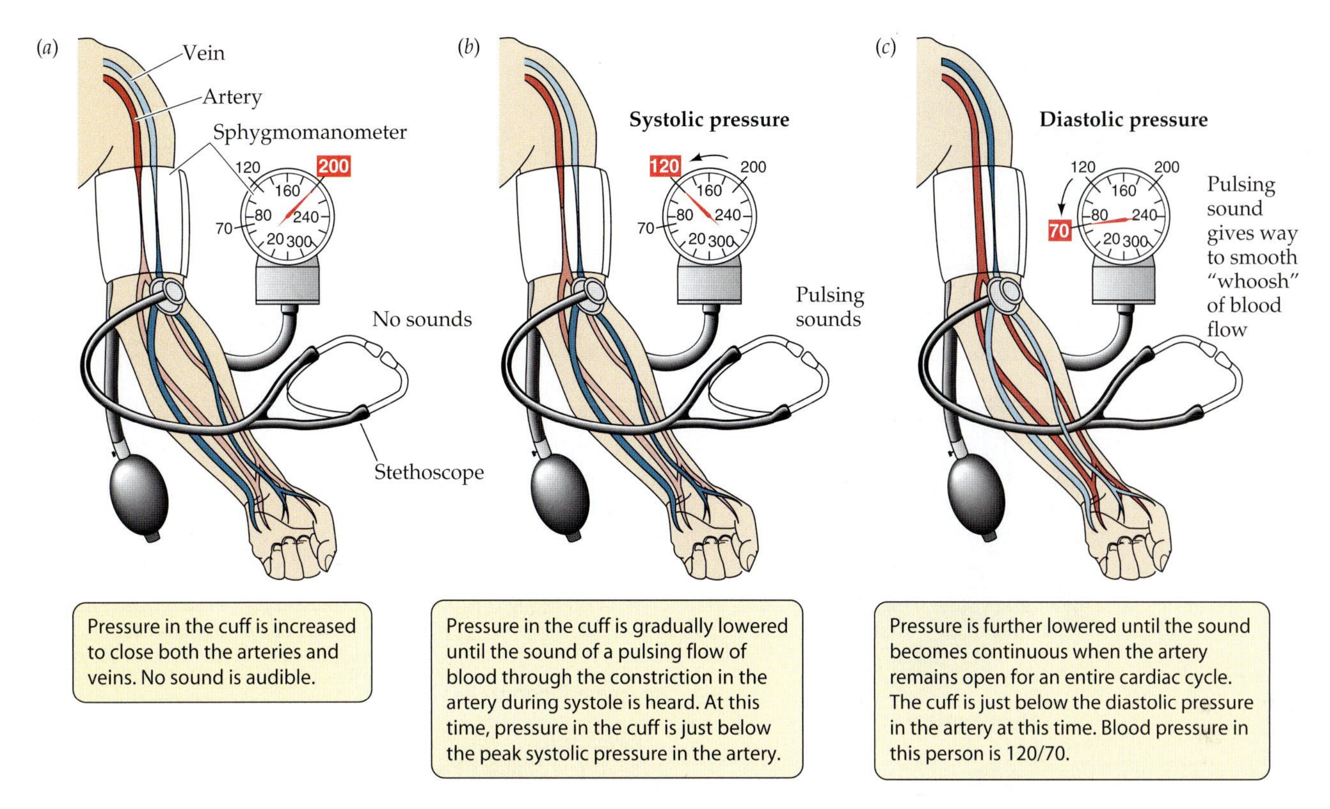

49.6 Measuring Blood Pressure
Blood pressure in the major artery of the arm can be measured with an inflatable pressure cuff called a sphygmomanometer.

Second, some cardiac muscle cells have the ability to initiate action potentials without stimulation from the nervous system. These cells stimulate neighboring cells to contract, thereby acting as pacemakers. The important characteristic of a pacemaker cell is that its resting membrane potential gradually becomes less negative until it reaches the threshold voltage for initiating an action potential (Figure 49.7).

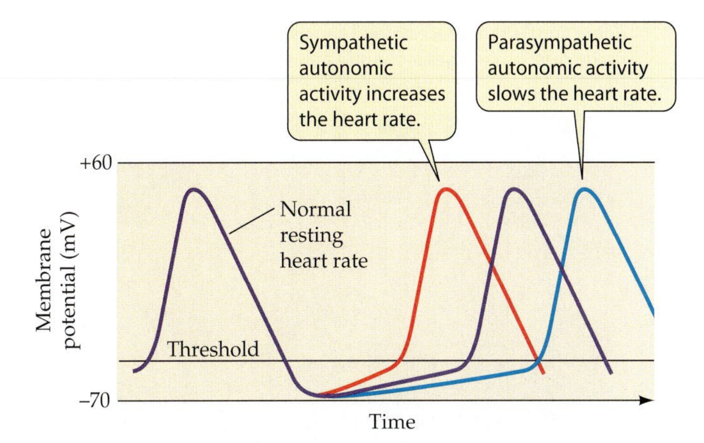

49.7 The Autonomic Nervous System Controls Heart Rate
The resting potentials of the plasma membranes of pacemaker cells spontaneously depolarize to threshold and fire action potentials. Signals from the two divisions of the autonomic nervous system raise and lower the heart rate, respectively.

These action potentials look different from the neuronal action potentials you saw in Chapter 41 because the depolarization is due primarily to the opening of voltage-gated calcium channels rather than voltage-gated sodium channels.

Like neurons, cardiac muscle repolarizes in part by opening potassium channels. The potassium channels in cardiac pacemaker cells, however, are unique. After an action potential, they open, causing the membrane potential to fall to its most negative level. Then they *gradually* close, and as they do so, the membrane potential becomes less negative—it *slowly* depolarizes. Sodium and calcium channels contribute to this gradual depolarization between action potentials. When membrane potential reaches threshold for the voltage-gated calcium channels, another action potential occurs.

The nervous system controls the heartbeat (speeds it up or slows it down) by influencing the rate at which pacemaker cells gradually depolarize between action potentials. Acetylcholine released by parasympathetic nerve endings onto the pacemaker cells slows their rate of depolarization and thereby slows the heart rate. Norepinephrine released by sympathetic nerve endings onto the pacemaker cells increases their rate of depolarization and thereby speeds the heart rate (see Figure 49.7).

Under normal circumstances, the heartbeat originates from pacemaker cells located at the junction of the superior vena cava and right atrium, in the **sinoatrial node** (Figure 49.8). An action potential spreads from the sinoatrial node across the atrial walls, causing the two atria to contract in unison. Since there are no gap junctions between the atria and the ventricles, however, this depolarization does not

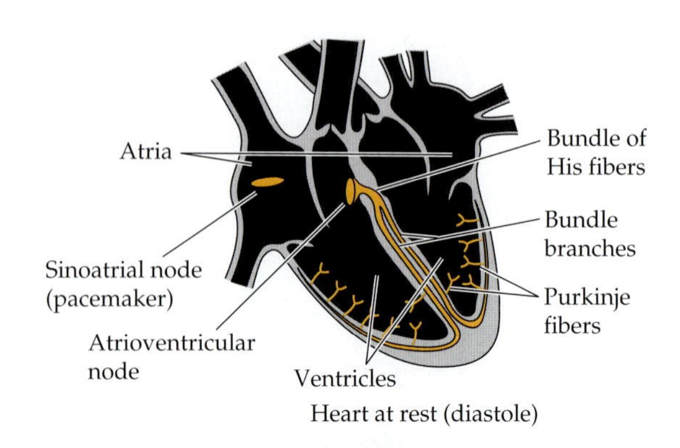

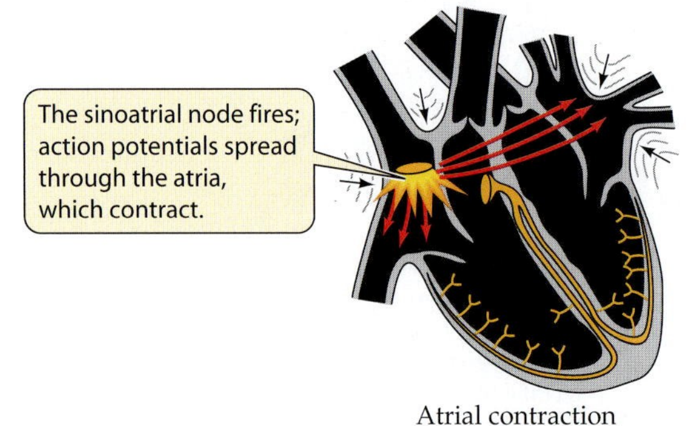

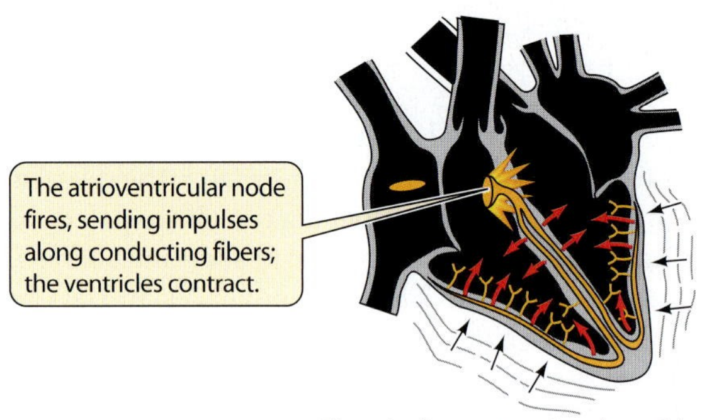

49.8 *The Heartbeat*
Pacemaker cells in the sinoatrial node initiate action potentials that spread through the walls of the atria, causing them to contract.

flow directly to the ventricles, and the ventricles do not contract in unison with the atria.

The action potential initiated in the atria passes to the ventricles through another node of modified cardiac muscle cells, the **atrioventricular node**. The atrioventricular node passes the action potential on to the ventricles via modified muscle fibers called the **bundle of His**. The bundle of His divides into right and left branches, which connect with **Purkinje fibers** that branch throughout the ventricular muscle.

The timing of the spread of the action potential from atria to ventricles is important. The atrioventricular node imposes a short delay in the spread of the action potential from atria to ventricles. Then the action potential spreads very rapidly throughout the ventricles, causing them to contract. Thus the atria contract before the ventricles do, so the blood passes progressively from the atria to the ventricles to the arteries.

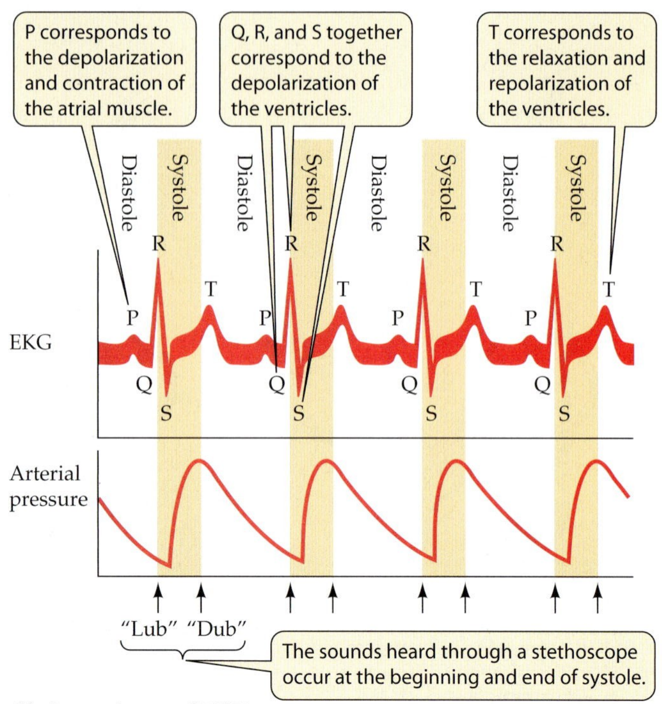

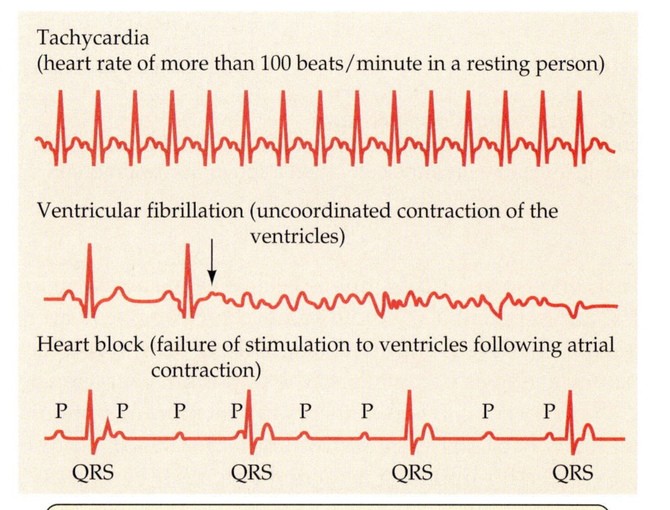

(c)

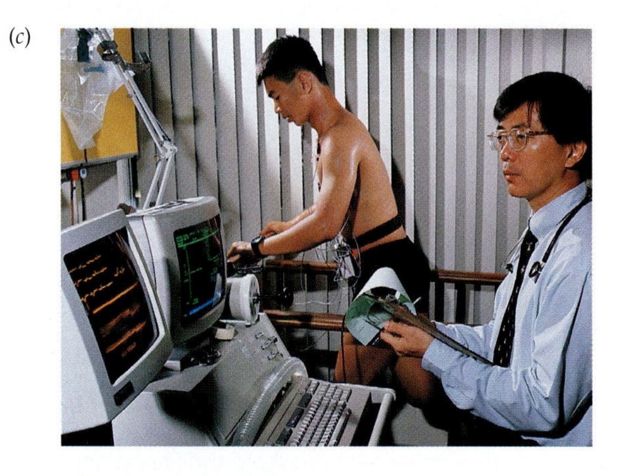

49.9 *The Electrocardiogram*
An EKG can be used to monitor heart function.

The EKG records the electrical activity of the heart

Electrical events in the cardiac muscle during the cardiac cycle can be recorded by electrodes placed on the surface of the body. Such a recording is called an **electrocardiogram**, or **EKG** ("EKG" because the Greek word for heart is *kardia*, but "ECG" is also used). The EKG is an important tool for diagnosing heart problems.

The action potentials that sweep through the muscles of the atria and the ventricles before they contract are such massive, localized electrical events that they cause electric currents to flow outward from the heart to all parts of the body. Electrodes placed on the surface of the body at different locations—usually on the wrists and ankles—detect those electric currents at different times and therefore register a voltage difference. The appearance of the EKG depends on the exact placement of the electrodes used for the recording. Electrodes placed on the right wrist and left ankle produced the normal EKG shown in Figure 49.9*a*. The waves of the EKG are designated P, Q, R, S, and T, each letter representing a particular event in the cardiac muscle.

The EKG is used by cardiologists (heart specialists) to diagnose heart problems. Figure 49.9*b* shows some abnormal EKGs that result from different problems. For patients who have had heart attacks, it is possible to determine which region of the heart has been damaged by placing electrodes at different locations on the chest. Comparing EKGs from the different electrodes tells the cardiologist which region of the heart is behaving abnormally.

The Vascular System: Arteries, Capillaries, and Veins

Blood circulates throughout the body in a system of blood vessels: arteries, capillaries, and veins. Arteries receive blood from the heart; accordingly, they are built to withstand high pressures. Arteries are important in controlling blood pressure and in the distribution of blood to different organs . Veins return blood to the heart at low pressures and serve as a blood reservoir. The capillaries are the site of all exchanges between the blood and the internal environment. In this section, we see how the structure of each of these vessel types supports its functions. In addition to arteries, capillaries, and veins, we consider another set of vessels, the lymphatic vessels, which return tissue fluid to the blood.

Arteries and arterioles have abundant elastic and muscle fibers

The walls of the large arteries have many elastic fibers that enable them to withstand high pressures (Figure 49.10). These elastic fibers have another important function as

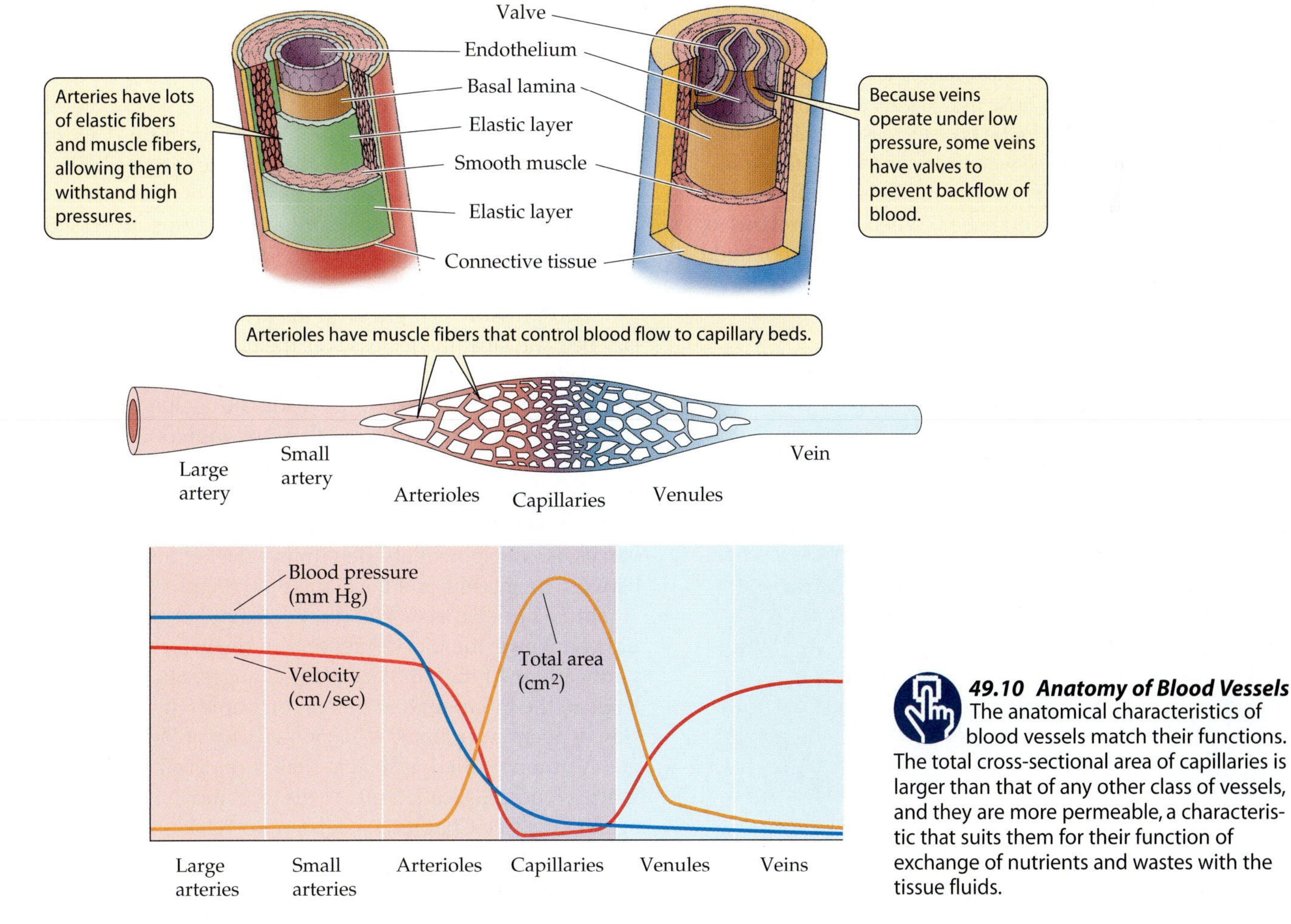

49.10 Anatomy of Blood Vessels The anatomical characteristics of blood vessels match their functions. The total cross-sectional area of capillaries is larger than that of any other class of vessels, and they are more permeable, a characteristic that suits them for their function of exchange of nutrients and wastes with the tissue fluids.

well: During systole, they are stretched, and thereby store some of the energy imparted to the blood by the heart. During diastole, they return this energy by squeezing the blood and pushing it forward. As a result, even though the flow of blood through the arterial system pulsates, it is smoother than it would be through a system of rigid pipes.

Smooth muscle cells in the arteries and arterioles make the diameter of those vessels variable. When their diameter changes, their resistance to blood flow changes as well, and the amount of blood flowing through them changes as a result. By influencing the contraction of the smooth muscle in the walls of arteries and arterioles, neural and hormonal mechanisms can control the distribution of blood to the different tissues of the bodyas well as central blood pressure. The arteries and arterioles are referred to as the *resistance vessels* because their resistance varies.

Materials are exchanged between blood and tissue fluid in the capillaries

Beds of capillaries lie between arterioles and venules. No cell of the body is more than a couple of cell diameters away from a capillary. The needs of the cells are served by the exchange of materials between blood and tissue fluid across the capillary walls. Capillaries have thin, permeable walls, and blood flows through them slowly, facilitating this exchange (Figure 49.11).

To anyone who has played with a garden hose, it may seem strange that blood flows through the large arteries rapidly at high pressures, but when it reaches the small capillaries, the pressure and rate of flow decrease. When you restrict the diameter of a garden hose by placing your thumb over the opening, the pressure in the hose increases, which in turn increases the velocity of the water spraying out of the hose.

This puzzle is solved by two more pieces of information. First, arterioles are highly branched. When one is restricted, blood flows into other branches, so pressure does not build up quickly. Second, each arteriole gives rise to many, many capillaries. Even though each capillary has a diameter so small that red blood cells must pass through in single file (see Figure 49.11), there are so many capillaries that their total cross-sectional area is much greater than that of any other class of vessel. As a result, all of the capillaries together have a much greater capacity for blood than do the arterioles. Returning to our garden hose analogy, if we connect the hose to many junctions leading to small irrigation tubes, the pressure and the flow in each of the irrigation tubes will be quite low.

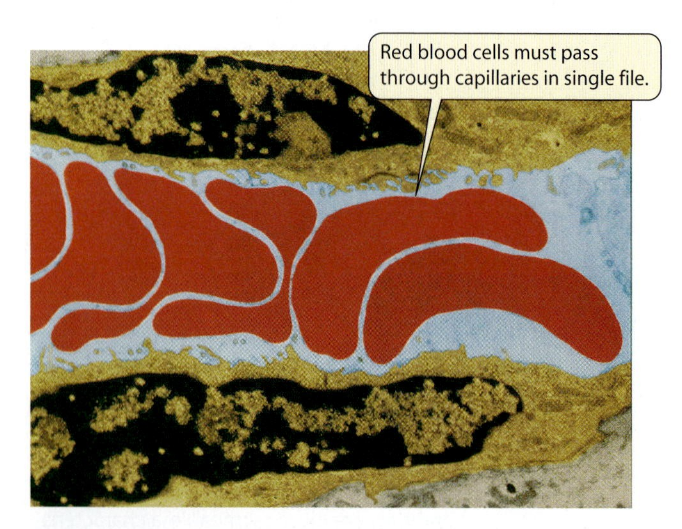

49.11 A Narrow Lane
Capillaries have a very small diameter, and blood flows through them slowly.

Materials are exchanged in capillary beds by filtration, osmosis, and diffusion

The walls of capillaries are made of a single layer of thin endothelial cells. In most tissues of the body other than the brain, these tubes of endothelial cells have tiny holes called *fenestrations*. Surrounding the endothelial cells is a very permeable basal lamina. So, capillaries are leaky. They are permeable to water, to some ions, and to some small molecules, but not to large molecules such as proteins. Blood pressure therefore tends to squeeze water and some solutes out of the capillaries and into the surrounding intercellular spaces. This process is called **filtration**. The large molecules that cannot cross the capillary wall create a difference in osmotic potential (also called osmotic pressure) between the plasma and the tissue fluid, which tends to draw water back into the capillary.

Recent research suggests that bicarbonate ions in the capillary plasma are an important contributor to the osmotic attraction of water back into the capillary. As we saw in Chapter 48, CO_2 diffuses into the plasma as the blood flows through the capillary. The conversion of this CO_2 to bicarbonate ions is catalyzed by the enzyme carbonic anhydrase. Therefore, there is a rise in bicarbonate concentration as blood flows through the capillary, and this increased concentration contributes to the resorption of water from the tissue fluid.

Blood pressure is highest on the arterial side of a capillary bed and steadily decreases as the blood flows to the venous side. Therefore, more water is squeezed out of the capillaries on the arterial side of the bed. The osmotic potential pulling water back into the capillary rises as blood flows toward the venous side. Gradually, osmotic potential becomes the dominant force, pulling water back into the plasma. The interaction of these two opposing forces—blood pressure versus osmotic potential—determines the net flow of water between the plasma and the tissue fluid (Figure 49.12).

The balance between blood pressure and osmotic potential changes if the blood pressure in the arterioles or the permeability of the capillary walls changes. Such a change leads to the inflammation that accompanies injuries to the skin or allergic reactions. A major mediator of inflammation is a hormone called *histamine* that is released mainly by white blood cells, called mast cells, that move to the damaged tissue (see Chapter 41). Histamine relaxes the smooth muscle of the arterioles, thus increasing blood flow to the damaged tissue and increasing pressure in the capillaries.

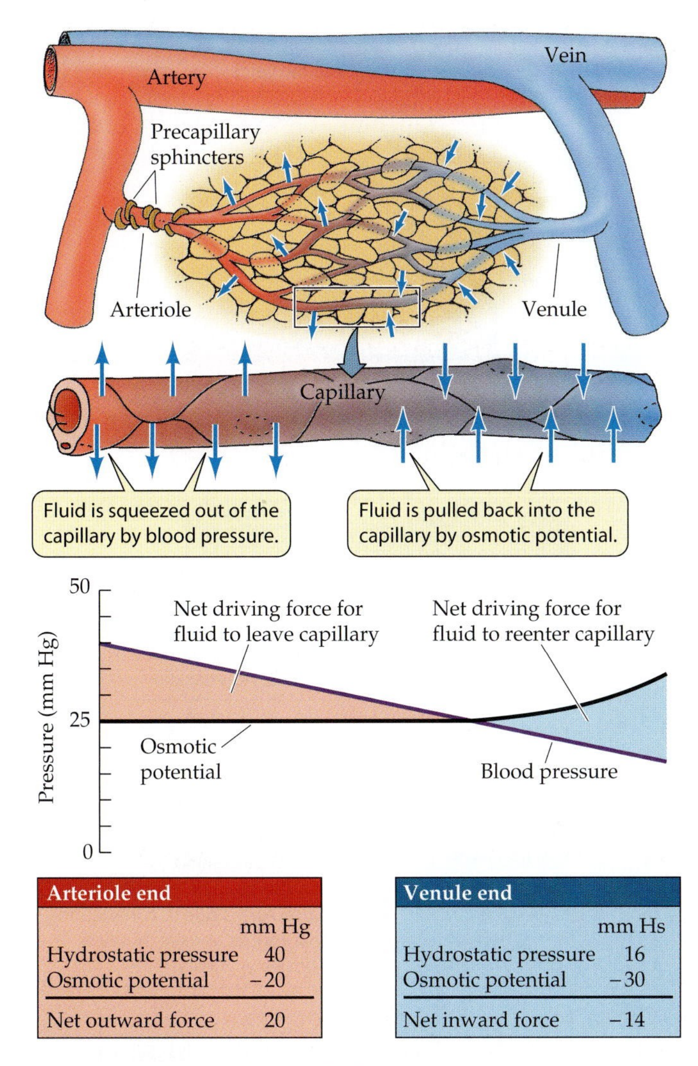

49.12 A Balance of Opposite Forces
Blood pressure and osmotic attraction (both expressed as millimeters of mercury, mm HG) control the exchange of fluids between blood vessels and intercellular space.

Histamine also increases the permeability of the capillaries, so that more water leaves the vessels. The accumulation of fluid in the intercellular spaces causes the tissue to swell, a condition known as **edema**. The use of drugs called *antihistamines* can alleviate inflammation and allergic reactions.

The loss of water from capillaries increases if the protein content of the blood decreases, as is seen in cases of liver failure due to alcoholism or liver disease. The liver is the major producer of blood proteins, and when it fails, blood protein levels fall. With a lower protein concentration in the plasma, there is less of an osmotic potential to pull water back into the capillaries. The result is that tissue fluid builds up, swelling the abdomen and the extremities.

Which specific small molecules can cross a capillary wall depends on the architecture of the capillary, the type of substance, and the concentration gradient of the substance between the blood and the tissue fluid. Capillary walls consist of the plasma membranes of endothelial cells and, as mentioned, may have actual holes (fenestrations) in them. Therefore, lipid-soluble substances and many small solute molecules can pass through them from an area of higher concentration to one of lower concentration (see Chapter 5).

The capillaries in different tissues, however, are differentially selective as to the sizes of molecules that can pass through them. In all capillaries, O_2, CO_2, glucose, lactate, and small ions such as Na^+ and Cl^- can cross. The capillaries of the brain do not have fenestrations, and therefore not much else can pass through them unless it is a lipid-soluble substance such as alcohol. This high selectivity of brain capillaries is known as the *blood–brain barrier* (see Chapter 44).

In other tissues, the capillaries are much less selective. Such capillaries are found in the digestive tract, where nutrients are absorbed, and in the kidneys, where wastes are filtered. Some capillaries have large gaps that permit the movement of even larger substances, such as red blood cells. These capillaries are found in the bone marrow, spleen, and liver. Substances move across many capillary walls by endocytosis (see Chapter 5).

Lymphatic vessels return tissue fluid to the blood

The tissue fluid that accumulates outside the capillaries contains water and small molecules, but no red blood cells, and less protein than there is in blood. A separate system of vessels—the **lymphatic system**—returns tissue fluid to the blood.

After entering the lymphatic vessels, the tissue fluid is called **lymph**. Fine lymphatic capillaries merge progressively into larger and larger vessels and end in a major vessel—the **thoracic duct**—that empties into the superior vena cava, returning blood to the heart (see Figure 19.1). Lymphatic vessels have one-way valves that keep the lymph flowing toward the thoracic duct. The force propelling the lymph is pressure on the lymphatic vessels from the contractions of nearby skeletal muscles.

Mammals and birds have *lymph nodes* along the major lymphatic vessels. Lymph nodes are an important component of the defensive machinery of the body (see Chapter 19). They are a major site of lymphocyte production and of the phagocytic action that removes microorganisms and other foreign materials from the circulation. The lymph nodes also act as mechanical filters. Particles become trapped there and are digested by the phagocytes that are abundant in the nodes.

Lymph nodes swell during infection. Some of them, particularly those on the sides of the neck or in the armpits, become noticeable when they swell. The nodes also trap metastasized cancer cells—that is, those that have broken free of the original tumor. Because such cells may start additional tumors, surgeons often remove the neighboring lymph nodes when they excise a malignant tumor.

Blood flows back to the heart through veins

The pressure of the blood flowing from capillaries to venules is extremely low, and is insufficient to propel blood back to the heart. Blood tends to accumulate in veins, and the walls of veins are more expandable than the walls of arteries. As much as 80 percent of the total blood volume may be in the veins at any one time. Because of their high capacity to store blood, veins are called *capacitance vessels*.

Blood must be returned from the veins to the heart so that circulation can continue. If the veins are above the level of the heart, gravity helps blood flow, but below the level of the heart, blood must be moved against the pull of gravity. If too much blood remains in the veins, then too little blood returns to the heart, and thus too little blood is pumped to the brain; a person may faint as a result. Fainting is self-correcting: A fainting person falls, thereby moving out of the position in which gravity caused blood to accumulate in the lower body. But means other than fainting also move blood from the tissues back to the heart.

The most important of the forces that propel venous and lymphatic return from the regions of the body below the heart is the squeezing of the vessels by the contractions of surrounding skeletal muscles. As muscles contract, the vessels are compressed, and the blood is squeezed through them. Blood flow might be temporarily obstructed during a prolonged muscle contraction, but with relaxation of the muscles the blood is free to move again.

One-way valves within the veins prevent the backflow of blood. Thus whenever a vein is squeezed, blood is propelled forward toward the heart (Figure 49.13). As we have already noted, the lymphatic vessels have similar one-way valves.

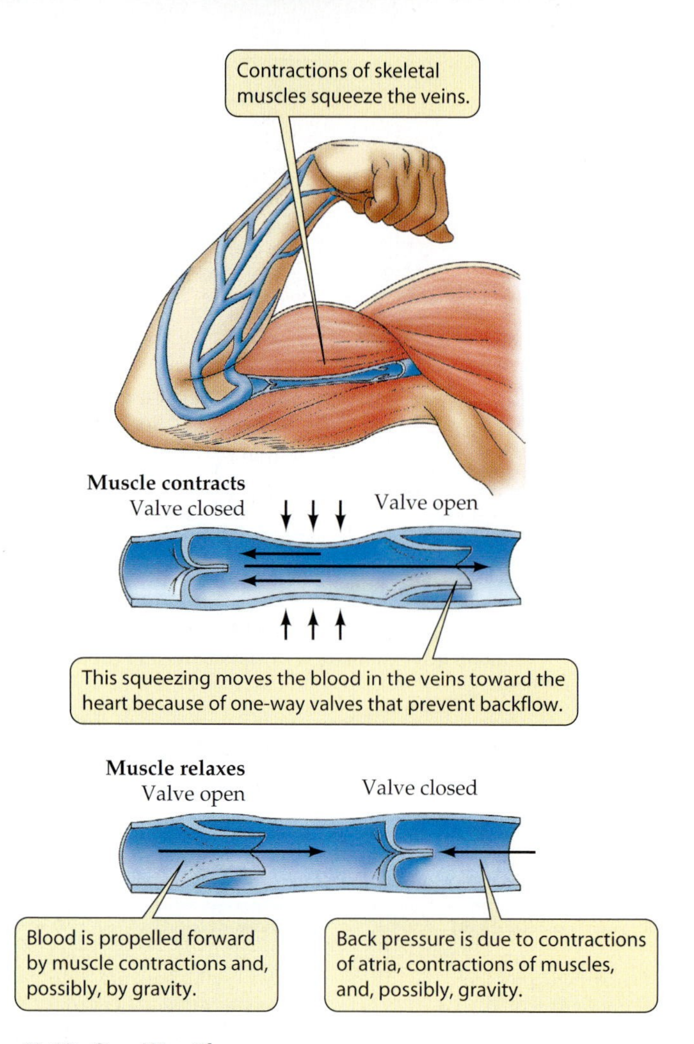

49.13 One-Way Flow
Veins have valves that prevent blood from flowing backward.

Gravity causes blood accumulation in veins and edema. The back pressure that builds up in the capillaries when blood accumulates in the veins shifts the balance between blood pressure and osmotic potential so that there is a net movement of fluid into the intercellular spaces. That is why you have trouble putting your shoes back on after you sit for a long time with your shoes off, such as on an airline flight. In persons with very expandable veins, the veins may become so stretched that the valves can no longer prevent backflow. This condition produces *varicose* (swollen) veins. Draining of these veins is highly desirable and can be aided by wearing support hose and periodically elevating the legs above the level of the heart.

During exercise, the squeezing action of muscles on veins speeds blood toward the heart to be pumped to the lungs and then to the respiring tissues. As an animal runs, its legs act as auxiliary vascular pumps, returning blood to the heart from the veins of the lower body. As a greater volume of blood is returned to the heart, the heart contracts more forcefully, and its pumping action becomes more effective. This strengthening of the heartbeat is due to a property of cardiac muscle cells referred to as the **Frank–Starling law**: If the cells are stretched, as they are when the volume of returning blood increases, they contract more forcefully. This principle holds (within a certain range) whenever venous return increases, by any mechanism.

The actions of breathing also help return venous blood to the heart. The ventilatory muscles create suction that pulls air into the lungs (see Chapter 48), and this suction also pulls blood and lymph toward the chest, increasing venous return to the right atrium.

Some smooth muscles in the walls of veins move venous blood back to the heart by constricting the veins and moving the blood forward. These muscles are rare in most veins and are totally absent from lymphatic vessels in humans. They do not play a major role in venous return. However, in the largest veins closest to the heart, contraction of smooth muscles at the onset of exercise can suddenly increase venous return and stimulate the heart in accord with the Frank–Starling law, thus increasing cardiac output.

Will you die of cardiovascular disease?

Cardiovascular disease is by far the largest single killer in the United States and Europe; it is responsible for about half of all deaths each year. The immediate cause of most of these deaths is heart attack or stroke, but those events are the end result of a disease called **atherosclerosis** (hardening of the arteries) that begins many years before symptoms are detected. Hence atherosclerosis is called the "silent killer." What is atherosclerosis, and how can it be prevented?

Healthy arteries have a smooth internal lining of endothelial cells (Figure 49.14*a*). This lining can be damaged by chronic high blood pressure, smoking, a high-fat diet, or microorganisms. Deposits called **plaque** begin to form at sites of endothelial damage. First the endothelial cells at the damaged site swell and proliferate; then they are joined by smooth muscle cells migrating from below. Lipids, espe-

(a)

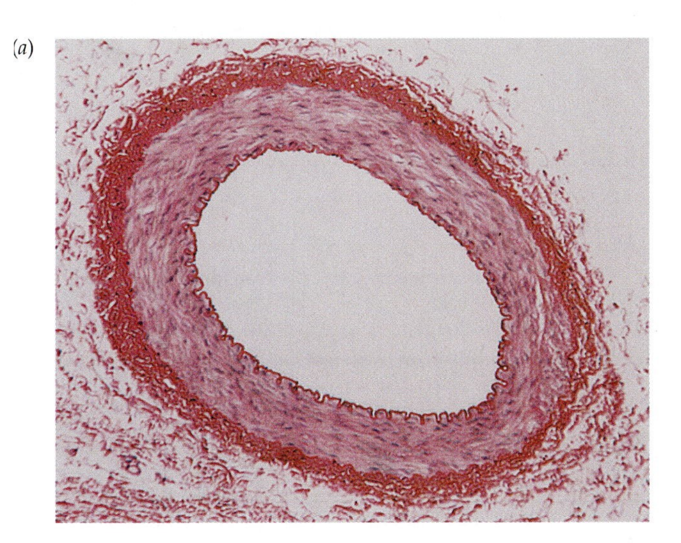

(b)

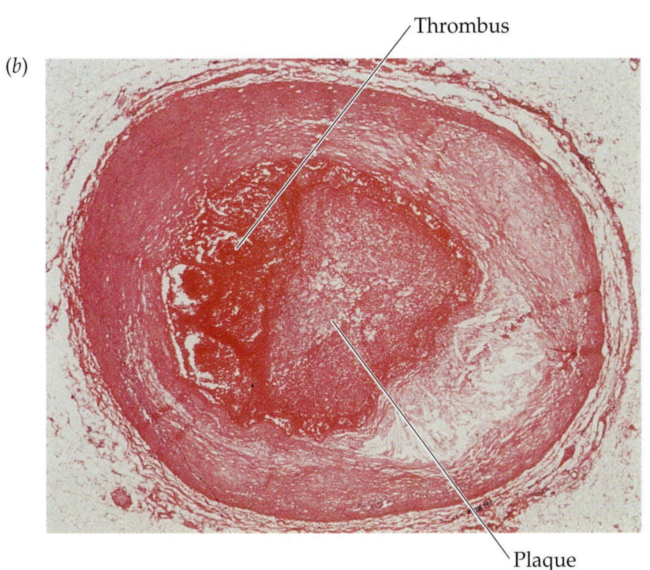

49.14 Atherosclerotic Plaque
(*a*) A healthy, clear artery. (*b*) An atherosclerotic artery, clogged with plaque and a thrombus.

cially cholesterol, are deposited in these cells, so the plaque becomes fatty. Fibrous connective tissue invades the plaque and, along with deposits of calcium, makes the artery wall less elastic; this process is what gives us the terms "atherosclerosis" and "hardening of the arteries." The growing plaque narrows the artery and causes turbulence in the blood flowing over it. Blood platelets (discussed later in this chapter) stick to the plaque and initiate the formation of a blood clot, called a **thrombus**, which further blocks the artery (Figure 49.14*b*).

The blood supply to the heart itself flows through the **coronary arteries**. These arteries are highly susceptible to atherosclerosis; as they narrow, blood flow to the heart muscles decreases. Chest pains and shortness of breath during mild exertion are symptoms of this condition. A person with atherosclerosis is at high risk of forming a thrombus in a coronary artery. This condition, called *coronary thrombosis*, can totally block the vessel, causing a heart attack, or *coronary infarction*.

A piece of a thrombus that breaks loose, called an *embolus*, is likely to travel to and become lodged in a vessel of smaller diameter, blocking its flow (a condition referred to as an *embolism*). Arteries already narrowed by plaque formation are likely places for an embolus to lodge. An embolism in an artery in the brain causes the cells fed by that artery to die. This event is called a *stroke*. The specific damage resulting from a stroke, such as memory loss, speech impairment, or paralysis, depends on the location of the blocked artery.

The most important approach to cardiovascular disease is prevention, not treatment. Probably the most important determinant of whether or not you will get atherosclerosis is your genetic predisposition. Environmental risk factors also play a large role, however, and if you do have a genetic predisposition to atherosclerosis, it is even more important to minimize environmental risk factors. These factors include high-fat and high-cholesterol diets, smoking, a sedentary lifestyle, hypertension (high blood pressure), obesity, and certain medical conditions such as diabetes. Changes in diet and behavior can prevent and reverse atherosclerosis and help fend off the silent killer.

Blood: A Fluid Tissue

Blood is classified as a connective tissue: it has cellular elements suspended in an extracellular matrix of complex, yet specific, composition. The unusual feature of blood is that the extracellular matrix is a liquid, so blood is a fluid tissue.

The cells of the blood can be separated from the fluid matrix, called **plasma**, by centrifugation (Figure 49.15). If a 100-ml sample of blood is spun in a centrifuge, all the cells move to the bottom of the tube, leaving the straw-colored, clear plasma on top. The **packed-cell volume**, or **hematocrit**, is the percentage of the blood volume made up by cells. Normal hematocrit is about 38 percent for women and 46 percent for men, but these values can vary considerably. They are usually higher, for example, in people who live and work at high altitudes because the low oxygen concentrations at high altitudes stimulate the production of more red blood cells.

In this section, we consider two classes of cellular elements in blood: the red blood cells and the platelets, which are pinched-off fragments of cells. We have already studied the other important class of blood cells—white blood cells, or *leucocytes*—in Chapter 19. Finally, we take a closer look at the content of plasma.

Red blood cells transport respiratory gases

Most of the cells in the blood are **erythrocytes**, or red blood cells. Mature red blood cells are biconcave, flexible discs packed with hemoglobin. Their function is to transport the respiratory gases. Their shape gives them a large surface area for gas exchange, and their flexibility enables them to squeeze through narrow capillaries. There are about 5 to 6 million red blood cells per milliliter of blood.

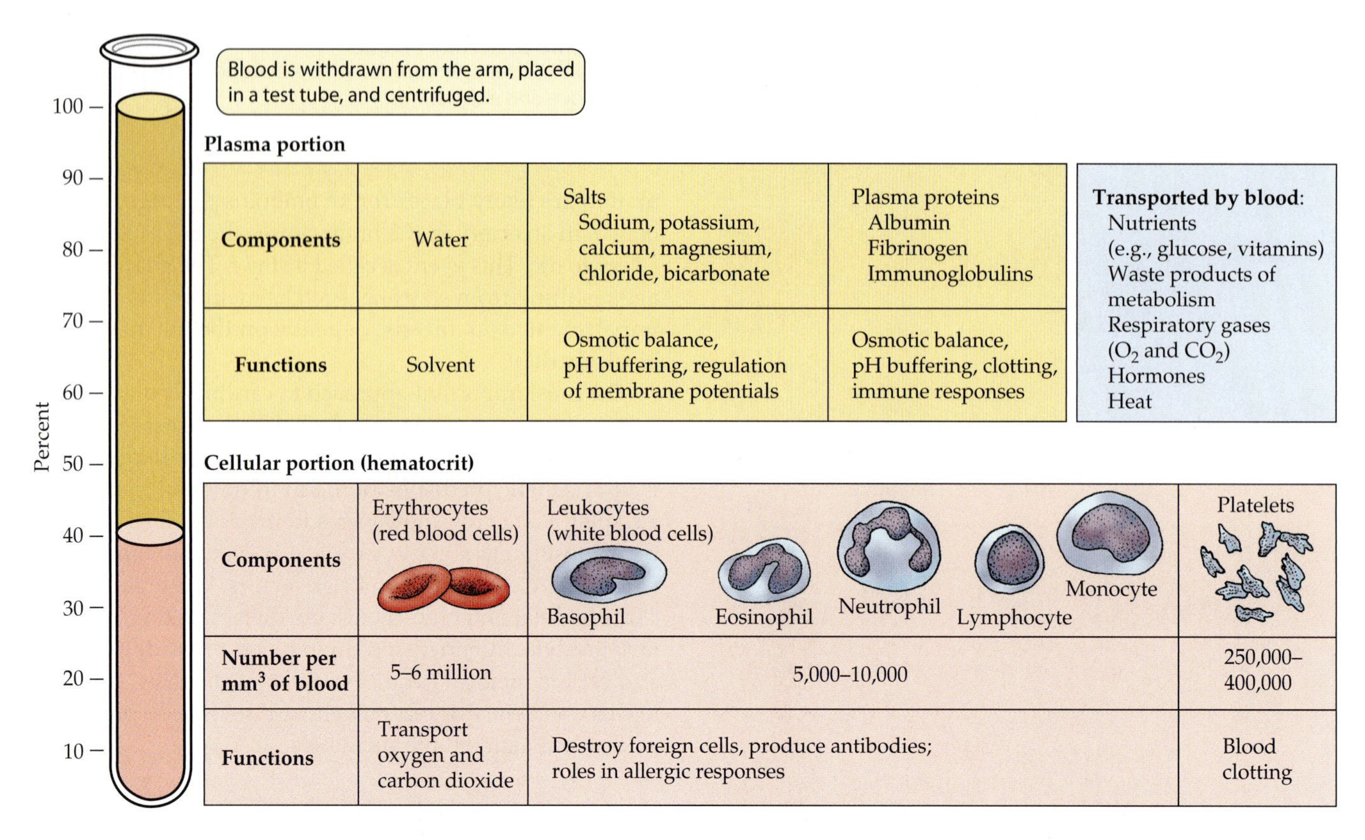

49.15 The Composition of Blood
Blood consists of a complex aqueous solution, numerous cell types, and cell fragments.

Red blood cells are generated by special cells in the bone marrow called **stem cells**, particularly in the bone marrow of the ribs, breastbone, pelvis, and vertebrae. Red blood cell production is controlled by a hormone, **erythropoietin**, which is released by cells in the kidney in response to insufficient oxygen (hypoxia) in the tissues (Figure 49.16*a*). Many tissues respond to hypoxia by expressing a transcription factor called *hypoxia-inducible factor 1* (*HIF-1*). In the kidney, HIF-1 activates the gene encoding erythropoietin. Cells in the kidney produce erythropoietin, which stimulates the stem cells to produce red blood cells.

Under normal conditions, your bone marrow produces about 2 million red blood cells every second. The developing, immature red blood cells divide many times while still in the bone marrow, and during this time they produce hemoglobin. When the hemoglobin content of a red blood cell approaches about 30 percent, its nucleus, endoplasmic reticulum, Golgi apparatus, and mitochondria begin to break down. This process is almost complete when the new red blood cell squeezes between the endothelial cells of capillaries in the bone marrow and enters the circulation.

Each red blood cell circulates for about 120 days and then breaks down. As it gets older, its membrane becomes less flexible and more fragile. Therefore, old red blood cells frequently rupture as they bend to fit through narrow capillaries. One place where they are really squeezed is in the **spleen**, an organ that sits near the stomach in the upper left side of the abdominal cavity. The spleen has many venous cavities, or sinuses, that serve as a reservoir for red blood cells, but to get into the sinuses, the red blood cells must squeeze between spleen cells. Old red blood cells are likely to be ruptured by this squeezing, and when they are, their remnants are broken down by macrophages.

Platelets are essential for blood clotting

Besides producing red blood cells, the stem cells in the bone marrow produce the leucocytes and cells called *megakaryocytes*. Megakaryocytes are large cells that remain in the bone marrow and continually break off cell fragments called **platelets** (Figure 49.16*b*). A platelet is just a tiny fragment of a cell, but it is packed with enzymes and chemicals necessary for its function: sealing leaks in blood vessels and initiating blood clotting.

Damage to a blood vessel exposes collagen fibers. When a platelet encounters collagen fibers, it is activated. It swells, becomes irregularly shaped and sticky, and releases chemicals that activate other platelets and initiate the clotting of blood. The sticky platelets form a plug at the damaged site, and the subsequent clotting forms a stronger patch on the vessel.

The clotting of blood requires many steps and many **clotting factors**. The absence of any one of these factors can impair clotting and cause excessive bleeding. Because the liver produces most of the clotting factors, liver diseases such as hepatitis and cirrhosis can result in excessive

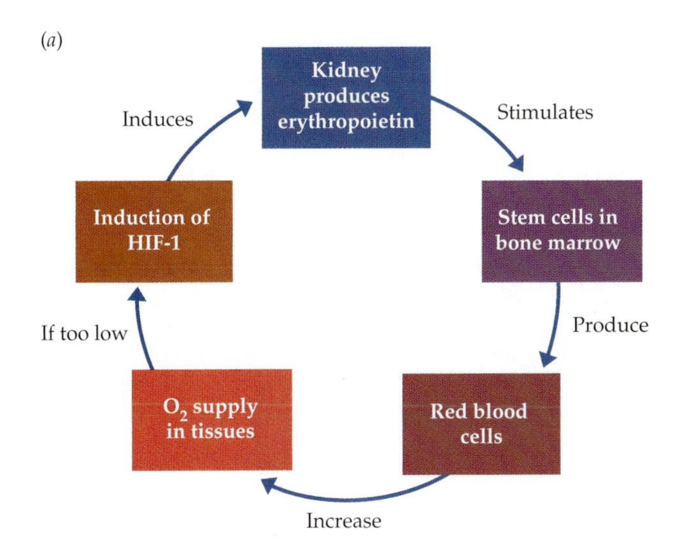

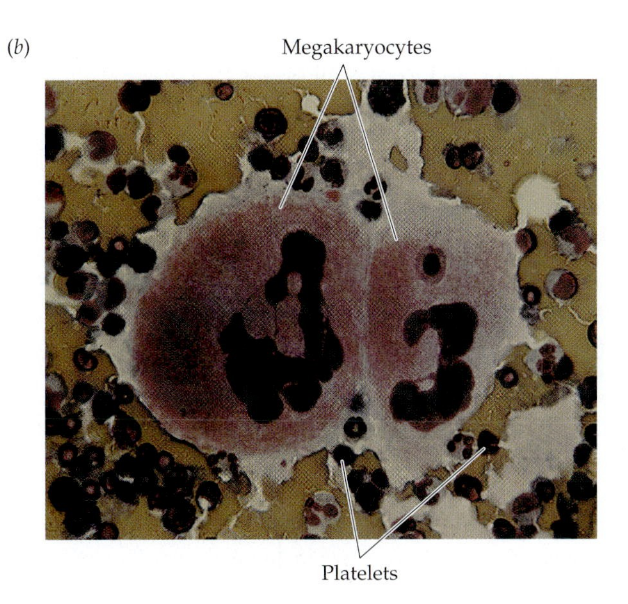

49.16 *Formation of Red Blood Cells and Platelets*
(*a*) Erythropoietin stimulates stem cells in the bone marrow to produce red blood cells. (*b*) In this micrograph, platelets can be seen breaking away from the edges of two megakaryocytes.

bleeding. The sex-linked trait hemophilia (see Chapter 10) is an example of a genetic inability to produce one of the clotting factors.

Blood clotting factors participate in a cascade of steps that activate other substances circulating in the blood. The cascade begins with cell damage and platelet activation and leads to the conversion of an inactive circulating enzyme, **prothrombin**, to its active form, **thrombin**. Thrombin causes molecules of a plasma protein called **fibrinogen** to polymerize and form **fibrin** threads. The fibrin threads form the meshwork that clots the blood, seals the vessel, and provides a scaffold for the formation of scar tissue (Figure 49.17).

Plasma is a complex solution

Plasma, the clear straw-colored liquid portion of the blood, contains gases, ions, nutrient molecules, proteins, and other molecules, such as nonprotein hormones. Most of the ions are Na^+ and Cl^- (hence the salty taste of blood), but many other ions are also present. Nutrient molecules in plasma include glucose, amino acids, lipids, cholesterol, and lactic acid.

The circulating proteins in plasma have many functions. We have just noted proteins that function in blood clotting; others of interest include albumin, which is partly responsible for the osmotic potential in capillaries that prevents a massive loss of water from plasma to intercellular spaces; antibodies (the immunoglobulins); hormones; and various carrier molecules, such as *transferrin*, which carries iron from the gut to where it is stored or used.

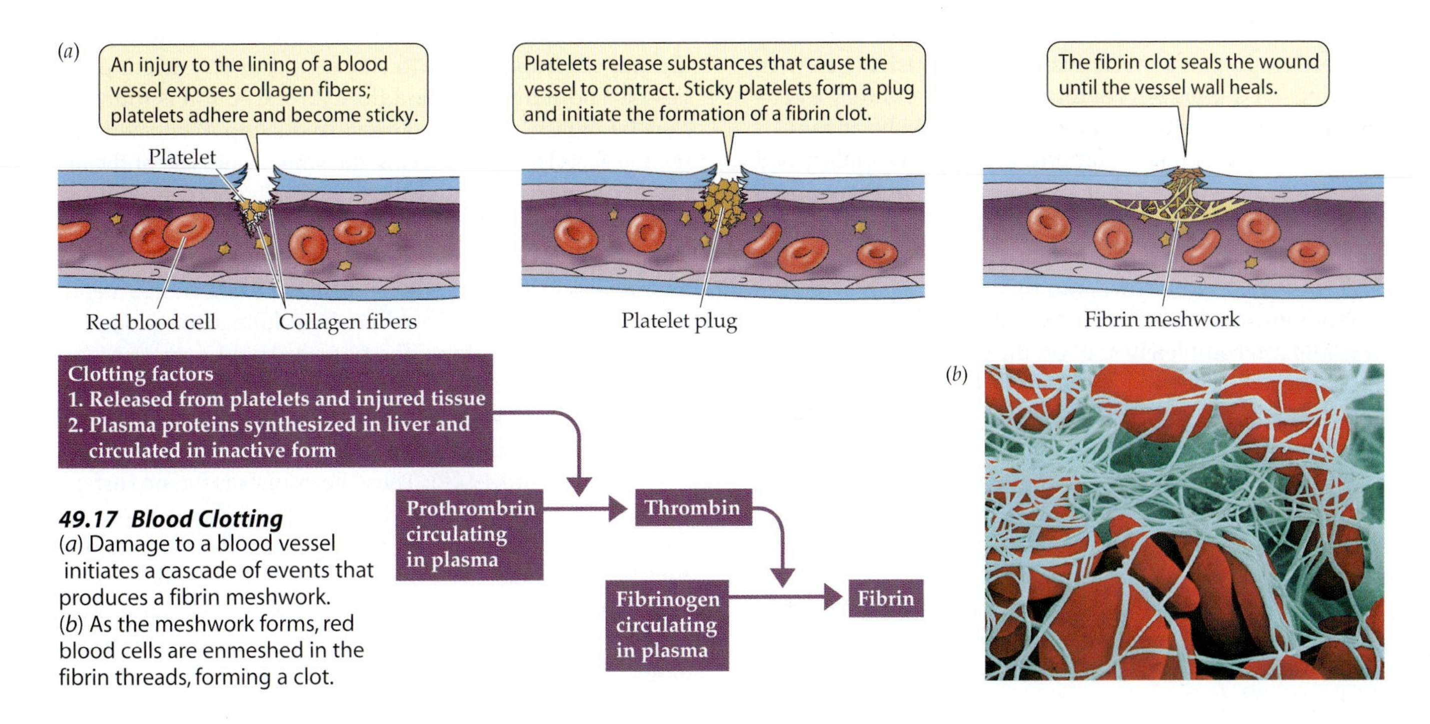

49.17 *Blood Clotting*
(*a*) Damage to a blood vessel initiates a cascade of events that produces a fibrin meshwork.
(*b*) As the meshwork forms, red blood cells are enmeshed in the fibrin threads, forming a clot.

Plasma is very similar to tissue fluid in composition, and most of its components move readily between these two fluid compartments of the body. The main difference between the two fluids is the higher concentration of proteins in the plasma.

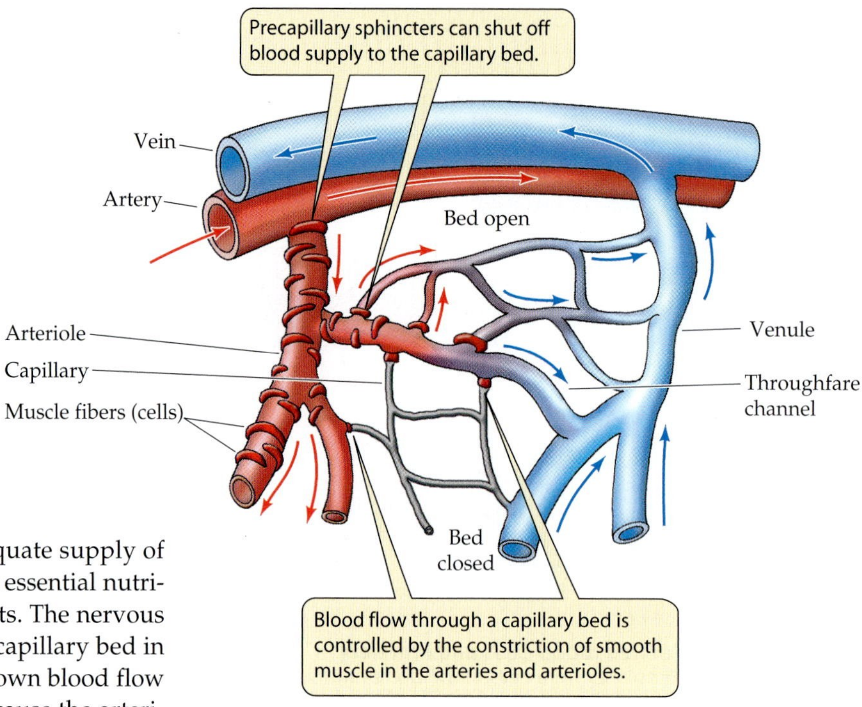

49.18 Local Control of Blood Flow
Low O_2 concentrations or high levels of metabolic by-products cause the smooth muscle of the arteries and arterioles to relax, thus increasing the supply of blood to the capillary bed.

Control and Regulation of Circulation

The circulatory system is controlled and regulated by neural and hormonal mechanisms at both the local and systemic levels. Every tissue requires an adequate supply of blood that is saturated withoxygen, carries essential nutrients, and is relatively free of waste products. The nervous system cannot monitor and control every capillary bed in the body. Instead, each tissue regulates its own blood flow through **autoregulatory mechanisms** that cause the arterioles supplying the tissue to constrict or dilate.

The autoregulatory actions of every capillary bed in every tissue influence the pressure and composition of the arterial blood leaving the heart. If many arterioles suddenly dilate, for example, allowing blood to flow through many more capillary beds, arterial blood pressure falls. If all the newly filled capillary beds contribute metabolic waste products to the blood at one time, the concentration of wastes in the blood returning to the heart increases. Thus events in all the capillary beds throughout the body produce combined effects on arterial blood pressure and blood composition. The nervous and endocrine systems respond to these changes by changing breathing, heart rate, and blood distribution to match the metabolic needs of the body.

Autoregulation matches local flow to local need

The autoregulatory mechanisms that adjust the flow of blood to a tissue are part of the tissue itself, but they can be influenced by the nervous system and certain hormones.

The amount of blood that flows through a capillary bed is controlled by the degree of contraction of the smooth muscle of the arteries and arterioles feeding that bed. The flow of blood in a typical capillary bed is diagrammed in Figure 49.18. Blood flows into the bed from an arteriole. Smooth muscle "cuffs," or **precapillary sphincters**, on the arteriole can completely shut off the supply of blood to the capillary bed. When the precapillary sphincters are relaxed and the arteriole is open, the arterial blood pressure pushes blood into the capillaries.

Autoregulation depends on the sensitivity of the smooth muscle to its chemical environment. Low O_2 concentrations and high CO_2 concentrations cause the smooth muscle to relax, thus increasing the supply of blood, which brings in more O_2 and carries away CO_2. Increases in other by-products of metabolism, such as lactate, hydrogen ions, potassium, and adenosine, promote increased blood flow through the same mechanism. Hence, activities that increase the metabolism of a tissue also increase blood flow to that tissue.

Arterial pressure is controlled and regulated by hormonal and neural mechanisms

The same smooth muscle of arteries and arterioles that responds to autoregulatory stimuli also responds to signals from the endocrine and central nervous systems. Most arteries and arterioles are innervated by the autonomic nervous system, particularly the sympathetic division. Most sympathetic neurons release norepinephrine, which causes the smooth muscle cells to contract, thus constricting the vessels and reducing blood flow. An exception is found in skeletal muscle, in which specialized sympathetic neurons release acetylcholine, causing the smooth muscle of the arterioles to relax and the vessels to dilate, increasing blood to flow to the muscle.

Hormones also can cause arterioles to constrict. Epinephrine, which has actions similar to those of norepinephrine, is released from the adrenal medulla during massive sympathetic activation—the fight-or-flight response. *Angiotensin*, produced when blood pressure in the kidneys falls, causes arterioles to constrict. *Vasopressin*, released by the posterior pituitary when blood pressure falls, has similar effects (Figure 49.19). These hormones influence arterioles located for the most part in peripheral tissues (extremities) or in tissues whose functions need not be maintained continuously (such as the gut). By reducing blood flow in those arterioles, the hormones increase central blood pressure and blood flow to essential organs such as the heart, brain, and kidneys.

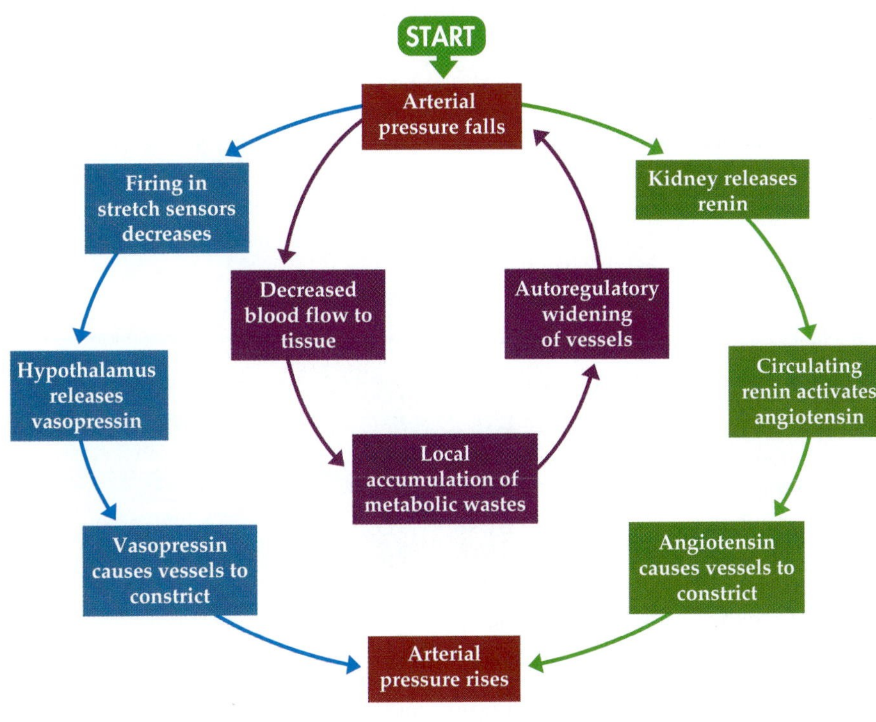

49.19 Hormonal Control of Blood Pressure through Vascular Resistance A drop in arterial pressure reduces blood flow to tissues, resulting in local accumulation of metabolic wastes. This change in the extracellular environment stimulates autoregulatory opening of the arteries, which would lead to a further decrease in central blood pressure if this were not prevented by the negative feedback mechanisms shown in this diagram, which work by promoting the constriction of arteries in less essential tissues.

The autonomic nervous system activity that controls heart rate and constriction of blood vessels originates in cardiovascular centers in the medulla of the brain stem. Many inputs converge on this central integrative network and influence the commands it issues via parasympathetic and sympathetic nerves (Figure 49.20). Of special importance is information about changes in blood pressure from stretch receptors in the walls of the great arteries leading to the brain—the aorta and the carotid arteries.

Increased activity in the stretch receptors indicates rising blood pressure and inhibits sympathetic nervous system output. As a result, the heart slows, and arterioles in peripheral tissues dilate. If pressure in the great arteries falls, the activity of the stretch receptors decreases, stimulating sympathetic output. As a result, the heart beats faster, and the arterioles in peripheral tissues constrict. When arterial pressure falls, the change in stretch receptor activity also causes the hypothalamus to release vasopressin, which helps increase blood pressure by stimulating peripheral arterioles to constrict.

You experience the action of the aortic and carotid stretch receptors when you get up each morning. While you are lying down, your blood pressure is rather evenly distributed from head to toe, but when you get up, gravity pulls blood to the lower part of your body. Blood return to the heart decreases, and therefore cardiac output decreases. As a result, the pressure in the aorta and the carotid arteries falls. This change is detected by the carotid and aortic stretch receptors, which stimulate corrective responses within two heartbeats. Now imagine the change in blood pressure detected by the carotid stretch receptors in the giraffe on the cover of this book when it raises and lowers its head. The giraffe has a much larger heart than would be expected for a mammal of its size because of the need to generate blood pressure sufficient to pump blood against gravity to a height more than 3 m above its heart.

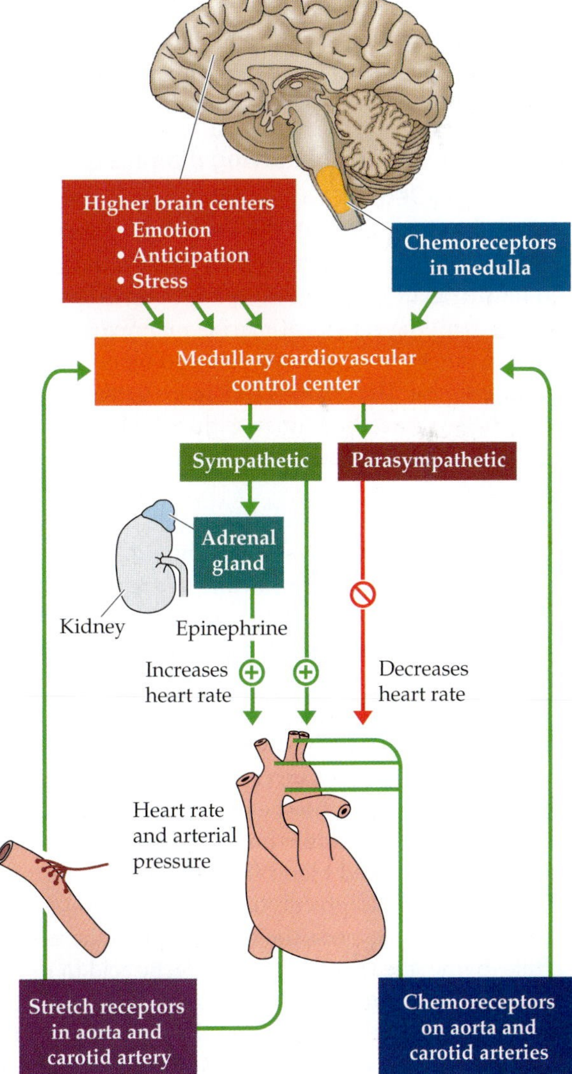

49.20 Regulating Blood Pressure The autonomic nervous system controls heart rate in response to information about blood pressure and blood composition that is integrated by regulatory centers in the medulla.

Other information that causes the medullary regulatory system to increase heart rate and blood pressure comes from chemoreceptors in the carotid and aortic bodies. These nodules of modified smooth muscle tissue respond to inadequate O_2 supply. If arterial blood flow slows or the O_2 content of the arterial blood falls drastically, these receptors are activated and send signals to the regulatory center. The regulatory center also receives input from other brain areas. Emotions or the anticipation of intense activity, as at the start of a race, can cause the center to increase heart rate and blood pressure.

Control and regulation in the cardiovascular system begins with the local autoregulatory mechanisms that cause dilation of local arterioles and precapillary sphincters when a tissue needs more oxygen or has accumulated wastes. As more blood flows into the tissues, the central blood pressure falls, and the composition of the blood returning to the heart reflects the exchanges that are occuring in the tissues. Changes in central blood pressure and blood composition are sensed, and both endocrine and central nervous system responses are activated to return blood pressure and composition to normal. Thus circulatory functions are matched to the regional and overall needs of the body.

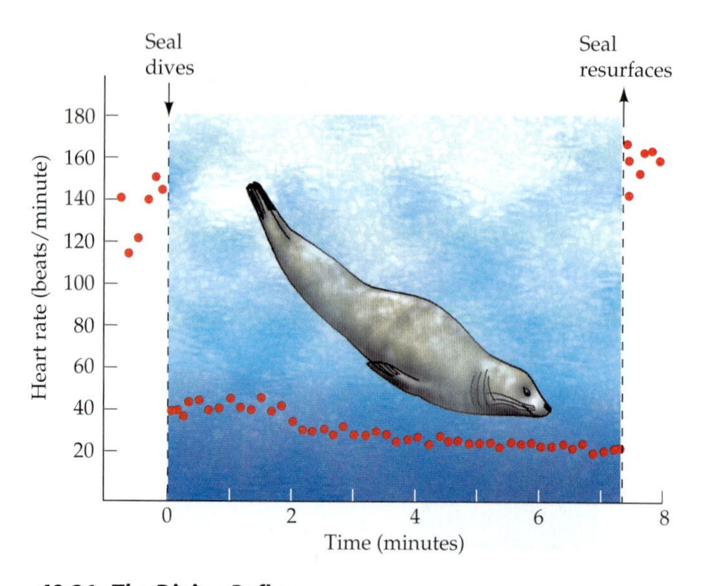

49.21 The Diving Reflex
When a marine mammal dives, its heart rate slows and the arteries to most of its organs constrict, so almost all blood flow and available oxygen goes to the animal's heart and brain. These adaptations enable some seals to remain underwater for up to an hour.

Cardiovascular control in diving mammals conserves oxygen

We began this chapter with the observation that when a seal begins underwater activity, its heart rate slows and blood flow to all of its tissues except its brain drops dramatically. This "diving reflex" of marine mammals is in stark contrast to our increase in heart rate and blood flow when we begin exercise. The obvious difference between the situation of the seal and the human is that the human has access to atmospheric oxygen during exercise and the diving seal does not.

The adaptations of the seal that enable it to remain underwater for a long time are several. The seal's oxygen storage capacity is about twice ours due to greater blood volume, the greater oxygen carrying capacity of its blood, and more myoglobin in its muscles. That is not sufficient, however, to explain dives of half an hour or more. The most important adaptation is the *diving reflex*, which is a slowing of the heart (Figure 49.21) and a constriction of major blood vessels going to all tissues except certain critical ones such as the nervous system, the heart, and the eyes. Central blood pressure remains high, but blood flow to the tissues decreases. This reduced blood flow has two effects: one is to switch the tissue to glycolytic (anaerobic) metabolism, and the other is to suppress the metabolism of the tissue.

While diving, the seal accumulates lactic acid in its muscles, which constitutes an "oxygen debt" to be paid back through elevated metabolism after the dive ends. But the total metabolic "payback" is much less than the metabolism that would have occurred over the same period of time had the seal not dived. The diving reflex caused the seal to be *hypometabolic* (below the basal metabolic rate) during the dive. Hypometabolism, increased oxygen stores, and a high capacity for anaerobic metabolism make it possible for the seal to perform amazing feats.

Chapter Summary

Circulatory Systems: Pumps, Vessels, and Blood

▶ The metabolic needs of the cells of small aquatic animals are met by direct exchange of materials with the external medium. The metabolic needs of the cells of larger animals are met by a circulatory system that transports nutrients, respiratory gases, and metabolic wastes throughout the body. **Review Figure 49.1**

▶ In open circulatory systems the blood or tissue fluid leaves vessels and percolates through tissues. **Review Figure 49.2**

▶ In closed circulatory systems the blood is contained in a system of vessels. **Review Figure 49.3**

Vertebrate Circulatory Systems

▶ The circulatory systems of vertebrates consist of a heart and a closed system of vessels containing blood that is separate from the tissue fluid. Arteries and arterioles carry blood from the heart; capillaries are the site of exchange between blood and tissue fluid; venules and veins carry blood back to the heart.

▶ The vertebrate heart evolved from two chambers in fishes to three in amphibians and reptiles and four in crocodilians, mammals, and birds. This evolutionary progression has led to an increasing separation of blood flow to the gas exchange organs and blood flow to the rest of the body. **Review Pages 868–870**

▶ In birds and mammals, blood circulates through two circuits: the pulmonary circuit and the systemic circuit.

The Human Heart: Two Pumps in One

▶ The human heart has four chambers. Valves in the heart prevent the backflow of blood. **Review Figure 49.4**

▶ The cardiac cycle has two phases: systole, in which the ventricles contract; and diastole, in which the ventricles relax. The sequential heart sounds ("lub-dub") are made by the closing of the heart valves. **Review Figure 49.5**

▶ Blood pressure can be measured using a sphygmomanometer and a stethoscope. **Review Figure 49.6**

▶ The autonomic nervous system controls heart rate. Sympathetic activity increases heart rate, and parasympathetic activity decreases it. These actions are due to the effects of norepinephrine and acetylcholine on the rate of depolarization of the membranes of pacemaker cells. **Review Figure 49.7**

▶ The sinoatrial node controls the cardiac cycle by initiating a wave of depolarization in the atria, which is conducted to the ventricles through the atrioventricular node. **Review Figure 49.8**

▶ The EKG records electric potentials resulting from the contraction and relaxation of the cardiac muscles. **Review Figure 49.9**

The Vascular System: Arteries, Capillaries, and Veins

▶ Arteries and arterioles have many elastic fibers that enable them to withstand high pressures. Abundant smooth muscle cells allow these vessels to contract and expand, altering their resistance and thus blood flow. **Review Figure 49.10**

▶ Capillary beds are the site of exchange of materials between blood and tissue fluid.

▶ The exchange of fluids between blood and tissues is determined by the balance between blood pressure and osmotic potential in the capillaries. **Review Figure 49.12**

▶ The ability of a specific molecule to cross a capillary wall depends on the architecture of the capillary, the type of substance, and the concentration gradient between the blood and the tissue fluid.

▶ A separate system of vessels, the lymphatic system, returns the tissue fluid to the blood.

▶ Veins have a high capacity for storing blood. Aided by gravity, by contractions of skeletal muscle, and by the actions of breathing, they carry blood back to the heart. **Review Figure 49.13**

▶ Cardiovascular disease is responsible for about half of all deaths in the United States and Europe. Atherosclerosis and thrombus formation can lead to potentially fatal conditions such as heart attack and stroke. Diet and behavior are the keys to good cardiovascular health. **Review Figure 49.14**

Blood: A Fluid Tissue

▶ Blood can be divided into a plasma portion (water, salts, and proteins) and a cellular portion (red blood cells, white blood cells, and platelets). All of the cellular components are produced from stem cells in the bone marrow. **Review Figure 49.15**

▶ Red blood cells transport respiratory gases. Their production in the bone marrow is stimulated by erythropoietin, which is produced in response to hypoxia in the tissues. **Review Figure 49.16**

▶ Platelets, along with circulating proteins, are involved in clotting responses. **Review Figure 49.17**

▶ Plasma is a complex solution that contains gases, ions, nutrient molecules, proteins, and other molecules.

Control and Regulation of Circulation

▶ Blood flow through capillary beds is controlled by local autoregulation mechanisms, hormones, and the autonomic nervous system. **Review Figure 49.18**

▶ Blood pressure is controlled in part by the hormones vasopressin and angiotensin, which stimulate contraction of blood vessels. **Review Figure 49.19**

▶ Heart rate is controlled by the autonomic nervous system, which responds to information about blood pressure and blood composition that is integrated by regulatory centers in the brain. **Review Figure 49.20**

▶ Diving mammals conserve blood oxygen stores by slowing the heart rate during dives. **Review Figure 49.21**

For Discussion

1. At the beginning of a race, cardiac output increases immediately before there is any change in blood oxygen or carbon dioxide concentrations. Explain two factors that contribute to this effect. Include the Frank–Starling law in your answer.
2. Explain how the hearts of crocodilians have the advantages of mammalian hearts during exercise but the efficiency of reptilian hearts during rest.
3. A sudden and massive loss of blood results in a decrease in blood pressure. Describe several mechanisms that help return blood pressure to normal.
4. You can describe the cycle of events in a ventricle of the heart by a graph that plots the pressure in the ventricle on the y axis and the volume of blood in the ventricle on the x axis. What would such a graph look like? Where would the heart sounds be on this graph? How would the graph differ for the left and the right ventricles?
5. If the major arteries become clogged with plaque and become less elastic because of calcification, the left ventricle must work harder and harder to pump an adequate supply of blood to the body. As a result, the left ventricle can become weakened and begin to fail even though the right ventricle is healthy. A heart attack primarily affecting the left ventricle can have the same effect. This condition is known as congestive heart failure, and commonly leads to fatal pulmonary edema. Explain how left ventricular failure can result in pulmonary edema, and why is it said that this condition creates a vicious circle that makes itself worse rapidly.

50 Animal Nutrition

THE CENTER FOR DISEASE CONTROL HAS recently determined that one out of every five people in the United States is obese—at least 30 percent over recommended body mass. This epidemic level of obesity reflects a dramatic increase: Less than 10 years ago, the number was one in eight. The CDC has placed obesity just below smoking as the second largest preventable cause of death. Yet, as a nation, we seem to be more health-conscious than ever before, and at the time of this writing, six diet books are on the *New York Times* bestseller list. So what is wrong?

Lifestyle changes have played a role. The consumption of "fast food" with a high fat content has risen, snack foods are more prevalent, and people have become more sedentary overall in spite of an increase in planned exercise. Yet different individuals exposed to the same routines and stimuli differ in their propensity to gain weight. Like all other physiological functions, food intake is under genetic and regulatory influences that we are just beginning to understand. In humans, we know that identical twins reared together or apart rarely differ in body mass by more than a few percent. In animals, and now in humans, we know of single-gene mutations that can cause excessive food intake and obesity, and we know of tiny brain areas that when damaged can cause increased or decreased food intake. Can knowledge of nutrition and the physiology of food intake help us fight the epidemic of obesity?

In this chapter we review the nutrients that organisms require for energy, for molecular building blocks, and for specific biochemical functions. We examine briefly the diversity of adaptations for acquiring and ingesting food. Most of the chapter is devoted to how food is digested and absorbed. Then we learn how the body regulates its traffic in metabolic fuels, and return to the quandaries we have just posed about the regulation of food intake and body mass . Last, we raise the issue of environmental toxicology. By taking in nutrients as food, animals also take in compounds that can be toxic. We briefly consider how animals deal with toxic compounds and how human activities that contribute new and highly dangerous toxic compounds to the environment are affecting human health and other organisms in the environment.

A Sizable Problem
Obesity is often caused or compounded by lifestyle. In the United States today, 1 in 5 adults is obese (30 percent or more above recommended weight limits) and liable to have resulting health problems.

Nutrient Requirements

Animals must eat to stay alive. They eat other organisms—both plants and animals. Since they derive their nutrition from other organisms, they are called **heterotrophs**. In contrast, **autotrophs** (most plants, some bacteria, and some protists) trap solar energy through photosynthesis and use that energy to synthesize all of their components. Directly or indirectly, heterotrophs take advantage of—indeed, depend on—the organic synthesis carried out by autotrophs.

Heterotrophs have evolved an enormous diversity of adaptations to exploit, directly or indirectly, the resources made available through the actions of autotrophs (Figure 50.1). In this section we discover how animals use nutrients for energy and to build more complex molecules. We also examine mineral nutrients, such as iron and calcium, and molecules called vitamins that animals need in small quantities.

(*a*) *Castor canadensis*

(*b*) *Cynanthus latirostris*

(*c*) *Eubalaena australis*

(*d*) *Ursus maritimus*

50.1 The Consumers
Heterotrophs have evolved a range of adaptations for exploiting sources of energy. (*a*) The staple food of the beaver, a mammalian herbivore, is bark. Trees and shrubs felled in the autumn are stored underwater in their lodges (also constructed of felled tree branches), insuring a constant supply of food during the winter. (*b*) The long bill and hovering flight pattern of the hummingbird, a fluid feeder, enable it to harvest nectar from individual flowers. (*c*) Filter-feeding right whales consume massive quantities of tiny phytoplankton by filtering ocean water through baleen plates in their mouths. (*d*) The carnivorous polar bear is a fearsome predator, preying mainly on marine mammals such as seals.

Energy can be measured in calories

In Chapters 6 and 7, we learned that energy in the chemical bonds of food molecules is transferred to the high-energy phosphate bonds of ATP. ATP provides the energy for cellular work. Each conversion of energy from food molecules to ATP and from ATP to cellular work is inefficient; in fact, most of the energy that was in the food is lost as heat. Even the useful energy conversions eventually are reduced to heat, as molecules that were synthesized are broken down and the energy of movement is dissipated by friction.

In time, all the energy that is transferred to ATP from the chemical bonds of food molecules is released to the environment as heat. Therefore, we can talk about the energy requirements of animals and the energy content of food in terms of a measure of heat energy: the **calorie**. A calorie is the amount of heat necessary to raise the temperature of 1 g of water 1°C. Since this value is a tiny amount of energy in comparison to the energy requirements of many animals, physiologists commonly use the **kilocalorie** (kcal) as a unit of measure (1 kcal = 1,000 calories). Nutritionists also use the kilocalorie as a standard unit of energy, but they traditionally refer to it as the **Calorie**, which is always capitalized to distinguish it from the single calorie. (Scientists are gradually abandoning the calorie as an energy unit as they switch to the International System of Units. In this system, the basic unit of energy is the joule: 1 calorie = 4.184 joules.)

The *metabolic rate* of an animal (see Chapter 40) is a measure of the overall energy needs that must be met by the animal's ingestion and digestion of food. The components of food that provide energy are fats, carbohydrates, and proteins. Fats yield 9.5 kcal/g, carbohydrates 4.2 kcal/g, and proteins about 4.1 kcal/g. The *basal metabolic rate* of a human

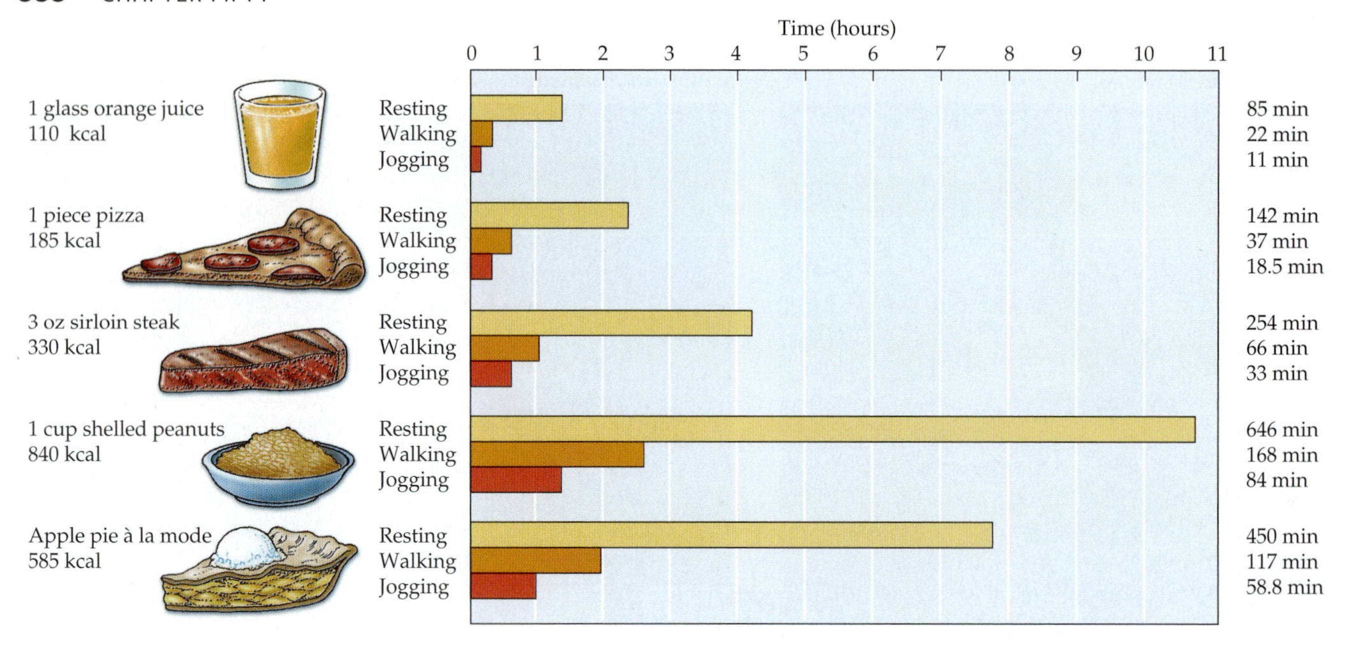

50.2 Food Energy and How Fast We Burn It
The energy in kilocalories contained in several common food items is shown at the left. The graph indicates about how long it would take a person with a basal metabolic rate of about 1,800 kcal/day to utilize the equivalent amount of energy while involved in various activities.

(the metabolic rate resulting from all of the essential physiological functions of a resting person) is about 1,300–1,500 kcal/day for an adult female and 1,600–1,800 kcal/day for an adult male. Physical activity adds to this basal energy requirement. For a person doing sedentary work, about 30 percent of total energy consumption is due to skeletal muscle activity, and for a person doing heavy physical labor, 80 percent or more of the total caloric expenditure is due to skeletal muscle activity. Some equivalencies of food, energy, and exercise are shown in Figure 50.2.

Sources of energy can be stored in the body

Although the cells of the body use energy continuously, most animals do not eat continuously. Humans generally eat several meals a day, a lion may eat once in several days, a boa constrictor may eat once a month, and hibernating animals may go 5 to 6 months without eating. Therefore, animals must store fuel molecules that can be released as needed between meals.

Carbohydrates are stored in liver and muscle cells as glycogen ("animal starch," see Chapter 3), but the total glycogen store is usually not more than the equivalent of a day's energy requirements. Fat is the most important form of stored energy in the bodies of animals. Not only does fat have the highest energy content per gram, but it can be stored with little associated water, making it more compact. If migrating birds had to store energy as glycogen rather than fat to fuel their long flights, they would be too heavy to fly! Protein is not used to store energy, although body protein can be metabolized as an energy source of last resort.

If an animal takes in too little food to meet its needs for metabolic energy, it is **undernourished**, and must make up the shortfall by metabolizing some of the molecules of its own body. Consumption of self for fuel begins with the storage compounds glycogen and fat. Protein loss is minimized for as long as possible, but eventually a starving animal uses its own proteins for fuel. The breakdown of body proteins impairs body functions and eventually leads to death. Blood proteins are among the first to go, resulting in loss of fluid to the intercellular spaces (edema; see Chapter 49). Muscles atrophy (waste away), and eventually even brain protein is lost. Figure 50.3 shows the course of starvation.

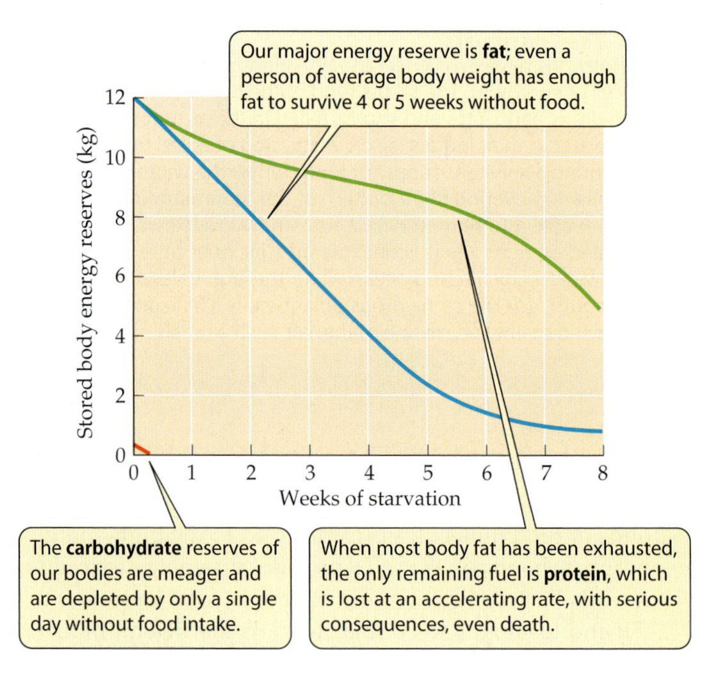

50.3 The Course of Starvation
In a person subjected to undernutrition, the energy reserves of the body are eventually depleted. Body fat is our major defense against starvation.

Undernourishment is rampant among people in nonindustrialized and war-torn nations, and a billion people—one-sixth of the world's population—are undernourished. (Ironically, one cause of life-threatening undernourishment in Western nations is a self-imposed starvation syndrome called *anorexia nervosa* that results from a psychological aversion to body fat.)

When an animal consistently takes in more food than it needs to meet its energy demands, it is **overnourished**. The excess nutrients are stored as increased body mass. First, glycogen reserves build up; then additional dietary carbohydrates, fats, and proteins are converted to body fat. In some species, such as hibernators, seasonal overnutrition is an important adaptation for surviving periods when food is unavailable. In humans, however, overnutrition can be a serious health hazard, increasing the risk of high blood pressure, heart attack, diabetes, and other disorders.

Food provides carbon skeletons for biosynthesis

Every animal requires certain basic organic molecules (*carbon skeletons*) that it cannot synthesize for itself, but needs to build its own complex organic molecules. An example of a required carbon skeleton is the acetyl group (Figure 50.4). Animals cannot make acetyl groups from carbon, oxygen, and hydrogen molecules; they obtain acetyl groups by metabolizing carbohydrates, fats, or proteins.

The acetyl group can be derived from the metabolism of almost any food. It is unlikely ever to be in short supply for an adequately nourished animal. Other carbon skeletons, however, are derived from more limited sources, and an animal can suffer a deficiency of these materials even if its caloric intake is adequate. This state of deficiency is called **malnutrition**.

Amino acids, the building blocks of proteins, are a good example of carbon skeletons that can be in short supply. Humans obtain amino acids by breaking down proteins from food and absorbing the resulting amino acids. Another source of amino acids is the breakdown of existing body proteins, which are in constant turnover as the tissues of the body undergo normal remodeling and renewal. From these amino acids and ones from food, the body synthesizes its own protein molecules as specified by its DNA.

Animals can synthesize some of their own amino acids by taking carbon skeletons synthesized from acetyl or other groups and transferring to them amino groups (—NH_2) derived from other amino acids. But most animals cannot synthesize all the amino acids they need. Each species has certain **essential amino acids** that must be obtained from food. Different species have different essential amino acids, and in general, herbivores have fewer essential amino acids than carnivores have.

If an animal does not take in one of its essential amino acids, its protein synthesis is impaired. Think of protein synthesis as using a keyboard to write a story. If one letter on the keyboard doesn't function, the story either comes to a stop or has an error in it wherever that letter is needed. In protein synthesis, the story usually comes to a stop, and a functional protein is not produced.

There are eight essential amino acids that humans must obtain from their food: isoleucine, leucine, lysine, methionine, phenylalanine, threonine, tryptophan, and valine. All eight are available in milk, eggs, or meat, but no plant food contains all eight. A strict vegetarian runs a risk of protein malnutrition. An appropriate dietary *mixture* of plant foods, however, supplies all eight essential amino acids (Figure 50.5). In general, grains are complemented by legumes or by milk products; legumes are complemented by grains, seeds, and nuts. Long before the chemical basis for this complementarity was understood, societies with little access to meat developed healthy dietary practices. Many

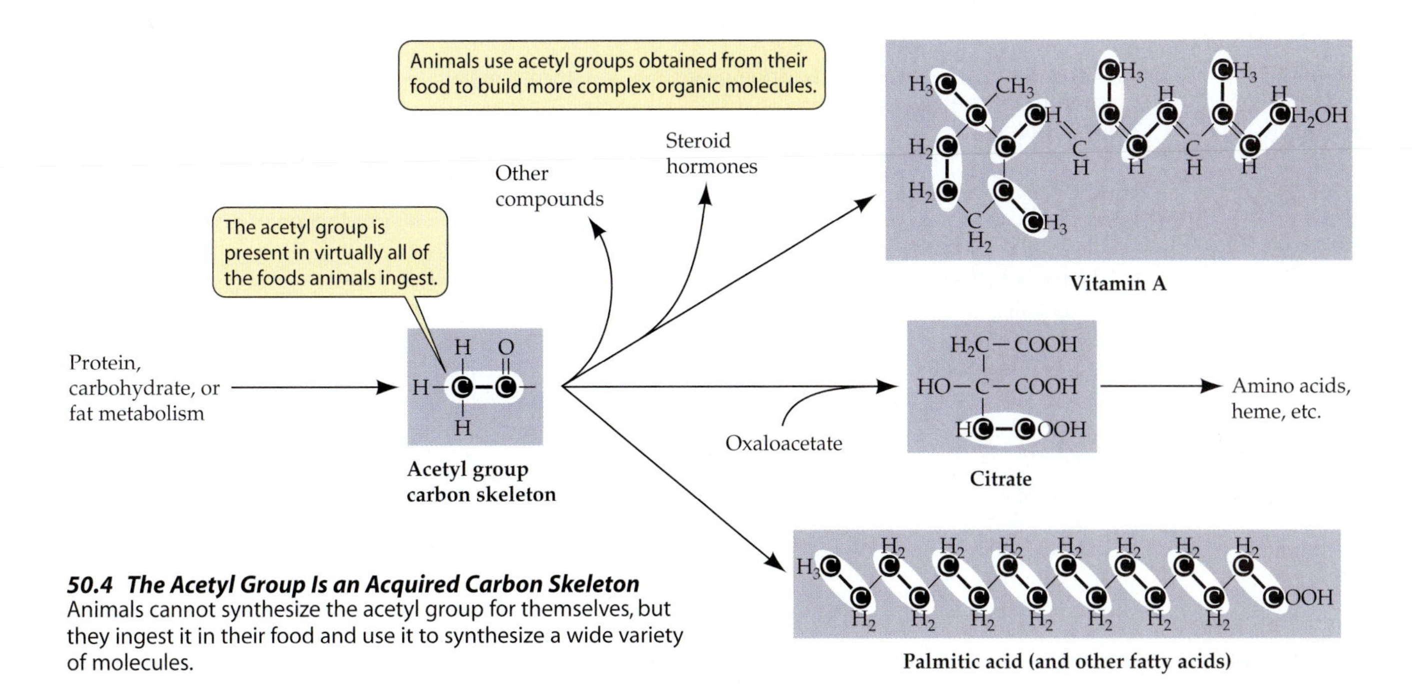

50.4 The Acetyl Group Is an Acquired Carbon Skeleton
Animals cannot synthesize the acetyl group for themselves, but they ingest it in their food and use it to synthesize a wide variety of molecules.

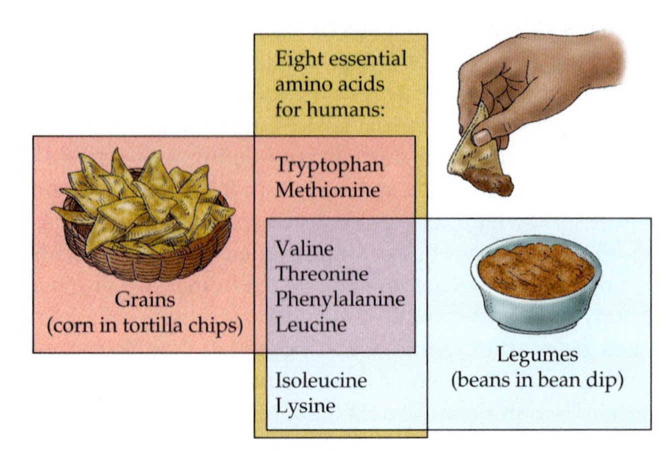

50.5 A Strategy for Vegetarians
By combining cereal grains and legumes, a vegetarian can obtain all eight essential amino acids.

Central and South American peoples traditionally eat beans with corn, and the native peoples of North America complemented their beans with squash.

Why are dietary proteins completely digested to their constituent amino acids before being used by the body? Wouldn't it be more energy-efficient to reuse some dietary proteins directly? There are several reasons why ingested proteins are not used "as is."

- Macromolecules such as proteins are not readily taken up through plasma membranes, but their constituent monomers (such as amino acids) are readily transported.
- Protein structure and function are highly species-specific. A protein that functions optimally in one species might not function well in another species.
- Foreign proteins entering the body directly from the gut would be recognized as invaders and would be attacked by the immune system.

50.1 *Mineral Elements Required by Animals*

ELEMENT	SOURCE IN HUMAN DIET	MAJOR FUNCTIONS
MACRONUTRIENTS		
Calcium (Ca)	Dairy foods, eggs, green leafy vegetables, whole grains, legumes, nuts	Found in bones and teeth; blood clotting; nerve and muscle action; enzyme activation
Chlorine (Cl)	Table salt (NaCl), meat, eggs, vegetables, dairy foods	Water balance; digestion (as HCl); principal negative ion in tissue fluid
Magnesium (Mg)	Green vegetables, meat, whole grains, nuts, milk, legumes	Required by many enzymes; found in bones and teeth
Phosphorus (P)	Dairy foods, eggs, meat, whole grains, legumes, nuts	Found in nucleic acids, ATP, and phospholipids; bone formation; buffers; metabolism of sugars
Potassium (K)	Meat, whole grains, fruits, vegetables	Nerve and muscle action; protein synthesis; principal positive ion in cells
Sodium (Na)	Table salt, dairy foods, meat, eggs, vegetables	Nerve and muscle action; water balance; principal positive ion in tissue fluid
Sulfur (S)	Meat, eggs, dairy foods, nuts, legumes	Found in proteins and coenzymes; detoxification of harmful substances
MICRONUTRIENTS		
Chromium (Cr)	Meat, dairy foods, whole grains, dried beans, peanuts, brewer's yeast	Glucose metabolism
Cobalt (Co)	Meat, tap water	Found in vitamin B_{12}; formation of red blood cells
Copper (Cu)	Liver, meat, fish, shellfish, legumes, whole grains, nuts	Found in active site of many redox enzymes and electron carriers; production of hemoglobin; bone formation
Fluorine (F)	Most water supplies	Resistance to tooth decay
Iodine (I)	Fish, shellfish, iodized salt	Found in thyroid hormones
Iron (Fe)	Liver, meat, green vegetables, eggs, whole grains, legumes, nuts	Found in active sites of many redox enzymes and electron carriers, hemoglobin, and myoglobin
Manganese (Mn)	Organ meats, whole grains, legumes, nuts, tea, coffee	Activates many enzymes
Molybdenum (Mo)	Organ meats, dairy foods, whole grains, green vegetables, legumes	Found in some enzymes
Selenium (Se)	Meat, seafood, whole grains, eggs, chicken, milk, garlic	Fat metabolism
Zinc (Zn)	Liver, fish, shellfish, and many other foods	Found in some enzymes and some transcription factors; insulin physiology

Most animals avoid these problems by digesting food proteins extracellularly and then absorbing the resulting amino acids into the body, where they synthesize new proteins that will function correctly and be recognized as "self" by the immune system.

Using acetyl groups obtained from food, humans can synthesize almost all the lipids required by the body, but we must have a dietary source of **essential fatty acids**—notably linoleic acid. Linoleic acid is an unsaturated fatty acid needed by mammals to synthesize other unsaturated fatty acids such as arachidonic acid, which in turn produces several signaling molecules, including prostaglandins. A deficiency of linoleic acid can lead to problems such as infertility and impaired lactation. Essential fatty acids are also necessary components of membrane phospholipids.

Animals need mineral elements in different amounts

The principal mineral elements required by animals are listed in Table 50.1. Elements required in large amounts are called **macronutrients**; elements required in only tiny amounts are called **micronutrients**. Some essential elements are required in such minute amounts that deficiencies are never observed, but these elements are nevertheless essential.

Calcium is an example of a macronutrient. It is the fifth most abundant element in the body; a 70-kg person contains about 1,200 g of calcium. Calcium phosphate is the principal structural material in bones and teeth. Muscle contraction, neuronal function, and many other intracellular functions in animals require calcium. The turnover of calcium in the extracellular fluid is quite high, as bones are constantly being remodeled and calcium is constantly entering and leaving cells. Calcium is lost from the body in urine, sweat, and feces, so it must be replaced regularly from the diet. Humans require about 800 to 1,000 mg of calcium per day in the diet.

Iron is an example of a micronutrient. Iron is found everywhere in the body because it is the oxygen-binding atom in hemoglobin and myoglobin and is a component of enzymes in the electron transport chain. Nevertheless, the total amount of iron in a 70-kg person is only about 4 g, and since iron is recycled so efficiently in the body and is not lost in the urine, we require only about 15 mg per day in ourfood. In spite of the small amount required, insufficient iron is the most common nutrient deficiency in the world today.

Animals must obtain vitamins from food

Another group of essential nutrients is the **vitamins**. Like essential amino acids and fatty acids, vitamins are carbon compounds that an animal requires for its normal growth and metabolism, but cannot synthesize for itself. Most vitamins function as coenzymes or parts of coenzymes and are required in very small amounts, compared with the essential amino acids and fatty acids, which have structural roles.

The list of vitamins varies from species to species. Most mammals, for example, can make their own ascorbic acid (vitamin C). However, primates (including humans) do not

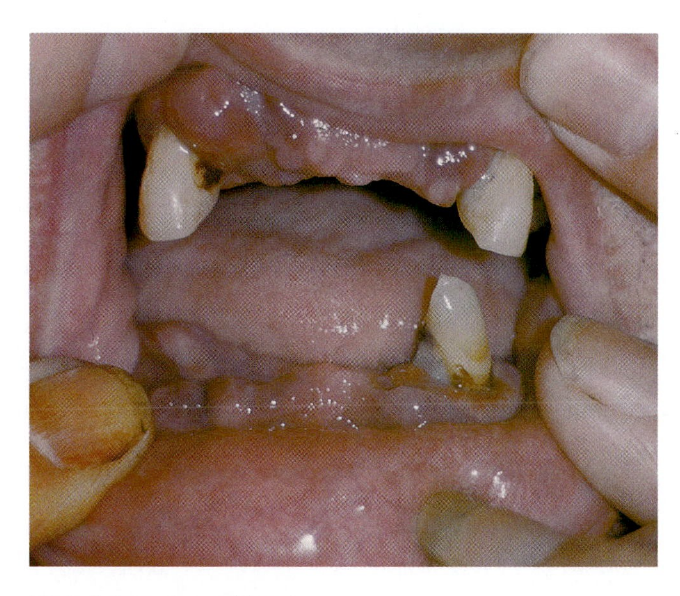

50.6 A Symptom of Scurvy
Because a vitamin C deficiency weakens connective tissue, the gums bleed, teeth fall out, and blood vessels under the skin break.

have this ability, so for primates ascorbic acid is a vitamin. If we do not get vitamin C in our food, we develop a disease known as *scurvy*, characterized by bleeding gums, loss of teeth, subcutaneous hemorrhages, and slow wound healing (Figure 50.6). Scurvy was a serious and frequently fatal problem for sailors on long voyages until a Scottish physician, James Lind, discovered that the disease could be prevented if the sailors ate fresh greens or fresh fruit. Eventually the British Admiralty made limes standard provisions for its ships, and ever since British sailors have been called "limeys." The active ingredient in limes was named ascorbic ("without scurvy") acid.

For humans, there are 13 vitamins. They are divided into two groups: water-soluble vitamins and fat-soluble vitamins. Table 50.2 presents these vitamins, their dietary sources, and their functions.

The fat-soluble vitamin D (calciferol), which is essential for the absorption and metabolism of calcium, is a special case because the body can make it. Certain lipids present in the human body can be converted into vitamin D by the action of ultraviolet light on the skin. Thus vitamin D must be obtained in the diet only by individuals with inadequate exposure to the sun, such as people who live in cold climates where clothing usually covers most of the body and where the sun may not shine for long periods of time.

The need for vitamin D may have been an important factor in the evolution of skin color. Human races that are adapted to equatorial and low latitudes have dark skin pigmentation as a protection against the damaging effects of ultraviolet radiation. These peoples generally have extensive skin areas exposed to the sun on a regular basis, so their skin synthesizes adequate amounts of vitamin D. Most human races that became adapted to higher latitudes lost dark skin pigmentation. Presumably, lighter skin facili-

50.2 Vitamins in the Human Diet

VITAMIN	SOURCE	FUNCTION	DEFICIENCY SYMPTOMS
WATER-SOLUBLE			
B_1, thiamin	Liver, legumes, whole grains, yeast	Coenzyme in cellular respiration	Beriberi, loss of appetite, fatigue
B_2, riboflavin	Dairy foods, organ meats, eggs, green leafy vegetables	Coenzyme in cellular respiration (in FAD and FMN)	Lesions in corners of mouth, eye irritation, skin disorders
Niacin (nicotinamide, nicotinic acid)	Meat, fowl, liver, yeast	Coenzyme in cellular metabolism (in NAD and NADP)	Pellagra, skin disorders, diarrhea, mental disorders
B_6, pyridoxine	Liver, whole grains, dairy foods	Coenzyme in amino acid metabolism	Anemia, slow growth, skin problems, convulsions
Pantothenic acid	Liver, eggs, yeast	Found in acetyl CoA	Adrenal problems, reproductive problems
Biotin	Liver, yeast, bacteria in gut	Found in coenzymes	Skin problems, loss of hair
B_{12}, cobalamin	Liver, meat, dairy foods, eggs	Coenzyme in formation of nucleic acids and proteins, and in red blood cell formation	Pernicious anemia
Folic acid	Vegetables, eggs, liver, whole grains	Coenzyme in formation of heme and nucleotides	Anemia
C, ascorbic acid	Citrus fruits, tomatoes, potatoes	Aids formation of connective tissues; prevents oxidation of cellular constituents	Scurvy, slow healing, poor bone growth
FAT-SOLUBLE			
A, retinol	Fruits, vegetables, liver, dairy foods	Found in visual pigments	Night blindness, damage to mucous membranes
D, calciferol	Fortified milk, fish oils, sunshine	Absorption of calcium and phosphorus	Rickets
E, tocopherol	Meat, dairy foods, whole grains	Muscle maintenance, prevents oxidation of cellular components	Anemia
K, menadione	Intestinal bacteria, liver	Blood clotting	Blood-clotting problems (in newborns)

tates vitamin D production in the relatively small areas of skin exposed to sunlight during the short days of winter. An exception to this correlation between latitude and skin pigmentation is the Inuit peoples of the Arctic. These dark-skinned people obtain plenty of vitamin D from the large amounts of meat and fish oils in their diet; for them, exposure to sunlight is not necessary for obtaining this vitamin.

When water-soluble vitamins are ingested in excess of bodily needs, they are simply eliminated in the urine. (This is the fate of much of the vitamin C that people take in excessive doses.) The fat-soluble vitamins, however, accumulate in body fat and may build up to toxic levels in the liver if taken in excess.

Nutrient deficiency diseases

A chronic shortage of any nutrient produces a characteristic deficiency disease. If the deficiency is not remedied, death may follow. An example is *kwashiorkor*, a disease that results from protein deficiency. It causes swelling of the extremities, distension of the abdomen (Figure 50.7), breakdown of the immune system, degeneration of the liver, mental retardation, and other problems.

A shortage of any of the vitamins results in specific deficiency symptoms (see Table 50.2). We have already described scurvy, which results from a lack of vitamin C. Another deficiency disease, *beriberi*, was directly involved in the discovery of vitamins. *Beriberi* means "extreme weakness." It became prevalent in Asia in the nineteenth century, after it became standard practice to mill rice to a high, white polish and discard the hulls that are present in brown rice. A critical observation was that birds—chickens and pigeons—developed beriberi-like symptoms when fed only polished rice. In 1912, Casimir Funk cured pigeons of beriberi by feeding them the discarded hulls.

At the time of Funk's discovery, all diseases were thought to be either caused by microorganisms or inherited. Funk suggested the radical idea that beriberi and some other diseases are dietary in origin and result from deficiencies in specific substances. Funk coined the term "vitamines" because he mistakenly thought that all these substances were amines (compounds with amino groups) vital for life. In 1926, thiamin (vitamin B_1)—the substance lost in the rice milling process—was the first vitamin to be isolated in pure form.

50.7 Kwashiorkor, "The Rejected One"
The swollen abdomen, face, hands, and feet due to edema (fluid retention), as well as the spindly limbs of these children, are hallmarks of serious protein starvation. These symptoms are a result of the body breaking down blood proteins and muscle tissue to obtain needed amino acids.

Deficiency diseases can also result from an inability to absorb or process an essential nutrient even if it is present in the diet. Vitamin B_{12} (cobalamin), for example, is present in all foods of animal origin. Since plants neither use nor produce vitamin B_{12}, a strictly vegetarian diet (not supplemented by vitamin pills) can lead to a B_{12} deficiency disease called *pernicious anemia*, characterized by a failure of red blood cells to mature. The most common cause of pernicious anemia, however, is not a lack of vitamin B_{12} in the diet, but an inability to absorb it. Normally, cells in the stomach lining secrete a peptide called *intrinsic factor*, which binds to vitamin B_{12} and makes it possible for it to be absorbed in the ileum of the small intestine. Conditions that damage the stomach lining can therefore cause anemia.

Inadequate mineral nutrition can also lead to deficiency diseases. Iodine, for example, is a constituent of the hormone thyroxine, which is produced in the thyroid gland. If insufficient iodine is obtained in the diet, the thyroid gland grows larger in an attempt to compensate for the inadequate production of thyroxine (see Figure 41.9). The swelling of the neck that results is called a *goiter*. The introduction of iodized table salt has greatly reduced the incidence of goiter in the United States.

Adaptations for Feeding

Heterotrophic organisms can be classified by how they acquire their nutrition. **Saprobes** (also called saprotrophs or decomposers) are mostly protists and fungi that absorb nutrients from dead organic matter. **Detritivores**, such as earthworms and crabs, actively feed on dead organic material. Animals that feed on living organisms are **predators**. **Herbivores** prey on plants, **carnivores** prey on animals, and **omnivores** prey on both. **Filter feeders**, such as clams and blue whales, prey on small organisms by filtering them from the environment. **Fluid feeders** include mosquitoes, aphids, and leeches, as well as birds that feed on plant nectar. The anatomical adaptations that enable a species to exploit a particular source of nutrition are usually quite obvious, but physiological and biochemical adaptations can be just as important.

The food of herbivores is often low in energy and hard to digest

Vegetation is frequently coarse and difficult to break down physically, but herbivores must process large amounts of it, since its energy content is low. Most herbivores spend a great deal of their time feeding. Many have striking adaptations for feeding, such as the trunk (a flexible, gripping nose) of the elephant or the long neck of the giraffe. Many types of grinding, rasping, cutting, and shredding mouthparts have evolved in invertebrates for ingesting plant material, and the teeth of herbivorous vertebrates have been shaped by selection to process coarse plant matter. The digestive processes of herbivores can also be quite specialized. The Australian koala, for example, eats nothing but the leaves of eucalyptus trees. Eucalyptus leaves are tough, low in nutrient content, and loaded with pungent, toxic compounds that evolved to protect the trees from predators. Yet the koala's gut can digest and detoxify the leaves and absorb all the nutrients the animal needs from this highly specialized and formidable diet.

Carnivores must detect, capture, and kill prey

The predatory behaviors of many carnivores are legendary. One need only call to mind the hunting skills of hawks, wolves, or any member of the cat family. Carnivores have evolved stealth, speed, power, large jaws, sharp teeth, and strong gripping appendages. Carnivores also have evolved remarkable means of detecting prey. Bats use echolocation, pit vipers sense infrared radiation from the warm bodies of their prey, and certain fishes detect electric fields created in the water by their prey (see Chapter 45).

Adaptations for killing and ingesting prey are diverse and highly specialized. These adaptations can be especially important when the prey are capable of inflicting damage on the predator. A snake may strike with poisonous fangs,

(a)

(b)

50.8 Adaptations for Feeding
(*a*) Snakes such as this Texas rat snake (*Elaphe obsoleta*) can ingest large prey (in this case a lizard) by dislocating their jaws. (*b*) This sea star is eating a clam. While it holds the clam with its arms, enzymes from its everted stomach digests it.

using its venom to immobilize its prey before ingesting it. To swallow large prey, a snake disengages its lower jaw from its joint with the skull (Figure 50.8*a*). The tentacles of jellyfishes, corals, squid, and octopuses, the long, sticky tongues of frogs and chameleons, and the webs of spiders are other fascinating examples of adaptations for capturing and immobilizing prey.

Because some prey items are impossible to ingest, some predators digest their prey externally. Sea stars evert their stomachs (turn them inside out) and digest their molluscan prey while they are still in their shells (Figure 50.8*b*). Spiders usually prey on insects with indigestible exoskeletons. The spider injects its prey with digestive enzymes and then sucks out the liquefied contents, leaving behind the empty exoskeletons frequently seen in old spider webs.

Vertebrate species have distinctive teeth

Teeth are adapted for the acquisition and initial processing of specific types of foods. Because they are one of the hardest structures of the body, an animal's teeth remain in the environment long after it dies. Paleontologists use teeth to identify animals that lived in the distant past and to deduce what their feeding behavior might have been.

All mammalian teeth have a general structure consisting of three layers (Figure 50.9*a*). An extremely hard material called **enamel**, composed principally of calcium phosphate, covers the crown of the tooth. Both the crown and the root contain a layer of bony material called **dentine**, inside of which is a **pulp cavity** containing blood vessels, nerves, and the cells that produce the dentine.

The shapes and organization of mammalian teeth, however, can be very different, since they are adapted to specific diets (Figure 50.9*b*). In general, incisors are used for cutting, chopping, or gnawing; canines are used for stabbing, ripping, and shredding; and molars and premolars (the cheek teeth) are used for shearing, crushing, and grinding. The highly varied diet of humans is reflected by our multipurpose set of teeth, as is common among omnivores.

Digestion

Most animals digest their food *extracellularly*. Animals take food into a body cavity that is continuous with the outside environment, into which they secrete digestive enzymes. The enzymes act on the food, reducing it to nutrient molecules that can be absorbed by the cells lining the cavity. Only after they are absorbed by the cells are the nutrients *within* the body of the animal.

The simplest digestive systems are gastrovascular cavities that connect to the outside world through a single opening. An example is the cnidarians, which capture prey using stinging nematocysts and cram it into their gastrovascular cavitues with tentacles (see Figures 31.7 and 49.1*a*). Enzymes in the gastrovascular cavity partly digest the prey. Cells lining the cavity take in small food particles by endocytosis. The vesicles that are created by endocytosis fuse with lysosomes containing digestive enzymes, and intracellular digestion completes the breakdown of the food. Nutrients are released to the cytoplasm as the vesicle breaks down.

Tubular guts have an opening at each end

The guts of most animals are *tubular*. A mouth takes in food; molecules are digested and absorbed throughout the length of the gut; and solid digestive wastes are excreted through an anus. Different regions in the tubular gut are specialized for particular functions (Figure 50.10). These functions must be coordinated so that they occur in proper sequence and at appropriate rates to maximize the efficiency of digestion and absorption of nutrients. Keep in mind as we discuss these regions that all locations within the tubular gut are really *outside* the body of the animal. Only by crossing the plasma membranes of cells lining the gut do nutrients enter the body.

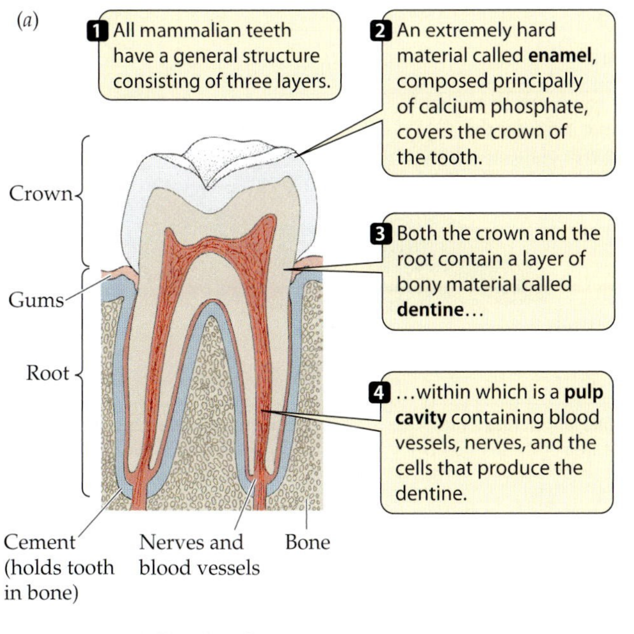

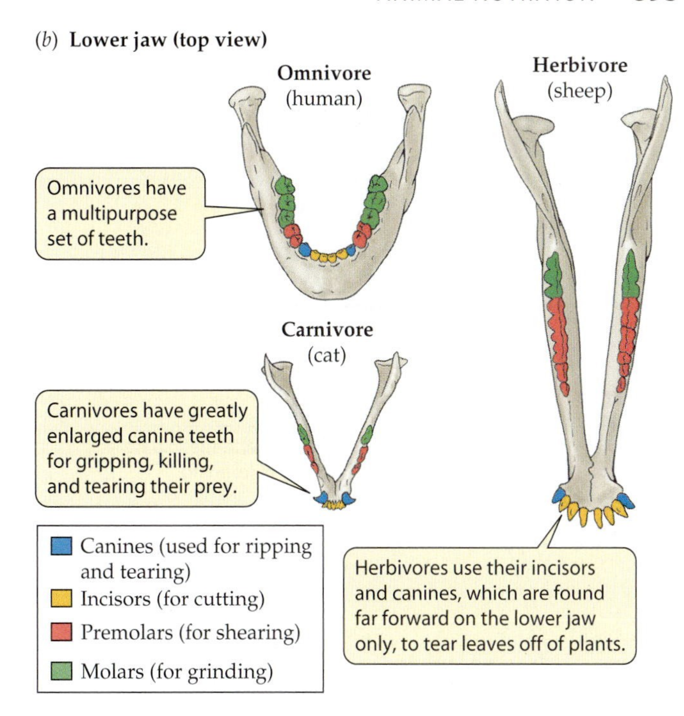

50.9 Mammalian Teeth
(*a*) A mammalian tooth has three layers: enamel, dentine, and pulp cavity. (*b*) The teeth of different mammalian species are specialized for different diets.

At the anterior end of the gut are the **mouth** (the opening itself) and **buccal cavity** (mouth cavity). Food may be broken up by teeth (in some vertebrates), by the radula (in snails), or by mandibles (in insects), or somewhat farther along the gut by structures such as the *gizzards* of birds and earthworms, where muscular contractions of the gut grind the food together with small stones. Some animals, such as snakes, simply ingest large chunks of food, with little or no fragmentation.

Stomachs and **crops** are storage chambers that enable animals to ingest relatively large amounts of food and digest it at leisure. In these storage chambers, food may be further fragmented and mixed, but digestion may or may not occur there, depending on the species. In any case, food delivered into the next section of gut, the **midgut** or **intestine**, is well minced and well mixed.

Most materials are digested and absorbed in the midgut. Specialized glands secrete some digestive enzymes into the intestine, and the gut wall itself secretes other digestive enzymes. The **hindgut** recovers water and ions and stores undigested wastes, or **feces**, so that they can be released to the environment at an appropriate time or place. A muscular **rectum** near the anus assists in the expulsion of feces, the process of **defecation**.

Within the hindguts of many species are colonies of endosymbiotic bacteria. These bacteria obtain their own nutrition from the food passing through the host's gut while contributing to the digestive processes of the host. Members of the leech genus *Hirudo*, for example, produce no enzymes that can digest the proteins in the blood they suck from vertebrates. A colony of gut bacteria produces the enzymes necessary to break down those proteins into amino acids, which are subsequently used by both the leech and

50.10 Compartments for Digestion and Absorption
Most animals have tubular guts that begin with a mouth that takes in food and end in an anus that excretes wastes. Between these two structures are specialized regions for digestion and nutrient absorption; these structures vary from species to species.

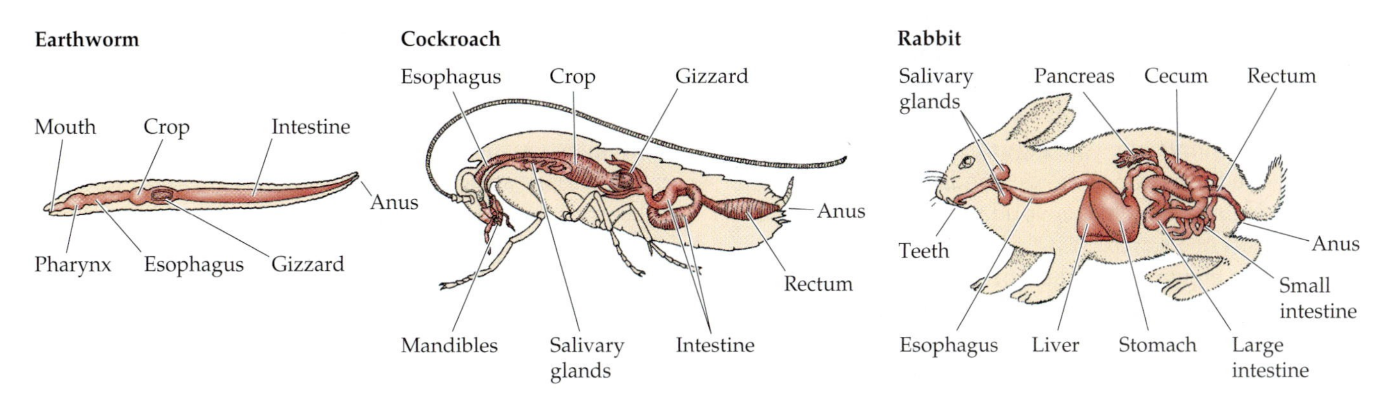

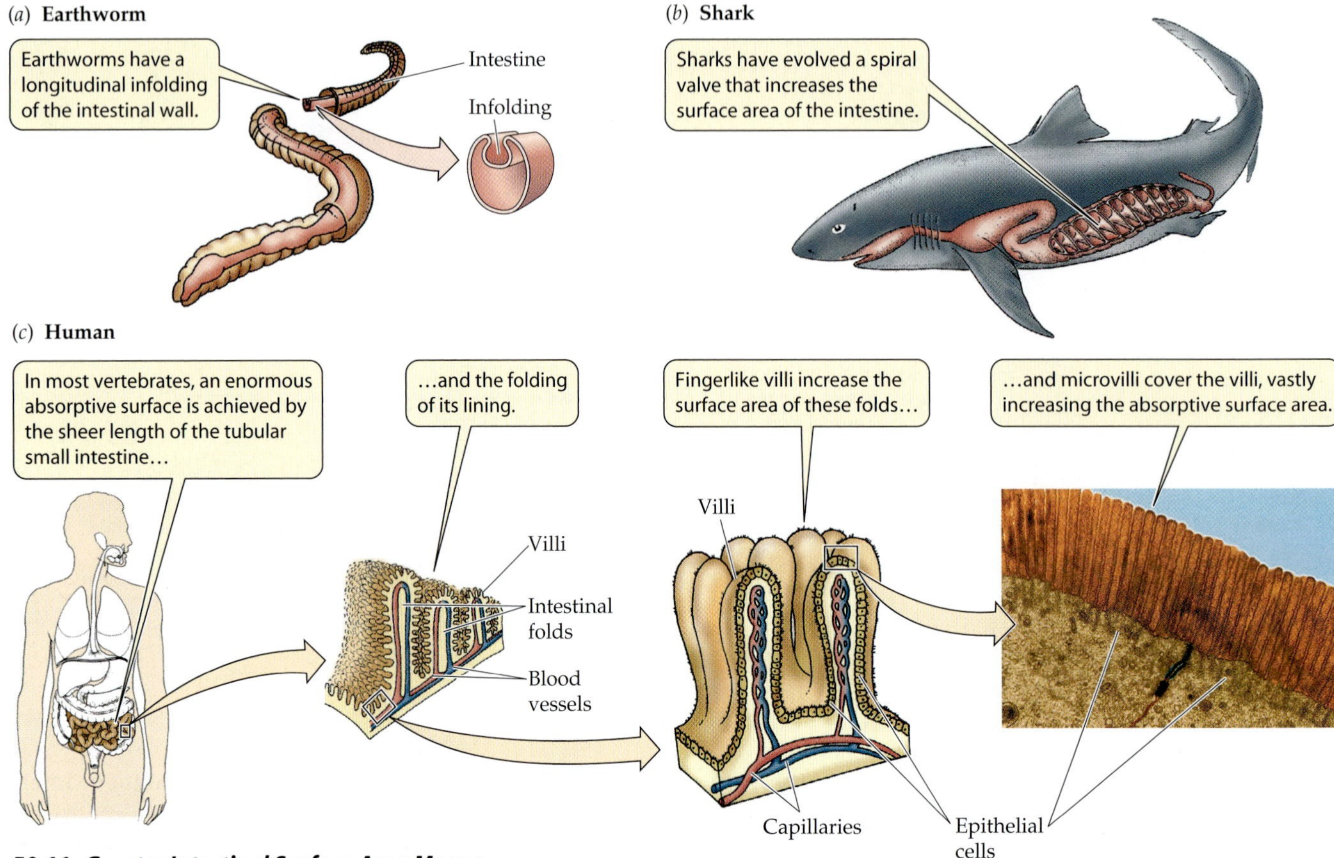

50.11 Greater Intestinal Surface Area Means More Nutrient Absorption
The guts of most animals have evolved to maximize their surface area.

the bacteria. Some animals rely on microorganisms in their guts to supply them with vitamins.

In many animals, the parts of the gut that absorb nutrients have evolved extensive surface areas (Figure 50.11*a, b*). In vertebrates, the wall of the gut is richly folded, with the individual folds bearing legions of tiny fingerlike projections called **villi** (Figure 50.11*c*). The cells that line the surfaces of the villi, in turn, have microscopic projections, called microvilli. The microvilli present an enormous surface area for the absorption of nutrients.

Digestive enzymes break down complex food molecules

Protein, carbohydrate, and fat macromolecules are broken down into their simplest monomeric units by hydrolytic enzymes. All of these enzymes cleave the chemical bonds of macromolecules through hydrolysis, a reaction that adds a water molecule (see Figure 3.2). Examples of hydrolysis are the breaking of the bonds between adjacent amino acids of a protein or peptide and between adjacent glucose units of a starch.

Digestive enzymes are classified according to the substances they hydrolyze: carbohydrases hydrolyze carbohydrates; proteases, proteins; peptidases, peptides; lipases, fats; and nucleases, nucleic acids. The prefixes *exo-* ("outside") and *endo-* ("within") indicate where the enzyme cleaves the molecule. An endoprotease hydrolyzes a protein at an internal site along the polypeptide chain, and an exoprotease snips away amino acids at the ends of the molecule.

How can an organism produce enzymes that hydrolyze biological macromolecules without digesting itself? Most digestive enzymes are produced in an inactive form, known as a **zymogen**, so that they cannot act on the cells that produce them. When secreted into the gut, zymogens are activated by another enzyme or by conditions in the gut (which, as you will remember, is outside the body). The lining of the gut is not digested because it is protected by a covering of mucus.

Structure and Function of the Vertebrate Gut

The digestive tract of vertebrates is a tubular gut that runs from mouth to anus (Figure 50.12). Different segments of the gut are specialized for different functions associated with digestion and absorption. In addition, there are several accessory structures that produce and export into the gut compounds that contribute to the digestive process.

Similar tissue layers are found in all regions of the vertebrate gut

The cellular architecture of the vertebrate gut follows a common plan throughout. Four major layers of different

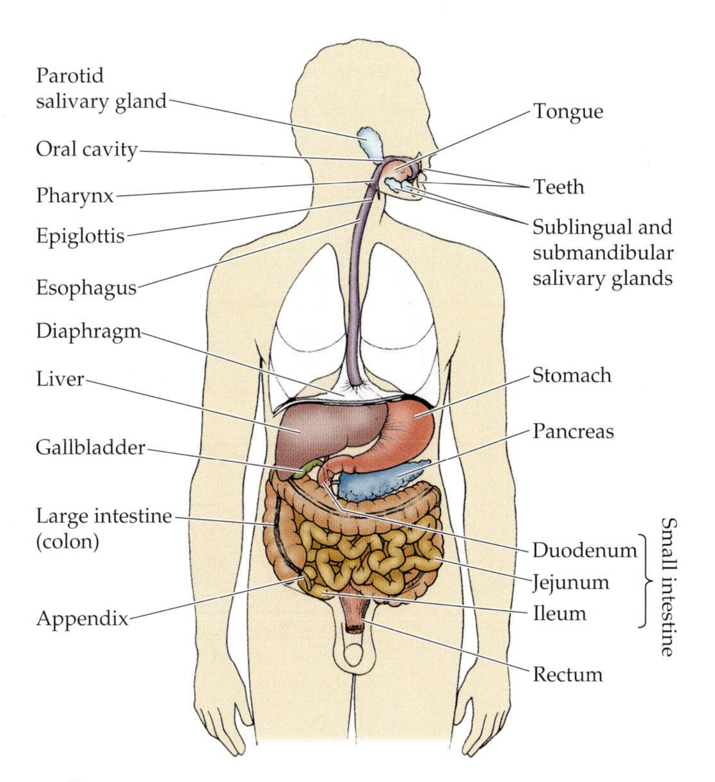

50.12 The Human Digestive System
Different compartments within the long tubular gut specialize in digesting food, absorbing nutrients, and storing and expelling wastes. Accessory organs contribute digestive juices containing enzymes and other molecules.

cell types form the wall of the gut (Figure 50.13). These layers differ somewhat from compartment to compartment, but they are always present.

Starting in the cavity, or **lumen**, of the gut, the first tissue layer is the **mucosa**. Mucosal cells have secretory and absorptive functions. Some secrete mucus, which lubricates and protects the walls of the gut. Others secrete digestive enzymes, and still others in the stomach secrete hydrochloric acid (HCl). In some regions of the gut, nutrients are absorbed across the plasma membranes of the mucosal cells; the plasma membranes of these absorptive cells have many folds that increase their surface area (see Figure 50.11*c*)

At the base of the mucosa are some smooth muscle cells, and just outside the mucosa is the second layer of cells, the **submucosa**. Here we find the blood and lymph vessels that carry absorbed nutrients to the rest of the body. The submucosa also contains a network of nerves; these neurons are both sensory (responsible for stomach aches) and regulatory (controlling the various secretory functions of the gut).

External to the submucosa are two layers of smooth muscle cells responsible for the movements of the gut. Innermost is the **circular muscle layer** with its cells oriented *around* the gut. Outermost is the **longitudinal muscle layer** with its cells oriented along the length of the gut. The circular muscles constrict the gut, and the longitudinal muscles shorten the gut. Between the two layers of muscle is another network of nerves that controls the movements of the gut, coordinating the movements of different regions with one another.

Surrounding the gut is a fibrous coat called the **serosa**. Like other abdominal organs, the gut is also covered and supported by a tissue called the **peritoneum**.

Peristalsis moves food through the gut

Food entering the mouth of most vertebrates is chewed and mixed with the secretions of salivary glands. A muscular tongue then pushes chewed food toward the back of the buccal cavity. By making contact with the soft tissue at the back of the mouth, the bolus of food initiates a complex series of autonomic reflex actions known as *swallowing*. Stand in front of a mirror and gently touch this tissue at the back of your mouth with a cotton swab. You may gag slightly, but you will also experience an uncontrollable urge to swallow. Swallowing involves many muscles doing a variety of jobs that propel the food through the **pharynx** (where the mouth cavity and the nasal passages join) and into the **esophagus** (the food tube). A structure called the *epiglottis*

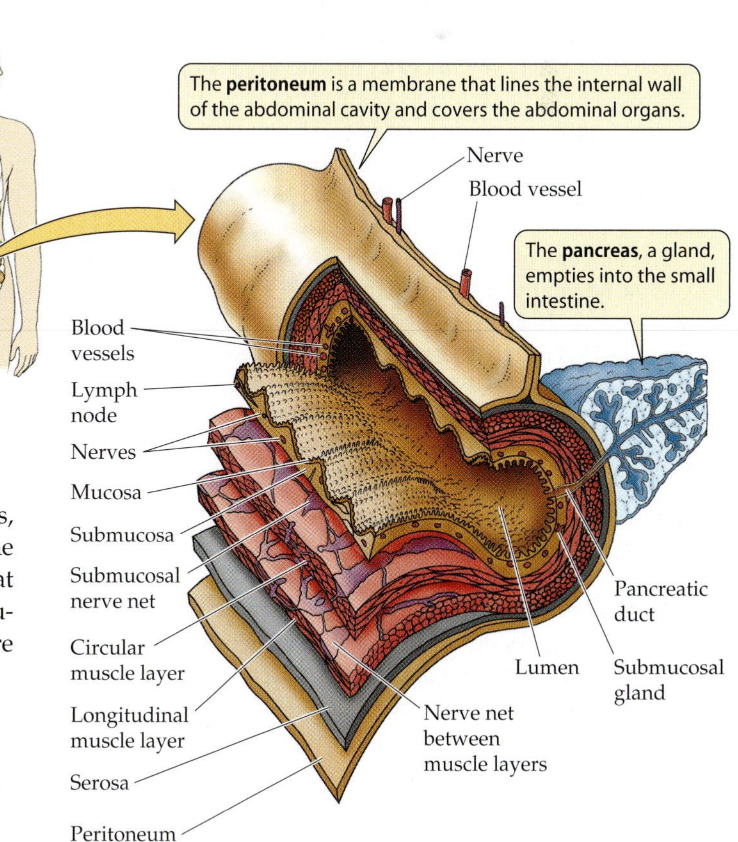

50.13 Tissue Layers of the Vertebrate Gut
In all compartments of the gut, the organization of the tissue layers is the same, but specialized adaptations of specific tissues characterize different regions.

(*a*) **Swallowing**

Brain stem reflex center and nerves controlling swallowing
Food
Tongue
Pharynx
Trachea (windpipe)
Esophagus
Soft palate
Larynx closes
Epiglottis

(*b*) **Peristalsis**

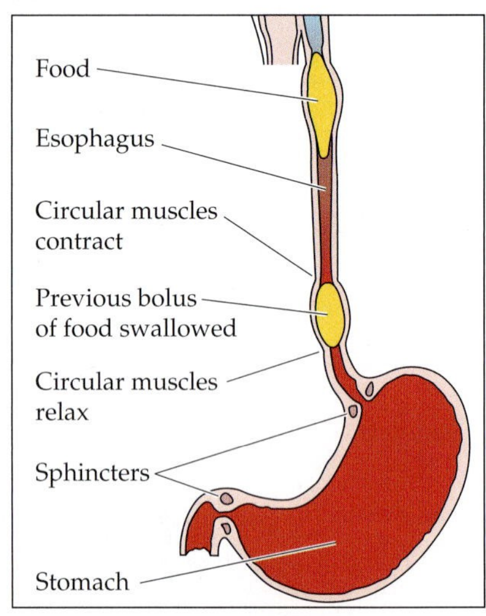

1 Food is chewed and the tongue pushes the bolus of food to the back of the mouth. Sensory nerves initiate the swallowing reflex.

2 The soft palate is pulled up as the vocal cords close the larynx.

3 The larynx is pulled up and forward and is covered by the epiglottis; the bolus of food enters the esophagus.

4 Peristaltic contractions propel the food to the stomach.

50.14 Swallowing and Peristalsis
Food pushed to the back of the mouth triggers the swallowing reflex. Once food enters the esophagus, peristalsis propels it through the gut.

prevents the food from entering the trachea (windpipe) or nasal passages (Figure 50.14).

Once the food is in the esophagus, peristalsis takes over and pushes it toward the stomach. **Peristalsis** is a wave of smooth muscle contraction that moves progressively down the gut from the pharynx toward the anus. The smooth muscle of the gut contracts in response to being stretched. Swallowing a bolus of food stretches the upper end of the esophagus, and this stretching initiates a wave of contraction that slowly pushes the contents of the gut toward the anus.

The movement of food from the stomach into the esophagus is normally prevented by a thick ring of circular smooth muscle at the junction of the esophagus and the stomach. This ring of muscle, the *esophageal sphincter*, is normally constricted. Waves of peristalsis cause it to relax enough to let food pass from the esophagus into the stomach. Sphincter muscles are found elsewhere in the digestive tract as well. The *pyloric sphincter* governs the passage of stomach contents into the intestine. Another important sphincter surrounds the anus.

Digestion begins in the mouth and the stomach

Food is chewed in the mouth, and carbohydrate digestion begins there. The enzyme **amylase** is secreted with saliva and mixed with the food as it is chewed. Amylase hydrolyzes the bonds between the glucose monomers that make up starch molecules. The action of amylase is what makes a piece of bread or cracker taste sweet if you hold it in your mouth long enough.

Most vertebrates can rapidly consume a large volume of food, but digesting that food is a long, slow process. The stomach stores the food consumed during a meal. The secretions of the stomach kill microorganisms that are taken in with the food and begin the digestion of proteins.

The major enzyme produced by the stomach is an endopeptidase called **pepsin**. Pepsin is secreted as a zymogen called **pepsinogen** by cells in the **gastric glands**—deep folds in the stomach lining (Figure 50.15). Other cells in the gastric glands produce hydrochloric acid, and still others near the openings of the gastric glands and throughout the stomach mucosa secrete mucus.

Hydrochloric acid (HCl) maintains the stomach fluid (the gastric juice) at a pH between 1 and 3. This low pH activates the conversion of pepsinogen to pepsin, which is achieved by the cleavage of a masking sequence of 44 amino acids from the N-terminal end of the pepsinogen molecule. The conversion is amplified as the newly formed pepsin activates other pepsinogen molecules, a process called **autocatalysis**. Hydrochloric acid also provides the right pH for the enzymatic action of pepsin. The low pH also helps dissolve the intercellular substances holding the ingested tissues together. Breakdown of the ingested tissues exposes more food surface area to the action of pepsin and eventually other digestive enzymes in the small intestine.

Mucus secreted by the stomach mucosa coats the walls of the stomach and protects them from being eroded and digested by HCl and pepsin. Sometimes, however, the walls of the stomach are exposed to HCl and pepsin; the resulting damage is called an *ulcer*. It was previously thought that ulcers were mostly due to stress and oversecretion of digestive juices. In recent years, however, it has been discovered that the basis for most ulcers is an infectious bacterium called *Helicobacter pylori*, which has the remarkable

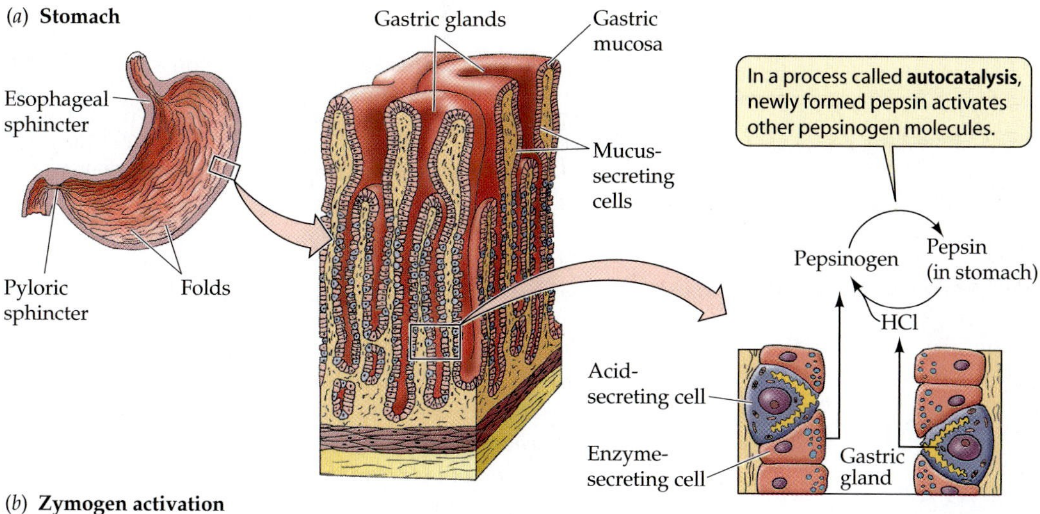

50.15 The Stomach
(*a*) The human stomach stores and breaks down ingested food. (*b*) Cells in the gastric glands secrete hydrochloric acid and the proteolytic enzyme pepsin. Both the glands and gastric mucosa secrete mucus that protects the stomach. (*c*) Pepsin is secreted as an inactive zymogen, pepsinogen, that is activated by low pH through the cleavage of a masking sequence of amino acids. Active pepsin also activates pepsinogen through autocatalysis.

ability to live in the highly acidic environment of the stomach. Lesions started by the bacterial infection are made worse by HCl and pepsin.

Contractions of the muscles in the walls of the stomach churn its contents, thoroughly mixing them with the stomach secretions. The acidic, fluid mixture of gastric juice and partly digested food in the stomach is called **chyme**. A few substances can be absorbed from the chyme across the stomach wall, including alcohol (hence its rapid effects), aspirin, and caffeine, but even these substances are absorbed in rather small quantities in the stomach.

Peristaltic contractions of the stomach walls push the chyme toward the bottom end of the stomach. These waves of peristalsis cause the pyloric sphincter to relax briefly so that little squirts of the chyme can enter the first region of the intestine. The human stomach empties itself gradually over a period of approximately 4 hours. This slow passage of food enables the intestine to work on a little material at a time and extends the digestive and absorptive processes throughout much of the time between meals.

The small intestine is the major site of digestion

In the **small intestine**, the digestion of carbohydrates and proteins continues, and the digestion of fats begins and the absorption of nutrients begins. Although the small intestine takes its name from its diameter, it is a very large organ. The small intestine of an adult human is more than 6 m long; its coils fill much of the lower abdominal cavity (see Figure 50.12). Because of its length, and because of the folds, villi, and microvilli of its lining, its inner surface area is enormous: about 550 m^2, or roughly the size of a tennis court. Across this surface the small intestine absorbs all the nutrient molecules derived from food. The small intestine has three sections. The initial section—the **duodenum**—is the site of most digestion; the **jejunum** and the **ileum** carry out 90 percent of the absorption of nutrients.

Digestion requires many specialized enzymes, as well as several other secretions. Two accessory organs that are not part of the digestive tract—the liver and the pancreas—provide many of these enzymes and secretions.

The liver synthesizes a substance called **bile** from cholesterol. Bile secreted from the liver flows through the **hepatic duct** to the gallbladder and to the duodenum. Bile reaches the gallbladder through a side branch of the hepatic duct (Figure 50.16). It is stored in the gallbladder until it is needed to assist in fat digestion. When fat enters the duodenum, a hormonal signal causes the walls of the gallbladder to contract rhythmically, squeezing bile back out toward the hepatic duct. Below the branch point to the gallbladder, the hepatic duct is called the **common bile duct**. Bile from the gallbladder flows down the common bile duct to the duodenum.

To understand the role of bile in fat digestion, think of the oil in salad dressing: it is not soluble in water (it is hydrophobic), and it tends to aggregate together in large globules. The enzymes that digest fat, the **lipases**, are water-soluble and must do their work in an aqueous medium. Bile stabilizes tiny droplets of fat so that they cannot aggregate into large globules. One end of each bile molecule is soluble in fat (it is lipophilic, or hydrophobic); the other end is soluble in water (it is hydrophilic, or lipophobic). Bile molecules bury their lipophilic ends in fat droplets, leaving their

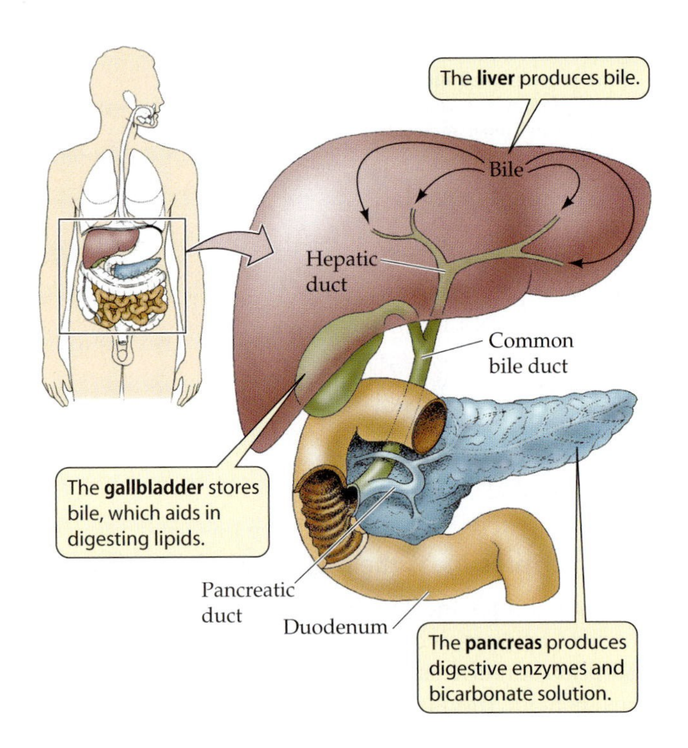

50.16 The Ducts of the Gallbladder and Pancreas
Bile produced in the liver leaves the liver via the hepatic duct. Branching off this duct is the gallbladder, which stores bile. Below the gallbladder, the hepatic duct is called the common bile duct and is joined by the pancreatic duct before entering the duodenum.

lipophobic ends sticking out. As a result, they prevent the fat droplets from sticking together. These very small fat particles are called **micelles**, and their small size maximizes the surface area exposed to lipase action (see Figure 50.17).

The **pancreas** is a large gland that lies just beneath the stomach (see Figures 50.12 and 50.16). It functions as both an endocrine (secreting hormones without ducts to the blood and tissue fluid) and an exocrine (secreting other substances through ducts to the outside of the body) gland. Here we will consider its exocrine products, which are delivered to the gut through the pancreatic duct. The pancreatic duct joins the common bile duct just before it enters the duodenum.

The pancreas produces a host of digestive enzymes, including lipases (Table 50.3). As in the stomach, these enzymes are released as zymogens; otherwise they would digest the pancreas and its ducts before they ever reached the duodenum. Once in the duodenum, one of these inactive enzymes, **trypsinogen**, is activated by *enterokinase*, which is produced by cells lining the duodenum. This process is similar to the activation of pepsinogen by low pH (see Figure 50.15). Active **trypsin** can cleave other trypsinogen molecules to release even more active trypsin (another example of autocatalysis). Similarly, trypsin acts on the other zymogens secreted by the pancreas and releases their active enzymes.

The mixture of zymogens produced by the pancreas can be very dangerous if the pancreatic duct is blocked or if the pancreas is injured by an infection or a severe blow to the abdomen. A few trypsinogen molecules spontaneously converting to trypsin can initiate a chain reaction of enzyme activation that digests the pancreas in a very short period of time, destroying both its endocrine and exocrine functions.

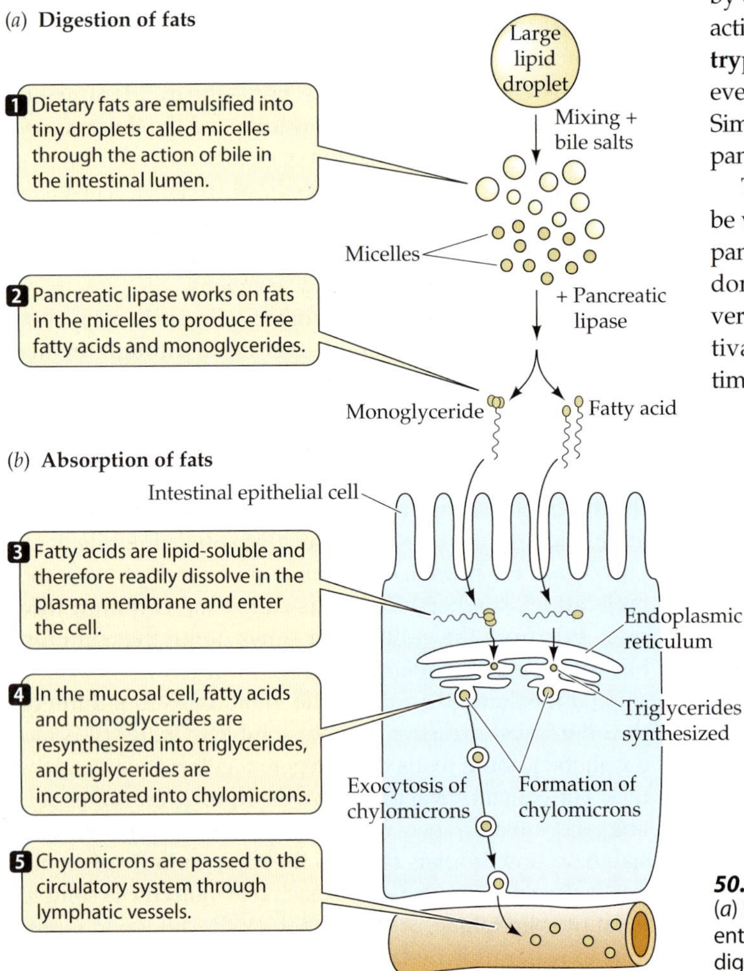

The pancreas produces, in addition to digestive enzymes, a secretion rich in bicarbonate ions (HCO_3^-). Bicarbonate ions neutralize the pH of the chyme that enters the duodenum from the stomach. This neutralization is essential because intestinal enzymes function best at a neutral or slightly alkaline pH.

Nutrients are absorbed in the small intestine

Only the smallest products of digestion can be absorbed through the mucosa of the small intestine and passed on to the blood and lymphatic vessels that lie in the submucosa. The final digestion of pro-

50.17 The Digestion and Absorption of Fats
(*a*) Dietary fats are broken up by bile into small micelles that present a large surface area to lipase action. (*b*) The products of fat digestion are absorbed by intestinal mucosal cells, where they are resynthesized into triglycerides and exported to lymphatic vessels.

50.3 **Sources and Functions of the Major Digestive Enzymes of Humans**

ENZYME	SOURCE	ACTION	SITE OF ACTION
Salivary amylase	Salivary glands	Starch → Maltose	Mouth
Pepsin	Stomach	Proteins → Peptides; autocatalysis	Stomach
Pancreatic amylase	Pancreas	Starch → Maltose	Small intestine
Lipase	Pancreas	Fats → Fatty acids and glycerol	Small intestine
Nuclease	Pancreas	Nucleic acids → Nucleotides	Small intestine
Trypsin	Pancreas	Proteins → Peptides; activation of zymogens	Small intestine
Chymotrypsin	Pancreas	Proteins → Peptides	Small intestine
Carboxypeptidase	Pancreas	Peptides → Peptides and amino acids	Small intestine
Aminopeptidase	Small intestine	Peptides → Peptides and amino acids	Small intestine
Dipeptidase	Small intestine	Dipeptides → Amino acids	Small intestine
Enterokinase	Small intestine	Trypsinogen → Trypsin	Small intestine
Nuclease	Small intestine	Nucleic acids → Nucleotides	Small intestine
Maltase	Small intestine	Maltose → Glucose	Small intestine
Lactase	Small intestine	Lactose → Galactose and glucose	Small intestine
Sucrase	Small intestine	Sucrose → Fructose and glucose	Small intestine

teins and carbohydrates that produces these absorbable products takes place among the microvilli. The mucosal cells with microvilli produce peptidases, which cleave larger peptides into tripeptides, dipeptides, and individual amino acids that the cells can absorb. These cells also produce the enzymes maltase, lactase, and sucrase, which cleave the common disaccharides into their constituent, absorbable monosaccharides—glucose, galactose, and fructose.

Many humans stop producing the enzyme *lactase* around the age of 4 years and thereafter have difficulty digesting lactose, which is the sugar in milk. Lactose is a disaccharide and cannot be absorbed without being cleaved into its constituent units, glucose and galactose. If a substantial amount of lactose remains unabsorbed and passes into the large intestine, its metabolism by bacteria in the large intestine causes abdominal cramps, gas, and diarrhea.

The mechanisms by which the cells lining the intestine absorb nutrient molecules and inorganic ions are diverse and not completely understood. Many inorganic ions are actively transported into the cells. Carrier proteins exist for sodium, calcium, and iron. Carriers also exist for certain classes of amino acids and for glucose and galactose, but their activity is much reduced if active sodium transport is blocked.

Sodium diffuses from the gut contents into the mucosal cells and is then actively transported from the mucosal cells into the submucosa. To diffuse into a mucosal cell, a sodium ion binds to a symport in the mucosal cell membrane. The symport also binds a nutrient molecule such as glucose or an amino acid. The diffusion of the sodium ion, driven by a concentration gradient, therefore drives the absorption of the nutrient molecule. This mechanism is called *sodium cotransport*.

The absorption of the products of fat digestion does not involve carrier proteins (Figure 50.17). Lipases break down fats into fatty acids and monoglycerides, which are lipid-soluble and are thus able to pass through the plasma membranes of the microvilli and diffuse into the mucosal cells. Once in the cells, the fatty acids and monoglycerides are resynthesized into triglycerides, combined with cholesterol and phospholipids, and coated with protein to form water-soluble **chylomicrons**, which are really little particles of fat. Rather than entering the blood directly, the chylomicrons pass into the lymphatic vessels in the submucosa. They then flow through the lymphatic system and enter the bloodstream through the thoracic duct. After a meal rich in fats, the chylomicrons can be so abundant in the blood that they give it a milky appearance.

The bile that emulsifies the fats is not absorbed along with the monoglycerides and the fatty acids, but shuttles back and forth between the gut contents and the microvilli. In the ileum, bile is actively resorbed and returned to the liver via the bloodstream. As noted earlier, bile is synthesized in the liver from cholesterol. Cholesterol comes from food, but it is also synthesized by liver cells and gut cells.

As we learned in Chapter 49, high cholesterol levels contribute to arterial plaque formation and therefore to cardiovascular disease. The body has no way of breaking down excess cholesterol, so high dietary intake or high levels of synthesis create problems. One major way that cholesterol leaves the body is through the elimination of unresorbed bile in the feces. The rationale for including certain kinds of fiber in the diet is that the fiber binds bile, decreases its resorption in the ileum, and thus helps to lower body cholesterol levels.

Water and ions are absorbed in the large intestine

Peristalsis gradually pushes the contents of the small intestine into the large intestine, or **colon**. The rate of peristalsis is controlled so that food passes through the small intestine slowly enough for digestion and absorption to be complete, but quickly enough to ensure an adequate supply of nutrients for the body. Most of the available nutrients have been removed from the material that enters the colon, but the material contains a lot of water and inorganic ions.

The colon absorbs water and ions, producing semisolid feces from the slurry of indigestible materials it receives from the small intestine. Absorption of too much water in the colon can cause constipation. The opposite condition, diarrhea, results if too little water is absorbed or if water is secreted into the colon. (Both constipation and diarrhea can be induced by toxins from certain microorganisms.) Feces are stored in the last segment of the colon and are periodically excreted.

Immense populations of bacteria live within the colon. One of the resident species is *Escherichia coli*, the bacterium that is so popular among researchers in biochemistry, genetics, and molecular biology. This inhabitant of the colon lives on matter indigestible to humans and produces some products useful to its host. Vitamin K and biotin, for example, are synthesized by *E. coli* and absorbed across the wall of the colon. Excessive or prolonged intake of antibiotics can lead to vitamin deficiency because the antibiotics kill the normal intestinal bacteria at the same time they are killing the disease-causing organisms for which they are intended. The intestinal bacteria produce gases such as methane and hydrogen sulfide as by-products of their largely anaerobic metabolism. Humans expel gas after eating beans because the beans contain certain carbohydrates that bacteria—but not humans—can break down.

The large intestine of humans has a small, fingerlike pouch called the **appendix**, which is best known for the trouble it causes when it becomes infected. The human appendix plays no essential role in digestion, but it does contribute to immune system function. It can be surgically removed without serious consequences. The part of the gut that forms the appendix in humans forms the much larger cecum in herbivores (see Figure 50.10), where it functions, as we will see, in cellulose digestion. As our primate ancestors evolved to exploit diets less rich in indigestible cellulose, the cecum no longer served an essential function and gradually became *vestigial* (reduced to a trace).

Herbivores have special adaptations to digest cellulose

Cellulose is the principal organic compound in the diets of herbivores. Most herbivores, however, cannot produce **cellulases**, the enzymes that hydrolyze cellulose. Exceptions include silverfish (well known for eating books and stored papers), earthworms, and shipworms. Other herbivores, from termites to cattle, rely on microorganisms living in their digestive tracts to digest cellulose for them.

The digestive tracts of **ruminants** (cud chewers) such as cattle, goats, and sheep are specialized to maximize the benefits of their endosymbiotic microorganisms. In place of the usual mammalian stomach, ruminants have a large, four-chambered organ (Figure 50.18). The first two chambers, the *rumen* and the *reticulum*, are packed with anaerobic microorganisms that break down cellulose. The ruminant periodically regurgitates the contents of the rumen (the cud) into the mouth for rechewing. When the more thoroughly ground-up vegetable fibers are swallowed again, they present more surface area to the microorganisms for their digestive actions.

The microorganisms in the rumen and reticulum metabolize cellulose and other nutrients to simple fatty acids, which become nutrients for the host. In addition, the microorganisms themselves provide an important source of protein for the host. The plant materials ingested by a ruminant are a poor source of protein, but they contain inorganic nitrogen that the microorganisms use to synthesize their own amino acids. A cow can derive more than 100 g of protein per day from digestion of its endosymbiotic microorganisms.

Carbon dioxide and methane are by-products of the fermentation of cellulose carried out by microorganisms. A single cow can produce and belch 400 liters of methane a day. Methane is the second most abundant of the "greenhouse gases" whose concentration in the atmosphere is increasing,

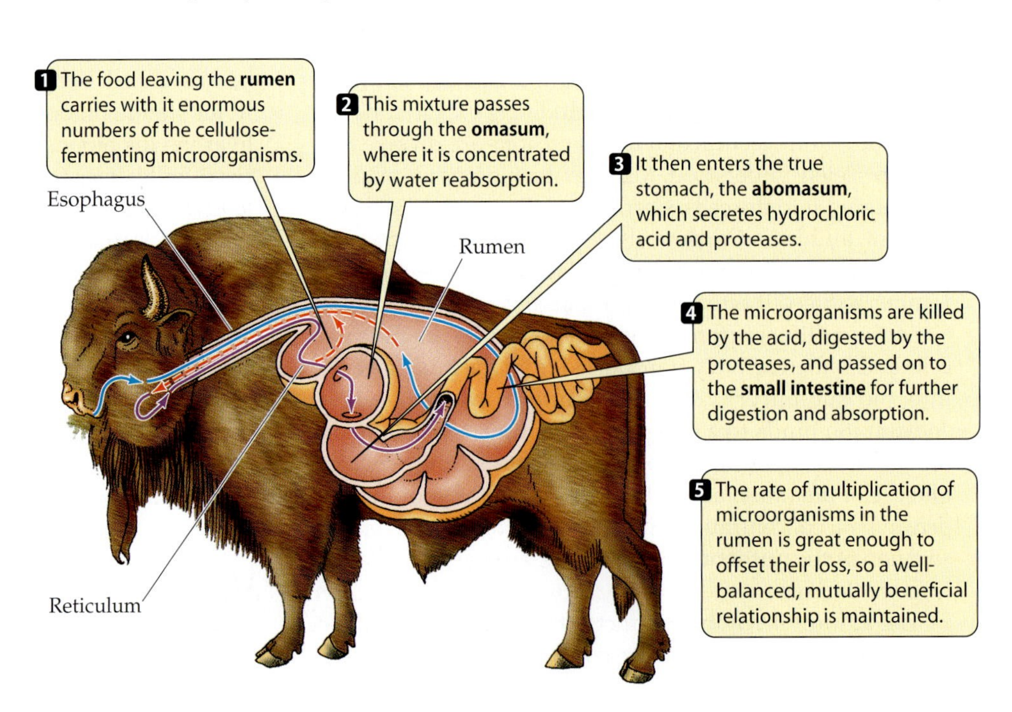

50.18 The Ruminant Stomach
Four specialized stomach compartments enable ruminants to digest and subsist on protein-poor plant material.

and domesticated ruminants are second only to industry as a source of methane emitted into the atmosphere.

The food leaving the rumen carries with it enormous numbers of the cellulose-fermenting microorganisms. This mixture passes through the *omasum*, where it is concentrated by water absorption. It then enters the true stomach, the *abomasum*, which secretes hydrochloric acid and proteases. The microorganisms are killed by the acid, digested by the proteases, and passed on to the small intestine for further digestion and absorption. The rate of multiplication of microorganisms in the rumen is great enough to offset their loss, so a well-balanced, mutually beneficial relationship is maintained.

Some mammalian herbivores other than ruminants have microbial farms and cellulose fermentation vats in a branch off the large intestine called the **cecum**. Rabbits and hares are good examples. Since the cecum empties into the large intestine, the absorption of the nutrients produced by the microorganisms is inefficient and incomplete. Therefore, some of these animals reingest some of their own feces, a behavior known as **coprophagy**. Coprophagous species usually produce two kinds of feces, one consisting of pure waste (which they discard), and one consisting mostly of cecal material, which they reingest directly from the anus. As this cecal material passes through the stomach and small intestine, the nutrients it contains are digested and absorbed.

Control and Regulation of Digestion

The vertebrate gut is an assembly line in reverse—a disassembly line. As with a standard assembly line, control and coordination of sequential processes is critical. Both neural and hormonal controls govern gut functions.

Autonomic reflexes coordinate functions in different regions of the gut

Everyone has experienced salivation stimulated by the sight or smell of food. That response is an autonomic reflex, as is the act of swallowing following tactile stimulation at the back of the mouth. Many autonomic reflexes coordinate activities in different regions of the digestive tract. Loading the stomach with food, for example, stimulates increased activity in the colon, which can lead to a bowel movement.

The digestive tract is unusual in that it has an intrinsic (that is, its own) nervous system. In addition to autonomic reflexes involving the CNS, such as salivation and swallowing, neural messages can travel from one region of the digestive tract to another without being processed by the CNS.

Hormones control many digestive functions

Several hormones control the activities of the digestive tract and its accessory organs (Figure 50.19). The first hormone ever discovered came from the duodenum; it was called **secretin** because it caused the pancreas to secrete digestive juices. We now know that secretin is one of several hormones that control pancreatic secretion; specifically, secretin stimulates the pancreas to secrete a solution rich in bicarbonate ions.

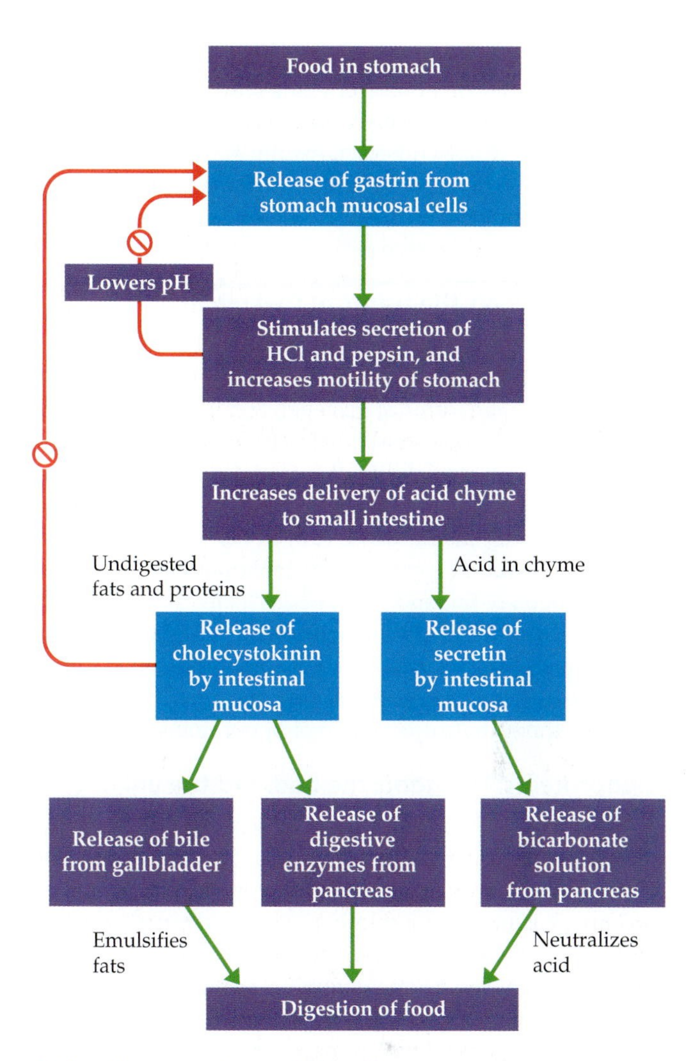

50.19 Hormones Control Digestion
Several hormones are involved in feedback loops that control the sequential processing of food in the digestive tract.

In response to the presence of fats and proteins in the chyme, the mucosa of the small intestine secretes **cholecystokinin**, a hormone that stimulates the gallbladder to release bile and the pancreas to release digestive enzymes. Cholecystokinin and secretin also slow the movements of the stomach, which slows the delivery of chyme into the small intestine.

The stomach secretes a hormone called **gastrin** into the blood. Cells in the lower region of the stomach release gastrin when they are stimulated by the presence of food. Gastrin circulates in the blood until it reaches cells in the upper areas of the stomach wall, where it stimulates the secretions and movements of the stomach. Gastrin release is inhibited when the stomach contents become too acidic—another example of negative feedback.

Control and Regulation of Fuel Metabolism

Most animals do not eat continuously. When they do eat, food is present in the gut and nutrients are being absorbed for some period of time after the meal, called the **absorp-**

tive period. Once the stomach and small intestine are empty, nutrients are no longer being absorbed. During this **postabsorptive period**, the continuous processes of energy metabolism and biosynthesis must run on internal reserves. Nutrient traffic must be controlled so that reserves accumulate during the absorptive period are used appropriately during the postabsorptive period.

The liver directs the traffic of fuel molecules

The liver directs the traffic of the nutrients that fuel metabolism. When nutrients are abundant in the blood, the liver stores them in the forms of glycogen and fats. The liver also synthesizes blood plasma proteins from circulating amino acids. When the availability of fuel molecules in the blood declines, the liver delivers glucose and fats back to the blood.

The liver has an enormous capacity to interconvert fuel molecules. Liver cells can convert monosaccharides into either glycogen or fat, and vice versa. The liver can also convert certain amino acids and some other molecules, such as pyruvate and lactate, into glucose—a process called **gluconeogenesis**. The liver is also the major controller of fat metabolism through its production of lipoproteins.

Lipoproteins: The good, the bad, and the ugly

In the intestine, bile solves the problem of processing hydrophobic fats in an aqueous medium. The transportation of fats in the circulatory system presents the same problem, and lipoproteins are the solution. A **lipoprotein** is a particle made up of a core of fat and cholesterol and a covering of protein that makes it water-soluble. The largest lipoprotein particles are the chylomicrons produced by the mucosal cells of the intestine, which transport dietary fat and cholesterol into the circulation (see Figure 50.17). As the chylomicrons circulate through the liver and to adipose (fat) tissues throughout the body, receptors on the capillary walls recognize their protein coats, and lipases begin to hydrolyze the fats, which are then absorbed into liver or fat cells. Thus the protein coat of the lipoprotein both makes it water-soluble and serves as an "address" that directs it to a specific tissue.

Lipoproteins other than chylomicrons originate in the liver and are classified according to their density. Fat has a low density (it floats in water), so the more fat a lipoprotein contains, the lower its density.

- **Very-low-density lipoproteins** (**VLDL**) produced by the liver contain mostly triglyceride fats that are being transported to fat cells in tissues around the body.
- **Low-density lipoproteins** (**LDL**) consist of about 50 to 60 percent cholesterol, which they transport to tissues around the body for use in biosynthesis and for storage.
- **High-density lipoproteins** (**HDL**) serve as acceptors of cholesterol (they consist of about 25 percent cholesterol) and are believed to remove cholesterol from tissues and carry it to the liver, where it can be used to synthesize bile.

Because of their differing functions in cholesterol regulation, LDL is sometimes called "bad cholesterol" and HDL "good cholesterol"—designations that are somewhat controversial. However, we do know that a high ratio of LDL to HDL in a person's blood is a risk factor for atherosclerotic heart disease. Cigarette smoking lowers HDL levels, and regular exercise increases them.

Fuel metabolism is controlled by hormones

During the absorptive period, blood glucose levels are high as carbohydrates are digested and absorbed. During this time, the liver takes up glucose from the blood and converts it to glycogen and fat, fat cells take up glucose from the blood and convert it to stored fat, and the cells of the body preferentially use glucose as their metabolic fuel.

During the postabsorptive period these processes are reversed. The liver breaks down glycogen to supply glucose to the blood, the liver and the adipose tissues supply fatty acids to the blood, and most of the cells of the body preferentially use fatty acids as their metabolic fuel.

One tissue that does not switch fuel sources during the postabsorptive period is the nervous system. The cells of the nervous system require a constant supply of glucose. Even though the nervous system can use other fuels to a limited extent, its overall dependence on glucose is the reason it is so important for other cells of the body to shift to fat metabolism during the postabsorptive period. This shift preserves the available glucose and glycogen stores for the nervous system for as long as possible.

What directs the traffic in fuel molecules? Insulin and glucagon, two hormones produced and released by the pancreas, are responsible for controlling the metabolic directions that fuel molecules take (Figure 50.20). The most

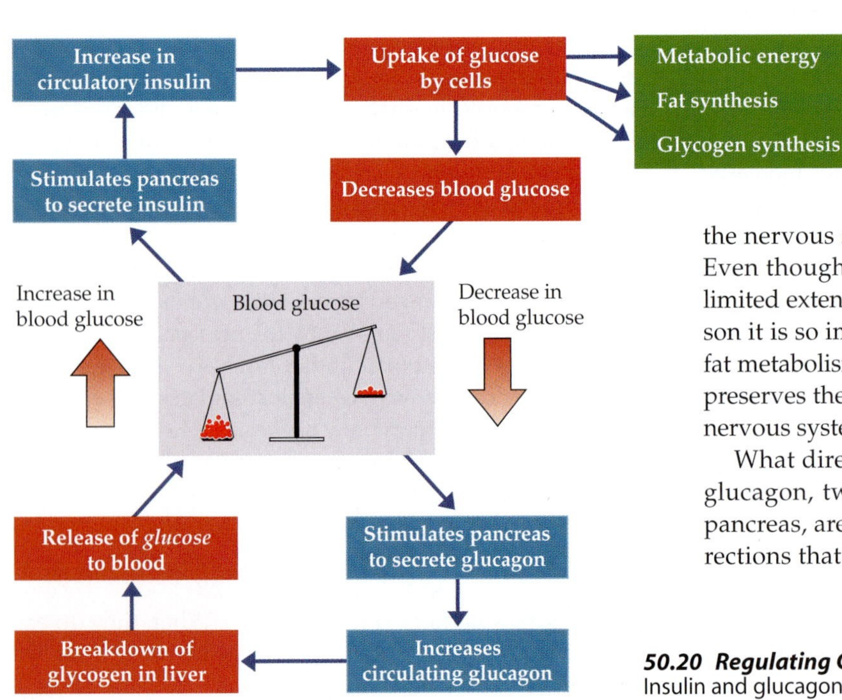

50.20 Regulating Glucose Levels in the Blood
Insulin and glucagon maintain the homeostasis of circulatory glucose.

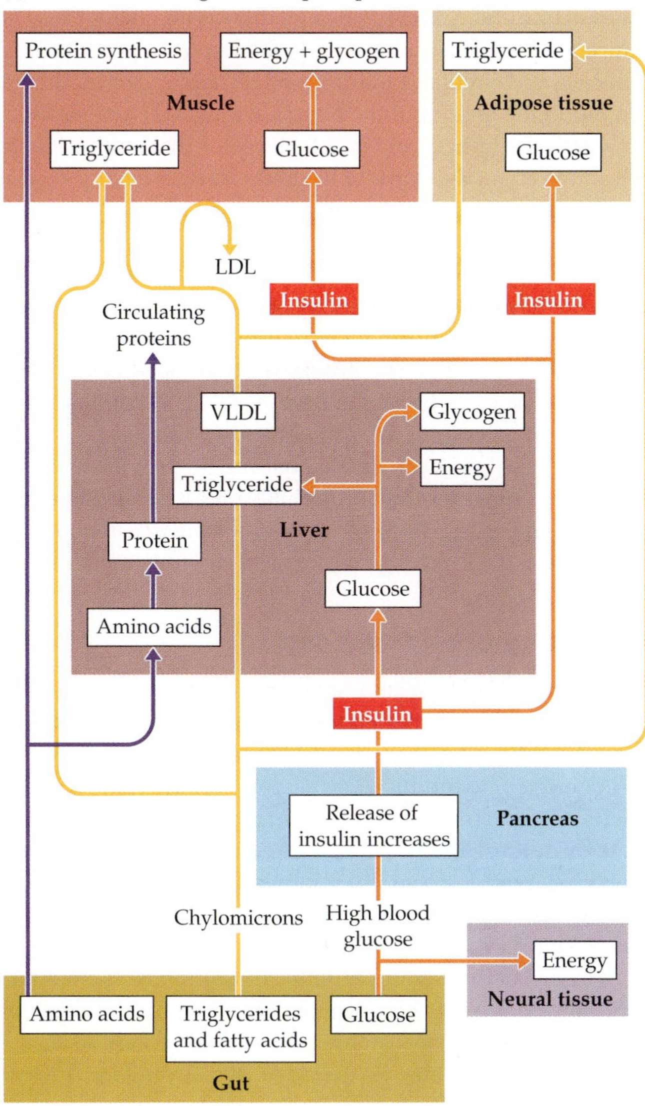

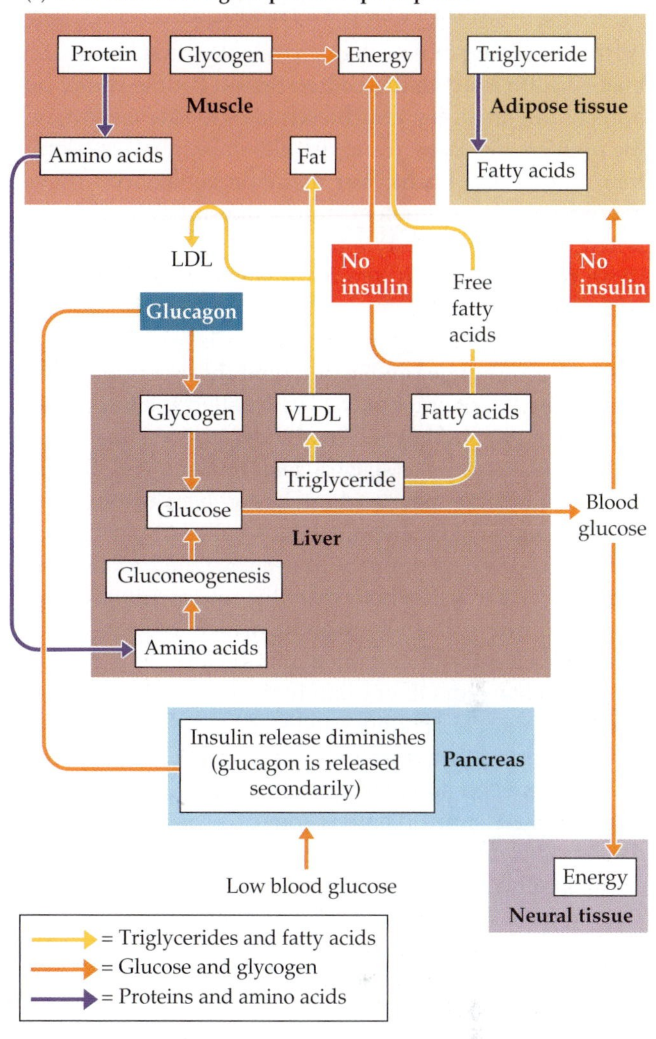

50.21 Fuel Molecule Traffic during the Absorptive and Postabsorptive Periods
Insulin promotes glucose uptake by liver, muscle, and fat cells during the absorptive period. During the postabsorptive period, the lack of insulin blocks glucose uptake by these same tissues and promotes fat and glycogen breakdown to supply metabolic fuel.

important of these hormones is **insulin**, which is produced in response to high blood glucose levels.

The pancreas releases insulin into the circulatory system when blood glucose rises above the normal postabsorptive level. Insulin facilitates the entry of glucose into most cells of the body. When insulin is present, most cells burn glucose as their metabolic fuel, fat cells use glucose to make fat, and liver cells convert glucose to glycogen and fat.

As soon as blood glucose falls back to postabsorptive levels, insulin release diminishes rapidly, and the entry of glucose into cells other than those of the nervous system is inhibited. Without a supply of glucose, cells switch to using glycogen and fat as their metabolic fuels. In the absence of insulin, the liver and fat cells stop synthesizing glycogen and fat and begin breaking them down. As a result, the liver supplies glucose to the blood rather than taking it from the blood, and both the liver and the adipose tissues supply fatty acids to the blood.

The pancreas releases **glucagon** when the blood glucose concentration falls below the normal postabsorptive level. Glucagon has the opposite effect of insulin: it stimulates liver cells to break down glycogen and to carry out gluconeogenesis. Thus, under the influence of glucagon, the liver produces glucose and releases it into the blood.

The traffic of fuel molecules during the absorptive and postabsorptive periods is summarized in Figure 50.21, which indicates the steps controlled by insulin and glucagon. During the absorptive period, all fuel molecules move toward storage, and glucose is the preferred energy source for all cells. During the postabsorptive period, most cells switch to metabolizing fat so that blood glucose reserves are saved for the nervous system. The level of circulating glucose is maintained through glycogen breakdown and gluconeogenesis.

The Regulation of Food Intake

At the beginning of this chapter we noted that obesity is a major health issue in the United States. People spend billions of dollars every year on schemes to lose weight, but the problem increases. A simple rule—take in fewer calories than your body burns, but maintain a balanced diet—should solve the problem, but it doesn't. Why? As we noted, social and lifestyle factors play a major role, but these factors play out against a genetic and regulatory background.

The amount of food an animal eats is governed by its sensations of hunger and satiety, and these sensations are influenced by a region of the brain called the hypothalamus. If a region in the middle of the hypothalamus of rats, called the *ventromedial hypothalamus*, is damaged, the animals will increase their food intake and become obese. If a different region of the hypothalamus, called the *lateral hypothalamus*, is damaged, rats will decrease their food intake and become skinny. In both cases the rats eventually reach a new equilibrium body weight, which they maintain. Thus, regulation is maintained, but the level of regulation has been changed. Other brain regions have also been implicated in control of hunger and satiety.

In Chapter 40 we learned that regulation involves feedback information and a means of comparing that information with a set point. There is some evidence that cells in the hypothalamus and in the liver are sensitive to the levels of blood glucose and insulin with high levels stimulating satiety and low levels stimulating hunger. There is even stronger evidence, however, that signals from fat metabolism influence hunger and satiety.

A single-gene mutation in mice, when present in the homozygous condition, results in mice that eat enormous amounts of food and become obese (Figure 50.22). Using genetics terms, these mice are called *ob/ob* mice due to their double dose of the recessive "obese" gene. The wild-type *ob* gene codes for a protein that has been named **leptin** (from the Greek *leptos*, "thin"). When leptin was injected into *ob/ob* mice, they ate less and lost body fat. Leptin is produced by fat cells and circulates in the blood. Receptors for leptin are found in the regions of the hypothalamus that are involved in control of hunger and satiety. It seems that leptin signals the brain about the status of the body fat reserves.

50.22 A Single-Gene Mutation Leads to Obesity in Mice
Leptin serves as a negative feedback signal to the brain to limit food intake. The fat cells of the *ob/ob* mouse (left) do not produce leptin. The wild-type mouse (right) does produce leptin and does not become obese when kept under the same conditions as the *ob/ob* mouse.

Could leptin be used to reduce human obesity? In a very few cases, obese humans do not produce the hormone leptin, and injections of leptin can curb their appetites and enable them to lose body mass. Most obese humans, however, have higher than normal circulating levels of leptin. It is likely that they have receptors with reduced sensitivity. Leptin appears to be one important feedback signal in the regulation of food intake. Understanding the actions of leptin in normal and obese individuals might provide a partial answer as to why some individuals find it easier than others to avoid excessive food intake and increases in body fat. Additional feedback signals are most certainly involved.

Toxic Compounds in Food

Plant and animal tissues contain nutrients, but as we have seen, they can also contain toxic compounds (Figure 50.23). Some mushrooms, for example, contain poisons and hallucinogens; some mollusks, fishes, and amphibians contain neurotoxins; some plants contain compounds that stimulate or depress the heart; and of course, tobacco contains nicotine, poppies contain opium, and marijuana contains tetrahydrocannabinol. Ingesting many plant and animal tissues, therefore, can be dangerous.

Human activities add millions of tons of synthetic toxic compounds to our environment every year, making the problem worse. Many of these compounds enter the air we breathe and the water we drink, as well as the food we eat. A whole new field, called **environmental toxicology**, has developed to address the problems of poisons in the environment.

Some toxins are retained and concentrated in organisms

The physical and chemical properties of a toxic compound affect its retention within a biological system. If a compound can dissolve in water, it may be quickly metabolized (and thus detoxified) because it is easily accessible to the wide variety of enzymes that can break down complex molecules in food.

In addition to being broken down or metabolized, many water-soluble compounds can be filtered out of the blood by the kidneys, and therefore do not accumulate in the body. That is why urine tests are used to detect illegal drug use by athletes and other individuals. However, some potentially dangerous water-soluble compounds can be incorporated into the body and disrupt normal functions. An example is lead, which can replace iron in blood and calcium in bone.

Lipid-soluble compounds are usually metabolized more slowly than water-soluble compounds, and they are often

(a)

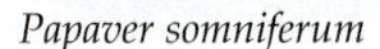

Papaver somniferum

(b)

Takifugu rubripes

50.23 Toxins Occur Naturally
Many animals and plants contain dangerous, potentially deadly toxins. (*a*) Opium poppies are the source of morphine and heroin. (*b*) The puffer fish is a delicacy in Japanese sushi restaurants; restaurants that serve it have special licenses affirming their chef's ability to prepare the fish so that its neurotoxins do not endanger diners.

stored in the body for a long time because they dissolve in the lipids of membranes and adipose tissues. Lipid-soluble compounds can accumulate in the body and reach very high concentrations. Even compounds that are beneficial or neutral to the body at low concentrations, such as fat-soluble vitamins, can become toxic at high concentrations.

Some toxins can bioaccumulate in the environment

Some lipid-soluble toxins, including many pesticides, can **bioaccumulate** in the environment; that is, they can become more and more concentrated in predators that eat contaminated prey. The pesticide load is passed up the food chain from prey to predator, growing increasingly concentrated in the tissues of each consumer in turn. In the top predator, the pesticide may be concentrated thousands or millions of times. Many top predators show high levels of pesticides and other synthetic chemicals in their tissues. Long-lived species, such as eagles or bears, are particularly at risk for heavy **body burdens** of accumulated pesticides because they have many years to accumulate them. Bioaccumulated toxins may be responsible for high rates of cancer and infertility in some wildlife populations.

A well-documented case was the effect of the pesticide DDT on predatory bird populations. Bioaccumulated DDT caused extreme thinning of eggshells and therefore high mortality of embryos (Figure 50.24). As a result, many species of predatory birds, such as ospreys, eagles, and falcons, became endangered. Since the banning of DDT use in the United States, these species have been recovering. Scientists and policy-makers are now setting standards for and researching alternatives to such pesticides to decrease the amounts of synthetic toxins bioaccumulating within natural systems. Since humans are at the top of the food chain, eating at all levels, we must consider the risks of our own exposure to these toxins.

The body cannot metabolize many synthetic toxins

How does the way that the body handles synthetic chemicals differ from the way it handles natural chemicals? In many cases, the detoxification systems that metabolize natural chemicals can also metabolize synthetic chemicals, breaking them apart and eliminating them through the urine. Enzymes called **cytochrome P450s** are responsible for much of this detoxification.

P450s are less specific in their abilities to bind substrates than are most enzymes. Thus, each P450 can catalyze reactions with a wide range of compounds, and there are many P450s. Phase I P450s make small chemical changes to the substrate, such as adding a —OH group or a —SO_3 group, which prepares the substrate for a second reaction. Phase II P450s use the —OH or —SO_3 group to attach a hydrophilic group onto the substrate, which facilitates the elimination of that substrate from the body. Few natural compounds can escape the P450s, even when the body encounters them for the first time.

Some synthetic chemicals, however, fall outside the range of structures that P450s and other enzymes can metabolize. When such chemicals are lipophilic, they bioaccumulate, and any biological effect they have is greatly magnified. If a synthetic chemical that cannot be metabolized is structurally similar to a hormone, that synthetic chemical may activate the hormonal signaling pathway within target cells. Whereas the natural hormonal signal can be turned off, the synthetic hormone cannot be, and control of function is lost.

50.24 DDT Affects Bird Eggs
Before its use was banned, the bioaccumulation of DDT in birds resulted in severe thinning of the eggshells of many species, with resulting population declines. This brown pelican egg cracked open long before the embryo inside was ready to hatch.

One example of a class of synthetic chemicals that appear to mimic hormones in animals is the polychlorinated biphenyls, **PCBs**. PCBs were produced extensively for use as an insulating fluid in electrical transformers from the 1930s until recently. PCBs are chemically stable, lipophilic, and are now found throughout the environment. They have been shown to bioaccumulate, reaching dangerously high levels in fish from contaminated waters such as the Great Lakes.

The biological response to exposure to PCBs in the diet varies among species. In rhesus monkeys, PCBs altered reproductive cycles, reduced weight gain in infants, depressed immune system responsiveness, and increased the incidence of death in developing embryos. In communities around the Great Lakes, studies have indicated cognitive impairment in children of mothers with a high body burden of PCBs, probably from eating fish caught in the Great Lakes. Studies of animals and of humans that have been exposed accidentally to high levels of PCBs have shown that effects are slow to reverse, lasting from several months to a year.

The risks of PCBs and DDT are now clear, but it is usually difficult to make a causal connection between a toxin in the environment and specific health effects in a population. Environmental toxicologists must be able to study large populations, use powerful statistics, and do controlled laboratory studies to obtain evidence that will support policy changes to stop and reverse the effects of synthetic environmental toxins.

Chapter Summary

Nutrient Requirements

▶ Animals are heterotrophs that derive their energy and structural building blocks from food, and therefore ultimately from autotrophs.

▶ Carbohydrates, fats, and proteins in food supply animals with metabolic energy. A measure of the energy content of food is the calorie. Excess caloric intake is stored as glycogen and fat. **Review Figure 50.2**

▶ An animal with insufficient caloric intake is undernourished and must metabolize its stored glycogen, fat, and finally its own protein for energy. Overnutrition in humans can be a serious health hazard. **Review Figure 50.3**

▶ For many animals, food provides essential carbon skeletons that they cannot synthesize themselves. **Review Figure 50.4**

▶ Humans require eight essential amino acids in their diet. All are available in milk, eggs, or meat, but not in all vegetables. Thus, vegetarians must eat a mix of foods. **Review Figure 50.5**

▶ Different animals need mineral elements in different amounts. Macronutrients, such as calcium, phosphorus, sodium chloride, and iron, are needed in large amounts. Micronutrients, such as iron, copper, magnesium, and zinc, are needed in small amounts. **Review Table 50.1**

▶ Vitamins are organic molecules that must be obtained in food. **Review Table 50.2**

▶ Malnutrition results when any essential nutrient is lacking from the diet. Lack of any essential nutrient causes a deficiency disease. **Review Table 50.2**

Adaptations for Feeding

▶ Animals can be characterized by how they acquire nutrition: Saprotrophs and detritivores depend on dead organic matter, filter feeders strain the environment for small food items, herbivores eat plants, and carnivores eat animals.

▶ Behavioral and anatomical adaptations reflect feeding types. In vertebrates, teeth have evolved to match the diet. **Review Figure 50.9**

Digestion

▶ Digestion involves the breakdown of complex food molecules into monomers that can be absorbed and utilized by cells. In most animals, digestion is extracellular and external to the body, taking place in a tubular gut that has different regions specialized for different digestive functions. **Review Figure 50.10**

▶ Absorptive areas of the gut are characterized by a large surface area. **Review Figure 50.11**

▶ Hydrolytic enzymes break down proteins, carbohydrates, and fats into their monomeric units. To prevent the organism itself from being digested, these enzymes are released as inactive zymogens, which become activated when secreted into the gut.

Structure and Function of the Vertebrate Gut

▶ The cells and tissues of the vertebrate gut are organized in the same way throughout its length. The innermost tissue layer, the mucosa, is the secretory and digestive surface. The submucosa contains secretory cells and glands, blood and lymph vessels, and nerves. External to the submucosa are two smooth muscle layers (circular and longitudinal) that move food through the gut. Between the two muscle layers is a nerve network that controls the movements of the gut. **Review Figure 50.13**

▶ Swallowing is a reflex that pushes food into the esophagus. Waves of smooth muscle contraction and relaxation called peristalsis move food from the beginning of the esophagus through the entire length of the gut. Sphincters block the gut at certain locations, but they relax as a wave of peristalsis approaches. **Review Figure 50.14**

▶ Enzymatic digestion begins in the mouth, where amylase is secreted with the saliva. Protein digestion begins in the stomach with pepsin and HCl secreted by the stomach mucosa. The mucosa also secretes mucus, protects the tissues of the gut. **Review Figure 50.15**

▶ In the the duodenum, pancreatic enzymes carry out most of the digestion of the food. Bile from the liver and gallbladder assists in the digestion of fats by breaking them into micelles. Bicarbonate ions from the pancreas neutralize the pH of the chyme entering from the stomach to produce an environment conducive to the actions of pancreatic enzymes. **Review Figure 50.16, Table 50.3**

▶ Final enzymatic cleavage of peptides and disaccharides occurs on the surface of the cells of the intestinal mucosa. Amino acids, monosaccharides, and many inorganic ions are absorbed by the microvilli of the mucosal cells. In many cases specific carrier proteins in the membranes of these cells transport nutrients into the cells. Sodium cotransport is a common mechanism for actively absorbing nutrient molecules and ions.

▶ Fats are absorbed mostly as monoglycerides and fatty acids, which are the product of lipase action on triglycerides in food. These products pass through the membranes of mucosal cells and are then resynthesized into triglycerides within the cells. The triglycerides are combined with cholesterol and coated with protein to form chylomicrons, which pass out of the mucosal cells and into lymphatic vessels in the submucosa. **Review Figure 50.17**

▶ Water and ions are absorbed in the large intestine so that waste matter is consolidated into feces, which are periodically excreted.

▶ In herbivores such as rabbits and ruminants, some compartments of the gut have large populations of microorganisms that aid in digesting molecules that otherwise would be indigestible to their host. **Review Figure 50.18**

Control and Regulation of Digestion

▶ The processes of digestion are coordinated and controlled by neural and hormonal mechanisms. Salivation and swallowing are autonomic reflexes. Actions of the stomach and small intestine are largely controlled by the hormones gastrin, secretin, and cholecystokinin. **Review Figure 50.19**

Control and Regulation of Fuel Metabolism

▶ The liver interconverts fuel molecules and plays a central role in directing their traffic. When food is being absorbed from the gut, the liver takes up and stores fats and carbohydrates, converting monosaccharides to glycogen or fat. The liver also takes up amino acids and uses them to produce blood plasma proteins.

▶ Fat and cholesterol are shipped out of the liver as low-density lipoproteins. High-density lipoproteins act as acceptors of cholesterol and are believed to bring fat and cholesterol back to the liver.

▶ Fuel metabolism during the absorptive period is controlled largely by the hormone insulin, which promotes glucose uptake and utilization by most cells of the body, as well as fat synthesis in adipose tissue. During the postabsorptive period, the lack of insulin blocks the uptake and utilization of glucose by most cells of the body except neurons. If blood glucose levels fall, the hormone glucagon is secreted, stimulating the liver to break down glycogen to release glucose to the blood. **Review Figures 50.20, 50.21**

The Regulation of Food Intake

▶ Food intake is governed by sensations of hunger and satiety that are determined by brain mechanisms. When one hypothalamic region is damaged, rats eat more and become obese; when another region is damaged, they eat less and become thin. A number of molecules, such as circulating insulin and glucose, provide feedback information to these brain areas.

▶ Leptin is a hormone produced by fat cells that inhibits food intake.

Toxic Compounds in Food

▶ Even natural plant and animal foods can contain toxic compounds in addition to nutrients. Human activities such as the use of pesticides and the release of pollutants into the environment have made the problem of toxins in food even worse.

▶ An organism can accumulate toxic compounds in its body, especially if those compounds are lipid-soluble or take the structural place of a natural molecule.

▶ Toxins such as PCBs and DDT that accumulate in the bodies of prey are transferred to and further concentrated in the bodies of their predators. This bioaccumulation produces high concentrations of toxins in animals high up the food chain.

For Discussion

1. Several current popular diet books recommend high fat and protein intake and low carbohydrate intake as a means of losing body mass. What could the rationale of a high-fat and high-protein diet be, and what health issues should be considered when someone considers going on such a diet?
2. Carnivores generally have more dietary vitamin requirements than herbivores do. Why?
3. It is said that the most important hormonal control of fuel metabolism in the postabsorptive period is the *lack* of insulin. Explain.
4. Why is obstruction of the common bile duct so serious? Consider in your answer the multiple functions of the pancreas and the way in which digestive enzymes are processed.
5. Trace the history of a fatty acid molecule from a piece of buttered toast to a plaque on a coronary artery. What possible forms and structures could it have passed through in the body? Describe a direct and an indirect route it could have taken.

51 Salt and Water Balance and Nitrogen Excretion

BLOOD, SWEAT, AND TEARS TASTE SALTY BEcause they reflect the composition of the tissue fluid that bathes the cells of the body. The volume and the composition of the tissue fluid must remain within certain limits and must be kept relatively free of wastes. Maintaining homeostasis of the tissue fluid can be challenging. Consider the problems of vampires—not the horror film kind, but the bat kind.

Vampire bats are small, tropical mammals that feed on the blood of other mammals, such as cattle. The bat lands on an unsuspecting (usually sleeping) victim, bites into a vein, and drinks blood—a high-protein, liquid food. The bat has only a short time to feed before the victim wakes and shakes it off. To maximize the volume of blood it can ingest, it eliminates water from its food as fast as it can by producing a lot of very dilute urine. The warm trickle down the neck of the victim is not blood!

Once feeding ends, however, this high rate of water loss cannot continue. Now the vampire bat is digesting protein and must excrete large amounts of nitrogenous breakdown products while conserving its body water. Within minutes, the excretory system of the vampire bat switches from producing lots of very dilute urine to producing a tiny amount of highly concentrated urine. Within minutes, the vampire bat is able to switch from an excretory physiology typical of an animal living in fresh water to an excretory physiology typical of an animal living in an arid desert.

In this chapter we discover how the vampire bat and other species accomplish the various feats of salt and water balance and excretion of wastes that adapt them to many different environments. We begin by discussing the challenges presented by different environments; we use some invertebrate examples to illustrate the basic mechanisms used in the excretory systems of all animals.

Turning to vertebrates, we show that the common anatomical unit that accomplishes all of these tasks is the nephron. The nephron evolved from a structure that enabled animals living in fresh water to excrete water to a structure that enabled animals living in dry terrestrial habitats to conserve water. Finally, we present the mechanisms that control and regulate salt and water balance in mammals, giving the vampire bat and other species their remarkable abilities to exploit unusual diets and extreme environments.

Tissue Fluids and Water Balance

Life evolved in the seas, and seawater is the extracellular environment for the cells of the simplest marine animals. More complex marine animals have an internal environment consisting of extracellular or tissue fluid, which is isolated from seawater but is similar to it in composition and osmotic concentration. Marine vertebrates and terrestrial animals maintain tissue fluids whose concentration and composition differ considerably from that of seawater (see Chapter 49). The concentration of the tissue fluid determines the water balance of the cells, and its composition influences the health and functions of the cells. Recall, for example, the importance of ionic gradients between the tissue fluid and the cytoplasm of nerve and muscle cells (see Chapter 44).

Blood as a Fast Food
The vampire bat, *Desmodus rotundus*, is able to adjust its excretory physiology from water-excreting to water-conserving, depending on whether it is ingesting or digesting its blood meal.

To understand what is meant by water balance, recall that cell plasma membranes are permeable to water and that the movement of water across membranes depends on differences in solute potential. (See the discussion of osmosis in Chapter 5.) If the solute potential (osmolarity) of the tissue fluid is less negative (that is, the fluid contains fewer solutes) than that of the intracellular fluids, water moves into the cells, causing them to swell and possibly burst. If the solute potential of the tissue fluid is more negative (the fluid contains more solutes) than that of the intracellular fluids, the cells lose water and shrink. The solute potential of tissue fluid determines both the volume and the solute potential of the intracellular environment.

Excretory organs control the solute potential of tissue fluid

Excretory organs control the solute potential and the volume of tissue fluid by excreting solutes that are in excess (such as NaCl when we eat lots of salty food) and conserving solutes that are valuable or in short supply (such as glucose and amino acids). In terrestrial organisms, these excretory organs also eliminate the waste products of nitrogen metabolism. The output of the excretory organs is called **urine**.

The functions of the excretory organs of a species, and therefore the composition of its urine, depend on the environment in which it lives. We will examine excretory systems that maintain salt and water balance and eliminate nitrogen in marine, freshwater, and terrestrial habitats. In spite of the evolutionary diversity of the anatomical and physiological details, all these systems obey a common rule: *there is no active transport of water*. Water must be moved either by pressure or by a difference in solute potential. Also, in spite of this evolutionary diversity, there are common mechanisms used by excretory systems.

Mechanisms used by excretory systems include filtration, secretion, and resorption

The excretory systems of many species filter the tissue fluid and then process the filtrate through a system of tubules to produce urine. This **filtration** process is usually carried out on blood plasma driven across capillary walls in the excretory organ by blood pressure. The filtrate then enters a system of tubules. The cells of the tubules change the composition of the filtrate by active **secretion** and **resorption** of specific solute molecules. These three mechanisms are used in the excretory systems of freshwater species, which excrete water and conserve salts, as well as in the systems of marine and terrestrial species, which conserve water and excrete salts.

Distinguishing Environments and Animals in Terms of Salt and Water

The salt concentration, or **osmolarity**, of ocean water is about 1,070 milliosmoles/liter (mosm/l), and fresh water is generally between 1 and 10 mosm/l. Aquatic environments grade continuously from fresh to extremely salty. Consider a place where a river enters the sea through a bay or a marsh. Aquatic environments within that bay or marsh range in osmolarity from that of the fresh water of the river to that of the open sea. Evaporating tide pools can reach an even greater osmolarity than seawater. Animals live in all these environments. Some species, called **osmoconformers**, allow the osmolarity of their tissue fluids to equilibrate with their environment. Others, called **osmoregulators**, maintain the osmolarity of their tissue fluids at a constant level as the environment changes.

Most marine invertebrates are osmoconformers

Over a wide range of environmental osmolarities, marine invertebrates simply equilibrate the osmolarity of their tissue fluid with that of the environment. There are limits to osmoconformity, however. No animal could survive if its tissue fluid had the osmolarity of fresh water; nor could animals survive with internal osmolarities as high as those that may be reached in an evaporating tide pool. Such solute concentrations cause proteins to denature.

Osmoregulators regulate the concentration of their tissue fluids

All animals have some solutes in their tissue fluids. Therefore, in fresh water, osmosis will cause water to invade the bodies of animals, so osmoregulation is essential. To osmoregulate in fresh water, animals must excrete water and conserve solutes; hence they produce large amounts of dilute urine. In salt water, the opposite problem exists. For animals that maintain the osmolarity of their tissue fluids below that of the environment, osmosis will cause a loss of water. To osmoregulate in salt water, animals must conserve water and excrete salts; thus they tend to produce small amounts of very concentrated urine.

Even animals that osmoconform over a wide range of environmental osmolarities must osmoregulate in extreme environments. An excellent example is the brine shrimp *Artemia* (Figure 51.1*a*), which lives in environments of almost any salinity. *Artemia* are found in huge numbers in the most saline environments known, such as Great Salt Lake in Utah or coastal evaporation ponds where salt is concentrated for commercial purposes (see Figure 26.23). The osmolarity of such water reaches 2,500 mosm/l. At these high environmental osmolarities, *Artemia* is capable of maintaining its tissue fluid osmolarity considerably below that of the environment, and therefore acts as a **hypotonic osmoregulator**. Very few organisms can survive in the crystallizing brine in which *Artemia* thrives. The main mechanism this small crustacean uses for hypotonic osmoregulation is the active transport of NaCl from its tissue fluids out across its gill membranes to the environment.

Artemia cannot survive in fresh water, but they can live in dilute seawater, in which they maintain the osmolarity of their tissue fluids above that of the environment. Under these conditions, *Artemia* behaves as a **hypertonic osmoregulator**; that is, it maintains the osmolarity of its tissue fluid above the osmolarity of the environment (Figure 51.1*b*).

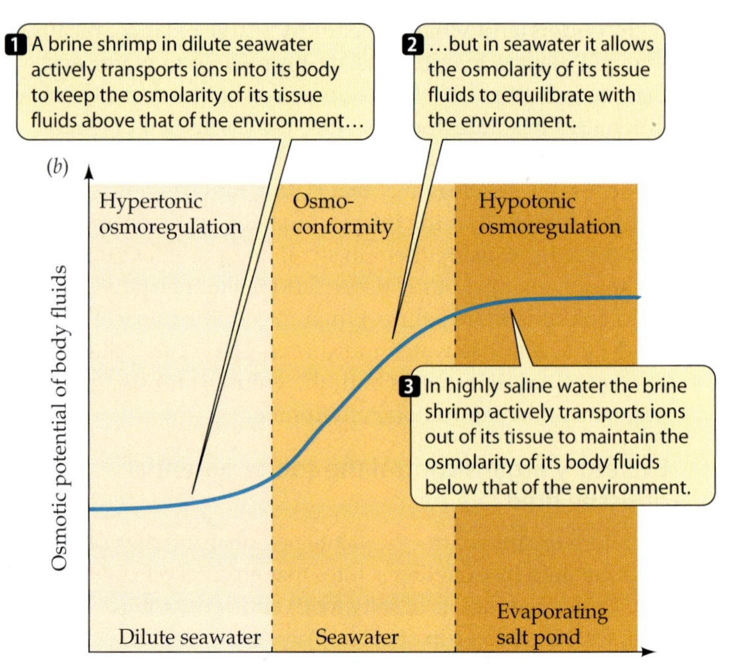

51.1 Environments Can Vary Greatly in Salt Concentration
(*a*) Brine shrimp are exposed to a range of very different salinities. (*b*) Animals like the brine shrimp that live at the extremes of environmental osmolarities display flexible osmoregulatory abilities. They become hypertonic osmoregulators in very dilute water, or hypotonic osmoregulators in very saline water.

The composition of tissue fluids can be regulated

Osmoconformers can be **ionic conformers**, allowing the ionic composition, as well as the osmolarity, of their tissue fluids to match that of the environment. Most osmoconformers, however, are **ionic regulators** to some degree: They employ active transport mechanisms to maintain specific ions in their tissue fluid at concentrations different from those in the environment.

The terrestrial environment presents problems of salt and water balance that are entirely different from those faced by aquatic organisms. Because the terrestrial environment is extremely desiccating (drying), most terrestrial animals must conserve water. (Exceptions are animals such as muskrats and beavers that spend most of their time in water.)

Terrestrial animals obtain their salts from their food. But plants generally have low concentrations of sodium, so most herbivores must conserve sodium ions. Some terrestrial herbivores travel long distances to naturally occurring salt licks. By contrast, birds that feed on marine animals must excrete the large excess of sodium they ingest with their food. Their **nasal salt glands** excrete a concentrated solution of sodium chloride via a duct that empties into the nasal cavity. Birds, such as penguins and seagulls, that have nasal salt glands can be seen frequently sneezing or shaking their heads to get rid of the very salty droplets that form (Figure 51.2).

Excreting Nitrogen

The end products of the metabolism of carbohydrates and fats are water and carbon dioxide, which are not difficult to eliminate. Proteins and nucleic acids, however, contain nitrogen, so their metabolism produces nitrogenous wastes in addition to water and carbon dioxide. The most common nitrogenous waste is ammonia (NH_3), which is highly toxic. Ammonia must be excreted continuously to prevent its accumulation, or it must be detoxified by conversion into other molecules for excretion. Those molecules are principally **urea** and **uric acid** (Figure 51.3).

Aquatic animals excrete ammonia

Continuous excretion of ammonia is relatively simple for aquatic animals. Ammonia diffuses in and is highly soluble in water. Animals that breathe water continuously lose ammonia from their blood to the environment by diffusion across their gill membranes. Animals, such as aquatic invertebrates and bony fishes, that excrete ammonia are said to be **ammonotelic**.

Many terrestrial animals and some fishes excrete urea

Ammonia is a dangerous metabolite for terrestrial animals that have limited access to water. In mammals, ammonia is lethal when it reaches only 5 mg/100 ml of blood. Therefore, terrestrial (and some aquatic) animals convert ammonia into either urea or uric acid. **Ureotelic** animals, such as mammals, amphibians, and cartilaginous fishes (sharks and rays), excrete urea as their principal nitrogenous waste product.

Urea is quite soluble in water, but excretion of urea solutions at low concentrations could result in a large loss of water that many terrestrial animals can ill afford. As we will see later in this chapter, mammals have evolved excretory systems that can conserve water by producing concentrated urea solutions. The cartilaginous fishes are another story. These marine species keep their body fluids almost isotonic to the marine environment by retaining high concentrations of urea.

Some terrestrial animals excrete uric acid

Animals that conserve water by excreting nitrogenous wastes as uric acid are said to be **uricotelic**. Insects, reptiles, birds, and some amphibians are uricotelic. Uric acid is very insoluble in water and is excreted as a semisolid (for example, the whitish material in bird droppings). Therefore, a uricotelic animal loses very little water as it disposes of its nitrogenous wastes.

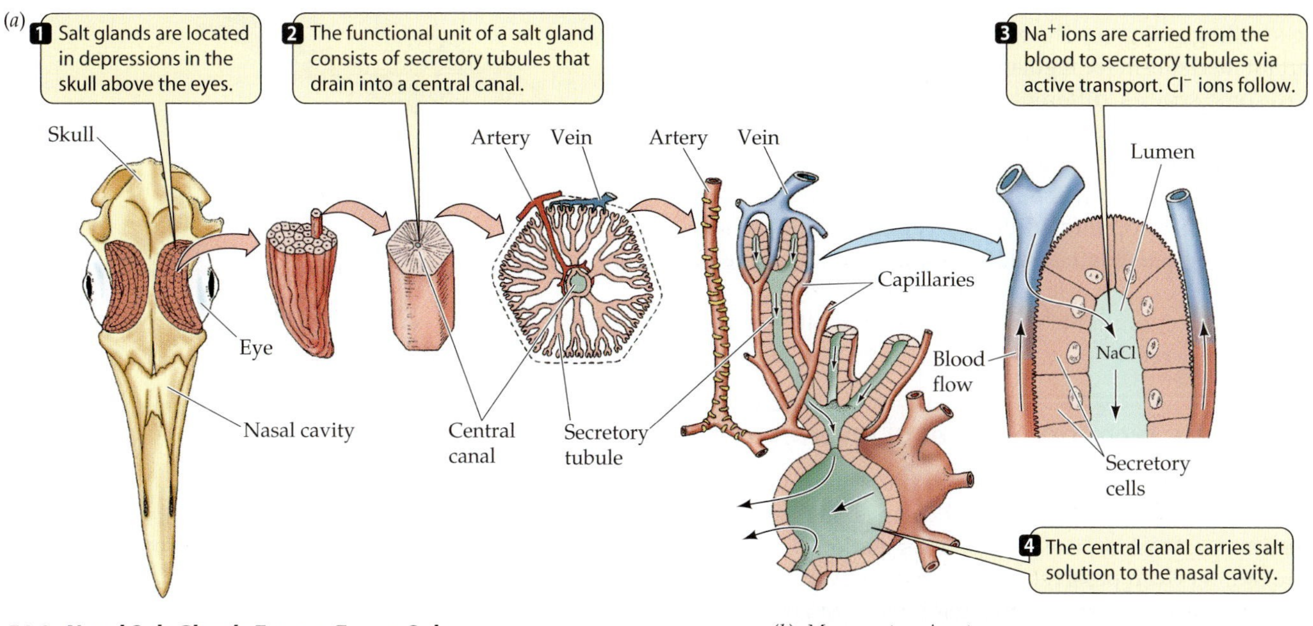

51.2 Nasal Salt Glands Excrete Excess Salt
(*a*) Marine birds have nasal salt glands adapted to excrete the excess salt from the seawater they consume with their food. (*b*) This giant petrel has returned from a feeding trip at sea and is excreting salt through its nasal salt gland. Note the drop of excreted salt at the tip of the bird's beak.

Most species produce more than one nitrogenous waste

Humans are ureotelic, yet we also excrete uric acid and ammonia. The uric acid in human urine comes largely from the metabolism of nucleic acids and caffeine. In the condition known as gout, uric acid levels in the tissue fluid increase and uric acid precipitates in the joints and elsewhere, caus-

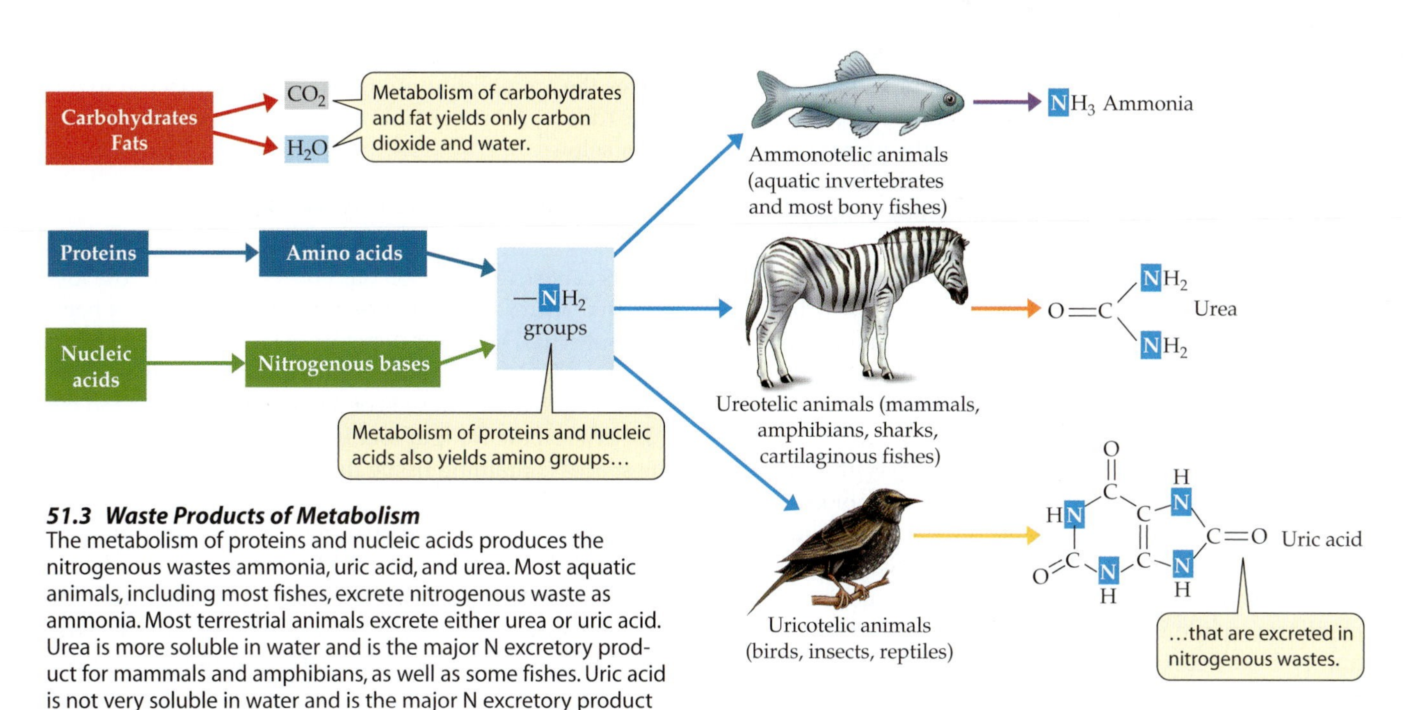

51.3 Waste Products of Metabolism
The metabolism of proteins and nucleic acids produces the nitrogenous wastes ammonia, uric acid, and urea. Most aquatic animals, including most fishes, excrete nitrogenous waste as ammonia. Most terrestrial animals excrete either urea or uric acid. Urea is more soluble in water and is the major N excretory product for mammals and amphibians, as well as some fishes. Uric acid is not very soluble in water and is the major N excretory product for birds, reptiles, and insects.

ing swelling and pain. The excretion of ammonia is an important mechanism for regulating the pH of the tissue fluid.

In some species, different developmental forms live in quite different habitats and have different forms of nitrogen excretion. The tadpoles of frogs and toads, for example, excrete ammonia across their gill membranes, but when they develop into adult frogs or toads, they generally excrete urea. Some adult frogs and toads that live in arid habitats excrete uric acid. These examples show the considerable evolutionary flexibility in how nitrogenous wastes are excreted.

The Diverse Excretory Systems of Invertebrates

Most marine invertebrates are osmoconformers and have few adaptations for salt and water balance other than active transport mechanisms for ionic regulation. To excrete nitrogen, they can passively lose ammonia by diffusion to the seawater. Freshwater and terrestrial invertebrates, however, have a wide variety of adaptations for maintaining salt and water balance and excreting nitrogen. All of these adaptations are based on the same set of mechanisms: filtration of body fluids and active secretion and resorption of specific ions.

Protonephridia excrete water and conserve salts

Many flatworms, such as *Planaria*, live in fresh water. These animals excrete water through an elaborate network of tubules running throughout their bodies. The tubules end in *flame cells*, so called because each tubule has a tuft of cilia beating inside it, giving the appearance of a flickering flame (Figure 51.4). A flame cell and a tubule together form a **protonephridium** (plural protonephridia; from the Greek *proto*, "before," and *nephros*, "kidney").

Tissue fluid enters the tubules (how it does so is not entirely clear), and the beating of the cilia causes this fluid to flow through the tubules toward the animal's excretory pore. As it flows, the cells of the tubules modify the fluid. As the modified tubule fluid (urine) leaves the planarian, it is less concentrated than the animal's tissue fluid, so ions are conserved and water is excreted by the protonephridium.

Metanephridia process coelomic fluid

Filtration of body fluids and modification of urine by tubules are highly developed processes in annelid worms, such as the earthworm. Recall that annelids are segmented and have a fluid-filled body cavity, called a coelom, in each segment (see Figure 31.23). Annelids have a closed circulatory system through which blood is pumped under pressure (see Figure 49.3). The pressure causes the blood to be filtered across the thin, permeable capillary walls into the coelom. This process is called *filtration* because the cells and large protein molecules of the blood stay behind in the capillaries while water and small molecules leave them and enter the coelom. In addition, some waste products, such as ammonia, diffuse directly from the tissues into the coelom. But where does this coelomic fluid go?

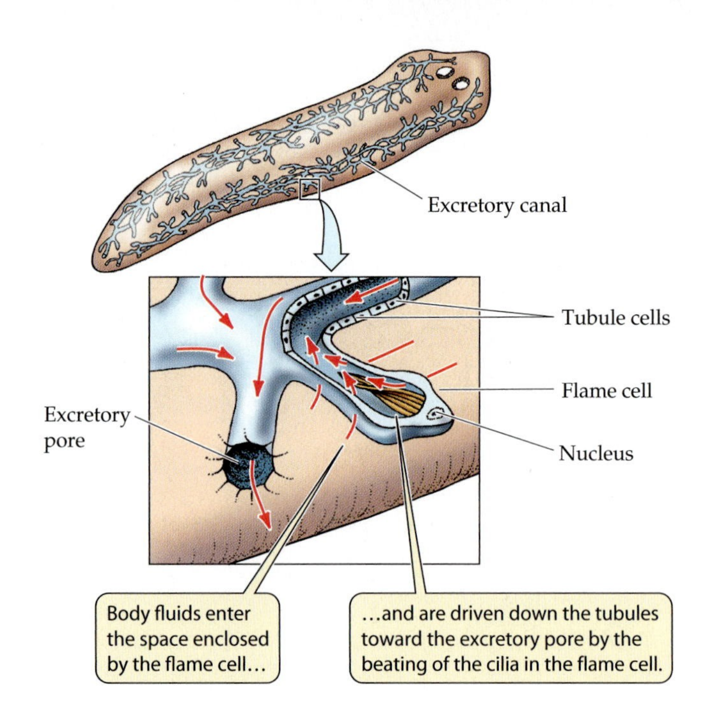

51.4 Protonephridia in Flatworms
The protonephridia of the flatworm *Planaria* consist of tubules ending in flame cells. The tubule cells modify the composition of the fluid passing through them.

Each segment of the earthworm contains a pair of **metanephridia** (singular metanephridium; from the Greek *meta*, akin to, and *nephros*, kidney). Each metanephridium begins in one segment as a ciliated, funnel-like opening in the coelom called a *nephrostome*, which leads into a tubule in the next segment. The tubule ends in a pore called the *nephridiopore*, which opens to the outside of the animal (Figure 51.5). Coelomic fluid enters the metanephridia through the nephrostomes. As the fluid passes through the tubules, the cells of the tubules actively resorb certain molecules from it and actively secrete other molecules into it. What leaves the animal through the nephridiopores is a hypotonic (dilute) urine containing nitrogenous wastes, among other solutes.

Malpighian tubules are the excretory organs of insects

Insects can excrete nitrogenous wastes with very little loss of water. Therefore, some species can live in the driest habitats on Earth. The insect excretory system consists of blind tubules called **Malpighian tubules**. An individual insect has from 2 to more than 100 of these tubules attached to the gut between the midgut and hindgut and projecting into the spaces containing tissue fluid (recall that insects have open circulatory systems) (Figure 51.6).

The cells of the Malpighian tubules actively transport uric acid, potassium ions, and sodium ions from tissue fluid into the tubules. As these solutes are secreted into the tubules, water follows because of the difference in solute potential. The walls of the Malpighian tubules have muscle fibers that contract to help move the contents of the tubules toward the hindgut.

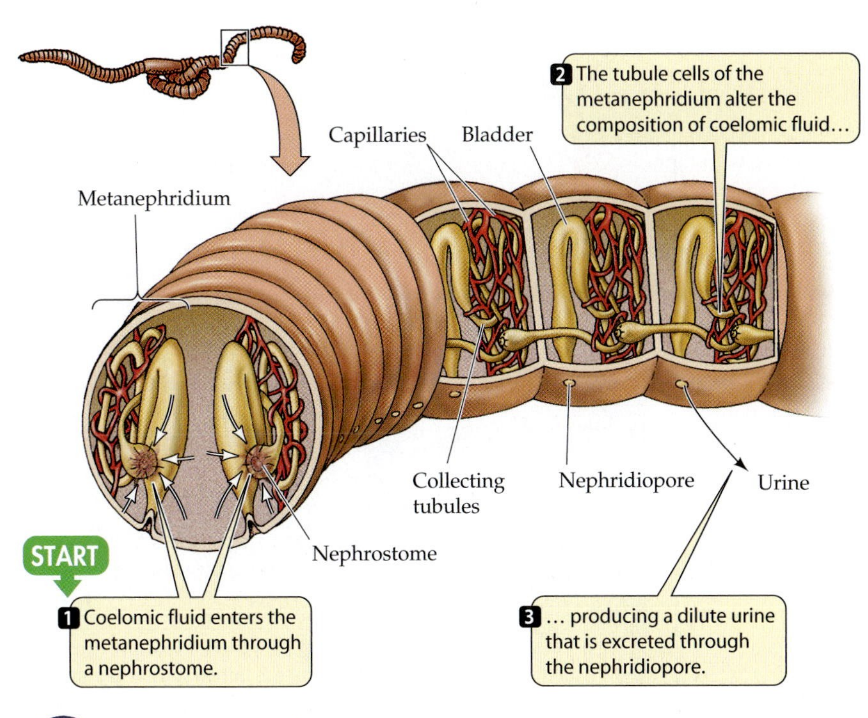

51.5 Metanephridia in Earthworms
The metanephridia of annelids are arranged segmentally. The cross section (left) shows a pair of metanephridia. Longitudinal sections (right) show only one metanephridium of the two in each segment.

The tubule fluid changes in composition while it is in the hindgut. The contents of the hindgut are more acidic than the tubule fluid; as a result, uric acid becomes less soluble and precipitates out of solution as it approaches and enters the rectum. The epithelial cells of the hindgut and rectum actively transport sodium and potassium ions from the gut contents back into the tissue fluid. Because the uric acid molecules have precipitated out of solution, water is free to follow the resorbed salts back into the tissue fluid through osmosis. Remaining in the rectum are crystals of uric acid mixed with undigested food; this dry matter is what the insect eliminates. The Malpighian tubule system is a highly effective mechanism for excreting nitrogenous wastes and some salts without giving up a significant fraction of the animal's precious water supply.

Vertebrate Excretory Systems Are Built of Nephrons

The major excretory organ of vertebrates is the **kidney**. The functional unit of the kidney is the **nephron**. Each human kidney has about a million nephrons. All vertebrate kidneys consist of nephrons, yet the kidneys of different species can serve opposite functions to maintain water and salt balance. The kidneys of freshwater fishes, for example, excrete water, but the kidneys of most mammals conserve water.

To understand how the kidney can fulfill opposite functions in different animals, we need to understand how the different parts of the nephron work and the different ways in which they can work together to influence the composition of the urine. The nephron has three main parts:

- A ball of capillaries called the glomerulus that filters the plasma
- Renal tubules that receive and modify the filtrate
- Peritubular capillaries that serve the tubules

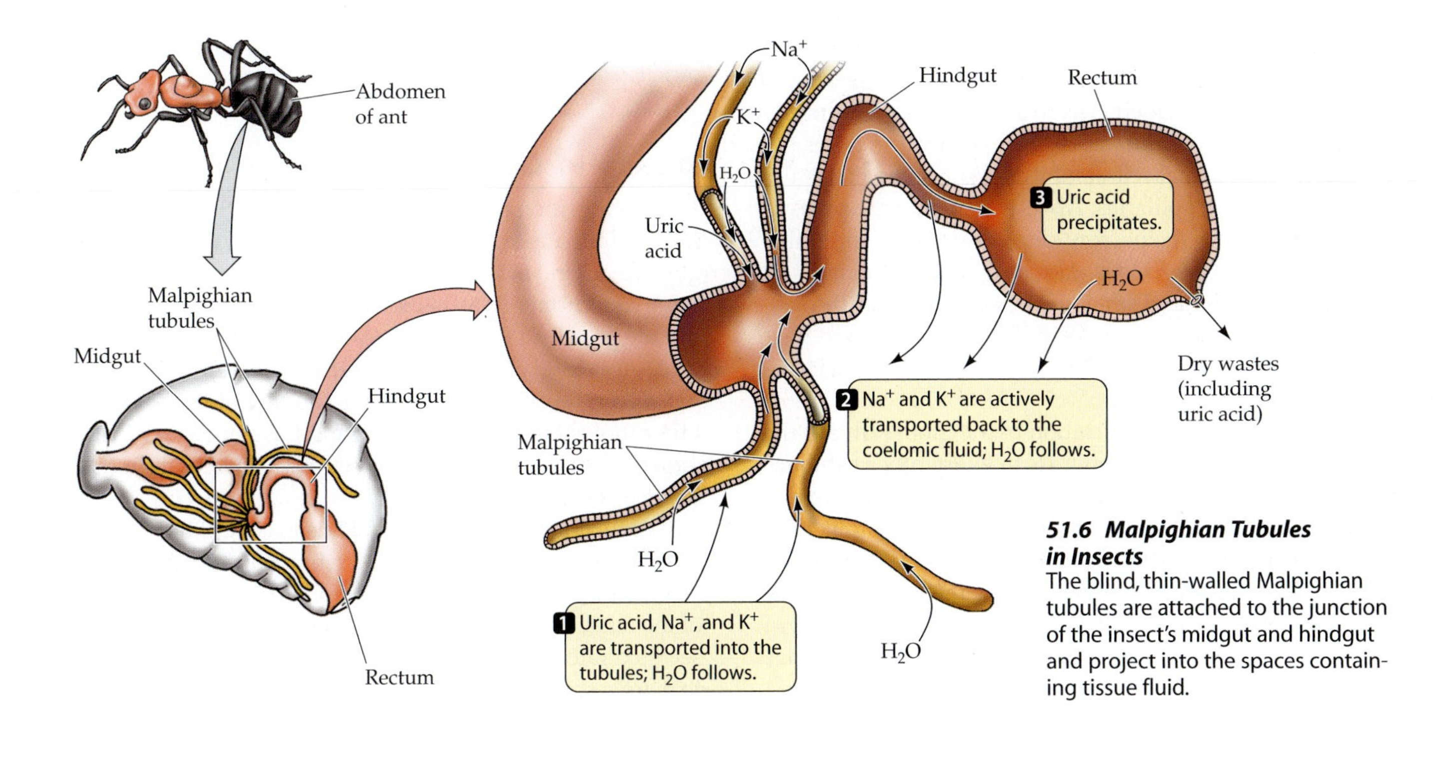

51.6 Malpighian Tubules in Insects
The blind, thin-walled Malpighian tubules are attached to the junction of the insect's midgut and hindgut and project into the spaces containing tissue fluid.

Blood is filtered in the glomerulus

Each nephron has both vascular and tubule components (Figure 51.7). The vascular component is unusual in that it consists of two capillary beds that lie between the arteriole that supplies it and the venule that drains it. The first capillary bed is a dense knot of very permeable vessels called the **glomerulus** (plural glomeruli) (Figure 51.8*a*). Blood enters the glomerulus through an **afferent arteriole** and exits through an **efferent arteriole**. The efferent arteriole gives rise to the second set of capillaries, the **peritubular capillaries**, which surround the tubule component of the nephron (see Figure 51.7).

The tubule component of the nephron, called a **renal tubule**, begins with **Bowman's capsule**, which encloses the glomerulus. The glomerulus appears to be pushed into Bowman's capsule much like a fist pushed into an inflated balloon. Together, the glomerulus and its surrounding Bowman's capsule are called the **renal corpuscle**. The cells of the capsule that come into direct contact with the glomerular capillaries are called **podocytes** (see Figure 51.7). These highly specialized cells have numerous armlike extensions, each with hundreds of fine, fingerlike projections. The podocytes wrap around the capillaries so that their fingerlike projections interdigitate and cover the capillaries completely (Figure 51.8*b*).

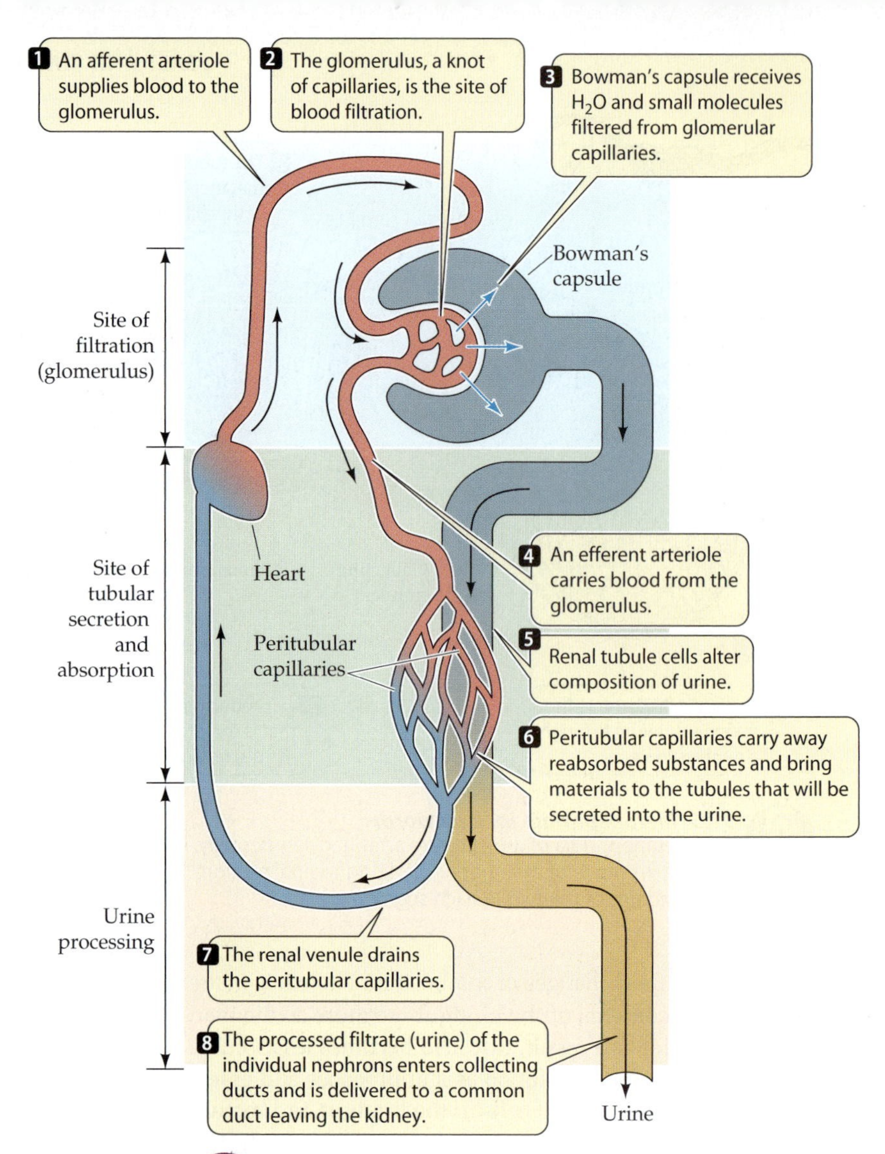

51.7 The Vertebrate Nephron
The vertebrate nephron consists of a renal tubule closely associated with a system of blood vessels. The end of the renal tubule system envelops the glomerulus so that the filtrate from the glomerular capillaries enters the tubules. The tubules change the composition of the filtrate by active absorption and secretion of solutes.

The glomerulus filters the blood to produce a tubule fluid that lacks cells and large molecules. The walls of the capillaries, the basal lamina of the capillary endothelium, and the podocytes of Bowman's capsule all participate in filtration. The endothelial walls of the capillaries have pores that allow water and small molecules to leave, but are too small to permit red blood cells to pass through. The meshwork of the basal lamina is even finer than the pores between the endothelial cells, and it prevents large molecules from leaving the capillaries. Also smaller than the pores in the capillaries are the narrow slits between the fingerlike projections of the podocytes. As a result of these anatomical adaptations, water and small molecules pass from the capillary blood and enter the renal tubule of the nephron (Figure 51.8*c*), but red blood cells and proteins remain in the capillaries.

The force that drives filtration in the glomerulus is the pressure of the arterial blood. As in every other capillary bed, the pressure of the blood entering the permeable capillaries causes the filtration of water and small molecules. The glomerular filtration rate is high because glomerular capillary blood pressure is unusually high, and because the capillaries of the glomerulus, along with their covering of podocytes, are much more permeable than other capillary beds in the body.

The renal tubules convert glomerular filtrate to urine

The composition of the filtrate that enters the nephron is similar to that of the blood plasma. This filtrate contains glucose, amino acids, ions, and nitrogenous wastes in the same concentrations as in the blood plasma, but it lacks the plasma proteins. As this fluid passes down the renal tubule, its composition changes as the cells of the tubule actively resorb certain molecules from the tubule fluid and secrete other molecules into it. When the tubule fluid leaves the kidney as urine, its composition is very different from that of the original filtrate.

The function of the renal tubules is to control the composition of the urine by actively secreting and resorbing specific molecules. The peritubular capillaries serve the needs of the renal tubules by bringing to them the molecules to be secreted into the tubules and carrying away the molecules that are resorbed from the tubules.

(*a*)

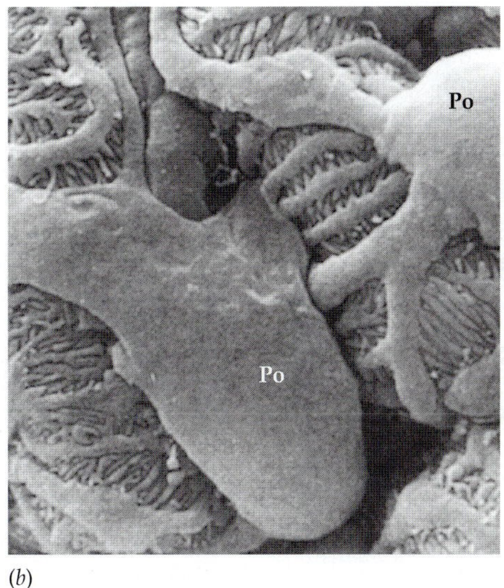

(*b*)

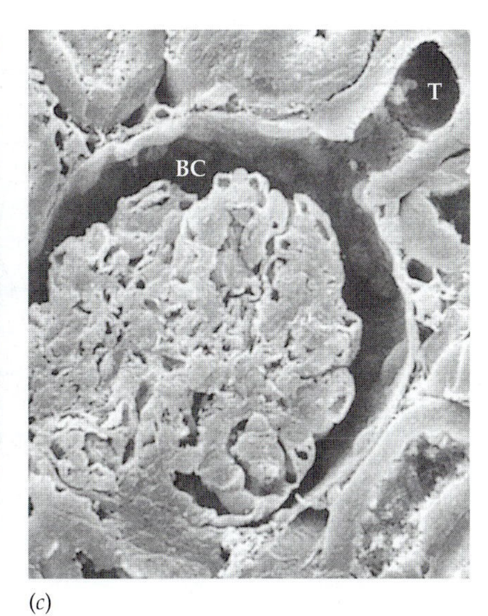

(*c*)

51.8 An SEM Tour of the Nephron
These scanning electron micrographs show the anatomical bases for kidney function. (*a*) The blood vessels in the kidney, showing the knots of capillaries that form the glomeruli. Each glomerulus (Gl) has an afferent and an efferent arteriole (Ar). Peritubular capillaries (Pt) are looser networks surrounding the tubules of the nephron. (*b*) Those cells that are in direct contact with the capillaries are the podocytes (Po). Each podocyte has hundreds of tiny fingerlike projections that create filtration slits between them. Anything passing from the glomerular capillaries into the tubule of the nephron must pass through these slits. (*c*) A cross section of a glomerulus shows that it is surrounded by the tubule cells that form Bowman's capsule (BC), which collects the filtrate and funnels it into the tubule (T) of the nephron. The relationship between the glomerulus and Bowman's capsule is like that of a fist punched into a balloon. Therefore, some of the renal tubule cells are in direct contact with the glomerular capillaries.

Both marine and terrestrial vertebrates must conserve water

Since the vertebrate nephron evolved as a structure for excreting water while conserving salts and essential small molecules, how have vertebrates adapted to environments where water must be conserved and salts excreted? The answer to this question differs for each vertebrate group. Even among the marine fishes, the adaptations of bony fishes are different from those of cartilaginous fishes.

MARINE BONY FISHES. Marine bony fishes cannot produce urine more concentrated than their tissue fluid, but unlike most marine animals, they osmoregulate their tissue fluid to only one-fourth to one-third the solute potential of seawater. They prevent excessive loss of water by producing very little urine. Their urine production is low because their kidneys have fewer glomeruli than do the kidneys of freshwater fishes. In some species of marine bony fishes, the kidneys have no glomeruli at all. Even though the glomeruli are reduced or absent, renal tubules with closed ends are retained for active excretion of ions and certain molecules.

Marine bony fishes take in seawater with their food, which results in a large salt load. The fish handle these salt loads by simply not absorbing some ions (such as Mg^{2+} or SO_4^{2-}) from their guts and by actively excreting others (such as Na^+ and Cl^-) from the gill membranes and from the renal tubules. Nitrogenous wastes are lost as ammonia from the gill membranes.

CARTILAGINOUS FISHES. Cartilaginous fishes are osmoconformers, but not ionic conformers. Unlike marine bony fishes, cartilaginous fishes convert nitrogenous wastes to urea and another compound called tri-methyl amine oxide and retain large amounts of these compounds in their tissue fluids. As a result, their tissue fluids have an osmolarity close to that of seawater. These species have adapted to a concentration of urea in the body fluids that would be fatal to other vertebrates.

Sharks and rays still have the problem of excreting the large amounts of salts they take in with their food. They have several sites of active secretion of NaCl, but the major one is a salt-secreting *rectal gland*.

AMPHIBIANS. Most amphibians live in or near fresh water and stay in humid habitats when they venture from the water. Like freshwater fishes, most amphibian species produce large amounts of dilute urine and conserve salts. Some amphibians, however, have adapted to habitats that require water conservation.

51.9 Burrowing Frogs
The banjo frog of the Australian desert survives long droughts by burrowing deep in the sand and entering estivation, a state of low metabolic activity. These frogs store water in the form of dilute urine in their enormous bladders.

Amphibians living in very dry terrestrial environments have reduced the water permeability of their skin. Some secrete a waxy substance that they spread over the skin to waterproof it. Several species of frogs that live in arid regions of Australia burrow deep into the ground, where they remain during long dry periods (Figure 51.9). There they enter *estivation,* a state of very low metabolic activity and therefore low water turnover. When it rains, these frogs come out of estivation, feed, and reproduce. But their most interesting adaptation is that they have enormous urinary bladders. Before entering estivation, they fill their bladders with dilute urine, which can amount to one-third of their body weight. This dilute urine serves as a water reservoir that they use gradually during the long period of estivation. Australian aboriginal peoples dig up estivating frogs as an emergency source of drinking water.

REPTILES. Reptiles occupy habitats ranging from aquatic to extremely hot and dry. Three major adaptations have freed the reptiles from maintaining the close association with water that is necessary for most amphibians. First, reptiles do not need fresh water to reproduce, because they employ internal fertilization and lay eggs with shells that retard evaporative water loss. Second, they have scaly, dry skins that retard evaporative water loss. Third, they excrete nitrogenous wastes as uric acid solids, therefore losing little water in the process.

BIRDS. Birds have the same adaptations for water conservation that reptiles have: internal fertilization, shelled eggs, skin that retards water loss, and uric acid as the nitrogenous waste product. In addition, some birds can produce a urine that is more concentrated than their tissue fluids. This last ability is most developed in mammals.

The Mammalian Excretory System

The adaptations of mammals and birds for producing urine hypertonic to their tissue fluids were an important step in vertebrate evolution. These adaptations enabled the excretory system to conserve water while still excreting excess salts and nitrogenous wastes. Mammals and birds have high body temperatures and high metabolic rates, and therefore have the potential for a high rate of water loss. Being able to minimize water loss from their excretory systems made it possible for these highly active species to occupy arid habitats.

We have seen how the nephron originally evolved to excrete water; now we will see how it can serve as the basic structural unit of an organ that is able to conserve water. To understand this evolutionary change of function, it is necessary to understand the structure and function of the nephron in the context of the overall anatomy of the kidney. First, however, let's look at the mammalian excretory system as a whole.

Kidneys produce urine, which the bladder stores

We will use humans here as our example of the mammalian excretory system. Humans have two kidneys just under the dorsal wall of the abdominal cavity in the mid-back region (Figure 51.10). Each kidney filters blood, processes the fil-

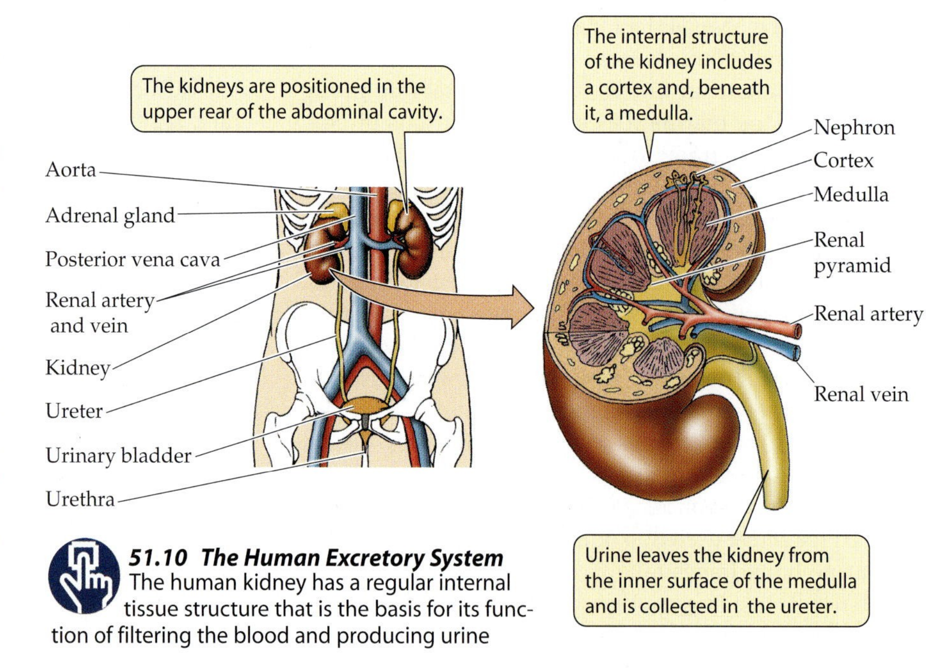

51.10 The Human Excretory System
The human kidney has a regular internal tissue structure that is the basis for its function of filtering the blood and producing urine

trate into urine, and releases that urine into a duct called the **ureter**. The ureter of each kidney leads to the **urinary bladder**, where the urine is stored until it is excreted through the urethra. The **urethra** is a short tube that opens to the outside at the end of the penis in males or just anterior to the vaginal opening in females.

Two sphincter muscles surrounding the base of the urethra control the timing of urination. One of these sphincters is a smooth muscle and is controlled by the autonomic nervous system. When the bladder is full, a spinal reflex relaxes this sphincter. This reflex is the only control of urination in infants, but the reflex gradually comes under the influence of higher centers in the nervous system as a child grows older. The other sphincter is a skeletal muscle and is controlled by the voluntary, or conscious, nervous system. When the bladder is *very* full, only serious concentration prevents urination.

Nephrons have a regular arrangement in the kidney

The kidney is shaped like a kidney bean; when cut down its long axis and split open as a bean splits open, its important anatomical features are revealed (see Figure 51.10). The ureter and the **renal artery** and **renal vein** enter the kidney on its concave (punched-in) side. The ureter divides into several branches, the ends of which envelop kidney tissues called **renal pyramids**. The renal pyramids make up the internal core, or **medulla**, of the kidney. The medulla is surrounded by tissue with a different appearance, called the **cortex**. The renal artery and vein give rise to many arterioles and venules in the region between the cortex and the medulla.

Each human kidney contains about a million nephrons, and their organization within the kidney is very regular. All of the glomeruli are located in the cortex. The initial segment of a renal tubule is called the **proximal convoluted tubule**—"proximal" because it is close to its glomerulus and "convoluted" because it is twisted (Figure 51.11). All the proximal convoluted tubules are also located in the cortex.

At a certain point, the renal tubule takes a dive directly down into the medulla. The portion of the tubule in the medulla is called the **loop of Henle**. It is called a loop because it runs straight down into the medulla, makes a hairpin turn, and comes straight back to the cortex. Where the ascending limb of the loop of Henle reaches the cortex, it becomes the distal convoluted tubule—"distal" because it is farther from its glomerulus than the proximal tubule is. The distal convoluted tubules of many nephrons join a common **collecting duct** in the cortex. The collecting ducts then run in parallel with the loops of Henle down through the medulla and empty into the ureter at the tips of the renal pyramids.

Blood vessels also have a regular arrangement in the kidney

The organization of the blood vessels of the kidney closely parallels the organization of the nephrons (see Figure 51.11). Arterioles branch from the renal arteriy and radiate into the cortex. An *afferent* arteriole carries blood to each

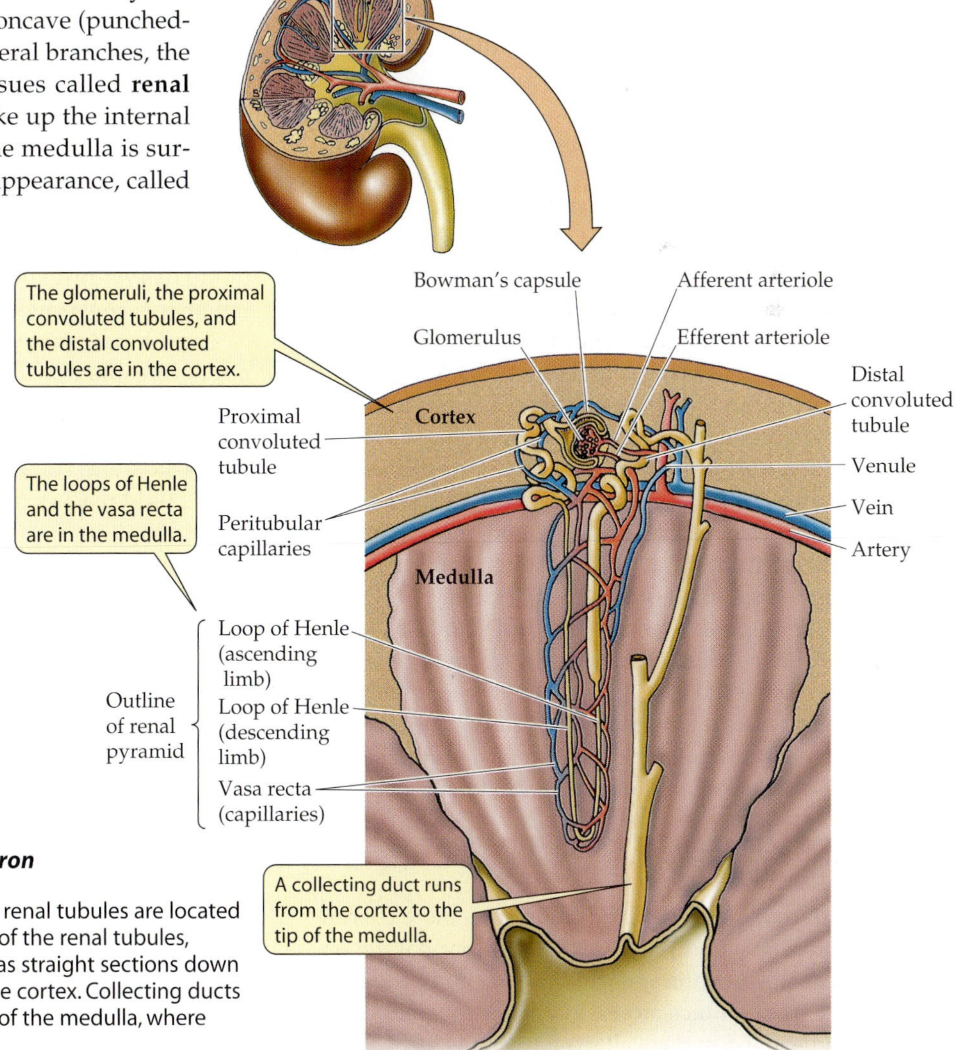

51.11 The Organization of the Nephron within the Mammalian Kidney
The glomeruli and major portions of the renal tubules are located in the cortex of the kidney, but portions of the renal tubules, called the loops of Henle, run in parallel as straight sections down into the renal medulla and back up to the cortex. Collecting ducts run from the cortex to the inner surface of the medulla, where they open into the ureter.

glomerulus. Draining each glomerulus is an *efferent* arteriole that gives rise to the peritubular capillaries, most of which surround the cortical portions of the tubules.

A few peritubular capillaries run into the medulla in parallel with the loops of Henle and the collecting ducts. These capillaries form the **vasa recta**. All the peritubular capillaries from a nephron join back together into a venule that joins with venules from other nephrons and eventually leads to the renal vein, which takes blood from the kidney.

The volume of glomerular filtration is greater than the volume of urine

Most of the water and solutes filtered in the glomerulus are resorbed and do not appear in the urine. We can reach this conclusion by comparing the rate of filtration by the glomeruli with the rate of urine production. The kidneys receive about 1 liter of blood per minute, or more than 1,400 liters of blood per day. How much of this huge volume is filtered in the glomeruli? The answer is about 12 percent. This is still a large volume—180 liters per day! Since we normally urinate 2 to 3 liters per day, about 98 to 99 percent of the fluid volume that is filtered in the glomerulus is resorbed into the blood. Where and how is this enormous fluid volume resorbed?

Most filtrate is resorbed by the proximal convoluted tubule

The proximal convoluted tubule is responsible for most of the resorption of water and solutes from the glomerular filtrate. The cells of this section of the renal tubule are cuboidal, and their surfaces facing into the tubule have thousands of microvilli, which greatly increase their surface area for resorption. These cells have lots of mitochondria—an indication that they are biochemically active. They actively transport Na^+ (with Cl^- following passively) and other solutes, such as glucose and amino acids, out of the tubule fluid. Almost all glucose and amino acid molecules that are filtered from the blood are actively resorbed by these cells and transported back into the tissue fluid. The active transport of solutes into the tissue fluid causes water to follow osmotically. The water and solutes moved into the tissue fluid are taken up by the peritubular capillaries and returned to the venous blood leaving the kidney.

Despite the large volume of water and solutes resorbed by the proximal convoluted tubule, the overall concentration, or osmolarity, of the fluid that enters the loop of Henle is not different from that of the blood plasma, although its composition is quite different. How, then, does the kidney produce urine that is hypertonic to the blood plasma?

The loop of Henle creates a concentration gradient in the surrounding tissue

Humans can produce urine that is four times more concentrated than their blood plasma. The vampire bat we encountered at the beginning of this chapter can produce a urine that is twenty times more concentrated than its blood plasma, and some desert-dwelling animals, as we will see, can produce even greater concentrations. This concentrating ability of the mammalian kidney is due to the loops of Henle, which function as a **countercurrent multiplier system**. The term "countercurrent" refers to the fact that tubule fluid in the descending limb of the loop flows in the opposite direction from that in the ascending limb. "Multiplier" refers to the ability of this system to create a concentration gradient in the renal medulla. The loops of Henle do not themselves produce a concentrated urine; rather, they increase the solute potential of the surrounding tissue fluid.

The segments of the loop of Henle differ anatomically and functionally. Cells of the descending limb and the initial cells of the ascending limb are flat, with no microvilli and few mitochondria. They are not specialized for transport. Partway up the ascending limb, the cells become specialized for active transport. They are cuboidal and have lots of mitochondria. Accordingly, the loop of Henle is divided into the *thin descending limb*, the *thin ascending limb*, and the *thick ascending limb*. To understand the countercurrent multiplier mechanism, it is easiest to move backward through the renal tubule, starting with the thick ascending limb (Figure 51.12).

The thick ascending limb actively resorbs Cl^- (with Na^+ following passively) from the tubule fluid and moves it into the surrounding tissue fluid. The thick ascending limb is not permeable to water, so the resorption of Na^+ and Cl^- raises the concentration of these solutes in the surrounding tissue fluid.

The thin descending limb, in contrast, is rather permeable to water, but not very permeable to Na^+ and Cl^-. Since the surrounding tissue fluid has been made more concentrated by the Na^+ and Cl^- resorbed from the neighboring thick ascending limb, water is withdrawn osmotically from the tubule fluid in the descending limb. Therefore, the fluid in the descending limb becomes more and more concentrated as it flows toward the bottom of the renal medulla.

The thin ascending limb, like the thick ascending limb, is not permeable to water. It is, however, permeable to Na^+ and Cl^-. As the concentrated tubule fluid flows up the thin ascending limb, it is more concentrated than the surrounding tissue fluid, so Na^+ and Cl^- diffuse out of it. When the tubule fluid reaches the thick ascending limb, active transport continues to move Na^+ and Cl^- from the tubule fluid to the tissue fluid, as we saw above.

As a result of the processes described above, the tubule fluid reaching the distal convoluted tubule is less concentrated than the blood plasma, and the solutes that have been left behind in the renal medulla have created a concentration gradient in the surrounding tissue fluid. The tissue fluid of the renal medulla becomes more and more concentrated as we move from the border with the cortex down to the tips of the renal pyramids.

Urine is concentrated in the collecting ducts

As Na^+ and Cl^- are transported out of the tubule fluid, urea and other waste products make up a greater proportion of its total solute content as it flows toward the collecting duct. Therefore, the tubule fluid entering the collecting duct is at the same *concentration* as the blood plasma, but its *composition* is considerably different from that of the plasma.

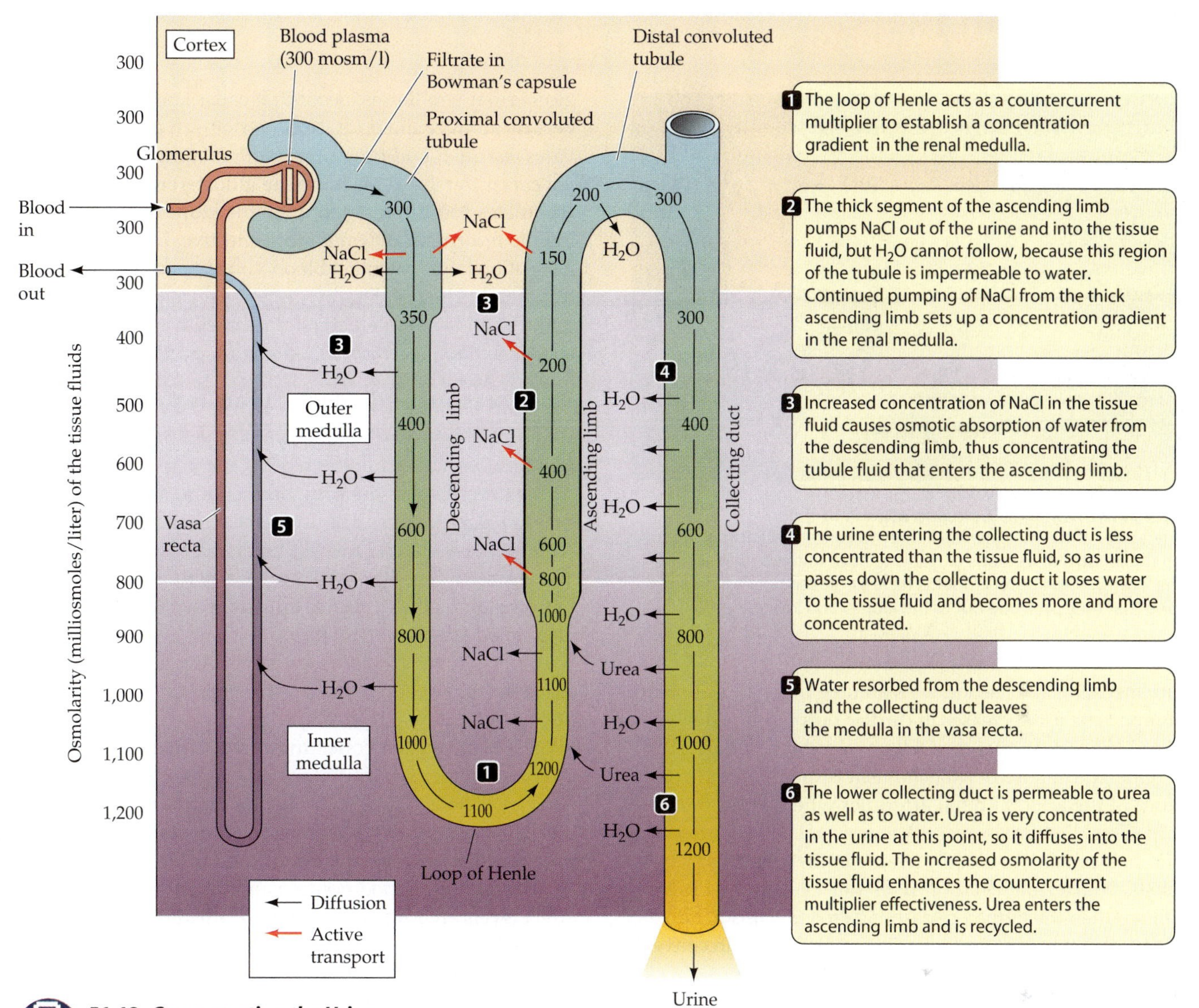

51.12 Concentrating the Urine
The countercurrent multiplier mechanism enables the kidney to produce urine that is far more concentrated than mammalian blood plasma.

The tubule fluid entering the distal convoluted tubule loses water osmotically as it flows toward the collecting duct.

The concentration gradient established in the renal medulla by the loops of Henle enables the urine to be concentrated in the collecting ducts. The collecting ducts begin in the renal cortex and run through the renal medulla before emptying into the ureter at the tips of the renal pyramids. As the solute concentration of the surrounding tissue fluid increases, more and more water is absorbed from the urine in the collecting duct. By the time it reaches the ureter, the urine has been greatly concentrated.

It follows from the process we have just described that the ability of a mammal to concentrate its urine will be determined by the maximum concentration gradient it can establish in its renal medulla. One way to increase the concentration gradient is to increase the lengths of the loops of Henle. That is precisely the adaptation we find in mammals that live in extremely arid habitats. The desert gerbil, for example, has such extremely long loops of Henle that its renal pyramid (each of its kidneys has only one, in contrast to ours) extends far out of the concave surface of the kidney and into the ureter (Figure 51.13). These animals are so effective in conserving water that they can survive on the water released by the metabolism of their dry food; they do not need to drink!

Control and Regulation of Kidney Functions

Control and regulatory mechanisms act on the kidneys to maintain blood osmolarity and blood pressure. We will discuss these various mechanisms separately, but keep in mind that they are always working together.

The kidneys act to maintain the glomerular filtration rate

If the kidneys stop filtering blood, they cannot accomplish any of their functions. The *glomerular filtration rate* (*GFR*) depends on an adequate blood supply to the kidneys at an adequate blood pressure. Therefore, the kidneys have mechanisms to maintain their blood supply and blood pressure regardless of what is happening elsewhere in the body. Because these adaptations of the kidney support the maintenance of kidney function, they are called *autoregula-*

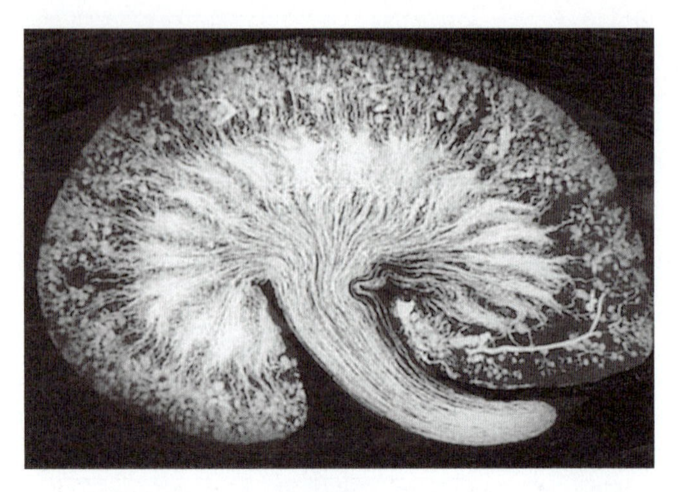

51.13 The Ability to Concentrate
The ability of the mammalian kidney to concentrate urine depends on the lengths of its loops of Henle relative to the overall size of the kidney. Some desert rodents have single renal pyramids so long that they protrude out of the kidney and into the ureter.

tory mechanisms. The kidney's autoregulatory adjustments compensate for decreases in cardiac output or decreases in blood pressure so that the GFR remains high (Figure 51.14).

One autoregulatory mechanism is the dilation (expansion) of the afferent renal arterioles when blood pressure falls. This dilation decreases the resistance in the arterioles and helps maintain blood pressure in the glomerular capillaries. If arteriole dilation does not keep the GFR from falling, then the kidney releases an enzyme, **renin**, into the blood. Renin acts on a circulating protein to begin converting it into an active hormone called **angiotensin**.

Angiotensin has several effects that help restore the GFR to normal. First, angiotensin causes the efferent renal arterioles to constrict, which elevates blood pressure in the glomerular capillaries. Second, it causes peripheral blood vessels all over the body to constrict—an action that elevates central blood pressure. Third, it stimulates the adrenal cortex to release the hormone **aldosterone**. Aldosterone stimulates sodium resorption by the kidney, thereby making the resorption of water more effective. Enhanced water resorption helps maintain blood volume and therefore central blood pressure. Finally, angiotensin acts on structures in the brain to stimulate thirst. Increased water intake in response to thirst increases blood volume and blood pressure.

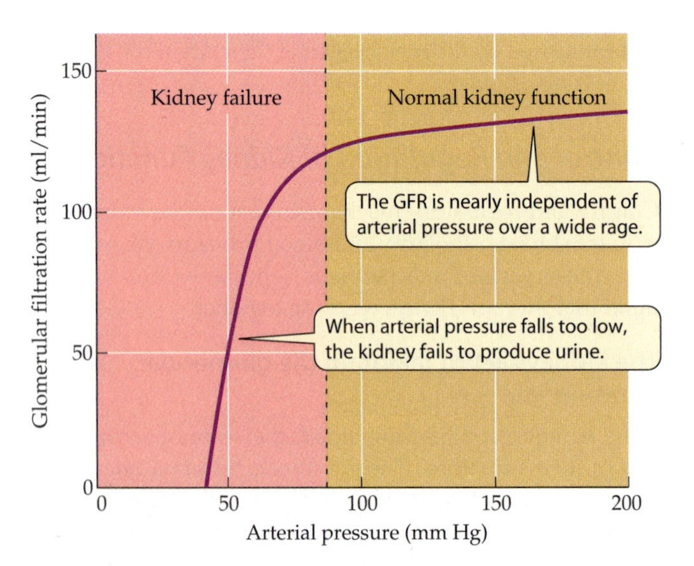

51.14 Maintaining the Glomerular Filtration Rate
Glomerular filtration is driven by arterial pressure, but autoregulatory mechanisms prevent rises and falls in glomerular filtration rate (GFR) over a wide range of pressures.

Blood pressure and osmolarity are regulated by ADH

When you lose blood volume, your blood pressure tends to fall. Besides activating the kidney autoregulatory mechanisms described in the previous section, a drop in blood pressure decreases the activity of the stretch receptors in the walls of the aorta and the carotid arteries (see Chapter 49). These stretch receptors provide information to cells in the hypothalamus that produce **antidiuretic hormone** (**ADH**, also called *vasopressin*) and send it down their axons to the posterior pituitary gland (see Chapter 41). As stretch receptor activity decreases, the production and release of this hormone increases (Figure 51.15).

ADH acts on the collecting ducts of the kidney to increase their permeability to water. When the circulating level of ADH is high, the collecting ducts are very permeable to water, more water is resorbed from the urine, and only small quantities of concentrated urine are produced, thus conserving blood volume and blood pressure. When ADH levels are low, water is not resorbed from the collecting ducts, and lots of dilute urine is produced.

ADH controls the permeability of the collecting ducts by stimulating the production and activity of membrane proteins that form water channels. These proteins, called **aquaporins**, are found in many tissues that are permeable to water—for example, the capillary endothelium, red blood cells, and the proximal convoluted tubules of the kidney. Differences among tissues in water permeability can be related to the presence or absence of aquaporins. Aquaporins are expressed in the descending limb of the loop of Henle, for example, but not in the ascending limb. One particular aquaporin is found in collecting duct cells and is controlled by ADH on both a long-term and a short-term basis. Over the long term, ADH levels influence the expression of the gene for this aquaporin; over the short term, ADH controls the insertion of the aquaporin into the cell membranes.

ADH also helps regulate blood osmolarity. Sensory cells in the hypothalamus monitor the solute potential of the blood. If blood osmolarity increases, these **osmoreceptors** stimulate increased release of ADH to enhance water resorption from the kidneys. The osmoreceptors also stimulate thirst. The resulting water retention and water intake dilutes the blood as it expands blood volume.

The heart produces a hormone that influences kidney function

When blood pressure becomes abnormally high, or when a weakened heart cannot pump blood effectively, the atria of

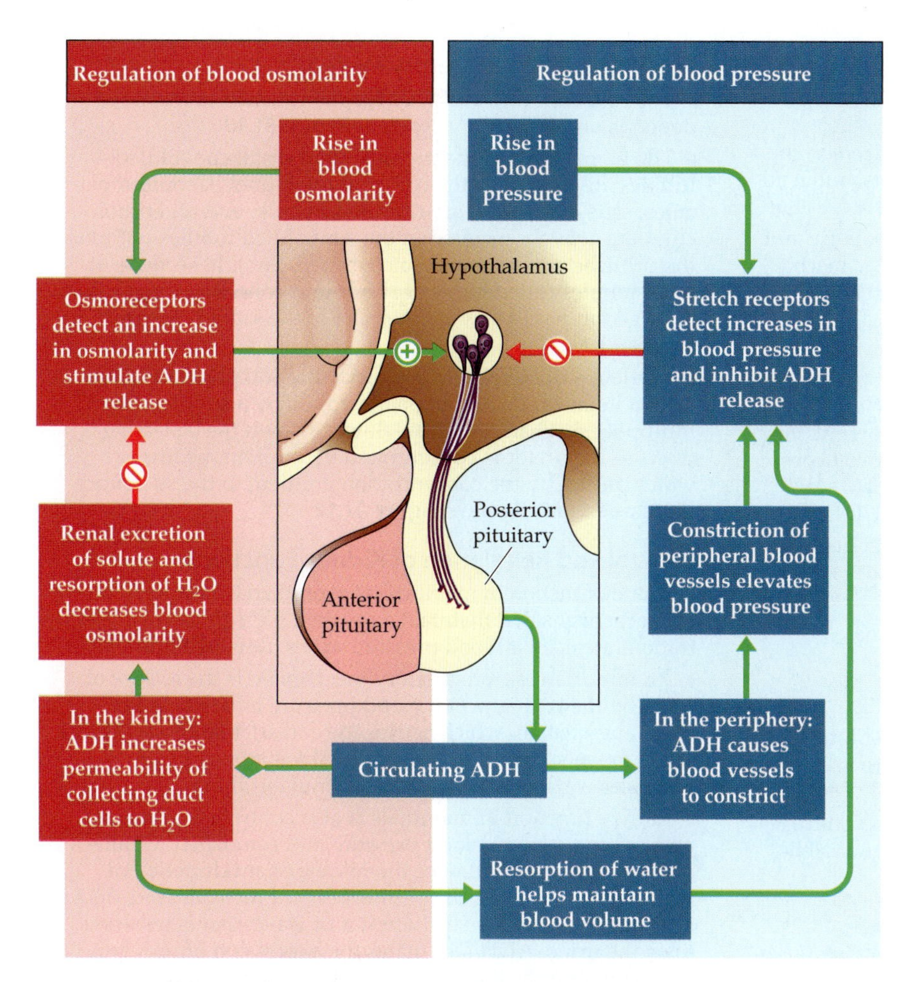

51.15 Antidiuretic Hormone Increases Blood Pressure and Promotes Water Resorption
ADH is produced by neurons in the hypothalamus and released from their axons in the posterior pituitary. The release of ADH is stimulated by hypothalamic osmoreceptors and inhibited by stretch receptors in the great arteries.

the heart become stretched. When the atrial muscle fibers are stretched too much, they release a peptide hormone called **atrial natriuretic hormone**. This hormone enters the circulation, and when it reaches the kidney, it decreases the resorption of sodium. The result is an increased loss of sodium and water, which has the effect of lowering blood volume and blood pressure.

The kidneys help regulate acid–base balance

Besides salt and water balance and nitrogen excretion, the kidneys have another important role in regulating the hydrogen ion concentration (the pH) of the blood. pH is a critical variable because it influences the structure and therefore the function of proteins. One way to minimize pH changes in a chemical solution is to add a *buffer*—a substance that can either absorb excess hydrogen ions or supply hydrogen ions (see Chapter 2). The major buffer in the blood is the bicarbonate ion, HCO_3^-, which is formed from the disassociation of carbonic acid, which in turn is formed by the hydration of CO_2 according to the following equilibrium reactions (see Chapter 48):

$$CO_2 + H_2O \rightleftharpoons H_2CO_3 \rightleftharpoons H^+ + HCO_3^-$$

You can see that if excess H^+ ions are added to this reaction mix, the reaction will move to the left and absorb the excess H^+. On the other hand, if H^+ ions are removed from the reaction mix, the reaction will move to the right and supply more H^+ ions.

The bicarbonate buffer system is important for controlling the pH of the blood because the reactions can be pushed and pulled physiologically. The lungs control the levels of CO_2 in the blood, and the kidneys control the levels of H^+ and HCO_3^- ions in the blood. The renal tubules secrete H^+ and resorb HCO_3^- (Figure 51.16). The kidney has other buffering systems as well, and together they greatly enhance the ability of the kidney to eliminate acid.

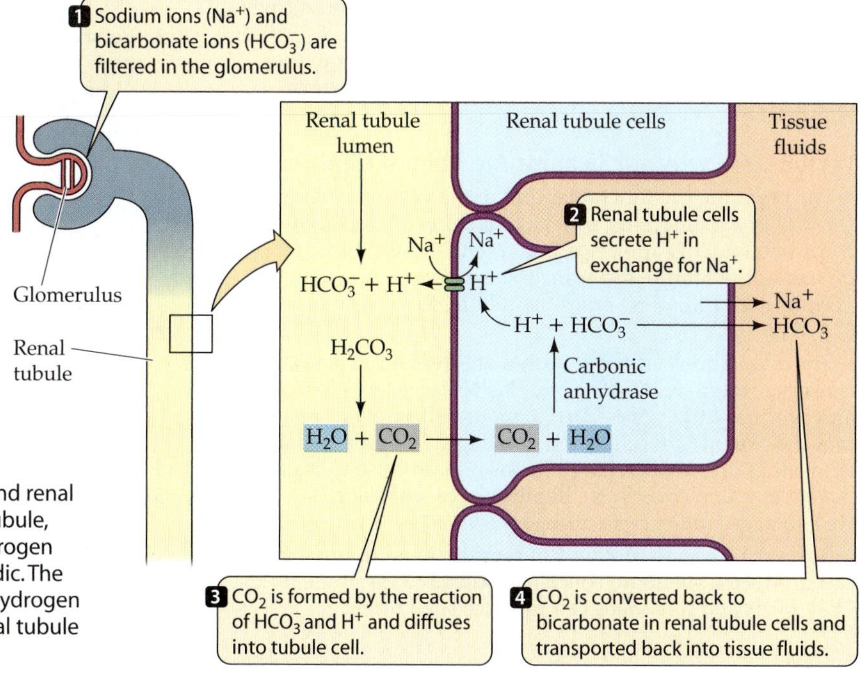

51.16 The Kidney Excretes Acids and Conserves Bases
Bicarbonate ions are filtered in the glomerulus, and renal tubule cells secrete hydrogen ions. In the renal tubule, the filtered bicarbonate buffers the secreted hydrogen ions and keeps the urine from becoming too acidic. The CO_2 formed by the reaction of bicarbonate and hydrogen ions is converted back to bicarbonate by the renal tubule cells and transported back into the tissue fluids.

Chapter Summary

Tissue Fluids and Water Balance

▶ The problems of salt and water balance and nitrogen excretion that animals face depend on their environments, but in all animal excretory systems, there is no active transport of water.

▶ All adaptations for maintaining salt and water balance and for excreting nitrogen wastes employ the same basic mechanisms: filtration of body fluids and active secretion and resorption of specific ions.

Distinguishing Environments and Animals in Terms of Salt and Water

▶ Marine animals can be osmoconformers or osmoregulators. Freshwater animals must be osmoregulators and must continually excrete water and conserve salts. All animals are ionic regulators to some degree. **Review Figure 51.1**

▶ On land, water conservation is essential, and diet determines whether salts must be conserved or excreted. Marine birds excrete excess salt through nasal salt glands. **Review Figure 51.2**

Excreting Nitrogen

▶ Aquatic animals can eliminate nitrogenous wastes such as ammonia by diffusion across their gill membranes. Terrestrial animals must detoxify ammonia by converting it to urea or uric acid for excretion. **Review Figure 51.3**

▶ Depending on the form in which they excrete their nitrogenous waste products, animals are classified as ammonotelic, ureotelic, or uricotelic.

The Diverse Excretory Systems of Invertebrates

▶ The protonephridia of flatworms consist of flame cells and excretory tubules. Tissue fluid is filtered into the tubules, which process the filtrate to produce a dilute urine. **Review Figure 51.4**

▶ In annelid worms, blood pressure causes filtration of the blood across capillary walls. The filtrate enters the coelomic cavity, where it is taken up by open-ended tubules called metanephridia. As the filtrate passes through the tubules to the outside, its composition is changed by active transport mechanisms. **Review Figure 51.5**

▶ The Malpighian tubules of insects receive ions and nitrogenous wastes by active transport across the tubule cells. Water follows by osmosis. Ions and water are resorbed from the rectum, so the insect excretes semisolid wastes. **Review Figure 51.6**

Vertebrate Excretory Systems Are Built of Nephrons

▶ The nephron, the functional unit of the vertebrate kidney, consists of a glomerulus, in which blood is filtered across the walls of a knot of capillaries, and a renal tubule, which processes the filtrate into urine to be excreted. A system of peritubular capillaries serves the tubule. **Review Figures 51.7, 51.8**

▶ The adaptations of marine fishes and terrestrial animals to conserve water are diverse. Bony fishes have few glomeruli and produce little urine. Cartilaginous fishes retain urea so that the osmotic concentration of their body fluids remains above that of seawater. Amphibians remain close to water or have waxy skin coverings . Reptiles have scaly skin, lay shelled eggs, and excrete nitrogenous wastes as uric acid.

▶ Birds share the adaptations of reptiles; in addition, they can produce urine more concentrated than their tissue fluids. Only birds and mammals can produce such urine.

The Mammalian Excretory System

▶ The concentrating ability of the mammalian kidney depends on its anatomy. **Review Figure 51.10**

▶ The glomeruli and the proximal and distal convoluted tubules are located in the cortex of the kidney. Certain molecules, salts, and water are resorbed in bulk, and other molecules are actively secreted in the convoluted tubules without the urine becoming more concentrated. Straight sections of renal tubules called loops of Henle and collecting ducts are arranged in parallel in the medulla of the kidney. **Review Figure 51.11**

▶ The loops of Henle create a concentration gradient in the extracellular fluids of the renal medulla by a countercurrent multiplier mechanism. Urine flowing down the collecting ducts to the ureter is concentrated by the osmotic loss of water caused by the concentration gradient in the surrounding tissue fluid. **Review Figure 51.12**

Control and Regulation of Kidney Functions

▶ Kidney function in mammals is controlled by autoregulatory mechanisms that maintain a constant high glomerular filtration rate even if blood pressure varies. **Review Figure 51.14**

▶ An important autoregulatory mechanism is the release of renin by the kidney when blood pressure falls. Renin activates angiotensin, which causes the constriction of peripheral blood vessels, causes the release of aldosterone (which enhances water resorption), and stimulates thirst.

▶ Kidney function in mammals is also controlled by mechanisms responsive to blood pressure and osmolarity. Changes in these variables influence the release of antidiuretic hormone, which controls the permeability of the collecting duct to water and therefore the amount of water that is resorbed from the urine. ADH stimulates the expression of proteins called aquaporins that serve as water channels in the membranes of collecting duct cells. **Review Figure 51.15**

▶ Hydrogen ions secreted by renal tubules are buffered in the urine by bicarbonate and other buffering systems. **Review Figure 51.16**

For Discussion

1. Why is it said that the oceans are a physiological desert? For what animals would this apply?
2. Persons with uncontrolled diabetes mellitus can have very high levels of glucose in their blood. Why do such individuals have a high level of urine production?
3. Inulin is a molecule that is filtered in the glomerulus, but is not secreted or resorbed by the renal tubules. If you injected inulin into an animal and after a brief time measured the concentration of inulin in its blood and urine, how could you determine the animal's glomerular filtration rate? Assume that the rate of urine production is 1 ml per minute.
4. After you did the inulin experiment to measure glomerular filtration rate, how could you use that information to determine whether another substance is secreted or resorbed by the renal tubules? Assume you can measure the concentration of that substance in the blood and in the urine. Urine production is still 1 ml per minute.
5. Explain what would happen with respect to control and regulation of your salt and water balance if you went to a movie and ate a lot of very salty popcorn.

52 Animal Behavior

A TROOP OF JAPANESE MACAQUES LIVING ON AN island was being studied by scientists, who fed the monkeys by throwing pieces of sweet potatoes onto the beach from a passing boat. The monkeys tried to brush the sand off the sweet potatoes, but they were still gritty. One day a young female monkey began taking her sweet potatoes to the water and washing them. Soon her siblings and other juveniles in her play group imitated her new behavior. Next their mothers began washing their potatoes. No adult males imitated the behavior of the juveniles or the adult females, but young males learned the behavior from their mothers and their siblings.

The scientists were fascinated by the way the creative, insightful behavior of one juvenile female spread through the population, so they presented the monkeys with a new challenge: They threw wheat onto the beach. Picking grains of wheat out of the sand was tedious and difficult. The same juvenile female came up with a solution: She carried handfuls of sand and grain to the water and threw them in. The sand sank, and the grain floated, enabling her to skim it off the surface and eat it. This behavior spread through the population in the same way potato washing did, first to other juveniles, then to mothers, and then from mothers to both their male and female offspring.

The macaques now routinely wash their food. They play in the water, which they did not do before, and they have added some marine items to their diet. Clearly, this population of monkeys has invented new behaviors that have spread by imitative learning and have become traditions in the population. One could say that they have acquired a *culture*: a set of behaviors shared by the population and transmitted by learned traditions.

The reason this study of macaques is so interesting is that it erodes what seemed to be a clear distinction between human behavior and the behavior of other animals. The behavior of most animals is largely determined by heredity, with learning playing a relatively minor role. In contrast, most human behaviors are acquired through cultural traditions and learning. The fact that other primates can invent novel behaviors and pass them on culturally shows that there is no absolute dividing line between human and animal behavior.

We begin this chapter with descriptions of some classic studies of behaviors that are largely shaped by inheritance, but to varying degrees are modified by experience. Then we explore how hormones influence the development and expression of behavior.

Next we discuss animal communication, showing how this behavior has been shaped by natural selection. Then we look at studies of biological rhythms and navigation to see how the mechanisms underlying these behaviors have been investigated. Throughout the chapter, we hope you will use what you read to raise your own questions about human behavior, to which we will return at the end of the chapter.

Learned Behaviors Shared by a Population Become a Culture
In the space of only a few generations, a population of Japanese macaques (*Macaca fuscata*) learned and transmitted a set of behaviors that included washing food, playing in the water, and eating marine food items—a new "culture" of water-related behaviors.

What, How, and Why Questions

Most species of animals can be identified by their behaviors. Behavior is highly visible and shows us what an animal does: how it gets food, how it avoids dangers in the environment, and how it reproduces. Behavior is highly adaptive, and it is therefore not surprising that many behaviors are shaped by natural selection, and are highly species-specific. On the other hand, flexibility of behavior can be extremely valuable to an animal that has to deal with changing conditions and complex situations, as in social interactions. Therefore, to varying degrees, behavior is modifiable by learning.

In studying any behavior, we can ask *what*, *how*, and *why* questions. *What* questions focus on the details of behavior, including the *proximate cause* of the behavior—in other words, what stimuli cause the animal to express the behavior. *How* questions are about the mechanisms of behavior—the underlying neural, hormonal, and anatomical mechanisms that we have been studying in Part Six. *How* questions can also focus on the means by which an animal acquires a behavior—the relative roles of genetically determined mechanisms and experience. Most behaviors involve complex interactions of inherited anatomical and physiological mechanisms and the ability to alter behavior through learning.

Why questions have to do with the *ultimate causes* of behavior—the selective pressures that shaped its evolution. In this chapter we will frequently discuss the adaptive nature of behavior, but the evolution of behavior will be the major focus of the next chapter.

Argiope aurantia

52.1 Spider Web Designs Are Species-Specific
Each spider performs a stereotypic sequence of movements typical of its species that results in a species-specific web design.

Behavior Shaped by Inheritance

Much of the behavior of many animals is highly **stereotypic** (it is performed in the same way every time) and **species-specific** (there is little variation in the way different individuals of the same species perform it). We can identify species of spiders, for example, by their web designs (Figure 52.1). Web spinning requires thousands of movements performed in just the right sequence, and for a given species, most of that sequence is performed the same way every time. Different spider species spin webs of different designs, using different sequences of movements.

Web spinning by spiders is also an example of a complex behavior that requires no learning or prior experience. When juvenile spiders hatch, their mother is already dead, and they disperse immediately (remember *Charlotte's Web* by E. B. White?). They have no experience of their mother's web. Yet, when they construct their own webs, they do it perfectly without the benefit of experience or a model to copy. In fact, their web spinning behavior is actually rather resistant to modification by learning. When confronted experimentally with challenges to web construction, young spiders appear incapable of learning how to modify the design of their webs.

Many classic studies of stereotypic and species-specific behaviors were performed by scientists who studied the behavior of animals in nature—a field called **ethology**. The early ethologists asked to what extent such behaviors are determined by inheritance and to what extent they are modifiable by experience. Two experimental approaches were used to test whether behaviors are hereditary: (1) depriving animals of opportunities to learn and (2) studying the behavior of the offspring of two parents that differ in their behavior.

1. Tail shake
2. Head flick
3. Tail shake
4. Bill shake
5. Grunt whistle
6. Tail shake

Deprivation and hybridization experiments test whether a behavior is inherited

In a *deprivation experiment,* an animal is reared so that it is deprived of all experience relevant to the behavior under study. In one such experiment, a tree squirrel was reared in isolation, on a liquid diet, and in a cage without soil or other particulate matter. When the young squirrel was given a nut, it put the nut in its mouth and ran around the cage. Eventually it made stereotypic digging movements in the corner of its cage, placed the nut in the imaginary hole, went through the motions of refilling the hole, and ended by tamping the nonexistent soil with its nose. The squirrel had never handled a food object and had never experienced soil, yet the stereotypic behavior of a squirrel burying a nut was fully expressed.

In a *hybridization experiment,* closely related species are interbred and the behavior of their offspring observed. Closely related species frequently show distinct differences in certain kinds of behavior. When such species can be interbred, it is possible to to see whether their offspring have inherited elements of the behavior of one or both parents.

Konrad Lorenz, a pioneer in the field of ethology, conducted hybridization experiments on ducks to investigate the genetic determinants of their elaborate courtship displays. Dabbling duck species such as mallards, teals, pintails, and gadwalls are closely related to one another and can interbreed, but because of the specificity of their courtship displays, they rarely do so in nature. Each male duck performs a carefully choreographed water ballet that is typical of his species (Figure 52.2), and a female is not likely to accept his advances unless the entire display is successfully and correctly completed.

When Lorenz crossbred duck species, he found that the hybrid offspring expressed some components of the courtship displays of each parent species, but expressed them in new combinations. Of particular interest was his observation that the hybrids sometimes showed display components that were not in the repertoire of either parent species, but were characteristic of the displays of other species. Lorenz's hybridization studies clearly demonstrated that the motor patterns of the courtship displays were inherited. The fact that natural selection was shaping these genetically determined behaviors was evidenced by the fact that females were not interested in males performing hybrid displays.

52.2 Courtship Ballet of the Mallard
The courtship display of the male mallard duck contains about ten elements. The displays of closely related duck species contain some of the same ten elements, but have other elements not displayed by mallards. The elements of the courtship display and their sequence are species-specific and act to prevent hybridization.

52.3 A Releaser of Aggressive Behavior
Red feathers serve as a releaser of aggressive behavior in male European robins.

Simple stimuli can trigger behaviors

If a behavior is not expressed during a deprivation experiment, it may nonetheless have genetic determinants. The right conditions may not have been available to stimulate the behavior during the experiment. The squirrel described above, for example, had to be given a nut for its digging and burying behaviors to be triggered. Specific stimuli are required to elicit the expression of many inherited behaviors. Two pioneering ethologists, Konrad Lorenz and Niko Tinbergen, who conducted classic studies of the nature of the stimuli that elicit such behaviors, called such stimuli **releasers**.

Releasers are usually a simple subset of all the sensory information available to an animal. Adult male European robins, for example, have red feathers on their breasts, which serve as releasers of aggressive behavior in other males. During the breeding season, the sight of an adult male robin stimulates another male robin to sing, perform aggressive displays, and attack the intruder if he does not heed these warnings. An immature male robin, whose feathers are all brown, does not elicit this aggressive behavior. A tuft of red feathers on a stick, however, is a sufficient releaser for male aggressive behavior in robins (Figure 52.3).

Tinbergen and A. C. Perdeck carefully examined the releasers involved in the interactions between herring gulls and their chicks during feeding. An adult herring gull has a

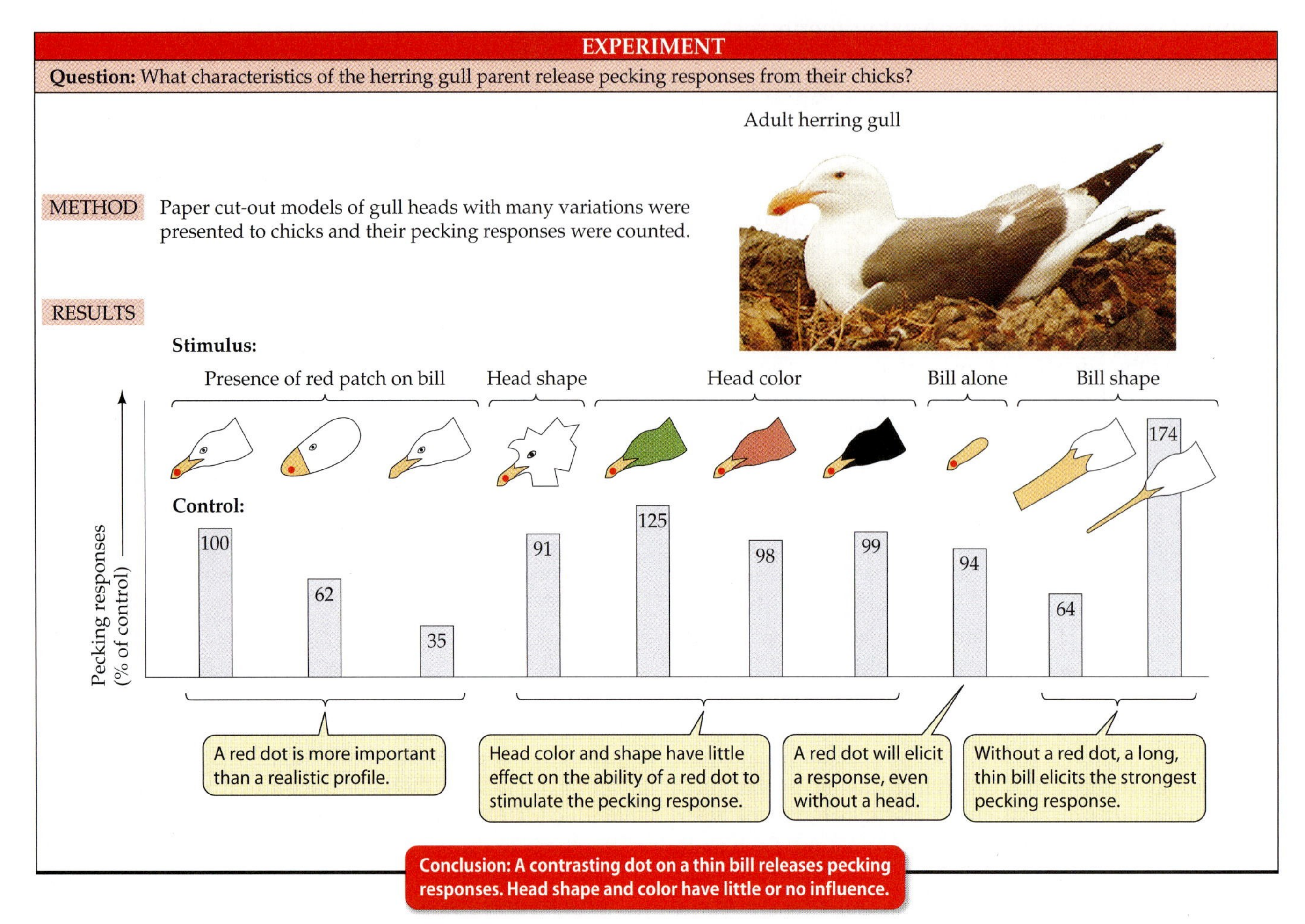

52.4 Releasing the Pecking Response
A series of experiments rated the pecking responses of herring gull chicks to artificial models of gull heads to discover which features of the parent were releasers of this behavior.

red dot at the end of its bill (Figure 52.4). When the gull returns to its nest with food, the chicks peck at the red dot, thereby stimulating the adult to regurgitate the food for the chicks to eat.

Tinbergen and Perdeck hypothesized that the red dot was a releaser for the chicks' begging behavior. To test their hypothesis, they made paper cutout models of gull heads and bills, varying the colors and the shapes. Then they rated each model according to how many pecks it received from naive, newly hatched chicks (Figure 52.4). The shape or color of the model head made no difference. In fact, a head was not even necessary; the chicks responded just as well to models of bills alone—as long as they had the red dot. Surprisingly, the most effective releaser for chick pecking behavior was a long, thin object with a dark tip that bore no resemblance to an adult herring gull. Clearly the chicks had inherited the ability to recognize a simple stimulus and respond to it with their also inherited begging behavior. To the ethologists, this represented an excellent example of a behavior that was genetically determined rather than learned.

Learning also shapes behavior

For their very significant contributions to our understanding of animal behavior, three ethologists, Lorenz, Tinbergen, and Karl von Frisch (whose work on honeybees you will encounter later in this chapter) shared a Nobel Prize in 1973. New generations of behavioral biologists, however, have moved beyond the ethologists' focus on inherited behavior to show that most behavior actually involves an interaction between inheritance and learning. The begging behavior of gull chicks is a case in point. Although newly hatched chicks respond maximally to simplistic artificial releasers, they gradually learn to discriminate between models and real gull heads, and they eventually beg only from their own parents. Thus, the inherited ability to recognize a simple releaser is subsequently refined by learning.

The early ethologists did not ignore learning or deny that it took place; in fact, they pioneered the study of learning. Tinbergen performed an early study of *spatial learning*, by which an animal learns to recognize features in its environment. In a classic experiment, he placed objects such as pine cones near the entrance of a nest dug by a female digger wasp. After the wasp left her nest, he moved the objects a short distance away. Upon returning, the wasp oriented to the moved objects and could not find her nest entrance (Figure 52.5). She had learned to recognize objects in the en-

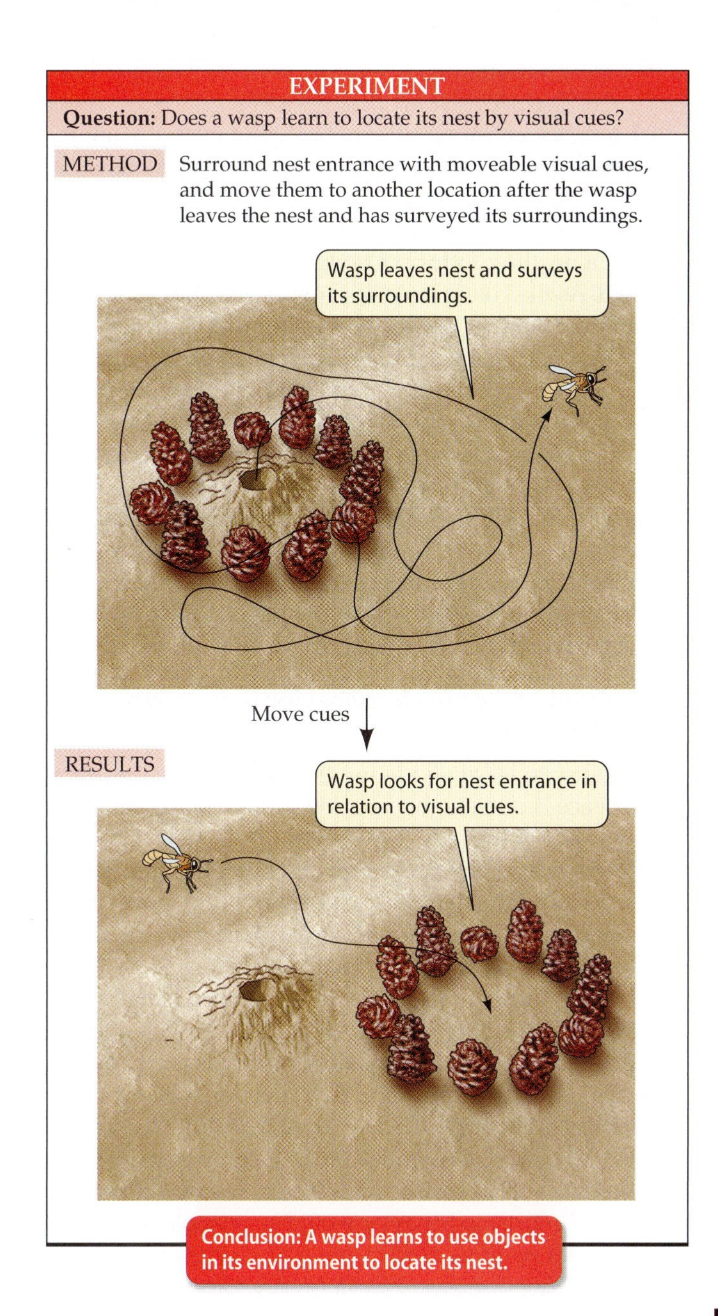

52.5 Spatial Learning
Tinbergen's classic experiment showed that a female digger wasp learns the positions of objects in her environment.

vironment to use as orientation cues. More recently, spatial learning has been studied in animals such as squirrels, chickadees, and jays that cache food items in hundreds and even thousands of locations. Their capacity for learning and remembering where their food is cached is phenomenal.

Imprinting is the learning of a complex releaser

Releasers are generally simple subsets of the available information because there are limits to what can be programmed genetically. A type of learning called **imprinting** makes it possible to learn, during a limited **critical period**, a complex set of stimuli that can later serve as a releaser. The classic example is the imprinting of offspring on their parents and parents on their offspring to ensure individual recognition even in a crowded situation such as a colony or a herd.

When Lorenz incubated goose eggs in an incubator, and he was the first thing the goslings saw when they hatched, they imprinted on him, following him everywhere and interacting with him as if he were their parent (Figure 52.6). When the experiment was repeated by his assistants, each wearing boots with a different design, the goslings imprinted on the boots, and would follow only a person wearing the boots they first saw when they hatched.

The critical period for imprinting is determined by a developmental or hormonal state and can be quite brief. If a mother goat, for example, does not nuzzle and lick her newborn within 5 to 10 minutes after birth, she will not recognize it as her own later. In this case, imprinting depends on olfactory cues, and the critical period is determined by the high levels of the hormone oxytocin circulating in the mother at the time of birth.

Inheritance and learning interact to produce bird song

Many behavior patterns are intricate interactions of inheritance and learning. One example that has been the subject of some elegant experiments is bird song. Adult male songbirds use a species-specific song in territorial displays and courtship. A few species, such as the song sparrow, express their species-specific song even during deprivation experiments, but most species do not. For most species, such as the white-crowned sparrow, learning is an essential step in the acquisition of song.

If the eggs of white-crowned sparrows are hatched in an incubator and the young male birds are reared in isolation, their adult songs will be unusual assemblages of sounds, not the typical species-specific song. This species cannot express its species-specific song without being imprinted on

52.6 Imprinting Enables an Animal to Learn a Complex Releaser
When Konrad Lorenz was the first thing newly hatched goslings saw, they imprinted on him, interacted with him as if he were their parent.

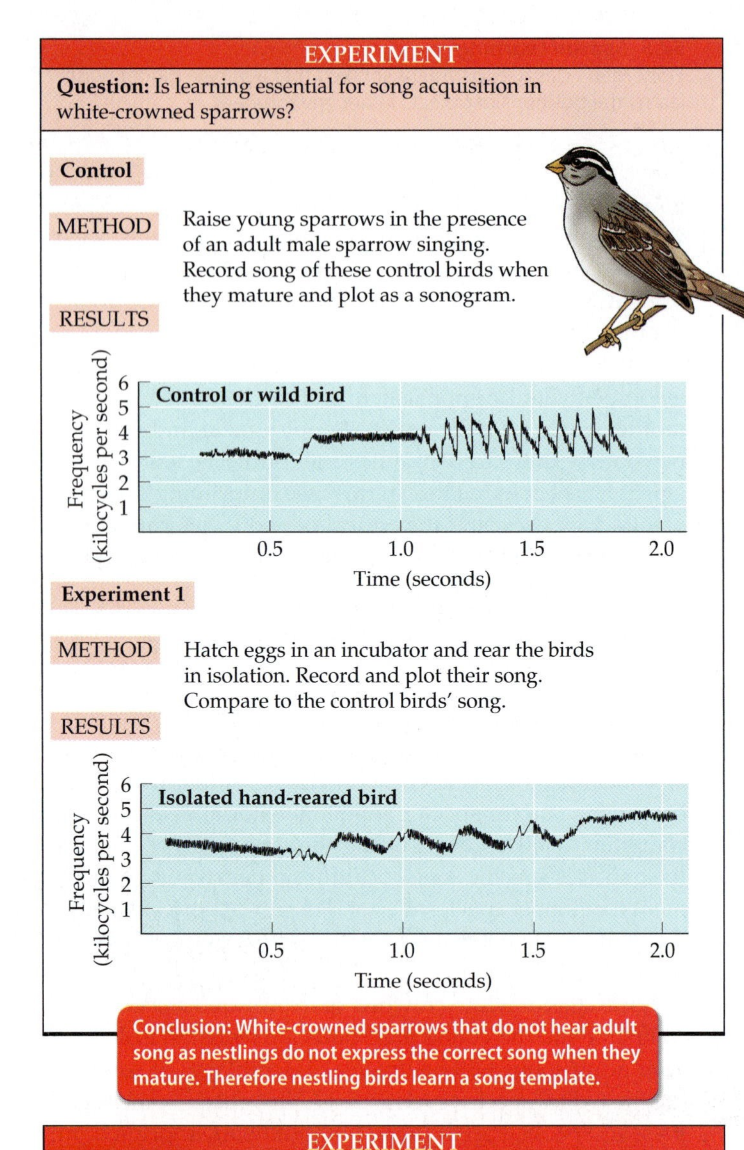

52.7 Two Critical Periods for Song Learning
To sing his species-specific song as an adult, a male white-crowned sparrow must acquire a song memory by hearing the song as a nestling, and must be able to hear himself as he attempts to match his singing to that memory.

that song as a nestling (Figure 52.7). But even though the male white-crowned sparrow must hear the song of his own species as a nestling to sing it as an adult, he does not sing it as a juvenile. Instead, he uses his auditory imprinting as a nestling to form a song memory in his nervous system. As the young male sparrow approaches sexual maturity the following spring, he tries to sing, and eventually he matches his imprinted song memory through trial and error. If a bird that has heard his species-specific song as a juvenile is deafened before he begins to express his song, he will not develop his species-specific song (see Figure 52.7). The bird must be able to hear himself to match his song memory. If he is deafened *after* he expresses his correct song, he will continue to sing like a normal bird. Two periods of learning are essential: the first in the nestling stage, the second as the bird approaches sexual maturity.

Genetically determined behavior is adaptive under certain conditions

The ability to learn and to modify behavior as a result of experience is often highly adaptive. Most human behavior is the result of learning. Why, then, are so many behavior patterns in so many species genetically determined? We have already touched on one answer to this question: If role models and opportunities to learn are not available—as in species with nonoverlapping generations, such as spiders—then there is no alternative to inherited behavior.

Inherited behaviors are also adaptive when mistakes are costly or dangerous. Mating with a member of the wrong species is a costly mistake; thus the function of much courtship behavior, such as that of dabbling ducks, is to guarantee species recognition. In an environment in which incorrect as well as correct models exist, learning the wrong pattern of courtship behavior would be possible.

Behavior patterns used to avoid predators or capture of dangerous prey allow no room for mistakes. If the behavior is not performed promptly and accurately the first time, there may not be a second chance (Figure 52.8).

Thus, *inherited behavior is highly adaptive for species that have little opportunity to learn, for species that might learn the wrong behavior, and in situations in which mistakes are costly or dangerous.*

52.8 Some Things Can't Be Learned by Trial and Error
In total darkness, the sound of a striking rattlesnake triggers an automatic escape jump in a kangaroo rat. The rat does not have to learn this behavior.

Hormones and Behavior

All behavior depends on the nervous system for initiation, coordination, and execution. Frequently, however, it is the endocrine system, through its controlling influences on the development and the physiological state of the animal, that determines when a particular behavior is performed, and even when certain behaviors can be learned. In this section we will present two complex cases in which hormones control the development, learning, and expression of behavior: sexual behavior in rats and maturation of the brain regions required for song learning and expression in birds.

Sex steroids determine the development and expression of sexual behavior in rats

Differences in the behavior of males and females of a species are clear examples of genes influencing the development and expression of behavior. Such sex differences in behavior are the result of actions of the sex steroids on the brain.

Rats, like most other animals, have stereotypic sexual behaviors. A female rat in estrus (receptive to males) responds to a tactile stimulus of her hindquarters by assuming a mating posture called *lordosis*. A male rat encountering a female in estrus engages in stereotypic copulatory behavior. The roles of genes and sex steroids in the development and expression of lordosis and male copulatory behavior have been investigated through experiments that manipulated the exposure of the developing and adult rat brain to sex steroids.

Experiments such as those shown in Figure 52.9 led to three conclusions:

- Sex steroids are necessary for adult rats to express sexual behavior. Moreover, the male sex steroid, testosterone, has an effect only in males, and the female sex steroid, estradiol, has an effect only in females.

52.9 *Hormonal Control of Sexual Behavior* In newborn rats of both sexes whose reproductive organs (ovaries or testes) have been removed, the presence of testosterone establishes male behavior patterns (no lordosis), and its absence establishes female patterns (lordosis).

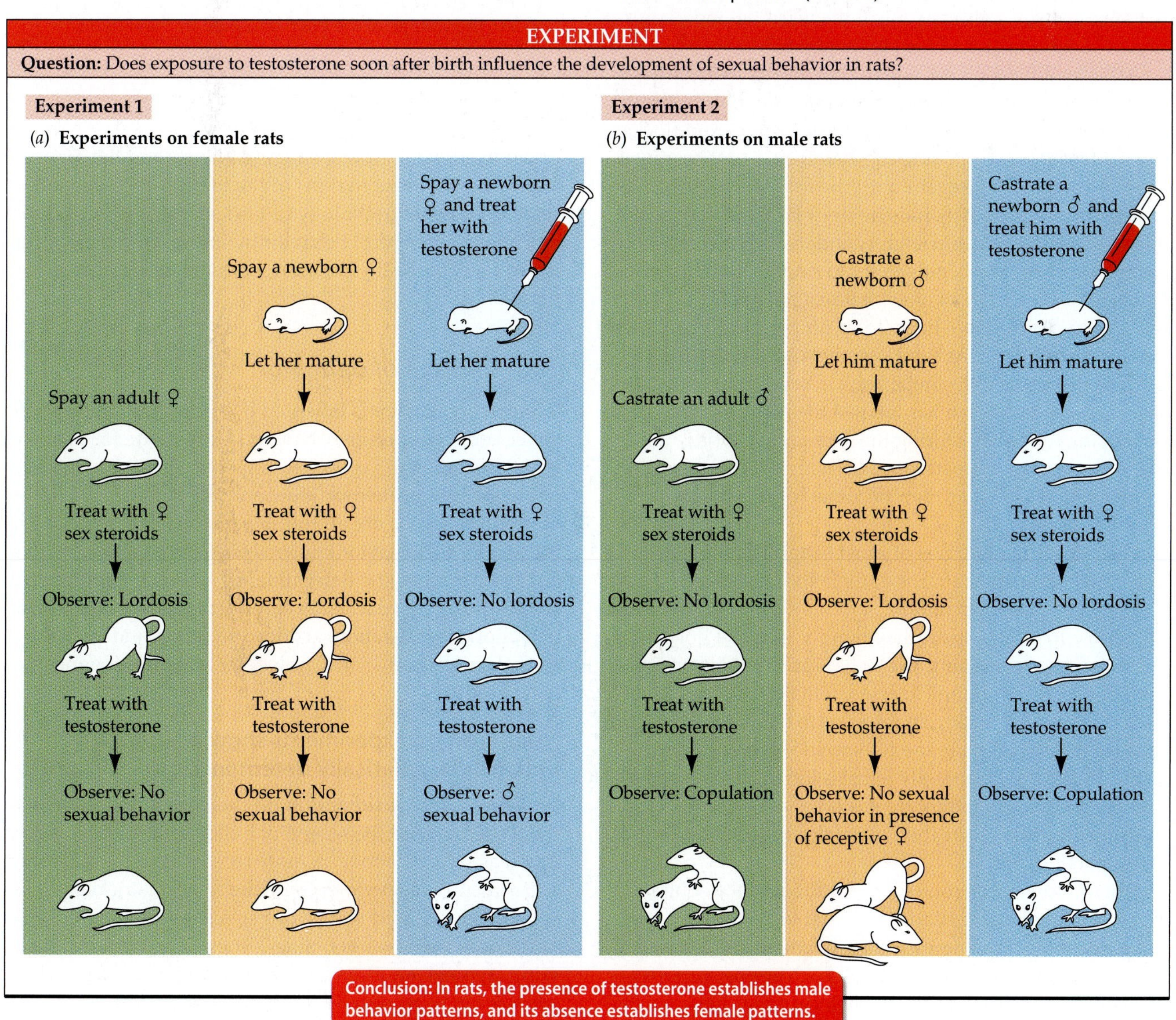

- Development of male sexual behavior requires the brain of the newborn rat to be exposed to testosterone, but development of female sexual behavior does not require the neonatal brain to be exposed to estradiol.
- Neonatal exposure to testosterone masculinizes the nervous systems of both genetic males and females so that they express male sexual behavior as adults.

Thus, the sex steroids that are present during development determine which pattern of sexual behavior develops, and the sex steroids that are present in adulthood determine whether that pattern is expressed.

Testosterone affects the development of the brain regions responsible for song in birds

As we saw above, learning is essential for the acquisition of bird song. Both male and female birds hear their species-specific song as nestlings, but only the males of most songbird species sing as adults. Male birds use song to claim territory, compete with other males, and declare dominance. They also use song to attract females, which suggests that the females know the song of their species even if they do not sing. Do sex steroids control the learning and expression of song in male and female songbirds?

After leaving the nest where they heard their father's song, young songbirds from temperate and arctic habitats migrate and associate with other species in mixed flocks. During this time they do not sing, and they do not hear their species-specific song again until the following spring. As that spring approaches and the days become longer, the young male's testes begin to grow and mature. As his testosterone level rises, he begins to try to sing. Even if he is isolated at this time from all other males of his species, his song will gradually improve until it is a proper rendition of his species-specific song. At that point the song is **crystallized**—the bird expresses it in similar form every spring thereafter. The young male's brain has learned the pattern of the song by hearing his father. During the subsequent spring, under the influence of testosterone, he learns to express that song—a behavior that then becomes rigidly fixed in his nervous system.

Why don't the females of most songbird species sing? Can't they learn the patterns of their species-specific song? Do they lack the muscular or nervous system capabilities necessary to sing? Or do they simply lack the hormonal stimulus for developing the behavior? To answer these questions, investigators injected female songbirds with testosterone in the spring. In response to these injections, the females developed their species-specific song and sang just as the males did. Apparently females learn the song pattern of their species when they are nestlings and have the capability to express it, but they normally lack the hormonal stimulation.

What does testosterone do to the brain of the songbird? A remarkable discovery revealed that testosterone causes the parts of the brain necessary for learning and expressing song to grow larger (Figure 52.10). Each spring, certain regions of the males' brains grow. Individual neurons increase in size and grow longer extensions, and the numbers of neurons in those regions of the brain increase. Such research on the neurobiology of bird song has revealed that hormones can control behavior by influencing brain structure as well as brain function, both developmentally and seasonally.

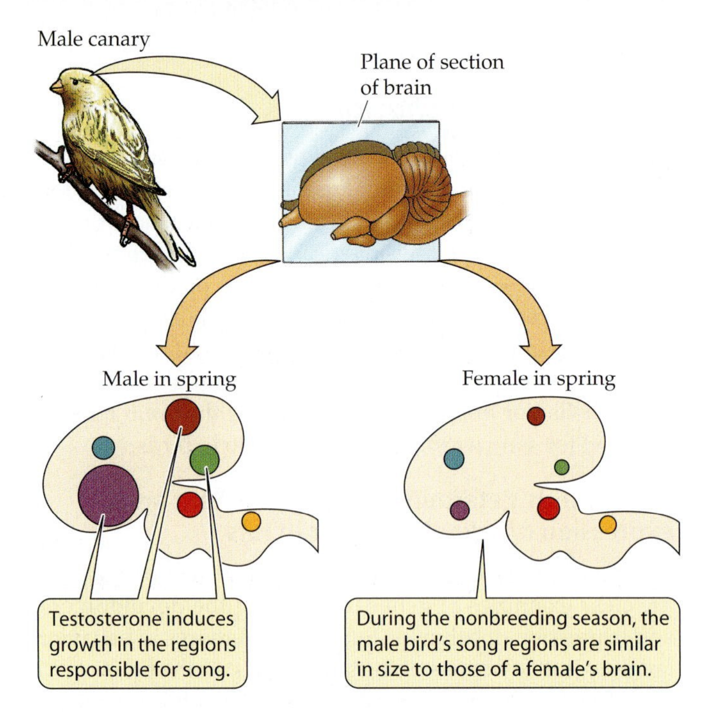

52.10 Effects of Testosterone on Bird Brains
In spring, rising testosterone levels in the male cause the song regions of the brain to develop. The size of each circle is proportional to the volume of the brain occupied by that region.

The Genetics of Behavior

To say that behavior is inherited does not mean that specific genes code for specific behaviors. Genes code for proteins, and there are many complex steps between the expression of a gene as a protein product and the expression of a behavior. In no case are all the steps between a gene and its influence on a behavior known. Nevertheless, it is clear that behavior has genetic determinants. In this section we will look at three approaches to investigating how genes affect behavior: hybridization, artificial selection and crossing of the selected strains, and molecular analysis of genes and gene products.

Hybridization experiments show whether a behavior is genetically determined

The effects of hybridization on the courtship displays of duck species were the subject of a classic ethological experiment, as we saw above. A more recent set of hybridization experiments was performed on the songs of crickets. Crickets songs, like bird songs, are species-specific, and as in birds, only male crickets "sing." They do so by rubbing one wing against another that has a serrated edge. These sounds can be recorded and analyzed quantitatively.

When two species of crickets were crossed, their offspring (the F_1 generation) expressed songs that had features of the songs of the two parental species. Backcrosses of F_1 individuals with the parental species produced individuals that had songs closer to the parental species used in the backcross. Clearly the genetic background determined the song pattern. What was amazing, however, was the demonstration that female preferences for male songs were under similar genetic control. Given a choice, females from each parental species preferred the calls of males from their own species, but hybrid females preferred the calls of hybrid males.

These genetic differences between the cricket species and the hybrids were reflected in the properties of their nervous systems. When specific neurons in the crickets' brains were stimulated, songs were expressed that reflected the genotypes of the crickets.

Artificial selection and crossbreeding experiments reveal the genetic complexity of behaviors

Domesticated animals provide abundant evidence that artificial selection of mating pairs on the basis of their behavior can result in strains with distinct behavioral as well as anatomical characteristics. Among dogs, consider retrievers, pointers, and shepherds. Each has a particular behavioral tendency that can be honed to a fine degree by training. However, dogs and other large animals are not the best subjects for genetic studies. Most artificial selection experiments in behavioral genetics have been done on more convenient laboratory animals with short life cycles and large numbers of offspring.

A favorite subject for behavioral genetic studies has been the fruit fly (*Drosophila*). Artificial selection has been successful in shaping a variety of behavior patterns in fruit flies, especially aspects of their courtship and mating behavior. Crossing of these artificially selected strains reveals that most of these behavioral differences are due to multiple genes that probably influence the behavior indirectly by altering general properties of the nervous system. Some single-gene effects, however, can be isolated. One example is the gene *per* (short for "period"), which alters the frequency of the wing vibrations that are part of the male's courtship display. The *per* gene is not a courtship behavior gene, however. It has subsequently been found that this gene codes for a transcription factor that plays an important role in the generation of daily rhythms of rest and activity, as we will see below. How it alters the development of wingbeat frequency is not clear.

Few behavioral genetic studies reveal simple Mendelian segregation of behavioral traits. An exception is nest-cleaning behavior in honeybees. One genetic strain of honeybees practices nest-cleaning, or *hygienic*, behavior, which makes them resistant to a bacterium that infects and kills the larvae of honeybees. When a larva dies, workers uncap its brood cell and remove the carcass from the hive. Another strain of honeybees does not show this hygienic behavior and therefore is more susceptible to the spread of the disease (Figure 52.11).

When these two strains of honeybees were crossed, the results indicated that the hygienic behavior was controlled by two recessive genes. All members of the F_1 generation were nonhygienic, indicating that the behavior is controlled by recessive genes. Backcrossing the F_1 with the hygienic strain produced the typical 3:1 ratio expected for a two-gene trait (see Chapter 10). The behavior of the nonhygienic hybrid individuals was very interesting. One-third of them showed no hygienic behavior at all; one-third uncapped the cells of dead larvae but did not remove them; and one-third did not uncap cells, but did remove carcasses if the cells were open.

52.11 Genes and Hygienic Behavior in Honeybees
Some honeybee strains remove the carcasses of dead larvae from their nests. This behavior seems to have two components: uncapping the larval cell (*u*) and removing the carcass (*r*), each of which is under the control of a recessive gene.

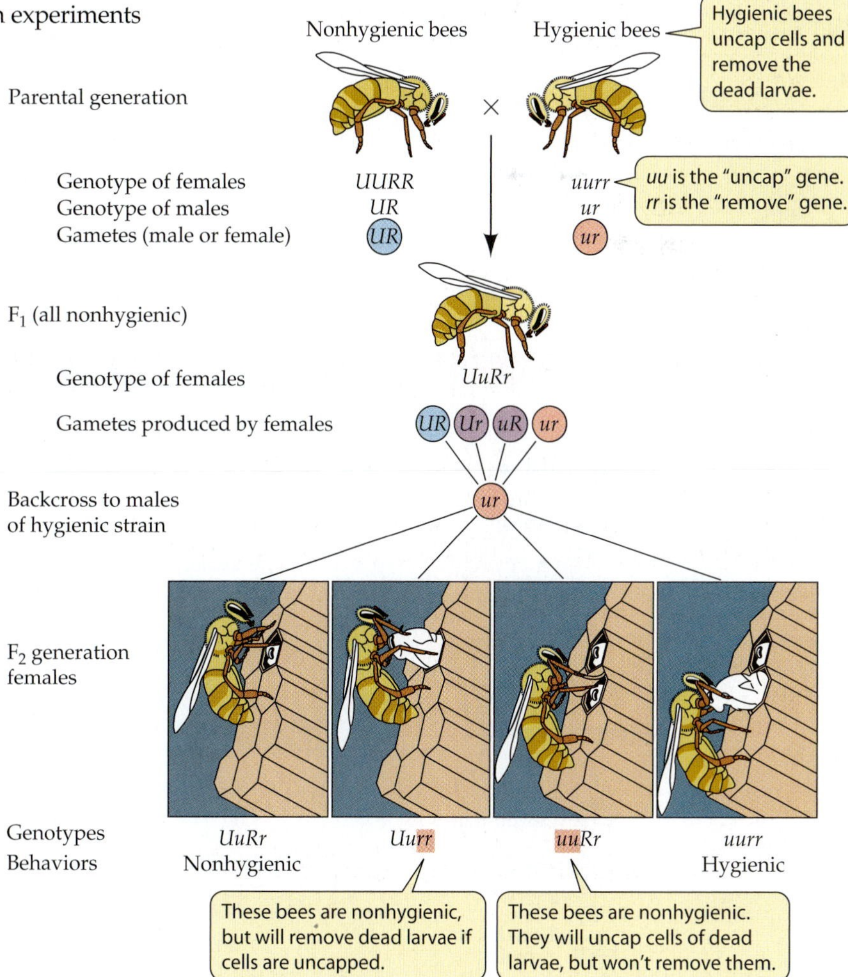

Even though these results appear to indicate a gene for uncapping and a gene for removal, these behavior patterns are complex. They involve sensory mechanisms, orientation movements, and motor patterns, each of which depends on multiple properties of many cells. The genetic deficits of nonhygienic bees could influence very small, specific, yet critical properties of some cells. If a single critical property, such as a crucial synapse or a particular sensory receptor, were lacking, the whole behavior would not be expressed. The responsible gene, then, is not a specific gene that codes for the entire behavior.

Molecular genetics techniques reveal specific genes that influence behavior

Molecular geneticists are investigating specific genes that influence behaviors. Male courtship behavior in fruit flies (*Drosophila*) is a subject of many such studies. This behavior is stereotypic, species-specific, and requires no learning. Males recognize potential mates, follow them, tap the female's body with their forelegs, extend and vibrate one wing, and lick the female's genitals. If the female is receptive, the male copulates with her (Figure 52.12*a*). Research in molecular genetics has now shown that most of this male courtship behavior is controlled by a single gene.

In fruit flies with two X chromosomes (females), a gene called *sex-lethal* (*sxl*) is expressed. This gene is at the top of a genetic hierarchy that determines all aspects of sexual differentiation and behavior (Figure 52.12*b*). The Sxl protein causes another gene called *transformer* (*tra*) to produce the female-specific Tra protein. Fruit flies without the *tra* gene develop into males anatomically and behaviorally, regardless of how many X chromosomes they have. But it is still another gene in the sex determination hierarchy that is responsible for male behavior.

The Tra protein controls two additional genes called *doublesex* (*dsx*) and *fruitless* (*fru*). The *dsx* gene mostly controls the anatomical differentiation of males, and *fru* causes the formation of a nervous system that expresses male courtship behavior. Mutations of the *fru* gene do not affect male body form, but they disrupt male courtship behavior. We don't know all of the actions that the male-specific Fru protein has in the development of the fruit fly nervous system, but this is about as close as we can get at present to identifying a gene that controls a complex behavior.

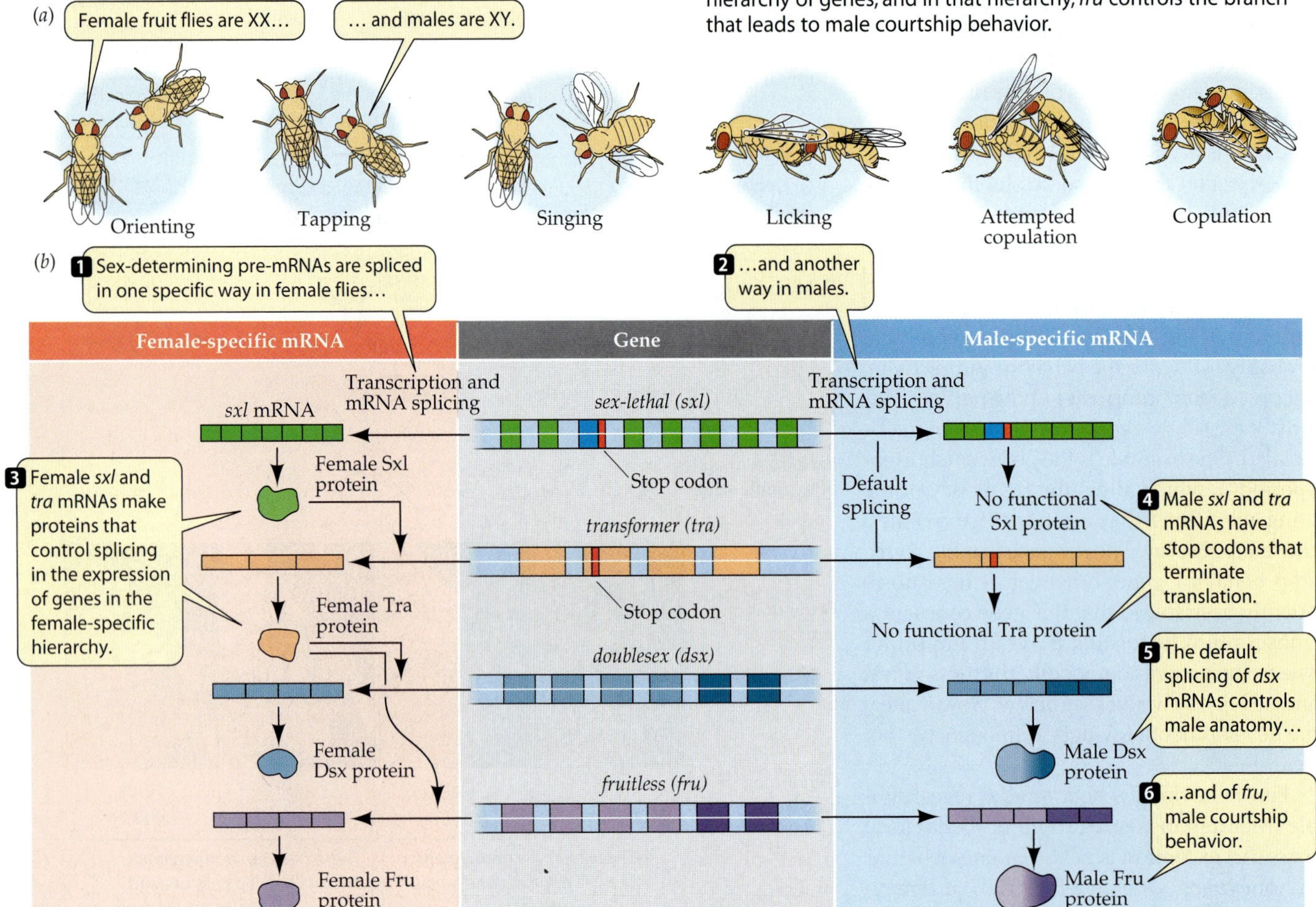

52.12 The* fruitless *Gene Controls Male Courtship Behavior in Fruit Flies
(*a*) Male fruit flies display stereotypic, species-specific courtship behavior. (*b*) Sexual differentiation in *Drosophila* is controlled by a hierarchy of genes, and in that hierarchy, *fru* controls the branch that leads to male courtship behavior.

Communication

Communication is behavior that influences the actions of other individuals. It consists of **displays** or **signals** that can be perceived by other individuals and which convey information to them. Natural selection shapes displays or signals into systems of communication if the transmission of information benefits both the sender and the receiver. Thus, the ultimate cause of communication is the selective advantage it gives to individuals that engage in it. The courtship displays of a male, for example, benefit the male if they attract females, and they benefit the female if they allow her to assess whether the male is of the right species and whether he is strong, vigorous, and has other attributes that will make him a good father. A common mutual benefit of communication is the reduction of uncertainty about the status or intentions of the signaler. Even in aggressive interactions, reducing uncertainty helps both sender and receiver to avoid physical harm.

Studies of communication can be complex because they must take into account the sender, the receiver, and the environment. The displays or signals that an animal can generate depend on its physiology and anatomy. Likewise, an animal's ability to perceive displays or signals depends on its sensory physiology and on the environment through which the display or signal must be transmitted.

In Chapter 45, we learned how sensory systems function in chemosensation, tactile sensation, audition, vision, and electrosensation. These are the channels of animal communication. In the discussion that follows, we will explore each of these five channels in turn.

Chemical signals are durable but inflexible

Molecules used for chemical communication between individual animals are called **pheromones**. Because of the diversity of their molecular structures, pheromones can communicate very specific messages that contain a great deal of information. The mate attraction pheromone of the female silkworm moth is a good example (see Figure 45.4). Male moths as far as several kilometers downwind are informed by these molecules that a female of their species is sexually receptive. By orienting to the wind direction and following the concentration gradient of the molecules, they can find her.

Territory marking is another example in which detailed information is conveyed by chemical communication (Figure 52.13). Pheromonal messages left by mammals such as cats and dogs, for example, can reveal a great deal of information about the animal: species, individual identity, reproductive status, size (indicated by the height of the message), and when the animal was last in the area (indicated by the strength of the scent).

An important feature of pheromones is that once they are released, they remain in the environment for a long time. By contrast, vocal or visual displays disappear as soon as the animal stops signaling or displaying. The durability of pheromonal signals enables them to be used to mark trails, as ants do, or to indicate directionality, as in the case of the moth sex attractant. However, it also means that the message cannot be changed rapidly. This inflexibility makes pheromonal communication unsuitable for a rapid exchange of information.

Panthera tigris

52.13 Many Animals Communicate with Pheromones
To mark her territory, this female tiger is spraying pheromonal secretions from a scent gland in her hindquarters onto a tree. Other tigers passing the spot will know that the area is "claimed," and they will know something about the animal who claimed it.

The chemical nature and the size of the pheromonal molecule determine its speed of diffusion. The greater the speed of diffusion, the more rapidly the message gets out and the farther it will reach, but the sooner it will disappear. Trail-marking and territory-marking pheromones tend to be relatively large molecules that diffuse slowly; sex attractants tend to be small molecules that diffuse rapidly.

Visual signals are rapid and versatile but are limited by directionality

Visual signals are easy to produce, come in an endless variety, can be changed very rapidly, and clearly indicate the position of the signaler. However, the extreme directionality of visual signals means that they are not the best means of getting the attention of a receiver. The receptors of the receiver must be focused on the signaler, or the message will be missed. Most animals are sensitive to light and can therefore receive visual signals, but sharpness of vision limits the detail that can be transmitted. The complexity of the environment also limits visual communication.

Because visual communication requires light, it is not useful at night or in environments that lack light, such as caves and the ocean depths. Some species have surmounted this constraint on visual communication by evolving their own light-emitting mechanisms. Fireflies use a enzymatic mechanism to create flashes of light. By emitting flashes in species-specific patterns, fireflies can advertise for mates at night.

Fireflies also illustrate how some species can exploit the communication systems of other species. There are predatory species of fireflies that mimic the mating flashes of other species. When an eager suitor approaches the signaling individual, it is eaten. Thus, deception can be part of animal communication systems, just as it is part of human use of language.

Auditory signals communicate well over a distance

Compared with visual communication, auditory communication has advantages and disadvantages. Sound can be used at night and in dark environments. It can go around objects that would interfere with visual signals, so it can be used in complex environments like forests. It is better than visual signals at getting the attention of a receiver because the receiver does not have to be focused on the signaler for the message to be received. Like visual signals, sound can provide directional information, as long as the receiver has at least two receptors spaced somewhat apart. By maximizing or minimizing the features of the sounds they emit, animals can make their location easier or more difficult to determine.

Sound is useful for communicating over long distances. Even though the intensity of sound decreases with distance from the source, loud sounds can be used to communicate over distances much greater than those possible with visual signals. An extreme example is the communication of whales. Some whales, such as the humpback, have very complex songs. When these sounds are produced at a certain depth (around 1,000 m), they can be heard hundreds of kilometers away. In this way, humpback whales can locate each other over vast areas of ocean.

Auditory signals cannot convey complex information as rapidly as visual signals can, as the expression "A picture is worth a thousand words" implies. When individuals are in visual contact, an enormous amount of information is exchanged instantaneously (for example, species, sex, individual identity, reproductive status, level of motivation, dominance, vigor, alliances with other individuals, and so on). Coding that amount of information, with all of its subtleties, as auditory signals would take considerable time, thus increasing the possibility that the communicators could be located by predators.

The animal world is relatively silent. Most invertebrates do not produce sound; cicadas and crickets are marvelous exceptions. Many amphibians, most fishes, and most reptiles produce no sound.

Tactile signals can communicate complex messages

Communication by touch is extremely common, although not always obvious. Animals in close contact use tactile interactions extensively, especially under conditions that do not favor visual communication. When eusocial insects such as ants, termites, or bees meet, they contact one another with their antennae and front legs. One of the best-studied uses of tactile communication, beginning with the pioneering work of ethologist Karl von Frisch, is the dance of honeybee. When a forager bee finds food, she returns to the hive and communicates her discovery to her hivemates by dancing in the dark on the vertical surface of the honeycomb. The dance is monitored by other bees, who follow and touch the dancer to interpret the message.

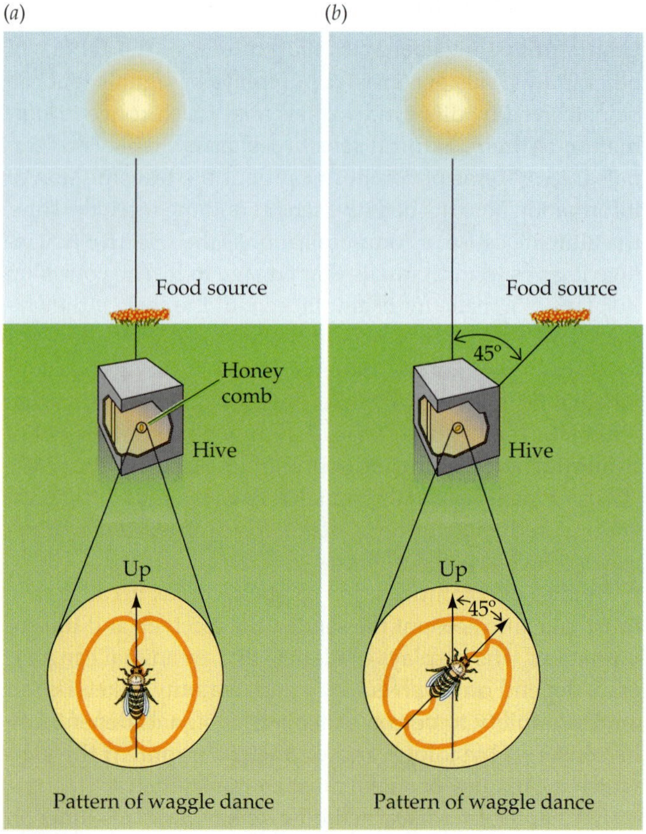

If the food is less than 80–100 meters from the hive, the forager performs a *round dance*, running rapidly in a circle and reversing her direction after each circumference. The odor on her body indicates the flower to be looked for, but the dance contains no information about the direction to go—only that it is within 100 meters of the hive.

If the food source is farther than 80–100 meters, the bee performs a *waggle dance*, which conveys information about both the distance and the direction of the food source. The bee repeatedly traces out a figure-eight pattern as she runs on the vertical surface. She alternates half-circles to the left and right with vigorous wagging of her abdomen in the short, straight run between turns. The angle of the straight run indicates the direction of the food source relative to the direction of the sun (Figure 52.14). The speed of the dancing

52.14 The Waggle Dance of the Honeybee
(*a*) By running straight up on the surface of the honeycomb in a dark hive, a honeybee tells her hivemates that there is a food source in the direction of the sun and at least 80 meters from the hive. The intensity of the waggle indicates exactly how far the food source is. If the food source were in the opposite direction from the sun, she would orient her waggle runs straight down. (*b*) When her waggle runs at an angle from the vertical, the other bees know that the same angle separates the direction of the food source from the direction of the sun.

indicates the distance to the food source: The farther away it is, the slower the waggle run.

Electric signals can also communicate messages

Some species of fish have evolved the ability to generate electric fields in the water around them by emitting a series of electric pulses (see Chapter 45). These trains of electric pulses can be used for sensing objects in the immediate surroundings, and they can also be used for communication.

An electrode connected to an amplifier and a speaker can be used to "listen" to the signals generated by glass knife fish in a tank. Each individual fish emits a pulse at a different frequency, and the frequency each fish uses relates to its status in the population. Males emit lower frequencies than females. The most dominant male has the lowest frequency, and the most dominant female has the highest frequency. When a new individual is introduced into the tank, the other individuals adjust their frequencies so that they do not overlap, and the signal of the new individual indicates its position in thehierarchy. In their natural environment—the murky waters of tropical rainforests—these fish can tell the identity, sex, and social position of another fish by its electric signals.

Communication has been a very fruitful area for investigating the ultimate causes of behavior and how the resulting adaptations have been shaped by the environment. Next we will return to some studies of proximate causes of behavior to see some examples of how "how" questions can be addressed.

The Timing of Behavior: Biological Rhythms

Among the important proximate causes of behavior are those that determine its organization through time. The study of biological rhythms has led to major discoveries about brain mechanisms down to the molecular level that enable animals to organize their behavior in time. In the discussion that follows, we will examine two types of biological rhythms: circadian rhythms and circannual rhythms.

Circadian rhythms control the daily cycle of behavior

Our planet turns on its axis once every 24 hours, creating a cycle of environmental conditions that has existed throughout the evolution of life. Daily cycles are characteristic of almost all organisms. What is surprising, however, is that this daily rhythmicity does not depend on the 24-hour cycle of light and dark.

If animals are kept in constant darkness, at a constant temperature with food and water available all the time, they still demonstrate daily cycles of activity, sleeping, eating, drinking, and just about anything else that can be measured. This persistence of the daily cycle in the absence of changes between light and dark suggests that animals have an endogenous (internal) clock. Without time cues from the environment, however, these daily cycles are not exactly 24 hours long. They are therefore called **circadian rhythms** (from the Latin *circa*, "about," and *dies*, "day").

To discuss biological rhythms, we must introduce some terminology. A rhythm can be thought of as a series of cycles, and the length of one of those cycles is the **period** of the rhythm. Any point on the cycle is a **phase** of that cycle:

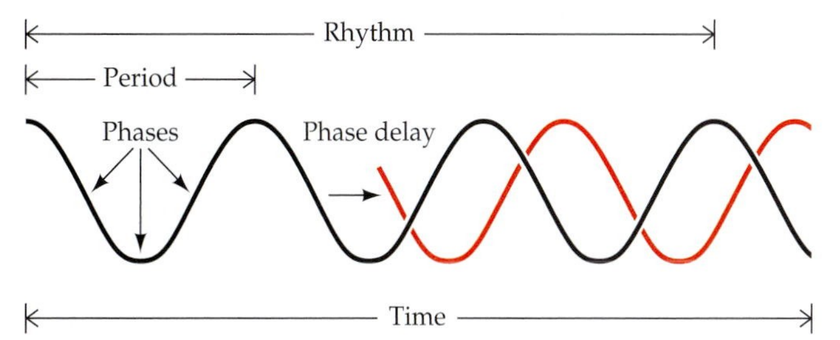

Hence, when two rhythms completely match, they are *in phase*, and if a rhythm is shifted (as in the resetting of a clock), it is *phase-advanced* or *phase-delayed*. Since the period of a circadian rhythm is not exactly 24 hours, it must be phase-advanced or phase-delayed every day to remain in phase with the daily cycle of the environment.

ENTRAINMENT. The process of resetting of the circadian rhythm by environmental cues is called **entrainment**. An animal kept in constant conditions will not be entrained to the 24-hour cycle of the environment, and its circadian clock will run according to its natural period—it will be **free-running**. If its period is less than 24 hours, the animal will begin its activity a little earlier each day (see the middle panel of Figure 52.15).

Animals with free-running circadian rhythms can be used in experiments to investigate the stimuli that phase-shift or entrain the circadian clock. Under natural conditions, environmental cues, such as the onset of light or dark, entrain the free-running rhythm to the 24-hour cycle of the real world. In the laboratory, it is possible to entrain circadian rhythms in free-running animals with short pulses of light or dark administered every 24 hours (bottom panel of Figure 52.15).

When you fly across several time zones, your circadian clock is out of phase with the real world at your destination; the result is jet lag. Gradually your endogenous rhythm synchronizes itself with the real world as it is reentrained by environmental cues. Since your endogenous rhythm cannot be shifted by more than 30 to 60 minutes each day, it takes several days to reentrain your clock to real time in your new location. This period of reentrainment is the time during which you experience jet lag, because your endogenous rhythm is waking you up, making you sleepy, initiating activities in your digestive tract, and stimulating many other physiological functions at inappropriate times of the day.

THE CIRCADIAN CLOCK. Where is the clock that controls the circadian rhythm? In mammals, the master circadian clock is located in two tiny groups of cells just above the optic chiasm, the place where the two optic nerves cross. These structures are called the **suprachiasmatic nuclei** (**SCN**). If

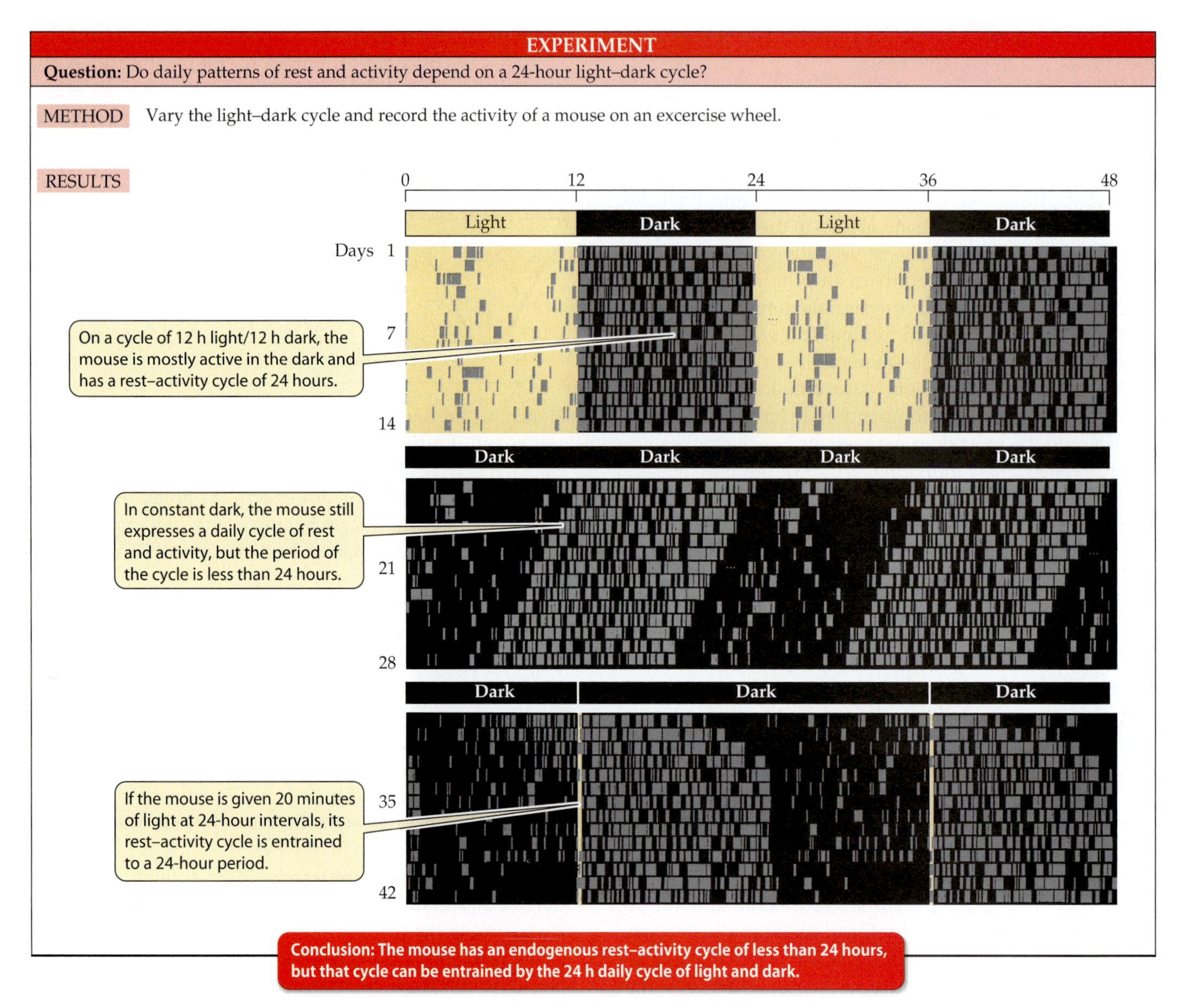

52.15 Circadian Rhythms

The marks indicate times when a mouse is running on an activity wheel. Two days of activity are recorded on each horizontal line, such that the data for each day are plotted twice, once on the right half of a line and again on the left half of the next line below; this double plotting makes patterns easier to see. The schedule of light and dark exposure is indicated by the solid bars running across the figure. First the mouse experiences 12 hours of light and 12 hours of dark every day (top panel), then it is placed in constant darkness (middle panel), and finally it is given a 20-minute exposure to light each day (bottom panel). In constant darkness, the circadian rhythm is free-running, but a 20-minute flash of light at 24-hour intervals can entrain it.

the SCN are destroyed, the animal loses circadian rhythmicity. Under constant conditions, the animal is equally likely to be active or asleep at any time of day (Figure 52.16).

Recent experiments have shown that circadian rhythms of rest and activity can be restored in an animal whose SCN have been destroyed if it receives a transplant of those nuclei from another animal. In no other known case can a brain tissue transplant restore such a complex behavior. Since the restored rhythm has the period of the animal that donates the tissue, the transplant clearly controls the recipient's behavior.

Circadian rhythms are found in every animal group, as well as in protists, plants, and fungi, but only vertebrates have SCN. Thus, natural selection has produced a variety of circadian clocks. In the mollusk *Bulla*, for example, the cells driving circadian behavior are in the eyes. Birds do have SCN, but the master clock of at least some species resides in the pineal gland, a mass of tissue between the cerebral hemispheres that produces the hormone melatonin. If the pineal gland of a bird is removed, the bird loses its circadian rhythm. In protists and fungi, circadian rhythmicity is a property of individual cells, and the individual cells of many multicellular animals can generate circadian rhythms. What are the molecular mechanisms of these circadian clocks?

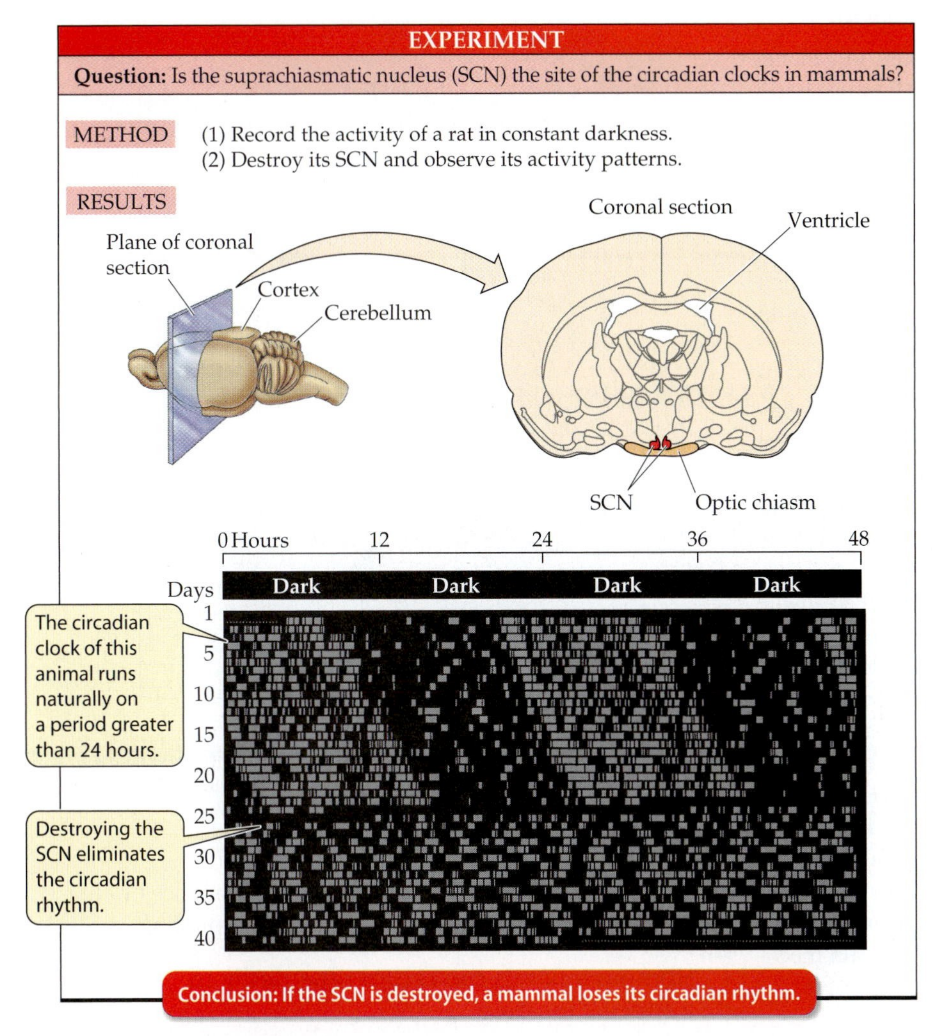

52.16 Where the Clock Is
The circadian clock of mammals is in the suprachiasmatic nuclei (SCN) of the brain. If its suprachiasmatic nuclei are destroyed, a mammal loses its circadian rhythm.

CLOCK GENES. Enormous progress has been made in recent years toward discovering the molecular basis of circadian rhythms. The surprise is that there is a high degree of homology in the genes involved across a very wide range of organisms, from bread molds to humans.

The story begins with a gene called *per* that was discovered in fruit flies, as mentioned above. Mutations of this gene cause flies to have either short or long circadian periods. Mutations of another circadian gene, called *tim* (short for "timeless"), cause a loss of circadian rhythms in fruit flies. The presence of mRNA for *per* and *tim* shows a daily cycle, as does the presence of the Per and Tim proteins. Thus, the transcription and translation of these two genes shows a circadian rhythm. But what controls the rhythm? The Per and Tim proteins dimerize in the cytoplasm, and the resulting heterodimer is translocated into the nucleus, where it acts as a transcription factor inhibiting the transcription of the *per* and *tim* genes (Figure 52.17). These two genes could thus be the wheels of a circadian clock.

It is now known that the mechanism is not so simple, and that there are a number of other genes involved. What is interesting, however, is that homologies have been found between these clock genes in fungi, insects, mice, and humans, indicating how fundamental molecular clock mechanisms are to living organisms on Earth.

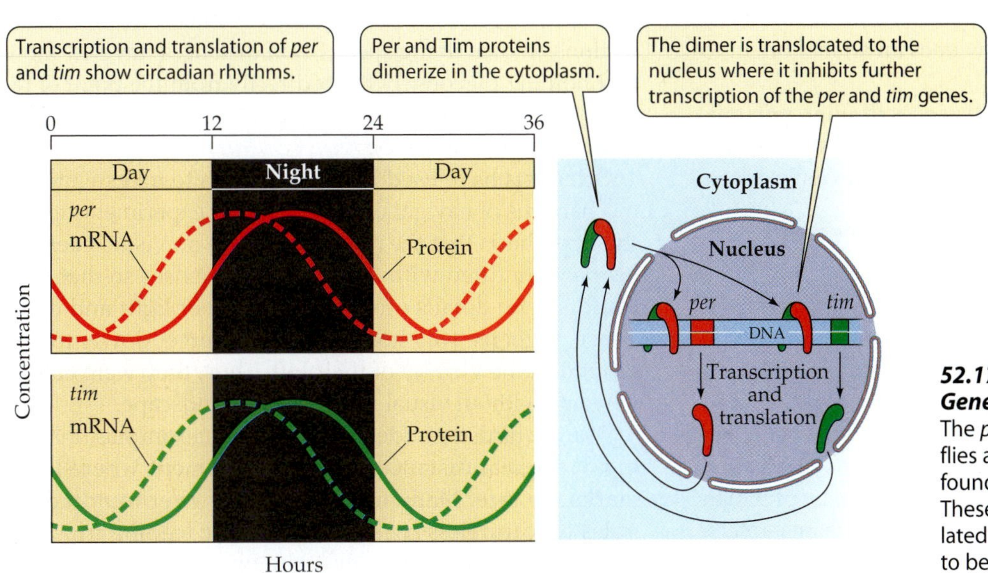

52.17 Circadian Rhythms May Be Generated by a Molecular Clock
The *per* and *tim* genes discovered in fruit flies are homologous to "clock genes" found in a wide range of organisms. These genes are transcribed and translated on a circadian rhythm that seems to be controlled by positive feedback.

Circannual rhythms control seasonal behaviors

In addition to turning on its axis every 24 hours, our planet revolves around the sun once every 365 days. Because Earth is tilted on its axis, its revolution around the sun results in seasonal changes in day length at all locations except the equator. These changes secondarily create seasonal changes in temperature, rainfall, and other variables. Because the behavior of animals must adapt to these seasonal changes, animals must be able to anticipate the seasons and adjust their behavior accordingly. Most animals, for example, should not produce young in the winter.

For many species, change in day length, or *photoperiod*, is a reliable indicator of seasonal changes to come. If day length has a direct effect on the physiology and behavior of a species, that species is said to be **photoperiodic**. If male deer, for example, are held in captivity and subjected to two cycles of change in day length in one year, they will grow and drop their antlers twice during that year.

For some animals, change in day length is not a reliable cue. Hibernators spend long months in dark burrows underground, away from any indicators of day length, but have to be physiologically prepared to breed almost as soon as they emerge in the spring. A bird overwintering in the tropics cannot use changes in photoperiod as a cue to time its migration north to the breeding grounds. Hibernators and equatorial migrants have endogenous annual rhythms, called **circannual rhythms**. In other words, their nervous systems have a built-in calendar. Just as circadian rhythms are not exactly 24 hours long, circannual rhythms are not exactly 365 days long, but usually shorter. The brain mechanisms of circannual rhythms are completely unknown.

Finding Their Way: Orientation and Navigation

Within a local environment, finding your way is not a problem. Like the wasps Tinbergen studied, you remember landmarks and organize your behavior spatially with respect to those reference points. Such **orientation** is a very common animal behavior. But what if the destination is a considerable distance away? How does an animal orient to it and find its way?

Many animals navigate long distances through unfamiliar territory. In this section we describe modes of navigation and examine some of their underlying mechanisms.

Piloting animals orient themselves by means of landmarks

In most cases an animal find its way using simple means: It knows and remembers the structure of its environment. It uses landmarks to find its nest, a safe hiding place, or a food source. Navigating by means of landmarks is called **piloting**. Gray whales, for example, migrate seasonally between the Bering Sea and the coastal lagoons of Mexico. They find their way by following the west coast of North America (Figure 52.18). Coastlines, mountain chains, rivers, water currents, and wind patterns can all serve as piloting cues. But some remarkable cases of long-distance orientation and movement cannot be explained by piloting.

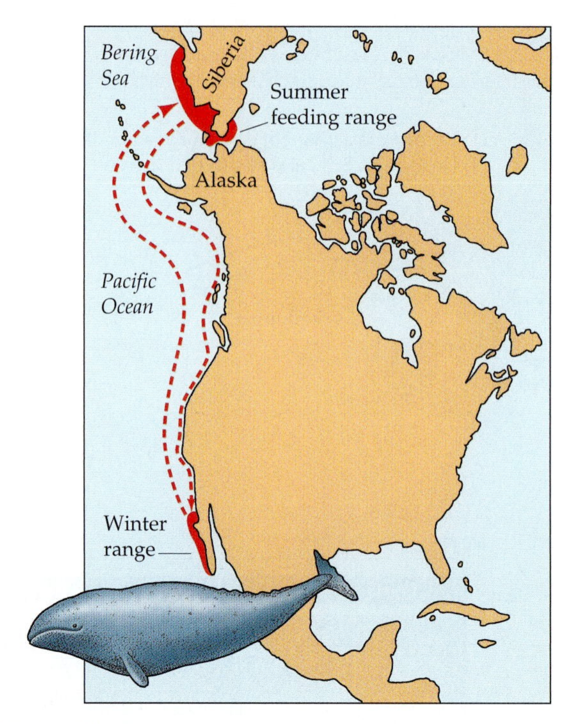

52.18 Piloting
Gray whales migrate south in winter from the Bering Sea to the coast of Baja California by following the coast of North America.

Homing animals can return repeatedly to a specific location

The ability of an animal to return to a nest site, burrow, or other specific location is called **homing**. In most cases, homing is merely piloting in a known environment, but some animals are capable of much more sophisticated homing.

People who breed and race homing pigeons take the pigeons from their home loft and release them at a remote site where they have never been before. The first pigeon that reaches its home wins the race. Data on departure directions, known flying speeds, and distances traveled show that homing pigeons fly fairly directly from the point of release to home. They do not randomly search until they encounter familiar territory.

Scientists have used homing pigeons to investigate the mechanisms of navigation. One series of experiments tested the hypothesis that the pigeons depend on visual cues. Pigeons were fitted with frosted contact lenses so that they could see no details other than degree of light and dark. These pigeons still homed and fluttered down to the ground in the vicinity of their loft. Thus, they were able to navigate without visual images of the landscape.

Marine birds provide many dramatic examples of homing over great distances in an environment where landmarks are rare. Many marine birds fly over hundreds of miles of featureless ocean on their daily feeding trips and then return directly to a nest site on a tiny island. Alba-

Diomedea melanophris

52.19 Coming Home
A pair of black-browed albatrosses engage in courtship display over their partially completed mud nest. Many albatrosses return to the site of their own birth to find a mate, and will return to that site year after year.

trosses display remarkable feats of homing. When a young albatross first leaves its nest on an oceanic island, it flies widely over the southern oceans for 8 or 9 years before it reaches reproductive maturity. At that time, it flies back to the island where it was raised to select a mate and build a nest (Figure 52.19). After the first mating season, the pair separate, and each bird resumes its solitary wanderings. The next year they return to the same nest site at the same time, reestablish their pair bond, and breed. Thereafter they return to the nest to breed every other year, spending many months in between at sea.

Migrating animals travel great distances with remarkable accuracy

For as long as humans have inhabited temperate and subpolar latitudes, they must have been aware that whole populations of animals, especially birds, disappear and reappear seasonally—that is, they **migrate**. Not until the early nineteenth century, however, were patterns of migration established by marking individual birds with identification bands around their legs. Being able to identify individual birds in a population made it possible to demonstrate that the same birds and their offspring returned to the same breeding grounds year after year, and that these same birds were found during the nonbreeding season at locations hundreds or even thousands of kilometers from the breeding grounds.

How do migrants find their way over such great distances? A reasonable hypothesis is that young birds on their first migration follow experienced birds and learn the landmarks by which they will pilot in subsequent years. However, adult birds of many species leave the breeding grounds before the young have finished fattening and are ready to begin their first migration. These naive birds must be able to navigate accurately on their own, and with little room for mistakes.

Navigation is based on internal and environmental cues

Since many homing and migrating species are able to take direct routes to their destinations through areas they have never experienced, they must have mechanisms of navigation other than piloting. Humans use two systems of navigation: distance-and-direction navigation and bicoordinate navigation. **Distance-and-direction navigation** requires knowing the direction to the destination and how far away that destination is. With a compass to determine direction and a means of measuring distance, humans can navigate. **Bicoordinate navigation**, also known as *true navigation*, requires knowing the latitude and longitude (the map coordinates) of both the current position and the destination. From that information, a route can be plotted to the destination.

DISTANCE-AND-DIRECTION NAVIGATION. Researchers conducted an experiment with European starlings to determine their method of navigation. These birds migrate between their breeding grounds in the Netherlands and northern Germany and their wintering grounds to the southwest, in southern England and western France (Figure 52.20). The researchers captured birds on their breeding grounds, marked them, transported them to Switzerland—south of their breeding grounds—and released them. The researchers expected that if the starlings were using distance-and-direction navigation, the marked birds would be recovered in France and Spain, to the southwest of where they were released. Naive juvenile starlings did use distance-and-direction navigation, but experienced adult birds were less disrupted by their geographic displacement.

How do animals determine distance and direction? In many instances, determining distance is not a problem as long as the animal recognizes its destination. Homing animals recognize landmarks and can pilot once they reach familiar areas. Evidence suggests that biological rhythms play a role in determining migration distances for some species. Birds kept in captivity display increased and oriented activity at the time of year when they would normally migrate. Such *migratory restlessness* has a definite duration, which corresponds to the usual duration of migration for the species. Since distance is determined by how long an animal moves in a given direction, the duration of migratory restlessness could set the distance for its migration.

Two obvious means of determining direction are the sun and the stars. During the day, the sun is an excellent compass, as long as the time of day is known. In the Northern Hemisphere, the sun rises in the east, sets in the west, and

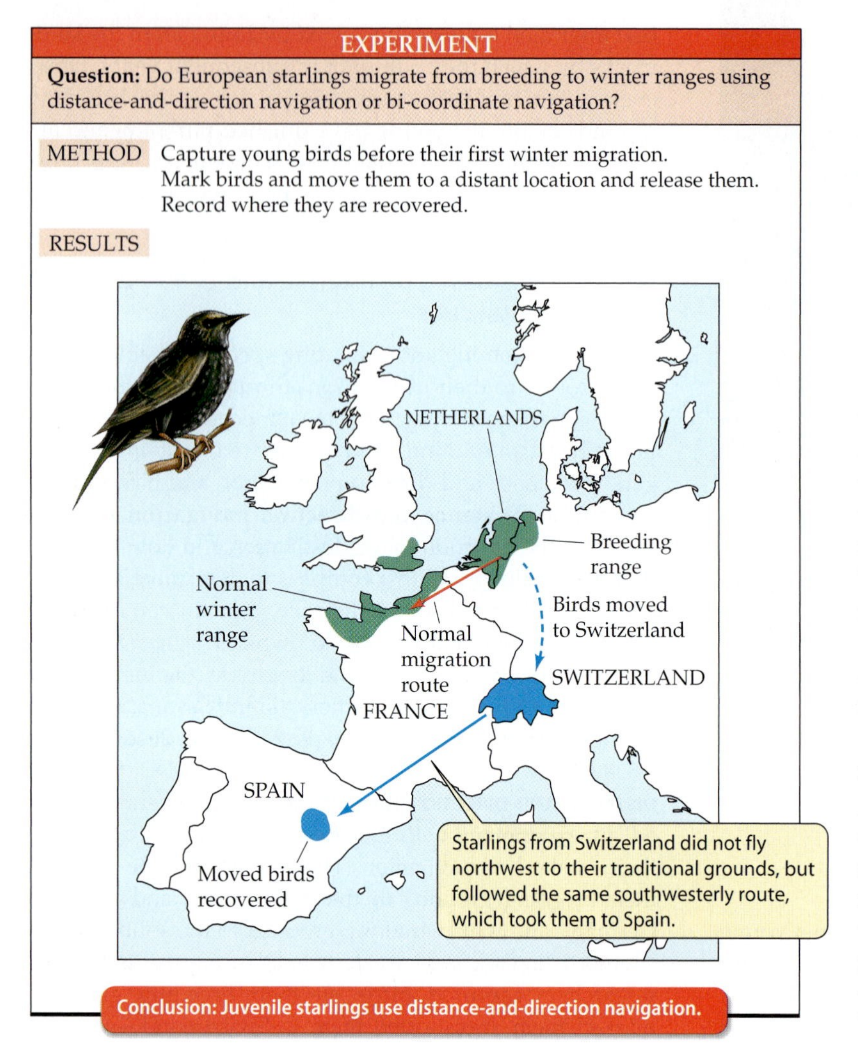

52.20 Distance-and-Direction Navigation European starlings normally make a short winter migration in a southwesterly direction, from the Netherlands to coastal France and southern England (red arrow). Experimental populations of starlings moved to a site in Switzerland did not fly northwest to their traditional wintering grounds, but followed the same southwesterly route (blue arrow), which took them to Spain.

points south at noon. As we have seen, animals can tell the time of day by means of their circadian clocks. Clock-shifting experiments have demonstrated that animals use their circadian clocks to determine direction from the position of the sun.

Researchers placed birds in a circular cage that enabled them to see the sun and sky, but no other visual cues (Figure 52.21). Food bins were arranged around the sides of the cage, and the birds were trained to expect food in the bin at one particular direction—south, for example. After training, no matter what time they were fed, and even if the cage was rotated between feedings, the birds always went to the bin at the southern end of the cage for food, even if that bin contained no food.

Next, the birds were placed in a room with a controlled light cycle, and their circadian rhythms were phase-shifted by turning the lights on at midnight and off at noon. After about 2 weeks, the birds' circadian clocks had been phase-advanced by 6 hours. Then the birds were returned to the circular cage under natural light conditions, with sunrise at 6:00 A.M. Because of the shift in their circadian rhythms, their endogenous clocks were indicating noon at the time the sun came up.

If food was always in the south bin, and it was sunup, the birds should have looked for food 90 degrees to the right of the direction of the sun. But since their circadian clocks were telling them it was noon, they looked for food in the direction of the sun—in the east bin. The 6-hour phase shift in their circadian clocks resulted in a 90-degree error in their orientation. These kinds of experiments on many species have shown that animals can orient by means of a *time-compensated solar compass.*

Many animals are normally active at night; in addition, many day-active bird species migrate at night and thus cannot use the sun to determine direction. The stars offer two sources of information about direction: moving constellations and a fixed point. The positions of constellations change because Earth is rotating. With a star map and a clock, direction can be determined from any constellation. But one point that does not change position during the night is the point directly over the axis on which Earth turns. In the Northern Hemisphere, a star called Polaris, or the North Star, lies in that position and always indicates north.

Stephen Emlen at Cornell University investigated whether birds use these sources of directional information from the stars. He raised young birds in a planetarium, where star patterns are projected on the ceiling of a large, domed room. The star patterns in the planetarium could be slowly rotated to simulate the rotation of Earth. If the star patterns were not rotated, birds caught in the wild could orient well in the planetarium, but birds raised in the planetarium under a nonmoving sky could not. If the star patterns in the planetarium were rotated each night as the young birds matured, they were able to orient in the planetarium, showing that birds can learn to use star patterns for orientation if the sky rotates (Figure 52.22).

These experiments provided no evidence that the birds used their circadian clocks to derive directional information from the star patterns. Experienced birds were not confused by a still sky, or by a sky that rotated faster than normal. These birds were orienting to the fixed point in the sky, the North Star. Young birds raised under a sky that rotated around a different star imprinted on that star and oriented to it as if it were the North Star. These studies showed that

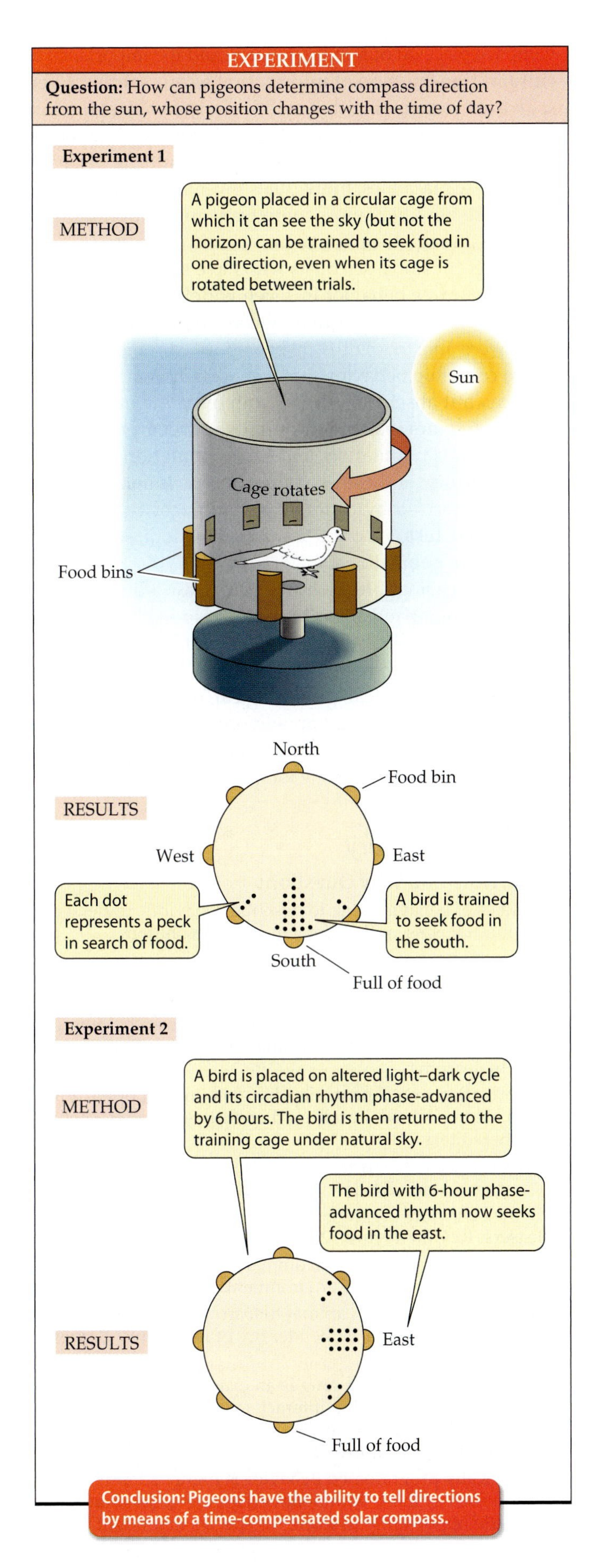

52.21 The Time-Compensated Solar Compass Birds whose circadian rhythms were phase-shifted forward by 6 hours oriented as though the dawn sun was at its noon position. These results showed that birds are capable of using their circadian clocks to determine direction from the position of the sun.

birds raised in the Northern Hemisphere learn a star map that they can use for orientation at night by imprinting on the fixed point in the sky.

Animals cannot use sun and star compasses when the sky is overcast, yet they still home and migrate under such conditions. Do other sources provide information they can use for orientation? There appears to be considerable redundancy in animals' abilities to sense direction. Pigeons are able to home as well on overcast days as on clear days, but this ability is severely impaired if small magnets are attached to their heads—evidence that the birds use a magnetic sense. Cells have been found in birds that contain small particles of the magnetic mineral magnetite, but the neurophysiology of the magnetic sense is largely unknown. Another possible cue is the plane of polarization of light, which can give directional information even under heavy cloud cover. Very low frequencies of sound can provide information about coastlines and mountain chains. Weather patterns can also provide considerable directional information.

BICOORDINATE NAVIGATION. Bicoordinate navigation involves knowing where you are (in longitude and latitude) and where you want to go, and plotting an appropriate course. Longitude can be determined by the position of the sun and the time of day: If the sun comes up earlier than expected, you must be east of where you want to be, and if the sun comes up later than expected, then you are west of

52.22 Star Patterns Can Be Altered in a Planetarium This scientist has placed birds in a planetarium. By changing the positions or movements of the stars projected on the planetarium ceiling, he can investigate what information the birds use for orientation.

where you want to be. Time and sun position can give information about latitude as well. At a given time of day in the Northern Hemisphere, a sun position higher in the sky than expected indicates you are farther south than you want to be, and if the sun is lower in the sky than expected, you are north of where you want to be. Information about longitude and latitude can also come from sensing Earth's magnetic lines of force and from the positions of the stars.

In spite of the remarkable navigational abilities of animals such as albatrosses, there is currently no evidence that animals use bicoordinate navigation. But, of course, it is not easy to do experiments on animals such as albatrosses!

Human Behavior

As we saw early in this chapter, the behavior of an animal is a mixture of components that are inherited and components that can be molded by learning. However, even some aspects of learned behavior patterns—such as what can be learned and when it can be learned—have genetic determinants. Thus natural selection shapes not only the physiology and morphology of a species, but also its behavior. In some situations natural selection favors inherited behaviors; in others, learned behaviors. In many cases, the optimal adaptation is a mixture of inherited and learned behavioral components. Given these considerations, how would we characterize human behavior?

An important characteristic of human behavior is the extent to which it can be modified by experience. The transmission of learned behavior from generation to generation—**culture**—is the hallmark of humans. Nevertheless, the structures and many functions of our brains are inherited, including drives, limits to and propensities for learning, and even some motor patterns. Biological drives such as hunger, thirst, sexual desire, and sleepiness are inherent in our nervous systems. Is it reasonable, therefore, to expect that emotions such as anger, aggression, fear, love, hate, and jealousy are solely the consequences of learning?

Our sensory systems enable us to use certain subsets of information from the environment; similarly, the structure of our nervous systems makes it more or less possible to process certain types of information. Consider, for example, how basic and simple it is for an infant to learn spoken language, yet how many years that same child must struggle to master reading and writing. Verbal communication is deeply rooted in our evolutionary past, whereas reading and writing are relatively recent products of human culture.

Some motor patterns seem to be programmed into our nervous systems. Studies of diverse human cultures from around the world reveal basic similarities of facial expressions and body language among human populations that have had little or no contact with one another. Infants born blind still smile, frown, and show other facial expressions at appropriate times, even though they have never observed such expressions in others.

Acknowledging that aspects of our behavior have been shaped through evolution in no way detracts from the value we place on our ability to learn and the importance of cultural transmission of information to our species. Even so, we are recognizing that culture, in its simplest form, is not uniquely human. In the introduction to this chapter we saw what has been characterized as pre-cultural behavior in Japanese macaques. Individuals invented new behaviors, and those new behaviors were transmitted by imitative learning through the population.

In a recent study, scientists who have spent years studying chimpanzee behavior in seven widely separated areas of Africa compared their findings on chimpanzee behavior. They were able to identify 39 behaviors, ranging from tool use to courtship behavior, that were common in some populations but absent in others. Moreover, the variation in these behaviors was much greater between populations than within a population, and each population had a distinct repertoire of these behaviors. Just as human societies are characterized by different assemblages of culturally transmitted customs or customary behaviors, so are these chimpanzee populations.

It is increasingly more difficult to draw a line between human behavior and animal behavior, especially that of our closest primate relatives. But why should we expect such a line to exist? We do not expect such a lack of continuity in molecular, biochemical, physiological, or anatomical characteristics. Similarly, human and animal behavior is on a continuum, and the challenge is to understand the common mechanisms and the reasons for quantitative differences.

Chapter Summary

What, How, and Why Questions

▶ Studies of behavior seek to describe behaviors, understand their mechanisms, and understand their evolution.

Behavior Shaped by Inheritance

▶ Many behaviors of many species are stereotypic and species-specific, and are thus largely determined by inheritance. They do not require learning and are minimally modifiable by learning.

▶ Deprivation experiments deprive an animal of opportunities to learn a behavior and can therefore reveal that a behavior is hereditary.

▶ Hybridization experiments can also reveal genetic influences on behavior. **Review Figure 52.2**

▶ Some behaviors are triggered by simple stimuli called releasers. **Review Figure 52.3, 52.4**

▶ Spatial learning enables an animal to learn and use information about its physical environment. **Review Figure 52.5**

▶ Imprinting enables an animal to learn the features of a complex releaser, such as the identity of its parents. **Review Figure 52.6**

▶ The acquisition of bird song is an example in which genetic determinants and learning interact, enabling an animal to learn a behavior by focusing on the correct stimuli at the correct times. **Review Figure 52.7**

▶ Genetically programmed behavior is highly adaptive for species, such as those with nonoverlapping generations, that have little opportunity to learn, for species that might learn the wrong behavior, and in situations in which mistakes are costly or dangerous.

Hormones and Behavior

▶ In rats, the sex steroids present during development determine what sexual behavior patterns develop, and the sex steroids present in the adult control the expression of those patterns. **Review Figure 52.9**

▶ In birds, testosterone determines a bird's ability to sing by causing the brain regions responsible for song to develop. **Review Figure 52.10**

The Genetics of Behavior

▶ There are many complex steps between the expression of a gene as a protein product and the expression of a behavior. Several types of experiments help reveal how genes affect behavior.

▶ Artificial selection and crossbreeding can produce individuals with particular behavioral traits that are inherited. **Review Figure 52.11**

▶ The techniques of molecular genetics can reveal the functions of specific genes that influence behavior. **Review Figure 52.12**

Communication

▶ Communication consists of displays or signals, that can be perceived by other individuals and which influence their behavior. Natural selection favors communication systems when both sender and receiver benefit from the exchange of information.

▶ The evolution of communication signals is constrained by the anatomical and physiological characteristics of a species that are available to be shaped by natural selection.

▶ Many animals communicate by emitting pheromones into the environment and by sensing the pheromones of other animals. Pheromonal messages can last a long time, but they cannot be changed quickly.

▶ Visual communication is easy, versatile, and rapid, but it is limited by its directionality, by the visual acuity of the receiver, and by environmental conditions such as darkness. Many animals communicate via visual signals.

▶ Auditory signals can be used at night, can go around objects that would interfere with visual communication, can easily get the receiver's attention, can provide directional information, and can travel long distances. Compared with visual communication, however, auditory communication is slow. Few animals communicate with auditory signals.

▶ Tactile signals can communicate complex messages, as the dance of the honeybee demonstrates. **Review Figure 52.14**

▶ The electric signals generated by some fishes can be used for communication.

The Timing of Behavior: Biological Rhythms

▶ Animal behaviors are expressed in daily cycles called circadian rhythms. A circadian rhythm is an endogenous rhythm with a period not equal to 24 hours. To remain in phase with the 24-hour daily cycle of the environment, a circadian rhythm must be phase-shifted every day. Phase-shifting cues such as the onset of light and dark entrain circadian rhythms to the natural 24-hour period. **Review Figure 52.15**

▶ In mammals, the clock that controls the circadian rhythm is located in the suprachiasmatic nuclei of the brain. In other animals, different structures function as the circadian clock. **Review Figure 52.16**

▶ Two genes have been identified that are involved in the clock mechanism in a variety of species. **Review Figure 52.17**

▶ Circannual rhythms ensure that animals, such as hibernators and equatorial migrants, that cannot rely on changes in day length as seasonal cues perform the appropriate behaviors at the appropriate times of year.

Finding Their Way: Orientation and Navigation

▶ Piloting animals find their way by orienting to landmarks. **Review Figure 52.18**

▶ Homing animals find their way through unfamiliar territory to specific locations. Migrating animals travel long distances with remarkable accuracy.

▶ Animals that navigate by distance and direction determine distance in part by recognizing landmarks in the vicinity of their destination and in part by biological rhythms timing how far they travel.

▶ Sources of directional information include a time-compensated solar compass and an ability to locate the fixed point in the nocturnal sky. **Review Figures 52.20, 52.21**

▶ The long-distance movements of some species are difficult to explain by distance-and-direction navigation mechanisms. Information for bicoordinate navigation is available from the physical environment, but there is no evidence that any species uses such information.

Human Behavior

▶ Like that of all other animals, human behavior consists of genetically determined and learned components. What sets humans apart from other animals is the extent to which we can modify our behavior on the basis of experience.

For Discussion

1. An oysstercatcher is a bird that normally lays a clutch of two eggs. If you place an artificial nest with either three artificial but normal-sized eggs or one very large artificial egg near the oystercatcher's nest, the oystercatcher will abandon its own two eggs and attempt to incubate the artificial eggs. How can you explain this behavior?
2. Cowbirds are nest parasites. A female cowbird lays her eggs in the nest of another species, which then incubates the eggs and raises the young. What do you think would characterize the acquisition of song in cowbirds? In a given area, cowbirds tend to parasitize the nests of particular bird species. How do you think female cowbirds learn this behavior? How would you test your hypothesis?
3. The short-tailed shearwater is a bird that winters in Antarctica and summers in the Arctic. What problems would this species have in using either the sun or the stars for navigation? What is the most likely means it uses to find its way to its summer and its winter feeding grounds?
4. Male dogs lift a hind leg when they urinate; female dogs squat. If a male puppy receives an injection of estrogen when it is a newborn, it will never lift its leg to urinate for the rest of its life; it will always squat. How might this result be explained?
5. If you were able to be the first person to visit a human population that had never been in contact with another culture, how could you use that opportunity to explore whether there were any human behaviors that were genetically determined?

Appendix: Some Measurements Used in Biology

QUANTITY	NAME OF UNIT	SYMBOL	DEFINITION
Length	meter (*also* metre)	m	A base unit. 1 m = 100 cm = 39.37 inches
	kilometer	km	1 km = 1000 m = 10^3 m
	centimeter	cm	1 cm = $\frac{1}{100}$ m = 10^{-2} m
	millimeter	mm	1 mm = $\frac{1}{1000}$ m = 10^{-3} m
	micrometer	µm	1 µm = $\frac{1}{1000}$ mm = 10^{-6} m
	nanometer	nm	1 nm = $\frac{1}{1000}$ µm = 10^{-9} m
Area	square meter	m^2	Area encompassed by a square, each side of which is 1 m in length
	hectare	ha	1 ha = 10,000 m^2 = 10^4 m^2 (2.47 acres)
	square centimeter	cm^2	1 cm^2 = $\frac{1}{10,000}$ m^2 = 10^{-4} m^2
Volume	liter (*also* litre)	l	1 l = $\frac{1}{1000}$ m^3 = 10^{-3} m^3 (1.057 qts)
	milliliter	ml	1 ml = $\frac{1}{1000}$ l = 10^{-3} l = 1 cm^3 = 1 cc
	microliter	µl	1 µl = $\frac{1}{1000}$ ml = 10^{-3} ml = 10^{-6} l
Mass	kilogram	kg	A basic unit. 1 kg = 1000 g = 2.20 lbs
	gram	g	1 g = $\frac{1}{1000}$ kg = 10^{-3} kg
	milligram	mg	1 mg = $\frac{1}{1000}$ g = 10^{-3} g = 10^{-6} kg
Time	second	s	A basic unit. 1 s = $\frac{1}{60}$ min
	minute	min	1 min = 60 s
	hour	h	1 h = 60 min = 3,600 s
	day	d	1 d = 24 h = 86,400 s
Temperature	kelvin	K	A basic unit. 0 K = –273.15°C = absolute zero
	degree Celsius	°C	0°C = 273.15 K = melting point of ice
Heat, work	calorie	cal	1 cal = heat necessary to raise 1 gram of pure water from 14.5°C to 15.5°C = 4.184 J
	kilocalorie	kcal	1 kcal = 1000 cal = 10^3 cal = (in nutrition) 1 Calorie
	joule	J	1 J = 0.2389 cal (The joule is now the accepted unit of heat in most sciences.)
Electric potential	volt	V	A unit of potential difference or electromotive force
	millivolt	mV	1 mV = $\frac{1}{1000}$ V = 10^{-3} V

Glossary

Abdomen (ab′ duh mun) [L.: belly] • In arthropods, the posterior portion of the body; in mammals, the part of the body containing the intestines and most other internal organs, posterior to the thorax.

Abscisic acid (ab sighs′ ik) [L. *abscissio*: breaking off] • A plant growth substance having growth-inhibiting action. Causes stomata to close.

Abscission (ab sizh′ un) [L. *abscissio*: breaking off] • The process by which leaves, petals, and fruits separate from a plant.

Absolute temperature scale • Also known as the Kelvin scale. A temperature scale in which zero is the state of no molecular motion. This "absolute zero" is –273° on the Celsius scale.

Absorption • (1) Of light: complete retention, without reflection or transmission. (2) Of liquids: soaking up (taking in through pores or cracks).

Absorption spectrum • A graph of light absorption versus wavelength of light; shows how much light is absorbed at each wavelength.

Abyssal zone (uh biss′ ul) [Gr. *abyssos*: bottomless] • That portion of the deep ocean floor where no light penetrates.

Accessory pigments • Pigments that absorb light and transfer energy to chlorophylls for photosynthesis.

Acetylcholine • A neurotransmitter substance that carries information across vertebrate neuromuscular junctions and some other synapses. **Acetylcholinesterase** is an enzyme that breaks down acetylcholine.

Acetyl CoA (acetyl coenzyme A) • Compound that reacts with oxaloacetate to produce citrate at the beginning of the citric acid cycle; a key metabolic intermediate in the formation of many compounds.

Acid [L. *acidus*: sharp, sour] • A substance that can release a proton in solution. (Contrast with base.)

Acid precipitation • Precipitation that has a lower pH than normal as a result of acid-forming precursors introduced into the atmosphere by human activities.

Acidic • Having a pH of less than 7.0 (a hydrogen ion concentration greater than 10^{-7} molar).

Acoelomate • Lacking a coelom.

Acquired Immune Deficiency Syndrome • See AIDS.

Acrosome (a′ krow soam) [Gr. *akros*: highest or outermost + *soma*: body] • The structure at the forward tip of an animal sperm which is the first to fuse with the egg membrane and enter the egg cell.

ACTH (adrenocorticotropin) • A pituitary hormone that stimulates the adrenal cortex.

Actin [Gr. *aktis*: a ray] • One of the two major proteins of muscle; it makes up the thin filaments. Forms the microfilaments found in most eukaryotic cells.

Action potential • An impulse in a neuron taking the form of a wave of depolarization or hyperpolarization imposed on a polarized cell surface.

Activating enzymes (also called aminoacyl-tRNA synthetases) • These enzymes catalyze the addition of amino acids to their appropriate tRNAs.

Activation energy (E_a) • The energy barrier that blocks the tendency for a set of chemical substances to react.

Active site • The region on the surface of an enzyme where the substrate binds, and where catalysis occurs.

Active transport • The transport of a substance across a biological membrane against a concentration gradient—that is, from a region of low concentration (of that substance) to a region of high concentration. Active transport requires the expenditure of energy and is a saturable process. (Contrast with facilitated diffusion, free diffusion; see primary active transport, secondary active transport.)

Adaptation (a dap tay′ shun) • In evolutionary biology, a particular structure, physiological process, or behavior that makes an organism better able to survive and reproduce. Also, the evolutionary process that leads to the development or persistence of such a trait.

Adenine (a′ den een) • A nitrogen-containing base found in nucleic acids, ATP, NAD, etc.

Adenosine triphosphate • See ATP.

Adenylate cyclase • Enzyme catalyzing the formation of cyclic AMP from ATP.

Adrenal (a dree′ nal) [L. *ad-*: toward + *renes*: kidneys] • An endocrine gland located near the kidneys of vertebrates, consisting of two glandular parts, the cortex and medulla.

Adrenaline • See epinephrine.

Adrenocorticotropin • See ACTH.

Adsorption • Binding of a gas or a solute to the surface of a solid.

Aerobic (air oh′ bic) [Gr. *aer*: air + *bios*: life] • In the presence of oxygen, or requiring oxygen.

Afferent (af′ ur unt) [L. *ad*: to + *ferre*: to bear] • To or toward, as in a neuron that carries impulses to the central nervous system, or a blood vessel that carries blood to a structure. (Contrast with efferents.)

AIDS (acquired immune deficiency syndrome) • Condition caused by a virus (HIV) in which the body's helper T lymphocytes are reduced, leaving the victim subject to opportunistic diseases.

Aldehyde (al′ duh hide) • A compound with a –CHO functional group. Many sugars are aldehydes. (Contrast with ketone.)

Aldosterone (al dahs′ ter own) • A steroid hormone produced in the adrenal cortex of mammals. Promotes secretion of potassium and reabsorption of sodium in the kidney.

Alga (al′ gah) (plural: algae) [L.: seaweed] • Any one of a wide diversity of protists belonging to the phyla Pyrrophyta, Chrysophyta, Phaeophyta, Rhodophyta, and Chlorophyta.

Allele (a leel′) [Gr. *allos*: other] • The alternate forms of a genetic character found at a given locus on a chromosome.

Allele frequency • The relative proportion of a particular allele in a specific population.

Allergy [Ger. *allergie*: altered reaction] • An overreaction to an antigen in amounts that do not affect most people; often involves IgE antibodies.

Allometric growth • A pattern of growth in which some parts of the body of an organism grow faster than others, resulting in a change in body proportions as the organism grows.

Allopatric speciation (al′ lo pat′ rick) [Gr. *allos*: other + *patria*: fatherland] • Also called geographical speciation, this is the formation of two species from one when reproductive isolation occurs because of the the interposition of (or crossing of) a physical geographic barrier such as a river. (Contrast with parapatric speciation, sympatric speciation.)

Allopolyploid • A polyploid in which the chromosome sets are derived from more than one species.

Allostery (al′ lo steer′ y) [Gr. *allos*: other + *stereos*: structure] • Regulation of the activity of a protein by the binding of an effector molecule at a site other than the active site.

Alpha helix • Type of protein secondary structure; a right-handed spiral.

Alternation of generations • The succession of haploid and diploid phases in some sexually reproducing organisms, notably plants.

Altruistism • A behavior whose performance harms the actor but benefits other individuals.

Alveolus (al ve′ o lus) (plural: alveoli) [L. *alveus*: cavity] • A small, baglike cavity, especially the blind sacs of the lung.

Amensalism (a men′ sul ism) • Interaction in which one animal is harmed and the other is unaffected. (Contrast with commensalism, mutualism.)

Amine • An organic compound with an amino group (see Amino acid).

Amino acid • An organic compound of the general formula H_2N–CH*R*–COOH, where *R* can be one of 20 or more different side groups. An amino acid is so named because it has both a basic amine group, $-NH_2$, and an acidic carboxyl group, –COOH. Proteins are polymers of amino acids.

Ammonotelic (am moan′ o teel′ ic) [Gr. *telos*: end] • Describes an organism in which the final product of breakdown of nitrogen-containing compounds (primarily proteins) is ammonia. (Contrast with ureotelic, uricotelic.)

Amniocentesis • A medical procedure in which cells from the fetus are obtained from the amniotic fluid. The genetic material of the cells is then examined. (Contrast with chorionic villus sampling.)

Amniote • An organism that lays eggs that can be incubated in air (externally) because the embryo is enclosed by a fluid-filled sac. Birds and reptiles are amniotes.

Amphipathic (am′ fi path′ ic) [Gr. *amphi*: both + *pathos*: emotion] • Of a molecule, having both hydrophilic and hydrophobic regions.

Amylase (am′ ill ase) • Any of a group of enzymes that digest starch.

Anabolism (an ab′ uh liz′ em) [Gr. *ana*: up, throughout + *ballein*: to throw] • Synthetic reactions of metabolism, in which complex molecules are formed from simpler ones. (Contrast with catabolism.)

Anaerobic (an ur row′ bic) [Gr. *an*: not + *aer*: air + *bios*: life] • Occurring without the use of molecular oxygen, O_2.

Anagenesis • Evolutionary change in a single lineage over time.

Analogy (a nal′ o jee) [Gr. *analogia*: resembling] • A resemblance in function, and often appearance as well, between two structures which is due to convergence in evolution rather than to common ancestry. (Contrast with homology.)

Anaphase (an′ a phase) [Gr. *ana*: indicating upward progress] • The stage in nuclear division at which the first separation of sister chromatids (or, in the first meiotic division, of paired homologues) occurs. Anaphase lasts from the moment of first separation to the time at which the moving chromosomes converge at the poles of the spindle.

Anaphylactic shock • A precipitous drop in blood pressure caused by loss of fluid from capillaries because of an increase in their permeability stimulated by an allergic reaction.

Ancestral trait • Trait shared by a group of organisms as a result of descent from a common ancestor.

Androgens (an′ dro jens) • The male sex steroids.

Aneuploidy (an′ you ploy dee) • A condition in which one or more chromosomes or pieces of chromosomes are either lacking or present in excess.

Angiosperm (an′ jee oh spurm) [Gr. *angion*: vessel + *sperma*: seed] • One of the flowering plants; literally, one whose seed is carried in a "vessel," which is the fruit. (See fruit.)

Angiotensin (an′ jee oh ten′ sin) • A peptide hormone that raises blood pressure by causing peripheral vessels to constrict; maintains glomerular filtration by constricting efferent glomerular vessels; stimulates thirst; and stimulates the release of aldosterone.

Animal [L. *animus*: breath, soul] • A member of the kingdom Animalia. In general, a multicellular eukaryote that obtains its food by ingestion.

Animal hemisphere • The metabolically active upper portion of some animal eggs, zygotes, and embryos, which does *not* contain the dense nutrient yolk. The **animal pole** refers to the very top of the egg or embyro. (Contrast with vegetal hemisphere.)

Anion (an′ eye one) • An ion with one or more negative charges. (Contrast with cation.)

Anisogamy (an′ eye sog′ a mee) [Gr. *aniso*: unequal + *gamos*: marriage] • The existence of two dissimilar gametes (egg and sperm).

Annual • Referring to a plant whose life cycle is completed in one growing season. (Contrast with biennial, perennial.)

Anterior pituitary • The portion of the vertebrate pituitary gland that derives from gut epithelium and produces tropic hormones.

Anther (an′ thur) [Gr. *anthos*: flower] • A pollen-bearing portion of the stamen of a flower.

Antheridium (an′ thur id′ ee um) (plural: antheridia) [Gr. *antheros*: blooming] • The multicellular structure that produces the sperm in bryophytes and ferns.

Antibody • One of millions of proteins, produced by the immune system, that specifically recognizes a foreign substance and initiates its removal from the body.

Anticodon • A "triplet" of three nucleotides in transfer RNA that is able to pair with a complementary triplet (a codon) in messenger RNA, thus aligning the transfer RNA on the proper place on the messenger. The codon (and, reciprocally, the anticodon) codes for a specific amino acid.

Antidiuretic hormone • A hormone that controls water reabsorption in the mammalian kidney. Also called vasopressin.

Antigen (an′ ti jun) • Any substance that stimulates the production of an antibody or antibodies in the body of a vertebrate.

Antigen processing • The breakdown of antigenic proteins into smaller fragments, which are then presented on the cell surface, along with MHC proteins, to T cells.

Antigenic determinant • A specific region of an antigen, which is recognized by and binds to a specific antibody.

Antiport • A membrane transport process that carries one substance in one direction and another in the opposite direction. (Contrast with symport.)

Antisense nucleic acid • A single-stranded RNA or DNA complementary to and thus targeted against the mRNA transcribed from a harmful gene such as an oncogene.

Anus (a′ nus) • Opening through which digestive wastes are expelled, located at the posterior end of the gut.

Aorta (a or′ tuh) [Gr. *aorte*: aorta] • The main trunk of the arteries leading to the systemic (as opposed to the pulmonary) circulation.

Apex (a′ pecks) • The tip or highest point of a structure, as the apex of a growing stem or root.

Apical (a′ pi kul) • Pertaining to the apex, or tip, usually in reference to plants.

Apical dominance • Inhibition by the apical bud of the growth of axillary buds.

Apical meristem • The meristem at the tip of a shoot or root; responsible for the plant's primary growth.

Apomixis (ap oh mix′ is) [Gr. *apo*: away from + *mixis*: sexual intercourse] • The asexual production of seeds.

Apoplast (ap′ oh plast) • in plants, the continuous meshwork of cell walls and extracellular spaces through which material can pass without crossing a plasma membrane. (Contrast with symplast.)

Apoptosis (ay′ pu toh sis) • A series of genetically programmed events leading to cell death.

Aquaporin • A transport protein in plant and animals cells through which water passes in osmosis.

Archegonium (ar′ ke go′ nee um) [Gr. *archegonos*: first of a kind] • The multicellular structure that produces eggs in bryophytes, ferns, and gymnosperms.

Archenteron (ark en′ ter on) [Gr. *archos*: beginning + *enteron*: bowel] • The earliest primordial animal digestive tract.

Arteriosclerosis • See atherosclerosis.

Artery • A muscular blood vessel carrying oxygenated blood away from the heart to other parts of the body. (Contrast with vein.)

Ascus (ass' cuss) [Gr. *askos*: bladder] • In fungi belonging to the phylum Ascomycota (the sac fungi), the club-shaped sporangium within which spores (ascospores) are produced by meiosis.

Asexual • Without sex.

Assortative mating • A breeding system in which mates are selected on the basis of a particular trait or group of traits.

Atherosclerosis (ath' er oh sklair oh' sis) • A disease of the lining of the arteries characterized by fatty, cholesterol-rich deposits in the walls of the arteries. When fibroblasts infiltrate these deposits and calcium precipitates in them, the disease become arteriosclerosis, or "hardening of the arteries."

Atmosphere • The gaseous mass surrounding our planet. Also: a unit of pressure, equal to the normal pressure of air at sea level.

Atom [Gr. *atomos*: indivisible] • The smallest unit of a chemical element. Consists of a nucleus and one or more electrons.

Atomic mass (also called atomic weight) • The average mass of an atom of an element on the amu scale. (The average depends upon the relative amounts of different isotopes of an element on Earth.)

Atomic number • The number of protons in the nucleus of an atom, also equal to the number of electrons around the neutral atom. Determines the chemical properties of the atom.

ATP (adenosine triphosphate) • A compound containing adenine, ribose, and three phosphate groups. When it is formed, useful energy is stored; when it is broken down (to ADP or AMP), energy is released to drive endergonic reactions. ATP is an energy storage compound.

ATP synthase • An integral membrane protein that couples the transport of proteins with the formation of ATP.

Atrium (a' tree um) • A body cavity, as in the hearts of vertebrates. The thin-walled chamber(s) entered by blood on its way to the ventricle(s). Also, the outer ear.

Autoimmune disease • A disorder in which the immune system attacks the animal's own antigens.

Autonomic nervous system • The system (which in vertebrates comprises sympathetic and parasympathetic subsystems) that controls such involuntary functions as those of guts and glands.

Autosome • Any chromosome (in a eukaryote) other than a sex chromosome.

Autotroph (au' tow trow' fik) [Gr. *autos*: self + *trophe*: food] • An organism that is capable of living exclusively on inorganic materials, water, and some energy source such as sunlight or chemically reduced matter. (Contrast with heterotroph.)

Auxin (awk' sin) [Gr. *auxein*: increase] • In plants, a substance (indoleacetic acid) that regulates growth and various aspects of development.

Auxotroph (awks' o trofe) [Gr. *auxanein*: to grow + *trophe*: food] • A mutant form of an organism that requires a nutrient or nutrients not required by the wild type, or reference, form of the organism. (Contrast with prototroph.)

Axon [Gr.: axle] • Fiber of a neuron which can carry action potentials. Carries impulses away from the cell body of the neuron; releases a neurotransmitter substance.

Axon hillock • The junction between an axon and its cell body; where action potentials are generated.

Axon terminals • The endings of an axon; they form synapses and release neurotransmitter.

Axoneme (ax' oh neem) • The complex of microtubules and their crossbridges that forms the motile apparatus of a cilium.

Bacillus (buh sil' us) [L.: little rod] • Any of various rod-shaped bacteria.

Bacteriophage (bak teer' ee o fayj) [Gr. *bakterion*: little rod + *phagein*: to eat] • One of a group of viruses that infect bacteria and ultimately cause their disintegration.

Bacteria (bak teer' ee ah) (singular: bacterium) [Gr. *bakterion*: little rod] • Prokaryote in the Domain Bacteria. The chromosomes of bacteria are not contained in nuclear envelopes.

Balanced polymorphism [Gr. *polymorphos*: having many forms] • The maintenance of more than one form, or the maintenance at a given locus of more than one allele, at frequencies of greater than one percent in a population. Often results when heterozygotes are superior to both homozygotes.

Bark • All tissues outside the vascular cambium of a plant.

Baroreceptor [Gr. *baros*: weight] • A pressure-sensing cell or organ.

Barr body • In mammals, an inactivated X chromosome.

Basal body • Centriole found at the base of a eukaryotic flagellum or cilium.

Basal metabolic rate • The minimum rate of energy turnover in an awake (but resting) bird or mammal that is not expending energy for thermoregulation.

Base • (1) A substance which can accept a proton (hydrogen ion; H^+) in solution. (Contrast with acid.) (2) In nucleic acids, a nitrogen-containing molecule that is attached to each sugar in the backbone. (See purine; pyrimidine.)

Base pairing • See complementary base pairing.

Basic • having a pH greater than 7.0 (having a hydrogen ion concentration lower than 10^{-7} molar).

Basidium (bass id' ee yum) • In fungi of the class Basidiomycetes, the characteristic sporangium in which four spores are formed by meiosis and then borne externally before being shed.

Batesian mimicry • Mimicry by a relatively harmless kind of organism of a more dangerous one, by which the mimic enjoys protection from predators that mistake it for the dangerous model. (Contrast with Müllerian mimicry.)

B cell • A type of lymphocyte involved in the humoral immune response of vertebrates. Upon recognizing an antigenic determinant, a B cell develops into a plasma cell, which secretes an antibody. (Contrast with a T cell.)

Benefit • An improvement in survival and reproductive success resulting from a behavior. (Contrast with cost.)

Benign (be nine') • A tumor that grows to a certain size and then stops, uaually with a fibrous capsule surrounding the mass of cells. Benign tumors do not spread (metastasize) to other organs.

Benthic zone [Gr. *benthos*: bottom of the sea] • The bottom of the ocean. (Contrast with pelagic zone.)

Beta-pleated sheet • Type of protein secondary structure; results from hydrogen bonding between polypeptide regions running antiparallel to each other.

Biennial • Referring to a plant whose life cycle includes vegetative growth in the first year and flowering and senescence in the second year. (Contrast with annual, perennial.)

Bilateral symmetry • The condition in which only the right and left sides of an organism, divided exactly down the back, are mirror images of each other. (Contrast with biradial symmetry.)

Bile • A secretion of the liver delivered to the small intestine via the common bile duct. In the intestine, bile emulsifies fats.

Binocular cells • Neurons in the visual cortex that respond to input from both retinas; involved in depth perception.

Binomial (bye nome' ee al) • Consisting of two names; for example, the binomial nomenclature of biology which gives the name of the genus followed by the name of the species.

Biodiversity crisis • The current high rate of loss of species, caused primarily by human activities.

Biogeochemical cycles • Movement of elements through living organisms and the physical environment.

Biogeography • The scientific study of the geographic distribution of organisms.

Biogeographic region • A continental-scale part of Earth that has a biota distinct from that of other such regions.

Biological species concept • The view that a species is most usefully defined as a population or series of populations within which there is a significant amount of gene flow under natural conditions, but which is genetically isolated from other populations.

Bioluminescence • The production of light by biochemical processes in an organism.

Biomass • The total weight of all the living organisms, or some designated group of living organisms, in a given area.

Biome (bye' ome) • A major division of the ecological communities of Earth; characterized by distinctive vegetation.

Biota (bye oh′ tah) • All of the organisms, including animals, plants, fungi, and microorganisms, found in a given area.

Biotechnology • The use of cells to make medicines, foods and other products useful to humans.

Biradial symmetry • Radial symmetry modified so that only two planes can divide the animal into similar halves.

Blastocoel (blass′ toe seal) [Br. *blastos*: sprout + *koilos*: hollow] • The central, hollow cavity of a blastula.

Blastodisc (blass′ toe disk) • A disk of cells forming on the surface of a large yolk mass, comparable to a blastula, but occurring in animals such as birds and reptiles, in which the massive yolk restricts cleavage to one side of the egg only.

Blastomere • A cell produced by the division of a fertilized egg.

Blastopore • The opening from the archenteron to the exterior of a gastrula.

Blastula (blass′ chu luh) [Gr. *blastos*: sprout] • An early stage in animal embryology; in many species, a hollow sphere of cells surrounding a central cavity, the blastocoel. (Contrast with blastodisc.)

Blood–brain barrier • A property of the blood vessels of the brain that prevents most chemicals from diffusing from the blood into the brain.

Body plan • A basic structural design that includes an entire animal, its organ systems, and the integrated functioning of its parts. Phylogenetic groups of organisms are classified in part on the basis of a shared body plan.

Bowman's capsule • An elaboration of kidney tubule cells that surrounds a know of capillaries (the glomerulus). Blood is filtered across the walls of these capillaries and the filtrate is collected into Bowman's capsule.

Brain stem • The portion of the vertebrate brain between the spinal cord and the forebrain.

Brassinosteroids • Plant steroid hormones that promote the elongation of stems and pollen tubes.

Bronchus (plural: bronchi) • The major airway(s) branching off the trachea into the vertebrate lung.

Brown fat • Fat tissue in mammals that is specialized to produce heat. It has many mitochondria and capillaries, and a protein that uncouples oxidative phosphorylation.

Browser • An animal that feeds on the tissues of woody plants.

Bryophyte (bri′ uh fite′) [Gr. *bruon*: moss + *phyton*: plant] • A moss. Formerly was often used to refer to all the nontracheophyte plants.

Budding • Asexual reproduction in which a more or less complete new organism simply grows from the body of the parent organism and eventually detaches itself.

Buffering • A process by which a system resists change—particularly in pH, in which case added acid or base is partially converted to another form.

C_3 photosynthesis • The form of photosynthesis in which 3-phosphoglycerate is the first stable product, and ribulose bisphosphate is the CO_2 receptor.

C_4 photosynthesis • The form of photosynthesis in which oxaloacetate is the first stable product, and phosphoenolpyruvate is the CO_2 acceptor. C_4 plants also perform the reactions of C_3 photosynthesis.

Calcitonin • A hormone produced by the thyroid gland; it lowers blood calcium and promotes bone formation. (Contrast with parathormone.)

Calmodulin (cal mod′ joo lin) • A calcium-binding protein found in all animal and plant cells; mediates many calcium-regulated processes.

calorie [L. *calor*: heat] • The amount of heat required to raise the temperature of one gram of water by one degree Celsius (1°C) from 14.5°C to 15.5°C. In nutrition studies, "Calorie" (spelled with a capital C) refers to the kilocalorie (1 kcal = 1,000 cal).

Calvin–Benson cycle • The stage of photosynthesis in which CO_2 reacts with RuBP to form 3PG, 3PG is reduced to a sugar, and RuBP is regenerated, while other products are released to the rest of the plant.

Calyx (kay′ licks) [Gr. *kalyx*: cup] • All of the sepals of a flower, collectively.

CAM • See crassulacean acid metabolism.

Cambium (kam′ bee um) [L. *cambiare*: to exchange] • A meristem that gives rise to radial rows of cells in stem and root, increasing them in girth; commonly applied to the vascular cambium which produces wood and phloem, and the cork cambium, which produces bark.

cAMP (cyclic AMP) • A compound, formed from ATP, that mediates the effects of numerous animal hormones. Also needed for the transcription of catabolite-repressible operons in bacteria. Used for communication by cellular slime molds.

Canopy • The leaf-bearing part of a tree. Collectively the aggregate of the leaves and branches of the larger woody plants of an ecological community.

Capillaries [L. *capillaris*: hair] • Very small tubes, especially the smallest blood-carrying vessels of animals between the termination of the arteries and the beginnings of the veins.

Capsid • The protein coat of a virus.

Carbohydrates • Organic compounds with the general formula $C_nH_{2m}O_m$. Common examples are sugars, starch, and cellulose.

Carboxylic acid (kar box sill′ ik) • An organic acid containing the carboxyl group, –COOH, which dissociates to the carboxylate ion, $-COO^-$.

Carcinogen (car sin′ oh jen) • A substance that causes cancer.

Cardiac (kar′ dee ak) [Gr. *kardia*: heart] • Pertaining to the heart and its functions.

Carnivore [L. *carn*: flesh + *vovare*: to devour] • An organism that feeds on animal tissue. (Contrast with detritivore, herbivore, omnivore.)

Carotenoid (ka rah′ tuh noid) [L. *carota*: carrot] • A yellow, orange, or red lipid pigment commonly found as an accessory pigment in photosynthesis; also found in fungi.

Carpel (kar′ pel) [Gr. *karpos*: fruit] • The organ of the flower that contains one or more ovules.

Carrier • (1) In facilitated diffusion, a membrane protein that binds a specific molecule and transports it through the membrane. (2) In respiratory and photosynthetic electron transport, a participating substance such as NAD that exists in both oxidized and reduced forms. (3) In genetics, a person heterozygous for a recessive trait.

Carrying capacity • In ecology, the largest number of organisms of a particular species that can be maintained indefinitely in a given part of the environment.

Cartilage • In vertebrates, a tough connective tissue found in joints, the outer ear, and elsewhere. Forms the entire skeleton in some animal groups.

Casparian strip • A band of cell wall containing suberin and lignin, found in the endodermis. Restricts the movement of water across the endodermis.

Catabolism [Ge. *kata*: down + *ballein*: to throw] • Degradational reactions of metabolism, in which complex molecules are broken down. (Contrast with anabolism.)

Catalyst (cat′ a list) [Gr. *kata-*, implying the breaking down of a compound] • A chemical substance that accelerates a reaction without itself being consumed in the overall course of the reaction. Catalysts lower the activation energy of a reaction. Enzymes are biological catalysts.

Cation (cat′ eye on) • An ion with one or more positive charges. (Contrast with anion.)

Caudal [L. *cauda*: tail] • Pertaining to the tail, or to the posterior part of the body.

cDNA • See complementary DNA.

Cecum (see′ cum) [L. *caecus*: blind] • A blind branch off the large intestine. In many nonruminant mammals, the cecum contains a colony of microorganisms that contribute to the digestion of food.

Cell adhesion molecules • Molecules on animal cell surfaces that affect the selective association of cells during development of the embryo.

Cell cycle • The stages through which a cell passes between one division and the next. Includes all stages of interphase and mitosis.

Cell division • The reproduction of a cell to produce two new cells. In eukaryotes, this process involves nuclear division (mitosis) and cytoplasmic division (cytokinesis).

Cell theory • The theory, well established, that organisms consist of cells, and that all cells come from preexisting cells.

Cell wall • A relatively rigid structure that encloses cells of plants, fungi, many protists, and most bacteria. The cell wall gives these cells their shape and limits their expansion in hypotonic media.

Cellular immune system • That part of the immune system that is based on the activities of T cells. Directed against parasites, fungi, intracellular viruses, and foreign tissues (grafts). (Contrast with humoral immune system.)

Cellular respiration • See respiration.

Cellulose (sell′ you lowss) • A straight-chain polymer of glucose molecules, used by plants as a structural supporting material.

Central dogma • The statement that information flows from DNA to RNA to polypeptide (in retroviruses, there is also information flow from RNA to cDNA).

Central nervous system • That part of the nervous system which is condensed and centrally located, e.g., the brain and spinal cord of vertebrates; the chain of cerebral, thoracic and abdominal ganglia of arthropods.

Centrifuge [L. *fugere*: to flee] • A device in which a sample can be spun around a central axis at high speed, creating a centrifugal force that mimics a very strong gravitational force. Used to separate mixtures of suspended materials.

Centriole (sen′ tree ole) • A paired organelle that helps organize the microtubules in animal and protist cells during nuclear division.

Centromere (sen′ tro meer) [Gr. *centron*: center + *meros*: part] • The region where sister chromatids join.

Centrosome (sen′ tro soam) • The major microtubule organizing center of an animal cell.

Cephalization (sef′ uh luh zay′ shun) [Gr. *kephale*: head] • The evolutionary trend toward increasing concentration of brain and sensory organs at the anterior end of the animal.

Cerebellum (sair′ uh bell′ um) [L.: diminutive of *cerebrum*: brain] • The brain region that controls muscular coordination; located at the anterior end of the hindbrain.

Cerebral cortex • The thin layer of gray matter (neuronal cell bodies) that overlays the cerebrum.

Cerebrum (su ree′ brum) [L.: brain] • The dorsal anterior portion of the forebrain, making up the largest part of the brain of mammals. In mammals, the chief coordination center of the nervous system; consists of two **cerebral hemispheres**.

Cervix (sir′ vix) [L.: neck] • The opening of the uterus into the vagina.

cGMP (cyclic guanosine monophosphate) • An intracellular messenger that is part of signal transmission pathways involving G proteins. (See G protein.)

Channel • A membrane protein that forms an aqueous passageway though which specific solutes may pass by simple diffusion; some channels are gated: they open and close in response to binding of specific molecules.

Chaperone protein • A protein that assists a newly forming protein in adopting its appropriate tertiary structure.

Chemical bond • An attractive force stably linking two atoms.

Chemiosmotic mechanism • The formation of ATP in mitochondria and chloroplasts, resulting from a pumping of protons across a membrane (against a gradient of electrical charge and of pH), followed by the return of the protons through a protein channel with ATPase activity.

Chemoautotroph • An organism that uses carbon dioxide as a carbon source and obtains energy by oxidizing inorganic substances from its environment. (Contrast with chemoheterotroph, photoautotroph, photoheterotroph.)

Chemoheterotroph • An organism that must obtain both carbon and energy from organic substances. (Contrast with chemoautotroph, photoautotroph, photoheterotroph.)

Chemoreceptor • A cell or tissue that senses specific substances in its environment.

Chemosynthesis • Synthesis of food substances, using the oxidation of reduced materials from the environment as a source of energy.

Chiasma (kie az′ muh) (plural: chiasmata) [Gr.: cross] • An X-shaped connection between paired homologous chromosomes in prophase I of meiosis. A chiasma is the visible manifestation of crossing over between homologous chromosomes.

Chitin (kye′ tin) [Gr. *kiton*: tunic] • The characteristic tough but flexible organic component of the exoskeleton of arthropods, consisting of a complex, nitrogen-containing polysaccharide. Also found in cell walls of fungi.

Chlorophyll (klor′ o fill) [Gr. *kloros*: green + *phyllon*: leaf] • Any of a few green pigments associated with chloroplasts or with certain bacterial membranes; responsible for trapping light energy for photosynthesis.

Chloroplast [Gr. *kloros*: green + *plast*: a particle] • An organelle bounded by a double membrane containing the enzymes and pigments that perform photosynthesis. Chloroplasts occur only in eukaryotes.

Choanocyte (cho′ an oh cite) • The collared, flagellated feeding cells of sponges.

Cholecystokinin (ko′ lee sis to kai nin) • A hormone produced and released by the lining of the duodenum when it is stimulated by undigested fats and proteins. It stimulates the gallbladder to release bile and slows stomach activity.

Chorion (kor′ ee on) [Gr. *khorion*: afterbirth] • The outermost of the membranes protecting mammal, bird, and reptile embryos; in mammals it forms part of the placenta.

Chorionic villus sampling • A medical procedure that extracts a portion of the chorion from a pregnant woman to enable genetic and biochemical analysis of the embryo. (Contrast with amniocentesis.)

Chromatid (kro′ ma tid) • Each of a pair of new sister chromosomes from the time at which the molecular duplication occurs until the time at which the centromeres separate at the anaphase of nuclear division.

Chromatin • The nucleic acid–protein complex found in eukaryotic chromosomes.

Chromatophore (krow mat′ o for) [Gr. *kroma*: color + *phoreus*: carrier] • A pigment-bearing cell that expands or contracts to change the color of the organism.

Chromosome (krome′ o sowm) [Gr. *kroma*: color + *soma*: body] • In bacteria and viruses, the DNA molecule that contains most or all of the genetic information of the cell or virus. In eukaryotes, a structure composed of DNA and proteins that bears part of the genetic information of the cell.

Chylomicron (ky low my′ cron) • Particles of lipid coated with protein, produced in the gut from dietary fats and secreted into the extracellular fluids.

Chyme (kime) [Gr. *kymus*, juice] • Created in the stomach; a mixture of ingested food with the digestive juices secreted by the salivary glands and the stomach lining.

Cilium (sil′ ee um) (plural: cilia) [L. *cilium*: eyelash] • Hairlike organelle used for locomotion by many unicellular organisms and for moving water and mucus by many multicellular organisms. Generally shorter than a flagellum.

Circadian rhythm (sir kade′ ee an) [L. *circa*: approximately + *dies*: day] • A rhythm in behavior, growth, or some other activity that recurs about every 24 hours under constant conditions.

Circannual rhythm (sir can′ you al) [L. *circa*: approximately + *annus*: year) • A rhythm of behavior, growth, or some other activity that recurs on a yearly basis.

Citric acid cycle • A set of chemical reactions in cellular respiration, in which acetyl CoA reacts with oxaloacetate to form citric acid, and oxaloacetate is regenerated. Acetyl CoA is oxidized to carbon dioxide, and hydrogen atoms are stored as NADH and $FADH_2$. Also called the Krebs cycle.

Class • In taxonomy, the category below the phylum and above the order; a group of related, similar orders.

Class I MHC molecules • These cell surface proteins participate in the cellular immune response directed against virus-infected cells.

Class II MHC molecules • These cell surface proteins participate in the cell-cell interactions (of helper T cells, macrophages, and B cells) of the humoral immune response.

Class switching • The process whereby a plasma cell changes the class of immunoglobulin that it synthesizes. This results from the deletion of part of the constant region of DNA, bringing in a new C segment. The variable region is the same as before, so that the new immunoglbulin has the same antigenic specificity.

Clathrin • A fibrous protein on the inner surfaces of animal cell membranes that strengthens coated vesicles and thus participates in receptor-mediated endocytosis.

Clay • A soil constituent comprising particles smaller than 2 micrometers in diameter.

Cleavages • First divisions of the fertilized egg of an animal.

Cline • A gradual change in the traits of a species over a geographical gradient.

Cloaca (klo ay' kuh) [L. *cloaca*: sewer] • In some invertebrates, the posterior part of the gut; in many vertebrates, a cavity receiving material from the digestive, reproductive, and excretory systems.

Clonal anergy • When a naive T cell encounters a self-antigen, the T cell may bind to the antigen but does not receive signals from an antigen-presenting cell. Instead of being activated, the T cell dies (becomes anergic). In this way, we avoid reacting to our own tissue-specific antigens.

Clonal deletion • In immunology, the inactivation or destruction of lymphocyte clones that would produce immune reactions against the animal's own body.

Clonal selection • The mechanism by which exposure to antigen results in the activation of selected T- or B-cell clones, resulting in an immune response.

Clone [Gr. *klon*: twig, shoot] • Genetically identical cells or organisms produced from a common ancestor by asexual means.

Cnidocytes • The feeding cells of cnidarians, within which nematocysts are housed.

Coacervate (ko as' er vate) [L. *coacervare*: to heap up] • An aggregate of colloidal particles in suspension.

Coacervate drop • Drops formed when a mixture of large proteins and polysaccharides is shaken in water. The interiors of these drops, which are often very stable, contain most of the proteins and polysaccharides.

Coated vesicle • Vesicle, sometimes formed from a coated pit, with characteristic "bristly" surface; its membrane contains distinctive proteins, including clathrin.

Coccus (kock' us) [Gr. *kokkos*: berry, pit] • Any of various spherical or spheroidal bacteria.

Cochlea (kock' lee uh) [Gr. *kokhlos*: a land snail] • A spiral tube in the inner ear of vertebrates; it contains the sensory cells involved in hearing.

Codominance • A condition in which two alleles at a locus produce different phenotypic effects and both effects appear in heterozygotes.

Codon • A "triplet" of three nucleotides in messenger RNA that directs the placement of a particular amino acid into a polypeptide chain. (Contrast with anticodon.)

Coefficient of relatedness • The probability that an allele in one individual is an identical copy, by descent, of an allele in another individual.

Coelom (see' lum) [Gr. *koiloma*: cavity] • The body cavity of certain animals, which is lined with cells of mesodermal origin.

Coelomate • Having a coelom.

Coenocyte (seen' a sight) [Gr.: common cell] • A "cell" bounded by a single plasma membrane, but containing many nuclei.

Coenzyme • A nonprotein molecule that plays a role in catalysis by an enzyme. The coenzyme may be part of the enzyme molecule or free in solution. Some coenzymes are oxidizing or reducing agents.

Coevolution • Concurrent evolution of two or more species that are mutually affecting each other's evolution.

Cohort (co' hort) [L. *cohors*: company of soldiers] • A group of similar-age organisms, considered as it passes through time.

Collagen [Gr. *kolla*: glue] • A fibrous protein found extensively in bone and connective tissue.

Collecting duct • In vertebrates, a tubule that receives urine produced in the nephrons of the kidney and delivers that fluid to the ureter for excretion.

Collenchyma (cull eng' kyma) [Gr. *kolla*: glue + *enchyma*: infusion] • A type of plant cell, living at functional maturity, which lends flexible support by virtue of primary cell walls thickened at the corners. (Contrast with parenchyma, sclerenchyma.)

Colon [Gr. *kolon*: large intestine] • The large intestine.

Commensalism • The form of symbiosis in which one species benefits from the association, while the other is neither harmed nor benefited.

Common bile duct • A single duct that delivers bile from the gallbladder and secretions from the pancreas into the small intestine.

Communication • A signal from one organism (or cell) that alters the pattern of behavior in another organism (or cell) in an adaptive fashion.

Community • Any ecologically integrated group of species of microorganisms, plants, and animals inhabiting a given area.

Companion cell • Specialized cell found adjacent to a sieve tube member in flowering plants.

Comparative analysis • An approach to studying evolution in which hypotheses are tested by measuring the distribution of states among a large number of species.

Comparative genomics • Computer-aided comparison of DNA sequences between different organisms to reveal genes with related functions.

Compensation point • The light intensity at which the rates of photosynthesis and of cellular respiration are equal.

Competitive inhibitor • A substance, similar in structure to an enzyme's substrate, that binds the active site and thus inhibits a reaction.

Competition • In ecology, use of the same resource by two or more species, when the resource is present in insufficient supply for the combined needs of the species.

Competitive exclusion • A result of competition between species for a limiting resource in which one species completely eliminates the other.

Competitive inhibitor • A substance, similar in structure to an enzyme's substrate, that binds the active site and inhibits a reaction.

Complement system • A group of eleven proteins that play a role in some reactions of the immune system. The complement proteins are not immunoglobulins.

Complementary base pairing • The A–T (or A–U), T–A (or U–A), C–G and G–C pairing of bases in double-stranded DNA, in transcription, and between tRNA and mRNA.

Complementary DNA (cDNA) • DNA formed by reverse transcriptase acting with an RNA template; essential intermediate in the reproduction of retroviruses; used as a tool in recombinant DNA technology; lacks introns.

Complete metamorphosis • A change of state during the life cycle of an organism in which the body is almost completely rebuilt to produce an individual with a very different body form. Characteristic of insects such as butterflies, moths, beetles, ants, wasps, and flies.

Compound • (1) A substance made up of atoms of more than one element. (2) Made up of many units, as the compound eyes of arthropods (as opposed to the simple eyes of the same group of organisms).

Condensation reaction • A reaction in which two molecules become connected by a covalent bond and a molecule of water is released. ($AH + BOH \rightarrow AB + H_2O$.)

Cones • (1) In the vertebrate retina: photoreceptors responsible for color vision. (2) In gymnosperms: reproductive structures consisting of many sporophylls packed relatively tightly.

Conidium (ko nid' ee um) [Gr. *konis*: dust] • An asexual fungus spore borne singly or in chains either apically or laterally on a hypha.

Conifer (kahn' e fer) [Gr. *konos*: cone + *phero*: carry] • One of the cone-bearing gymnosperms, mostly trees, such as pines and firs.

Conjugation (kahn' jew gay' shun) [L. *conjugare*: yoke together] • The close approximation of two cells during which they exchange genetic material, as in *Paramecium* and other ciliates, or during which DNA passes from one to the other through a tube, as in bacteria.

Connective tissue • An animal tissue that connects or surrounds other tissues; its cells are embedded in a collagen-containing matrix.

Connexon • In a gap junction, a protein channel linking adjacent animal cells.

Consensus sequences • Short stretches of DNA that appear, with little variation, in many different genes.

Constant region • The constant region in an immunoglobulin is encoded by a single exon and determines the function, but not the specificity, of the molecule. The constant region of the T cell receptor anchors the protein to the plasma membrane.

Constitutive enzyme • An enzyme that is present in approximately constant amounts in a system, whether its substrates are present or absent. (Contrast with inducible enzyme.)

Consumer • An organism that eats the tissues of some other organism.

Continental drift • The gradual drifting apart of the world's continents that has occurred over a period of billions of years.

Convergent evolution • The evolution of similar features independently in unrelated taxa from different ancestral structures.

Cooperative act • Behavior in which two or more individuals interact to their mutual benefit. No conscious awareness by the actors of the effects of their behavior is implied.

Cooption • The act of capturing something for a particular use. In ecology refers to the diversion of ecological production for human use. Such production is said to be coopted.

Copulation • Reproductive behavior that results in a male depositing sperm in the reproductive tract of a female.

Corepressor • A low molecular weight compound that unites with a protein (the repressor) to prevent transcription in a repressible operon.

Cork • A waterproofing tissue in plants, with suberin-containing cell walls. Produced by a cork cambium.

Corolla (ko role' lah) [L.: diminutive of *corona*: wreath, crown] • All of the petals of a flower, collectively.

Coronary (kor' oh nair ee) • Referring to the blood vessels of the heart.

Corpus luteum (kor' pus loo' tee um) [L. *corpus*: body + *luteum*: yellow] A structure formed from a follicle after ovulation; it produces hormones important to the maintenance of pregnancy.

Cortex [L.: bark or rind] • (1) In plants, the tissue between the epidermis and the vascular tissue of a stem or root. (2) In animals, the outer tissue of certain organs, such as the adrenal cortex and cerebral cortex.

Corticosteroids • Steroid hormones produced and released by the cortex of the adrenal gland.

Cost • See energetic cost, opportunity cost, risk cost.

Cotyledon (kot' ul lee' dun) [Gr. *kotyledon*: a hollow space] • A "seed leaf." An embryonic organ which stores and digests reserve materials; may expand when seed germinates.

Countercurrent exchange • An adaptation that promotes maximum exchange of heat or any diffusible substance between two fluids by the fluids flow in opposite directions through parallel tubes in close approximation to each other. An example is countercurrent heat exchange between arterioles and venules in the extremities of some animals.

Covalent bond • A chemical bond that arises from the sharing of electrons between two atoms. Usually a strong bond.

Crassulacean acid metabolism (CAM) • A metabolic pathway enabling the plants that possess it to store carbon dioxide at night and then perform photosynthesis during the day with stomata closed.

Crista (plural: cristae) • A small, shelflike projection of the inner membrane of a mitochondrion; the site of oxidative phosphorylation.

Critical night length • In the photoperiodic flowering response of short-day plants, the length of night above which flowering occurs and below which the plant remains vegetative. (The reverse applies in the case of long-day plants.)

Critical period • The age during which some particular type of learning must take place or during which it occurs much more easily than at other times. Typical of song learning among birds.

Cross section (also called a transverse section) • A section taken perpendicular to the longest axis of a structure.

Crossing over • The mechanism by which linked markers undergo recombination. In general, the term refers to the reciprocal exchange of corresponding segments between two homologous chromatids.

CRP • The cAMP receptor protein that interacts with the promoter to enhance transcription; a lowered cAMP concentration results in catabolite repression.

Crustacean (crus tay' see an) • A member of the phylum Crustacea, such as a crab, shrimp, or sowbug.

Cryptic appearance [Gr. *kryptos*: hidden] • The resemblance of an animal to some part of its environment, which helps it to escape detection by predators.

Cryptochromes [Gr. *kryptos*: hidden + *kroma*: color] • Photoreceptors mediating some blue-light effects in plants and animals.

Culture • (1) A laboratory association of organisms under controlled conditions. (2) The collection of knowledge, tools, values, and rules that characterize a human society.

Cuticle • A waxy layer on the outer surface of a plant or an insect, tending to retard water loss.

Cyanobacteria (sigh an' o bacteria) [Gr. *kuanos*: the color blue] • A division of photosynthetic bacteria, formerly referred to as blue-green algae; they lack sexual reproduction, and they use chlorophyll *a* in their photosynthesis.

Cyclic AMP • See cAMP.

Cyclins • Proteins that activate cyclin-dependent kinases, bringing about transitions in the cell cycle.

Cyclin-dependent kinase (cdk) • A kinase is an enzyme that catalzyes the addition of phosphate groups from ATP to target molecules. Cdk's target proteins involved in transitions in the cell cycle and are active only when complexed to additional protein subunits, cyclins.

Cyst (sist) [Gr. *kystis*: pouch] • (1) A resistant, thick-walled cell formed by some protists and other organisms. (2) An abnormal sac, containing a liquid or semisolid substance, produced in response to injury or illness.

Cytochromes (sy' toe chromes) [Gr. *kytos*: container + *chroma*: color] • Iron-containing red proteins, components of the electron-transfer chains in photophosphorylation and respiration.

Cytokinesis (sy' toe kine ee' sis) [Gr. *kytos*: container + *kinein*: to move] • The division of the cytoplasm of a dividing cell. (Contrast with mitosis.)

Cytokinin (sy' toe kine' in) [Gr. *kytos*: container + *kinein*: to move] • A member of a class of plant growth substances playing roles in senescence, cell division, and other phenomena.

Cytoplasm • The contents of the cell, excluding the nucleus.

Cytoplasmic determinants • In animal development, gene products whose spatial distribution may determine such things as embryonic axes.

Cytosine (site' oh seen) • A nitrogen-containing base found in DNA and RNA.

Cytoskeleton • The network of microtubules and microfilaments that gives a eukaryotic cell its shape and its capacity to arrange its organelles and to move.

Cytosol • The fluid portion of the cytoplasm, excluding organelles and other solids.

Cytotoxic T cells • Cells of the cellular immune system that recognize and directly eliminate virus-infected cells. (Contrast with helper T cells, suppressor T cells.)

Decomposer • See detritivore.

Degeneracy • The situation in which a single amino acid may be represented by any of two or more different codons in messenger RNA. Most of the amino acids can be represented by more than one codon.

Degradative succession • Ecological succession occuring on the dead remains of the bodies of plants and animals, as when leaves or animal bodies rot.

Deletion (genetic) • A mutation resulting from the loss of a continuous segment of a gene or chromosome. Such mutations never revert to wild type. (Contrast with duplication, point mutation.)

Deme (deem) [Gr. *demos*: common people] • Any local population of individuals belonging to the same species that interbreed with one another.

Demographic processes • The events—such as births, deaths, immigration, and emigration—that determine the number of individuals in a population.

Demographic stochasticity • Random variations in the factors influencing the size, density, and distribution of a population.

Demography • The study of dynamical changes in the sizes, densities, and distributions of populations.

Denaturation • Loss of activity of an enzyme or nucleic acid molecule as a result of structural changes induced by heat or other means.

Dendrite [Gr. *dendron*: a tree] • A fiber of a neuron which often cannot carry action potentials. Usually much branched and relatively short compared with the axon, and commonly carries information to the cell body of the neuron.

Denitrification • Metabolic activity by which inorganic nitrogen-containing ions are reduced to form nitrogen gas and other products; carried on by certain soil bacteria.

Density dependence • Change in the severity of action of agents affecting birth and death rates within populations that are directly or inversely related to population density.

Density independence • The state where the severity of action of agents affecting birth and death rates within a population does not change with the density of the population.

Deoxyribonucleic acid • See DNA.

Depolarization • A change in the electric potential across a membrane from a condition in which the inside of the cell is more negative than the outside to a condition in which the inside is less negative, or even positive, with reference to the outside of the cell. (Contrast with hyperpolarization.)

Derived trait • A trait found among members of a lineage that was not present in the ancestors of that lineage.

Dermal tissue system • The outer covering of a plant, consisting of epidermis in the young plant and periderm in a plant with extensive secondary growth. (Contrast with ground tissue system and vascular tissue system.)

Desmosome (dez′ mo sowm) [Gr. *desmos*: bond + *soma*: body] • An adhering junction between animal cells.

Determination • Process whereby an embryonic cell or group of cells becomes fixed into a predictable developmental pathway.

Detritivore (di try′ ti vore) [L. *detritus*: worn away + *vorare*: to devour] • An organism that obtains its energy from the dead bodies and/or waste products of other organisms.

Deuterostome • A major evolutionary lineage in animals, characterized by radial cleavage, enterocoelous development, and other traits. (Compare with protostome.)

Development • Progressive change, as in structure or metabolism; in most kinds of organisms, development continues throughout the life of the organism.

Diaphragm (dye′ uh fram) [Gr. *diaphrassein*, to barricade] • (1) A sheet of muscle that separates the thoracic and abdominal cavities in mammals; responsible for the action of breathing. (2) A method of birth control in which a sheet of rubber is fitted over the woman's cervix, blocking the entry of sperm.

Diastole (dye ahs′ toll ee) [Gr.: dilation] • The portion of the cardiac cycle when the heart muscle relaxes. (Contrast with systole.)

Dicot (short for dicotyledon) [Gr. *di*: two + *kotyledon*: a hollow space] • This term, not used in this book, formerly referred to all angiosperms other than the monocots. (See eudicot, monocot.)

Differentiation • Process whereby originally similar cells follow different developmental pathways. The actual expression of determination.

Diffusion • Random movement of molecules or other particles, resulting in even distribution of the particles when no barriers are present.

Digestibility-reducing chemicals • Defensive chemicals produced by plants that make the plant's tissued difficult to digest.

Digestion • Enzyme-catalyzed process by which large, usually insoluble, molecules (foods) are hydrolyzed to form smaller molecules of soluble substances.

Dihybrid cross • A mating in which the parents differ with respect to the alleles of two loci of interest.

Dikaryon (di care′ ee ahn) [Gr. *dis*: two + *karyon*: kernel] • A cell or organism carrying two genetically distinguishable nuclei. Common in fungi.

Dioecious (die eesh′ us) [Gr.: two houses] • Organisms in which the two sexes are "housed" in two different individuals, so that eggs and sperm are not produced in the same individuals. Examples: humans, fruit flies, oak trees, date palms. (Contrast with monoecious.)

Diploblastic • Having two cell layers. (Contrast with triploblastic.)

Diploid (dip′ loid) [Gr. *diploos*: double] • Having a chromosome complement consisting of two copies (homologues) of each chromosome. A diploid individual (or cell) usually arises as a result of the fusion of two gametes, each with just one copy of each chromosome. Thus, the two homologues in each chromosome pair in a diploid cell are of separate origin, one derived from the female parent and one from the male parent.

Directional selection • Selection in which phenotypes at one extreme of the population distribution are favored. (Contrast with disruptive selection; stabilizing selection.)

Disaccharide • A carbohydrate made up of two monosaccharides (simple sugars).

Dispersal stage • Stage in its life history at which an organism moves from its birthplace to where it will live as an adult.

Displacement activity • Apparently irrelevant behavior performed by an animal under conflict situations, especially when tendencies to attack and escape are closely balanced.

Display • A behavior that has evolved to influence the actions of other individuals.

Disruptive selection • Selection in which phenotypes at both extremes of the population distribution are favored. (Contrast with directional selection; stabilizing selection.)

Distal • Away from the point of attachment or other reference point. (Contrast with proximal.)

Disturbance • A short-term event that disrupts populations, communities, or ecosystems by changing the environment.

Diverticulum (di ver tic′ u lum) [L. *divertere*: turn away] • A small cavity or tube that connects to a major cavity or tube.

Division • A term used by some microbiologists and formerly by botanists, corresponding to the term phylum.

DNA (deoxyribonucleic acid) • The fundamental hereditary material of all living organisms. In eukaryotes, stored primarily in the cell nucleus. A nucleic acid using deoxyribose rather than ribose.

DNA chip • A small glass or plastic square onto which thousands of single-stranded DNA sequences are fixed. Hybridization of cell-derived RNA or DNA to the target sequences can be performed. (See DNA hybridization.)

DNA hybridization • A process by which DNAs from two species are mixed and heated so that interspecific double helixes are formed.

DNA ligase • Enzyme that unites Okazaki fragments of the lagging strand during DNA replication; also mends breaks in DNA strands. It connects pieces of a DNA strand and is used in recombinant DNA technology.

DNA methylation • Addition of methyl groups to DNA; plays role in regulation of gene expression; protects a bacterium's DNA against its restriction endonucleases.

DNA polymerase • Any of a group of enzymes that catalyze the formation of DNA strands from a DNA template.

Domain • The largest unit in the current taxonomic nomenclature. Members of the three domains (Bacteria, Archaea, and Eukarya) are believed to have been evolving independently of each other for at least a billion years.

Dominance • In genetic terminology, the ability of one allelic form of a gene to determine the phenotype of a heterozygous individual, in which the homologous chromosome carries both it and a different allele. For example, if *A* and *a* are two allelic forms of a gene, *A* is said to be dominant to *a* if *AA* diploids and *Aa* diploids are phenotypically identical and are distinguishable from *aa* diploids. The *a* allele is said to be **recessive**.

Dominance hierarchy • In animal behavior, the set of relationships within a group of animals, usually established and maintained by aggression, in which one individual has precedence over all others in eating, mating, and other activities.

Dormancy • A condition in which normal activity is suspended, as in some seeds and buds.

Dorsal [L. *dorsum*: back] • Pertaining to the back or upper surface. (Contrast with ventral.)

Double fertilization • Process virtually unique to angiosperms in which one sperm nucleus combines with the egg to produce a zygote, and the other sperm nucleus combines with the two polar nuclei to produce the first cell of the triploid endosperm.

Double helix • Of DNA: molecular structure in which two complementary polynucleotide strands, antiparallel to each other, form a right-handed spiral.

Duodenum (doo′ uh dee′ num) • The beginning portion of the vertebrate small intestine. (Contrast with ileum, jejunum.)

Duplication (genetic) • A mutation resulting from the introduction into the genome of an extra copy of a segment of a gene or chromosome. (Contrast with deletion, point mutation.)

Dynein [Gr. *dunamis*: power] • A protein that undergoes conformational changes and thus plays a part in the movement of eukaryotic flagella and cilia.

Ecdysone (eck die' sone) [Gr. *ek*: out of + *dyo*: to clothe] • In insects, a hormone that induces molting.

Ecological biogeography • The study of the distributions of organisms from an ecological perspective, usually concentrating on migration, dispersal, and species interactions.

Ecological community • The species living together at a particular site.

Ecological niche (nitch) [L. *nidus*: nest] • The functioning of a species in relation to other species and its physical environment.

Ecological succession • The sequential replacement of one population assemblage by another in a habitat following some disturbance. Succession sometimes ends in a relatively stable ecosystem.

Ecology [Gr. *oikos*: house + *logos*: discourse, study] • The scientific study of the interaction of organisms with their environment, including both the physical environment and the other organisms that live in it.

Ecoregion • A large geographic unit characterized by a typical climate and a widespread assemblage of similar species.

Ecosystem (eek' oh sis tum) • The organisms of a particular habitat, such as a pond or forest, together with the physical environment in which they live.

Ecto- (eck' toh) [Gr.: outer, outside] • A prefix used to designate a structure on the outer surface of the body. For example, ectoderm. (Contrast with endo- and meso-.)

Ectoderm [Gr. *ektos*: outside + *derma*: skin] • The outermost of the three embryonic tissue layers first delineated during gastrulation. Gives rise to the skin, sense organs, nervous system, etc.

Ectotherm [Gr. *ektos*: outside + *thermos*: heat] • An animal unable to control its body temperature. (Contrast with endotherm.)

Edema (i dee' mah) [Gr. *oidema*: swelling] • Tissue swelling caused by the accumulation of fluid.

Edge effect • The changes in ecological processes in a community caused by physical and biological factors originating in an adjacent community.

Effector • Any organ, cell, or organelle that moves the organism through the environment or else alters the environment to the organism's advantage. Examples include muscle, bone, and a wide variety of exocrine glands.

Effector cell • A lymphocyte that performs a role in the immune system without further differentiation.

Effector phase • In this phase of the immune response, effector T cells called cytotoxic T cells attack virus-infected cells, and effector helper T cells assist B cells to differentiate into plasma cells, which release antibodies.

Efferent [L. *ex*: out + *ferre*: to bear] • Away from, as in neurons that conduct action potentials out from the central nervous system, or arterioles that conduct blood away from a structure. (Contrast with afferent.)

Egg • In all sexually reproducing organisms, the female gamete; in birds, reptiles, and some other vertebrates, a structure witin which early embryonic development occurs.

Elasticity • The property of returning quickly to a former state after a disturbance.

Electrocardiogram (EKG) • A graphic recording of electrical potentials from the heart.

Electroencephalogram (EEG) • A graphic recording of electrical potentials from the brain.

Electromyogram (EMG) • A graphic recording of electrical potentials from muscle.

Electron (e lek' tron) [L. *electrum*: amber (associated with static electricity), from Gr. *slektor*: bright sun (color of amber)] • One of the three most important fundamental particles of matter, with mass approximately 0.00055 amu and charge –1.

Electronegativity • The tendency of an atom to attract electrons when it occurs as part of a compound.

Electrophoresis (e lek' tro fo ree' sis) [L. *electrum*: amber + Gr. *phorein*: to bear] • A separation technique in which substances are separated from one another on the basis of their electric charges and molecular weights.

Electrotonic potential • In neurons, a hyperpolarization or small depolarization of the membrane potential induced by the application of a small electric current. (Contrast with action potential, resting potential.)

Elemental substance • A substance composed of only one type of atom.

Embolus (em' buh lus) [Gr. *embolos*: inserted object; stopper] • A circulating blood clot. Blockage of a blood vessel by an embolus or by a bubble of gas is referred to as an **embolism**. (Contrast with thrombus.)

Embryo [Gr. *en-*: in + *bryein*: to grow] • A young animal, or young plant sporophyte, while it is still contained within a protective structure such as a seed, egg, or uterus.

Embryo sac • In angiosperms, the female gametophyte. Found within the ovule, it consists of eight or fewer cells, membrane bounded, but without cellulose walls between them.

Emergent property • A property of a complex system that is not exhibited by its individual component parts.

Emigration • The deliberate and usually oriented departure of an organism from the habitat in which it has been living.

3′ End (3-prime) • The end of a DNA or RNA strand that has a free hydroxyl group at the 3′-carbon of the sugar (deoxyribose or ribose).

5′ End (5-prime) • The end of a DNA or RNA strand that has a free phosphate group at the 5′-carbon of the sugar (deoxyribose or ribose).

Endemic (en dem' ik) [Gr. *endemos*: dwelling in a place] • Confined to a particular region, thus often having a comparatively restricted distribution.

Endergonic reaction • One for which energy must be supplied. (Contrast with exergonic reaction.)

Endo- [Gr.: within, inside] • A prefix used to designate an innermost structure. For example, endoderm, endocrine. (Contrast with ecto-, meso-.)

Endocrine gland (en' doh krin) [Gr. *endon*: inside + *krinein*: to separate] • Any gland, such as the adrenal or pituitary gland of vertebrates, that secretes certain substances, especially hormones, into the body through the blood.

Endocrinology • The study of hormones and their actions.

Endocytosis • A process by which liquids or solid particles are taken up by a cell through invagination of the plasma membrane. (Contrast with exocytosis.)

Endoderm [Gr. *endon*: within + *derma*: skin] • The innermost of the three embryonic tissue layers first delineated during gastrulation. Gives rise to the digestive and respiratory tracts and structures associated with them.

Endodermis [Gr. *endon*: within + *derma*: skin] • In plants, a specialized cell layer marking the inside of the cortex in roots and some stems. Frequently a barrier to free diffusion of solutes.

Endomembrane system • Endoplasmic reticulum plus Golgi apparatus plus, when present, lysosomes; thus, a system of membranes that exchange material with one another.

Endoplasmic reticulum [Gr. *endon*: within + L. *plasma*: form; L. *reticulum*: little net] • A system of membrane-bounded tubes and flattened sacs found in the cytoplasm of eukaryotes. Exists as rough ER, studded with ribosomes; and smooth ER, lacking ribosomes.

Endorphins • Naturally occurring, opiate-like substances in the mammalian brain.

Endoskeleton [Gr. *endon*: within + *skleros*: hard] • A skeleton covered by other, soft body tissues. (Contrast with exoskeleton.)

Endosperm [Gr. *endon*: within + *sperma*: seed] • A specialized triploid seed tissue found only in angiosperms; contains stored food for the developing embryo.

Endosymbiosis [Gr. *endon*: within + *syn*: together + *bios*: life] • The living together of two species, with one living inside the body (or even the cells) of the other.

Endosymbiotic theory • Theory that the eukaryotic cell evolved from a prokaryote that contained other, endosymbiotic prokaryotes.

Endotherm [Gr. *endon*: within + *thermos*: hot] • An animal that can control its body temperature by the expenditure of its own metabolic energy. (Contrast with ectotherm.)

Endotoxins [Gr. *endon*: within + L. *toxicum*: poison] • Lipopolysaccharides released by the lysis of some Gram-negative bacteria that cause fever and vomiting in a host organism.

Energetic cost • The difference between the energy an animal would have expended had it rested, and that expended in performing a behavior.

Energy • The capacity to do work.

Enhancer • In eukaryotes, a DNA sequence, lying on either side of the gene it regulates, that stimulates a specific promoter.

Enterocoelous development • A pattern of development in which the coelum is formed by an outpocketing of the embryonic gut (enteron).

Enterokinase (ent uh row kine' ase) • An enzyme secreted by the mucosa of the duodenum. It activates the zymogen trypsinogen to create the active digestive enzyme trypsin.

Entrainment • With respect to circadian rhythms, the process whereby the period is adjusted to match the 24-hour environmental cycle.

Entropy (en' tro pee) [Gr. *en*: in + *tropein*: to change] • A measure of the degree of disorder in any system. A perfectly ordered system has zero entropy; increasing disorder is measured by positive entropy. Spontaneous reactions in a closed system are always accompanied by an increase in disorder and entropy.

Environment • An organism's surroundings, both living and nonliving; includes temperature, light intensity, and all other species that influence the focal organism.

Environmental toxicology • The study of the distribution and effects of toxic compounds in the environment.

Enzyme (en' zime) [Gr. *en*: in + *zyme*: yeast] • A protein, on the surface of which are chemical groups so arranged as to make the enzyme a catalyst for a chemical reaction.

Epi- [Gr.: upon, over] • A prefix used to designate a structure located on top of another; for example: epidermis, epiphyte.

Epicotyl (epp' i kot' il) [Gr. *epi*: upon + *kotyle*: something hollow] • That part of a plant embryo or seedling that is above the cotyledons.

Epidermis [Gr. *epi*: upon + *derma*: skin] • In plants and animals, the outermost cell layers. (Only one cell layer thick in plants.)

Epididymis (epuh did' uh mus) [Gr. *epi*: upon + *didymos*: testicle] • Coiled tubules in the testes that store sperm and conduct sperm from the seiminiferous tubules to the vas deferens.

Epinephrine (ep i nef' rin) [Gr. *epi*: upon + *nephros*: a kidney] • The "fight or flight" hormone. Produced by the medulla of the adrenal gland, it also functions as a neurotransmitter. Also known as adrenaline.

Epiphyte (ep' e fyte) [Gr. *epi*: upon + *phyton*: plant] • A specialized plant that grows on the surface of other plants but does not parasitize them.

Episome • A plasmid that may exist either free or integrated into a chromosome. (See plasmid.)

Epistasis • An interaction between genes, in which the presence of a particular allele of one gene determines whether another gene will be expressed.

Epithelium • In animals, a layer of cells covering or lining an external surface or a cavity.

Equilibrium • (1) In biochemistry, a state in which forward and reverse reactions are proceeding at counterbalancing rates, so there is no observable change in the concentrations of reactants and products. (2) In evolutionary genetics, a condition in which allele and genotype frequencies in a population are constant from generation to generation.

Erythrocyte (ur rith' row sight) [Gr. *erythros*: red + *kytos*: hollow vessel] • A red blood cell.

Esophagus (i soff' i gus) [Gr. *oisophagos*: gullet] • That part of the gut between the pharynx and the stomach.

Ester linkage • A condensation (water-releasing) reaction in which the carboxyl group of a fatty acid reacts with the hydroxyl group of an alcohol. Lipids are formed in this way.

Estivation (ess tuh vay' shun) [L. *aestivalis*: summer] • A state of dormancy and hypometabolism that occurs during the summer; usually a means of surviving drought and/or intense heat. Contrast with hibernation.

Estrogen • Any of several steroid sex hormones, produced chiefly by the ovaries in mammals.

Estrus (es' truss) [L. *oestrus*: frenzy] • The period of heat, or maximum sexual receptivity, in some female mammals. Ordinarily, the estrus is also the time of release of eggs in the female.

Ethylene • One of the plant hormones, the gas $H_2C{=}2CH_2$.

Euchromatin • Chromatin that is diffuse and non-staining during interphase; may be transcribed. (Contrast with heterochromatin.)

Eudicots (yew di' kots) [Gr. *eu*: true + *di*: two + kotyledon: a cup-shaped hollow] • Members of the angiosperm class Eudicotyledones, flowering plants in which the embryo produces two cotyledons prior to germination. Leaves of most eudicots have major veins arranged in a branched or reticulate pattern.

Eukaryotes (yew car' ry otes) [Gr. *eu*: true + *karyon*: kernel or nucleus] • Organisms whose cells contain their genetic material inside a nucleus. Includes all life other than the viruses, Archaebacteria, and Eubacteria.

Eusocial • Term applied to insects, such as termites, ants, and many bees and wasps, in which individuals cooperate in the care of offspring, there are sterile castes, and generations overlap.

Eutrophication (yoo trofe' ik ay' shun) [Gr. *eu-*: well + *trephein*: to flourish] • The addition of nutrient materials to a body of water, resulting in changes to species composition therein.

Evolution • Any gradual change. Organic evolution, often referred to as evolution, is any genetic and resulting phenotypic change in organisms from generation to generation.

Evolutionary agent • Any factor that influences the direction and rate of evolutionary changes.

Evolutionarily conservative • Traits of organisms that evolve very slowly.

Evolutionary innovations • Major changes in body plans of organisms; these have been very rare during evolutionary history.

Evolutionary radiation • The proliferation of species within a single evolutionary lineage.

Evolutionary reversal • The reappearance of the ancestral state of a trait in a lineage in which that trait had acquired a derived state.

Excision repair • The removal and damaged DNA and its replacement by the appropriate nucleotides.

Excitatory postsynaptic potential (EPSP) • A change in the resting potential of a postsynaptic membrane in a positive (depolarizing) direction. (Contrast with inhibitory postsynaptic potential.)

Excretion • Release of metabolic wastes by an organism.

Exergonic reaction • A reaction in which free energy is released. (Contrast with endergonic reaction.)

Exo- (eks' oh) • Same as ecto-.

Exocrine gland (eks' oh krin) [Gr. *exo*: outside + *krinein*: to separate] • Any gland, such as a salivary gland, that secretes to the outside of the body or into the gut.

Exocytosis • A process by which a vesicle within a cell fuses with the plasma membrane and releases its contents to the outside. (Contrast with endocytosis.)

Exon • A portion of a DNA molecule, in eukaryotes, that codes for part of a polypeptide. (Contrast with intron.)

Exoskeleton (eks' oh skel' e ton) [Gr. *exos*: outside + *skleros*: hard] • A hard covering on the outside of the body to which muscles are attached. (Contrast with endoskeleton.)

Exotoxins • Highly toxic proteins released by living, multiplying bacteria.

Experiment • A scientific method in which particular factors are manipulated while other factors are held constant so that the potential influences of the manipulated factors can be determined.

Exponential growth • Growth, especially in the number of organisms in a population, which is a simple function of the size of the growing entity: the larger the entity, the faster it grows. (Contrast with logistic growth.)

Expression vector • A DNA vector, such as a plasmid, that carries a DNA sequence that includes the adjacent sequences for its expression into mRNA and protein in a host cell.

Expressivity • The degree to which a genotype is expressed in the phenotype— may be affected by the environment.

Extensor • A muscle the extends an appendage.

Extinction • The termination of a lineage of organisms.

Extrinsic protein • A membrane protein found only on the surface of the membrane. (Contrast with intrinsic protein.)

F_1 generation • The immediate progeny of a parental (P) mating; the first filial generation.

F_2 generation • The immediate progeny of a mating between members of the F_1 generation.

Facilitated diffusion • Passive movement through a membrane involving a specific carrier protein; does not proceed against a concentration gradient. (Contrast with active transport, free diffusion.)

Family • In taxonomy, the category below the order and above the genus; a group of related, similar genera.

Fat • A triglyceride that is solid at room temperature. (Contrast with oil.)

Fatty acid • A molecule with a long hydrocarbon tail and a carboxyl group at the other end. Found in many lipids.

Fauna (faw' nah) • All of the animals found in a given area. (Contrast with flora.)

Feces [L. *faeces*: dregs] • Waste excreted from the digestive system.

Feedback control • Control of a particular step of a multistep process, induced by the presence or absence of a product of one of the later steps. A thermostat regulating the flow of heating oil to a furnace in a home is a negative feedback control device.

Fermentation (fur men tay' shun) [L. *fermentum*: yeast] • The degradation of a substance such as glucose to smaller molecules with the extraction of energy, without the use of oxygen (i.e., anaerobically). Involves the glycolytic pathway.

Fertilization • Union of gametes. Also known as syngamy.

Fertilization membrane • A membrane surrounding an animal egg which becomes rapidly raised above the egg surface within seconds after fertilization, serving to prevent entry of a second sperm.

Fetus • The latter stages of an embryo that is still contained in an egg or uterus; in humans, the unborn young from the eighth week of pregnancy to the moment of birth.

Fiber • An elongated and tapering cell of flowering plants, usually with a thick cell wall. Serves a support function.

Fibrin • A protein that polymerizes to form long threads that provide structure to a blood clot.

Filter feeder • An organism that feeds upon much smaller organisms, that are suspended in water or air, by means of a straining device.

Filtration • In the excretory physiology of some animals, the process by which the initial urine is formed; water and most solutes are transferred into the excretory tract, while proteins are retained in the blood or hemolymph.

First law of thermodynamics • Energy can be neither created nor destroyed.

Fission • Reproduction of a prokaryote by division of a cell into two comparable progeny cells.

Fitness • The contribution of a genotype or phenotype to the composition of subsequent generations, relative to the contribution of other genotypes or phenotypes. (See inclusive fitness.)

Fixed action pattern • A behavior that is genetically programmed.

Flagellum (fla jell' um) (plural: flagella) [L. *flagellum*: whip] • Long, whiplike appendage that propels cells. Prokaryotic flagella differ sharply from those found in eukaryotes.

Flexor • A muscle that flexes an appendage.

Flora (flore' ah) • All of the plants found in a given area. (Contrast with fauna.)

Florigen • A plant hormone (not yet isolated) involved in the conversion of a vegetative shoot apex to a flower.

Flower • The total reproductive structure of an angiosperm; its basic parts include the calyx, corolla, stamens, and carpels.

Fluorescence • The emission of a photon of visible light by an excited atom or molecule.

Follicle [L. *folliculus*: little bag] • In female mammals, an immature egg surrounded by nutritive cells.

Follicle-stimulating hormone • A gonadotropic hormone produced by the anterior pituitary.

Food chain • A portion of a food web, most commonly a simple sequence of prey species and the predators that consume them.

Food web • The complete set of food links between species in a community; a diagram indicating which ones are the eaters and which are consumed.

Forb • Any broad-leaved (dicotyledonous), herbaceous plant. Especially applied to such plants growing in grasslands.

Fossil • Any recognizable structure originating from an organism, or any impression from such a structure, that has been preserved over geological time.

Fossil fuel • A fuel (particularly petroleum products) composed of the remains of organisms that lived in the remote past.

Founder effect • Random changes in allele frequencies resulting from establishment of a population by a very small number of individuals.

Fovea [L. *fovea*; a small pit] • The area, in the vertebrate retina, of most distinct vision.

Frame-shift mutation • A mutation resulting from the addition or deletion of a single base pair in the DNA sequence of a gene. As a result of this, mRNA transcribed from such a gene is translated normally until the ribosome reaches the point at which the mutation has occurred. From that point on, codons are read out of proper register and the amino acid sequence bears no resemblance to the normal sequence. (Contrast with missense mutation, nonsense mutation, synonymous mutation.)

Free energy • That energy which is available for doing useful work, after allowance has been made for the increase or decrease of disorder. Designated by the symbol G (for Gibbs free energy), and defined by: $G = H - TS$, where H = heat, S = entropy, and T = absolute (Kelvin) temperature.

Frequency-dependent selection • Selection that changes in intensity with the proportion of individuals having the trait.

Fruit • In angiosperms, a ripened and mature ovary (or group of ovaries) containing the seeds. Sometimes applied to reproductive structures of other groups of plants, and includes any adjacent parts which may be fused with the reproductive structures.

Fruiting body • A structure that bears spores.

Fundamental niche • The range of condition under which an organism could survive if it were the only one in the environment. (Contrast with realized niche.)

Fungus (fung' gus) • A member of the kingdom Fungi, a (usually) multicellular eukaryote with absorptive nutrition.

G_1 phase • In the cell cycle, the gap between the end of mitosis and the onset of the S phase.

G_2 phase • In the cell cycle, the gap between the S (synthesis) phase and the onset of mitosis.

G protein • A membrane protein involved in signal transduction; characterized by binding guanyl nucleotides. The activation of certain receptors activates the G protein, which in turn activates adenylate cyclase. G protein activation involves binding a GTP molecule in place of a GDP molecule.

Gametangium (gam i tan' gee um) [Gr. *gamos*: marriage + *angeion*: vessel or reservoir] • Any plant or fungal structure within which a gamete is formed.

Gamete (gam' eet) [Gr. *gamete*: wife, *gametes*: husband] • The mature sexual reproductive cell: the egg or the sperm.

Gametocyte (ga meet' oh site) [Gr. *gamete*: wife, *gametes*: husband + *kytos*: cell] • The cell that gives rise to sex cells, either the eggs or the sperm. (See oocyte and spermatocyte.)

Gametogenesis (ga meet' oh jen' e sis) [Gr. *gamete*: wife, *gametes*: husband + *genesis*: source] • The specialized series of cellular divisions that leads to the production of sex cells (gametes). (Contrast with oogenesis and spermatogenesis.)

Gametophyte (ga meet' oh fyte) • In plants and photosynthetic protists with alternation of generations, the haploid phase that produces the gametes. (Contrast with sporophyte.)

Ganglion (gang' glee un) [Gr.: tumor] • A group or concentration of neuron cell bodies.

Gap junction • A 2.7-nanometer gap between plasma membranes of two animal cells, spanned by protein channels. Gap junctions allow chemical substances or electrical signals to pass from cell to cell.

Gas exchange • In animals, the process of taking up oxygen from the environment and releasing carbon dioxide to the environment.

Gastrovascular cavity • Serving for both digestion (gastro) and circulation (vascular); in particular, the central cavity of the body of jellyfish and other cnidarians.

Gastrula (gas' true luh) [Gr. *gaster*: stomach] • An embryo forming the characteristic three cell layers (ectoderm, endoderm, and mesoderm) which will give rise to all of the major tissue systems of the adult animal.

Gastrulation • Development of a blastula into a gastrula.

Gated channel • A channel (membrane protein) that opens and closes in response to binding of specific molecules or to changes in membrane potential.

Gel electrophoresis (jel ul lec tro for' eesis) • A semisolid matrix suspended in a salty buffer in which molecules can be separated on the basis of their size and change when current is passed through the gel.

Gene [Gr. *gen*: to produce] • A unit of heredity. Used here as the unit of genetic function which carries the information for a single polypeptide.

Gene amplification • Creation of multiple copies of a particular gene, allowing the production of large amounts of the RNA transcript (as in rRNA synthesis in oocytes).

Gene cloning • Formation of a clone of bacteria or yeast cells containing a particular foreign gene.

Gene family • A set of identical, or once-identical, genes, derived from a single parent gene; need not be on the same chromosomes; classic example is the globin family in vertebrates.

Gene flow • The exchange of genes between different species (an extreme case referred to as hybridization) or between different populations of the same species caused by migration following breeding.

Gene pool • All of the genes in a population.

Gene therapy • Treatment of a genetic disease by providing patients with cells containing wild type alleles for the genes that are nonfunctional in their bodies.

Generative nucleus • In a pollen tube, a haploid nucleus that undergoes mitosis to produce the two sperm nuclei that participate in double fertilization. (Contrast with tube nucleus.)

Genet • The genetic individual of a plant that is composed of a number of nearly identical but repeated units.

Genetic drift • Changes in gene frequencies from generation to generation in a small population as a result of random processes.

Genetic stochasticity • Variation in the frequencies of alleles and genotypes in a population over time.

Genetics • The study of heredity.

Genetic structure • The frequencies of alleles and genotypes in a population.

Genome (jee' nome) • The genes in a complete haploid set of chromosomes.

Genotype (jean' oh type) [Gr. *gen*: to produce + *typos*: impression] • An exact description of the genetic constitution of an individual, either with respect to a single trait or with respect to a larger set of traits. (Contrast with phenotype.)

Genus (jean' us) (plural: genera) [Gr. *genos*: stock, kind] • A group of related, similar species.

Geotropism • See gravitropism.

Germ cell • A reproductive cell or gamete of a multicellular organism.

Germination • The sprouting of a seed or spore.

Gestation (jes tay' shun) [L. *gestare*: to bear] • The period during which the embryo of a mammal develops within the uterus. Also known as **pregnancy**.

Gibberellin (jib er el' lin) [L. *gibberella*: hunchback (refers to shape of a reproductive structure of a fungus that produces gibberellins)] • One of a class of plant growth substances playing roles in stem elongation, seed germination, flowering of certain plants, etc. Named for the fungus *Gibberella*.

Gill • An organ for gas exchange in aquatic organisms.

Gill arch • A skeletal structure that supports gill filaments and the blood vessels that supply them.

Gizzard (giz' erd) [L. *gigeria*: cooked chicken parts] • A very muscular port of the stomach of birds that grinds up food, sometimes with the aid of fragments of stone.

Gland • An organ or group of cells that produces and secretes one or more substances.

Glans penis • Sexually sensitive tissue at the tip of the penis.

Glia (glee' uh) [Gr.: glue] • Cells, found only in the nervous system, which do not conduct action potentials.

Glomerulus (glo mare' yew lus) [L. *glomus*: ball] • Sites in the kidney where blood filtration takes place. Each glomerulus consists of a knot of capillaries served by afferent and efferent arterioles.

Glucocorticoids • Steroid hormones produced by the adrenal cortex. Secreted in response to ACTH, they inhibit glucose uptake by many tissues in addition to mediating other stress responses.

Glucagon • A hormone produced and released by cells in the islets of Langerhans of the pancreas. It stimulates the breakdown of glycogen in liver cells.

Gluconeogenesis • The biochemical synthesis of glucose from other substances, such as amino acids, lactate, and glycerol.

Glucose (glue' kose) [Gr. *gleukos*: sweet wine mash for fermentation] • The most common sugar, one of several monosaccharides with the formula $C_6H_{12}O_6$.

Glycerol (gliss' er ole) • A three-carbon alcohol with three hydroxyl groups, the linking component of phospholipids and triglycerides.

Glycogen (gly' ko jen) • A branched-chain polymer of glucose, similar to starch (which is less branched and may be of lower molecular weight). Exists mostly in liver and muscle; the principal storage carbohydrate of most animals and fungi.

Glycolysis (gly kol' li sis) [from glucose + Gr. *lysis*: loosening] • The enzymatic breakdown of glucose to pyruvic acid. One of the oldest energy-yielding machanisms in living organisms.

Glycosidic linkage • The connection in an oligosaccharide or polysaccharide chain, formed by removal of water during the linking of monosaccharides.by root pressure.

Glyoxysome (gly ox' ee soam) • An organelle found in plants, in which stored lipids are converted to carbohydrates.

Golgi apparatus (goal' jee) • A system of concentrically folded membranes found in the cytoplasm of eukaryotic cells. Plays a role in the production and release of secretory materials such as the digestive enzymes manufactured in the pancreas. First described by Camillo Golgi (1844–1926).

Gonad (go' nad) [Gr. *gone*: seed, that which produces seed] • An organ that produces sex cells in animals: either an ovary (female gonad) or testis (male gonad).

Gonadotropin • A hormone that stimulates the gonads.

Gondwana • The large southern land mass that existed from the Cambrian (540 mya) to the Jurassic (138 mya). Present-day South America, Africa, India, Australia, and Antarctica.

Gram stain • A differential stain useful in characterizing bacteria.

Granum • Within a chloroplast, a stack of thylakoids.

Gravitropism • A directed plant growth response to gravity.

Grazer • An animal that eats the vegetative tissues of herbaceous plants.

Green gland • An excretory organ of crustaceans.

Greenhouse effect • The heating of Earth's atmosphere by gases that are transparent to sunlight but opaque to radiated heat.

Gross primary production • The total energy captured by plants growing in a particular area.

Ground meristem • That part of an apical meristem that gives rise to the ground tissue system of the primary plant body.

Ground tissue system • Those parts of the plant body not included in the dermal or vascular tissue systems. Ground tissues function in storage, photosynthesis, and support.

Group transfer • The exchange of atoms between molecules.

Growth • Irreversible increase in volume (probably the most accurate definition, but at best a dangerous oversimplification).

Growth factors • A group of proteins that circulate in the blood and trigger the normal growth of cells. Each growth factor acts only on certain target cells.

Guanine (gwan'een) • A nitrogen-containing base found in DNA, RNA and GTP.

Guard cells • In plants, paired epidermal cells which surround and control the opening of a stoma (pore).

Gut • An animal's digestive tract.

Guttation • The extrusion of liquid water through openings in leaves, caused by root pressure.

Gymnosperm (jim' no sperm) [Gr. *gymnos*: naked + *sperma*: seed] • A plant, such as a pine or other conifer, whose seeds do not develop within an ovary (hence, the seeds are "naked").

Gyrus (plural: gyri) • The raised or ridged portion of the convoluted surface of the brain. (Contrast to sulcus.)

Habit • The form or pattern of growth characteristic of an organism.

Habitat • The environment in which an organism lives.

Habituation (ha bich' oo ay shun) • The simplest form of learning, in which an animal presented with a stimulus without reward or punishment eventually ceases to respond.

Hair cell • A type of mechanoreceptor in animals.

Half-life • The time required for half of a sample of a radioactive isotope to decay to its stable, nonradioactive form.

Halophyte (hal' oh fyte) [Gr. *halos*: salt + *phyton*: plant] • A plant that grows in a saline (salty) environment.

Haploid (hap' loid) [Gr. *haploeides*: single] • Having a chromosome complement consisting of just one copy of each chromosome. This is the normal "ploidy" of gametes or of asexual spores produced by meiosis or of organisms (such as the gametophyte generation of plants) that grow from such spores without fertilization.

Hardy–Weinberg equililbrium • The percentages of diploid combinations expected from a knowledge of the proportions of alleles in the population if no agents of evolution are acting on the population.

Haustorium (haw stor' ee um) [L. *haustus*: draw up] • A specialized hypha or other structure by which fungi and some parasitic plants draw food from a host plant.

Haversian systems • Units of organization in compact bone that reflect the action of intercommunicating osteoblasts.

Heat-shock proteins • Chaperone proteins expressed in cells exposed to high temperatures or other forms of environmental stress.

Helper T cells • T cells that participate in the activation of B cells and of other T cells; targets of the HIV-I virus, the agent of AIDS. (Contrast with cytotoxic T cells, suppressor T cells.)

Hematocrit (heme at o krit) [Gr. *haima*: blood + *krites*: judge] • The proportion of 100 cc of blood that consists of red blood cells.

Hemizygous(hem' ee zie' gus) [Gr. *hemi*: half + *zygotos*: joined] • In a diploid organism, having only one allele for a given trait, typically the case for X-linked genes in male mammals and Z-linked genes in female birds. (Contrast with homozygous, heterozygous.)

Hemoglobin (hee' mo glow' bin) [Gr. *haima*: blood + L. *globus*: globe] • The colored protein of vertebrate blood (and blood of some invertebrates) which transports oxygen.

Hepatic (heh pat' ik) [Gr. *hepar*: liver] • Pertaining to the liver.

Hepatic duct • The duct that conveys bile from the liver to the gallbladder.

Herbicide (ur' bis ide) • A chemical substance that kills plants.

Herbivore [L. *herba*: plant + *vorare*: to devour] • An animal which eats the tissues of plants. (Contrast with carnivore, detritivore, omnivore.)

Heritable • Able to be inherited; in biology usually refers to genetically determined traits.

Hermaphroditism (her maf' row dite' ism) [Gr. *hermaphroditos*: a person with both male and female traits] • The coexistence of both female and male sex organs in the same organism.

Hertz (abbreviated as Hz) • Cycles per second.

Hetero- [Gr.: other, different] • A prefix used in biology to mean that two or more different conditions are involved; for example, heterotroph, heterozygous.

Heterochromatin • Chromatin that retains its coiling during interphase; generally not transcribed. (Contrast with euchromatin.)

Heterocyst • A large, thick-walled cell in the filaments of certain cyanobacteria; performs nitrogen fixation.

Heterogeneous nuclear RNA (hnRNA) • The product of transcription of a eukaryotic gene, including transcripts of introns.

Heteromorphic (het' er oh more' fik) [Gr. *heteros*: different + *morphe*: form] • having a different form or appearance, as two heteromorphic life stages of a plant. (Contrast with isomorphic.)

Heterosporous (het' er os' por us) • Producing two types of spores, one of which gives rise to a female megaspore and the other to a male microspore. Heterosporous plants produce distinct female and male gametophytes. (Contrast with homosporous.)

Heterotherm • An animal that regulates its body temperature at a constant level at some times but not others, such as a hibernator.

Heterotroph (het' er oh trof) [Gr. *heteros*: different + *trophe*: food] • An organism that requires preformed organic molecules as food. (Contrast with autotroph.)

Heterozygous (het' er oh zie' gus) [Gr. *heteros*: different + *zygotos*: joined] • Of a diploid organism having different alleles of a given gene on the pair of homologues carrying that gene. (Contrast with homozygous.)

Hibernation [L. *hibernus*: winter] • The state of inactivity of some animals during winter; marked by a drop in body temperature and metabolic rate.

Highly repetitive DNA • Short DNA sequences present in millions of copies in the genome, next to each other (in tandem). In a In a reassociation experiment, denatured highly repetitive DNA reanneals very quickly.

Hippocampus • A part of the forebrain that takes part in long-term memory formation.

Histamine (hiss; tah meen) • A substance released within a damaged tissue by a type of white blood cell. Histamines are responsible for aspects of allergice reactions, including the increased vascular permeability that leads to edema (swelling).

Histology • The study of tissues.

Histone • Any one of a group of basic proteins forming the core of a nucleosome, the structural unit of a eukaryotic chromosome. (See nucleosome.)

hnRNA • See heterogeneous nuclear RNA.

Homeobox • A 180-base-pair segment of DNA found in a few genes (called **Hox genes**), perhaps regulating the expression of other genes and thus controlling large-scale developmental processes.

Homeostasis (home' ee o sta' sis) [Gr. *homos*: same + *stasis*: position] • The maintenance of a steady state, such as a constant temperature or a stable social structure, by means of physiological or behavioral feedback responses.

Homeotherm (home' ee o therm) [Gr. *homos*: same + *therme*: heat] • An animal which maintains a constant body temperature by virtue of its own heating and cooling mechanisms. (Contrast with heterotherm, poikilotherm.)

Homeotic genes (home' ee ott' ic) • Genes that determine what entire segments of an animal become. Drastic mutations in these genes cause the transformation of body segments in *Drosophila*. Homeotic genes studied in the plant *Arabidopsis* are called organ identity genes.

Homolog (home' o log') [Gr. *homos*: same + *logos*: word] • One of a pair, or larger set, of chromosomes having the same overall genetic composition and sequence. In diploid organisms, each chromosome inherited from one parent is matched by an identical (except for mutational changes) chromosome—its homolog—from the other parent.

Homology (ho mol' o jee) [Gr. *homologi(a)*: agreement] • A similarity between two structures that is due to inheritance from a common ancestor. The structures are said to be homologous. (Contrast with analogy.)

Homoplasy (home' uh play zee) [Gr. *homos*: same + *plastikos*: to mold] • The presence in several species of a trait not present in their most common ancestor. Can result from convergent evolution, reverse evolution, or parallel evolution.

Homosporous • Producing a single type of spore that gives rise to a single type of gametophyte, bearing both female and male reproductive organs. (Contrast with heterosporous.)

Homozygous (home' o zie' gus) [Gr. *homos*: same + *zygotos*: joined] • Of a diploid organism having identical alleles of a given gene on both homologous chromosomes. An organism may be a "homozygote" with respect to one gene and, at the same time, a "heterozygote" with respect to another. (Contrast with heterozygous.)

Hormone (hore' mone) [Gr. *hormon*: excite, stimulate] • A substance produced in one part of a multicellular organism and transported to another part where it exerts its specific effect on the physiology or biochemistry of the target cells.

Host • An organism that harbors a parasite and provides it with nourishment.

Host–parasite interaction • The dynamic interaction between populations of a host and the parasites that attack it.

Hox genes • See homeobox.

Humoral immune system • The part of the immune system mediated by B cells; it is mediated by circulating antibodies and is active against extracellular bacterial and viral infections.

Humus (hew' muss) • The partly decomposed remains of plants and animals on the surface of a soil. Its characteristics depend primarily upon climate and the species of plants growing on the site.

Hyaluronidase (hill yew ron' uh dase) • An enzyme that digests proteoglycans. Found in sperm cells, it helps digest the coatings surrounding an egg so the sperm can penetrate the egg cell membrane.

Hybrid (high' brid) [L. *hybrida*: mongrel] • The offspring of genetically dissimilar parents. In molecular biology, a double helix formed of nucleic acids from different sources.

Hybridoma • A cell produced by the fusion of an antibody-producing cell with a myeloma cell; it produces monoclonal antibodies.

Hybrid zone • A narrow zone where two populations interbreed, producing hybrid individuals.

Hydrocarbon • A compound containing only carbon and hydrogen atoms.

Hydrogen bond • A chemical bond which arises from the attraction between the slight positive charge on a hydrogen atom and a slight negative charge on a nearby fluorine, oxygen, or nitrogen atom. Weak bonds, but found in great quantities in proteins, nucleic acids, and other biological macromolecules.

Hydrological cycle • The sum total of movement of water from the oceans to the atmosphere, to the soil, and back to the oceans. Some water is cycled many times within compartments of the system before completing one full circuit.

Hydrolyze (hi' dro lize) [Gr. *hydro*: water + *lysis*: cleavage] • To break a chemical bond, as in a peptide linkage, with the insertion of the components of water, –H and –OH, at the cleaved ends of a chain. The digestion of proteins is a hydrolysis.

Hydrophilic [Gr. *hydro*: water + *philia*: love] • Having an affinity for water. (Contrast with hydrophobic.)

Hydrophobic [Gr. *hydro*: water + *phobia*: fear] • Molecules and amino acid side chains, which are mainly hydrocarbons (compounds of C and H with no charged groups or polar groups), have a lower energy when they are clustered together than when they are distributed through an aqueous solution. Because of their attraction for one another and their reluctance to mix with water they are called "hydrophobic." Oil is a hydrophobic substance; phenylalanine is a hydrophobic animo acid in a protein. (Contrast with hydrophilic.)

Hydrostatic skeleton • The incompressible internal liquids of some animals that transfer forces from one part of the body to another when acted upon by the surrounding muscles.

Hydroxyl group • The —OH group, characteristic of alcohols.

Hyperpolarization • A change in the resting potential of a membrane so the inside of a cell becomes more electronegative. (Contrast with depolarization.)

Hypersensitive response • A defensive response of plants to microbial infection; it results in a "dead spot."

Hypertension • High blood pressure.

Hypertonic [Gk. *hyper*: above, over] • Having a greater solute concentration. Said of one solution in comparing it to another. (Contrast with hypotonic, isotonic.)

Hypha (high' fuh) (plural: hyphae) [Gr. *hyphe*: web] • In the fungi, any single filament. May be multinucleate (zygomycetes, ascomycetes) or multicellular (basidiomycetes).

Hypocotyl [Gk. *hypo*: beneath, under + *kotyledon*: hollow space] • That part of the embryonic or seedling plant shoot that is below the cotyledons.

Hypothalamus • The part of the brain lying below the thalamus; it coordinates water balance, reproduction, temperature regulation, and metabolism.

Hypothesis • A tentative answer to a question, from which testable predictions can be generated. (Contrast with theory.)

Hypothetico-deductive method • A method of science in which hypotheses are erected, predictions are made from them, and experiments and observations are performed to test the predictions.

Hypotonic [Gk. *hypo*: beneath, under] • Having a greater solute concentration. Said of one solution in comparing it to another. (Contrast with hypotonic, isotonic.)

Imaginal disc • In insect larvae, groups of cells that develop into specific adult organs.

Immune system [L. *immunis*: exempt] • A system in mammals that recognizes and eliminates or neutralizes either foreign substances or self substances that have been altered to appear foreign.

Immunization • The deliberate introduction of antigen to bring about an immune response.

Immunoglobulins • A class of proteins, with a characteristic structure, active as receptors and effectors in the immune system.

Immunological memory • Certain clones of immune system cells made to respond to an antigen persist. This leads to a more rapid and massive response of the immune system to any subsequenct exposure to that antigen.

Immunological tolerance • A mechanism by which an animal does not mount an immune response to the antigenic determinants of its own macromolecules.

Imprinting • (1) In genetics, the differential modification of a gene depending on whether it is present in a male or a female. (2) In animal behavior, a rapid form of learning in which an animal comes to make a particular response, which is maintained for life, to some object or other organism.

Inclusive fitness • The sum of an individual's own fitness (the effect of producing its own offspring: the individual selection component) plus its influence on fitness in relatives other than direct descendants (the kin selection component).

Incomplete dominance • Condition in which the heterozygous phenotype is intermediate between the two homozygous phenotypes.

Incomplete metamorphosis • Insect development in which changes between instars are gradual.

Incus (in' kus) [L. *incus*: anvil] • The middle of the three bones that conduct movements of the eardrum to the oval window of the inner ear. (See malleus, stapes.)

Independent assortment • The random separation during meiosis of nonhomologous chromosomes and of genes carried on nonhomologous chromosomes.

Individual fitness • That component of inclusive fitness that results from an organism producing its own offspring. (Contrast with kin selection component.)

Indoleacetic acid • See auxin.

Inducer • (1) In enzyme systems, a small molecule which, when added to a growth medium, causes a large increase in the level of some enzyme. (2) In embryology, a substance that causes a group of target cells to differentiate in a particular way.

Inducible enzyme • An enzyme that is present in much larger amounts when a particular compound (the inducer) has been added to the system. (Contrast with constitutive enzyme.)

Inflammation • A nonspecific defense against pathogens; characterized by redness, swelling, pain, and increased temperature.

Inflorescence • A structure composed of several flowers.

Inhibitor • A substance which binds to the surface of an enzyme and interferes with its action on its substrates.

Inhibitory postsynaptic potential • A change in the resting potential of a postsynaptic membrane in the hyperpolarizing (negative) direction.

Initiation complex • Combination of a ribosomal light subunit, an mRNA molecule, and the tRNA charged with the first amino acid coded for by the mRNA; formed at the onset of translation.

Initiation factors • Proteins that assist in forming the translation initiation complex at the ribosome.

Inositol triphosphate (IP3) • An intracellular second messenger derived from membrane phospholipids.

Instar (in' star) [L.: image, form] • An immature stage of an insect between molts.

Insulin (in' su lin) [L. *insula*: island] • A hormone, synthesized in islet cells of the pancreas, that promotes the conversion of glucose to the storage material, glycogen.

Integrase • An enzyme that integrates retroviral cDNA into the genome of the host cell.

Integrated pest management • A method of control of pests in which natural predators and parasites are used in conjunction with sparing use of chemical methods to achieve control of a pest without causing serious adverse environmental side effects.

Integument [L. *integumentum*: covering] • A protective surface structure. In gymnosperms and angiosperms, a layer of tissue around the ovule which will become the seed coat. Gymnosperm ovules have one integument, angiosperm ovules two.

Intercalary meristem • A meristematic region in plants which occurs not apically, but between two regions of mature tissue. Intercalary meristems occur in the nodes of grass stems, for example.

Intercostal muscles • Muscles between the ribs that can augment breathing movements by elevating and suppressing the rib cage.

Interferon • A glycoprotein produced by virus-infected animal cells; increases the resistance of neighboring cells to the virus.

Interkinesis • The phase between the first and second meiotic divisions.

Interleukins • Regulatory proteins, produced by macrophages and lymphocytes, that act upon other lymphocytes and direct their development.

Intermediate filaments • Fibrous proteins that stabilize cell structure and resist tension.

Internode • Section between two nodes of a plant stem.

Interphase • The period between successive nuclear divisions during which the chromosomes are diffuse and the nuclear envelope is intact. It is during this period that the cell is most active in transcribing and translating genetic information.

Interspecific competition • Competition between members of two or more species.

Intertropical convergence zone • The tropical region where the air rises most strongly; moves north and south with the passage of the sun overhead.

Intraspecific competition • Competition among members of a single species.

Intrinsic protein • A membrane protein that is embedded in the phospholipid bilayer of the membrane. (Contrast with extrinsic protein.)

Intrinsic rate of increase • The rate at which a population can grow when its density is low and environmental conditions are highly favorable.

Intron • A portion of a DNA molecule that, because of RNA splicing, is not involved in coding for part of a polypeptide molecule. (Contrast with exon.)

Invagination • An infolding.

Inversion (genetic) • A rare mutational event that leads to the reversal of the order of genes within a segment of a chromosome, as if that segment had been removed from the chromosome, turned 180°, and then reattached.

Invertebrate • Any animal that is not a vertebrate, that is, whose nerve cord is not enclosed in a backbone of bony segments.

In vitro [L.: in glass] • In a test tube, rather than in a living organism. (Contrast with in vivo.)

In vivo [L.: in the living state] • In a living organism. Many processes that occur in vivo can be reproduced in vitro with the right selection of cellular components. (Contrast with in vitro.)

Ion (eye' on) [Gr.: wanderer] • An atom or group of atoms with electrons added or removed, giving it a negative or positive electrical charge.

Ion channel • A membrane protein that can let ions pass across the membrane. The channel can be ion-selective, and it can be voltage-gated or ligand-gated.

Ionic bond • A chemical bond which arises from the electrostatic attraction between positively and negatively charged ions. Usually a strong bond.

Iris (eye' ris) [Gr. *iris*: rainbow] • The round, pigmented membrane that surrounds the pupil of the eye and adjusts its aperture to regulate the amount of light entering the eye.

Irruption • A rapid increase in the density of a population. Often followed by massive emigration.

Islets of Langerhans • Clusters of hormone-producing cells in the pancreas.

Iso- [Gr.: equal] • Prefix used to denote two separate but similar or identical states of a characteristic. (See isomers, isomorphic, isotope.)

Isolating mechanism • Geographical, physiological, ecological, or behavioral mechanisms that lead to a reduction in the frequency of hybrid matings.

Isomers • Molecules consisting of the same numbers and kinds of atoms, but differing in the way in which the atoms are combined.

Isomorphic (eye' so more' fik) [Gr. *isos*: equal + *morphe*: form] • having the same form or appearance, as two isomorphic life stages. (Contrast with heteromorphic.)

Isotonic • Having the same solute concentration; said of two solutions. (Contrast with hypertonic, hypotonic.)

Isotope (eye' so tope) [Gr. *isos*: equal + *topos*: place] • Two isotopes of the same chemical element have the same number of protons in their nuclei, but differ in the number of neutrons.

Jasmonates • Plant hormones that trigger defenses against pathogens and herbivores.

Jejunum (jih jew' num) • The middle division of the small intestine, where most absorption of nutrients occurs. (See duodenum, ileum.)

Joule (jool, or jowl) • A unit of energy, equal to 0.24 calories.

Juvenile hormone • In insects, a hormone maintaining larval growth and preventing maturation or pupation.

Karyotype • The number, forms, and types of chromosomes in a cell.

Kelvin temperature scale • See absolute temperature scale.

Keratin (ker' a tin) [Gr. *keras*: horn] • A protein which contains sulfur and is part of such hard tissues as horn, nail, and the outermost cells of the skin.

Ketone (key' tone) • A compound with a C==O group attached to two other groups, neither of which is an H atom. Many sugars are ketones. (Contrast with aldehyde.)

Keystone species • A species that exerts a major influence on the composition and dynamics of the community in which it lives.

Kidneys • A pair of excretory organs in vertebrates.

Kin selection • The component of inclusive fitness resulting from helping the survival of relatives containing the same alleles by descent from a common ancestor.

Kinase (kye' nase) • An enzyme that transfers a phosphate group from ATP to another molecule. Protein kinases transfer phosphate from ATP to specific proteins, playing important roles in cell regulation.

Kinesis (ki nee' sis) [Gr.: movement] • Orientation behavior in which the organism does not move in a particular direction with reference to a stimulus but instead simply moves at an increasing or decreasing rate until it ends up farther from the object or closer to it. (Contrast with taxis.)

Kinetochore (kin net′ oh core) [Gr. *kinetos*: moving + *khorein*: to move] • Specialized structure on a centromere to which microtubules attach.

Koch's posulates • Four rules for establishing that a particular microorganism causes a particular disease.

Krebs cycle • See citric acid cycle.

Lactic acid • The end product of fermentation in vertebrate muscle and some microorganisms.

Lagging strand • In DNA replication, the daughter strand that is synthesized discontinuously.

Lamella • Layer.

Larynx (lar′ inks) • A structure between the pharynx and the trachea that includes the vocal cords.

Larva (plural: larvae) [L.: ghost, early stage] • An immature stage of any invertebrate animal that differs dramatically in appearance from the adult.

Lateral • Pertaining to the side.

Lateral gene transfer • The movement of genes from one prokaryotic species to another.

Lateral meristems • The vascular cambium and cork cambium, which give rise to secondary tissue in plants.

Laterization (lat′ ur iz ay shun) • The formation of a nutrient-poor soil that is rich in insoluble iron and aluminum compounds.

Law of independent assortment • The random separation during meiosis of nonhomologous chromosomes and of genes carried on nonhomologous chromosomes. Mendel's second law.

Law of segregation • Alleles segregate from one another during gamete formation, Mendel's first law.

Leader sequence • A sequence of amino acids at the N-terminal end of a newly synthesized protein, determining where the protein will be placed in the cell.

Leading strand • In DNA replication, the daughter strand that is synthesized continuously.

Lenticel • Spongy region in a plant's periderm, allowing gas exchange.

Leukocyte (loo′ ko sight) [Gr. *leukos*: clear + *kutos*: hollow vessel] • A white blood cell.

Lichen (lie′ kun) [Gr. *leikhen*: licker] • An organism resulting from the symbiotic association of a true fungus and either a cyanobacterium or a unicellular alga.

Life cycle • The entire span of the life of an organism from the moment of fertilization (or asexual generation) to the time it reproduces in turn.

Life history • The stages an individual goes through during its life.

Life table • A table showing, for a group of equal-aged individuals, the proportion still alive at different times in the future and the number of offspring they produce during each time interval.

Ligament • A band of connective tissue linking two bones in a joint.

Ligand (lig′ and) • A molecule that binds to a receptor site of another molecule.

Lignin • The principal noncarbohydrate component of wood, a polymer that binds together cellulose fibrils in some plant cell walls.

Limbic system • A group of primitive vertebrate forebrain nuclei that form a network and are involved in emotions, drives, instinctive behaviors, learning, and memory.

Limiting resource • The required resource whose supply most strongly influences the size of a population.

Linkage • Association between genetic markers on the same chromosome such that they do not show random assortment and seldom recombine; the closer the markers, the lower the frequency of recombination.

Lipase (lip′ ase; lye′ pase) • An enzyme that digests fats.

Lipids (lip′ ids) [Gr. *lipos*: fat] • Substances in a cell which are easily extracted by organic solvents; fats, oils, waxes, steroids, and other large organic molecules, including those which, with proteins, make up the cell membranes. (See phospholipids.)

Litter • The partly decomposed remains of plants on the surface and in the upper layers of the soil.

Littoral zone • The coastal zone from the upper limits of tidal action down to the depths where the water is thoroughly stirred by wave action.

Liver • A large digestive gland. In vertebrates, it secretes bile and is involved in the formation of blood.

Lobes • Regions of the human cerebral hemispheres; includes the temporal, frontal, parietal, and occipital lobes.

Locus • In genetics, a specific location on a chromosome. May be considered to be synonymous with "gene."

Logistic growth • Growth, especially in the size of an organism or in the number of organisms that constitute a population, which slows steadily as the entity approaches its maximum size. (Contrast with exponential growth.)

Loop of Henle (hen′ lee) • Long, hairpin loop of the mammalian renal tubule that runs from the cortex down into the medulla, and back to the cortex. Creates a concentration gradient in the interstitial fluids in the medulla.

Lophophore • A U-shaped fold of the body wall with hollow, ciliated tentacles that encircles the mouth of animals in several different phyla. Used for filtering prey from the surrounding water.

Lordosis (lor doe′ sis) [Gk. *lordosis*: curving forward] • A posture assumed by females of some mammalian species (especially rodents) to signal sexual receptivity.

Lumen (loo′ men) [L.: light] • The cavity inside any tubular part of an organ, such as a piece of gut or a kidney tubule.

Lungs • A pair of saclike chambers within the bodies of some animals, functioning in gas exchange.

Luteinizing hormone • A gonadotropin produced by the anterior pituitary. It stimulates the gonads to produce sex hormones.

Lymph [L. *lympha*: water] • A clear, watery fluid that is formed as a filtrate of blood; it contains white blood cells; it collects in a series of special vessels and is returned to the bloodstream.

Lymph nodes • Specialized tissue regions that act as filters for cells, bacteria and foreign matter.

Lymphocyte • A major class of white blood cells. Includes T cells, B cells, and other cell types important in the immune response.

Lysis (lie′ sis) [Gr.: a loosening] • Bursting of a cell.

Lysogenic • The condition of a bacterium that carries the genome of a virus in a relatively stable form. (Contrast with lytic.)

Lysosome (lie′ so soam) [Gr. *lysis*: a loosening + *soma*: body] • A membrane-bounded inclusion found in eukaryotic cells (other than plants). Lysosomes contain a mixture of enzymes that can digest most of the macromolecules found in the rest of the cell.

Lysozyme (lie′ so zyme) • An enzyme in saliva, tears, and nasal secretions that attacks bacterial cell walls, as one of the body's nonspecific defense mechanisms.

Lytic • Condition in which a bacterium lyses shortly after infection by a virus; the viral genome does not become stabilized within the bacterial cell. (Contrast with lysogenic.)

Macro- (mack′ roh) [Gr. *makros*: large, long] • A prefix commonly used to denote something large. (Contrast with micro-.)

Macroevolution • Evolutionary changes occurring over long time spans and usually involving changes in many traits. (Contrast with microevolution.)

Macromolecule • A giant polymeric molecule. The macromolecules are proteins, polysaccharides, and nucleic acids.

Macronutrient • A mineral element required by plant tissues in concentrations of at least 1 milligram per gram of their dry matter.

Macrophage (mac′ roh faj) • A type of white blood cell that endocytoses bacteria and other cells.

Major histocompatibility complex (MHC) • A complex of linked genes, with multiple alleles, that control a number of immunological phenomena; it is important in graft rejection.

Malignant tumor • A tumor whose cells can invade surrounding tissues and spread to other organs.

Malleus (mal′ ee us) [L. *malleus*: hammer] • The first of the three bones that conduct movements of the eardrum to the oval window of the inner ear. (See incus, stapes.)

Malpighian tubule (mal pee′ gy un) • A type of protonephridium found in insects.

Mammal [L. *mamma*: breast, teat] • Any animal of the class Mammalia, characterized by the production of milk by the female mammary glands and the possession of hair for body covering.

Mantle • A sheet of specialized tissues that covers most of the viscera of mollusks; provides protection to internal organs and secretes the shell.

Map unit • In eukaryotic genetics, one map unit corresponds to a recombinant frequency of 0.01.

Mapping • In genetics, determining the order of genes on a chromosome and the distances between them.

Marine [L. *mare*: sea, ocean] • Pertaining to or living in the ocean. (Contrast with aquatic, terrestrial.)

Marsupial (mar soo' pee al) • A mammal belonging to the subclass Metatheria, such as opossums and kangaroos. Most have a pouch (marsupium) that contains the milk glands and serves as a receptacle for the young.

Mass extinctions • Geological periods during which rates of extinction were much higher than during intervening times.

Mass number • The sum of the number of protons and neutrons in an atom's nucleus.

Mast cells • Typically found in connective tissue, mast cells can be provoked by antigens or inflammation to release histamine.

Maternal effect genes • These genes code for morphogens that determine the polarity of the egg and larva in the fruit fly, *Drosophila melanogaster.*

Maternal inheritance (cytoplasmic inheritance) • Inheritance in which the phenotype of the offspring depends on factors, such as mitochondria or chloroplasts, that are inherited from the female parent through the cytoplasm of the female gamete.

Maturation • The automatic development of a pattern of behavior, which becomes increasingly complex or precise as the animal matures. Unlike learning, the development does not require experience to occur.

Mechanoreceptor • A cell that is sensitive to physical movement and generates action potentials in response.

Medulla (meh dull' luh) [L.: narrow] • (1) The inner, core region of an organ, as in the adrenal medulla (adrenal gland) or the renal medulla (kidneys). (2) The portion of the brain stem that connects to the spinal cord.

Mega- [Gr. *megas*: large, great] • A prefix often used to denote something large. (Contrast with micro-.)

Megaspore [Gr. *megas*: large + *spora*:seed] • In plants, a haploid spore that produces a female gametophyte.

Meiosis (my oh' sis) [Gr.: diminution] • Division of a diploid nucleus to produce four haploid daughter cells. The process consists of two successive nuclear divisions with only one cycle of chromosome replication.

Membrane potential • The difference in electrical charge between the inside and the outside of a cell, caused by a difference in the distribution of ions.

Memory cells • Long-lived lymphocytes produced by exposure to antigen. They persist in the body and are able to mount a rapid response to subsequent exposures to the antigen.

Mendelian population • A local population of individuals belonging to the same species and exchanging genes with one another.

Menopause • The time in a human female's life when the ovarian and menstrual cycles cease.

Menstrual cycle • The monthly sloughing off of the uterine lining if fertilization does not occur in the female. Occurs between puberty and menopause.

Meristem [Gr. *meristos*: divided] • Plant tissue made up of actively dividing cells.

Mesenchyme (mez' en kyme) [Gr. *mesos*: middle + *enchyma*: infusion] • Embryonic or unspecialized cells derived from the mesoderm.

Meso- (mez' oh) [Gr.: middle] • A prefix often used to designate a structure located in the middle, or a stage that appears at some intermediate time. For example, mesoderm, Mesozoic.

Mesoderm [Gr. *mesos*: middle + *derma*: skin] • The middle of the three embryonic tissue layers first delineated during gastrulation. Gives rise to skeleton, circulatory system, muscles, excretory system, and most of the reproductive system.

Mesophyll (mez' a fill) [Gr. *mesos*: middle + *phyllon*: leaf] • Chloroplast-containing, photosynthetic cells in the interior of leaves.

Mesosome (mez' o soam') [Gr. *mesos*: middle + *soma*: body] • A localized infolding of the plasma membrane of a bacterium.

Messenger RNA (mRNA) • A transcript of one of the strands of DNA, it carries information (as a sequence of codons) for the synthesis of one or more proteins.

Meta- [Gr.: between, along with, beyond] • A prefix used in biology to denote a change or a shift to a new form or level; for example, as used in metamorphosis.

Metabolic compensation • Changes in biochemical properties of an organism that render it less sensitive to temperature changes.

Metabolic pathway • A series of enzyme-catalyzed reactions so arranged that the product of one reaction is the substrate of the next.

Metabolism (meh tab' a lizm) [Gr. *metabole*: to change] • The sum total of the chemical reactions that occur in an organism, or some subset of that total (as in "respiratory metabolism").

Metamorphosis (met' a mor' fo sis) [Gr. *meta*: between + *morphe*: form, shape] • A radical change occurring between one developmental stage and another, as for example from a tadpole to a frog or an insect larva to the adult.

Metaphase (met' a phase) [Gr. *meta*: between] • The stage in nuclear division at which the centromeres of the highly supercoiled chromosomes are all lying on a plane (the metaphase plane or plate) perpendicular to a line connecting the division poles.

Metapopulation • A population divided into subpopulations, among which there are occasional exchanges of individuals.

Metastasis (meh tass' tuh sis) • The spread of cancer cells from their original site to other parts of the body.

Methanogen • Any member of a group of Archaebacteria that release methane as a metabolic product. This group is considered to be an extremely ancient one.

MHC • See major histocompatibility complex.

Micro- (mike' roh) [Gr. *mikros*: small] • A prefix often used to denote something small. (Contrast with macro-, mega-.)

Microbiology [Gr. *mikros*: small + *bios*: life + *logos*: discourse] • The scientific study of microscopic organisms, particularly bacteria, unicellular algae, protists, and viruses.

Microevolution • The small evolutionary changes typically occurring over short time spans; generally involving a small number of traits and minor genetic changes. (Contrast with macroevolution.)

Microfilament • Minute fibrous structure generally composed of actin found in the cytoplasm of eukaryotic cells. They play a role in the motion of cells.

Micronutrient • A mineral element required by plant tissues in concentrations of less than 100 micrograms per gram of their dry matter.

Micropyle (mike' roh pile) [Gr. *mikros*: small + *pyle*: gate] • Opening in the integument(s) of a seed plant ovule through which pollen grows to reach the female gametophyte within.

Microspores [Gr. *mikros*: small + *spora*: seed] • In plants, a haploid spore that produces a male gametophyte.

Microtubules • Minute tubular structures found in centrioles, spindle apparatus, cilia, flagella, and other places in the cytoplasm of eukaryotic cells. These tubules play roles in the motion and maintenance of shape of eukaryotic cells.

Microvilli (singular: microvillus) • The projections of epithelial cells, such as the cells lining the small intestine, that increase their surface area.

Middle lamella • A layer of derivative polysaccharides that separates plant cells; a common middle lamella lies outside the primary walls of the two cells.

Migration • The regular, seasonal movements of animals between breeding and nonbreeding ranges.

Mimicry (mim' ik ree) • The resemblance of one kind of organism to another, or to some inanimate object; serves the function of making the organism difficult to find, of discouraging potential enemies or of attracting potential prey. (See Batesian mimicry and Müllerian mimicry.)

Mineral • An inorganic substance other than water.

Mineralocorticoid • A hormone produced by the adrenal cortex that influences mineral ion balance; aldosterone.

Mismatch repair • When a single base in DNA is changed into a different base, or the wrong base inserted during DNA replication, there is a mismatch in base pairing with the base on the opposite strand. A repair system removes the incorrect base and inserts the proper one for pairing with the opposite strand.

Missense mutation • A nonsynonymous mutation, or one that changes a codon for one amino acid to a codon for a different amino acid. (Contrast with frame-shift mutation, nonsense mutation, synonymous mutation.)

Mitochondrial matrix • The fluid interior of the mitochondrion, enclosed by the inner mitochondrial membrane.

Mitochondrion (my' toe kon' dree un) (plural: mitochondria) [Gr. *mitos*: thread + *chondros*: cartilage, or grain] • An organelle that occurs in eukaryotic cells and contains the enzymes of the ctric acid cycle, the respiratory chain, and oxidative phosphorylation. A mitochondrion is bounded by a double membrane.

Mitosis (my toe' sis) [Gr. *mitos*: thread] • Nuclear division in eukaryotes leading to the formation of two daughter nuclei each with a chromosome complement identical to that of the original nucleus.

Mitotic center • Cellular region that organizes the microtubules for mitosis. In animals a centrosome serves as the mitotic center.

Moderately repetitive DNA • DNA sequences that appear hundreds to thousands of times in the genome. They include the DNA sequences coding for rRNAs and tRNAs, as well as the DNA at telomeres.

Modular organism • An organism which grows by producing additional units of body construction (modules) that are very similar to the units of which it is already composed.

Mole • A quantity of a compound whose weight in grams is numerically equal to its molecular weight expressed in atomic mass units. Avogadro's number of molecules: 6.023×10^{23} molecules.

Molecular clock • The theory that macromolecules diverge from one another over evolutionary time at a constant rate, and that discovering this rate gives insight into the phylogenetic relationships of organisms.

Molecular weight • The sum of the atomic weights of the atoms in a molecule.

Molecule • A particle made up of two or more atoms joined by covalent bonds or ionic attractions.

Molting • The process of shedding part or all of an outer covering, as the shedding of feathers by birds or of the entire exoskeleton by arthropods.

Mono- [Gr. *monos*: one] • Prefix denoting a single entity. (Contrast with poly.)

Monoclonal antibody • Antibody produced in the laboratory from a clone of hybridoma cells, each of which produces the same specific antibody.

Monocot (short for monocotyledon) [Gr. *monos*: one + *kotyledon*: a cup-shaped hollow] • Any member of the angiosperm class Monocotyledones, plants in which the embryo produces but a single cotyledon (seed leaf). Leaves of most monocots have their major veins arranged parallel to each other.

Monocytes • White blood cells that produce macrophages.

Monoecious (mo nee' shus) [Gr.: one house] • Organisms in which both sexes are "housed" in a single individual, which produces both eggs and sperm. (In some plants, these are found in different flowers within the same plant.) Examples: corn, peas, earthworms, hydras. (Contrast with dioecious, perfect flower.)

Monohybrid cross • A mating in which the parents differ with respect to the alleles of only one locus of interest.

Monomer [Gr.: one unit] • A small molecule, two or more of which can be combined to form oligomers (consisting of a few monomers) or polymers (consisting of many monomers).

Monophyletic (mon' oh fih leht' ik) [Gk. *monos*: single + *phylon*: tribe] • Being descended from a single ancestral stock.

Monosaccharide • A simple sugar. Oligosaccharides and polysaccharides are made up of monosaccharides.

Monosynaptic reflex • A neural reflex that begins in a sensory neuron and makes a single synapse before activating a motor neuron.

Morphogens • Diffusible substances whose concentration gradients determine patterns of development in animals and plants.

Morphogenesis (more' fo jen' e sis) [Gr. *morphe*: form + *genesis*: origin] • The development of form. Morphogenesis is the overall consequence of determination, differentiation, and growth.

Morphology (more fol' o jee) [Gr. *morphe*: form + *logos*: discourse] • The scientific study of organic form, including both its development and function.

Mosaic development • Pattern of animal embryonic development in which each blastomere contributes a specific part of the adult body. (Contrast with regulative development.)

Motor end plate • The modified area on a muscle cell membrane where a synapse is formed with a motor neuron.

Motor neuron • A neuron carrying information from the central nervous system to an effector such as a muscle fiber.

Motor unit • A motor neuron and the set of muscle fibers it controls.

mRNA • (See messenger RNA.)

Mucosa (mew koh' sah) • An epithelial membrane containing cells that secrete mucus. The inner cell layers of the digestive and respiratory tracts.

Müllerian mimicry • The resemblance of two or more unpleasant or dangerous kinds of organisms to each other.

Multicellular [L. *multus*: much + *cella*: chamber] • Consisting of more than one cell, as for example a multicellular organism. (Contrast with unicellular.)

Muscle • Contractile tissue containing actin and myosin organized into polymeric chains called microfilaments. In vertebrates, the tissues are either cardiac muscle, smooth muscle, or striated (skeletal) muscle.

Muscle fiber • A single muscle cell. In the case of striated muscle, a syncitial, multinucleate cell.

Muscle spindle • Modified muscle fibers encased in a connective sheat and functioning as stretch receptors.

Mutagen (mute' ah jen) [L. *mutare*: change + Gr. *genesis*: source] • Any agent (e.g., chemicals, radiation) that increases the mutation rate.

Mutation • An inherited change along a very narrow portion of the nucleic acid sequence.

Mutation pressure • Evolution (change in gene proportions) by different mutation rates alone.

Mutualism • The type of symbiosis, such as that exhibited by fungi and algae or cyanobacteria in forming lichens, in which both species profit from the association.

Mycelium (my seel' ee yum) [Gr. *mykes*: fungus] • In the fungi, a mass of hyphae.

Mycorrhiza (my' ka rye' za) [Gr. *mykes*: fungus + *rhiza*: root] • An association of the root of a plant with the mycelium of a fungus.

Myelin (my' a lin) • A material forming a sheath around some axons. It is formed by Schwann cells that wrap themselves about the axon. It serves to insulate the axon electrically and to increase the rate of transmission of a nervous impulse.

Myofibril (my' oh fy' bril) [Gr. *mys*: muscle + L. *fibrilla*: small fiber] • A polymeric unit of actin or myosin in a muscle.

Myogenic (my oh jen' ik) [Gr. *mys*: muscle + *genesis*: source] • Originating in muscle.

Myoglobin (my' oh globe in) [Gr. *mys*: muscle + L. *globus*: sphere] • An oxygen-binding molecule found in muscle. Consists of a heme unit and a single globin chain, and carries less oxygen than hemoglobin.

Myosin [Gr. *mys*: muscle] • One of the two major proteins of muscle, it makes up the thick filaments. (See actin.)

NAD (nicotinamide adenine dinucleotide) • A compound found in all living cells, existing in two interconvertible forms: the oxidizing agent NAD^+ and the reducing agent NADH.

NADP (nicotinamide adenine dinucleotide phosphate) • Like NAD, but possessing another phosphate group; plays similar roles but is used by different enzymes.

Natural selection • The differential contribution of offspring to the next generation by various genetic types belonging to the same population. The mechanism of evolution proposed by Charles Darwin.

Necrosis (nec roh' sis) • Tissue damage resulting from cell death.

Negative control • The situation in which a regulatory macromolecule (generally a repressor) functions to turn off transcription. In the absence of a regulatory macromolecule, the structural genes are turned on.

Nekton [Gr. *nekhein*: to swim] • Animals, such as fish, that can swim against currents of water. (Contrast with plankton.)

Nematocyst (ne mat' o sist) [Gr. *nema*: thread + *kystis*: cell] • An elaborate, thread-like structure produced by cells of jellyfish and other cnidarians, used chiefly to paralyze and capture prey.

Nephridium (nef rid' ee um) [Gr. *nephros*: kidney] • An organ which is involved in excretion, and often in water balance, involving a tube that opens to the exterior at one end.

Nephron (nef' ron) [Gr. *nephros*: kidney] • The basic component of the kidney, which is made up of numerous nephrons. Its form varies in detail, but it always has at one end a device for receiving a filtrate of blood, and then a tubule that absorbs selected parts of the filtrate back into the bloodstream.

Nephrostome (nef' ro stome) [Gr. *nephros*: kidney + *stoma*: opening] An opening in a nephridium through which body fluids can enter.

Nerve • A structure consisting of many neuronal axons and connective tissue.

Net primary production • Total photosynthesis minus respiration by plants.

Neural plate • A thickened strip of ectoderm along the dorsal side of the early vertebrate embryo; gives rise to the central nervous system.

Neural tube • An early stage in the development of the vertebrate nervous system consisting of a hollow tube created by two opposing folds of the dorsal ectoderm along the anterior–posterior body axis.

Neuromuscular junction • The region where a motor neuron contacts a muscle fiber, creating a synapse.

Neuron (noor' on) [Gr. *neuron*: nerve, sinew] • A cell derived from embryonic ectoderm and characterized by a membrane potential that can change in response to stimuli, generating action potentials. Action potentials are generated along an extension of the cell (the axon), which makes junctions (synapses) with other neurons, muscle cells, or gland cells.

Neurotransmitter • A substance, produced in and released by one neuron, that diffuses across a synapse and excites or inhibits the postsynaptic neuron.

Neurula (nure' you la) [Gr. *neuron*: nerve] • Embryonic stage during formation of the dorsal nerve cord by two ectodermal ridges.

Neutral allele • An allele that does not alter the functioning of the proteins for which it codes.

Neutral theory • A view of molecular evolution that postulates that most mutations do not affect the amino acid being coded for, and that such mutations accumulate in a population at rates driven by genetic drift and mutation rates.

Neutron (new' tron) [E.: neutral] • One of the three most fundamental particles of matter, with mass approximately 1 amu and no electrical charge.

Nicotinamide adenine dinucleotide • (See NAD.)

Nicotinamide adenine dinucleotide phosphate • (See NADP.)

Nitrification • The oxidation of ammonia to nitrite and nitrate ions, performed by certain soil bacteria.

Nitrogenase • In nitrogen-fixing organisms, an enzyme complex that mediates the stepwise reduction of atmospheric N_2 to ammonia.

Nitrogen fixation • Conversion of nitrogen gas to ammonia, which makes nitrogen available to living things. Carried out by certain prokaryotes, some of them free-living and others living within plant roots.

Node [L. *nodus*: knob, knot] • In plants, a (sometimes enlarged) point on a stem where a leaf is or was attached.

Node of Ranvier • A gap in the myelin sheath covering an axons, where the axonal membrane can fire action potentials.

Noncompetitive inhibitor • An inhibitor that binds the enzyme at a site other than the active site. (Contrast with competitive inhibitor.)

Nondisjunction • Failure of sister chromatids to separate in meiosis II or mitosis, or failure of homologous chromosomes to separate in meiosis I. Results in aneuploidy.

Nonpolar molecule • A molecule whose electric charge is evenly balanced from one end of the molecule to the other.

Nonsense (chain-terminating) mutation • Mutations that change a codon for an amino acid to one of the codons (UAG, UAA, or UGA) that signal termination of translation. The resulting gene product is a shortened polypeptide that begins normally at the amino-terminal end and ends at the position of the altered codon. (Contrast with frame-shift mutation, missense mutation, synonymous mutation.)

Nonspecific defenses • Immunologic responses directed against most or all pathogens, generally without reference to the pathogens' antigens. These defenses include the skin, normal flora, lysozyme, the acidic stomach, interferon, and the inflammatory response.

Nonsynonymous mutation • A nucleotide substitution that that changes the amino acid specified (i.e., AGC → AGA, or serine → arginine). (Compare with frame-shift mutation, missense mutation, nonsense mutation.)

Nonsynonymous substitution • The situation when a nonsynonymous mutation becomes widespread in a population. Typically influenced by natural selection. (Contrast with synonymous substitution.)

Nontracheophytes • Those plants lacking well-developed vascular tissue; the liverworts, hornworts, and mosses. (Contrast with tracheophytes.)

Normal flora • The bacteria and fungi that live on animal body surfaces without causing disease.

Norepinephrine • A neurotransmitter found in the central nervous system and also at the postganglionic nerve endings of the sympathetic nervous system. Also called noradrenaline.

Notochord (no' tow kord) [Gr. *notos*: back + *chorde*: string] • A flexible rod of gelatinous material serving as a support in the embryos of all chordates and in the adults of tunicates and lancelets.

Nuclear envelope • The surface, consisting of two layers of membrane, that encloses the nucleus of eukaryotic cells.

Nucleic acid (new klay' ik) [E.: nucleus of a cell] • A long-chain alternating polymer of deoxyribose or ribose and phosphate groups, with nitrogenous bases—adenine, thymine, uracil, guanine, or cytosine (A, T, U, G, or C)—as side chains. DNA and RNA are nucleic acids.

Nucleoid (new' klee oid) • The region that harbors the chromosomes of a prokaryotic cell. Unlike the eukaryotic nucleus, it is not bounded by a membrane.

Nucleolar organizer (new klee' o lar) • A region on a chromosome that is associated with the formation of a new nucleolus following nuclear division. The site of the genes that code for ribosomal RNA.

Nucleolus (new klee' oh lus) [from L. diminutive of *nux*: little kernel or little nut] • A small, generally spherical body found within the nucleus of eukaryotic cells. The site of synthesis of ribosomal RNA.

Nucleoplasm (new' klee o plazm) • The fluid material within the nuclear envelope of a cell, as opposed to the chromosomes, nucleoli, and other particulate constituents.

Nucleosome • A portion of a eukaryotic chromosome, consisting of part of the DNA molecule wrapped around a group of histone molecules, and held together by another type of histone molecule. The chromosome is made up of many nucleosomes.

Nucleotide • The basic chemical unit (monomer) in a nucleic acid. A nucleotide in RNA consists of one of four nitrogenous bases linked to ribose, which in turn is linked to phosphate. In DNA, deoxyribose is present instead of ribose.

Nucleus (new' klee us) [from L. diminutive of *nux*: kernel or nut] • (1) In chemistry, the dense central portion of an atom, made up of protons and neutrons, with a positive charge. Surrounded by a cloud of negatively charged electrons. (2) In cells, the centrally located chamber of eukaryotic cells that is bounded by a double membrane and contains the chromosomes. The information center of the cell.

Null hypothesis • The assertion that an effect proposed by its companion hypothesis does not in fact exist.

Nutrient • A food substance; or, in the case of mineral nutrients, an inorganic element required for completion of the life cycle of an organism.

Oil • A triglyceride that is liquid at room temperature. (Contrast with fat.)

Okazaki fragments • Newly formed DNA strands making up the lagging strand in DNA replication. DNA ligase links the Okazaki fragments to give a continuous strand.

Olfactory • Having to do with the sense of smell.

Oligomer [Gr.: a few units] • A compound molecule of intermediate size, made up of two to a few monomers. (Contrast with monomer, polymer.)

Oligosaccharins • Plant hormones, derived from the plant cell wall, that trigger defenses against pathogens.

Ommatidium [Gr. *omma*: an eye] • One of the units which, collected into groups of up to 20,000, make up the compound eye of arthropods.

Omnivore [L. *omnis*: all, everything + *vorare*: to devour] • An organism that eats both animal and plant material. (Contrast with carnivore, detritivore, herbivore.)

Oncogenic (ong' co jen' ik) [Gr. *onkos*: mass, tumor + *genes*: born] • Causing cancer.

Oocyte (oh' eh site) [Gr. *oon*: egg + *kytos*: cell] • The cell that gives rise to eggs in animals.

Oogenesis (oh' eh jen e sis) [Gr. *oon*: egg + *genesis*: source] • Female gametogenesis, leading to production of the egg.

Oogonium (oh' eh go' nee um) • In some algae and fungi, a cell in which an egg is produced.

Operator • The region of an operon that acts as the binding site for the repressor.

Operon • A genetic unit of transcription, typically consisting of several structural genes that are transcribed together; the operon contains at least two control regions: the promoter and the operator.

Opportunity cost • The sum of the benefits an animal forfeits by not being able to perform some other behavior during the time when it is performing a given behavior.

Opsin (op' sin) [Gr. *opsis*: sight] • The protein protion of the visual pigment rhodopsin. (See rhodopsin.)

Optic chiasm • Stucture on the lower surface of the vertebrate brain where the two optic nerves come together.

Optical isomers • Isomers that differ in the configuration of the four different groups attached to a single carbon atom; so named because solutions of the two isomers rotate the plane of polarized light in opposite directions. The two isomers are mirror images of one another.

Optimality models • Models developed to determine the structures or behaviors that best solve particular problems faced by organisms.

Order • In taxonomy, the category below the class and above the family; a group of related, similar families.

Organ • A body part, such as the heart, liver, brain, root, or leaf, composed of different tissues integrated to perform a distinct function for the body as a whole.

Organ identity genes • Plant genes that specify the various parts of the flower. See homeotic genes.

Organ of Corti • Structure in the inner ear that transforms mechanical forces produced from pressure waves ("sound waves") into action potentials that are sensed as sound.

Organelles (or' gan els') [L.: little organ] • Organized structures that are found in or on cells. Examples: ribosomes, nuclei, mitochrondria, chloroplasts, cilia, and contractile vacuoles.

Organic • Pertaining to any aspect of living matter, e.g., to its evolution, structure, or chemistry. The term is also applied to any chemical compound that contains carbon.

Organism • Any living creature.

Organizer, embryonic • A region of an embryo which directs the development of nearby regions. In amphibian early gastrulas, the dorsal lip of the blastopore.

Origin of replication • A DNA sequence at which helicase unwinds the DNA double helix and DNA polymerase binds to initiate DNA replication.

Osmoregulation • Regulation of the chemical composition of the body fluids of an organism.

Osmoreceptor • A neuron that converts changes in the osmotic potential of interstial fluids into action potentials.

Osmosis (oz mo' sis) [Gr. *osmos*: to push] • The movement of water through a differentially permeable membrane from one region to another where the water potential is more negative. This is often a region in which the concentration of dissolved molecules or ions is higher, although the effect of dissolved substances may be offset by hydrostatic pressure in cells with semi-rigid walls.

Ossicle (ah' sick ul) [L. *os*: bone] • The calcified construction unit of echinoderm skeletons.

Osteoblasts • Cells that lay down the protein matrix of bone.

Osteoclasts • Cells that dissolve bone.

Otolith (oh' tuh lith) [Gk.*otikos*: ear + *lithos*: stone[• Structures in the vertebrate vestibular apparatus that mechanically stimulate hair cells when the head moves or changes position.

Outgroup • A taxon that separated from another taxon, whose lineage is to be inferred, before the latter underwent evolutionary radiation.

Oval window • The flexible membrane which, when moved by the bones of the middle ear, produces pressure waves in the inner ear

Ovary (oh' var ee) • Any female organ, in plants or animals, that produces an egg.

Oviduct [L. *ovum*: egg + *ducere*: to lead] • In mammals, the tube serving to transport eggs to the uterus or to outside of the body.

Oviparous (oh vip' uh rus) • Reproduction in which eggs are released by the female and development is external to the mother's body. (Contrast with viviparous.)

Ovulation • The release of an egg from an ovary.

Ovule (oh' vule) [L. *ovulum*: little egg] • In plants, an organ that contains a gametophyte and, within the gametophyte, an egg; when it matures, an ovule becomes a seed.

Ovum (oh' vum) [L.: egg] • The egg, the female sex cell.

Oxidation (ox i day' shun) • Relative loss of electrons in a chemical reaction; either outright removal to form an ion, or the sharing of electrons with substances having a greater affinity for them, such as oxygen. Most oxidation, including biological ones, are associated with the liberation of energy. (Contrast with reduction.)

Oxidative phosphorylation • ATP formation in the mitochondrion, associated with flow of electrons through the respiratory chain.

Oxidizing agent • A substance that can accept electrons from another. The oxidizing agent becomes reduced; its partner becomes oxidized.

P generation • Also called the parental generation. The individuals that mate in a genetic cross. Their immediate offspring are the F_1 generation.

Pacemaker • That part of the heart which undergoes most rapid spontaneous contraction, thus setting the pace for the beat of the entire heart. In mammals, the sinoatrial (SA) node. Also, an artificial device, implanted in the heart, that initiates rhythmic contraction of the organ.

Pacinian corpuscle • A sensory neuron surrounded by sheaths of connective tissue. Found in the deep layers of the skin, where it senses touch and vibration.

Pair rule genes • Segmentation genes that divide the *Drosophila* larva into two segments each.

Paleomagnetism • The record of the changing direction of Earth's magnetic field as stored in lava flows. Used to accurately date extremely ancient events.

Paleontology (pale' ee on tol' oh jee) [Gr. *palaios*: ancient, old + *logos*: discourse] • The scientific study of fossils and all aspects of extinct life.

Pancreas (pan' cree us) • A gland, located near the stomach of vertebrates, that secretes digestive enzymes into the small intestine and releases insulin into the bloodstream.

Pangaea (pan jee' uh) [Gk. *pan*: all, every] • The single land mass formed when all the continents came together in the Permian period. (Contrast with Gondwana.)

Parabronchi • Passages in the lungs of birds through which air flows.

Paradigm • A general framework within which a scientific or philosophical discipline is viewed and within which questions are asked and hypotheses are developed. Scientific revolutions usually involve major paradigm changes. (Contrast with hypothesis, theory.)

Parallel evolution • Evolutionary patterns that exist in more than one lineage. Often the result of underlying developmental processes.

Parapatric speciation [Gr. *para*: beside + *patria*: fatherland] • Development of reproductive isolation when the barrier is not geographic but is a difference in some other physical condition (such as soil nutrient content) that prevents gene flow between the subpopulations. (Contrast with allopatric speciation, sympatric speciation.)

Paraphyletic taxon • A taxon that includes some, but not all, of the descendants of a single ancestor.

Parasite • An organism that attacks and consumes parts of an organism much larger than itself. Parasites sometimes, but not always, kill the host.

Parasitoid • A parasite that is so large relative to its host that only one individual or at most a few individuals can live within a single host.

Parasympathetic nervous system • A portion of the autonomic (involuntary) nervous system. Activity in the parasympathetic nervous system produces effects such as decreased blood pressure and decelerated heart beat. (Contrast with sympathetic nervous system.)

Parathormone • Hormone secreted by the parathyroid glands. Stimulates osteoclast activity and raises blood calcium levels.

Parathyroids • Four glands on the posterior surface of the thyroid that produce and release parathormone.

Parenchyma (pair eng' kyma) [Gr. *para*: beside + *enchyma*: infusion] • A plant tissue composed of relatively unspecialized cells without secondary walls.

Parental investment • Investment in one offspring or group of offspring that reduces the ability of the parent to assist other offspring.

Parsimony • The principle of preferring the simplest among a set of plausible explanations of a phenomenon. Commonly employed in evolutionary and biogeographic studies.

Parthenocarpy • Formation of fruit from a flower without fertilization.

Parthenogenesis (par' then oh jen' e sis) [Gr. *parthenos*: virgin + *genesis*: source] • The production of an organism from an unfertilized egg.

Partial pressure • The portion of the barometric pressure of a mixture of gases that is due to one component of that mixture. For example, the partial pressure of oxygen at sea level is 20.9% of barometric pressure.

Patch clamping • A technique for isolating a tiny patch of membrane to allow the study of ion movement through a particular channel.

Pathogen (path' o jen) [Gr. *pathos*: suffering + *gignomai*: causing] • An organism that causes disease.

Pattern formation • In animal embryonic development, the organization of differentiated tissues into specific structures such as wings.

Pedigree • The pattern of transmission of a genetic trait in a family.

Pelagic zone (puh ladj' ik) [Gr. *pelagos*: the sea] • The open waters of the ocean.

Penetrance • Of a genotype, the proportion of individuals with that genotype who show the expected phenotype.

PEP carboxylase • The enzyme that combines carbon dioxide with PEP to form a 4-carbon dicarboxylic acid at the start of C_4 photosynthesis or of Crassulacean acid metabolism (CAM).

Pepsin [Gr. *pepsis*: digestion] • An enzyme, in gastric juice, that digests protein.

Peptide linkage • The connecting group in a protein chain, –CO–NH–, formed by removal of water during the linking of amino acids, –COOH to $-NH_2$.

Peptidoglycan • The cell wall material of many prokaryotes, consisting of a single enormous molecule that surrounds the entire cell.

Perennial (per ren' ee al) [L. *per*: through + *annus*: a year] • Referring to a plant that lives from year to year. (Contrast with annual, biennial.)

Perfect flower • A flower with both stamens and carpels, therefore hermaphroditic.

Pericycle [Gr. *peri*: around + *kyklos*: ring or circle] • In plant roots, tissue just within the endodermis, but outside of the root vascular tissue. Meristematic activity of pericycle cells produces lateral root primordia.

Periderm • The outer tissue of the secondary plant body, consisting primarily of cork.

Period • (1) A minor category in the geological time scale. (2) The duration of a cyclical event, such as a circadian rhythm.

Peripheral nervous system • Neurons that transmit information to and from the central nervous system and whose cell bodies reside outside the brain or spinal cord.

Peristalsis (pair' i stall' sis) [Gr. *peri*: around + *stellein*: place] • Wavelike muscular contractions proceeding along a tubular organ, propelling the contents along the tube.

Peritoneum • The mesodermal lining of the coelom among coelomate animals.

Permease • A membrane protein that specifically transports a compound or family of compounds across the membrane.

Peroxisome • An organelle that houses reactions in which toxic peroxides are formed. The peroxisome isolates these peroxides from the rest of the cell.

Petal • In an angiosperm flower, a sterile modified leaf, nonphotosynthetic, frequently brightly colored, and often serving to attract pollinating insects.

Petiole (pet' ee ole) [L. *petiolus*: small foot] • The stalk of a leaf.

pH • The negative logarithm of the hydrogen ion concentration; a measure of the acidity of a solution. A solution with pH = 7 is said to be neutral; pH values higher than 7 characterize basic solutions, while acidic solutions have pH values less than 7.

Phage (fayj) • Short for bacteriophage.

Phagocyte • A white blood cell that ingests microorganisms by endocytosis.

Phagocytosis [Gr.: *phagein* to eat; cell-eating] • A form of endocytosis, the uptake of a solid particle by forming a pocket of plasma membrane around the particle and pinching off the pocket to form an intracellular particle bounded by membrane. (Contrast with pinocytosis.)

Pharynx [Gr.: throat] • The part of the gut between the mouth and the esophagus.

Phenotype (fee' no type) [Gr. *phanein*: to show] • The observable properties of an individual as they have developed under the combined influences of the genetic constitution of the individual and the effects of environmental factors. (Contrast with genotype.)

Phenotypic plasticity • The fact that the phenotype of an organism is determined by a complex series of developmental processes that are affected by both its genotype and its environment.

Pheromone (feer' o mone) [Gr. *phero*: carry + *hormon*: excite, arouse] • A chemical substance used in communication between organisms of the same species.

Phloem (flo' um) [Gr. *phloos*: bark] • In vascular plants, the food-conducting tissue. It consists of sieve cells or sieve tubes, fibers, and other specialized cells.

Phosphate group • The functional group $-OPO_3H_2$; the transfer of energy from one compound to another is often accomplished by the transfer of a phosphate group.

Phosphodiester linkage • The connection in a nucleic acid strand, formed by linking two nucleotides.

Phospholipids • Cellular materials that contain phosphorus and are soluble in organic solvents. An example is lecithin (phosphatidyl choline). Phospholipids are important constituents of cellular membranes. (See lipids.)

Phosphorylation • The addition of a phosphate group.

Photoautotroph • An organism that obtains energy from light and carbon from carbon dioxide. (Contrast with chemoautotroph, chemoheterotroph, photoheterotroph.)

Photoheterotroph • An organism that obtains energy from light but must obtain its carbon from organic compounds. (Contrast with chemoautotroph, chemoheterotroph, photoautotroph.)

Photon (foe' tohn) [Gr. *photos*: light] • A quantum of visible radiation; a "packet" of light energy.

Photoperiod (foe' tow peer' ee ud) • The duration of a period of light, such as the length of time in a 24-hour cycle in which daylight is present. The regulation of processes such as flowering by the changing length of day (or of night) is known as photoperiodism.

Photoreceptor • (1) A protein (pigment) that triggers a physiological response when it absorbs a photon. (2) A cell that senses and responds to light energy.

Photorespiration • Light-driven uptake of oxygen and release of carbon dioxide, the carbon being derived from the early reactions of photosynthesis.

Photosynthesis (foe tow sin' the sis) [literally, "synthesis out of light"] • Metabolic processes, carried out by green plants, by which visible light is trapped and the energy used to synthesize compounds such as ATP and glucose.

Phototropin • A yellow protein that is the photoreceptor responsible for phototropism.

Phototropism [Gr. *photos*: light + *trope*: a turning] • A directed plant growth response to light.

Phylogenetic tree • Graphic representation of lines of descent among organisms.

Phylogeny (fy loj' e nee) [Gr. *phylon*: tribe, race + *genesis*: source] • The evolutionary history of a particular group of organisms; also, the diagram of the "family tree" that shows genetic linkages between ancestors and descendants.

Phylum (plural: phyla) [Gr. *phylon*: tribe, stock] • In taxonomy, a high-level category just beneath kingdom and above the class; a group of related, similar classes.

Physiology (fiz' ee ol' o jee) [Gr. *physis*: natural form + *logos*: discourse, study] • The scientific study of the functions of living organisms and the individual organs, tissues, and cells of which they are composed.

Phytoalexins • Substances toxic to fungi, produced by plants in response to fungal infection.

Phytochrome (fy' tow krome) [Gr. *phyton*: plant + *chroma*: color] • A plant pigment regulating a large number of developmental and other phenomena in plants; can exist in two different forms, one of which is active and the other is not. Different wavelengths of light can drive it from one form to the other.

Phytoplankton (fy' tow plangk' ton) [Gr. *phyton*: plant + *planktos*: wandering] • The autotrophic portion of the plankton, consisting mostly of algae.

Pigment • A substance that absorbs visible light.

Pilus (pill' us) [Lat. *pilus*: hair] • A surface appendage by which some bacteria adhere to one another during conjugation.

Pinocytosis [Gr.: drinking cell] • A form of endocytosis; the uptake of liquids by engulfing a sample of the external medium into a pocket of the plasma membrane followed by pinching off the pocket to form an intracellular vesicle. (Contrast with phagocytosis and endocytosis.)

Pistil [L. *pistillum*: pestle] • The female structure of an angiosperm flower, within which the ovules are borne. May consist of a single carpel, or of several carpels fused into a single structure. Usually differentiated into ovary, style, and stigma.

Pith • In plants, relatively unspecialized tissue found within a cylinder of vascular tissue.

Pituitary • A small gland attached to the base of the brain in vertebrates. Its hormones control the activities of other glands. Also known as the hypophysis.

Placenta (pla sen' ta) [Gr. *plax*: flat surface] • The organ found in most mammals that provides for the nourishment of the fetus and elimination of the fetal waste products.

Placental (pla sen' tal) • Pertaining to mammals of the subclass Eutheria, a group characterized by the presence of a placenta; contains the majority of living species of mammals.

Plankton [Gr. *planktos*: wandering] • The free-floating organisms of the sea and fresh water that for the most part move passively with the water currents. Consisting mostly of microorganisms and small plants and animals. (Contrast with nekton.)

Plant • A member of the kingdom Plantae. Multicellular, gaining its nutrition by photosynthesis.

Planula (plan' yew la) [L. *planum*: something flat] • The free-swimming, ciliated larva of the cnidarians.

Plaque (plack) [Fr.: a metal plate or coin] • (1) A circular clearing in a turbid layer (lawn) of bacteria growing on the surface of a nutrient agar gel. Produced by successive rounds of infection initiated by a single bacteriophage. (2) An accumulation of prokaryotic organisms on tooth enamel. Acids produced by the metabolism of these microorganisms can cause tooth decay.

Plasma (plaz' muh) [Gr. *plassein*: to mold] • The liquid portion of blood, in which blood cells and other particulates are suspended.

Plasma cell • An antibody-secreting cell that developed from a B cell. The effector cell of the humoral immune system.

Plasma membrane • The membrane that surrounds the cell, regulating the entry and exit of molecules and ions. Every cell has a plasma membrane.

Plasmid • A DNA molecule distinct from the chromosome(s); that is, an extrachromosomal element. May replicate independently of the chromosome.

Plasmodesma (plural: plasmodesmata) [Gr. *plasma*: formed or molded + *desmos*: band] • A cytoplasmic strand connecting two adjacent plant cells.

Plasmolysis (plaz mol' i sis) • Shrinking of the cytoplasm and plasma membrane away from the cell wall, resulting from the osmotic outflow of water. Occurs only in cells with rigid cell walls.

Plastid • Organelle in plants that serves for food manufacture (by photosynthesis) or food storage; bounded by a double membrane.

Platelet • A membrane-bounded body without a nucleus, arising as a fragment of a cell in the bone marrow of mammals. Important to blood-clotting action.

Pleiotropy (plee' a tro pee) [Gr. *pleion*: more] • The determination of more than one character by a single gene.

Pleural membrane [Gk. *pleuras*: rib, side] • The membrane lining the outside of the lungs and the walls of the thoracic cavity. Inflammation of these membranes is a condition known as *pleurisy*.

Podocytes • Cells of Bowman's capsule of the nephron that cover the capillaries of the glomerulus, forming filtration slits.

Poikilotherm (poy' kill o therm) [Gr. *poikilos*: varied + *therme*: heat] • An animal whose body temperature tends to vary with the surrounding environment. (Contrast with homeotherm, heterotherm.)

Point mutation • A mutation that results from a small, localized alteration in the chemical structure of a gene. Such mutations can give rise to wild-type revertants as a result of reverse mutation. In genetic crosses, a point mutation behaves as if it resided at a single point on the genetic map. (Contrast with deletion.)

Polar body • A nonfunctional nucleus produced by meiosis, accompanied by very little cytoplasm. The meiosis which produces the mammalian egg produces in addition three polar bodies.

Polar molecule • A molecule in which the electric charge is not distributed evenly in the covalent bonds.

Polarity • In development, the difference between one end and the other. In chemistry, the property that makes a polar molecule.

Pollen [L.: fine powder, dust] • The fertilizing element of seed plants, containing the male gametophyte and the gamete, at the stage in which it is shed.

Pollination • Process of transferring pollen from the anther to the receptive surface (stigma) of the ovary in plants.

Poly- [Gr. *poly*: many] • A prefix denoting multiple entities.

Polygamy [Gr. *poly*: many + *gamos*: marriage] • A breeding system in which an individual acquires more than one mate. In polyandry, a female mates with more than one male, in polygyny, a male mates with more than one female.

Polygenes • Multiple loci whose alleles increase or decrease a continuously variable phenotypic trait.

Polymer • A large molecule made up of similar or identical subunits called monomers. (Contrast with monomer, oligomer.)

Polymerase chain reaction (PCR) • A technique for the rapid production of millions of copies of a particular stretch of DNA.

Polymerization reactions • Chemical reactions that generate polymers by means of condensation reactions.

Polymorphism (pol′ lee mor′ fiz um) [Gr. *poly*: many + *morphe*: form, shape] • (1) In genetics, the coexistence in the same population of two distinct hereditary types based on different alleles. (2) In social organisms such as colonial cnidarians and social insects, the coexistence of two or more functionally different castes within the same colony.

Polyp • The sessile, asexual stage in the life cycle of most cnidarians.

Polypeptide • A large molecule made up of many amino acids joined by peptide linkages. Large polypeptides are called proteins.

Polyphyletic group • A group containing taxa, not all of which share the most recent common ancestor.

Polyploid (pol′ lee ploid) • A cell or an organism in which the number of complete sets of chromosomes is greater than two.

Polysaccharide • A macromolecule composed of many monosaccharides (simple sugars). Common examples are cellulose and starch.

Polysome • A complex consisting of a threadlike molecule of messenger RNA and several (or many) ribosomes. The ribosomes move along the mRNA, synthesizing polypeptide chains as they proceed.

Polytene (pol′ lee teen) [Gr. *poly*: many + *taenia*: ribbon] • An adjective describing giant interphase chromosomes, such as those found in the salivary glands of fly larvae. The characteristic, reproducible pattern of bands and bulges seen on these chromosomes has provided a method for preparing detailed chromosome maps of several organisms.

Pons [L. *pons*: bridge] • Region of the brain stem anterior to the medulla.

Population • Any group of organisms coexisting at the same time and in the same place and capable of interbreeding with one another.

Population density • The number of individuals (or modules) of a population in a unit of area or volume.

Population genetics • The study of genetic variation and its causes within populations.

Population structure • The proportions of individuals in a population belonging to different age classes (age structure). Also, the distribution of the population in space.

Portal vein • A vein connecting two capillary beds, as in the hepatic portal system.

Positive control • The situation in which a regulatory macromolecule is needed to turn transcription of structural genes on. In its absence, transcription will not occur.

Positive cooperativity • Occurs when a molecule can bind several ligands and each one that binds alters the conformation of the molecule so that it can bind the next ligand more easily. The binding of four molecules of O_2 by hemoglobin is an example of positive cooperativity.

Postabsorptive period • When there is no food in the gut and no nutrients are being absorbed.

Postsynaptic cell • The cell whose membranes receive the neurotransmitter released at a synapse.

Predator • An organism that kills and eats other organisms. Predation is usually thought of as involving the consumption of animals by animals, but it can also mean the eating of plants.

Presynaptic excitation/inhibition • Occurs when a neuron modifies activity at a synapse by releasing a neurotransmitter onto the presynaptic nerve terminal.

Prey [L. *praeda*: booty] • An organism consumed as an energy source.

Primary active transport • Form of active transport in which ATP is hydrolyzed, yielding the energy required to transport ions against their concentration gradients. (Contrast with secondary active transport.)

Primary growth • In plants, growth produced by the apical meristems. (Contrast with secondary growth.)

Primary producer • A photosynthetic or chemosynthetic organism that synthesizes complex organic molecules from simple inorganic ones.

Primary succession • Succession that begins in an areas initially devoid of life, such as on recently exposed glacial till or lava flows.

Primary structure • The specific sequence of amino acids in a protein.

Primary wall • Cellulose-rich cell wall layers laid down by a growing plant cell.

Primate (pry′ mate) • A member of the order Primates, such as a lemur, monkey, ape, or human.

Primer • A short, single-stranded segment of DNA serving as the necessary starting material for the synthesis of a new DNA strand, which is synthesized from the 3′ end of the primer.

Primitive streak • A line running axially along the blastodisc, the site of inward cell migration during formation of the three-layered embryo. Formed in the embryos of birds and fish.

Primordium [L. *primordium*: origin] • The most rudimentary stage of an organ or other part.

Principle of continuity • States that because life probably evolved from nonlife by a continuous, gradual process, all postulated stages in the evolution of life should be derivable from preexisting states. (Compare with signature principle.)

Pro- [L.: first, before, favoring] • A prefix often used in biology to denote a developmental stage that comes first or an evolutionary form that appeared earlier than another. For example, prokaryote, prophase.

Probe • A segment of single stranded nucleic acid used to identify DNA molecules containing the complementary sequence.

Procambium • Primary meristem that produces the vascular tissue.

Progesterone [L. *pro*: favoring + *gestare*: to bear] • A vertebrate female sex hormone that maintains pregnancy.

Prokaryotes (pro kar′ ry otes) [L. *pro*: before + Gk. *karyon*: kernel, nucleus] • Organisms whose genetic material is not contained within a nucleus. The bacteria. Considered an earlier stage in the evolution of life than the eukaryotes.

Prometaphase • The phase of nuclear division that begins with the disintegration of the nuclear envelope.

Promoter • The region of an operon that acts as the initial binding site for RNA polymerase.

Proofreading • The correction of an error in DNA replication just after an incorrectly paired base is added to the growing polynucleotide chain.

Prophage (pro′ fayj) • The noninfectious units that are linked with the chromosomes of the host bacteria and multiply with them but do not cause dissolution of the cell. Prophage can later enter into the lytic phase to complete the virus life cycle.

Prophase (pro′ phase) • The first stage of nuclear division, during which chromosomes condense from diffuse, threadlike material to discrete, compact bodies.

Prostaglandin • Any one of a group of specialized lipids with hormone-like functions. It is not clear that they act at any considerable distance from the site of their production.

Prosthetic group • Any nonprotein portion of an enzyme.

Protease (pro′ tee ase) • See proteolytic enzyme.

Protein (pro′ teen) [Gr. *protos*: first] • One of the most fundamental building substances of living organisms. A long-chain polymer of amino acids with twenty different common side chains. Occurs with its polymer chain extended in fibrous proteins, or coiled into a compact macromolecule in enzymes and other globular proteins.

Proteolytic enzyme • An enzyme whose main catalytic function is the digestion of a protein or polypeptide chain. The digestive enzymes trypsin, pepsin, and carboxypeptidase are all proteolytic enzymes (proteases).

Protist • Those eukaryotes not included in the kingdoms Animalia, Fungi, or Plantae.

Protobiont • Aggregates of abiotically produced molecules that cannot reproduce but do maintain internal chemical environments that differ from their surroundings.

Protoderm • Primary meristem that gives rise to epidermis.

Proton (pro′ ton) [Gr. *protos*: first] • One of the three most fundamental particles of matter, with mass approximately 1 amu and an electrical charge of +1.

Proto-oncogenes • The normal alleles of genes possessing oncogenes (cancer-causing genes) as mutant alleles. Proto-oncogenes encode growth factors and receptor proteins.

Protostome • One of the major lineages of animal evolution. Characterized by spiral, determinate cleavage of the egg, and by schizocoelous development. (Compare with deuterostome.)

Prototroph (pro′ tow trofe′) [Gr. *protos*: first + *trophein*: to nourish] • The nutritional wild type, or reference form, of an organism. Any deviant form that requires growth nutrients not required by the prototrophic form is said to be a nutritional mutant, or auxotroph.

Protozoa • A group of single-celled organisms classified by some biologists as a single phylum; includes the flagellates, amoebas, and ciliates. This textbook follows most modern classifications in elevating the protozoans to a distinct kingdom (Protista) and each of their major subgroups to the rank of phylum.

Proximal • Near the point of attachment or other reference point. (Contrast with distal.)

Pseudocoelom • A body cavity not surrounded by a peritoneum. Characteristic of nematodes and rotifers.

Pseudogene • A DNA segment that is homologous to a functional gene but contains a nucleotide change that prevents its expression.

Pseudoplasmodium [Gr. *pseudes*: false + *plasma*: mold or form] • In the cellular slime molds such as *Dictyostelium*, an aggregation of single amoeboid cells. Occurs prior to formation of a fruiting structure.

Pseudopod (soo′ do pod) [Gr. *pseudes*: false + *podos*: foot] • A temporary, soft extension of the cell body that is used in location, attachment to surfaces, or engulfing particles.

Pulmonary • Pertaining to the lungs.

Punctuated equiilibrium • An evolutionary pattern in which periods of rapid change are separated by longer periods of little or no change.

Pupa (pew′ pa) [L.: doll, puppet] • In certain insects (the Holometabola), the encased developmental stage that intervenes between the larva and the adult.

Pupil • The opening in the vertebrate eye through which light passes.

Purine (pure′ een) • A type of nitrogenous base. The purines adenine and guanine are found in nucleic acids.

Purkinje fibers • Specialized heart muscle cells that conduct excitation throughout the ventricular muscle.

Pyramid of biomass • Graphical representation of the total body masses at different trophic levels in an ecosystem.

Pyramid of energy • Graphical representation of the total energy contents at different trophic levels in an ecosystem.

Pyrimidine (peer im′ a deen) • A type of nitrogenous base. The pyrimidines cytosine, thymine, and uracil are found in nucleic acids.

Pyruvate • A three-carbon acid; the end product of glycolysis and the raw material for the citric acid cycle.

Q_{10} • A value that compares the rate of a biochemical process or reaction over a 10°C range of temperature. A process that is not temperature-sensitive has a Q_{10} of 1. Values of 2 or 3 mean the reaction speeds up as temperature increases.

Quantum (kwon′ tum) [L. *quantus*: how great] • An indivisible unit of energy.

Quaternary structure • Of aggregating proteins, the arrangement of polypeptide subunits.

R factor (resistance factor) • A plasmid that contains one or more genes that encode resistance to antibiotics.

Radial symmetry • The condition in which two halves of a body are mirror images of each other regardless of the angle of the cut, providing the cut is made along the center line. Thus, a cylinder cut lengthwise down its center displays this form of symmetry. (Contrast with biradial symmetry.)

Radioisotope • A radioactive isotope of an element. Examples are carbon-14 (^{14}C) and hydrogen-3, or tritium (^{3}H).

Radiometry • The use of the regular, known rates of decay of radioisotopes of elements to determine dates of events in the distant past.

Rain shadow • A region of low precipitation on the leeward side of a mountain range.

Ramet • The repeated morphological units of sessile, modular organisms. (Contrast with genet.)

Random genetic drift • Evolution (change in gene proportions) by chance processes alone.

Rate constant • Of a particular chemical reaction, a constant which, when multiplied by the concentration(s) of reactant(s), gives the rate of the reaction.

Reactant • A chemical substance that enters into a chemical reaction with another substance.

Reaction, chemical • A process in which atoms combine or change bonding partners.

Realized niche • The actual niche occupied by an organism; it differs from the fundamental niche because of the presence of other species.

Receptive field • Of a neuron, the area on the retina from which the activity of that neuron can be influenced.

Receptor potential • The change in the resting potential of a sensory cell when it is stimulated.

Recessive • See dominance.

Reciprocal altruism • The exchange of altruistic acts between two or more individuals. The acts may be separated considerably in time.

Reciprocal crosses • A pair of crosses, in one of which a female of genotype A mates with a male of genotype B and in the other of which a female of genotype B mates with a male of genotype A.

Recognition site (also called a restriction site) • A sequence of nucleotides in DNA to which a restriction enzyme binds and then cuts the DNA.

Recombinant • An individual, meiotic product, or single chromosome in which genetic materials originally present in two individuals end up in the same haploid complement of genes. The reshuffling of genes can be either by independent segragation, or by crossing over between homologous chromosomes. For example, a human may pass on genes from both parents in a single haploid gamete.

Recombinant DNA technology • The application of genetic tools (restriction endonucleases, plasmids, and transformation) to the production of specific proteins by biological "factories" such as bacteria.

Rectum • The terminal portion of the gut, ending at the anus.

Redox reaction • A chemical reaction in which one reactant becomes oxidized and the other becomes reduced.

Reducing agent • A substance that can donate electrons to another substance. The reducing agent becomes oxidized, and its partner becomes reduced.

Reduction (re duk′ shun) • Gain of electrons; the reverse of oxidation. Most reductions lead to the storage of chemical energy, which can be released later by an oxidation reaction. Energy storage compounds such as sugars and fats are highly reduced compounds. (Contrast with oxidation.)

Reflex • An automatic action, involving only a few neurons (in vertebrates, often in the spinal cord), in which a motor response swiftly follows a sensory stimulus.

Refractory period • Of a neuron, the time interval after an action potential, during which another action potential cannot be elicited.

Regulative development • A pattern of animal embryonic development in which the fates of the first blastomeres are not absolutely fixed. (Contrast with mosaic development.)

Regulatory gene • A gene that contains the information for making a regulatory macromolecule, often a repressor protein.

Releaser • A sensory stimulus that triggers a fixed action pattern.

Releasing hormone • One of several hypothalamic hormones that stimulates the secretion of anterior pituitary hormone.

REM sleep • A sleep state characterized by dreaming, skeletal muscle relaxation, and rapid eye movements.

Renal [L. *renes*: kidneys] • Relating to the kidneys.

Replication fork • A point at which a DNA molecule is replicating. The fork forms by the unwinding of the parent molecule.

Repressible enzyme • An enzyme whose synthesis can be decreased or prevented by the presence of a particular compound. A repressible opren often controls the syhthesis of such an enzyme.

Repressor • A protein coded by the regulatory gene. The repressor can bind to a specific operator and prevent transcription of the operon.

Reproductive isolating mechanism • Any trait that prevents individuals from two different populations from producing fertile hybrids.

Reproductive isolation • The condition in which a population is not exchanging genes with other populations of the same species.

Resolving power • Of an optical device such as a microscope, the smallest distance between two lines that allows the lines to be seen as separate from one another.

Resource • Something in the environment required by an organism for its maintenance and growth that is consumed in the process of being used.

Resource defense polygamy • A breeding system in which individuals of one sex (usually males) defend resources that are attractive to individuals of the other sex (usually females); individuals holding better resources attract more mates.

Respiration (res pi ra' shun) [L. *spirare*: to breathe] • (1) Cellular respiration; the oxidation of the end products of glycolysis with the storage of much energy in ATP. The oxidant in the respiration of eukaryotes is oxygen gas. Some bacteria can use nitrate or sulfate instead of O_2. (2) Breathing.

Respiratory chain • The terminal reactions of cellular respiration, in which electrons are passed from NAD or FAD, through a series of intermediate carriers, to molecular oxygen, with the concomitant production of ATP.

Resting potential • The membrane potential of a living cell at rest. In cells at rest, the interior is negative to the exterior. (Contrast with action potential, electrotonic potential.)

Restoration ecology • The science and practice of restoring damaged or degraded ecosystems.

Restriction endonuclease • Any one of several enzymes, produced by bacteria, that break foreign DNA molecules at very specific sites. Some produce "sticky ends." Extensively used in recombinant DNA technology.

Restriction map • A partial genetic map of a DNA molecule, showing the points at which particular restriction endonuclease recognition sites reside.

Reticular system • A central region of the vertebrate brain stem that includes complex fiber tracts conveying neural signals between the forebrain and the spinal cord, with collateral fibers to a variety of nuclei that are involved in autonomic functions, including arousal from sleep.

Retina (rett' in uh) [L. *rete*: net] • The light-sensitive layer of cells in the vertebrate or cephalopod eye.

Retinal • The light-absorbing portion of visual pigment molecules. Derived from β-carotene.

Retrovirus • An RNA virus that contains reverse transcriptase. Its RNA serves as a template for cDNA production, and the cDNA is integrated into a chromosome of the mammalian host cell.

Reverse transcriptase • An enzyme that catalyzes the production of DNA (cDNA), using RNA as a template; essential to the reproduction of retroviruses.

RFLP (Restriction fragment length polymorphism) • Coexistence of two or more patterns of restriction fragments (patterns produced by restriction enzymes), as revealed by a probe. The polymorphism reflects a difference in DNA sequence on homologous chromosomes.

Rhizoids (rye' zoids) [Gr. *rhiza*: root] • Hairlike extensions of cells in mosses, liverworts, and a few vascular plants that serve the same function as roots and root hairs in vascular plants. The term is also applied to branched, rootlike extensions of some fungi and algae.

Rhizome (rye' zome) [Gr. *rhizoma*: mass of roots] • A special underground stem (as opposed to root) that runs horizontally beneath the ground.

Rhodopsin • A photopigment used in the visual process of transducing photons of light into changes in the membrane potential of photoreceptor cells.

Ribonucleic acid • See RNA.

Ribosomal RNA (rRNA) • Several species of RNA that are incorporated into the ribosome. Involved in peptide bond formation.

Ribosome • A small organelle that is the site of protein synthesis.

Ribozyme • An RNA molecule with catalytic activity.

Ribulose 1,5-bisphosphate (RuBP) • The compound in chloroplasts which reacts with carbon dioxide in the first reaction of the Calvin-Benson cycle.

Risk cost • The increased chance of being injured or killed as a result of performing a behavior, compared to resting.

RNA (ribonucleic acid) • A nucleic acid using ribose. Various classes of RNA are involved in the transcription and translation of genetic information. RNA serves as the genetic storage material in some viruses.

RNA polymerase • An enzyme that catalyzes the formation of RNA from a DNA template.

RNA splicing • The last stage of RNA processing in eukaryotes, in which the transcripts of introns are excised through the action of small nuclear ribonucleoprotein particles (snRNP).

Rods • Light-sensitive cells (photoreceptors) in the retina. (Contrast with cones.)

Root cap • A thimble-shaped mass of cells, produced by the root apical meristem, that protects the meristem and that is the organ that perceives the gravitational stimulus in root gravitropism.

Root hair • A specialized epidermal cell with a long, thin process that absorbs water and minerals from the soil solution.

rRNA • See ribosomal RNA.

Rubisco (RuBP carboxylase) • Enzyme that combines carbon dioxide with ribulose bisphosphate to produce 3-phosphoglycerate, the first product of C_3 photosynthesis. The most abundant protein on Earth.

Rumen (rew' mun) • The first division of the ruminant stomach. It stores and initiates bacterial fermentation of food. Food is regurgitated from the rumen for further chewing.

Ruminant • An herbivorous, cud-chewing mammal such as a cow, sheep, or deer, having a stomach consisting of four compartments.

S phase • In the cell cycle, the stage of interphase during which DNA is replicated. (Contrast with G_1 phase, G_2 phase.)

Saprobe [Gr. *sapros*: rotten + *bios*: life] • An organism (usually a bacterium or fungus) that obtains its carbon and energy directly from dead organic matter.

Sarcomere (sark' o meer) [Gr. *sark*: flesh + *meros*: a part] • The contractile unit of a skeletal muscle.

Saturated hydrocarbon • A compound consisting only of carbon and hydrogen, with the hydrogen atoms connected by single bonds.

Schizocoelous development • Formation of a coelom during embryological development by a splitting of mesodermal masses.

Schwann cell • A glial cell that wraps around part of the axon of a peripheral neuron, creating a myelin sheath.

Sclereid [Gr. *skleros*: hard] • A type of sclerenchyma cell, commonly found in nutshells, that is not elongated.

Sclerenchyma (skler eng' kyma) [Gr. *skleros*: hard + *kymus*, juice] • A plant tissue composed of cells with heavily thickened cell walls, dead at functional maturity. The principal types of sclerenchyma cells are fibers and sclereids.

Secondary active transport • Form of active transport in which ions or molecules are transported against their concentration gradient using energy obtained by relaxation of a gradient of sodium ion concentration rather than directly from ATP. (Contrast with primary active transport.)

Secondary compound • A compound synthesized by a plant that is not needed for basic cellular metabolism. Typically has an antiherbivore or antiparasite function.

Secondary growth • In plants, growth produced by vascular and cork cambia, contributing to an increase in girth. (Contrast with primary growth.)

Secondary structure • Of a protein, localized regularities of structure, such as the α helix and the β pleated sheet.

Secondary succession • Ecological succession after a disturbance that does not eliminate all the organisms that originally lived on the site.

Secondary wall • Wall layers laid down by a plant cell that has ceased growing; often impregnated with lignin or suberin.

Second law of thermodynamics • States that in any real (irreversible) process, there is a decrease in free energy and an increase in entropy.

Second messenger • A compound, such as cyclic AMP, that is released within a target cell after a hormone or other "first messenger" has bound to a surface receptor on a cell; the second messenger triggers further reactions within the cell.

Secretin (si kreet' in) • A peptide hormone secreted by the upper region of the small intestine when acidic chyme is present. Stimulates the pancreatic duct to secrete bicarbonate ions.

Section • A thin slice, usually for microscopy, as a tangential section or a transverse section.

Seed • A fertilized, ripened ovule of a gymnosperm or angiosperm. Consists of the embryo, nutritive tissue, and a seed coat.

Seed crop • The number of seeds produced by a plant during a particular bout of reproduction.

Seedling • A young plant that has grown from a seed (rather than by grafting or by other means.)

Segmentation genes • In insect larvae, genes that determine the number and polarity of larval segments.

Segment polarity genes • Genes that determine the boundaries and front-to-back organization of the segments in the *Drosophila* larva.

Segregation (genetic) • The separation of alleles, or of homologous chromosomes, from one another during meiosis so that each of the haploid daughter nuclei produced by meiosis contains one or the other member of the pair found in the diploid mother cell, but never both.

Selective permeability • A characteristic of a membrane, allowing certain substances to pass through while other substances are excluded.

Selfish act • A behavioral act that benefits its performer but harms the recipients.

Semelparous organism • An organism that reproduces only once in its lifetime. (Contrast with iteroparous.)

Semen (see' men) [L.: seed] • The thick, whitish liquid produced by the male reproductive organ in mammals, containing the sperm.

Semicircular canals • Part of the vestibular system of mammals.

Semiconservative replication • The common way in which DNA is synthesized. Each of the two partner strands in a double helix acts as a template for a new partner strand. Hence, after replication, each double helix consists of one old and one new strand.

Seminiferous tubules • The tubules within the testes within which sperm production occurs.

Senescence [L. *senescere*: to grow old] • Aging; deteriorative changes with aging; the increased probability of dying with increasing age.

Sensory neuron • A neuron leading from a sensory cell to the central nervous system. (Contrast with motor neuron.)

Sepal (see' pul) • One of the outermost structures of the flower, usually protective in function and enclosing the rest of the flower in the bud stage.

Septum [L.: partition] • A membrane or wall between two cavities.

Sertoli cells • Cells in the seminiferous tubules that nuture the developing sperm.

Serum • That part of the blood plasma that remains after clots have formed and been removed.

Sessile (sess' ul) [L. *sedere*: to sit] • Permanently attached; not moving.

Set point • In a regulatory system, the threshold sensitivity to the feedback stimulus.

Sex chromosome • In organisms with a chromosomal mechanism of sex determination, one of the chromosomes involved in sex determination.

Sex linkage • The pattern of inheritance characteristic of genes located on the sex chromosomes of organisms having a chromosomal mechanism for sex determination.

Sexual selection • Selection by one sex of characteristics in individuals of the opposite sex. Also, the favoring of characteristics in one sex as a result of competition among individuals of that sex for mates.

Shoot • The aerial part of a vascular plant, consisting of the leaves, stem(s), and flowers.

Sieve tube • A column of specialized cells found in the phloem, specialized to conduct organic matter from sources (such as photosynthesizing leaves) to sinks (such as roots). Found principally in flowering plants.

Sieve tube member • A single cell of a sieve tube, containing cytoplasm but relatively few organelles, with highly specialized perforated end walls leading to elements above and below.

Sign stimulus • The single stimulus, or one out of a very few stimuli, by which an animal distinguishes key objects, such as an enemy, or a mate, or a place to nest, etc.

Signal sequence • The sequence of a protein that directs the protein through a particular cellular membrane.

Signal transduction pathway • The series of biochemical steps whereby a stimulus to a cell (such as a hormone or neurotransmitter binding to a receptor) is translated into a response of the cell.

Signature principle • States that because of continuity, prebiotic processes should leave some trace in contemporary biochemistry. (Compare with principle of continuity.)

Silencer • A sequence of eukaryotic DNA that binds proteins that inhibit the transcription of an associated gene.

Silent mutations • Genetic changes that do not lead to a phenotypic change. At the molecular level, these are DNA sequence changes that, because of the redundancy of the genetic code, result in the same amino acids in the resulting protein. See synonymous mutation.

Similarity matrix • A matrix to compare the structures of two molecules constructed by adding the number of their amino acids that are identical or different

Sinoatrial node (sigh' no ay' tree al) • The pacemaker of the mammalian heart.

Sinus (sigh' nus) [L. *sinus*: a bend, hollow] • A cavity in a bone, a tissue space, or an enlargement in a blood vessel.

Skeletal muscle • See striated muscle.

Sliding filament theory • A proposed mechanism of muscle contraction based on formation and breaking of crossbridges between actin and myosin filaments, causing them to slide together.

Small intestine • The portion of the gut between the stomach and the colon, consisting of the duodenum, the jejunum, and the ileum.

Small nuclear ribonucleoprotein particle (snRNP) • A complex of an enzyme and a small nuclear RNA molecule, functioning in RNA splicing.

Smooth muscle • One of three types of muscle tissue. Usually consists of sheets of mononucleated cells innervated by the autonomic nervous system.

Society • A group of individuals belonging to the same species and organized in a cooperative manner; in the broadest sense, includes parents and their offspring.

Sodium–potassium pump • The complex protein in plasma membranes that is responsible for primary active transport; it pumps sodium ions out of the cell and potassium ions into the cell, both against their concentration gradients.

Solute • A substance that is dissolved in a liquid (solvent).

Solute potential • A property of any solution, resulting from its solute contents; it may be zero or have a negative value.

Solution • A liquid (solvent) and its dissolved solutes.

Solvent • A liquid that has dissolved or can dissolve one or more solutes.

Somatic [Gr. *soma*: body] • Pertaining to the body, or body cells (rather than to germ cells).

Somite (so' might) • One of the segments into which an embryo becomes divided longitudinally, leading to the eventual segmentation of the animal as illustrated by the spinal column, ribs, and associated muscles.

Spatial summation • In the production or inhibition of action potentials in a postsynaptic neuron, the interaction of depolarizations and hyperpolarizations produced by several terminal boutons.

Spawning • The direct release of sex cells into the water.

Speciation (spee′ shee ay′ shun) • The process of splitting one population into two populations that are reproductively isolated from one another.

Species (spee′ shees) [L.: kind] • The basic lower unit of classification, consisting of a population or series of populations of closely related and similar organisms. The more narrowly defined "biological species" consists of individuals capable of interbreeding freely with each other but not with members of other species.

Species diversity • A weighted representation of the species of organisms living in a region; large and common species are given greater weight than are small and rare ones. (Contrast with species richness.)

Species richness • The number of species of organisms living in a region. (Contrast with species diversity.)

Specific heat • The amount of energy that must be absorbed by a gram of a substance to raise its temperature by one degree centigrade. By convention, water is assigned a specific heat of one.

Sperm [Gr. *sperma*: seed] • A male reproductive cell.

Spermatocyte (spur mat′ oh site) [Gr. *sperma*: seed + *kytos*: cell] • The cell that gives rise to the sperm in animals.

Spermatogenesis (spur mat′ oh jen′ e sis) [Gr. *sperma*: seed + *genesis*: source] • Male gametogenesis, leading to the production of sperm.

Spermatogonia • Undifferentiated germ cells that give rise to primary spermatocytes and hence to sperm.

Sphincter (sfingk′ ter) [Gr. *sphinkter*: that which binds tight] • A ring of muscle that can close an orifice, for example at the anus.

Spindle apparatus • An array of microtubules stretching from pole to pole of a dividing nucleus and playing a role in the movement of chromosomes at nuclear division. Named for its shape.

Spiracle (spy′ rih kel) [L. *spirare*: to breathe] • An opening of the treacheal respiratory system of terrestrial arthorpods.

Spiteful act • A behavioral act that harms both the actor and the recipient of the act.

Spliceosome • An RNA–protein complex that splices out introns from eukaryotic pre-mRNAs.

Splicing • The removal of introns and connecting of exons in eukaryotic pre-mRNAs.

Spontaneous generation • The idea that life is generated continually from nonliving matter. Usually distinguished from the current idea that life evolved from nonliving matter under primordial conditions at an early stage in the history of earth.

Spontaneous reaction • A chemical reaction which will proceed on its own, without any outside influence. A spontaneous reaction need not be rapid.

Sporangium (spor an′ gee um) [Gr. *spora*: seed + *angeion*: vessel or reservoir] • In plants and fungi, any specialized stucture within which one or more spores are formed.

Spore [Gr. *spora*: seed] • Any asexual reproductive cell capable of developing into an adult plant without gametic fusion. Haploid spores develop into gametophytes, diploid spores into sporophytes. In prokaryotes, a resistant cell capable of surviving unfavorable periods.

Sporophyte (spor′ o fyte) [Gr. *spora*: seed + *phyton*: plant] • In plants with alternation of generations, the diploid phase that produces the spores. (Contrast with gametophyte.)

Stabilizing selection • Selection against the extreme phenotypes in a population, so that the intermediate types are favored. (Contrast with disruptive selection.)

Stamen (stay′ men) [L.: thread] • A male (pollen-producing) unit of a flower, usually composed of an anther, which bears the pollen, and a filament, which is a stalk supporting the anther.

Starch [O.E. *stearc*: stiff] • An α-linked polymer of glucose; used by plants as a means of storing energy and carbon atoms.

Start codon • The mRNA triplet (AUG) that acts as signals for the beginning of translation at the ribosome. (Compare with stop codons. There are a few mnior exceptions to these codons.)

Stasis • Period during which little or no evolutionary change takes place within a lineage or groups of lineages.

Statocyst (stat′ oh sist) [Gk. *statos*: stationary + *kystos*: pouch] • An organ of equilibrium in some invertebrates.

Statolith (stat′ oh lith) [Gk. *statos*: stationary + *lithos*: stone] • A solid object that responds to gravity or movement and stimulates the mechanoreceptors of a statocyst.

Stele (steel) [Gr. *stele*: pillar] • The central cylinder of vascular tissue in a plant stem.

Stem cell • A cell capable of extensive proliferation, generating more stem cells and a large clone of differentiated progeny cells, as in the formation of red blood cells.

Step cline • A sudden change in one or more traits of a species along a geographical gradient.

Steroid • Any of numerous lipids based on a 17-carbon atom ring system.

Sticky ends • On a piece of two-stranded DNA, short, complementary, one-stranded regions produced by the action of a restriction endonuclease. Sticky ends allow the joining of segments of DNA from different sources.

Stigma [L.: mark, brand] • The part of the pistil at the apex of the style, which is receptive to pollen, and on which pollen germinates.

Stimulus • Something causing a response; something in the environment detected by a receptor.

Stolon • A horizontal stem that forms roots at intervals.

Stoma (plural: stomata) [Gr. *stoma*: mouth, opening] • Small opening in the plant epidermis that permits gas exchange; bounded by a pair of guard cells whose osmotic status regulates the size of the opening.

Stop codons • Triplets (UAG, UGA, UAA) in mRNA that act as signals for the end of translation at the ribosome. (See also start codon. There are a few mnior exceptions to these codons.)

Stratosphere • The part of the atmosphere above the troposphere; extends upward to approximately 50 kilometers above the surface of the earth; contains very little water.

Stratum (plural strata) • A layer or sedimentary rock laid down at a particular time in a past.

Striated muscle • Contractile tissue characterized by multinucleated cells containing highly ordered arrangements of actin and myosin microfilaments. Also known as skeletal muscle.

Stroma • The fluid contents of an organelle, such as a chloroplast.

Stromatolite • A composite, flat-to-domed structure composed of successive mineral layers. Some are known to be produced by the action of bacteria in salt or fresh water, and some ancient ones are considered to be evidence for early life on the earth.

Structural formula • A representation of the positions of atoms and bonds in a molecule.

Structural gene • A gene that encodes the primary structure of a protein.

Style [Gr. *stylos*: pillar or column] • In flowering plants, a column of tissue extending from the tip of the ovary, and bearing the stigma or receptive surface for pollen at its apex.

Sub- [L.: under] • A prefix often used to designate a structure that lies beneath another or is less than another. For example, subcutaneous, subspecies.

Submucosa • (sub mew koe′ sah) • The tissue layer just under the epithelial lining of the lumen of the digestive tract. (Contrast with mucosa.)

Substrate (sub′ strayte) • (1) The molecule or molecules on which an enzyme exerts catalytic action. (2) The base material on which an organism lives.

Substrate level phosphorylation • ATP formation resulting from direct transfer of a phosphate group to ADP from an intermediate in glycolysis. (Contrast with oxidative phosphorylation.)

Succession • In ecology, the gradual, sequential series of changes in species composition of a community following a disturbance.

Sulcus (plural: sulci) [L. *sulcare*: to plow] • The valleys or creases between the raised portions of the convoluted surface of the brain. (Contrast to gyrus.)

Sulfhydryl group • The —SH group.

Summation • The ability of a neuron to fire action potentials in response to numerous subthreshold postsynaptic potentials arriving simultaneously at differentiated places on the cell, or arriving at the same site in rapid succession.

Surface area-to-volume ratio • For any cell, organism, or geometrical solid, the ratio of surface area to volume; this is an important factor in setting an upper limit on the size a cell or organism can attain.

Surfactant • A substance that decreases the surface tension of a liquid. Lung surfactant, secreted by cells of the alveoli, is mostly phospholipid and decreases the amount of work necessary to inflate the lungs.

Symbiosis (sim' bee oh' sis) [Gr.: to live together] • The living together of two or more species in a prolonged and intimate ecological relationship. (See parasitism, commensalism, mutualism.)

Symmetry • In biology, the property that two halves of an object are mirror images of each other. (See bilateral symmetry and biradial symmetry.)

Sympathetic nervous system • A division of the autonomic (involuntary) nervous system. Its activities include increasing blood pressure and acceleration of the heartbeat. The neurotransmitter at the sympathetic terminals is epinephrine or norepinephrine. (Contrast with parasympathetic nervous system.)

Sympatric speciation (sim pat' rik) [Gr. *sym*: same + *patria*: homeland] • The occurrence of genetic reproduction isolation and the subsequent formation of new species without any physical separation of the subpopulation. (Contrast with allopatric speciation, parapatric speciation.)

Symplast • The continuous meshwork of the interiors of living cells in the plant body, resulting from the presence of plasmodesmata. (Contrast with apoplast.)

Symport • A membrane transport process that carries two substances in the same direction across the membrane. (Contrast with antiport.)

Synapse (sin' aps) [Gr. *syn*: together + *haptein*: to fasten] • The narrow gap between the terminal bouton of one neutron and the dendrite or cell body of another.

Synapsis (sin ap' sis) • The highly specific parallel alignment (pairing) of homologous chromosomes during the first division of meiosis.

Synaptic vesicle • A membrane-bounded vesicle, containing neurotransmitter, which is produced in and discharged by the presynaptic neuron.

Syngamy (sing' guh mee) [Gr. *sun-*: together + *gamos*: marriage] • Union of gametes. Also known as fertilization.

Synonymous mutation • A mutation that substitutes one nucleotide for another but does not change the amino acid specified (i.e., UUA → UUG, both specifying leucine). (Compare with frame-shift mutation, missense mutation, nonsense mutation.)

Synonymous substitution • The situation when a synonymous mutation becomes widespread in a population. Typically not influenced by natural selection, these substitutions can accumulate in a population. (Contrast with nonsynonymous substitution.)

Systematics • The scientific study of the diversity of organisms.

Systemic circulation • The part of the circulatory system serving those parts of the body other than the lungs or gills.

Systemin • The only polypeptide plant hormone; participates in response to tissue damage.

Systole (sis' tuh lee) [Gr.: contraction] • Contraction of a chamber of the heart, driving blood forward in the circulatory system.

T cell • A type of lymphocyte, involved in the cellular immune response. The final stages of its development occur in the thymus gland. (Contrast with B cell; see also cytotoxic T cell, helper T cell, suppressor T cell.)

T cell receptor • A protein on the surface of a T cell that recognizes the antigenic determinant for which the cell is specific.

T tubules • A system of tubules that runs throughout the cytoplasm of muscle fibers, through which action potentials spread.

Target cell • A cell with the appropriate receptors to bind and respond to a particular hormone or other chemical mediator.

Taste bud • A structure in the epithelium of the tongue that includes a cluster of chemoreceptors innervated by sensory neurons.

TATA box • An eight-base-pair sequence, found about 25 base pairs before the starting point for transcription in many eukaryotic promoters, that binds a transcription factor and thus helps initiate transcription.

Taxis (tak' sis) [Gr. *taxis*: arrange, put in order] • The movement of an organism in a particular direction with reference to a stimulus. A taxis usually involves the employment of one sense and a movement directly toward or away from the stimulus, or else the maintenance of a constant angle to it. Thus a positive phototaxis is movement toward a light source, negative geotaxis is movement upward (away from gravity), and so on.

Taxon • A unit in a taxonomic system.

Taxonomy (taks on' oh me) [Gr. *taxis*: arrange, classify] • The science of classification of organisms.

Telomeres (tee' lo merz) [Gr. *telos*: end] • Repeated DNA sequences at the ends of eukaryotic chromosomes.

Telophase (tee' lo phase) [Gr. *telos*: end] • The final phase of mitosis or meiosis during which chromosomes became diffuse, nuclear envelopes reform, and nucleoli begin to reappear in the daughter nuclei.

Template • In biochemistry, a molecule or surface upon which another molecule is synthesized in complementary fashion, as in the replication of DNA. In the brain, a pattern that responds to a normal input but not to incorrect inputs.

Template strand • In a stretch of double-stranded DNA, the strand that is transcribed.

Temporal summation • In the production or inhibition of action potentials in a postsynaptic neuron, the interaction of depolarizations or hyperpolarizations produced by rapidly repeated stimulation of a single point.

Tendon • A collagen-containing band of tissue that connects a muscle with a bone.

Terrestrial (ter res' tree al) [L. *terra*: earth] • Pertaining to the land. (Contrast with aquatic, marine.)

Territory • A fixed area from which an animal or group of animals excludes other members of the same species by aggressive behavior or display.

Tertiary structure • In reference to a protein, the relative locations in three-dimensional space of all the atoms in the molecule. The overall shape of a protein. (Contrast with primary, secondary, and quaternary structures.)

Test cross • A cross of a dominant-phenotype individual (which may be either heterozygous or homozygous) with a homozygous-recessive individual.

Testis (tes' tis) (plural: testes) [L.: witness] • The male gonad; that is, the organ that produces the male sex cells.

Testosterone (tes toss' tuhr own) • A male sex steroid hormone.

Tetanus [Gr. *tetanos*: stretched] • (1) In physiology, a state of sustained, maximal muscular contraction caused by rapidly repeated stimulation. (2) In medicine, an often-fatal disease ("lockjaw") caused by the bacterium *Clostridium tetani*.

Thalamus • A region of the vertebrate forebrain; involved in integration of sensory input.

Thallus (thal' us) [Gr.: sprout] • Any algal body which is not differentiated into root, stem, and leaf.

Theory • An explanation or hypothesis that is supported by a wide body of evidence. (Contrast with hypothesis, paradigm.)

Thermoneutral zone • The range of temperatures over which an endotherm does not have to expend extra energy to thermoregulate.

Thermoreceptor • A cell or structure that responds to changes in temperature.

Thoracic cavity • The portion of the mammalian body cavity bounded by the ribs, shoulders, and diaphragm. Contains the heart and the lungs.

Thorax • In an insect, the middle region of the body, between the head and abdomen. In mammals, the part of the body between the neck and the diaphragm.

Thrombin • An enzyme that converts fibrinogen to fibrin, thus triggering the formation of blood clots.

Thrombus (throm' bus) [Gk. *thrombos*: clot] • A blood clot that forms within a blood vessel and remains attached to the wall of the vessel. (Contrast with embolus.)

Thylakoid • A flattened sac within a chloroplast. The membranes of the numerous thylakoids contain all of the chlorophyll in a plant, in addition to the electron carriers of photophosphorylation. Thylakoids stack to form grana.

Thymine • A nitrogen-containing base found in DNA.

Thymus • A ductless, glandular portion of the lymphoid system, involved in development of the immune system of vertebrates.

Thyroid [Gr. *thyreos*: door-shaped] • A two-lobed gland in vertebrates. Produces the hormone thyroxin.

Thyrotropic hormone • A hormone that is produced in the pituitary gland of amphibia such as frogs and transported in the bloodstream to the thyroid gland, inducing the thyroid gland to produce the thyroid hormone that regulates metamorphosis from tadpole to adult frog.

Tight junction • A junction between epithelial cells, in which there is no gap whatever between the adjacent cells. Materials may get through a tight junction only by entering the epithelial cells themselves.

Tissue • A group of similar cells organized into a functional unit and usually integrated with other tissues to form part of an organ such as a heart or leaf.

Tonus • A low level of muscular tension that is maintained even when the body is at rest.

Totipotency • In a cell, the condition of possessing all the genetic information and other capacities necessary to form an entire individual.

Toxigenicity [L. *toxicum*: poison] • The ability of a bacterium to produce chemical substances injurious to the tissues of the host organism.

Trachea (tray′ kee ah) [Gr. *trakhoia*: a small tube] • A tube that carries air to the bronchi of the lungs of vertebrates, or to the cells of arthropods.

Tracheid (tray′ kee id) • A distinctive conducting and supporting cell found in the xylem of nearly all vascular plants, characterized by tapering ends and walls that are pitted but not perforated.

Tracheophytes [Gr. *trakhoia*: a small tube + *phyton*: plant] • Those plants with xylem and phloem, including psilophytes, club mosses, horsetails, ferns, gymnosperms, and angiosperms. (Contrast with nontracheophytes.)

Trait • One form of a character: Eye color is a character; brown eyes and blue eyes are traits.

Transcription • The synthesis of RNA, using one strand of DNA as the template.

Transcription factors • Proteins that assemble on a eukaryotic chromosome, allowing RNA polymerase II to perform transcription.

Transduction • (1) Transfer of genes from one bacterium to another, with a bacterial virus acting as the carrier of the genes. (2) In sensory cells, the transformation of a stimulus (e.g., light energy, sound pressure waves, chemical or electrical stimulants) into action potentials.

Transfection • Uptake, incorporation, and expression of recombinant DNA.

Transfer cell • A modified parenchyma cell that transports solutes from its cytoplasm into its cell wall, thus moving the solutes from the symplast into the apoplast.

Transfer RNA (tRNA) • A category of relatively small RNA molecules (about 75 nucleotides). Each kind of transfer RNA is able to accept a particular activated amino acid from its specific activating enzyme, after which the amino acid is added to a growing polypeptide chain.

Transformation • Mechanism for transfer of genetic information in bacteria in which pure DNA extracted from bacteria of one genotype is taken in through the cell surface of bacteria of a different genotype and incorporated into the chromosome of the recipient cell.

Transgenic • Containing recombinant DNA incorporated into its genetic material.

Translation • The synthesis of a protein (polypeptide). This occurs on ribosomes, using the information encoded in messenger RNA.

Translocation • (1) In genetics, a rare mutational event that moves a portion of a chromosome to a new location, generally on a nonhomologous chromosome. (2) In vascular plants, movement of solutes in the phloem.

Transpiration [L. *spirare*: to breathe] • The evaporation of water from plant leaves and stem, driven by heat from the sun, and providing the motive force to raise water (plus ions) from the roots.

Transposable element • A segment of DNA that can move to, or give rise to copies at, another locus on the same or a different chromosome.

Triglyceride • A simple lipid in which three fatty acids are combined with one molecule of glycerol.

Triplet • See codon.

Triplet repeat • Occurrence of repeated triplet of bases in a gene, often leading to genetic disease, as does excessive repetition of CGG in the gene responsible for fragile-X syndrome.

Triploblastic • Having three cell layers. (Contrast with diploblastic.)

Trisomic • Containing three, rather than two members of a chromosome pair.

tRNA • See transfer RNA.

Trochophore (troke′ o fore) [Gr. *trochos*: wheel + *phoreus*: bearer] • The free-swimming larva of some annelids and mollusks, distinguished by a wheel-like band of cilia around the middle, and indicating an evolutionary relationship between these two groups.

Trophic level • A group of organisms united by obtaining their energy from the same part of the food web of a biological community.

Tropic hormones • Hormones of the anterior pituitary that control the secretion of hormones by other endocrine glands.

Tropism [Gr. *tropos*: to turn] • In plants, growth toward or away from a stimulus such as light (phototropism) or gravity (gravitropism).

Tropomyosin (troe poe my′ oh sin) • A protein that, along with actin, constitutes the thin filaments of myofibrils. It controls the interactions of actin and myosin necessary for muscle contraction.

Troposphere • The atmospheric zone reaching upward approximately 17 km in the tropics and subtropics but only to about 10 km at higher latitudes. The zone in which virtually all the water vapor in the atmosphere is located.

Trypsin • A protein-digesting enzyme. Secreted by the pancreas in its inactive form (trypsinogen), it becomes active in the duodenum of the small intestine.

T-tubules • A set of transverse tubes that penetrates skeletal muscle fibers and terminates in the sarcoplasmic reticulum. The T-system transmits impulses to the sacs, which then release Ca^{2+} to initiate muscle contraction.

Tube nucleus • In a pollen tube, the haploid nucleus that does not participate in double fertilization. (Contrast with generative nucleus.)

Tubulin • A protein that polymerizes to form microtubules.

Tumor • A disorganized mass of cells, often growing out of control. Malignant tumors spread to other parts of the body.

Tumor suppressor genes • Genes which, when homozygous mutant, result in cancer. Such genes code for protein products that inhibit cell proliferation.

Twitch • A single unit of muscle contraction.

Tympanic membrane [Gr. *tympanum*: drum] • The eardrum.

Umbilical cord • Tissue made up of embryonic membranes and blood vessels that connects the embryo to the placenta in eutherian mammals.

Understory • The aggregate of smaller plants growing beneath the canopy of dominant plants in a forest.

Unicellular (yoon′ e sell′ yer ler) [L. *unus*: one + *cella*: chamber] • Consisting of a single cell; as for example a unicellular organism. (Contrast with multicellular.)

Uniport • A membrane transport process that carries a single substance. (Contrast with antiport, symport.)

Unsaturated hydrocarbon • A compound containing only carbon and hydrogen atoms. One or more pairs of carbon atoms are connected by double bonds.

Upwelling • The upward movement of nutrient-rich, cooler water from deeper layers of the ocean.

Urea • A compound serving as the main excreted form of nitrogen by many animals, including mammals.

Ureotelic • Describes an organism in which the final product of the breakdown of nitrogen-containing compounds (primarily proteins) is urea. (Contrast with ammonotelic, uricotelic.)

Ureter (your′ uh tur) [Gr. *ouron*: urine] • A long duct leading from the vertebrate kidney to the urinary bladder or the cloaca.

Urethra (you ree′ thra) [Gr. *ouron*: urine] • In most mammals, the canal through which urine is discharged from the bladder and which serves as the genital duct in males.

Uric acid • A compound that serves as the main excreted form of nitrogen in some animals, particularly those which must conserve water, such as birds, insects, and reptiles.

Uricotelic • Describes an organism in which the final product of the breakdown of nitrogen-containing compounds (primarily proteins) is uric acid. (Contrast with ammonotelic, ureotelic.)

Urinary bladder • A structure structure that receives urine from the kidneys via the ureter, stores it, and expels it periodically through the urethra.

Urine (you′ rin) [Gk. *ouron*: urine] • In vertebrates, the fluid waste product containing the toxic nitrogenous by-products of protein and amino acid metabolism.

Uterus (yoo′ ter us) [L.: womb] • The uterus or womb is a specialized portion of the female reproductive tract in certain mammals. It receives the fertilized egg and nurtures the embryo in its early development.

Vaccination • Injection of virus or bacteria or their proteins into the body, to induce immunization. The injected material is usually attenuated (weakened) before injection.

Vacuole (vac′ yew ole) [Fr.: small vacuum] • A liquid-filled cavity in a cell, enclosed within a single membrane. Vacuoles play a wide variety of roles in cellular metabolism, some being digestive chambers, some storage chambers, some waste bins, and so forth.

Vagina (vuh jine′ uh) [L.: sheath] • In female mammals, the passage leading from the external genital orifice to the uterus; receives the copulatory organ of the male in mating.

van der Waals interaction • A weak attraction between atoms resulting from the interaction of the electrons of one atom with the nucleus of the other atom. This attraction is about one-fourth as strong as a hydrogen bond.

Variable regions • The part of an immunoglobulin molecule or T-cell receptor that includes the antigen-binding site.

Vascular (vas′ kew lar) • Pertaining to organs and tissues that conduct fluid, such as blood vessels in animals and phloem and xylem in plants.

Vascular bundle • In vascular plants, a strand of vascular tissue, including conducting cells of xylem and phloem as well as thick-walled fibers.

Vascular ray • In vascular plants, radially oriented sheets of cells produced by the vascular cambium, carrying materials laterally between the wood and the phloem.

Vascular tissue system • The conductive system of the plant, consisting primarily of xylem and phloem. (Contrast with dermal tissue system, ground tissue system.)

Vasopressin • See antidiuretic hormone.

Vector • (1) An agent, such as an insect, that carries a pathogen affecting another species. (2) A plasmid or virus that carries an inserted piece of DNA into a bacterium for cloning purposes in recombinant DNA technology.

Vegetal hemisphere • The lower portion of some animal eggs, zygotes, and embryos, in which the dense nutrient yolk settles. The **vegetal pole** refers to the very bottom of the egg or embyro. (Contrast with animal hemisphere.)

Vegetative • Nonreproductive, or nonflowering, or asexual.

Vein [L. *vena*: channel] • A blood vessel that returns blood to the heart. (Contrast with artery.)

Ventral [L. *venter*: belly, womb] • Toward or pertaining to the belly or lower side. (Contrast with dorsal.)

Ventricle • A muscular heart chamber that pumps blood through the body.

Vernalization [L. *vernalis*: belonging to spring] • Events occurring during a required chilling period, leading eventually to flowering.

Vertebral column • The jointed, dorsal column that is the primary support structure of vertebrates.

Vertebrate • An animal whose nerve cord is enclosed in a backbone of bony segments, called vertebrae. The principal groups of vertebrate animals are the fishes, amphibians, reptiles, birds, and mammals.

Vessel [L. *vasculum*: a small vessel] • In botany, a tube-shaped portion of the xylem consisting of hollow cells (vessel elements) placed end to end and connected by perforations. Together with tracheids, vessel elements conduct water and minerals in the plant.

Vestibular apparatus (ves tib′ yew lar) [L. *vestibulum*: an enclosed passage] • Structures associated with the vertebrate ear; these structures sense changes in position or momentum of the head, affecing balance and motor skills.

Vestigial (ves tij′ ee al) [L. *vestigium*: footprint, track] • The remains of body structures that are no longer of adaptive value to the organism and therefore are not maintained by selection.

Vicariance (vye care′ ee unce) [L. *vicus*: change] • The splitting of the range of a taxon by the imposition of some barrier to dispersal of its members.

Vicariant distribution • A distribution resulting from the disruption of a formerly continuous range by a vicariant event.

Villus (vil′ lus) (plural: villi) [L.: shaggy hair] • A hairlike projection from a membrane; for example, from many gut walls.

Virion (veer′ e on) • The virus particle, the minimum unit capable of infecting a cell.

Viroid (vye′ roid) • An infectious agent consisting of a single-stranded RNA molecule with no protein coat; produces diseases in plants.

Virus [L.: poison, slimy liquid] • Any of a group of ultramicroscopic infectious particles constructed of nucleic acid and protein (and, sometimes, lipid) that can reproduce only in living cells.

Visceral mass • The major internal organs of a mollusk.

Vitamin [L. *vita*: life] • Any one of several structurally unrelated organic compounds that an organism cannot synthesize itself, but nevertheless requires in small quantity for normal growth and metabolism.

Viviparous (vye vip′ uh rus) [L. *vivus*: alive] • Reproduction in which fertilization of the egg and development of the embryo occur inside the mother's body. (Contrast with oviparous.)

Waggle dance • The running movement of a working honey bee on the hive, during which the worker traces out a repeated figure eight. The dance contains elements that transmit to other bees the location of the food.

Water potential • In osmosis, the tendency for a system (a cell or solution) to take up water from pure water, through a differentially permeable membrane. Water flows toward the system with a more negative water potential. (Contrast with osmotic potential, turgor pressure.)

Water vascular system • The array of canals and tubelike appendages that serves as the circulatory system, locomotory system, and food-capturing system of many echinoderms; is in direct connection with the surrounding sea water.

Wavelength • The distance between successive peaks of a wave train, such as electromagnetic radiation.

Wild type • Geneticists' term for standard or reference type. Deviants from this standard, even if the deviants are found in the wild, are said to be mutant.

Xanthophyll (zan′ tho fill) [Gr. *xanthos*: yellowish-brown + *phyllon*: leaf] • A yellow or orange pigment commonly found as an accessory pigment in photosynthesis, but found elsewhere as well. An oxygen-containing carotenoid.

X-linked (also called sex-linked) • A character that is coded for by a gene on the X chromosome.

Xerophyte (zee′ row fyte) [Gr. *xerox*: dry + *phyton*: plant] • A plant adapted to an environment with a limited water supply.

Xylem (zy′ lum) [Gr. *xylon*: wood] • In vascular plants, the woody tissue that conducts water and minerals; xylem consists, in various plants, of tracheids, vessel elements, fibers, and other highly specialized cells.

Yeast artificial chromosome • A laboratory-made DNA molecule containing sequences of yeast chromosomes (origin of replication, telomeres, centromere, and selectable markers) so that it can be used as a vector in yeast.

Yolk • The stored food material in animal eggs, usually rich in protein and lipid.

Z-DNA • A form of DNA in which the molecule spirals to the left rather than to the right.

Zooplankton (zoe′ o plang ton) [Gr. *zoon*: animal + *planktos*: wandering] • The animal portion of the plankton.

Zoospore (zoe′ o spore) [Gr. *zoon*: animal + *spora*: seed] • In algae and fungi, any swimming spore. May be diploid or haploid.

Zygote (zye′ gote) [Gr. *zygotos*: yoked] • The cell created by the union of two gametes, in which the gamete nuclei are also fused. The earliest stage of the diploid generation.

Zymogen • An inactive precursor of a digestive enzyme secreted into the lumen of the gut, where a protease cleaves it to form the active enzyme.

Illustration Credits

Authors' Photograph
Christopher Small

Table of Contents Photographs
Golgi: © Biophoto Associates/Science Source/Photo Researchers, Inc.
Fibroblast: © Dennis Kunkel, U. Hawaii.
Plasmid: © A. B. Dowsett/Science Photo Library/Photo Researchers, Inc.
Protein: © James King-Holmes/OCMS/Science Photo Library/Photo Researchers, Inc.
Fossils: © Tom and Therisa Stack/Tom Stack and Assoc.
Starfish: © Darrell Gulin/DRK PHOTO.
Fireweed: © Stephen J. Krasemann/DRK PHOTO.
Blueberries: © Pat O'Hara/DRK PHOTO.
Fetus: © Science Pictures, Ltd. OSF/DRK PHOTO.
Polar bears: © Mike and Lisa Husar/DRK PHOTO.
Giraffes: © BIOS/Peter Arnold, Inc.
Poppies: © Larry Ulrich/DRK PHOTO.
Zebras: © Art Wolfe.

Part-Opener Photographs
Part 1: © K. R. Porter/Science Source/Photo Researchers, Inc.
Part 2: © Conly L. Rieder/BPS.*
Part 3: © Tui de Roy/Minden Pictures.
Part 4: © Dan Dempster/Dembinsky Photo Assoc.
Part 5: © Gerry Ellis/Minden Pictures.
Part 6: © Michael Fogden/DRK PHOTO.
Part 7: © Doug Perrine/DRK PHOTO.

Chapter 1 *Opener*: © Enric Marti/Associated Press. 1.3: © T. Stevens and P. McKinley/Photo Researchers, Inc. 1.4: © Dennis Kunkel, U. Hawaii. 1.5: © Science Pictures Limited/CORBIS. 1.6 *larva*: © Valorie Hodgson/Visuals Unlimited. 1.6 *pupa*: © Dick Poe/Visuals Unlimited. 1.6 *butterfly*: © Bill Beatty/Visuals Unlimited. 1.7: © K. and K. Amman/Planet Earth Pictures. 1.8*a*: © Staffan Widstrand. 1.8*b*: © Joe MacDonald/Tom Stack & Assoc. 1.8*c*: © Steve Kaufman/Peter Arnold, Inc. 1.8*d*: © Luiz C. Marigo/Peter Arnold, Inc. 1.11: J. S. Bleakney, courtesy of J. S. Boates. 1.12: © T. Leeson/Photo Researchers, Inc. 1.13: © J. S. Boates.

*BPS = Biological Photo Service

Chapter 2 *Opener*: NASA. 2.3 *left*: © SIU/Visuals Unlimited. 2.3 *right*: © SIU/Visuals Unlimited. 2.14: © Art Wolfe. 2.16: © P. Armstrong/Visuals Unlimited.

Chapter 3 *Opener*: © Dennis Kunkel, U. Hawaii. 3.6*a,b,c*; 3.7: © Dan Richardson. 3.14*c left*: © Biophoto Associates/Photo Researchers, Inc. 3.14*c middle*: © W. F. Schadel/BPS. 3.14*c right*: © CNRI, Science World Enterprises/BPS. 3.15 *upper*: © Robert Brons/BPS. 3.15 *lower*: © Peter J. Bryant/BPS. 3.18: © Dan Richardson.

Chapter 4 *Opener*: © Dennis Kunkel, U. Hawaii. 4.1: After N. Campbell, 1990. *Biology*, 2nd Ed., Benjamin Cummings Publishing Co. 4.3 *upper row*: © David M. Phillips/Visuals Unlimited. 4.3 *middle row, left*: © Conly L. Rieder/BPS. 4.3 *middle row, center*: David Albertini, Tufts U. School of Medicine. 4.3 *middle row, right*: © M. Abbey/Photo Researchers, Inc. 4.3 *bottom row, left*: © D. P. Evenson/BPS. 4.3 *bottom row, center*: © BPS. 4.3 *bottom row, right*: © L. Andrew Staehelin, U. Colorado. 4.4: © J. J. Cardamone Jr. & B. K. Pugashetti/BPS. 4.5: © Stanley C. Holt/BPS. 4.6*a*: © J. J. Cardamone Jr./BPS. 4.6*c*: S. Abraham & E. H. Beachey, VA Medical Center, Memphis, TN. 4.7 *centrioles*: © Barry F. King/BPS. 4.7 *mitochondrion*: © K. Porter, D. Fawcett/Visuals Unlimited. 4.7 *rough ER*: © Don Fawcett/Science Source/Photo Researchers, Inc. 4.7 *plasma membrane*: J. David Robertson, Duke U. Medical Center. 4.7 *peroxisome*: © E. H. Newcomb & S. E. Frederick/BPS. 4.7 *nucleolus*: © Richard Rodewald/BPS. 4.7 *golgi apparatus*: © L. Andrew Staehelin, U. Colorado. 4.7 *smooth ER*: © Don Fawcett, D. Friend/Science Source/Photo Researchers, Inc. 4.7 *ribosome*: From Boublik et al., 1990, *The Ribosome*, p. 177. Courtesy of American Society for Microbiology. 4.7 *cell wall*: © David M. Phillips/Visuals Unlimited. 4.7 *chloroplast*: © W. P. Wergin, courtesy of E. H. Newcomb/BPS. 4.8 *upper*: © Richard Rodewald/BPS. 4.8 *lower*: © Don W. Fawcett/Photo Researchers, Inc. 4.9*a*: © Barry King, U. California, Davis/BPS. 4.9*b*: © Biophoto Associates/Science Source/Photo Researchers, Inc. 4.10: From Aebi, U., et al., 1986. *Nature* 323:560–564. © Macmillan Publishers Ltd. 4.11: © Don Fawcett/Visuals Unlimited. 4.12: © L. Andrew Staehelin, U. Colorado. 4.13: © K. G. Murti/Visuals Unlimited. 4.14: © K. Porter, D. Fawcett/Visuals Unlimited. 4.15: © W. P. Wergin, courtesy of E. H. Newcomb/BPS. 4.16*a*: © John Durham/Science Photo Library/Photo Researchers, Inc. 4.16*b*: © Ed Reschke/Peter Arnold, Inc. 4.16*c*: © Chuck Davis/Tony Stone Images. 4.17*a*: © Richard Shiell/Animals Animals. 4.17*a inset*: © Richard Green/Photo Researchers, Inc. 4.17*b*: © G. Büttner/Naturbild/OKAPIA/Photo Researchers, Inc. 4.17*b inset*: © R. R. Dute. 4.19: © E. H. Newcomb & S. E. Frederick/BPS. 4.20: © M. C. Ledbetter, Brookhaven National Laboratory. 4.21: Courtesy of Vic Small, Austrian Academy of Sciences, Salzburg. 4.22: © Nancy Kedersha/Immunogen/Science Photo Library/Custom Medical Stock Photo. 4.23: © N. Hirokawa. 4.24: © W. L. Dentler/BPS. 4.25: B. J. Schnapp et al., 1985. *Cell* 40:455. Courtesy of B. J. Schnapp, R. D. Vale, M. P. Sheetz, and T. S. Reese. 4.26: © Barry F. King/BPS. 4.27: © David M. Phillips/Visuals Unlimited. 4.28 *left*: Courtesy of David Sadava. 4.28 *upper right*: © J. Gross, Biozentrum/Science Photo Library/Photo Researchers, Inc. 4.28 *lower right*: © From J. A. Buckwalter and L. Rosenberg, 1983. *Coll. Rel. Res.* 3:489. Courtesy of L. Rosenberg.

Chapter 5 *Opener*: © Reuters Newmedia Inc./CORBIS. 5.2: After L. Stryer, 1981. *Biochemistry*, 2nd Ed., W. H. Freeman. 5.3: © L. Andrew Staehelin, U. Colorado. 5.6*a*: © D. S. Friend, U. California, San Francisco. 5.6*b*: © Darcy E. Kelly, U. Washington. 5.6*c*: Courtesy of C. Peracchia. 5.15: Courtesy of J. Casley-Smith. 5.16: From M. M. Perry, 1979. *J. Cell Sci.* 39:26.

Chapter 6 *Opener*: © Geoff Tompkinson/Science Photo Library/Photo Researchers, Inc. 6.2: © Jonathan Scott/Masterfile. 6.7: © Jeff J. Daly/Visuals Unlimited. 6.15: © The Mona Group. 6.17: © Clive Freeman, The Royal Institution/Photo Researchers, Inc. 6.19: © The Mona Group.

Chapter 7 *Opener*: © Catherine Ursillo/Photo Researchers, Inc. 7.13: © Ephraim Racker/BPS.

Chapter 8 *Opener*: © C. S. Lobban/BPS. 8.1: © C. G. Van Dyke/Visuals Unlimited. 8.15: © Lawrence Berkeley National Laboratory. 8.18, 8.20: © E. H. Newcomb & S. E. Frederick/BPS. 8.21 *left*: © Arthur R. Hill/Visuals Unlimited. 8.21 *right*: © David Matherly/Visuals Unlimited.

Chapter 9 *Opener*: © Nancy Kedersha/Science Photo Library/Photo Researchers, Inc. 9.1*a,c*: © John D. Cunningham, Visuals, Unlimited. 9.1*b*: © David M. Phillips/Visuals Unlimited. 9.2: © Ruth Kavenoff, Designergenes Ltd., P.O. Box 100, Del Mar, CA 90214. 9.3*b*: © John J. Cardamone Jr./BPS. 9.6: © G. F. Bahr/BPS. 9.7 *upper inset*: © A. L. Olins/BPS. 9.7 *lower inset*: © Biophoto Associates/Science Source/Photo Researchers, Inc. 9.8: © Andrew S. Bajer, U. Oregon. 9.9*b*: © Conly L. Rieder/BPS. 9.10*a*: © T. E. Schroeder/BPS. 9.10*b*: © B. A. Palevitz, U. Wisconsin, courtesy of E. H. Newcomb/BPS. 9.11: © Gary T. Cole/BPS. 9.12*a*: © Andrew Syred/Science Photo Library/Photo Researchers, Inc. 9.12*b*: © E. Webber/Visuals Unlimited. 9.12*c*: © Bill Kamin/Visuals Unlimited. 9.13: © Dr. Thomas Ried and Dr. Evelin Schröck, NIH. 9.14: © C. A. Hasenkampf/BPS. 9.15: © Klaus W. Wolf, U. West Indies. 9.19*b*: © Gopal Murti/Photo Researchers, Inc.

Chapter 10 *Opener*: © David H. Wells/CORBIS. 10.2: © R. W. Van Norman/Visuals Unlimited. 10.12: Courtesy the American Netherland Dwarf Rabbit Club. 10.15: © NCI/Photo Researchers, Inc. 10.16: Courtesy of Pioneer Hi-Bred International, Inc. 10.17: After N. Campbell, 1990. *Biology*, 2nd Ed., Benjamin Cummings Publishing Co. 10.26: © Science VU/Visuals Unlimited. *Bay scallops*: © Barbara J. Miller/BPS.

Chapter 11 *Opener*: © From coordinates provided by N. Geacintov, NYU. 11.2: © Lee D. Simon/Photo Researchers, Inc. 11.4: Courtesy of Prof. M. H. F. Wilkins, Dept. of Biophysics, King's College, U. London. 11.6*a*: © A. Barrington Brown/Photo Researchers, Inc. 11.6*b*: © Dan Richardson.

Chapter 12 *Opener*: © David Wrobel/Visuals Unlimited. 12.7: © Dan Richardson. 12.13*b*: Courtesy of J. E. Edström and *EMBO J*. 12.17*a*: © Stanley Flegler/Visuals Unlimited. 12.17*b*: © Stanley Flegler/Visuals Unlimited.

Chapter 13 *Opener*: © Rosenfeld Images LTD/Photo Researchers, Inc. 13.1*a*: © Dennis Kunkel, U. Hawaii. 13.1*b*: © E.O.S./Gelderblom/Photo Researchers, Inc. 13.1*c*: © Dennis Kunkel, U. Hawaii. 13.8: Courtesy of L. Caro and R. Curtiss. 13.21: Based on an illustration by Anthony R. Kerlavage, Institute for Genomic Research. *Science* 269: 449–604 (1995).

Chapter 14 *Opener*: © Andrew Syred/Tony Stone images. 14.8: © Tiemeier et al., 1978. *Cell* 14:237–246. 14.18: Courtesy of Murray L. Barr, U. Western Ontario. 14.19: Courtesy of O. L. Miller, Jr.

Chapter 15 *Opener*: © Victoria Blackie/Tony Stone Images. 15.3 *inset*: © Biophoto Associates/Photo Researchers, Inc. 15.4: © From de Vos et al., 1992. *Science* 255: 306–312. 15.14: © Stephen A. Stricker, courtesy of Molecular Probes, Inc.

Chapter 16 *Opener*: © Yorgos Nikas, Karolinska Institute. 16.4: © Roddy Field, the Roslin Institute. 16.5: Courtesy of T. Wakayama and R. Yanagimachi. 16.9: J. E. Sulston and H. R. Horvitz, 1977. *Dev. Bio.* 56:100. 16.10*b*; 16.12 *left*: Courtesy of J. Bowman. 16.12 *right*: Courtesy of Detlef Weigel. 16.13: Courtesy of W. Driever and C. Nüsslein-Vollhard. 16.20: Courtesy of F. R. Turner, Indiana U.

Chapter 17 *Opener*: Courtesy of Nexia Biotechnologies, Inc. 17.2: © Philippe Plailly/Photo Researchers, Inc. 17.7: Pamela Silver and Jason A. Kahana, courtesy of Chroma Technology. 17.16 *left*: © Custom Medical Stock Photography. 17.16 *right*: Courtesy of Ingo Potrykus, Swiss Federal Institute of Technology. 17.18: © Bettmann/CORBIS.

Chapter 18 *Opener*: Willard Centerwall, from Lyman, F. L. (ed.), 1963. *Phenylketonuria*. Charles C. Thomas, Springfield, IL. 18.5: C. Harrison et al., 1983. *J. Med. Genet.* 20:280. 18.10: Courtesy of Harvey Levy and Cecelia Walraven, New England Newborn Screening Program. 18.13: © P. P. H. Debruyn and Yongock Cho, U. Chicago/BPS.

Chapter 19 *Opener*: © Francis G. Mayer/CORBIS. 19.4: © Dennis Kunkel, U. Hawaii. 19.10: © Dr. Gopal Murti/Science Photo Library/Photo Researchers, Inc. 19.15: A. Liepins, Sloan-Kettering Research Inst. 19.17: David Phillips/Science Source/Photo Researchers, Inc.

Chapter 20 *Opener*: © Robert Fried/Tom Stack & Assoc. 20.1: © Richard Coomber/Planet Earth Pictures. 20.5: © François Gohier/The National Audubon Society Collection/Photo Researchers, Inc. 20.6: © W. B. Saunders/BPS. 20.9 *left*: © Ken Lucas/BPS. 20.9 *right*: © Stanley M. Awramik/BPS. 20.10: © Chip Clark. 20.11: © Tom McHugh/Field Museum, Chicago/Photo Researchers, Inc. 20.12: © Chase Studios, Cedarcreek, MO. 20.14: Transparency no. 5800 (3), photo by D. Finnin, painting by Robert J. Barber. Courtesy the Library, American Museum of Natural History.

Chapter 21 *Opener*: © Toshiyuki Yoshino/Nature Production. 21.1: © Science Photo Library/Photo Researchers, Inc. 21.2: Levi, W. 1965. *Encyclopedia of Pigeon Breeds*. T. F. H. Publications, Jersey City, NJ. (*a,b*: photos by R. L. Kienlen, courtesy of Ralston Purina Company; *c,d*: photos by Stauber.). 21.9: © Frank S. Balthis. 21.11: © Lincoln Nutting/The National Audubon Society Collection/Photo Researchers, Inc. 21.13*a*: © C. Allan Morgan/Peter Arnold, Inc. 21.16: © Based on drawings produced by the Net-Spinner Web Program by Peter Fuchs and Thiemo Krink. 21.17: After D. Futuyma, 1987. *Evolutionary Biology*, 2nd Ed., Sinauer Associates, Inc. 21.20: Courtesy P. Brakefield and S. Carroll, from Brakefield et al., *Nature* 372:458–461. © Macmillan Publishers Ltd. 21.21*a*: © Marilyn Kazmers/Dembinsky Photo Assoc. 21.21*b*: © Randy Morse/Tom Stack and Assoc.

Chapter 22 *Opener*: © Patti Murray/Animals Animals. 22.1*a*: © Gary Meszaros/Dembinsky Photo Assoc. 22.1*b*: © Lior Rubin/Peter Arnold, Inc. 22.7*a*: © Virginia P. Weinland/Photo Researchers, Inc. 22.7*b*: © José Manuel Sánchez de Lorenzo Cáceres. 22.8: © Reed/Williams/Animals Animals. 22.10 *upper, lower*: © Peter J. Bryant/BPS. 22.10 *center*: © Kenneth Y. Kaneshiro, U. Hawaii. 22.13 *left*: © Peter K. Ziminsky/Visuals Unlimited. 22.13 *center*: © Elizabeth N. Orians. 22.13 *right*: © Noble Proctor/The National Audubon Society Collection/Photo Researchers, Inc.

Chapter 23 *Opener*: © Gary Brettnacher/Adventure Photo & Film. 23.3 *left*: © Adam Jones/Dembinsky Photo Assoc. 23.3 *right*: © Brian Parker/Tom Stack & Assoc. 23.10*a*: © Michael Giannechini/Photo Researchers, Inc. 23.10*b*: © Helen Carr/BPS. 23.10*c*: © Skip Moody/Dembinsky Photo Assoc.

Chapter 24 *Opener*: © John Reader/Science Photo Library/Photo Researchers, Inc. 24.2, 24.5: © Richard Alexander, U. Pennsylvania. 24.8: Courtesy of E. B. Lewis.

Chapter 25 *Opener*: © Mehau Kulyk/Science Photo Library/Photo Researchers, Inc. 25.1: © Stanley M. Awramik/BPS. 25.2: © Roger Ressmeyer/CORBIS. 25.5*a*: © Tom & Therisa Stack/Tom Stack & Assoc. 25.5*b*: © Gary Bell/Planet Earth Pictures.

Chapter 26 *Opener*: Photo by Ferran Garcia Pichel, from the cover of *Science* 284 (no. 5413). 26.1: © Kari Lounatmaa/Photo Researchers, Inc. 26.3*a*: © David Phillips/Photo Researchers, Inc. 26.3*b*: © R. Kessel-G. Shih/Visuals Unlimited. 26.3*c*: © Stanley Flegler/Visuals Unlimited. 26.4: © T. J. Beveridge/BPS. 26.5*a*: © J. A. Breznak and H. S. Pankratz/BPS. 26.5*b*: © J. Robert Waaland/BPS. 26.6: © George Musil/Visuals Unlimited. 26.7*a left*: © S. C. Holt/BPS. 26.7*a center*: © David M. Phillips/Visuals Unlimited. 26.7*b left*: © Leon J. LeBeau/BPS. 26.7*b center*: © A. J. J. Cardamone, Jr./BPS. 26.8: © Alfred Pasieka/Photo Researchers, Inc. 26.9: © Wolfgang Baumeister/Science Photo Library/Photo Researchers, Inc. 26.13: © Phil Gates, U. Durham/BPS. 26.14: © S. C. Holt/BPS. 26.15*a*: © Paul W. Johnson/BPS. 26.15*b*: © H. S. Pankratz/BPS. 26.15*c*: © Bill Kamin/Visuals Unlimited. 26.16: © Science VU/Visuals Unlimited. 26.17: © Randall C. Cutlip/BPS. 26.18: © T. J. Beveridge/BPS.

26.19: © G. W. Willis/BPS. 26.20: © Science VU/Visuals Unlimited. 26.21: © Michael Gabridge/Visuals Unlimited. 26.23: © Krafft/Hoa-qui/Photo Researchers, Inc. 26.24: © Martin G. Miller/Visuals Unlimited.

Chapter 27 *Opener*: © Mike Abbey/Visuals Unlimited. 27.1*a*: © David Phillips/Visuals Unlimited. 27.1*b*: © J. Paulin/Visuals Unlimited. 27.1*c*: © Randy Morse/Tom Stack & Assoc. 27.7*a*: © Christian Gautier/Jacana/Photo Researchers, Inc. 27.7*b*: © Cabisco/Visuals Unlimited. 27.7*c*: © Alex Rakosy/Dembinsky Photo Assoc. 27.8: © David M. Phillips/Visuals Unlimited. 27.11: © Oliver Meckes/Photo Researchers, Inc. 27.12: © Sanford Berry/Visuals Unlimited. 27.14*a*: © Mike Abbey/Visuals Unlimited. 27.14*b*: © Dennis Kunkel, U. Hawaii. 27.14*c,d*: © Paul W. Johnson/BPS. 27.15*b*: © M. A. Jakus, NIH. 27.18*a*: © Manfred Kage/Peter Arnold, Inc. 27.18*b*: © Biophoto Associates/Photo Researchers, Inc. 27.20*a*: © Joyce Photographics/The National Audubon Society Collection/Photo Researchers, Inc. 27.20*b*: © J. Robert Waaland/BPS. 27.21*a*: © Jeff Foott/Tom Stack & Assoc. 27.21*b*: © J. N. A. Lott/BPS. 27.23: © James W. Richardson/Visuals Unlimited. 27.24*a*: © Maria Schefter/BPS. 27.24*b*: © J. N. A. Lott/BPS. 27.25*a*: © Cabisco/Visuals Unlimited. 27.25*b*: © Andrew J. Martinez/Photo Researchers, Inc. 27.25*c*: © Alex Rakosy/Dembinsky Photo Assoc. 27.31*a*: © Robert Brons/BPS. 27.31*b*: © A. M. Siegelman/Visuals Unlimited. 27.32*a*: © Barbara J. Miller/BPS. 27.32*b*: © Cabisco/Visuals Unlimited. 27.33*a*: © D. W. Francis, U. Delaware. 27.33*b*: © David Scharf/Peter Arnold, Inc.

Chapter 28 *Opener*: © Fred Bruemmer/DRK PHOTO. 28.1*a*: © Ron Dengler/Visuals Unlimited. 28.1*b*: © Larry Mellichamp/Visuals Unlimited. 28.4*a,b*: © J. Robert Waaland/BPS. 28.5*a*: © Rod Planck/Dembinsky Photo Assoc. 28.5*b*: © William Harlow/Photo Researchers, Inc. 28.5*c*: © Science VU/Visuals Unlimited. 28.6: © Dr. David Webb, U. Hawaii. 28.7*a*: © Brian Enting/Photo Researchers, Inc. 28.7*b*: © J. H. Troughton. 28.9: Figure information provided by Hermann Pfefferkorn, Dept. of Geology, U. Pennsylvania. Original oil painting by John Woolsey. 28.14*a*: © Ed Reschke/Peter Arnold, Inc. 28.14*b*: © Cabisco/Visuals Unlimited. 28.15*a*: © J. N. A. Lott/BPS. 28.15*b*: © David Sieren/Visuals Unlimited. 28.16: © W. Ormerod/Visuals Unlimited. 28.17*a*: © Rod Planck/Dembinsky Photo Assoc. 28.17*b*: © Nuridsany et Perennou/Photo Researchers, Inc. 28.17*c*: © Dick Keen/Visuals Unlimited. 28.18: © L. West/Photo Researchers, Inc.

Chapter 29 *Opener*: © Marty Cordano/DRK PHOTO. 29.3: © Phil Gates/BPS. 29.4*a*: © Roland Seitre/Peter Arnold, Inc. 29.4*b*: © Bernd Wittich/Visuals Unlimited. 29.4*c*: © M. Graybill/J. Hodder/BPS. 29.4*d*: © Louisa Preston/Photo Researchers, Inc. 29.7*a*: © Dick Poe/Visuals Unlimited. 29.7*b*: © Richard Shiell. 29.7*c*: © Richard Shiell/Dembinsky Photo Assoc. 29.8*a*: © Richard Shiell. 29.8*b*: © Noboru Komine/Photo Researchers, Inc. 29.11*a*: © Inga Spence/Tom Stack & Assoc. 29.11*b*: © Holt Studios/Photo Researchers, Inc. 29.11*c*: © Catherine M. Pringle/BPS. 29.11*d*: © Inga Spence/Tom Stack & Assoc. 29.12: © U. California, Santa Cruz, and UCSC Arboretum. 29.12 *inset*: © Sandra K. Floyd, U. Colorado. 29.14*a*: © Ken Lucas/Visuals Unlimited. 29.14*b*: © Ed Reschke/Peter Arnold, Inc. 29.14*c*: © Adam Jones/Dembinsky Photo Assoc. 29.15*a*: © Richard Shiell. 29.15*b*: © Adam Jones/Dembinsky Photo Assoc. 29.15*c*: © Alan & Linda Detrick/The National Audubon Society Collection/Photo Researchers, Inc.

Chapter 30 *Opener*: © S. Nielsen/DRK PHOTO. 30.1*a*: © Inga Spence/Tom Stack & Assoc. 30.1*b*: © L. E. Gilbert/BPS. 30.1*c*: © G. L. Barron/BPS. 30.2: © David M. Phillips/Visuals Unlimited. 30.4: © G. T. Cole/BPS. 30.5: © N. Allin and G. L. Barron/BPS. 30.7: © J. Robert Waaland/BPS. 30.8: © Gary R. Robinson/Visuals Unlimited. 30.9: © Tom Stack/Tom Stack & Assoc. 30.10: © John D. Cunningham/Visuals Unlimited. 30.11*a*: © Richard Shiell/Dembinsky Photo Assoc. 30.11*b*: © Matt Meadows/Peter Arnold, Inc. 30.12: © Andrew Syred/Science Photo Library/Photo Researchers, Inc. 30.14*a*: © Angelina Lax/Photo Researchers, Inc. 30.14*b*: © Manfred Danegger/Photo Researchers, Inc. 30.14*c*: © Stan Flegler/Visuals Unlimited. 30.15 *inset*: © Biophoto Associates/Photo Researchers, Inc. 30.16*a*: © R. L. Peterson/BPS. 30.16*b*: © Merton F. Brown/Visuals Unlimited. 30.17*a*: © Ed Reschke/Peter Arnold, Inc. 30.17*b*: © Gary Meszaros/Dembinsky Photo Assoc. 30.18*a*: © J. N. A. Lott/BPS.

Chapter 31 *Opener*: © Paolo Curto/The Image Bank. 31.5*a*: © Don Fawcett/Visuals Unlimited. 31.5*b*: © Christian Petron/Planet Earth Pictures. 31.5*c*: © Gillian Lythgoe/Planet Earth Pictures. 31.6*a*: © Robert Brons/BPS. 31.6*b*: © Tom & Therisa Stack/Tom Stack & Assoc. 31.6*c*: © Randy Morse/Tom Stack & Assoc. 31.7, 31.8, 31.9, 31.10: Adapted from Bayerand, F. M., and H. B. Owre, 1968. *The Free-Living Lower Invertebrates*, Macmillan Publishing Co. 31.11*a*: © G. Carleton Ray/Photo Researchers, Inc. 31.11*b*: © Fred Bavendam/Minden Pictures. 31.12: © David J. Wrobel/BPS. 31.13: From M. W. Martin, 2000. *Science* 288:841–845. 31.15*a*: © Fred McConnaughey/Photo Researchers, Inc. 31.17*b*: © James Solliday/BPS. 31.20*a*: © Chamberlain, MC/DRK PHOTO. 31.21: © David J. Wrobel/BPS. 31.22: © Jeff Mondragon. 31.24*a*: © Brian Parker/Tom Stack & Assoc. 31.24*b*: © Roger K. Burnard/BPS. 31.24*c*: © Stanley Breeden/DRK PHOTO. 31.24*d*: © R. R. Hessler, Scripps Institute of Oceanography. 31.26*a*: © Ken Lucas/Planet Earth Pictures. 31.26*b*: © Dave Fleetham/Tom Stack & Assoc. 31.26*c*: © Mike Severns/Tom Stack & Assoc. 31.26*d*: © Milton Rand/Tom Stack & Assoc. 31.26*e*: © Dave Fleetham/Tom Stack & Assoc. 31.26*f*: © A. Kerstitch/Visuals Unlimited.

Chapter 32 *Opener*: © John Mitchell/The National Audubon Society Collection/Photo Researchers, Inc. 32.2: © Dr. Rick Hochberg, U. New Hampshire. 32.4: © R. Calentine/Visuals Unlimited. 32.5*b,c*: © James Solliday/BPS. 32.7*a*: © Doug Wechsler. 32.7*b*: © Diane R. Nelson/Visuals Unlimited. 32.8: © Ken Lucas/Visuals Unlimited. 32.9*a*: © Joel Simon. 32.9*b*: © Fred Bruemmer/DRK PHOTO. 32.10*a*: © Peter J. Bryant/BPS. 32.10*b*: © David Maitland/Masterfile. 32.10*c*: © W. M. Beatty/Visuals Unlimited. 32.10*d*: © Robert Brons/BPS. 32.11*a*: © Henry W. Robison/Visuals Unlimited. 32.11*b*: © Stephen P. Hopkin/Planet Earth Pictures. 32.11*c*: © Peter David/Planet Earth Pictures. 32.11*d*: © A. Flowers & L. Newman/The National Audubon Society Collection/Photo Researchers, Inc. 32.13*a*: © Charles R. Wyttenbach/BPS. 32.13*b*: © William Leonard/DRK PHOTO. 32.15*a*: © David P. Maitland/Planet Earth Pictures. 32.15*b*: © Konrad Wothe/Minden Pictures. 32.15*c*: © Peter J. Bryant/BPS. 32.15*d*: © David Maitland/Masterfile. 32.15*e*: © Steve Nicholls/Planet Earth Pictures. 32.15*f*: © Brian Kenney/Planet Earth Pictures. 32.15*g*: © Simon D. Pollard/The National Audubon Society Collection/Photo Researchers, Inc. 32.15*h*: © L. West/The National Audubon Society Collection/Photo Researchers, Inc.

Chapter 33 *Opener*: © Norbert Wu/DRK PHOTO. 33.3*a*: © Hal Beral/Visuals Unlimited. 33.3*b*: © Randy Morse/Tom Stack & Assoc. 33.3*c*: © Mark J. Thomas/Dembinsky Photo Assoc. 33.3*d*: © Randy Morse/Tom Stack & Assoc. 33.3*e*: © John A. Anderson/Animals Animals. 33.4: © C. R. Wyttenbach/BPS. 33.5: © Gary Bell/Masterfile. 33.6*b*, 33.9: © Norbert Wu/DRK PHOTO. 33.11*a*: © Dave Fleetham/Tom Stack & Assoc. 33.11*b*: © Marty Snyderman/Masterfile. 33.12*a*: © Ken Lucas/Planet Earth Pictures. 33.12*b*: © Fred Bavendam/Minden Pictures. 33.12*c*: © Dave Fleetham/Visuals Unlimited. 33.12*d*: © Dr. Paul A. Zahl/The National Audubon Society Collection/Photo Researchers, Inc. 33.13: © Tom McHugh, Steinhart Aquarium/The National Audubon Society Collection/Photo Researchers, Inc. 33.15*a*: © Ken Lucas/BPS. 33.15*b*: © Nick Garbutt/Indri Images. 33.15*c*: © Art Wolfe. 33.19*a*: © Michael Fogden/DRK PHOTO. 33.19*b*: © Joe McDonald/Tom Stack & Assoc. 33.19*c*: © C. Alan Morgan/Peter Arnold, Inc. 33.19*d*: © Dave B. Fleetham/Tom Stack & Assoc. 33.19*e*: © Mark J. Thomas/Dembinsky Photo Assoc. 33.20*a*: Courtesy of Carnegie Museum of Natural History, Pittsburgh. 33.20*b*: Fossil from the Natural History Museum of Basel, photographed by Severino Dahint. 33.21*a*: © Joe McDonald/Tom Stack & Assoc. 33.21*b*: © John Shaw/Tom Stack & Assoc. 33.21*c*: © Skip Moody/Dembinsky

Photo Assoc. 33.22*a*: © Ed Kanze/Dembinsky Photo Assoc. 33.22*b*: © Dave Watts/Tom Stack & Assoc. 33.23*a*: © Art Wolfe. 33.23*b*: © Jany Sauvanet/Photo Researchers, Inc. 33.23*c*: © Hans & Judy Beste/Animals Animals. 33.24*a*: © Rod Planck/Dembinsky Photo Assoc. 33.24*b*: © Joe McDonald/Tom Stack & Assoc. 33.24*c*: © Doug Perrine/Planet Earth Pictures. 33.24*d*: © Erwin & Peggy Bauer/Tom Stack & Assoc. 33.26*a*: © Art Wolfe. 33.26*b*: © Gary Milburn/Tom Stack & Assoc. 33.27*a*: © Steve Kaufman/DRK PHOTO. 33.27*b*: © John Bracegirdle/Masterfile. 33.28*a*: © Art Wolfe. 33.28*b*: © Anup Shah/Dembinsky Photo Assoc. 33.28*c*: © Anup Shah/Dembinsky Photo Assoc. 33.28*d*: © Stan Osolinsky/Dembinsky Photo Assoc. 33.31*a*: © Dembinsky Photo Assoc. 33.31*b*: © Tim Davis/Photo Researchers, Inc. 33.31*c*: © John Downer/Planet Earth Pictures.

Chapter 34 *Opener*: © D. Cavagnaro/Visuals Unlimited. 34.3*a*: © Jan Tove Johansson/Planet Earth Pictures. 34.3*b*: © R. Calentine/Visuals Unlimited. 34.4*a*: © Joyce Photographics/Photo Researchers, Inc. 34.4*b*: © Renee Lynn/Photo Researchers, Inc. 34.4*c*: © C. K. Lorenz/The National Audubon Society Collection/Photo Researchers, Inc. 34.7: © Biophoto Associates/Photo Researchers, Inc. 34.9*a,b*: © Phil Gates, U. Durham/BPS. 34.9*c*: © Biophoto Associates/Photo Researchers, Inc. 34.9*d*: © Jack M. Bostrack/Visuals Unlimited. 34.9*e*: © John D. Cunningham/Visuals Unlimited. 34.9*f*: © J. Robert Waaland/BPS. 34.11*b*, 34.14: © J. Robert Waaland/BPS. 34.16*a*: © Jim Solliday/BPS. 34.16*b*: © Microfield Scientific LTD/Photo Researchers, Inc. 34.16*c*: © Ray F. Evert, U. Wisconsin, Madison. 34.16*d*: © John D. Cunningham/Visuals Unlimited. 34.18*a left*: © Cabisco/Visuals Unlimited. 34.18*a right*: © J. Robert Waaland/BPS. 34.18*b left*: © Cabisco/Visuals Unlimited. 34.18*b right*: © J. Robert Waaland/BPS. 34.20: © J. N. A. Lott/BPS. 34.21: © Jim Solliday/BPS. 34.22: © Phil Gates, U. Durham/BPS. 34.23*b*: © Thomas Eisner, Cornell U. 34.23*c*: © C. G. Van Dyke/Visuals Unlimited.

Chapter 35 *Opener*: © Patti Murray/Animals Animals. 35.5: Brentwood, B., and J. Cronshaw, 1978. *Planta* 140:111–120. 35.6: © Ed Reschke/Peter Arnold, Inc. 35.9*a*: © David M. Phillips/Visuals Unlimited. 35.13: © M. H. Zimmermann.

Chapter 36 *Opener*: © J. H. Robinson/The National Audubon Society Collection/Photo Researchers, Inc. 36.1: © Inga Spence/Tom Stack & Assoc. 36.4: © Kathleen Blanchard/Visuals Unlimited. 36.6: © Hugh Spencer/Photo Researchers, Inc. 36.8: © E. H. Newcomb and S. R. Tandon/BPS. 36.10: © Gilbert S. Grant/Photo Researchers, Inc. 36.11: © Milton Rand/Tom Stack & Assoc.

Chapter 37 *Opener*: © Jeremy Woodhouse/DRK PHOTO. 37.4: © Tom J. Ulrich/Visuals Unlimited. 37.5: © John Eastcott, Yva Momatiuk/DRK PHOTO. 37.6: © J. N. A. Lott/BPS. 37.8: © J. A. D. Zeevaart, Michigan State U. 37.13: © Ed Reschke/Peter Arnold, Inc. 37.16*a*: © Biophoto Associates/Photo Researchers, Inc. 37.19: © T. A. Wiewandt/DRK PHOTO. 37.22: Dr. Eva Huala, Carnegie Institution of Washington.

Chapter 38 *Opener*: © C. C. Lockwood/Animals Animals. 38.1 *lower*: © J. R. Waaland/BPS. 38.1 *upper*: © Jim Solliday/BPS. 38.2: © Oliver Meckes/Science Source/Photo Researchers, Inc. 38.3: © Stephen Dalton/The National Audubon Society Collection/Photo Researchers, Inc. 38.5: © Bowman, J. (ed.), 1994. *Arabiopsis: An Atlas of Morphology and Development*. Springer-Verlag, New York. Photo by S. Craig & A. Chaudhury. 38.9*a*: © C. P. George/Visuals Unlimited. 38.9*b*: © Tess & David Young/Tom Stack & Assoc. 38.17*a*: © Nigel Cattlin, Holt Studios International/Photo Researchers, Inc. 38.17*b*: © Jerome Wexler/The National Audubon Society Collection/Photo Researchers, Inc.

Chapter 39 *Opener*: Agricultural Research Service, USDA. 39.2: © D. Cavagnaro/Visuals Unlimited. 39.4: © Stan Osolinski/Dembinsky Photo Assoc. 39.7: © Thomas Eisner, Cornell U. 39.8: © Adam Jones/Dembinsky Photo Assoc. 39.9: © J. N. A. Lott/BPS. 39.10, 39.11: © Richard Shiell. 39.12: © Janine Pestel/Visuals Unlimited. 39.13: © Chip Isenhart/Tom Stack & Assoc. 39.14: © J. N. A. Lott/BPS. 39.15: © Robert & Linda Mitchell. 39.16: © Budd Titlow/Visuals Unlimited.

Chapter 40 *Opener*: © S. Asad/Peter Arnold, Inc. 40.3*a,b*: © Biophoto Associates/Science Source/Photo Researchers, Inc. 40.3*c*: © G. W. Willis/BPS. 40.4*a*: © Cabisco/Visuals Unlimited. 40.4*b*: © Biophoto Associates/Science Source/Photo Researchers, Inc. 40.4*c*: © Cabisco/Visuals Unlimited. 40.4*d*: © David M. Phillips/Visuals Unlimited. 40.10*a*: © B. & C. Alexander/Photo Researchers, Inc. 40.10*b*: © Timothy Ransom/BPS. 40.12: © Auscape (Parer-Cook)/Peter Arnold, Inc. 40.16: © G. W. Willis/BPS. 40.17*a*: © Stephen J. Kraseman/DRK PHOTO. 40.17*b*: © Jim Roetzel/Dembinsky Photo Assoc.

Chapter 41 *Opener*: © R. D. Fernald, Stanford U. 41.6*a*: © Associated Press Photo. 41.6*b*: © Bettman/CORBIS. 41.14*a*: Courtesy of Gerhard Heldmaier, Philipps University.

Chapter 42 *Opener*: © Nik Wheeler. 42.1*a*: © Biophoto Associates/Photo Researchers, Inc. 42.1*b*: © Brian Parker/Tom Stack & Assoc. 42.1*c*: © Thomas Eisner, Cornell U. 42.2: © Patricia J. Wynne. 42.3: © David M. Phillips/Science Source/Photo Researchers, Inc. 42.5: © Fred Bavendam/Minden Pictures. 42.6: © David T. Roberts, Nature's Images/The National Audubon Society Collection/Photo Researchers, Inc. 42.7*a*: © Mitsuaki Iwago/Minden Pictures. 42.7*b*: © Johnny Johnson/DRK PHOTO. 42.12 *inset*: © P. Bagavandoss/Photo Researchers, Inc. 42.16: © CC Studio/Photo Researchers, Inc.

Chapter 43 *Opener*: © Dave B. Fleetham/Tom Stack & Assoc. 43.5 *inset*: Courtesy of Richard Elinson, U. Toronto. 43.24*a*: © C. Eldeman/Photo Researchers, Inc. 43.24*b*: © Nestle/Photo Researchers, Inc. 43.26: © S. I. U. School of Med./Photo Researchers, Inc.

Chapter 44 *Opener*: © Associated Press Photo. 44.4: © C. Raines/Visuals Unlimited.

Chapter 45 *Opener*: Courtesy of Grace Sours, ATF. 45.4 *left*: © R. A. Steinbrecht. 45.4 *right*: © G. I. Bernard/Animals Animals. 45.6, 45.12: © P. Motta/Photo Researchers, Inc. 45.15*b*: © S. Fisher, U. California, Santa Barbara. 45.19*a*: © Dennis Kunkel, U. Hawaii. 45.22: © Omikron/Science Source/Photo Researchers, Inc. 45.26: © Joe McDonald/Tom Stack & Assoc.

Chapter 46 *Opener*: From Harlow, J. M., 1869. *Recovery from the passage of an iron bar through the head*. Boston: David Clapp & Son. 46.14: David Joel, courtesy of Bio-logic Systems Corp. 46.16: © Wellcome Dept. of Cognitive Neurology/Science Photo Library/Photo Researchers, Inc.

Chapter 47 *Opener*: © AFP/CORBIS. 47.2: © P. Motta/Photo Researchers, Inc. 47.5 *upper*: © CNRI/Photo Researchers, Inc. 47.5 *center*: © G. W. Willis/BPS. 47.5 *lower*: © Michael Abbey/Photo Researchers, Inc. 47.7: © Frank A. Pepe/BPS. 47.12: Courtesy of Jesper L. Andersen. 47.14: © Skip Moody/Dembinsky Photo Assoc. 47.18*a*: © G. Mili. 47.18*b*: © Robert Brons/BPS. 47.22*a*: © Ken Lucas/Visuals Unlimited. 47.22*b*: © Fred McConnaughey/The National Audubon Society Collection/Photo Researchers, Inc.

Chapter 48 *Opener*: © Darrell Gulin/Tony Stone Images. 48.1*a*: © Ed Robinson/Tom Stack & Assoc. 48.1*b*: © Robert Brons/BPS. 48.1*c*: © Tom McHugh/Photo Researchers, Inc. 48.3: © Eric Reynolds/Adventure Photo. 48.5*b*: © Skip Moody/Dembinsky Photo Assoc. 48.5*c*: © Thomas Eisner, Cornell U. 48.9: © Walt Tyler, U. California, Davis. 48.12 *left inset*: © Science Photo Library/Photo Researchers, Inc. 48.12 *right inset*: © P. Motta/Photo Researchers, Inc. 48.15: © Fred Bruemmer/DRK PHOTO.

Chapter 49 *Opener*: © Norbert Wu/DRK PHOTO. 49.9: © Geoff Tompkinson/Photo Researchers, Inc. 49.11: © Dennis Kunkel, U. Hawaii. 49.14*a*: © Chuck Brown/Science Source/Photo Researchers, Inc. 49.14*b*: © Biophoto Associates/Science Source/Photo Researchers, Inc. 49.15: After N. Campbell, 1990. *Biology*, 2nd Ed., Benjamin Cummings Publishing Co. 49.16*a*: © NYU Franklin Research Fund/Phototake. 49.17*b*: © CNRI/Photo Researchers, Inc.

Chapter 50 *Opener*: © Bettmann/CORBIS. 50.1*a*: © Mike Barlow/Dembinsky Photo Assoc. 50.1*b*: © Jim Battles/Dembinsky Photo Assoc. 50.1*c*: © James Watt/Animals Animals. 50.1*d*: © Tom Walker/Visuals Unlimited. 50.6: © Ken Greer/Visuals Unlimited. 50.7: © Carl Purcell/Photo Researchers, Inc. 50.8*a*: © David Roberts/Nature's Images/Photo Researchers, Inc. 50.8*b*: © Gary Milburn/Tom Stack & Assoc. 50.11: © Dennis Kunkel, U. Hawaii. 50.22: © Jackson/Visuals Unlimited. 50.23 *left*: © Richard Shiell/Animals Animals. 50.23 *right*: © Katsutoshi Ito/Nature Productions. 50.24: © L. Kiff/Visuals Unlimited.

Chapter 51 *Opener*: © Michael Fogden/DRK PHOTO. 51.1*a*: © Brian Kenney/Planet Earth Pictures. 51.2*b*: © Rod Planck/Photo Researchers, Inc. 51.8: From Kessel, R. G., and R. H. Kardon, 1979. *Tissues and Organs*. W. H. Freeman, San Francisco. 51.9: © John Cancalosi/DRK PHOTO. 51.13: © Lise Bankir, INSERM Unit, Hôpital Necker, Paris.

Chapter 52 *Opener*: © Frans de Waal, Emory U. 52.1: © Bill Beatty/Visuals Unlimited. 52.4: © Marc Chappell, U. California, Riverside. 52.6: © Nina Leen/TimePix. 52.13: © François Savigny/Animals Animals. 52.19: © Fritz Polking/Dembinsky Photo Assoc. 52.25: © Jonathan Blair/Woodfin, Camp & Assoc.

Chapter 53 *Opener*: © Anup and Monoj Shah/A Perfect Exposure. 53.4*b*: © Dominique Braud/Tom Stack & Assoc. 53.5: © John Alcock, Arizona State U. 53.6: Courtesy of Arild Johnsen, University of Oslo. 53.7: © Anup Shah/Planet Earth Pictures. 53.10: © Art Wolfe. 53.11: © D. Houston/Bruce Coleman, Inc. 53.12: © John Gerlach/Dembinsky Photo Assoc. 53.13: © Andrew J. Martinez/The National Audubon Society Collection/Photo Researchers, Inc. 53.14: © Art Wolfe. 53.16: © Jonathan Scott/Planet Earth Pictures.

Chapter 54 *Opener*: © Pat Anderson/Visuals Unlimited. 54.1*a*: © Breck P. Kent/Animals Animals. 54.1*b*: © John Gerlach/Visuals Unlimited. 54.2: © Ted Mead/Woodfall Wild Images. 54.9*a*: © Frans Lanting/Minden Pictures. 54.9*b*: © Art Wolfe. 54.10: © Adam Jones/Dembinsky Photo Assoc. 54.11*a*: © Art Wolfe/The National Audubon Society Collection/Photo Researchers, Inc. 54.11*b*: © Frans Lanting/Photo Researchers, Inc. 54.15: © John R. Hosking, NSW Agriculture, Australia.

Chapter 55 *Opener*: © Maximilian Stock Ltd./Science Photo Library/Photo Researchers, Inc. 55.5: © Tom Brakefield/DRK PHOTO. 55.8*a*: © Thomas Eisner and Daniel Aneshansley, Cornell U. 55.8*b*: © David J. Wrobel/BPS. 55.9, 55.10: © Lawrence E. Gilbert/BPS. 55.11: © Callanan/Visuals Unlimited. 55.12: © Stephen G. Maka/DRK PHOTO. 55.13: © Patti Murray/Animals Animals. 55.14: © Daniel Janzen, U. Pennsylvania. 55.15*a*: © Merlin D. Tuttle/Bat Conservation International/Photo Researchers, Inc. 55.15*b*: © Roger Wilmshurst/Photo Researchers, Inc. 55.16: © Dr. Edward S. Ross, California Academy of Sciences. 55.17*a*: © Darrell Gulin/Dembinsky Photo Assoc. 55.17*b*: © Gilbert Grant/Photo Researchers, Inc. 55.18: © Jan Tove Johansson/Planet Earth Pictures. 55.19: © Jeff Mondragon. 55.21: After Begon, M., J. Harper, and C. Townsend, 1986. *Ecology*, Blackwell Scientific Publications.

Chapter 56 *Opener*: © Tom Bean/DRK PHOTO. 56.9*a*: © Joe McDonald/Tom Stack & Assoc. 56.9*b*: © Inga Spence/Tom Stack & Assoc.

Chapter 57 *Opener*: © Dave Watts/Tom Stack & Assoc. 57.3: © Elizabeth N. Orians. 57.7: Courtesy of E. O. Wilson. *Tundra, upper*: © Tom & Pat Leeson/DRK PHOTO. *Tundra, lower*: © Elizabeth N. Orians. *Boreal forest, upper*: © Carr Clifton/Minden Pictures. *Boreal forest, lower*: © Ted Mead/Woodfall Wild Images. *Temperate deciduous forest*: © Paul W. Johnson/BPS. *Temperate grasslands, upper*: © Robert and Jean Pollock/BPS. *Temperate grasslands, lower*: © Elizabeth N. Orians. *Cold desert, upper*: © Edward Ely/BPS. *Cold desert, lower*: © Art Wolfe. *Hot desert, left*: © Terry Donnelly/Tom Stack & Assoc. *Hot desert, right*: © Ted Mead/Woodfall Wild Images. *Chaparral*: © Elizabeth N. Orians. *Thorn forest*: © Nick Garbutt/Indri Images. *Savanna*: © Tim Davis/The National Audubon Society Collection/Photo Researchers, Inc. *Tropical deciduous forest*: © Donald L. Stone. *Tropical evergreen forest*: © Elizabeth N. Orians. 57.14: © Gary Bell/Masterfile.

Chapter 58 *Opener*: © D. Pratt (Oil Painting), Bishop Museum. 58.4: © Kevin Schafer/Tom Stack & Assoc. 58.5*a*: © Elizabeth N. Orians. 58.8*a*: © Peter Bichier/Smithsonian Migratory Bird Center. 58.8*b*: © Inga Spence/Visuals Unlimited. 58.10: Richard Bierregaard, Courtesy of the Smithsonian Institution, Office of Environmental Awareness. 58.11*a*: © Frans Lanting/Minden Pictures. 58.11*b*: © John S. Dunning/The National Audubon Society Collection/Photo Researchers, Inc. 58.12: © Adrienne Gibson/Animals Animals. 58.13: © Ed Reschke/Peter Arnold, Inc. 58.14: © David Boynton. 58.15*a*: © John Gerlach/DRK PHOTO. 58.16*a*: © Richard P. Smith. 58.17: © Frans Lanting/Photo Researchers, Inc. 58.18: © Nick Garbutt/Indri Images.

Index

Numbers in ***boldface italic*** refer to information in an illustration, caption, or table.